T0091624

Advanced
Particle Physics
Volume I

Advanced
Particle Physics
Volume I
Particles, Fields, and
Quantum Electrodynamics

O. M. Boyarkin

CRC Press
Taylor & Francis Group
Boca Raton London New York

CRC Press is an imprint of the
Taylor & Francis Group, an **informa** business

CRC Press
Taylor & Francis Group
6000 Broken Sound Parkway NW, Suite 300
Boca Raton, FL 33487-2742

First issued in paperback 2017

© 2011 by Taylor and Francis Group, LLC
CRC Press is an imprint of Taylor & Francis Group, an Informa business

No claim to original U.S. Government works

ISBN-13: 978-1-4398-0414-8 (hbk)
ISBN-13: 978-1-138-11599-6 (pbk)

This book contains information obtained from authentic and highly regarded sources. Reasonable efforts have been made to publish reliable data and information, but the author and publisher cannot assume responsibility for the validity of all materials or the consequences of their use. The authors and publishers have attempted to trace the copyright holders of all material reproduced in this publication and apologize to copyright holders if permission to publish in this form has not been obtained. If any copyright material has not been acknowledged please write and let us know so we may rectify in any future reprint.

Except as permitted under U.S. Copyright Law, no part of this book may be reprinted, reproduced, transmitted, or utilized in any form by any electronic, mechanical, or other means, now known or hereafter invented, including photocopying, microfilming, and recording, or in any information storage or retrieval system, without written permission from the publishers.

For permission to photocopy or use material electronically from this work, please access www.copyright.com (http://www.copyright.com/) or contact the Copyright Clearance Center, Inc. (CCC), 222 Rosewood Drive, Danvers, MA 01923, 978-750-8400. CCC is a not-for-profit organization that provides licenses and registration for a variety of users. For organizations that have been granted a photocopy license by the CCC, a separate system of payment has been arranged.

Trademark Notice: Product or corporate names may be trademarks or registered trademarks, and are used only for identification and explanation without intent to infringe.

Visit the Taylor & Francis Web site at
http://www.taylorandfrancis.com

and the CRC Press Web site at
http://www.crcpress.com

To all
    those whom
I sweetly
    was deceived by.

Contents

Introduction

Among the natural sciences, only physics assumes the role of an all-embracing discipline because its investigative subject is the universe as a whole. Physics enables a unified approach to all objects of the universe from the elementary particles that constitute atoms to giant astronomical structures. The greatest discovery of the twentieth century is the realization that the physical world surrounding us was nonexistent at one time. There is no other problem of science that could be so challenging as an effort to explain the origin of the universe and to establish the reasons why and how the world arrangement has been regulated. Possibilities to develop the first realistic models for the evolution of the universe have been opened only with a breakthrough in the theory of elementary particles because this theory provides the principal way to understand the laws of nature. The greatest achievements in this section of physics are at first well-known and understandable only for a narrow group of the experts and later become cultural elements alongside with masterpieces of art. The latest particular discoveries in elementary particle physics enables one to describe all natural phenomena within the scope of a unified descriptive scheme and, hence, to establish a link between the macrocosm, where the galaxies and their aggregations are scattered like scarce particles of dust and the microcosm of elementary particles. There are two poles of the world: the giant universe, on the one hand, and the fundamental particles, invisible despite the use of any available microscope, on the other hand. And it has been found that a young universe possessed the properties of a microparticle, whereas some microobjects, for example, microscopic black holes, are liable to harbor whole galactic worlds.

Elementary particle physics is a relatively new sphere of human knowledge. Its birth year is 1897, when the first elementary particle, an electron, was discovered. De Broglie's hypothesis and its confirmation during the electron diffraction experiments have formed one of the principal statements of microscopic physics, wave–corpuscle dualism, that has provided the foundation for nonrelativistic quantum mechanics (NQM). Being the first theory in microworld physics, the NQM is sufficiently successful to explain the structure of matter at an atomic and molecular level.

Merging a special relativity theory (SRT) and the NQM has culminated in the derivation of quantum field theory (QFT) that is the foundation for elementary particle physics. At first, particle physics seemed to represent a set of the formulations spun out of thin air and of intuitive assumptions. Each form of the fundamental interactions was studied separately and almost independently of all others. The only feature in common that unified all the interactions was associated with the divergences present in series of the perturbation theory. A theory of electromagnetic interactions between electrons and positrons, quantum electrodynamics (QED), was a pleasant exception to this gloomy landscape. However, the situation had changed drastically by the 1970s. Because of a series of experiments, the chain of *molecules–atoms–nuclei–nucleons* was extended by the inclusion of a further layer in the structure of matter: *the quark-lepton level*. And it was found that the quark and lepton dynamics may be described using quantum theories with a local gauge symmetry. The use of the gauge symmetry group $SU(2)_{EW} \times U(1)_{EW}$ together with a spontaneous symmetry breaking hypothesis allows for the unification of electromagnetic and weak interactions. Then, the inclusion of a theory of strong interaction based on the gauge color group $SU(3)_c$ has resulted in the creation of a so-called standard model (SM). The SM perfectly illuminates not only the microworld phenomena, but numerous cosmological phenomena as well, for example, the Big Bang theory. However, by this time there are existing experimental results, which will possibly demand going out beyond the SM. And it is not improbable that, with the increase of accelerator energies, better accuracy of observations in γ- and neutrino-astronomy, the number of phenomena at variance with the SM will be increased resulting in the development of a new theory. Then, according to the correspondence principle reading *that every new and more exact theory incorporates, in the limiting case, an old, less exact theory*, all formulae of the SM could be derived from the appropriate expressions of a new theory in a certain limiting procedure. In other words, the SM might be a low-energy fragment of a new theory with the extended gauge group relative to the SM. Further advance to the unified picture of the world consists in the unification of strong and electroweak interactions, that is, in the framing of grand unification theory (GUT). We have every reason to believe that the solution will be again found in the searches of the gauge group including the gauge group of the SM as a subgroup. No doubt that for the formation of a unified field theory, that involves both the GUT and gravitational interaction theory, gauge symmetries will play a significant role.

This textbook is intended for anyone who wants to know how the world in which we live really works. Having gained this knowledge, he or she might as well add his or her own new page to this story. Particle physics is the main method to get to know laws of the nature. The way to this science is thorny and complicated. Particle physics is similar to the arts since it takes as much inspiration and hard work as literature, painting, and music.

Microworld physics is one of the "most challenging territories" of human knowledge. Using several textbooks under its study leads to the fact that the process of immersion into the subject noticeably loses its efficacy because of the changing presentation style, the variation of complexity of the mathematical apparatus, the passage to differing designations, and so on. And each textbook has its specific "language" and manner of presentation of the material under study. My experience as a lecturer in elementary particle physics shows that students or beginning post graduates are, as a rule, rarely willing to use several textbooks, rather preferring a single favorite. This textbook contains all the particle physics foundations and, as a result, the necessity of consulting any additional textbooks and monographs is minimized.

To ease mastering the computation procedures in a quantum field theory, the majority of the calculations have been performed without dropping the complex intermediate steps. Therefore, the "as is easy to see"-type phrases, which may be associated with the omission of the most subtle computational parts in some monographs, here denote the actual situation and should not give anxiety to the reader.

A physicist-theorist has not only to perform complicated and skilled calculations and to propose new theories, but he or she has to know experimental procedures and measuring methods as well. Because of this, I have tried, as far as possible, to follow the links between theory and experiment. Moreover, this textbook gives descriptions for modern devices used in the physics of the microworld: accelerators, detectors of elementary particles, and neutrino telescopes. Besides, I include such actual divisions of the modern particle physics as the SM extensions and physics of the massive neutrinos that are usually contained in monographs.

The textbook is constructed in the following manner. The first volume consists of four parts. Part I covers the mathematical basis of modern quantum field theory. The most necessary knowledge of the group theory are provided for, the Noether theorem is proved, and the major motion integrals to be connected with both space and internal symmetry are deduced. In Part II, fundamental interactions and the ways of their unifications are discussed. The main theoretical preconditions and experiments that allowed one to establish matter structure at the quark-lepton level are considered. In Part III, the secondary quantized theories of free fields with spin 0, 1/2, and 1 are investigated. Special attention is attached to the neutrino field. Part IV is devoted to the first successfully operating quantum field theory, quantum electrodynamics. In this part the methods of calculations with polarized and unpolarized particles with and without the inclusion of radiative corrections are yielded. Different renormalization schemes of the quantum field theory are discussed.

The second volume includes three parts. In Part V a quantum chromodynamics (QCD) is expounded. The quantization scheme with the help of functional integrals is considered and the renormalization problems are investigated. Cross sections of basic hard processes are calculated. Nonperturbative methods (lattice approach, QCD vacuum) are also presented. Part VI is devoted to electroweak interactions. The description of the Glashow–Weinberg–Salam theory is given. Composite models and a left-right symmetric model are viewed as the standard model extensions. In Part VII, massive neutrino physics is propounded.

Every part of the first volume contains problems that are an integral part of the text. The goal of a majority of them is mastering the calculation technique to be described in the textbook. There are also the problems that aim at generalizations of the results obtained.

This textbook could be used for teaching the following courses: (i) Introduction to Physics of Elementary Particles; (ii) Foundations of the Quantum Field Theory; (iii) Quantum Electrodynamics; (iv) Electroweak Interaction; (v) Quantum Chromodynamics; (vi) Physics beyond the SM; and (vii) Physics of Massive Neutrinos.

This two volume-textbook is mainly meant for students and post-graduate students who are specialized in particle physics. It will also be useful for post-doctors, researchers, and academics. In order to understand the textbook contents, the reader must have knowledge of nonrelativistic quantum mechanics, the special theory of relativity, and higher mathematics fundamentals only. The textbook will help any attentive reader to be confident, when sailing in the ocean of specialized literature devoted to the problems of strong- and electroweak-interaction physics, and to be adequately prepared to begin research activities.

In conclusion, I would like to thank all physicists for personal contacts that contributed greatly to my scientific horizons. Among them are Zygmunt Ajduk, James Daniel Bjorken, Alexander Pankov, Stefan Pokorski, Dieter Rein, Alexander Studenikin, Dieter Schild-knecht, Goran Senjanovic, and Victor Tikhomirov. I especially thank my wife and Dr. B. Abai for the constant support during the long course of writing this book, and for their inexhaustible belief that it would be completed.

Designations

Let us mark three-dimensional indices with Latin letters, four-dimensional indices running values 0,1,2,3 by Greek ones. All components of four-dimensional vectors are real numbers. We introduce two kinds of four-dimensional tensors. Thus, for four-dimensional coordinates by definition we have

$$x^\mu = (x^0, x^1, x^2, x^3) = (ct, x, y, z),$$

$$x_\mu = (x_0, x_1, x_2, x_3) = (ct, -x, -y, -z).$$

Four-dimensional vectors with upper (low) index we name contravariant (covariant) vectors. In the same way the difference is made for covariant and contravariant tensors with the rank higher than one. Let us define the metric tensor

$$g_{\mu\nu} = \begin{pmatrix} 1 & 0 & 0 & 0 \\ 0 & -1 & 0 & 0 \\ 0 & 0 & -1 & 0 \\ 0 & 0 & 0 & -1 \end{pmatrix}.$$

Since the determinant in this matrix is not equal to 0, there is its inverse matrix $g_{\mu\nu}$ for which takes place

$$g_{\mu\nu} = g^{\mu\nu}.$$

To raise and lower indices the metric tensor is employed. Thus, for example,

$$x_\mu = g_{\mu\nu} x^\nu, \qquad T^{\mu\nu} = g^{\mu\lambda} g^{\nu\sigma} T_{\lambda\sigma}, \qquad \text{etc.}$$

One calls twice-repeated indices dummy ones and on them summarizing is meant. Note, that only space components change its sign under transition from covariant to contravariant four-dimensional vectors. The scalar product of two four-dimensional vectors a_μ and b^μ is defined as follows:

$$a_\mu b^\mu = a^0 b^0 - \mathbf{ab},$$

where

$$\mathbf{ab} = a^k b^k = a_k b_k = a^1 b^1 + a^2 b^2 + a^3 b^3.$$

Four-dimensional vector of energy-momentum has the following form:

$$p^\mu = (E/c, \mathbf{p})$$

and for it

$$p^\mu p_\mu = m^2 c^2$$

is true.

Four-dimensional generalization for the Nabla operator is given by the expression:

$$\partial_\mu \equiv \frac{\partial}{\partial x^\mu} = (\partial_0, \nabla).$$

The symbol $\square$ is used for the D'Alembert operator:

$$\square = \partial_\mu \partial^\mu = \frac{1}{c^2} \frac{\partial^2}{\partial t^2} - \triangle,$$

where $\triangle$ is the Laplace operator. Quantity ε_{ijk} is a completely antisymmetric tensor:

$$\varepsilon_{ijk} = \begin{cases} 1, & n - \text{even} \\ -1, & n - \text{odd} \\ 0, & \text{two and more indices coinside} \end{cases}$$

where n is the number of transpositions that leads indices ijk to the sequence 123. Symbol $\varepsilon_{\mu\nu\lambda\sigma}$ denotes the four-dimensional generalization of tensor ε^{ijk} with $\varepsilon^{0123} = 1$ (while $\varepsilon_{0123} = -1$).

Upper signs $*$, T, and $\dagger$ mean operations of complex conjugation, transposition, and Hermitian conjugation respectively. A continuous line above spinors indicates the operation of the Dirac conjugation:

$$\overline{u} = u^\dagger \gamma_4,$$

where γ_μ are Dirac matrices.

For basic vectors of representations and state vectors we use the Dirac bra- $(|... >)$ and ket- $(< ...|)$ vectors. Thus, for example:

$$\Psi(\mathbf{p}, s_3) \equiv |\mathbf{p}, s_3 >, \qquad \text{and} \qquad \Psi^\dagger(\mathbf{p}, s_3) \equiv < \mathbf{p}, s_3|.$$

Let us mark the three-dimensional radius vector with $\mathbf{r}$, and its module with r, where $r = \sqrt{x_1^2 + x_2^2 + x_3^2}$.

In this book the Heaviside system of units are used, in which $e^2/(4\pi\hbar c) = \alpha_{em}$. This very normalization of electric charge is adopted in periodic literature on quantum field theory. In the Gauss system of units, the normalization $e^2/(\hbar c) = \alpha_{em}$ is used (the value of electric charge corresponding to it is usually given in tables of physical constants). In the Heaviside system of units, the equations of electromagnetic fields have a more convenient form, since the multiplier 4π does not enter there. The Coulomb law, however, in this system has the following form:

$$F_c = \frac{q_1 q_2}{4\pi r^2}.$$

In contrast to this, field equations in the Gauss system of units contain the factor 4π, and the Coulomb law has the simple form $F_c = q_1 q_2/r^2$. It is obvious that the value α_{em} is the same in all systems of units, while the magnitude of the elementary charge e takes different values.

Part I

Mathematical Prelude

1

Relativistic invariance

*Any constitution will lose its ground
against good accoutrements.*
Kozma Prutkov, "Thoughts about Democracy"

The classical physics description of all nature phenomena consists of two essentially different components—matter particles and fields. Movements of mass points, from which as it seemed at that time, one could build all the variety of objects existing in nature, were completely defined by laws of Newton mechanics. A classical particle is a small object localized in a restricted region of space. In the most general case, the position of the particle is defined by three spatial coordinates, that is, its maximum number of degrees of freedom equals three. Having set initial conditions and having solved the evolution equation (the second Newton law), we obtain exhaustive information about the particle.

Description of the electromagnetic field, which represented the sole field known by that time, was much more difficult, since it was necessary to indicate magnitudes and directions of electric and magnetic field strengths at any point in space and at every instant of time. Thus a field, unlike a mass point, had not three but an infinite number of degrees of freedom. Interference, dispersion, and diffraction constituted other important field characteristics, inaccessible by classical particles. Moreover, already at the time of Poisson (1811), the field concept to describe electric and magnetic phenomena was created as an alternative to the long-range interaction theory. So, since particles and fields are bearers of essentially different individual characteristics, the universe picture in classical physics possesses clear features of duality. However, the description of particles and fields has some things in common. First, for particles and fields classical (Laplace's) determinism operates. Second, dynamic variables, used for their description, are the ordinary c-numbers ($c_1 c_2 = c_2 c_1$).

Nonrelativistic quantum mechanics changed the dual scheme of classical physics for a partly more consecutive picture of wave-corpuscle dualism. According to the de Broglie hypothesis, all the matter particles possess not only corpuscular, but also wave properties. On the other hand, an electromagnetic field alongside with wave properties exhibits discrete characteristics, which before were ascribed only to particles. The wave-corpuscle dualism demanded reconsideration of both motion laws and every method of objects description. In nonrelativistic quantum mechanics dynamical observables are no longer associated with c-numbers, now they are associated with operators (q-numbers) satisfying definite commutation relations. The Schrödinger equation is the evolution equation of a quantum system:

$$i\hbar \frac{\partial}{\partial t} \Psi(\mathbf{r}, t) = H \Psi(\mathbf{r}, t), \qquad (1.1)$$

where H is a Hamiltonian operator, $\Psi(\mathbf{r}, t)$ is a wavefunction of the system. In the quantum theory a wavefunction is the main source of information about corpuscular and wave properties of a microobject. Employing the wavefunction the average values of dynamical observables are obtained and $|\Psi(\mathbf{r}, t)|^2 \, dV$ defines the probability of a particle location in the volume element dV. The description of states in classical and quantum regions is essentially different. Coordinates and momenta of classical particles are the quantities that are precisely and simultaneously measured by experiments. The wavefunction $\Psi(\mathbf{r}, t)$ describing the state of quantum particles cannot be directly measured by experiment, that is to say, it has no physical meaning (only the square of its module has physical meaning). From it follows that in the quantum theory the notion of determinism finds different meaning. In this case the links of cause and effect reveal itself only at the probability level, namely, the specifying of interaction form and wavefunction describing a system at some moment of time, must uniquely determine wavefunction in subsequent moments of time. The procedure of correlation between dynamical observables and operators, at which the wavefunction remains to be the c-number is called *formalism of primary quantization*. However, in quantum mechanics, an electromagnetic field had been keeping on to be apart. It was described by classical Maxwell equations, that is, it was considered to be a classical continuous field. Nonrelativistic quantum mechanics allows the description of movements of electrons, nucleons, mesons, and other particles but not their production and annihilation, that means, it is only applicable to systems with a constant number of particles. However, the universal mutual convertibility of particles is one of the most characteristic features of the microworld. Thus, the conservation law of the particles number is approximate, to be exact, it is valid in the low energy region (of the order of atomic electrons energy) only. To describe production and annihilation of particles, merging of quantum mechanics and the special theory of relativity was needed.

The necessity for further development of quantum theory was as well dictated by a more serious task—to quantize fields, corresponding to the known elementary particles, that is, to build a quantum theory for systems with infinite numbers of degrees of freedom. The solution of the above-mentioned problems led to the generalization of quantum mechanics, which is called a *quantum field theory*. Quantized wave field is a fundamental physical concept, logically reflecting the unified wave-corpuscle viewpoint at objects of nature. Each kind of elementary particles is correlated with the corresponding wave field. Wave field properties are determined by spin, charge, and other characteristics of these particles. To quantize a wave field means to set up a correspondence between field and discrete energy quanta (we call them "quanta of the given field"). Interactions between particles are carried out by exchange of quanta of a certain field. For example, electromagnetic quanta (photons) are particles, which describe interactions between electrically charged particles. Thus, within the quantum field theory properties of elementary particles and their interactions are formulated and described. A field is also characterized by a certain function $\psi(x)$ (x means plurality of four-dimensional coordinates), for which one uses the name a field function, and sometimes, the name a wavefunction just as in nonrelativistic quantum mechanics. In the quantum field theory field functions describe the population of particles, with the theory containing mutual conversion processes of particles in the explicit form. In correspondence with it, field functions acquire operator sense and fall into production and annihilation operators of particles, between which commutation relations are established. The procedure of imposing anticommutative relations (for fermions) and commutative ones (for bosons) on field functions derived the name "formalism of secondary quantization". Field functions are no longer classical functions as it used to be in quantum mechanics. They become operators acting upon a state vector $\mid \Phi_N >$, which characterizes completely a physical system consisting of N particles. The vector $\mid \Phi_N >$ is represented by a beam in the complex Hilbert space , that is, in the complex infinite-dimensional space consisting of a set of elements. In this space besides operations of addition and multiplication by a number, the scalar product (Φ, Ψ) is also specified, which meets the following demands:

$$(\Psi, \Psi) \geq 0; \qquad (\Psi, \Psi) = 0 \qquad \text{only when } \mid \Phi >= 0, \tag{1.2}$$

$$(\Phi, \Psi + \Theta) = (\Phi, \Psi) + (\Phi, \Theta), \tag{1.3}$$

$$(\Phi, \alpha\Psi) = \alpha(\Phi, \Psi), \qquad \alpha \in C^1, \tag{1.4}$$

$$(\Phi, \Psi) = (\Psi, \Phi)^*. \tag{1.5}$$

If N is known with assurance, then the module square $\mid \Phi_N >$ defining probability of the given state evidently turns into 1. It means, that a state vector with any fixed N is normalized to one. The form of an evolution equation for the system state vector depends on a representation used in the formulation of a theory. The most convenient representation is an interaction one, in which (as we demonstrate further) field functions satisfy free equations of motion and the evolution of a system state vector is defined by a Schrödinger-like equation:

$$i\hbar\frac{\partial}{\partial t} \mid \Phi_N(t) >= H_{int} \mid \Phi_N(t) >, \tag{1.6}$$

where H_{int} is an interaction Hamiltonian. While looking at (1.6) some reasons for anxiety comes into being. Einstein's relativity principle reads, that all physical laws must have the same form in all inertial frames of reference (IFR). Since the IFR are connected with each other through the Lorentz transformation, then the relativity principle demands invariance of a physical theory with respect to the Lorentz transformations, that is, relativistic covariance. The equation (1.6) as it has been written is not relativistically covariant (covariance occurs only when all the four coordinates enter an equation at the equal foots). However, it is possible to attach Eq. (1.6) the so-called "seeming" covariance [1,2]. The idea is to change the nonrelativistic consideration of the whole space at some moment of time by introducing the space-like arbitrary hypersurface σ. The hypersurface σ is defined by only one condition, namely, it must contain no points that could be connected by a light signal, that is, an interval between any two points x and y at σ is always a space-like one $((x - y)^2 < 0)$. Since the concept of space-like hypersurface is relativistically invariant, then setting of the system state vector as a functional on σ would be also relativistically invariant. A state vector can be determined by independent observations at all points of σ, because perturbations caused by observations cannot spread faster than light speed. Due to the aforesaid, we are not limited by a certain system of coordinates always working with regions where all the points are physically independent. Let us identify an element of volume dV with the projection of the space-like hypersurface element $d\sigma = (d\sigma_\mu) = (dx_1 dx_2 dx_3, cdx_2 dx_3 dt, cdx_1 dx_3 dt, cdx_1 dx_2 dt)$ on a hyperplane σ_t, for which t=const. The hypersurface in four-dimensional space is an ordinary two-dimensional plane analog in three-dimensional space, that is to say, it is some three-dimensional manifold in the four-dimensional world. In the three-dimensional space the volume of a parallelepiped, built on three vectors dr, dr', and dr'', is equal to a determinant of the third order composed from components of these vectors. In four-dimensional space the projections of the volume of a "parallelepiped" (that is, the area of the hyperplane), built on four-vectors dx^μ, $dx^{\mu'}$, and $dx^{\mu''}$, are expressed in the same way. They constitute an antisymmetric tensor of the third rank, whose components are defined by the expression:

$$d\sigma'^{\mu\nu\lambda} = \begin{vmatrix} dx^\mu & dx'^\mu & dx''^\mu \\ dx^\nu & dx'^\nu & dx''^\nu \\ dx^\lambda & dx'^\lambda & dx''^\lambda \end{vmatrix}.$$

As an element of integration over the hypersurface it is more convenient to use the four-vector $d\sigma^\mu$, which is dual to the tensor $d\sigma^{\nu\lambda\rho}$:

$$d\sigma^\mu = -\frac{1}{6}\varepsilon^{\mu\nu\lambda\rho}d\sigma_{\nu\lambda\rho}.$$

Geometrically $d\sigma^\mu$ is a four-vector that is equal in magnitude to the area of the hypersurface element and normal to this element (that is, $d\sigma^\mu$ is perpendicular to all lines drawn in the hypersurface element). It is possible to ascribe local time $t(\mathbf{r})$ to any point $\mathbf{r}$ of the space-like hypersurface σ. A unit normal vector to the hypersurface σ at any point x is a time-like vector, that is, $n^{\mu}(x)n_{\mu}(x) = 1$. The transition to the hyperplane σ_t takes place when $n_{\mu} = (1,0,0,0)$.

Now let us consider Eq. (1.6) as a result of infinite series of equations, obtained by introducing local time for every point at the hypersurface σ. Exhibit the interaction Hamiltonian as the sum over small three-dimensional cells ΔV, belonging to σ:

$$H_{int}(t) = \sum_{\sigma} \mathcal{H}_{int}(x)\Delta V. \tag{1.7}$$

The equation that is fulfilled in the small cell the around space-time point with coordinates $(t(\mathbf{r}), \mathbf{r})$ can be represented in the following form:

$$i\hbar\frac{\partial \mid \Phi_N(t(\mathbf{r})) >}{\partial t(\mathbf{r})} = \mathcal{H}_{int}(x)\Delta V \mid \Phi_N(t(\mathbf{r})) > . \tag{1.8}$$

The result obtained is quite obvious: since the change of $\mid \Phi(t) >$ at the infinitesimal replacement of the hyperplane σ_t as a whole is given by the integral:

$$\int_{\sigma_t} \mathcal{H}_{int}(x)d\mathbf{r},$$

then the variation of $\mid \Phi_N(t(\mathbf{r})) >$ at the point $(t(\mathbf{r}), \mathbf{r})$ must be defined by the density of the interaction Hamiltonian $\mathcal{H}_{int}(x)\Delta V$ in the infinitesimal vicinity of the point x. In Eq. (1.8) we shall switch over from the noninvariant differentiation with respect to time to the covariant differentiation with respect to σ. For this purpose we consider the value of the state vector $\mid \Phi_N(t(\mathbf{r})) >=\mid \Phi_N(\sigma) >$ at two space-like hypersurfaces σ and σ', which are distinguished by an infinitesimal only in the vicinity of the space-time point x (see Fig. 1.1).

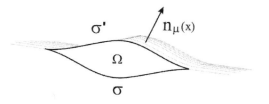

FIGURE 1.1
Two space-like hypersurfaces σ and σ'.

Then we have:

$$\frac{\delta}{\delta\sigma(x)}\mid \Phi_N(\sigma) >= \lim_{\Delta\Omega(x)\to 0}\frac{\mid \Phi_N(\sigma') > - \mid \Phi_N(\sigma) >}{\Omega(x)} =$$

$$= \lim_{\Delta t\Delta V\to 0}\frac{\mid \Phi_N(t(\mathbf{r}) + \Delta t(\mathbf{r})) > - \mid \Phi_N(t(\mathbf{r})) >}{c\Delta t(\mathbf{r})\int_{\Delta V}d\mathbf{r}} =$$

$$= \lim_{\Delta t\Delta V\to 0}\frac{\mid \Phi_N(\sigma') > - \mid \Phi_N(\sigma) >}{c\Delta t(\mathbf{r})\Delta V}, \tag{1.9}$$

where Ω is a space-time volume (an invariant quantity) compressed into the point x in the limit. Using Eqs. (1.8) and (1.9) one can arrive at the relativistically invariant evolution equation for the system state vector:

$$ i\hbar c \frac{\delta}{\delta\sigma(x)} \mid \Phi_N(\sigma) >= \mathcal{H}_{int}(x) \mid \Phi_N(\sigma) > . \tag{1.10}$$

The total Hamiltonian is connected with the total Lagrangian by the relation:

$$ \mathcal{H} = \dot{\psi}_j \frac{\delta\mathcal{L}}{\delta\dot{\psi}_j} - \mathcal{L}, \tag{1.11}$$

where $j = 1, 2, ..n$, $\dot{\psi}_j \equiv \partial_t\psi_j$, and n is the freedom degrees number of a quantum field system. Since the Lagrangian is a relativistically invariant quantity, then for the interaction Hamiltonian to be invariant, the derivatives with respect to four-coordinates must be absent in the part of the Lagrangian, which is responsible for the interaction. For such interactions we used the term "couplings without derivatives." In case of couplings without derivatives we simply have

$$ \mathcal{H}_{int} = -\mathcal{L}_{int}. \tag{1.12}$$

Thus, for example, the Lagrangian of interaction between particles with spin 1/2 and an electromagnetic field contains no derivatives, while the electromagnetic interaction between massive particles with spin 0 and 1 is an example of couplings with derivatives. However, paradoxical though it may seem, in the latter case the theory finds the relativistic covariance at the final stage. This transformation, to be similar to the metamorphosis of an ugly duckling into a beautiful swan, is caused by the following circumstances. For the solutions of Eq. (1.10) to exist, the integrability condition must be fulfilled:

$$ \frac{\delta}{\delta\sigma(x)} \frac{\delta}{\delta\sigma(x')} \mid \Phi_N(\sigma) >= \frac{\delta}{\delta\sigma(x')} \frac{\delta}{\delta\sigma(x)} \mid \Phi_N(\sigma) >, \tag{1.13}$$

where x and x' are two points on the hypersurface σ. In its turn, for the condition to be carried out, it is necessary, that:

$$ [\mathcal{H}_{int}(x : n), \mathcal{H}_{int}(x' : n)] = i\hbar c \left(\frac{\partial}{\partial\sigma(x)} \mathcal{H}_{int}(x' : n) - \frac{\partial}{\partial\sigma(x')} \mathcal{H}_{int}(x : n) \right), \tag{1.14}$$

where $\partial/(\partial\sigma(x))$ is the differentiation with respect to arguments, explicitly depending on σ and the designation $\mathcal{H}_{int}(x : n)$ emphasizes the dependence of the interaction Hamiltonian on the normal vector to the space-like hypersurface $n_\mu(x)$. By means of Eq. (1.14), which is the basis to define the interaction Hamiltonian in case of couplings with derivatives (see for details [3]), it is possible to demonstrate, that

$$ \mathcal{H}_{int}(x : n) = -\mathcal{L}_{int}(x) + W^{\mu\nu}(x)n_\mu(x)n_\nu(x) + ... \tag{1.15}$$

Besides $\mathcal{H}_{int}(x : n)$, Green functions of equations for vector and scalar particles also displayed dependence on $n_\mu(x)$. In quantum field theory we call these functions "propagators" and use them to describe the spreading of particles in virtual states. While calculating the experimentally observed quantities (for example, cross sections), the so-called $\sigma(x)$-dependence compensation procedure comes into play. It works as follows [4]. Terms, which depend on the space-like hypersurface, that is, on $n_\mu(x)$, appear in both an interaction Hamiltonian and in propagators. Further, the exact mutual cancellation of these terms in the S-matrix elements in all orders of the perturbation theory takes place. Thus, the correct computational procedure in the theory of electromagnetic interactions of scalar and vector bosons is to completely ignore the terms, which depend on the orientation of the space-like hypersurface both in an interaction Hamiltonian and in particle propagators. In the modern theory, which unifies strong, weak, and electromagnetic interactions, couplings with derivatives for fundamental particles occur for particles with spin 1 and 0. In this case such couplings are always contained for the vector particles and may be held for the scalar particles only in some theory modifications. Then, we can claim that the evolution equation for the state vector (1.10) is relativistically covariant in our picture of the world. The fulfillment of Eq. (1.12) instills pleasant confidence in us, that the refusal from the Hamiltonian formalism in favor of the Lagrangian one in the field theory is possible. Notice that the Hamiltonian formalism is not relativistically invariant due to the singled-out role of time. Let us underline that there is no proof that compensation of the $\sigma(x)$-dependence for an arbitrary spin theory, which contains couplings with derivatives, takes place. Consequently, if an experiment, being the last judge in the controversy, adds any new, say, charged particles with spin 3/2 to the existing fundamental particles, we would be obliged to check whether the compensation of the $\sigma(x)$-dependence takes place in the theory.

One may come to relativistic invariance of a theory from the other more constructive position, namely, from the viewpoint of investigating transformation properties of field functions under the Lorentz group transformations. To put this another way, it is necessary to determine the Lorentz group representation for field functions. As this takes place, unitary representations of the Lorentz group are of principal interest for us. The importance of obtaining such representations is due to the fact that they completely define a wave equation for a physical system. To carry out this program we must supplement our mathematical arsenal with the necessary information about the group theory. Let us make one remark concerning our terminology. We shall resort to the theory of secondary quantization only in Part III and in the first two Parts we shall work within the primary quantization formalism. It allows us to view wavefunctions as classical objects in Parts I and II.

2

Three-dimensional world

2.1   Orthogonal transformation group $O(3)$

Let us start our excursus in the group theory from studies of transformations of the three-dimensional Euclidean space. We consider transformations:

$$\mathbf{x}' = R\mathbf{x}, \tag{2.1}$$

leaving a scalar product $(\mathbf{x} \cdot \mathbf{y})$ invariant, for which the relation:

$$(\mathbf{x}' \cdot \mathbf{y}') = (\mathbf{x} \cdot \mathbf{y}) \tag{2.2}$$

is true. As follows from (2.2), these transformations are carried out through orthogonal matrices

$$RR^T = R^T R = 1.$$

Since:

$$\det(RR^T) = \det R^T \det R = (\det R)^2 = 1,$$

then $\det R = \pm 1$. In case $\det R = +1$ we have proper orthogonal transformations or rotations, while when $\det R = -1$ we obtain improper orthogonal transformations . The latter is exemplified by space inversion, which is represented by the matrix:

$$R_{-} = \begin{pmatrix} -1 & 0 & 0 \\ 0 & -1 & 0 \\ 0 & 0 & -1 \end{pmatrix}. \tag{2.3}$$

Any rotation can be represented as a product of three consecutive rotations around orthogonal axes x_1, x_2, x_3 about angles θ_1, θ_2, θ_3. Under the rotations around axis x_1 coordinates $\mathbf{x}$ are transformed in the following way:

$$x_1' = x_1 + 0 \times x_2 + 0 \times x_3,$$

$$x_2' = 0 \times x_1 + \cos\theta_1 x_2 + \sin\theta_1 x_3,$$

$$x_1' = 0 \times x_1 - \sin\theta_1 x_2 + \cos\theta_1 x_3.$$

In the matrix form it looks as follows:

$$\begin{pmatrix} x_1' \\ x_2' \\ x_3' \end{pmatrix} = R_{\theta_1} \begin{pmatrix} x_1 \\ x_2 \\ x_3 \end{pmatrix} = \begin{pmatrix} 1 & 0 & 0 \\ 0 & \cos\theta_1 & \sin\theta_1 \\ 0 & -\sin\theta_1 & \cos\theta_1 \end{pmatrix} \begin{pmatrix} x_1 \\ x_2 \\ x_3 \end{pmatrix}.$$

In the same way we can represent transformations of rotations around axes x_2 and x_3:

$$R_{\theta_2} = \begin{pmatrix} \cos\theta_2 & 0 & \sin\theta_2 \\ 0 & 1 & 0 \\ -\sin\theta_2 & 0 & \cos\theta_2 \end{pmatrix}, \qquad R_{\theta_3} = \begin{pmatrix} \cos\theta_3 & \sin\theta_3 & 0 \\ -\sin\theta_3 & \cos\theta_3 & 0 \\ 0 & 0 & 1 \end{pmatrix}.$$

The matrix of a rotation around an arbitrary axis is defined by the product of matrices: $R_{\theta_1} R_{\theta_2} R_{\theta_3}$. Pay attention to the fact that the final result depends on the order of the multiplication of R_θ matrices, that is to say, rotation transformations are not commutative. Totality of all the orthogonal transformations in the three-dimensional Euclidean space makes up a group, namely, an orthogonal group $O(3)$. Transformations with the determinant equal to 1 are called either "unimodular" or "special" (the name of a group is indicated by the letter S). Thus, the rotation group is the group $SO(3)$. Groups consisting of commutative elements are called *Abelian groups*, while ones with noncommutative elements are called *non-Abelian groups* . The group $SO(3)$ is a non-Abelian group. The group $SO(3)$, supplemented with reflections, makes up the group $O(3)$. It means, that $SO(3)$ is a subgroup of $O(3)$. At this point it is high time to introduce the rules of the game, adopted in the group theory.

A group is a set of elements G, which meets the following demands:

1. On the set G the group action is defined (call it "multiplication" conventionally), which puts in correlation to every pair of elements f and g a certain h element from the same set. The element h is a product of elements f and g, that is, $h = fg$. In the most general case $fg \neq gf$.

2. The set contains such a unit element e, for which the relation $ef = fe = f$ is true, where f is any element from the set G.

3. Alongside with any element f, the set G contains an inverse element f^{-1}, which possesses the following property:

$$ff^{-1} = f^{-1}f = e.$$

The group is finite (infinite) when the number of its elements is finite (infinite). The three-dimensional rotation group is infinite while the reflection group is finite. The group is called *discrete (continuous)* when its elements take discrete (continuous) values. The three-dimensional rotation group is continuous, while the reflection group is discrete. The order of group is a number of independent parameters, which defined a group. The three-dimensional rotation group is the group of the third order (its independent parameters are θ_1, θ_2, θ_3). The order of the reflection group is the same. The order of the transformation matrix (two-column, three-column, etc.) defines dimensionality of a group (d). The dimensionality of the three-dimensional rotation group and of the reflection group is three. The dimensionality of the $O(n)$-group is defined by the formula $d = n(n-1)/2$. Let us note, that the order of a group differs from its dimensionality. In the case of the three-dimensional orthogonal group this coincidence is accidental. Continuous groups of the finite order are Lie groups. The three-dimensional rotation group is an example of a Lie group.

Now our purpose is to define all representations of an orthogonal group. In general, representation of a certain G group is a mapping, which compares every element g of G with linear operator U_g, acting in some vector space V. In such a mapping a multiplication table for a group is maintained, and the unit element e of the group G is represented by identity transformation of I in V:

$$U_{g_1} U_{g_2} = U_{g_3} \qquad (g_3 = g_1 g_2), \qquad U_e = I.$$

The most simple representations, of which one can build all the rest of group representations by means of multiplication, are named fundamental. If the group representation is formed by the group generators themselves then that is regular or adjoint.

Subspace V_1 of the space V is called the *invariant subspace* with respect to representation U_g in case if all the vectors v in V_1 are transformed by any U_g into vectors v' belonging again to V_1. A representation is irreducible, if the only subspace V, which are invariant with respect to representation $g \to U_g$, are the whole space and subspace, consisting of one zero vector. Otherwise a representation is reducible.

Studies of representations could be limited to studies of irreducible representations. Moreover, only nonequivalent irreducible representations are of a main interest for us. Two representations U_g and U'_g are equivalent, if there is a unitary operator M, which ensures

$$U'_g = MU_gM^{-1} \qquad v' = Mv,$$

where v and v' are vectors of representation spaces. Thus, two equivalent representations could be seen as realizations of one and the same representation in terms of two different bases in a vector space.

All irreducible representations are finite-dimensional and any representation is a direct sum of irreducible finite-dimensional representations. The direct sum of two square matrices D_1 and D_2 is by definition a square matrix D_3 for which

$$D_3 = \begin{pmatrix} D_1 & 0 \\ 0 & D_2 \end{pmatrix}.$$

is true. Symbolically, it is set down as follows:

$$D_3 = D_1 \bigoplus D_2.$$

Further on, infinitesimal rotations and matrices, corresponding to them will play a fundamental role. Their importance is caused by the fact, that they produce one-parameter subgroups and any finite rotation can be composed of a sequence of infinitesimal rotations. To obtain matrices for infinitesimal rotations, we expand every element of the matrix of finite rotations into the Taylor series and constrain ourselves by the terms of the first order:

$$A_i = \frac{d}{d\theta_i} R(\theta_i)\bigg|_{\theta_i=0}, \qquad (i = 1, 2, 3) \tag{2.4}$$

The obtained matrices (2.4) will be called *generators of rotations* around i-axis. Their obvious form is given by

$$A_1 = \begin{pmatrix} 0 & 0 & 0 \\ 0 & 0 & 1 \\ 0 & -1 & 0 \end{pmatrix}, \qquad A_2 = \begin{pmatrix} 0 & 0 & -1 \\ 0 & 0 & 0 \\ 1 & 0 & 0 \end{pmatrix},$$

$$A_3 = \begin{pmatrix} 0 & 1 & 0 \\ -1 & 0 & 0 \\ 0 & 0 & 0 \end{pmatrix}. \tag{2.5}$$

From (2.5) it is clear, that infinitesimal rotations commute with each other. A rotation about a finite angle θ_i can be viewed as a result of n rotations about angle θ_i/n:

$$R(\theta_i) = \lim_{n\to\infty} \left(1 + \frac{\theta_i}{n} A_i\right)^n = \exp\left(A_i\theta_i\right). \tag{2.6}$$

By direct calculations it is possible to check that generators A_i satisfy the commutation relation:

$$[A_i, A_k] = -\varepsilon_{ikl} A_l \tag{2.7}$$

and commute with the reflection operator:

$$[A_i, R_-] = 0. \tag{2.8}$$

A set of A generators constitute the so-called Lie algebra of the three-dimensional rotation group.

The set of elements N is a Lie algebra, corresponding to a certain group, if it meets the following demands:

1. If X and Y are elements of the set N, then both the sum of $X + Y$ and the product αX (where α is an arbitrary number) belong to N once again.

2. Commutator of two elements X and Y of the set N is again expressed in terms of elements of the set N.

3. Commutators of elements of the set N satisfy the relation:

$$[X, Y] + [Y, X] = 0, \qquad [X, (Y + Z)] = [X, Y] + [X, Z],$$

$$[X, [Y, Z]] + [Y, [Z, X]] + [Z, [X, Y]] = 0, \qquad \text{(Jacobi identity)}. \qquad (2.9)$$

The representation of the Lie algebra is a correspondence $X \to T(X)$, which confronts every element X with the linear operator $T(X)$ acting in a vector space in such a way, that:

$$T(X + Y) = T(X) + T(Y), \qquad T(cX) = cT(X),$$

$$T([X, Y]) = [T(X), T(Y)].$$

By the example of the rotation group, we show that the Lie algebra representation uniquely determines the representation of the group itself. Infinitesimal rotation about the angle $\delta\theta$ is put down as follows:

$$R(\delta\theta) = I + A_i \delta\theta_i + O(\delta\theta^2).$$

The corresponding representation operator has the form:

$$U(\delta\theta) = I + M_i \delta\theta_i + O(\delta\theta^2), \qquad (2.10)$$

where M_i constitute the representation of the Lie algebra generators and obey the same commutation relations as A_i:

$$[M_i, M_k] = -\varepsilon_{ikl} M_l. \qquad (2.11)$$

Let us consider two rotations about the finite angles $a\theta$ and $b\theta$. Because rotations about one and the same axis commute:

$$R(a\theta) R(b\theta) = R((a + b)\theta)$$

then as a result:

$$U(a\theta) U(b\theta) = U((a + b)\theta). \qquad (2.12)$$

Differentiating (2.12) with respect to a, using (2.10) and taking into consideration that

$$\frac{d}{da} \leftrightarrow \frac{d}{db},$$

at $a = 0$ we obtain the equation

$$\frac{d}{da} U(a\theta) U(b\theta)|_{a=0} = \frac{d}{db} U(b\theta) = (M_i \theta_i) U(b\theta). \qquad (2.13)$$

To integrate Eq. (2.13) and use the initial condition $U(0) = 1$ produces the expression for operators of the representation of the three-dimensional rotation group:

$$U(\theta) = \exp(M_i \theta_i). \qquad (2.14)$$

Since in unitary representations operators U are unitary, then operators M_i are anti-Hermitian. Now we proceed to Hermitian operators $J_i = -iM_i$,, which satisfy ordinary permutation relations for angular moment operators:

$$[J_k, J_m] = i\varepsilon_{kmn}J_n. \tag{2.15}$$

The problem of finding all the irreducible rotation group representations is equivalent to that of finding all the possible sets of matrices J_1, J_2, J_3, complying with the permutation relations (2.15). The Shur lemma has the fundamental importance in representations theory of groups, realized by complex matrices. It reads: *"for representation to be irreducible, it is necessary and sufficient, that the only matrices, which commute with all the representation matrices are the matrices which are multiple of the identity matrix."* Such matrices being polynomials of generators, are called *Casimir operators* U_i^K. According to the Shur lemma, a representation is irreducible in the case of

$$U_i^K = \lambda I$$

and as a result

$$U_i^K \psi = \lambda\psi,$$

where ψ denotes the field function. Since the Casimir operators enter the total set* of commuting operators, then conserved physical quantities correspond to them. The number of the Casimir operators defines a group rank.

So, if we have found all the Casimir operators, have chosen one eigenvalue from each operator, and built up a representation that acts in space stretched on corresponding eigenfunctions, such a representation is irreducible, according to the Shur lemma. In other words, to classify irreducible representations of a group one must find the spectrum of eigenvalues and eigenfunctions of the Casimir operators.

Because the three-dimensional rotations group has one Casimir operator:

$$\mathbf{J}^2 = J_1^2 + J_2^2 + J_3^2, \tag{2.16}$$

then the rank of this group is equal to 1. In this case the Casimir operator is nothing else but the squared angular moment operator with eigenvalues $j(j+1)$, where $j = 0, 1/2, 1, 3/2, 2, ...$. Consequently, every irreducible representation of the three-dimensional rotations group is characterized by the positive integer or half-integer number j, which also defines a representation dimensionality by the formula $2j + 1$. From quantum mechanics we know that the squared angular moment operator commutes with a projection of the angular moment operator onto a certain direction singled out in space, for example, the x_3-axis. Thus, the basic representation functions are the eigenfunction of the operators $\mathbf{J}^2$ and J_3 and could be marked by their eigenvalues.

Passing on to the classification of irreducible representations of the orthogonal group we keep in mind that the linear operator U_- corresponding to the operation of space inversion R_- commutes with all the operators of the rotations group representation. According to the Shur lemma, in every irreducible representation U_- must be a multiple of the unit operator. Thus, the irreducible representation of the orthogonal group is classified by a pair of indices (j, η_p), where the latter is the eigenvalue of U_- corresponding to the given representation. Representations with integer j are called *one-valued* or *tensor representations*, in the case of half-integer j representations are called *two-valued* or *spinor representations*.

---

*It undoubtedly contains the Hamiltonian as well.

2.2   Tensor representations of the $SO(3)$-group

Since at integer values j the relation

$$(U_-)^2\psi = I\psi$$

takes place, then eigenvalues of the reflection operator are $+1$ and -1. Thus, there are two different representations of the orthogonal group, in the former $U_- = I$ and in the latter $U_- = -I$.

At $j = 0$ the representation is one-dimensional, each group element is expressed in terms of the unit operator, and generators are identically equal to zero. Let us call representation $\{0,1\}$ as a scalar and $\{0,-1\}$ as a pseudoscalar one. Quantities transformed according to (pseudo)scalar representation are (pseudo)tensors of zero-rank or (pseudo)scalars.

At $j = 1$ the representation is three-dimensional. One might use matrices of generators of the three-dimensional rotations group as the matrix representation of generators M_i. For the representations $\{1,-1\}$ and $\{1,1\}$ we use the terms vector representation and pseudovector representation, respectively. Three-dimensional quantities transformed with respect to (pseudo)vector representations are called *(pseudo)tensors of first rank* or *(pseudo)vectors*. Very often pseudovectors are called *axial vectors* and vectors are called *polar vectors*. All the representations with $j = 0, 1$ are irreducible, while the representations with $j \geq 2$ are reducible. To summarize the aforesaid, we can give the following definition: a three-dimensional tensor (pseudotensor) of the n-rank is a quantity, which is transformed under the representation $\{n, (-1)^n\}$ ($\{n, (-1)^{n+1}\}$) of the group $O(3)$. More convenient for practical use language, the tensor and pseudotensor of the n-rank are the quantities whose components, when rotating, are transformed according to the law:

$$\psi'_{\underbrace{ik..m}_{n}}(\mathbf{r}') = R_{ip}R_{kl}...R_{ms}\psi_{\underbrace{pl..s}_{n}}(\mathbf{r}). \tag{2.17}$$

However, under reflections we have:

$$\psi'_{\underbrace{ik..m}_{n}}(\mathbf{r}') = (-1)^n \psi_{\underbrace{ik..m}_{n}}(\mathbf{r}), \tag{2.18}$$

for a tensor

$$\psi'_{\underbrace{ik..m}_{n}}(\mathbf{r}') = (-1)^{n+1}\psi_{\underbrace{ik..m}_{n}}(\mathbf{r}), \tag{2.19}$$

and for a pseudotensor, respectively.

The ability to expand reducible representations in irreducible ones will make one more element of our mathematical culture. As an example, we consider a field function ψ_{ik}, which is a tensor of the second rank with respect to the three-dimensional rotations group and consequently it has nine components. A tensor of the second rank is a product of two three-dimensional vectors, which are transformed according to three-dimensional representations. Schematically it can be written in the form:

$$3 \bigotimes 3. \tag{2.20}$$

Let us represent ψ_{ik} as a sum of symmetric and antisymmetric tensors:

$$\psi_{ik}^s = \frac{1}{2}(\psi_{ik} + \psi_{ki}), \qquad \psi_{ik}^a = \frac{1}{2}(\psi_{ik} - \psi_{ki}).$$

By the direct check one can be sure, that

$$\psi_{ik}^{s\prime} = R_{im}^{(1)} R_{kn}^{(1)} \psi_{mn}^s \qquad \psi_{ik}^{a\prime} = R_{im}^{(1)} R_{kn}^{(1)} \psi_{mn}^a,$$

that is, under the rotations the components of the symmetric and antisymmetric tensors are not mixed with each other. In other words, under transformations of the group $SO(3)$ the nine components of the tensor ψ_{ik} fall into two independent totalities: three-dimensional one ψ_{ik}^a and six-dimensional one ψ_{ik}^s. Further we remember, that the scalar product of three-dimensional vectors is invariant under the rotations. Then the symmetric tensor could be also expanded into two independent totalities, namely, a scalar

$$\psi^0 = \frac{1}{3}\mathrm{Sp}\{\psi_{mn}\} = \frac{1}{3}\psi_{nn} = \frac{1}{3}(\psi_{11} + \psi_{22} + \psi_{33})$$

and five components, which constitute a matrix with a zero-spur:

$$\psi_{ik}^s = \psi_{ik}^s - \frac{1}{3}\delta_{ik}\psi_{nn}.$$

So, from the viewpoint of the rotation transformations, the second rank tensor ψ_{ik} is the sum of the three independent quantities: the one-dimensional ψ^0, the three-dimensional ψ_k^a ($k = 1, 2, 3$), and the five-dimensional ψ_l^s ($l = 1, 2, 3, 4, 5$). The corresponding representation of the rotations group consists of matrices of a less dimensionality and has the box-diagonal form:

$$U \begin{pmatrix} \psi_1^a \\ \psi_2^a \\ \psi_3^a \\ \psi_1^s \\ \psi_2^s \\ \psi_3^s \\ \psi_4^s \\ \psi_5^s \\ \psi^0 \end{pmatrix} = \begin{pmatrix} U_{11} & U_{12} & U_{13} & 0 & 0 & 0 & 0 & 0 & 0 \\ U_{21} & U_{22} & U_{23} & 0 & 0 & 0 & 0 & 0 & 0 \\ U_{31} & U_{32} & U_{33} & 0 & 0 & 0 & 0 & 0 & 0 \\ 0 & 0 & 0 & U_{44} & U_{45} & U_{46} & U_{47} & U_{48} & 0 \\ 0 & 0 & 0 & U_{54} & U_{55} & U_{56} & U_{57} & U_{58} & 0 \\ 0 & 0 & 0 & U_{64} & U_{65} & U_{66} & U_{67} & U_{68} & 0 \\ 0 & 0 & 0 & U_{74} & U_{75} & U_{76} & U_{77} & U_{78} & 0 \\ 0 & 0 & 0 & U_{84} & U_{85} & U_{86} & U_{87} & U_{88} & 0 \\ 0 & 0 & 0 & 0 & 0 & 0 & 0 & 0 & 1 \end{pmatrix} \begin{pmatrix} \psi_1^a \\ \psi_2^a \\ \psi_3^a \\ \psi_1^s \\ \psi_2^s \\ \psi_3^s \\ \psi_4^s \\ \psi_5^s \\ \psi^0 \end{pmatrix}. \qquad (2.21)$$

If a representation matrix can be written down in the box-diagonal form, the corresponding representation is reducible (otherwise it is irreducible). Then, Eq. (2.21) means

$$3 \bigotimes 3 = 9 = 5 \bigoplus 3 \bigoplus 1. \qquad (2.22)$$

Let us try to present the sequence of manipulations from Eqs. (2.20) to (2.22) in a more obvious form. In Fig. 2.1 we plot basis vectors of the representations entering the left-hand side of Eq. (2.22), where values of the angular momentum projection m_j are denoted with dots. To obtain the totality of the basis vectors, composing representations in the right-

FIGURE 2.1
The basis vectors of the representations 3 and 3.

hand side of Eq. (2.22), one might use the moments multiplication rule, which is well known from quantum mechanics. In order to get the products of two moments, we superimpose the center $m_j = 0$ of the second diagram with each of the three states $m_j = 1, 0, -1$ of the first diagram. Further, we should calculate frequency n_i with which every summation result m_j appears (in our example $m_j = +1$ occurs twice, $m_j = 0$ does triply). From the obtained states we single out diagrams with the same value of j. This procedure is displayed in Fig. 2.2. Then, if one takes into account that the states number with the angular momentum

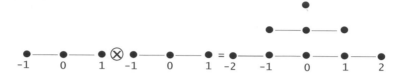

FIGURE 2.2
Graphical interpretation of the moments summation rule.

eigenvalue j is equal to $2j + 1$, that is,

$$(2j_1 + 1) \bigotimes (2j_2 + 1) \equiv j_1 \bigotimes j_2,$$

the generalization of the relation obtained takes the familiar form:

$$j_1 \bigotimes j_2 = \sum_{\oplus} j, \tag{2.23}$$

where every j-representation occurs once, $j = | j_1 - j_2 |, | j_1 - j_2 + 1 |, \dots | j_1 + j_2 |$ and subscript $\oplus$ indicates that the sum is direct, not normal.

§ 1.2.3. Spinor representations of the three-dimensional rotations group

At $j = 1/2$ representations of the rotations group is two-dimensional and generators $J_k^{(1/2)}$ might be chosen in the form of $i\sigma_k/2$, where σ_k are Pauli matrices

$$\sigma_1 = \begin{pmatrix} 0 & 1 \\ 1 & 0 \end{pmatrix}, \qquad \sigma_2 = \begin{pmatrix} 0 & -i \\ i & 0 \end{pmatrix}, \qquad \sigma_3 = \begin{pmatrix} 1 & 0 \\ 0 & -1 \end{pmatrix}, \tag{2.24}$$

that satisfy the relation:

$$\sigma_k \sigma_l = \delta_{kl} + i\varepsilon_{klm}\sigma_m. \tag{2.25}$$

The operator to transform a wavefunction under the rotations about θ angle around the axis with the unit vector $\mathbf{n} = (n_1, n_2, n_3)$, is written down as follows:

$$U^{(1/2)}(\mathbf{n}, \theta) = \exp\left[i(\mathbf{n} \cdot \boldsymbol{\sigma})\frac{\theta}{2}\right] = \sum_{k=0}^{\infty} \frac{1}{k!}\left(\frac{i}{2}(\mathbf{n} \cdot \boldsymbol{\sigma})\theta\right)^k = [1 - \frac{1}{2!}\left(\frac{\theta}{2}\right)^2 + \frac{1}{4!}\left(\frac{\theta}{2}\right)^2 -$$

$$\dots] + i(\mathbf{n} \cdot \boldsymbol{\sigma})[\frac{\theta}{2} - \frac{1}{3!}\left(\frac{\theta}{2}\right)^3 + \frac{1}{5!}\left(\frac{\theta}{2}\right)^5 - \dots] = I\cos\frac{\theta}{2} + i(\mathbf{n} \cdot \boldsymbol{\sigma})\sin\frac{\theta}{2}. \tag{2.26}$$

From (2.26) it follows that the rotation matrices are unitary and their determinant equals 1. Objects transformed on the representation $j = 1/2$ shall be called spinors of the first rank. As a basis in space of such spinors, one might choose eigenvectors of the matrices $(1/4)\boldsymbol{\sigma}^2$ (spin square) and $(1/2)\sigma_3$ (spin projection onto the x_3-axis) which have the form:

$$\omega_+ = \begin{pmatrix} 1 \\ 0 \end{pmatrix}, \qquad \omega_- = \begin{pmatrix} 0 \\ 1 \end{pmatrix}.$$

Since:

$$\frac{1}{4}\boldsymbol{\sigma}^2 \omega_\pm = \frac{3}{4}\omega_\pm$$

and

$$\frac{1}{2}\sigma_3 \omega_+ = +\frac{1}{2}\omega_+, \qquad \frac{1}{2}\sigma_3 \omega_- = -\frac{1}{2}\omega_-,$$

it becomes obvious, that spinors of the first rank are used to describe particles with the spin $1/2$.

The system rotation around the angle 2π causes quite unexpected results:

$$U^{(1/2)}(\mathbf{n}, \theta + 2\pi)\psi = -U^{(1/2)}(\mathbf{n}, \theta)\psi, \tag{2.27}$$

that is, the system does not return to the starting state. It means that observable quantities cannot be represented by a spinor, since spinors have no definite transformation properties under the rotations. From a physical point of view, it makes the main difference between the spinor and tensor representations. It should be noted that in the tensor representations observables can be connected with field functions directly (electromagnetic field strengths may serve as an example)

A Hermitian conjugation of a spinor is carried out in the ordinary way, that is, it consists of the transposition and the complex conjugation:

$$\xi^\dagger = (\xi_1^*, \xi_2^*).$$

An observable can be represented as a bilinear combination of spinor quantities. Needless to say, it will possess single-valued transformation properties. Thus, for example, under the rotations the quantity $\xi^\dagger \chi$ behaves as the scalar:

$$\xi'^\dagger \chi' = \xi^\dagger U^{(1/2)\dagger}(\mathbf{n}, \theta)U^{(1/2)}(\mathbf{n}, \theta)\chi = \xi^\dagger \chi.$$

It can be shown in the same way, that quantity $\xi^\dagger \boldsymbol{\sigma} \chi$ is a vector and so on.

From (2.27) it follows that the representation with the weight $j = 1/2$ is two-valued, namely,

$$R(\mathbf{n}, \theta) \to \pm U^{(1/2)}(\mathbf{n}, \theta). \tag{2.28}$$

It brings us to the idea, that from the point of view of the $SO(3)$-group, the existence of two kinds of spinors is possible.

Let us slightly alter the above-mentioned facts in form. In two-dimensional complex space $S^{(2)}$ (spinor space) we introduce basis e_1 and e_2 by which any spinor ξ can be expanded:

$$\xi = \xi^1 e_1 + \xi^2 e_2. \tag{2.29}$$

At the rotation about the angle θ around an axis with the unit vector $\mathbf{n}$ spinors belonging to $S^{(2)}$ are transformed according to the formula:

$$\xi'^\alpha = U^{(1/2)\alpha}{}_\beta \xi^\beta, \tag{2.30}$$

where α and β are spinor indices taking the values 1 and 2. The operation of raising and lowering the indices is carried out by means of an antisymmetric "metric tensor":

$$\epsilon_{\alpha\beta} = \epsilon^{\alpha\beta} = \begin{pmatrix} 0 & 1 \\ -1 & 0 \end{pmatrix} \tag{2.31}$$

that is, $\epsilon = i\sigma_2$.

Now let us examine another example of a spinor space $S^{(\dot{2})}$ whose spinors we shall consider to be different from ones of $S^{(2)}$. For any basis (e_1, e_2) of the spinor space $S^{(2)}$ one may uniquely determine the dual basis connected with the initial one by the relation:

$$e^{\dot{\alpha}}e_\beta = \delta^{\dot{\alpha}}_\beta = \begin{cases} 0, & \dot{\alpha} \neq \beta \\ 1, & \dot{\alpha} = \beta, \end{cases} \tag{2.32}$$

In (2.32) we have supplied the spinor indices of the space $S^{(\dot{2})}$ with dots to emphasize the fact that the transformation law of the spinors on these indices differs from that of the spinors of the space $S^{(2)}$. An arbitrary spinor η belonging to $S^{(\dot{2})}$ is expanding in dual basis in the following way:

$$\eta = \eta_{\dot{1}}e^{\dot{1}} + \eta_{\dot{2}}e^{\dot{2}}. \tag{2.33}$$

At the three-dimensional rotations the transformation law of the spinors belonging to the space $S^{(\dot{2})}$ is given by the expression:

$$\eta^{\dot{\alpha}\prime} = \tilde{U}^{(1/2)\dot{\alpha}}_{\dot{\beta}}\eta^{\dot{\beta}}. \tag{2.34}$$

A matrix of the transformation $\tilde{U}^{(1/2)}$ meets the choice of the rotation generators M_k in the form $-i\sigma_k^*/2$ to ensure the commutative relations fulfillment (2.11) as well. Passing to the spinors with the low indices by means of the metric tensor $\epsilon_{\dot{\alpha}\dot{\beta}} = \epsilon^{\dot{\alpha}\dot{\beta}} = \epsilon^{\alpha\beta}$ in Eq. (2.34) and with allowance made for the relation:

$$\sigma_2\boldsymbol{\sigma}^*\sigma_2 = -\boldsymbol{\sigma}. \tag{2.35}$$

we arrive at:

$$\eta'_{\dot{\sigma}} = \epsilon_{\dot{\sigma}\dot{\alpha}}\tilde{U}^{(1/2)\dot{\alpha}}_{\dot{\beta}}\epsilon^{\dot{\beta}\dot{\tau}}\eta_{\dot{\tau}} = U^{(1/2)\dot{\alpha}\dagger}_{\dot{\sigma}}\eta_{\dot{\alpha}}.$$

A two-component quantity

$$\xi^\alpha = \begin{pmatrix} \xi^1 \\ \xi^2 \end{pmatrix},$$

which is transformed by the law (2.30) at the three-dimensional rotations we call a *contravariant spinor of the first rank*. Spinors whose transformation law has the form:

$$\eta'_\alpha = U^{(1/2)\beta\dagger}_\alpha\eta_\beta \tag{2.36}$$

we call *covariant spinors*. In Eq. (2.36) we consciously omitted dots above the indices, but further on we shall always imply that the transformation law on low (covariant) indices is determined by the formula (2.36). Now it is possible to obtain a bilinear form from covariant and contravariant spinors:

$$(\xi, \eta) = \xi^\alpha\epsilon_{\alpha\beta}\eta^\beta = \xi^1\eta_1 + \xi^2\eta_2, \tag{2.37}$$

to be invariant under the three-dimensional rotations. We can interpret it as the scalar product.

Now we proceed to a discussion about spinor properties at the space inversion. Let us introduce an arbitrary plane P with the unit normal vector $\mathbf{n}$. Decompose the vector $\mathbf{x}$ into two components to be parallel and perpendicular to the vector $\mathbf{n}$:

$$\mathbf{x} = \mathbf{x}^\| + \mathbf{x}^\perp = (\mathbf{x} \cdot \mathbf{n})\mathbf{n} - [\mathbf{n} \times [\mathbf{n} \times \mathbf{x}]]. \tag{2.38}$$

We introduce the matrix-vector $\mathbf{R}$ whose components are three inversion matrices of the coordinate axes:

$$R_{-1} = \begin{pmatrix} -1 & 0 & 0 \\ 0 & 1 & 0 \\ 0 & 0 & 1 \end{pmatrix}, \qquad R_{-2} = \begin{pmatrix} 1 & 0 & 0 \\ 0 & -1 & 0 \\ 0 & 0 & 1 \end{pmatrix}, \qquad R_{-3} = \begin{pmatrix} 1 & 0 & 0 \\ 0 & 1 & 0 \\ 0 & 0 & -1 \end{pmatrix}.$$

Transition to a vector $\mathbf{x}'$

$$\mathbf{x} \to \mathbf{x}' = -(\mathbf{x} \cdot \mathbf{n})\mathbf{n} - [\mathbf{n} \times [\mathbf{n} \times \mathbf{x}]] = \mathbf{x} - 2\mathbf{n}(\mathbf{x} \cdot \mathbf{n}), \tag{2.39}$$

which is in fact a specular reflection of the vector $\mathbf{x}$ in the plane P, is carried out by the inversion matrix: $\mathbf{R}_-^{\parallel} = (\mathbf{R}_- \cdot \mathbf{n})\mathbf{n}$. By direct calculations it is possible to check that $\mathbf{R}_-^{\parallel}$ satisfies the following relations:

$$\{R_{-i}^{\parallel}, A_k^{\perp}\} = 0, \qquad [R_{-i}^{\parallel}, A_k^{\parallel}] = 0, \tag{2.40}$$

where $i, k = 1, 2, 3$. In the expression (2.40) braces denotes the anticommutator, and the rotation generators are also expanded into $\mathbf{A}_{\parallel}$ and $\mathbf{A}_{\perp}$:

$$\mathbf{A} = \mathbf{A}_{\parallel} + \mathbf{A}_{\perp} = (\mathbf{A} \cdot \mathbf{n})\mathbf{n} - [\mathbf{n} \times [\mathbf{n} \times \mathbf{A}]]. \tag{2.41}$$

In any representation permutation relations of an operator, corresponding to $\mathbf{R}_-^{\parallel}$, and generators of this representation must have the same form as the relation (2.40), that is,

$$\{T_{-i}^{\parallel}, M_k^{\perp}\} = 0, \qquad [T_{-i}^{\parallel}, M_k^{\parallel}] = 0, \tag{2.42}$$

Consistency of the conditions (2.42) and allowing for $(\mathbf{T}^{\parallel})^2 = 1$ result in the conclusion, that under the reflection (2.39) the spinor ξ is transformed by the law:

$$\xi^{\alpha} \to \xi'^{\alpha} = (\mathbf{n} \cdot \boldsymbol{\sigma})_{\beta}^{\alpha} \xi^{\beta} = N_{\beta}^{\alpha} \xi^{\beta}, \tag{2.43}$$

where from two possible signs in front of the reflection matrix N the sign "plus" was chosen arbitrarily (two reflections might be considered as the rotation about the angle 2π). As the result of matching the transformation laws for the spinors ξ and η at the space rotations (2.30) and (2.36), we obtain the relation:

$$\eta_{\beta} \rightsquigarrow \epsilon_{\beta\alpha} \xi^{\alpha*} \tag{2.44}$$

where ξ^* is a column with components ξ^{1*} and ξ^{2*}, and a sign $\rightsquigarrow$ denotes "transformed as follows." Using (2.43) and (2.44) we discover, that at the reflection in the plane P the following expression is true for the covariant spinor η:

$$\eta_{\beta} \to \eta'_{\beta} \rightsquigarrow \epsilon_{\beta\alpha} N_{\tau}^{\alpha*} \xi^{\tau*} = -\epsilon_{\beta\alpha} N_{\tau}^{\alpha*} \epsilon^{\tau\kappa} \epsilon_{\kappa\gamma} \xi^{\gamma*},$$

With (2.35) and (2.44) taken into account, we finally have:

$$\eta'_{\beta} = -N_{\beta}^{\alpha} \eta_{\alpha} \tag{2.45}$$

So, the transformation laws for the η- and ξ-spinors are different at the reflections. It once again proves the fact, that these two types of spinors may not be reduced to each other. To put it differently, a linear transformation that would transform the spinors of the space $S^{(\dot{2})}$ into ones of the space $S^{(2)}$ is absent, that is, such a transformation is identically equal to zero. Actually, if such a transformation D existed, the following relations would be true:

$$\eta = D\xi, \qquad \eta_N = D\xi_N, \tag{2.46}$$

where $\eta_N = N\eta$ and $\xi_N = N\xi$, while the matrix D had to anticommute with N. Then orienting the plane P sequentially along the coordinate axes x_1, x_2, and x_3 we would have obtained anticommutation of the matrix D with all the matrices σ_i. However, the Pauli matrices and the unit matrix I constitute a system of linear independent matrices (it means, that any 2×2-matrix can be expanded in terms of them). Hence, a 2×2-matrix that would commute with all the Pauli matrices is absent, that is, we arrive at

$$\{D, N\} = 0 \qquad \text{only at} \qquad D = 0.$$

Spinors of the highest ranks are built by analogy with the tensors theory. The set 2^n of complex numbers $\alpha_1, \alpha_2, ...\alpha_n$ transformed by the law:

$$\left(\psi^{\alpha_1' \alpha_2'..\alpha_n'}\right)' = U_{\alpha_1' \alpha_1} U_{\alpha_2' \alpha_2}...U_{\alpha_n' \alpha_n} \psi^{\alpha_1 \alpha_2..\alpha_n}. \tag{2.47}$$

is called a *contravariant spinor* of the rank n. Similarly, a covariant spinor of the rank n is defined through its transformation properties:

$$\left(\psi_{\beta_1' \beta_2'..\beta_n'}\right)' = U^{\beta_1' \beta_1 \dagger} U^{\beta_2' \beta_2 \dagger}...U^{\beta_n' \beta_n \dagger} \psi_{\beta_1 \beta_2..\beta_n}. \tag{2.48}$$

The set of quantities transformed by the rule:

$$\left(\psi^{\alpha_1' \alpha_2'..\alpha_n'}_{\beta_1' \beta_2'..\beta_l'}\right)' = U_{\alpha_1' \alpha_1} U_{\alpha_2' \alpha_2}...U_{\alpha_n' \alpha_n} U^{\beta_1' \beta_1 \dagger} U^{\beta_2' \beta_2 \dagger}...U^{\beta_l' \beta_l \dagger} \psi^{\alpha_1 \alpha_2..\alpha_n}_{\beta_1 \beta_2..\beta_l}. \tag{2.49}$$

is named a mixed spinor of the rank $n + l$. By means of simplification (summation) over covariant and contravariant indices, one can obtain spinors of lowest ranks.

All representations with $j > 1/2$ are reducible. Let us consider, for example, a field function, which is a mixed spinor of the second rank ψ^α_β, that is, it is transformed by the representation $j = 3/2$. Using the above-formulated diagram rule for expansion into irreducible representations, we obtain:

$$2 \bigotimes \overline{2} = 3 \bigoplus 1, \tag{2.50}$$

where 2 ($\overline{2}$) denotes covariant (contravariant) components of the rotations group representation. Hereinafter the term "group representation" is used in two meanings: as an operator transforming field functions, and as field functions transformed by a given representation.

Since the quantum field theory is a relativistic theory, it operates with quantities, defined in the four-dimensional Minkowski space, hence, we have to attach the four-dimensional aspect to the developed classification. The three-dimensional classification, however, will not prove to be useless, since it can be successfully used to study properties of tensors and spinors in abstract spaces (for example, in the space of an isotopic or color "spins").

3

The four-dimensional Minkowski space

3.1 The homogeneous Lorenz group

Let us consider homogeneous Lorenz transformations, that is, linear transformations:

$$x'^{\mu} = \Lambda^{\mu}_{\nu} x^{\nu}, \tag{3.1}$$

which leave the scalar product of four-dimensional vectors $x^{\mu} y_{\mu}$ invariant. Since the squares of three variables enter $x^{\mu} y_{\mu}$ with the plus sign and one variable square enters it with the minus sign, the Lorentz group is denoted by $O(3.1)$. All the transformation coefficients Λ^{μ}_{ν} are real. From the condition:

$$r^{\mu} y_{\mu} = \text{inv},$$

it results:

$$\Lambda^{\nu}_{\mu} \Lambda^{\mu}_{\sigma} = \Lambda^{\nu\mu} \Lambda_{\sigma\mu} = \delta^{\nu}_{\sigma}. \tag{3.2}$$

Thus, the following condition:

$$\det \Lambda = \pm 1 \tag{3.3}$$

must be true. A subgroup with the determinant to be equal to $+1$ is denoted by $SU(3,1)$. Eq. (3.2) means, in its turn, that there exists the reverse Lorenz transformation. Since the product of two Lorentz transformations is again a Lorentz transformation, then the set of these transformations composes the group.

Having set $\nu = \sigma = 0$ in Eq. (3.2), we obtain:

$$\left(\Lambda^0_0\right)^2 = 1 + \sum_{i=1}^{3} \left(\Lambda^i_0\right)^2 \geq 1. \tag{3.4}$$

From (3.4) it follows:

$$\mid \Lambda^0_0 \mid \geq 1.$$

By this fact the Lorentz transformations, depending upon the signs of $\det \Lambda$ and Λ^0_0, are divided into four categories:

1) proper orthochronous transformations $L^{\uparrow}_{+}$ with $\det \Lambda = 1$, $\Lambda^0_0 > 0$);
2) nonproper orthochronous transformations ($L^{\uparrow}_{-}$ with $\det \Lambda = -1$, $\Lambda^0_0 > 0$);
3) proper nonorthochronous transformations ($L^{\downarrow}_{+}$ with $\det \Lambda = 1$, $\Lambda^0_0 < 0$);
4) nonproper nonorthochronous transformations ($L^{\downarrow}_{-}$ with $\det \Lambda = -1$, $\Lambda^0_0 < 0$).

Besides continuous transformations (the rotations in the four-dimensional space), the Lorentz transformations also contain two discrete operations, namely, the space and time inversions (P and T). They are defined as follows:

$$P\begin{pmatrix} x^0 \\ x^1 \\ x^2 \\ x^3 \end{pmatrix} = \begin{pmatrix} 1 & 0 & 0 & 0 \\ 0 & -1 & 0 & 0 \\ 0 & 0 & -1 & 0 \\ 0 & 0 & 0 & -1 \end{pmatrix} \begin{pmatrix} x^0 \\ x^1 \\ x^2 \\ x^3 \end{pmatrix} = \begin{pmatrix} x^0 \\ -x^1 \\ -x^2 \\ -x^3 \end{pmatrix}, \tag{3.5}$$

$$T\begin{pmatrix} x^0 \\ x^1 \\ x^2 \\ x^3 \end{pmatrix} = \begin{pmatrix} -1 & 0 & 0 & 0 \\ 0 & 1 & 0 & 0 \\ 0 & 0 & 1 & 0 \\ 0 & 0 & 0 & 1 \end{pmatrix} \begin{pmatrix} x^0 \\ x^1 \\ x^2 \\ x^3 \end{pmatrix} = \begin{pmatrix} -x^0 \\ x^1 \\ x^2 \\ x^3 \end{pmatrix}. \tag{3.6}$$

Hereinafter to write down 4×4-matrices we follow the agreement:

$$\Lambda \equiv \Lambda^\mu_\nu,$$

where the superscript numerates a column, and the subscript indicates a row. The transformations $L^\downarrow_\pm$ involve the time reflections, while $L^\uparrow_\pm$ do not. Analogously, the transformations $L^{\downarrow\uparrow}_-$ contain the space inversion, while $L^{\downarrow\uparrow}_+$ do not. The homogeneous Lorentz group can be symbolically represented as the sum:

$$L = L^\uparrow_+ + PTL^\uparrow_+ + PL^\uparrow_+ + TL^\uparrow_+ \tag{3.7}$$

Since any Lorentz transformation can be presented as the product $L^\uparrow_+$, P, T, PT, we can limit ourselves to studying the space-space rotations (the rotations in the planes $[x^1x^2]$, $[x^1x^3]$ and $[x^2x^3]$) and the purely Lorentz transformations (the transformations connecting two inertial reference frames that move with the relative speed $\mathbf{v}^*$) only. The totality of these transformations constitutes a group of the limited homogeneous Lorentz transformations $L^\uparrow_+$ (the proper orthochronous Lorentz group or, simply, the proper Lorentz group).

If the speed of relative movement of the inertial reference frames (IRFs) under consideration is directed along the common axis x^1, then we have the following relation between their coordinates:

$$\begin{pmatrix} x^0 \\ x^1 \\ x^2 \\ x^3 \end{pmatrix}' = \begin{pmatrix} (1-\beta^2)^{-1/2} & -\beta(1-\beta^2)^{-1/2} & 0 & 0 \\ -\beta(1-\beta^2)^{-1/2} & (1-\beta^2)^{-1/2} & 0 & 0 \\ 0 & 0 & 1 & 0 \\ 0 & 0 & 0 & 1 \end{pmatrix} \begin{pmatrix} x^0 \\ x^1 \\ x^2 \\ x3 \end{pmatrix}, \tag{3.8}$$

where $\beta = v/c$. Setting:

$$(1-\beta^2)^{-1/2} = \cosh u, \qquad \beta(1-\beta^2)^{-1/2} = \sinh u,$$

we can interpret the Lorenz boost along the axis x^1 as a rotation in the plane x^1x^0 by the angle $u = \operatorname{arctanh} \beta$. Actually,

$$\begin{pmatrix} x^0 \\ x^1 \\ x^2 \\ x^3 \end{pmatrix}' = \Lambda(10; u) \begin{pmatrix} x^0 \\ x^1 \\ x^2 \\ x3 \end{pmatrix} = \begin{pmatrix} \cosh u & -\sinh u & 0 & 0 \\ -\sinh u & \cosh u & 0 & 0 \\ 0 & 0 & 1 & 0 \\ 0 & 0 & 0 & 1 \end{pmatrix} \begin{pmatrix} x^0 \\ x^1 \\ x^2 \\ x^3 \end{pmatrix}. \tag{3.9}$$

It is not out of space to emphasize the conventional character of such a interpretation of the boost. Since $\tan u = i\beta$, the rotation proceeds by the imaginary angle. This approach,

---

*We shall also use for these transformations the term "boosts."

however, allows one to interpret the limited homogeneous Lorenz transformations as the group, which consist only of the rotations in the four-dimensional Minkowski space.

Let us determine the explicit form for generators of the Lorentz transformations. A generator for the rotation in the plane $[x^1 x^0]$ is defined by the formula:

$$J^{10} = \frac{d}{du} \Lambda(10; u) \Big|_{u=0}$$

and has the form

$$J^{10} = \begin{pmatrix} 0 & -1 & 0 & 0 \\ -1 & 0 & 0 & 0 \\ 0 & 0 & 0 & 0 \\ 0 & 0 & 0 & 0 \end{pmatrix}. \tag{3.10}$$

Analogously, the generators J^{20} and J^{30}, corresponding to the boosts along the coordinate axes x^2 and x^3, are given by the expressions:

$$J^{20} = \begin{pmatrix} 0 & 0 & -1 & 0 \\ 0 & 0 & 0 & 0 \\ -1 & 0 & 0 & 0 \\ 0 & 0 & 0 & 0 \end{pmatrix}, \qquad J^{30} = \begin{pmatrix} 0 & 0 & 0 & -1 \\ 0 & 0 & 0 & 0 \\ 0 & 0 & 0 & 0 \\ -1 & 0 & 0 & 0 \end{pmatrix}. \tag{3.11}$$

The generators of the space rotations in the four-dimensional notations look as follows:

$$J^{12} = \begin{pmatrix} 0 & 0 & 0 & 0 \\ 0 & 0 & 1 & 0 \\ 0 & -1 & 0 & 0 \\ 0 & 0 & 0 & 0 \end{pmatrix}, \qquad J^{23} = \begin{pmatrix} 0 & 0 & 0 & 0 \\ 0 & 0 & 0 & 0 \\ 0 & 0 & 0 & 1 \\ 0 & 0 & -1 & 0 \end{pmatrix}, \tag{3.12}$$

$$J^{31} = \begin{pmatrix} 0 & 0 & 0 & 0 \\ 0 & 0 & 0 & -1 \\ 0 & 0 & 0 & 0 \\ 0 & 1 & 0 & 0 \end{pmatrix}. \tag{3.13}$$

Since the rotation in the plane $[x^\mu x^\nu]$ by the angle φ is identical to that in the plane $[x^\nu x^\mu]$ by the angle $-\varphi$, then the equality: $J^{\mu\nu} = -J^{\nu\mu}$ takes place. An arbitrary infinitesimal transformation of the homogeneous Lorentz group can be written down in the following way:

$$\Lambda(\omega) = I + \frac{1}{2} \omega_{\mu\nu} J^{\mu\nu}, \tag{3.14}$$

where $\omega_{\mu\nu} = -\omega_{\nu\mu}$. By means of direct calculations one can be convinced, that the generators $J^{\mu\nu}$ of the Lorentz group obey the permutation relations:

$$[J^{\mu\nu}, J^{\rho\sigma}] = g^{\mu\rho} J^{\nu\sigma} + g^{\nu\sigma} J^{\mu\rho} - g^{\mu\sigma} J^{\nu\rho} - g^{\nu\rho} J^{\mu\sigma}. \tag{3.15}$$

3.2 Classification of irreducible representations of the homogeneous Lorentz group

Suppose, there exists two IRFs: K and K' with the coordinates x and x', respectively. Let one and the same wave field be described by a K-frame observer with the help of functions

$\psi_i(x)$ and a K'-frame observer with the help of functions $\psi'_l(x')$. Then values of $\psi_i(x)$ completely define those of $\psi'_l(x')$, since x and x' correspond to one and the same event in the space-time. Thus, $\psi'_l(x')$ must be expressed in terms of $\psi_i(x)$. Consequently, under transition from the first observer to the second one, transformation of the fields, described by both, occurs according to the law:

$$\psi^{n\prime}(x') = \mathcal{D}^{nk}(\Lambda)\psi_k(x). \tag{3.16}$$

It should be emphasized, that the matrix $\mathcal{D}$ depends on the Lorentz transformation matrix and the indices n and k may have tensor dimensionality. For the sake of simplicity, the law of the field transformation (3.16) can be written down in a more convenient form:

$$\psi^{n\prime}(x) = \mathcal{D}^{nk}(\Lambda)\psi_k(\Lambda^{-1}x), \tag{3.17}$$

The totality of matrices (or, what is the same, operators) $\mathcal{D}(\Lambda)$, which specify transformations of field components, constitute the Lorentz group representation. The matrix of any representation of the limited homogeneous Lorenz group $\mathcal{D}(\Lambda)$ can be set down as follows:

$$\mathcal{D}(\Lambda) = I + \frac{1}{2}\omega^{\mu\nu}M_{\mu\nu}, \tag{3.18}$$

where $M^{\mu\nu}$ compose a representation for the Lee-algebra generators, satisfying permutation relations, which are similar to the permutation relations for the Lorentz group generators (3.15):

$$[M_{\mu\nu}, M_{\rho\sigma}] = g_{\mu\rho}M_{\nu\sigma} + g_{\nu\sigma}M_{\mu\rho} - g_{\mu\sigma}M_{\nu\rho} - g_{\nu\rho}M_{\mu\sigma}. \tag{3.19}$$

Finding representations of the limited Lorentz group is reduced to determining all the irreducible matrices, suiting to the relations (3.19). The group under consideration has finite-dimensional and infinite-dimensional irreducible representations. From the point of view of physical applications we are interested in classification of quantities with the finite number of components transformed according to a finite-dimensional representation of the homogeneous Lorentz group. However, one should not go so far to lose an interest in infinite-dimensional representations in general. Thus, unitary infinite-dimensional irreducible representations of the Poincare group deal with the classification of elementary particles. A thoughtful reader can guess at once, that in the last case we are dealing with the presence in a complete set, classifying the representation basis, of such operators that possess a continuous spectrum, not limited from above (as, for example, the momentum operator). In other words, basis vectors of a representation have the finite number of components, while the number of basis vectors as well as eigenvalues of an operator with a continuous spectrum is infinite.

Now we consider finite-dimensional representations of the limited homogeneous Lorentz group, namely the situation, when transformation matrices contain the finite number of rows (or columns, since matrices are square). Let us define the operators

$$\mathbf{M} = (M_{32}, M_{13}, M_{21}), \qquad \mathbf{N} = (M_{01}, M_{02}, M_{03}). \tag{3.20}$$

They obey the following permutation relations:

$$[M_i, M_j] = \varepsilon_{ijk}M_k, \qquad [N_i, N_j] = -\varepsilon_{ijk}M_k,$$

$$[M_i, N_j] = \varepsilon_{ijk}N_k. \tag{3.21}$$

Next, we pass on to new quantities according to the formulae:

$$\mathbf{X} = \frac{1}{2}(\mathbf{M} + i\mathbf{N}), \qquad \mathbf{Y} = \frac{1}{2}(\mathbf{M} - i\mathbf{N}). \tag{3.22}$$

It is easy to show that they satisfy commutation relations to be already quite clear for physical interpretation:

$$[X_i, X_j] = i\varepsilon_{ijk}X_k, \qquad [Y_i, Y_j] = i\varepsilon_{ijk}Y_k, \qquad [X_i, Y_j] = 0. \qquad (3.23)$$

Thus, we have demonstrated that infinitesimal operators of the Lorentz group $M^{\mu\nu}$ could be reduced to two independent vector operators $\mathbf{X}$ and $\mathbf{Y}$ to obey the permutation relations for the moment of momentum projection operators. Therefore, we can obtain all irreducible finite-dimensional representations of the homogeneous Lorentz group when the matrices for the moments are taken as $\mathbf{X}$ and $\mathbf{Y}$. Since the angular momentum square commutes with its every projection

$$[\mathbf{X}^2, X_i] = 0, \qquad [\mathbf{Y}^2, Y_i] = 0, \qquad (3.24)$$

then the operators $\mathbf{X}^2$, $\mathbf{Y}^2$ are the Casimir operators. It means, as we already know, that they are multiple to the unit operator:

$$\mathbf{X}^2 = j(j+1)I_{2j+1}, \qquad \mathbf{Y}^2 = k(k+1)I_{2k+1}, \qquad (3.25)$$

where j and k are arbitrary positive integer or half-integer numbers, I_{2j+1} (I_{2k+1}) is the identity matrix, containing $2j + 1$ ($2k + 1$) rows and columns. So, every irreducible representation of the limited homogeneous Lorentz group is characterized by two integer or half-integer numbers j and k. This representation denoted by (j, k), has the dimensionality $(2j + 1)(2k + 1)$, since it is stretched over the totality $(2j + 1)(2k + 1)$ of the basis vectors $| j, m_j; k, m_k >$ satisfying the eigenvalue equations:

$$\mathbf{X}^2 | j, m_j; k, m_k >= j(j+1) | j, m_j; k, m_k >, \qquad (3.26)$$

$$X_3 | j, m_j; k, m_k >= m_j | j, m_j; k, m_k >, \qquad (3.27)$$

$$\mathbf{Y}^2 | j, m_j; k, m_k >= k(k+1) | j, m_j; k, m_k >, \qquad (3.28)$$

$$Y_3 | j, m_j; k, m_k >= m_k | j, m_j; k, m_k > . \qquad (3.29)$$

Let us find out for what representations complex-conjugated field functions are transformed. Taking into account that

$$\mathbf{M}^* = \mathbf{M}, \qquad \mathbf{N}^* = \mathbf{N},$$

and making use of (3.22), we conclude that at the complex conjugation the replacement:

$$\mathbf{X} \to \mathbf{Y}, \qquad \mathbf{Y} \to \mathbf{X} \qquad (3.30)$$

takes place. So, if $| j, m_j; k, m_k >$ is transformed under representation (j, k), then its complex-conjugated field function is transformed under the mirror representation (k, j).

Under the space inversion the speed, entering into a boost, reverses its sign $\mathbf{v} \to -\mathbf{v}$. Then:

$$\mathbf{N} \to -\mathbf{N}, \qquad \mathbf{M} \to \mathbf{M}.$$

As a result we come to recognize that under the space inversion the transformations (j, k) and (k, j) are interchanged by their places. A similar phenomenon takes place for the time inversion. Thus, for the theory to be invariant under the P- or T-transformations, the representation must either have the form (j, j) or must be the direct product of the representations (j, k) and (k, j).

Depending on whether the number $j + k$ is integer or half-integer, irreducible representations (j, k) of the limited homogeneous Lorentz group may fall in two categories. In the former case, the space rotation by the angle 2π corresponds to the identical transformation I (one-valued, or the tensor representations), and in the latter case, a field function reverses its sign (two-valued, or the spinor representations).

3.3 Tensor representations of the limited homogeneous Lorentz group

Let us give a definition of four-dimensional tensors. Quantity, which is not changed under the transformations of the limited Lorentz group (3.1), is called a *zero-rank tensor* or a *scalar*. The representation $(0, 0)$ corresponds to the scalar. The totality of quantities which transformation law at (3.1) has the form:

$$a'^\mu = \Lambda^\mu_\nu x^\nu, \qquad (a'_\mu = \Lambda^\nu_\mu x_\nu), \tag{3.31}$$

where

$$\Lambda^\mu_\nu = \frac{\partial x'^\mu}{\partial x^\nu}.$$

is called a *contravariant (covariant) first rank tensor* or a *vector*. The representation $(1/2, 1/2)$ corresponds to the vector.

Now we are quite ready to introduce a mixed tensor, which is contravariant with respect to indices $\mu_1\mu_2...\mu_n$ and covariant with respect to indices $\nu_1\nu_2....\nu_k$. It is a totality of quantities with the following transformation properties:

$$T'^{\mu_1\mu_2...\mu_n}_{\nu_1\nu_2....\nu_k} = \Lambda^{\mu_1}_{\sigma_1}\Lambda^{\mu_2}_{\sigma_2}...\Lambda^{\mu_n}_{\sigma_n}\Lambda^{\rho_1}_{\nu_1}\Lambda^{\rho_2}_{\nu_2}...\Lambda^{\rho_k}_{\nu_k}T^{\sigma_1\sigma_2...\sigma_n}_{\rho_1\rho_2...\rho_k}. \tag{3.32}$$

The representations of the limited homogeneous Lorentz group (j, k) are irreducible. The same representations, but composed by direct products like $(j_1, k_1) \bigotimes (j_2, k_2) \bigotimes ...$ are reducible. The recipe of the expansion of such products follows from the obvious generalization of the formula (2.23):

$$(j_1, k_1) \bigotimes (j_2, k_2) = \sum_\oplus (j, k), \tag{3.33}$$

where $j =| j_1 - j_2 |, | j_1 - j_2 + 1 |, ... | j_1 + j_2 |$, $k =| k_1 - k_2 |, | k_1 - k_2 + 1 |, ... | k_1 + k_2 |$, and every representations (j, k) occurs once. Let us apply (3.33) to analyze a representation of a contravariant tensor of the second rank $T^{\mu\nu}$. This tensor is a product of two four-vectors and, consequently, is transformed by the representation $(1/2, 1/2) \bigotimes (1/2, 1/2)$. Since:

$$(1/2, 1/2) \bigotimes (1/2, 1/2) = (1, 1) \bigoplus (1, 0) \bigoplus (0, 1) \bigoplus (0, 0),$$

we could state, that the representation for $T^{\mu\nu}$ is reducible and falls into three irreducible ones: a symmetric tensor with zero spur $((1, 1))$, an antisymmetric tensor $((1, 0) \bigoplus (0, 1))$, and a scalar to be equal to the spur of the initial tensor $((0, 0))$.

3.4 Spinor representations of the homogeneous Lorentz group

Let us now turn to the two-valued or spinor representations. In case of the first rank spinors, the permutation relations for the generators M_k and N_k are satisfied both by the choice:

$$M_k = i\sigma_k/2, \qquad N_k = \sigma_k/2,$$

and by the choice:

$$M_k = -i\sigma_k^*/2, \qquad N_k = \sigma_k^*/2.$$

Consequently, as in the three-dimensional case, we can speak about two variants of a spinor space or about two kinds of spinors, which we denote by ξ and η, as in the case of the $SO(3)$-group. The operation of raising and lowering all spinor indices is again carried out by means of an antisymmetric "metric tensor" $\epsilon_{\alpha\beta}$. A spinor of the first kind will be a spinor ξ^α, which under the space rotations is transformed like the contravariant spinor under the three-dimensional rotations, that is, according to the law:

$$\xi'^\alpha = \left[\exp\left(\frac{i(\mathbf{n}\cdot\boldsymbol{\sigma})}{2}\theta\right)\right]^\alpha_\beta \xi^\beta, \tag{3.34}$$

and under the hyperbolic rotations by the angle φ in the plane $(x_0, \mathbf{n})$ (Lorentz boosts), its transformation law has the form:

$$\xi'^\alpha = \left[\exp\left(\frac{(\mathbf{n}\cdot\boldsymbol{\sigma})}{2}\varphi\right)\right]^\alpha_\beta \xi^\beta. \tag{3.35}$$

Thus, a choice of the generators for this kind of the spinors is as follows:

$$Y = 0, \qquad X \neq 0. \tag{3.36}$$

It means, in its turn, that they are transformed under representation $(1/2, 0)$. Choosing the generators M_k and N_k in the form:

$$M_k = -\frac{i\sigma_k^*}{2}, \qquad N_k = \frac{\sigma_k^*}{2}, \tag{3.37}$$

we obtain $Y \neq 0$ and $X = 0$, that is, the representation $(0, 1/2)$. Under the $SO(3,1)$-rotations the transformation law for spinors, belonging to this representation, has the shape:

$$\eta'^\alpha = \left(\exp\left[-i\frac{(\mathbf{n}\cdot\boldsymbol{\sigma}^*)}{2}\theta + \frac{(\mathbf{n}\cdot\boldsymbol{\sigma}^*)}{2}\varphi\right]\right)^\alpha_\beta \eta^\beta, \tag{3.38}$$

From Eq. (3.38) emerges, that:

$$\eta'_\alpha = \left(\exp\left[i\frac{(\mathbf{n}\cdot\boldsymbol{\sigma})}{2}\theta - \frac{(\mathbf{n}\cdot\boldsymbol{\sigma})}{2}\varphi\right]\right)^\beta_\alpha \eta_\beta, \tag{3.39}$$

that is, the transformation laws for the spinors ξ^α and η_α follow from each other at the Hermitian conjugation. The spinors, whose transformation properties at the $SO(3,1)$-transformations are given by the formula (3.39) hereinafter we call the *spinors of the second kind*. It is important to stress, that we deal with nonequivalent representations of the limited homogeneous Lorentz group, that is, there exists no matrix S, for which

$$\mathcal{N} = S\mathcal{M}S^{-1}$$

($\mathcal{N}$ and $\mathcal{M}$ are the transformation matrices for the first- and second-kind spinors under the $SO(3,1)$ rotations) is true.

At transformations of the space inversion the representations $(j, 0)$ and $(0, j)$ are interchanged by the places, and, as a result

$$\xi^\alpha \leftrightarrow \eta_\alpha.$$

The inversion inclusion means the transition from the proper Lorentz group $SO(3,1)$ to the Lorentz group $0(3,1)$. Because of this the simplest irreducible spinor representation of

$0(3, 1)$ is four-dimensional and composed by bispinor $\xi^\alpha \otimes \eta_\alpha$, usually written down in the form of a column matrix:

$$\Psi = \begin{pmatrix} \xi^\alpha \\ \eta_\alpha \end{pmatrix}.$$

Thus, under the homogeneous Lorentz transformations the four-dimensional spinor Ψ is transformed as follows:

$$\begin{pmatrix} \xi \\ \eta \end{pmatrix} \rightarrow \begin{pmatrix} \exp\left[i\dfrac{(\mathbf{n}\sigma)}{2}\theta + \dfrac{(\mathbf{n}\sigma)}{2}\varphi\right] & 0 \\ 0 & \exp\left[-i\dfrac{(\mathbf{n}\sigma)}{2}\theta + \dfrac{(\mathbf{n}\sigma)}{2}\varphi\right] \end{pmatrix} \begin{pmatrix} \xi \\ \eta \end{pmatrix}. \tag{3.40}$$

Since matrices of transformation of the Lorentz boosts are not unitary, then finite dimensional representations of the homogeneous Lorentz group are also nonunitary. It is a very sad circumstance, since only for unitary representations of symmetry groups a probability of a transition between two states does not depend on a reference frame, chosen for measurements. Thus, we are facing the conflict with the relativity principle. Unlike the three-dimensional rotations group, the proper Lorentz group is noncompact, which results in nonunitarity of its representations. For the rotations group θ is varying from 0 up to 2π, and these extreme points are identified in such a way, that the line is closed in the circle. Therefore, the group space of the $SO(3)$-group is finite. The Lorentz group noncompactness is caused by properties of the Lorentz boosts. Since the speeds, which are the Lorentz boost parameters, take values in the open interval from $v = 0$ up to $v = c$, then the group space of the $SO(3, 1)$-group is infinite and, as a consequence, this group is noncompact. Notice, all of this is true only for finite-dimensional representations of the homogeneous Lorentz group, while infinite-dimensional representations are unitary. There is something, however, that forces us to search for other groups of space-time symmetry under describing elementary particles. The Casimir operators and group generators that commute with each other, must fix a relativistic particle state. At the first glance on the homogeneous Lorentz group, we are convinced that there are no such quantities among its invariants and generators, which could define such important characteristics of elementary particles as mass, spin, and momentum.

A fundamental group in elementary particle physics is a nonhomogeneous Lorentz group (Poincare group) rather than the limited homogeneous Lorentz group. The Poincare group consists of the Lorentz boosts, the three-dimensional rotations, and space-time translations. We shall study it in the next subsection and now it is not out of place to dwell on terminology we are going to use at a later time. So, as we have learned, there are two kinds of tensor quantities in physics, namely, three- and four-dimensional ones. Three-dimensional tensors are defined in the three-dimensional Euclidian space $\mathbf{R^3}$ with respect to the rotations group $SO(3)$, while four-dimensional tensors are defined in the four-dimensional Minkowski space $\mathbf{M^4}$ with respect to the proper Lorentz group $SO(3, 1)$. In the same way, there are two different spinors, related to the groups $SO(3)$ and $SO(3, 1)$, respectively. If for $\mathbf{R^3}$ one uses the ordinary coordinate space, then tensor and spinor quantities, given in it, describe non-relativistic physics. One would think that it is more pertinent to call them "non-relativistic tensors and spinors" and use the terms "relativistic tensors and spinors" for the corresponding quantities, given in space-time continuum of Minkowski. However, such a terminology will result in misunderstandings if we work not in coordinate, but in some inner, abstract space. For the same reason we shall avoid the terms "relativistic" and "non-relativistic" in tensor and spinor algebra. Instead, we shall use the terms "tensors" and "spinors" in $\mathbf{R^3}$ or in $\mathbf{M^4}$.

3.5 Poincare group and its representations

It turns out that the nonhomogeneous Lorentz group, the Poincare group, to perform transformations between any IRFs, satisfies the relativity principle. Notice, that in quantum theory the IRF is connected with the totality of macroscopic devices (laboratory), necessary for the complete description of elementary particles system under study. The devices participate in the quantum mechanic description twice: at first, by means of some device a quantum object state is prepared and then measurements of objects are performed with another device. The Poincare group transformation may be assigned a double meaning. On the one hand, it can describe a change of space-time location of a physical system during two measurements in the same IRF, that is: difference in location of two identical physical systems with respect to the given IRF (active viewpoint). On the other hand, this transformation can characterize the difference between two IRFs, in which one and the same system is being studied (passive viewpoint). We shall stick to the active interpretation of the Poincare group transformation.

The Poincare transformations $\mathcal{P} = \{a, \Lambda\}$ leave the square of a distance between two four-dimensional vectors $(x - y)_\mu (x - y)^\mu$ invariant. $\mathcal{P}$ are defined by:

$$x'_\mu = (\mathcal{P}x)_\mu = \Lambda_\mu^\nu x_\nu + a_\mu, \qquad (3.41)$$

that is, as the product of operations of the shift by the real vector a_μ and of the homogeneous Lorentz transformation Λ. It is important that the shift takes place after the homogeneous Lorentz transformation. Product of two Poincare transformations $\mathcal{P}_1$ and $\mathcal{P}_2$ is equal to:

$$\{a_1, \Lambda_1\}\{a_2, \Lambda_2\} = \{a_1 + \Lambda_1 a_2, \Lambda_1 \Lambda_2\}. \qquad (3.42)$$

Under active interpretation of the Poincare transformation with the element $g = (a, \Lambda)$, every initial physical state $\Psi^{(in)}$ is compared with a transformed state $\Psi_g^{(in)}$, which differs from the initial state $\Psi^{(in)}$ by arrangement of preparing devices. At a creation of the state $\Psi_g^{(in)}$, a facility with preparing devices has been either shifted, turned, or is evenly moving with regard to the previous position (according to geometrical sense of g). In the same way, every state $\Psi^{(f)}$, registered by measuring devices, can be compared with a transformed state $\Psi_g^{(f)}$. To find the probability of a transition $\Psi_g^{(in)} \to \Psi_g^{(f)}$, which we denote by $W_{i_g \to f_g}$, we must repeat the experiment $\Psi^{(in)} \to \Psi^{(f)}$ (with the same or an identical physical system). In so doing, all the preparing and measuring devices must be transformed with regard to the initial state according to the geometrical sense of g. The existence of the space-time symmetry exhibits in the equality:

$$W_{i \to f} = W_{i_g \to f_g},$$

that is, in an invariance of the theory with respect to the Poincare group transformations. Analogously, as in the case of the homogeneous Lorentz transformation, the Poincare group can be divided in four classes differing according to signs of the quantities $\det \Lambda$ and Λ_0^0:

1) $\mathcal{P}_+^\uparrow$ ($\det \Lambda = 1$, $\Lambda_0^0 > 0$);
2) $\mathcal{P}_-^\uparrow$ ($\det \Lambda = -1$, $\Lambda_0^0 > 0$);
3) $\mathcal{P}_+^\downarrow$ ($\det \Lambda = 1$, $\Lambda_0^0 < 0$);
4) $\mathcal{P}_-^\downarrow$ ($\det \Lambda = -1$, $\Lambda_0^0 < 0$).

Thus, we can write down a symbolic equality:

$$\mathcal{P} = \mathcal{P}_+^\uparrow + PT\mathcal{P}_+^\uparrow + P\mathcal{P}_+^\uparrow + T\mathcal{P}_+^\uparrow. \tag{3.43}$$

Classification of elementary particle states deals, first of all, with studies of the proper orthochronous group $\mathcal{P}_+^\uparrow$. This group contains ten generators: three generators $\mathbf{M}$ correspond to the space-space rotations, three generators $\mathbf{N}$ denote boosts, and four generators P_μ correspond to the space-time translations. We have already obtained expressions for $\mathbf{M}$ and $\mathbf{N}$ in a matrix form. It turns out that to perform the same operation for translation operators we are forced to introduce a fifths fictitious coordinate without any physical meaning. It remains unchanged under the Poincare transformations. Then transformations (3.41) can be presented as follows:

$$\begin{pmatrix} x^0 \\ x^1 \\ x^2 \\ x^3 \\ 1 \end{pmatrix}' = \begin{pmatrix} \Lambda_0^0 & \Lambda_1^0 & \Lambda_2^0 & \Lambda_3^0 & a^0 \\ \Lambda_0^1 & \Lambda_1^1 & \Lambda_2^1 & \Lambda_3^1 & a^1 \\ \Lambda_0^2 & \Lambda_1^2 & \Lambda_2^2 & \Lambda_3^2 & a^2 \\ \Lambda_0^3 & \Lambda_1^3 & \Lambda_2^3 & \Lambda_3^3 & a^3 \\ 0 & 0 & 0 & 0 & 1 \end{pmatrix} \begin{pmatrix} x^0 \\ x^1 \\ x^2 \\ x^3 \\ 1 \end{pmatrix}. \tag{3.44}$$

In order to make the generators be Hermitian quantities we introduce the factor $-i$ into Eq. (2.4). Then, for shift generators we obtain 5×5-matrices with the elements:

$$(P_\mu)_\beta^\lambda = -i\delta_\mu^\lambda \delta_\beta^5. \tag{3.45}$$

By means of direct calculation one may be convinced that the following permutation relations:

$$[M_{\mu\nu}, M_{\rho\sigma}] = -i(g_{\mu\rho}M_{\nu\sigma} + g_{\nu\sigma}M_{\mu\rho} - g_{\mu\sigma}M_{\nu\rho} - g_{\nu\rho}M_{\mu\sigma}). \tag{3.46}$$

$$[P_\mu, P_\nu] = 0, \tag{3.47}$$

$$[P_\sigma, M_{\mu\nu}] = i(g_{\sigma\mu}P_\nu - g_{\sigma\nu}P_\mu) \tag{3.48}$$

are valid. Eqs. (3.47)–(3.48) bring us to the idea that a shift generator must be identified with a system momentum operator.

However, introduction of the fifths coordinate can cause vague anxiety in our readers: "Maybe we are dealing with hidden dimensions of the space-time?" For this reason we shall obtain Eqs. (3.46)–(3.48) in a less radical way. Recall about the existence of two versions of quantum mechanics, namely, the Schrödinger wave mechanics and the Heisenberg matrix mechanics. In the former case the differential form of operators is used while in the latter case operators are represented in the matrix form. Let us also turn to the differential form of writing the Poincare group generators. A generator, corresponding to parameter $\alpha^{(k)}$ ($k = 1, 2 \ldots, n$, where n is a number of parameters in a group), is defined by its action on the function $\psi(x)$:

$$L_{\alpha^{(k)}}\psi(x) = -i \lim_{\alpha^{(k)} \to 0} \left[\frac{\psi(x') - \psi(x)}{\alpha^{(k)}} \right],$$

(to introduce the factor i ensures the Hermitian character of generators). For an infinitesimal Poincare transformation we have:

$$x'_\mu = \left(\delta_\mu^\nu + \sum_{i=1}^{6} \epsilon^{(i)} \lambda_\mu^{(i)\nu} \right) x_\nu + \delta a_\mu, \tag{3.49}$$

where $\lambda_{\nu\mu}^{(k)} = -\lambda_{\mu\nu}^{(k)}$, and $\epsilon^{(i)}$ and δa_μ are infinitesimal parameters of the four-dimensional rotations and the four-dimensional translations, respectively. By means of (3.49) it is possible to obtain a working formula to define the generator $L_{\alpha^{(k)}}$ (by $\alpha^{(k)}$ we mean all ten parameters of the transformation (3.49)):

$$L_{\alpha^{(k)}} = -i\left[\left(\frac{\partial x'_1}{\partial\alpha^{(k)}}\right)_{\alpha^{(k)}=0}\frac{\partial}{\partial x_1} + \left(\frac{\partial x'_2}{\partial\alpha^{(k)}}\right)_{\alpha^{(k)}=0}\frac{\partial}{\partial x_2} + \right.$$

$$\left. + \left(\frac{\partial x'_3}{\partial\alpha^{(k)}}\right)_{\alpha^{(k)}=0}\frac{\partial}{\partial x_3} - \left(\frac{\partial x'_0}{\partial\alpha^{(k)}}\right)_{\alpha^{(k)}=0}\frac{\partial}{\partial x_0}\right] = i\left(\frac{\partial x'^\mu}{\partial\alpha^{(k)}}\right)_{\alpha^{(k)}=0}\frac{\partial}{\partial x^\mu}. \tag{3.50}$$

By using the obtained formula, we find the expression for generators of the Poincare group

$$M_k = -i\varepsilon_{klm}x^l\frac{\partial}{\partial x^m}, \qquad N_k = i(t\frac{\partial}{\partial x^k} + x^k\frac{\partial}{\partial t}), \qquad P_\mu = i\frac{\partial}{\partial x^\mu}. \tag{3.51}$$

From the relations derived it follows that $\hbar M_k$ is the orbital angular momentum projection operator whereas the translation operator, multiplied by $\hbar$, represents the four-dimensional energy-momentum vector p_μ. One can be easily convinced of the fact, that the operators $M_{\mu\nu}$ and P_μ satisfy the commutation relations (3.46)–(3.48).

Relations (3.47) and (3.48) demonstrate, that the generators **M** and **P** commute with the Hamiltonian P_0, what allows us to classify physical states according to eigenvalues of these generators. The eigenvalues of the Lorentz boosts generators **N** cannot be used for this purpose because they do not commute with P_0.

In case of the finite-dimensional Poincare transformations, the law of fields transformation has the form:

$$\psi'_l(x) = \mathcal{B}_l^i(\Lambda)\psi_i[(\Lambda^{-1}(x-a)],$$

where the set of matrices $\mathcal{B}(\Lambda)$ forms representations of the Poincare group. Once again to classify all irreducible unitary representations of the group one must find all the representations of the permutation relations (3.46)–(3.48) in the form of Hermitian operators. To achieve this goal we fix Casimir operators. Let us define the so-called Pauli-Lubanski pseudovector :

$$W_\sigma = \frac{1}{2}\varepsilon_{\sigma\mu\nu\lambda}\mathcal{M}^{\mu\nu}p^\lambda. \tag{3.52}$$

In Eq. (3.52) $\mathcal{M}^{\mu\nu}$ is a generator of representation of the proper Lorentz group, and consequently, it is already a sum of both orbital and spin moments of a wave field. In the three-dimensional vector notations W_σ has the form:

$$W^0 = \mathbf{p}\cdot\mathcal{M}, \qquad \mathbf{W} = p_0\mathcal{M} - [\mathbf{p}\times\mathcal{N}]. \tag{3.53}$$

It is obvious that the four-dimensional pseudovector W_σ is orthogonal to the four-dimensional vector p^σ

$$W_\sigma p^\sigma = 0. \tag{3.54}$$

The permutation relations for the Pauli-Lubanski pseudovector have the shape:

$$[\mathcal{M}_{\mu\nu}, W_\sigma] = i(g_{\nu\sigma}W_\mu - g_{\mu\sigma}W_\nu), \qquad [W_\mu, p_\nu] = 0,$$

$$[W_\mu, W_\nu] = i\varepsilon_{\mu\nu\sigma\lambda}W^\sigma p^\lambda. \tag{3.55}$$

Now it is possible to check, that the scalar operators:

$$p^\mu p_\mu = m^2, \qquad w = -W^\mu W_\mu, \tag{3.56}$$

commute with all the generators $\mathcal{M}_{\mu\nu}$ and p_μ, that is, they are the Casimir operators. Thus, their eigenvalues can be used to classify irreducible representations of the Poincare group.

At $m^2 \geq 0$ one more invariant quantity, composed of momenta, exists. It is an energy sign operator:

$$\epsilon = \frac{p_0}{|p_0|} \qquad (3.57)$$

whose eigenvalues are ± 1. Conservation of ϵ is caused by the fact that the proper orthochronous transformations of the Poincare group do not change a sign of a time component of a time-like vector. Thus, a complete set of states, composing basis for the Poincare group representations, breaks up on six different classes depending on the values of m^2 and sign of ϵ:

$$I. \qquad m^2 > 0, \qquad \epsilon > 0,$$
$$II. \qquad m^2 > 0, \qquad \epsilon < 0,$$
$$III. \qquad p^\mu \neq 0, \qquad m^2 = 0, \qquad \epsilon > 0,$$
$$IV. \qquad p^\mu \neq 0, \qquad m^2 = 0, \qquad \epsilon < 0,$$
$$V. \qquad p^\mu = 0,$$
$$VI. \qquad m^2 < 0.$$

The first and the third classes correspond to a massive and massless physical particles respectively, the fifth class—to vacuum, the sixth class—to virtual particles (which can possess space-like momentum). The remaining classes are probably not physical ones. In case of a massive particle we can perform a transformation to the reference frame where a particle is at rest in which $\mathbf{p} = 0$, $p_0 = m$, and

$$W^\mu = m(0, \mathcal{M}_{23}, \mathcal{M}_{31}, \mathcal{M}_{12}) = m(0, S_1, S_2, S_3).$$

It is easy to check, that the operators S_k $(k = 1, 2, 3)$ satisfy the ordinary commutation relations for the moment of momentum

$$[S_k, S_l] = i\varepsilon_{kln} S_n. \qquad (3.58)$$

The quantity W^2/m^2 is equal to the moment of momentum square at rest, that is, the spin square. Eigenvalues of $\mathbf{S}^2$ are $J(J+1)$ with $J = 0, 1/2, 1, 3/2, 2, ...$, and S_k are generators of the irreducible $(2J + 1)$-dimensional representation of the three-dimensional rotations group. Since W^2 is an invariant operator, then in any reference frame for the state $\mid \mathbf{p}, m >$, transformed by irreducible representation $\mathcal{P}_+^\uparrow$, the equality:

$$W^2 \mid \mathbf{p}, m >= m^2 J(J+1) \mid \mathbf{p}, m > .$$

will be true.

To establish covariant connection between W_μ and the spin operator S_k, we introduce space-like normals $n_\mu^{(k)}$, with properties: $n_\mu^{(k)} n^{(i)\mu} = \delta_{ki}$, $n_\mu^{(k)} p^\mu = 0$ $(k, i = 1, 2, 3)$. Together with the speed $v_\mu = p_\mu/m = n_\mu^{(0)}$ they make a complete system of the four-dimensional normals:

$$n_\mu^{(\lambda)} g^{\mu\nu} n_\nu^{(\sigma)} = g^{\lambda\sigma}. \qquad (3.59)$$

Since $W^\mu n_\mu^{(0)} = 0$, then there are only three independent space-like components of W^μ, that is

$$W_\mu = mS_i n_\mu^{(i)}. \qquad (3.60)$$

Multiplying (3.60) by $g^{\mu\nu} n_\nu^{(k)}$, we obtain an expression for the spin operator in the covariant form:

$$S_k = \frac{1}{m} W^\mu n_\mu^{(k)}. \qquad (3.61)$$

Besides the main invariants W^2, m^2, and ϵ (at $m^2 > 0$) there is one more operator in the Poincare group:

$$z = \exp\left(2\pi i M_3\right) = (-1)^{2M_3}, \tag{3.62}$$

which commutes with all the generators of the group $\mathcal{P}_+^\uparrow$. Eigenvalues z are equal to $(-1)^J$, that is, for particles with an integer spin value (we call them *bosons*) $z = +1$, and for the particles with a half-integer spin value (they are called *fermions*) $z = -1$. Basic vectors of representations with a different z value are mutually orthogonal, so that superposition of states with integer and half-integer spin value is impossible.

Thus, the irreducible unitary Poincare group representations are singled out by the values of mass m, spin J, and the energy sign (at $m^2 \geq 0$)). Basic vectors of an irreducible representation are also classified according to operator eigenvalues, entering one of the complete sets, which are composed in turn, out of the generators $M_{\mu\nu}$ and P_μ. For example, in the canonical basis the complete set of mutually commuting operators contains operators:

$$m^2, \qquad S^2, \qquad p_i, \qquad S_3 = \frac{1}{m} W^\mu n_\mu^{(3)}. \tag{3.63}$$

The vectors $|+, m, J; \mathbf{p}, J_3>$ constitute a basis of representation ("+" corresponds to the energy sign). At fixed value of $\mathbf{p}$ for every pair (m, J) the representation basis consists of $2J + 1$ vectors. The fact that irreducible representation is infinite-dimensional, means that any elementary system can take an infinitely large number of linearly independent states. For each pair (m, J) and a given energy sign there is one, and only one irreducible representation of the Poincare group, if we do not take into consideration unitary-equivalent representations. At integer J a representation is one-valued, while at half-integer J it is two-valued.

Any elementary particle can be defined as an elementary quantum mechanical system, all the possible states of which are governed by functions, which realize in their totality some invariant space of the Poincare group representations, in other words, by functions, which are transformed under an irreducible representation of this group. A particle is elementary in case it has no inner coordinate, which can divide its states in two or more groups in the invariant manner. So, to describe elementary particles one must choose only relativistic equations, based on the usage of the field functions, which realize irreducible representations of the Poincare group. Important to note, that it is a very strong statement, because it forces us to reject quite a wide class of equations. By way of example we cite Stückelberg equations for the neutral [5] and charged [6] particles:

$$\left.\begin{array}{l} \partial^\mu \psi_{\mu\nu} + m^2 \psi_\nu + \partial_\nu \psi = 0, \\ \psi_{\mu\nu} = \partial_\mu \psi_\nu - \partial_\nu \psi_\mu, \\ \psi = \partial^\mu \psi_\mu. \end{array}\right\} \tag{3.64}$$

These equations describe a particle which can dwell either in the state with spin 1 or with spin 0.

However, we still have some ambiguity in choosing motion equations, since the homogeneous Lorentz group is in the mean time the subgroup of the Poincare group. Thus, for example, a particle with spin 1 and mass m can be described by two different representations of the homogeneous Lorentz group, namely, by:

$$(0,1) \bigoplus (1,0) \bigoplus (1/2, 1/2) \qquad \text{or} \qquad (0,1) \bigoplus (1,0). \tag{3.65}$$

The well-known Proca equations [7]:

$$\left.\begin{array}{l} \partial^\mu \psi_{\mu\nu} + m^2 \psi_\nu = 0, \\ \psi_{\mu\nu} = \partial_\mu \psi_\nu - \partial_\nu \psi_\mu. \end{array}\right\} \tag{3.66}$$

correspond to the first choice. In the second case we arrive at equations [8]:

$$\left.\begin{array}{l} \partial^\mu \psi_{\mu\nu\lambda} + m^2 \psi_{\nu\lambda} + \partial_\nu \psi_\lambda - \partial_\lambda \psi_\nu = 0, \\ \psi_{\mu\nu\lambda} = \partial_\mu \psi_{\nu\lambda} - \partial_\nu \psi_{\mu\lambda} + \partial_\lambda \psi_{\mu\nu}, \\ \psi_\nu = \partial^\mu \psi_{\mu\nu}. \end{array}\right\} \tag{3.67}$$

As we can guess from these examples, under formulation of motion equations the homogeneous Lorentz group still plays an important role. The transformation law for field functions under Poincare group transformations has a nonlocal character (which is caused by the translation generator, the energy-momentum operator), while for the homogeneous Lorentz group this law is local. However, it is clear that we should like to deal with quantities transformed by the local law. It turns out that elimination of nonlocality in the transformation law could be fulfilled by means of separating the transformation matrix $\mathcal{B}_l^i(\Lambda)$ into three matrices, which already belong to representation of the homogeneous Lorentz group [9].

Below we give the traditional rules, which allow to further narrow the class of permissible equations of motion for quantum fields.

1. Under the Lorentz transformations the field functions $\Psi_k(x)$ must be transformed by finite-dimensional (and consequently nonunitary) representation of the homogeneous Lorentz group. In this case representation might be reducible and the number of components could exceed $2J + 1$.

2. The field functions must admit build-up of an invariant bilinear Hermitian form:

$$\Psi_k^\dagger(x)\eta^{kk'}\Psi_{k'}(x) = \Psi_k^\dagger(\Lambda x)\eta^{kk'}\Psi_{k'}(\Lambda x), \tag{3.68}$$

where elements of the Hermitian operator $\eta^{kk'}$ are constants.

3. The field functions $\Psi_k(x)$ must satisfy not only motion equations but must meet some additional demands as well (they follow from motion equations), which ensure reduction of independent components $\Psi_k(x)$ up to $2J + 1$.

The vector field described by Eqs. (3.68) contains six independent components. The doubling of field function components is caused by the fact that a particle can be in states with either positive or negative space parity. Then, the combination of states with the opposite space parity will correspond to a particle without definite space parity. For the first time the possibility, that such particles exist in the Nature, was mentioned in Ref. [10]. However, due to a very radical rule 3, such theories, similar to the one, described by Eqs. (3.68) further on will be out of our consideration.

Now let us switch over to the analysis of representations belonging to the IIIth class. In this case there are two different types of representations: 1) $w = 0$; 2) $w = a^2$, where a is a real number. To characterize the first type of representations, we must introduce a new quantum number, because $w = m^2 = 0$. Since the isotropic vectors W_μ and p_μ are orthogonal to each other, the relation

$$W^\mu = -\lambda p^\mu. \tag{3.69}$$

must be fulfilled. Multiplying the both sides of Eq. (3.69) by the vector $n_\mu^{(0)}$ being normal to arbitrary space-like hypersurface we arrive at the relation:

$$\lambda = -\frac{W^\mu n_\mu^{(0)}}{p^\mu n_\mu^{(0)}} = \frac{1}{2} \frac{\varepsilon^{\mu\nu\lambda\sigma} \mathcal{M}_{\nu\lambda} p_\sigma n_\mu^{(0)}}{p^\mu n_\mu^{(0)}} \tag{3.70}$$

If we choose the normal vector as $n_\mu^{(0)} = (1, 0, 0, 0)$, what corresponds to the transition to the hyperplane $t =$const, then Eq. (3.70) produces:

$$\lambda = \frac{\boldsymbol{\mathcal{M}} \cdot \mathbf{p}}{p_0} = \frac{\mathbf{S} \cdot \mathbf{p}}{|\mathbf{p}|}. \tag{3.71}$$

It follows from (3.71), that λ is a particle spin projection onto its motion direction, or in other words, it is a helicity. Physical states of a particle with zero mass are classified according to its helicity. One independent state corresponds to each value of helicity.

Under the space inversion P the helicity sign is reversed, because of

$$\mathbf{S} \xrightarrow{P} \mathbf{S}, \qquad \mathbf{p} \xrightarrow{P} -\mathbf{p}.$$

Then, in the case of the P-invariance of the theory, for any nonzero spin value a particle with zero mass has two polarization states (at $J = 0$, λ is also equal to zero). When the P-invariance does not take place, the number of polarization states can be either equal to two, or reduced to one. It is worthwhile to notice that the statements "the spin is parallel or antiparallel to motion direction" are relativistically covariant for zero-mass particles, since expression (3.71), determining λ, does not change its form at any Lorentz transformations. Spin moment of a particle with nonzero mass can be also either parallel or antiparallel to its velocity. These notions, however, are not Lorentz-invariant. If a velocity and a spin are parallel in one reference frame it does not necessarily mean that they are parallel to each other in an other reference frame. In particular, it is obvious, that in the very reference frame, in which a particle is at rest, the spin moment is not parallel to the velocity, since the latter is equal to zero. Any particle with a nonzero mass can be described in the reference frame, in which it rests. For such a particle a statement, that its spin is parallel to its velocity is not Lorentz-invariant. However, for a zero-mass particle moving with the light velocity there is no reference frame where it would be at rest. It explains the difference between zero-mass particles with only two polarization directions for $J \neq 0$, and particles with the nonzero mass, which have $2J + 1$ states of polarization for the spin J. For $m = 0$ and $w = a^2$ the Poincare group representations are infinite-dimensional on the spin variable. That is why, if corresponding particles existed, they would possess continuous spin. Even if such particles exist, nature is successfully hiding them from us.

In general, there are other invariant operators too (they are related to the invariance of the system with respect to the transformations in inner spaces, for example, operators of electric charge, operators of baryon and lepton charge, and so on). Such operators commute with the Casimir operators of the Poincare group and their eigenvalues characterize a physical system state as well. Therefore, the more general writing for the basis vectors of space of an irreducible representation is $| +, m, J; \mathbf{p}, J_3, \alpha >$, where α denotes eigenvalues of the Casimir operators of the groups, related to dynamic symmetries.

4

Lagrangian formulation of field theory

4.1 Principle of least action. Lagrange-Euler equations

Components of a field function are functions of coordinates and time and satisfy some definite differential equations, which are called *field equations*. Field equations can be given a very general variational or Lagrangian form, if one accepts, that a field as the dynamic system must be characterized by a definite Lagrangian function. For a mechanical system this function is written down as the sum over all the material points of the system. For a continuous system like a wave field, Lagrangian function is expressed by the space integral from Lagrangian function density:

$$L(t) = \int \mathcal{L} d^3 x. \tag{4.1}$$

However, the variational principle deals not with the Lagrangian function but with an action S obtained from it by means of integration over $x^0 = ct$

$$S = c \int L(t) dt. \tag{4.2}$$

Therefore, noncovariant expression (4.1) in Lagrangian formalism is in fact intermediate and it is sufficient to consider the Lagrangian function density $\mathcal{L}$, which depends on the four space-time coordinates x in an implicit way. Hereinafter, we shall simply call $\mathcal{L}$ by the *Lagrangian*. Analogously, the Hamiltonian function density $\mathcal{H}$ will be named the *Hamiltonian*. The Lagrangian depends on components of a field wavefunction ψ_k and on their first derivatives with respect to the space-time coordinates $\partial \psi_k / \partial x^\nu$. In so doing we assign the meaning of generalized field coordinates to ψ_k, and the meaning of generalized field velocities to $\partial \psi_k / \partial x^\nu$.

The demand of lacking derivatives with respect to field functions above the first order in the Lagrangian is caused by the following reason. Dynamics of all known quantum fields is described either by equations of the first or the second order. On the one hand, these equations follow from the principle of a least action:

$$\delta S = 0, \tag{4.3}$$

and under action variation the order of derivatives, entering $\mathcal{L}$, is raised by one. The Lagrangian must be a relativistically covariant quantity, namely, a relativistic scalar $\mathcal{L}'(x') = \mathcal{L}(x)$. Thanks to the invariance under the space-time translations, we conclude that the Lagrangian must not manifestly contain any space-time coordinates x. Since dynamical observables (four-dimensional vector of the energy-momentum, the moment of momentum tensor, and so on) are real quantities, then the Lagrangian must be also a real quantity. We shall also impose the locality condition on the Lagrangian, that is, we assume that the Lagrangian depends only on the state of fields in the infinitely small vicinity of the point

x. So:

$$\mathcal{L}(x) = \mathcal{L}(\psi_k, \frac{\partial \psi_k}{\partial x^\nu}) = \mathcal{L}\left(\psi_k(x), \psi_{k;\nu}(x)\right).$$

General variation of an action, related with varying both field wave functions and of integration domain boundaries, is equal to:

$$\delta S = \delta \int \mathcal{L}(x) dx = \int \mathcal{L}(x') dx' - \int \mathcal{L}(x) dx, \tag{4.4}$$

where $dx = dx^0 d\mathbf{r}$. For $\mathcal{L}'(x')$ we have the expression:

$$\mathcal{L}'(x') = \mathcal{L}\left(\psi_k'(x'), \psi_{k;\nu}'(x')\right) = \mathcal{L}(x) + \delta\mathcal{L}(x) = \mathcal{L}(x) + \bar{\delta}\mathcal{L}(x) + \frac{d\mathcal{L}}{dx^\lambda}\delta x^\lambda, \tag{4.5}$$

In Eq. (4.5) we denote a form variation of corresponding quantities by the symbol $\bar{\delta}$. This way $\bar{\delta}\mathcal{L}(x)$ represents varying the Lagrangian only on the account of the variation of field functions and field function derivatives:

$$\bar{\delta}\mathcal{L}(x) = \frac{\partial \mathcal{L}}{\partial \psi_k}\bar{\delta}\psi_k + \frac{\partial \mathcal{L}}{\partial \psi_{k;\nu}}\bar{\delta}\psi_{k;\nu}, \tag{4.6}$$

and $\bar{\delta}\psi_k(x)$ does a variation of a field function form:

$$\bar{\delta}\psi_k(x) = \psi_k'(x) - \psi_k(x).$$

Thus, we have:

$$\delta S = \int \left(\bar{\delta}\mathcal{L}(x) + \frac{d\mathcal{L}}{dx^\lambda}\delta x^\lambda\right) dx + \int \mathcal{L}(x) dx' - \int \mathcal{L}(x) dx. \tag{4.7}$$

For an infinitesimal coordinate transformation:

$$x^\lambda \to x'^\lambda = x^\lambda + \delta x^\lambda, \tag{4.8}$$

volume elements in the new and old coordinate systems are connected by the relation:

$$dx' = dx_0' dx_1' dx_2' dx_3' = \frac{\partial(x_0', x_1', x_2', x_3')}{\partial(x_0, x_1, x_2, x_3)} dx \approx \left(1 + \frac{\partial \delta x^\lambda}{\partial x^\lambda}\right) dx. \tag{4.9}$$

Then, in the final form the action variation becomes:

$$\delta S = \int [\bar{\delta}\mathcal{L}(x) + \frac{\partial}{\partial x^\lambda}\left(\mathcal{L}(x)\delta x^\lambda\right)] dx = \int [\frac{\partial \mathcal{L}}{\partial \psi_k} - \frac{\partial}{\partial x_\lambda}\frac{\partial \mathcal{L}}{\partial \psi_{k;\lambda}}]\bar{\delta}\psi_k dx +$$

$$+ \int \frac{\partial}{\partial x^\lambda}\left(\frac{\partial \mathcal{L}}{\partial \psi_{k;\lambda}}\bar{\delta}\psi_k + \mathcal{L}\delta x^\lambda\right) dx. \tag{4.10}$$

Let us consider the action variation at fixed integration boundaries, in the mean time assuming, that wavefunctions variation at these boundaries is equal to zero. With the help of the four-dimensional Gauss theorem the second integral in Eq. (4.10) can be rewritten as follows:

$$\int_\Omega \frac{\partial}{\partial x^\lambda}\mathcal{D}_{(n)}^\lambda(x) dx = \int_\Sigma \mathcal{D}_{(n)}^\lambda(x) d\sigma_\lambda, \tag{4.11}$$

where

$$\mathcal{D}_{(n)}^\lambda(x) = \frac{\partial \mathcal{L}}{\partial \psi_{k;\lambda}}\bar{\delta}\psi_k + \mathcal{L}\delta x^\lambda$$

and Σ is an arbitrary closed space-like hypersurface, restricting four-dimensional volume Ω. As integration domains one should choose space-like hypersurfaces Σ_i and Σ_f corresponding to the initial and final states, and, as a closing surface, one should take an arbitrary space-like hypersurface Σ_0, connecting Σ_i and Σ_f. Then, we obtain:

$$\int_\Sigma \mathcal{D}^\lambda_{(n)}(x)d\sigma_\lambda = \int_{\Sigma_i} \mathcal{D}^\lambda_{(n)}(x)d\sigma_\lambda^{(i)} + \int_{\Sigma_f} \mathcal{D}^\lambda_{(n)}(x)d\sigma_\lambda^{(f)} + \int_{\Sigma_0} \mathcal{D}^\lambda_{(n)}(x)d\sigma_\lambda. \qquad (4.12)$$

Taking into consideration arbitrariness in choice of hypersurface Σ_0, we carry it away on infinity. Then, for all points belonging to Σ_0 we have

$$x_\nu^2 = c^2 t^2 - \mathbf{r}^2 \to -\infty$$

or

$$\mid \mathbf{r} \mid \to \infty.$$

However, on infinity there are no fields, what means:

$$\psi(x) \mid_{x \in \Sigma_0} = 0, \qquad \text{or} \qquad \int_{\Sigma_0} \mathcal{D}^\lambda_{(n)}(x)d\sigma_\lambda = 0.$$

As a result Eq.(4.12) takes the form:

$$\int_\Sigma \mathcal{D}^\lambda_{(n)}(x)d\sigma_\lambda = \int_{\Sigma_i} \mathcal{D}^\lambda_{(n)}(x)d\sigma_\lambda^{(i)} + \int_{\Sigma_f} \mathcal{D}^\lambda_{(n)}(x)d\sigma_\lambda^{(f)}.$$

Since $\bar{\delta}\psi_k$ and δx^λ at the initial and final space-like hypersurfaces are equal to zero, the right-hand side of Eq.(4.11) turns into zero. Then the Lagrangian-Euler equations:

$$\frac{\partial \mathcal{L}}{\partial \psi_k} - \frac{\partial}{\partial x_\lambda}\frac{\partial \mathcal{L}}{\partial \psi_{k;\lambda}} = 0, \qquad (4.13)$$

follow from condition (4.3) and arbitrariness of integration domain. These equations are interpreted as field motion equations. Since physical properties of the system are governed by the action (4.3) then the Lagrangian $\mathcal{L}(x)$ is defined up to a four-dimensional divergence. It means, that Lagrangians:

$$\mathcal{L}(x) \qquad \text{and} \qquad \mathcal{L}(x) + \frac{\partial}{\partial x_\lambda} F^\lambda$$

result in the same field motion equations, since the variation of the term $\partial F^\lambda/\partial x_\lambda$ gives zero contribution due to the four-dimensional Gauss theorem.

4.2   Noether theorem and dynamic invariants

For the complete description of physical systems one must know not only the motion equations, but the main characteristics of these systems as well. As such characteristics it is convenient to use quantities, constant in time, which we call either *integrals of motion* or *dynamic invariants*. It turns out, that dynamic invariants are connected with symmetry of a physical system with respect to a definite transformation group. This connection is described by a theorem, proved by E. Noether [11].

Let us assume that there is some finite-parametric (depending on l constant parameters) continuous transformation of coordinates and field functions under which action variation vanishes. Then l dynamic invariants—l combinations of field functions and field function derivatives that are conserved in time—correspond to such a transformation.

To prove this, we consider infinitesimal coordinates transformation:

$$x^\lambda \to x'^\lambda = x^\lambda + \delta x^\lambda, \qquad \delta x^\lambda = X_n^\lambda \delta \omega^n = \left(I_{(n)}\right)_\nu^\lambda x^\nu \delta \omega^n. \tag{4.14}$$

Under (1.14) the field functions are transformed by the law:

$$\psi_i(x) \to \psi_i'(x') = \psi_i(x) + \delta \psi_i(x), \qquad \delta \psi_i(x) = Y_{i(n)} \delta \omega^n = \left(K_{(n)}\right)_i^j \psi_j \delta \omega^n. \tag{4.15}$$

In Eqs. (4.14) and (4.15) index n takes the values $1, 2, \dots l$, ω^n are transformation parameters, $I_{(n)}$ are generators of a coordinate transformations group, and $K_{(n)}$ are generators of representation of this group. Indices of field functions i and transformation parameters n can have either tensor or spinor contents (it will be specified in every particular case). Notice, that a transformation law of field function derivatives:

$$\psi_{i;k}(x) \to \psi_{i;k}'(x') = \psi_{i;k}(x) + \delta \psi_{i;k}(x) \tag{4.16}$$

contains variations $\delta \psi_{i;k}(x)$, which are not derivatives with respect to $\delta \psi_i$. In other words, the operations δ and $\partial/\partial x$ are not commutative. It is caused by the fact that $\delta \psi_i$ is the variation of a field function due to both change of its form and change of the argument. The form variation of a field function $\overline{\delta} \psi_i(x)$ up to infinitesimal of the second order can be presented as follows:

$$\overline{\delta} \psi_i(x) = \delta \psi_i(x) - \psi_{i;k} \delta x^k = \left(Y_{i(n)} - \psi_{i;k} X_{(n)}^k\right) \delta \omega^n. \tag{4.17}$$

By definition the operation $\overline{\delta}$ is commutative with that $\partial/\partial x$.

Now we turn to the general expression for the action variation. We consider Eq. (4.10) for true motion, that is, when the Lagrangian-Euler equations are fulfilled. Further we suppose, that an integration domain Ω together with a wave field as a whole is subjected either to infinitesimal translation or to infinitesimal rotation. Then, from (4.10) follows:

$$\delta S = -\delta \omega^n \int_\Omega \frac{\partial}{\partial x^\lambda} \mathcal{D}_{(n)}^\lambda(x) dx = 0, \tag{4.18}$$

where quantities:

$$\mathcal{D}_{(n)}^\lambda(x) = \frac{\partial \mathcal{L}(x)}{\partial \psi_{i;\lambda}} \left(\psi_{i;\nu} X_{(n)}^\nu - Y_{i(n)}\right) - \mathcal{L}(x) X_{(n)}^\lambda \tag{4.19}$$

hereinafter we will call the Noether currents. Using arbitrariness of integration domain over four-dimensional volume and Eq. (4.18), we obtain the continuity equation:

$$\frac{\partial}{\partial x^\lambda} \mathcal{D}_{(n)}^\lambda(x) = 0. \tag{4.20}$$

Integration of Eq. (4.20) over the volume V results in the expression:

$$\int_V \partial_0 \mathcal{D}_{(n)}^0(x) d\mathbf{r} + \int_V \partial_l \mathcal{D}_{(n)}^l(x) d\mathbf{r} = \partial_0 \int_V \mathcal{D}_{(n)}^0(x) d\mathbf{r} + \oint_S \mathcal{D}_{(n)}^l(x) dS_l =$$

$$= \partial_0 \int_V \mathcal{D}_{(n)}^0(x) d\mathbf{r} = 0, \tag{4.21}$$

where we take into account the three-dimensional Gauss theorem and the fact that any fields are absent at infinity.

In its turn, from (4.21) follows, that the integrals:

$$C_{(n)} = \int \mathcal{D}^0_{(n)}(x) d\mathbf{r} = \text{const} \qquad (4.22)$$

do not depend on time. It is possible to rewrite the expression (4.22) in manifestly relativistically invariant form. To reach this purpose let us use the definition of the functional derivative of the functional $F(\sigma)$ (see Eq.(1.9)). In the case when $F_\mu(\sigma)$ is represented by an integral over space-like hypersurface

$$F_\mu(\sigma) = \int_\sigma F(x) d\sigma_\mu,$$

calculations give:

$$\frac{\delta}{\delta\sigma(x)} F_\mu(\sigma) = \lim_{\Omega \to 0} \left[\int_\sigma - \int_{\sigma'} \right] \frac{F(x) d\sigma_\mu}{\Omega} = \partial_\mu F(x). \qquad (4.23)$$

Because of this the relation:

$$\frac{\delta}{\delta\sigma(x)} F(\sigma) = \partial_\mu F^\mu(x) \qquad (4.24)$$

is carried out for the functional:

$$F(\sigma) = \int_\sigma F^\mu(x) d\sigma_\mu$$

where $F^\mu(x)$ is an analytical function. Then, from Eq. (4.24) follows, that the functional $F(\sigma)$ does not depend on the choice of the hypersurface σ ($\delta F(\sigma)/\delta\sigma = 0$) if $F_\mu(x)$ satisfies the continuity equation. Since for $\mathcal{D}^\lambda_{(n)}(x)$ Eq.(4.20) is fulfilled, the equivalent form of writing (4.22) is

$$C_{(n)} = \int_\sigma \mathcal{D}^\mu_{(n)}(x) d\sigma_\mu = \text{const}, \qquad (4.25)$$

as due to arbitrariness of σ, it could be always chosen as hyperplane with $t=$const.

Thus, we proved, that l dynamic invariants $C_{(n)}$ ($n = 1, 2, ..l$) correspond to every continuous l-parametric transformation of coordinates (4.14) and field functions (4.15) that converts the action variation to zero. The relations (4.20) and (4.25) present mathematical writing of the conservation laws in the integral and differential forms, respectively. It is worthwhile to notice, that the Noether currents $\mathcal{D}^\lambda_{(n)}(x)$ are ambiguously defined. Actually, the continuity equations will be satisfied for both $\mathcal{D}^\lambda_{(n)}(x)$ and

$$\mathcal{D}'^\lambda_{(n)}(x) = \mathcal{D}^\lambda_{(n)}(x) + \frac{\partial}{\partial x^\nu} f^{[\nu\lambda]}_{(n)}(x), \qquad (4.26)$$

where $f^{[\nu\lambda]}_{(n)}(x)$ are arbitrary functions, which are antisymmetric in indices λ and ν. Let us proceed to the concrete definition of dynamic invariants and symmetry groups connected with them.

4.3   Energy-momentum tensor

The opportunity of selecting the natural laws among the chaos of phenomena surrounding us is based on two reasons. First, in most cases it is possible to determine the plurality of initial conditions amenable to adjustment that is important for the phenomenon under study. Such a definition is not a trivial one, it makes the foundation of art in experimental investigation of phenomenon. Second, under the same given initial conditions, the experimental result will be always the same, regardless of place and time of experiment. In the language of initial conditions it appears as follows: *absolute location and absolute time are never essential initial conditions.* In other words, the laws of physics do not depend on the choice of the origin of coordinates both on the space and on the time axes, that is, they are invariant under translations along the coordinates axes in four-dimensional space. If correlations between events of the same events sequence were changing day after day and were different at different points in space, then the existence of physics would be impossible. However, it may turn out that invariance with respect to the space-time translations, which we connect with homogeneity of the time and space, could bear an approximate character. The application of the translational invariance postulate on cosmic scales presumes the homogeneous and stationary character of the universe. Nowadays, the Big Bang theory succeeded the stationary universe model, which was keeping a delicate silence about the genesis of our world. The Big Bang theory does not rule out the possibility of homogeneity violation in the space-time continuum. At this point we should calm too impressionable readers, that such regions of the universe have not been discovered by now. Space and time homogeneity means the Lagrangian invariance under the space and time translations group. We shall demonstrate that the conservation law of the energy-momentum four-dimensional vector corresponds to such an invariance. As this takes place, the three-dimensional momentum conservation law follows from space homogeneity, while the energy conservation law follows from the time homogeneity.

The invariance of a theory with respect to the four-parametric group of the four-dimensional translations:

$$x'^{\nu} = x^{\nu} + a^{\nu},$$

will be ensured automatically if one takes into account that under these transformations field functions do not change their form, that is:

$$\psi_k(x) = \psi'_k(x') = \psi'_k(x + a). \tag{4.27}$$

In the case of infinitesimal translations:

$$x'^{\nu} = x^{\nu} + \delta a^{\nu} = x^{\nu} + \delta x^{\nu}, \tag{4.28}$$

we have

$$\delta\omega^{\mu} = \delta x^{\mu}, \qquad X^{\nu}_{\mu} = \delta^{\nu}_{\mu}; \qquad \delta\psi_k = 0, \qquad Y_{i(n)} = 0;$$

$$\psi_k(x) = \psi'_k(x) + \psi_{k;\nu}(x)\delta x^{\nu}, \qquad \overline{\delta}\psi_k = -\psi_{k;\nu}(x)\delta x^{\nu}.$$

Thus, in the case under consideration, the Noether current $\mathcal{D}^{\lambda}_{(n)}(x)$ proves to be the second rank tensor:

$$T'^{\nu}_{\mu} = \frac{\partial\mathcal{L}}{\partial\psi_{k;\nu}}\frac{\partial\psi_k}{\partial x^{\mu}} - \mathcal{L}\delta^{\nu}_{\mu}, \tag{4.29}$$

satisfying the continuity equation:

$$\partial_{\nu}T'^{\nu}_{\mu} = 0. \tag{4.30}$$

The corresponding dynamic invariant is:

$$P^\nu = \int T^{\nu 0} d\mathbf{x}. \tag{4.31}$$

In classical mechanics the time component of this vector represents the Hamiltonian function, that is, the energy. From reasons of the relativistic covariance it follows that the four-dimensional vector P^ν is the energy-momentum vector. The tensor defined by Eq. (4.29) shall be called the *canonical tensor of the energy-momentum*. In the most general case it is not symmetric. However, with allowance made for ambiguity in its definition (see Eq. (4.25)), it is always possible to make the energy-momentum tensor symmetric by adding a specially chosen quantity of the form: $\partial_\lambda f_\mu^{[\lambda\nu]}$.

4.4 Tensor of angular momentum

As stated before, it would be impossible to discover the laws of nature, if they were not invariant under the space-time translations. The same would be true in the case of lacking an invariance under the space rotations. As experiments show, three-dimensional space surrounding us is not only homogeneous, but also isotropic. Space isotropy means that properties of the space are equivalent in all directions, that is, properties of a closed system do not change under coordinate system rotations. Of course, it is difficult to check the space isotropy by direct experiment in conditions of the earth's gravitation or any other gravitation. In this case there exists obvious difference between "up" and "down" directions, as well as between "left" and "right" directions. Notwithstanding the fact, Newton formulated the gravitation theory in such a way that the motion equations were invariant under rotations. A conservation law, following from three-dimensional space isotropy, is known to us as a conservation law of three-dimensional angular momentum $\mathbf{L}$. In this case the connection established by the Noether theorem, has the following form

$$space\ isotropy \rightarrow SO(3)\ invariance \rightarrow \mathbf{L}-const$$

However, from the relativity principle it follows, that physical laws are also invariant under rotations in the space-time planes. Due to this reason the four-dimensional space of the universe, in which we have the luck to live, is isotropic or invariant with respect to the six-parametric group of the four-dimensional rotations, the $SO(3,1)$-group.

Under infinitesimal four-dimensional rotations the coordinate transformation law has the form:

$$x'^\nu = x^\nu + x_\mu \delta\omega^{[\mu\nu]} = x^\nu + \delta x^\nu, \tag{4.32}$$

where, as we see, the index (n) fell into indices μ and ν ($\mu < \nu$) denoting the plane, in which the rotation with the parameter $\delta\omega^{[\mu\nu]}$ takes place. δx^ν is given as:

$$\delta x^\nu = X^\nu_{[\mu\sigma]} \delta\omega^{[\mu\sigma]} = x_\mu \delta\omega^{[\mu\nu]} = x_\mu \delta\omega^{[\mu\sigma]} \delta^\nu_\sigma =$$

$$= \delta\omega^{[\mu\sigma]} (x_\mu \delta^\nu_\sigma - x_\sigma \delta^\nu_\mu). \tag{4.33}$$

From Eqs. (14.33) and (4.14) it follows

$$X^\nu_{[\mu\sigma]} = x_\mu \delta^\nu_\sigma - x_\sigma \delta^\nu_\mu, \qquad \left(I^{\nu\lambda}\right)_{\sigma\mu} = \delta^\lambda_\mu \delta^\nu_\sigma - \delta^\nu_\mu \delta^\lambda_\sigma. \tag{4.34}$$

So, now the Noether current $\mathcal{D}_{\mu\nu}^\lambda(x)$ is the third rank tensor having the form:

$$M_{\mu\nu}^\lambda(x) = \frac{\partial \mathcal{L}(x)}{\partial \psi_{i;\lambda}(x)}\{\psi_{i;\sigma}(x)[x_\nu \delta_\mu^\sigma - x_\mu \delta_\nu^\sigma] - (K_{\mu\nu})_i^j \psi_j(x)\} + \mathcal{L}(x)\left(x_\mu \delta_\nu^\lambda -\right.$$

$$\left. - x_\nu \delta_\mu^\lambda\right) = \left(x_\nu T_\mu^{\prime\lambda} - x_\mu T_\nu^{\prime\lambda}\right) - \frac{\partial \mathcal{L}(x)}{\partial \psi_{i;\lambda}(x)}(K_{\mu\nu})_i^j \psi_j(x). \tag{4.35}$$

For a scalar field $(K_{\mu\nu})_i^j = 0$ the relation between the tensor $M_{\mu\nu}^\lambda(x)$ and the energy-momentum tensor takes the same form as in the classical mechanics of the material point. For this reason the quantity:

$$L^{[\mu\nu]\lambda} = x^\nu T^{\prime\mu\lambda} - x^\mu T^{\prime\nu\lambda} = T^{\prime\sigma\lambda}(I^{\mu\nu})_{\sigma\tau} x^\tau \tag{4.36}$$

should be interpreted as the intrinsic angular moment of the field. Since in the case of a multicomponent field $(K_{\mu\nu})_i^j \neq 0$, the second quantity in the expression (4.35):

$$S_{[\mu\nu]}^\lambda(x) = -\frac{\partial \mathcal{L}(x)}{\partial \psi_{i;\lambda}(x)}(K_{\mu\nu})_i^j \psi_j(x). \tag{4.37}$$

is logically identified with the spin moment tensor of particles to be described by the wave field. Thus, the Lagrangian invariance with respect to the four-dimensional rotations group in the Minkowski space resulted in the conservation of the sum of the spin and orbital moments of the field:

$$M^{[\mu\nu]} = \int M^{[\mu\nu]0} d\mathbf{r} = \int \left(L^{[\mu\nu]0} + S^{[\mu\nu]0}\right) d\mathbf{r} = L^{[\mu\nu]} + S^{[\mu\nu]}. \tag{4.38}$$

Carrying out the convolution of space components of the spin moment tensor with the three-dimensional antisymmetric third rank tensor $\varepsilon_{\alpha\beta\gamma}$, we obtain components of the three-dimensional spin pseudovector:

$$S_\alpha = \frac{1}{2}\varepsilon_{\alpha\beta\gamma}S_{\beta\gamma}, \tag{4.39}$$

in which, as it should be expected, we recognize a classical analog of the spin operator, namely, the Pauli-Lubanski pseudovector.

As it follows from (4.36) for any field the orbital moment tensor is defined by the same Hermitian generators of the Lorentz group $(I^{\mu\nu})_{\sigma\tau}$, while the spin moment tensor is expressed by means of infinitesimal operators $(K_{\mu\nu})_i^j$ to form the Lorentz group representation for functions of a given field.

Now we should like to get rid of arbitrariness in definition of the Noether currents $\mathcal{D}_{(n)}^\lambda(x)$. For this purpose it would be very attractive to use some physical principles. Let us turn to the general relativity theory where one of the most important elements of the theory is the energy-momentum tensor $T_{\mu\nu}$. The Einstein equations have the form:

$$R_{\mu\nu}(\eta) - \frac{1}{2}R(\eta)\eta_{\mu\nu}(x) = \frac{8\pi G_N}{c^4}T_{\mu\nu}(\eta), \tag{4.40}$$

where $\eta_{\mu\nu}(x)$ is a metric tensor, $R_{\mu\nu}(\eta)$ is the Ricci tensor, $R(\eta) = R_{\mu\nu}(\eta)\eta^{\mu\nu}$ and G_N is the Newton constant. Since the tensors $R_{\mu\nu}(\eta)$ and $\eta_{\mu\nu}$ are symmetric, then the energy-momentum tensor must be of the same property. Further, remembering the fact that the energy-momentum tensor is automatically symmetric for a scalar field, we understand that the symmetrization procedure of this tensor for fields with the spin is connected with usage of the spin moment tensor combinations instead of $f_{[\mu\nu]}^\lambda$. Following these decisive statements, after straightforward calculations we arrive at the symmetric (metric) energy-momentum tensor:

$$T_{\mu\nu}^{(metr)} = T'_{\mu\nu} + \partial_\lambda f_{[\mu\lambda]\nu} = T_{\nu\mu}^{(metr)}, \qquad (4.41)$$

where

$$f_{[\mu\lambda]\nu} = \frac{1}{2}\left(S_{\mu[\nu\lambda]} + S_{\lambda[\mu\nu]} + S_{\nu[\mu\lambda]}\right).$$

Then connection between the total moment tensor and the energy-momentum tensor is of the same form as in material point mechanics:

$$M^{[\mu\nu]\lambda} = x^\nu T^{(metr)\mu\lambda} - x^\mu T^{(metr)\nu\lambda}. \qquad (4.42)$$

This writing allows us to interpret the obtained results as follows. The $SO(3.1)$-invariance of the theory results in conservation of the quantity:

$$M^{[\mu\nu]} = \int \left(x^\nu T^{(metr)\mu0} - x^\mu T^{(metr)\nu0}\right) d\mathbf{r}, \qquad (4.43)$$

where invariance with respect to the three-dimensional rotations is associated with conservation of the space components $M^{[12]}$, $M^{[23]}$, and $M^{[31]}$ (the total moment of momentum of the field), whereas invariance with respect to the Lorenz boosts leads to conservation of the space time components $M^{[01]}$, $M^{[02]}$, and $M^{[03]}$, describing motion of the system mass center.

4.5 Electromagnetic current vector and electric charge

Thus, we considered dynamic invariants relating with symmetry in the Minkowski space. Besides them, other dynamic invariants are introduced into particle physics, so-called inner quantum numbers, caused by symmetries of physical systems with respect to transformations in abstract spaces. Corresponding symmetries are called "nongeometrical, inner" or "dynamic". The electric charge conservation law is an excellent example illustrating the connection:

$$inner \; symmetry \rightarrow invariance \rightarrow conservation \; law.$$

Let us consider the arbitrary field, described by N-component complex functions $\psi_k(x)$ and $\psi_k^\dagger(x) = (\psi_k^*)^T$ (k=1,2,...,N). The demand for the field Lagrangian $\mathcal{L}(x)$ to be real means, that it contains only bilinear combinations of field functions and their first derivatives of the following form:

$$\psi_k^\dagger(x)\mathcal{A}^{kl}\psi_l(x), \qquad \partial^\mu\psi_k^\dagger(x)\mathcal{B}_{\mu\nu}^{kl}\partial^\nu\psi_l(x), \qquad (4.44)$$

$$\partial^\mu\psi_k^\dagger(x)\mathcal{C}_\mu^{kl}\psi_l(x), \qquad \psi_k^\dagger(x)\mathcal{C}_\nu^{\prime kl}\partial^\nu\psi_l(x), \qquad (4.45)$$

where $\mathcal{A}^{kl}$, $\mathcal{B}_{\mu\nu}^{kl}$, $\mathcal{C}_\mu^{kl}$, $\mathcal{C}_\nu^{\prime kl}$ are quantities that do not depend on x, and the indices k and l can have either tensor or matrix dimensionality. Thus, for example, for a free electron-positron field we have:

$$\mathcal{A}^{kl} = -m(\gamma^0)^{kl}, \qquad \mathcal{B}_{\mu\nu}^{kl} = 0, \qquad \mathcal{C}_\mu^{\prime kl} = -\mathcal{C}_\mu^{kl} = \frac{i}{2}(\gamma^0\gamma^\mu)^{kl},$$

where γ_μ is 4×4 Dirac matrices, satisfying relations of Clifford algebra:

$$\{\gamma^\mu, \gamma^\nu\} = 2g^{\mu\nu}, \qquad (4.46)$$

k and l are matrix indices. From Eqs. (4.44) and (4.45) it follows that the field functions are defined with a precision of an arbitrary phase factor. It means, that under transformation:

$$\left.\begin{array}{l} \psi_k(x) \to \psi_k'(x) = U(\alpha)\psi_k(x) = \exp{(i\alpha)}\psi_k(x), \\ \psi_k^\dagger(x) \to \psi_k'^\dagger(x) = U^\dagger(\alpha)\psi_k^\dagger(x) = \exp{(-i\alpha)}\psi_k^\dagger(x), \end{array}\right\} \qquad (4.47)$$

where α is an arbitrary constant, the system Lagrangian remains invariant. In other words, physical reality, corresponding to the description in terms of the old $(\psi_k(x))$ and new $(\psi_k'(x))$ field functions, is the same. Transformations $U(\alpha)$ make up one-parametric group of the global gauge transformations (transformations are global, since phase α does not depend on x). Sometimes such a group is called the *gauge transformation group of the first kind*. The $U(\alpha)$-group is unitary, that is:

$$U(\alpha)U^\dagger(\alpha) = I,$$

where I is a unit matrix. This group is Abelian as well, since all its elements commute with each other. Then, under infinitesimal transformation of field functions:

$$\psi_k'(x) = (1 + i\delta\alpha)\psi_k(x) = \psi_k(x) + \delta\psi_k(x), \qquad (4.48)$$

from the Noether theorem it follows the continuity equation for four-dimensional current vector:

$$\partial_\mu j^\mu(x) = 0, \qquad (4.49)$$

where

$$j^\mu(x) = i\left(\psi_k^\dagger(x)\frac{\partial\mathcal{L}(x)}{\partial[\partial_\mu\psi_k^\dagger(x)]} - \frac{\partial\mathcal{L}(x)}{\partial[\partial_\mu\psi_k(x)]}\psi_k(x)\right). \qquad (4.50)$$

It leads, in its turn, to conservation of the corresponding charge:

$$Q = \int j^0(x)d\mathbf{x} = \text{const.} \qquad (4.51)$$

It is obvious, that if the current j^μ satisfies the continuity equation, the same is true for the current to be equal to $const \times j^\mu$. In the method we used, a unit of measurement of the charge was not fixed. It can be fixed only by means of additional physical assumptions. Setting $\alpha = q\alpha_0$, where q is an electric charge of particles, corresponding to a wave field, we arrive at the electric charge conservation law:

$$Q_{em} = \int j_{em}^0(x)d\mathbf{r} = iq\int\left(\psi_k^\dagger(x)\frac{\partial\mathcal{L}(x)}{\partial[\partial_0\psi_k^\dagger(x)]} - \frac{\partial\mathcal{L}(x)}{\partial[\partial_0\psi_k(x)]}\psi_k(x)\right)d\mathbf{r} = \text{const.} \qquad (4.52)$$

The electric charge belongs to additive characteristics. The experimental test of the electric charge conservation law is based on a test of electron stability and the zero mass of a photon. Analysis of possible events of atmospheric electricity, which could appear as a result of electron decays in the atmosphere gives for the lowest limit of the electron life time $> 10^{21}$ years. The existence of a large-scale magnetic field in disk components of the Galaxy leads to the most strict upper limit on the photon mass $\leq 10^{-27}$ eV/c^2. Strict equality in the absolute value of the proton and electron charges also confirms the electric charge conservation law. All of these facts make rather problematic the existence of theories to admit violation of (4.52) and give us all the grounds to consider the electric charge conservation law to be exact. That means that any processes, which are proceeding with its violation, are forbidden.

It is worthwhile to note, that in the case of the real fields

$$\psi_k^\dagger(x) = \psi_k^T(x),$$

and the expression for $j^\mu(x)$ turns into zero. In the heat of the moment one may state that complex field functions describe charged particles, while real functions are only used for neutral particles. What actually happens is that the real functions are only used for true neutral particles, that is, in the case when a particle is equivalent to its antiparticle (photon, π^0 meson, and so on). Clearly, simply neutral particles generate no electric currents. However, such particles possess another sorts of "currents" and "charges" related with these currents. We shall turn back to this topic later.

For further analysis it is useful to give geometrical form to the gauge transformation (4.47). For the sake of simplicity, let us consider one-component field $\phi(x)$ (such fields describe spinless particles). The field functions $\phi(x)$ and $\phi^*(x)(\text{x})$ can be presented in the form:

$$\phi(x) = \frac{\phi_1(x) + i\phi_2(x)}{\sqrt{2}}, \qquad \phi^*(x) = \frac{\phi_1(x) - i\phi_2(x)}{\sqrt{2}}, \tag{4.53}$$

where $\phi_1(x)$ and $\phi_2(x)$ are real quantities. Then the gauge transformations (4.47) will look like:

$$\left.\begin{array}{l} \phi_1'(x) + i\phi_2'(x) = \exp(i\alpha)[\phi_1(x) + i\phi_2(x)], \\ \phi_1'(x) - i\phi_2'(x) = \exp(-i\alpha)[\phi_1(x) - i\phi_2(x)]. \end{array}\right\} \tag{4.54}$$

Since Eqs. (4.54) could be shaped to the form:

$$\begin{pmatrix} \phi_1(x) \\ \phi_2(x) \end{pmatrix}' = \begin{pmatrix} \cos\alpha & \sin\alpha \\ -\sin\alpha & \cos\alpha \end{pmatrix} \begin{pmatrix} \phi_1(x) \\ \phi_2(x) \end{pmatrix}, \tag{4.55}$$

then it is obvious, that the gauge transformations (4.47) may be interpreted as rotations of the vector $\phi(x) = (\phi_1(x), \phi_2(x))$ about the angle α. In group theory slang the above said means, that the group $U(1)$ is locally isomorphic to the orthogonal rotations group $SO(2)$ in two-dimensional space of the real functions $\phi_1(x)$ and $\phi_2(x)$. As $\alpha = \text{const}$, this transformation must be the same for all points of the space-time continuum, that is, it is the "global" gauge transformation. To put this another way, when a rotation in one point by the angle α in the inner space of the field $\phi(x)$ is made, the same rotation in all other points must be simultaneously carried out. We would have no reason to worry, if a conserved quantity related to invariance under this transformation were not the source of physical field. The electric charge, however, is producing the electromagnetic field in space and the electromagnetic interaction is spreading at the finite speed, not instantly. To avoid conflict with the short-range interaction theory (or the field concept), we are forced to localize the gauge transformation $U(1)$. It means that we consider the phase α to be different for various space-time points, that is, $\alpha = \alpha(x)$. The corresponding transformation is called the "local" gauge transformation or the *gauge transformation of the second kind*.

Let us investigate the physical consequences of theory invariance under the local gauge transformations. As an example, we refer to the electron-positron field $\psi(x)$, for which the Lagrangian function density is given by the expression:

$$\mathcal{L} = \frac{i}{2}\left[\overline{\psi}(x)\gamma^\mu\partial_\mu\psi(x) - \partial_\mu\overline{\psi}(x)\gamma^\mu\psi(x)\right] - m\overline{\psi}(x)\psi(x). \tag{4.56}$$

where the line above the field function means the Dirac conjugation, that is: $\overline{\psi}(x) = (\psi^*(x))^T\gamma^0$. The field function transformation law under the local gauge transformations of the $U(1)$-group has the form:

$$\psi(x) \to \psi'(x) = \exp[i\alpha(x)]\psi(x). \tag{4.57}$$

Due to the presence of the derivative the Lagrangian (4.56) is not invariant under these transformations, since:

$$\partial_\mu \psi(x) \rightarrow \exp\left[i\alpha(x)\right][\partial_\mu \psi(x) + i\psi(x)\partial_\mu \alpha(x)]. \tag{4.58}$$

For the Lagrangian to be invariant, we must introduce a new derivative in such a way that a derivative of the field function is transformed in the same manner as the field function itself:

$$D_\mu \psi(x) \rightarrow \exp\left[i\alpha(x)\right]D_\mu \psi(x). \tag{4.59}$$

The relation (4.59) will be fulfilled under the condition that

$$D_\mu = \partial_\mu - igA_\mu(x), \tag{4.60}$$

where under the local transformations (4.57) the introduced vector field $A_\mu(x)$ must behave by the following way:

$$A_\mu(x) \rightarrow A_\mu(x) + g^{-1}\partial_\mu \alpha(x). \tag{4.61}$$

We call new derivative D_μ a *covariant derivative*. Since, besides the fermions field our system contains the vector field as well, the Lagrangian (4.56) must be supplemented with a free Lagrangian of the field $A_\mu(x)$, which must not violate the local gauge invariance and be relativistically covariant at the same time. If we call for fulfillment of the superposition principle, then only a quantity like:

$$aF_{\mu\nu}(x)F^{\mu\nu}(x) + b\tilde{F}_{\mu\nu}(x)\tilde{F}^{\mu\nu}(x), \tag{4.62}$$

where

$$F_{\mu\nu}(x) = \partial_\mu A_\nu(x) - \partial_\nu A_\mu(x), \qquad \tilde{F}_{\mu\nu}(x) = \frac{1}{2}\varepsilon_{\mu\nu\lambda\sigma}F^{\lambda\sigma}(x),$$

is possible. Now, to obtain the Lagrangian of quantum electrodynamics (the theory of electromagnetic interaction between electrons and positrons) it is sufficient to identify g with the electric charge of the electron e, $A_\mu(x)$—with the four-dimensional potential of the electromagnetic field and choose the coefficients in (4.62) as follows:

$$a = -1/4, \qquad b = 0.$$

The corresponding Noether current (the electromagnetic current) and the conserved charge (the electric charge) are defined by the expressions (4.51) and (4.52) multiplied by e. Thus, the Lagrangian invariance with respect to the local gauge transformation group not only ensures the electric charge conservation, but results in harmony with the special relativity theory as well. The appearance of the gauge boson related to the gauge group $U(1)$, the electromagnetic interaction carrier, which is called a *photon*, is one more consequence of the locality of this group. Thus, if one distracts from some arbitrariness in the choice of the free field Lagrangians, then the aforesaid is a strong argument in favor of the fact that the local gauge invariance represents a fundamental principle laying in the basis of a theory of any interaction.

4.6   Isotopic spin

It is known from the experiment, that if one neglects electromagnetic interaction, then it is possible to unify some elementary particles in families, the so-called isotopic multiplets.

Particles entering these multiplets have identical properties and approximately the same masses. The well-known examples are the isotopic doublet of nucleons (p, n) and the isotopic triplet of pions (π^+, π^0, π^-). The observed mass spectrum for members of the above-listed multiplets has the form:

$$m_p = 938.27231 \pm 0.00028 \text{ MeV}/c^2, \qquad m_n = 939.56563 \pm 0.00028 \text{ MeV}/c^2$$

$$m_{\pi^\pm} = 139.56995 \pm 0.00035 \text{ MeV}/c^2, \qquad m_{\pi^0} = 134.9764 \pm 0.0006 \text{ MeV}/c^2.$$

Believing, that under the absence of electromagnetic interaction masses of members of the same multiplet coincide, we can consider particles of a multiplet as different states of one and the same particle, every state being marked by a definite charge value. By analogy with the ordinary spin, we confront every multiplet with so-called isotopic spin, whose number of projections is equal to the number of particles in the multiplet; we also confront every projection with a particle of definite charge. We also introduce abstract (inner) three-dimensional space, which we call the *isospace*.

Since the isospin operators must satisfy the same commutation relations as the ordinary spin operators, then the isospin square and one of its projections, for example, the third one, can be simultaneously measured. Further on we shall have to introduce a wide variety of operators, similar to the spin operator, namely, a unitary spin operator, a color spin operator, and a weak isospin operator. For spin-like operators let us use the symbol S with a superscript defining an operator class ($S^{(is)}$, $S^{(un)}$, $S^{(col)}$, and S^W). As far as operator eigenvalues are concerned, we use notations, adopted by "Review of Particle Physics." Thus, eigenvalues of the isotopic spin S^{is} (the ordinary spin S) are marked by I (J). So, a field function of particles of a given multiplet must depend not only on the coordinates and the ordinary spin, but also on the isospin and its projection, that is, it must be a product of an isotopic function, a coordinate function, and a spin function. As in the case of an ordinary spin, an isospin dependence is described by multicomponent isotopic wavefunctions. For example, the nucleon doublet (p, n) is related with an isospinor ($I = 1/2$), while the pion triplet is associated to an isovector ($I = 1$). A correlation of the sign of the projection I_3 with the charge sign is arbitrary. A common choice is the sign's coincidence of the electric charge and I_3. Thus, for the proton $I_3 = 1/2$, for the neutron $I_3 = -1/2$, for the π^+, π^0, π^--mesons $I_3 = +1, 0, -1$, respectively. If the electromagnetic interaction is switched off, then multiplet particles are identical, in other words, the system of isomultiplet particles is degenerate on the electric charge. In mathematical language this degeneracy is formulated as an invariance with respect to the three-dimensional rotations group in the isospace. The degree of coincidence of particle masses of the same multiplet could be viewed as validity criteria for isotopic invariance. For the pions:

$$\frac{m_{\pi^\pm} - m_{\pi^0}}{\tilde{m}_\pi} = \delta_\pi \approx 3\%, \tag{4.63}$$

that is, much more than the corresponding value for nucleons. In the case of the proton and neutron, the mass difference is caused not only by the Coulomb forces, but also by magnetic ones (magnetic moments of nucleons are not equal to zero), that is, certain compensation takes place. The mass difference inside of the pion triplet is caused only by the Coulomb forces since both the pion spins and the magnetic moments are equal to zero. Obviously, the quantity δ_i ($i = N, \pi, ...$) represents the limit of a precision that should be waited under the fulfillment of predictions based on the isotopic invariance within one or an other multiplet. As we shall demonstrate later, the isotopic invariance (or charge independence) leads to the isotopic spin conservation law. However, before we analyze the consequences of this invariance, we must broaden our outlook a little in the group theory.

Let us address the simplest one-parametric unitary group $U(1)$ again. In the n-dimensional vector space of complex functions $\psi_k(x)$ one may define the set of linear homogeneous transformations:

$$\psi'_k(x) = U\psi_k(x), \qquad \psi'^\dagger_k(x) = (U\psi_k(x))^\dagger = \psi^\dagger_k(x)U^\dagger, \tag{4.64}$$

which leave a scalar product invariant and satisfy the unitary demand:

$$UU^\dagger = U^\dagger U = I. \tag{4.65}$$

The set of all the square $n \times n$-matrices U with the above-mentioned properties constitutes a unitary group $U(n)$. To impose on unitary matrices U the unimodular condition:

$$\det U = 1, \tag{4.66}$$

will single out from $U(n)$ its subgroup, namely, a special unitary group $SU(n)$. In this case

$$U(n) = U(1) \bigotimes SU(n), \tag{4.67}$$

that is, the transformations $U \in U(n)$ differ from those belonging to $SU(n)$ by means of a phase factor of the form $\exp(i\alpha) \in U(1)$. The theory of the $SU(n)$-groups is in many ways similar to that of the $SO(n)$-groups. The $SU(n)$-group goes over into the $SO(n)$-group if one passes from a complex n-dimensional space into a real one. The $SU(n)$-group is a finite-parametric continuous Lie group. The relations (4.65) and (4.66) defining its elements impose $n^2 + 1$ conditions on $2n^2$ real parameters. Therefore, the number of independent parameters of the $SU(n)$-group is $n^2 - 1$. The simplest example of a special unitary group is a group $SU(2)$. Recall that any unitary unimodular $n \times n$-matrix can be written down in an exponential form:

$$U = \exp(iH),$$

where due to demands (4.65) and (4.66) a $n \times n$-matrix H satisfies the conditions:

$$H^\dagger = H, \qquad \mathrm{Sp}\,\{H\} = 0. \tag{4.68}$$

It is an easy matter to check that at $n = 2$ a special unitary group is formed by the already familiar matrices:

$$U(\theta) = \exp[i(\boldsymbol{\sigma} \cdot \boldsymbol{\theta})/2], \tag{4.69}$$

which realize the spinor representation of the three-dimensional rotations group. In principle, we could have guessed this from the very beginning. Since the generators of the three-dimensional rotations group and generators of any corresponding representation satisfy the same relations, then the $SO(3)$- and $SU(2)$-groups are locally isomorphic to each other. Thus, we came to the conclusion, that to solve a problem of constructing irreducible representations of the $SU(2)$-group and its Lie algebra is equivalent to the solution of a similar problem for the $SO(3)$-group. It is therefore concluded that both the generators T_α ($\alpha = 1, 2, 3$) of an arbitrary representation for the $SU(2)$-group and the generators K_α of the $SO(3)$-group representation are governed by the same permutation relations, that is,

$$[T_\alpha, T_\beta] = i\varepsilon_{\alpha\beta\gamma}T_\gamma, \qquad (\alpha, \beta, \gamma = 1, 2, 3). \tag{4.70}$$

Now we are quite ready to study the consequences of the isotopic invariance. Assume that the Lagrangian is invariant with respect to the $SU(2)$-group. Ordinary coordinates x^μ are not changed under the isotopic transformations $X^\lambda_n = 0$, and the law of the infinitesimal transformation of the wavefunction has the form:

$$\psi'_k(x) = \psi_k(x) + \delta\psi_k(x), \qquad \delta\psi_k(x) = (T_\gamma)^j_k\,\psi_j\delta\theta^\gamma, \tag{4.71}$$

where for the sake of simplicity we turned from $\delta\omega^{[\alpha\beta]}$ to element $\delta\theta^\gamma$, which represents an infinitesimal angle of rotation about an axis γ in the isotopic space defined by:

$$\delta\theta^\gamma = \varepsilon^{\alpha\beta\gamma}\delta\omega^{[\alpha\beta]}.$$

Since $(T_\gamma)^j_k = (K_\gamma)^j_k$, then the Noether current corresponding to the isotopic invariance is completely equivalent to the three-dimensional part of the spin moment tensor:

$$N^\lambda_\gamma(x) = -\frac{\partial\mathcal{L}(x)}{\partial\psi_{i;\lambda}(x)}\,(T_\gamma)^j_i\,\psi_j(x). \tag{4.72}$$

Spatial integrals

$$S^{(is)}_\gamma = \int N^0_\gamma d\mathbf{r}, \tag{4.73}$$

determining an isospin pseudovector, are constant in time. The total isotopic spin of a particles system is given by a vector sum of the isotopic spin vectors of all the particles (in full analogy with the rules for composition of moments). The isotopic invariance means the conservation of $S^{(is)2}$ and $S^{(is)}_3$. Worthwhile to note, when the electromagnetic interaction is switched on, the theory loses invariance with respect to arbitrary rotations in the isospace. In this case the theory remains invariant only under rotations around the x_3-axis. As a consequence, $S^{(is)2}$ ceases to be motion integral and only $S^{(is)}_3$ is conserved, that is, the electric charge remains the motion integral.

5

Discrete symmetry operations

5.1 Spatial inversion

Rotations are the examples of continuous transformations, since a rotation by any finite angle can be viewed as an infinite sequence of infinitesimal rotations. There exist also discrete transformations, which cannot be presented as a series of infinitesimal steps. The space inversion $\mathbf{r} \rightarrow -\mathbf{r}$ is an example of such a transformation. The corresponding group is finite and contains only two elements: the unit element I and an element P, satisfying a condition $P^2 = 1$ (to use the space inversion transformation twice is equivalent to the identical transformation).

Laws of classical physics are invariant under the space inversion. It means, that a mirror image of any process, which obeys laws of classical physics and takes place in nature, is subject to the same laws and its existence is physically possible. In other words, symmetry with respect to the space inversion means, that two physicists, one of which is using a left-coordinate system (a lefthander) while the other is using a right-coordinate system (a righthander), describing the corresponding law of the nature, will arrive at the same expression. In classical physics, however, the existence of mirror symmetry does not produce any conserved quantity.

In quantum theory strong, electromagnetic, and gravitation interactions are conserving an invariance with respect to the mirror reflection, while a weak interaction does not possess such a property. Distracting for a while from weak interaction, we show that the mirror symmetry results in a conserved quantum number, a so-called space parity. First of all we demonstrate, that when the theory is invariant with respect to the operation: $\mathbf{r} \rightarrow -\mathbf{r}$ (Hamiltonian commutes with the operator P), then it is always possible to choose stationary states of a physical system in such a way, that they are orthogonal eigenstates of the operator P (without degeneration this result is obvious). For the sake of simplicity we restrict ourselves to a field obeying to superposition principle. In the most general case a wave field of an arbitrary spin is described by a system of linear equations. These equations can be always represented in the form of a matrix equation of the first order:

$$\left(i\Lambda_\mu \frac{\partial}{\partial x_\mu} - \frac{mc}{\hbar} \right) \psi(x) = 0, \tag{5.1}$$

where the algebra of square matrices Λ_μ is defined by a choice of a representation according to which the multicomponent wavefunction $\psi(x)$ is transformed. Such a conclusion follows from the fact that derivatives of any order

$$\frac{\partial^k}{(\partial x_\mu)^k} \psi_j(x)$$

can be always reduced to derivatives of the first order, if one sequentially introduces new functions viewing them as additional components of the initial wavefunctions:

$$\psi'_{j;\mu} = \frac{\partial}{\partial x_\mu} \psi_j(x), \qquad \psi''_{j;\mu,\nu} = \frac{\partial}{\partial x_\nu} \psi'_{j;\mu}(x), \qquad \ldots$$

In its turn, any equation of the first order can be always represented in the Schrödinger-like form:

$$i\hbar \frac{\partial \psi(x)}{\partial t} = \mathcal{H}\psi(x). \tag{5.2}$$

Applying the P-operator to Eq. (5.2) and taking into account

$$[\mathcal{H}, P] = 0, \tag{5.3}$$

one obtains:

$$i\hbar \frac{\partial \psi'(x)}{\partial t} = \mathcal{H}\psi'(x), \tag{5.4}$$

where $\psi'(x) = P\psi(x)$. Consequently, combinations:

$$\psi_\pm(x) = \frac{1}{2}\left(\psi(x) \pm P\psi(x)\right),$$

are also solutions of the Schrödinger equation. Furthermore, these combinations are orthogonal eigenstates of the P-operator with or without degeneration in the Hamiltonian spectrum

$$P\psi_\pm(x) = \pm\psi_\pm(x).$$

Under $\mathbf{r} \to -\mathbf{r}$ the momentum reverses its sign, in other words, the relation:

$$\{P, \mathbf{p}\} = 0$$

takes place. Therefore, the space inversion operator commutes with the moment of momentum operator $\mathbf{L}$:

$$[P, \mathbf{L}] = 0, \tag{5.5}$$

that is, the eigenvalues of these operators are simultaneously and therewith exactly measurable.

Let us define the eigenvalues of the space inversion operator (space parity) for a particle with the given moment of momentum $\mathbf{L}$ and zero-spin. Recall, that an eigenfunction of the operator $\mathbf{L}^2$ is the spherical function $Y_l^{|m|}(\theta, \varphi)$, having the form

$$Y_l^m(\theta, \varphi) = \sqrt{\frac{(2l+1)(l-|m|)!}{4\pi(l+|m|)!}} \sin^{|m|}\theta P_l^{|m|}(\cos\theta)e^{im\varphi}, \tag{5.6}$$

where $P_l^{|m|}(\cos\theta)$ are the generalized Legendre polynomials

$$P_l^{|m|}(x) = \frac{1}{2^l l!} \frac{d^{|m|+l}}{dx^{|m|+l}}(x^2 - 1)^l.$$

Allowing for that under the space inversion the Decart coordinates:

$$x = r\sin\theta\cos\varphi, \qquad y = r\sin\theta\sin\varphi, \qquad z = r\cos\theta,$$

reverse their signs we find without trouble:

$$r \xrightarrow{P} r, \qquad \theta \xrightarrow{P} \pi - \theta, \qquad \varphi \xrightarrow{P} \varphi + \pi.$$

Then:

$$Y_l^m(\theta, \varphi) \xrightarrow{P} (-1)^l Y_l^m(\theta, \varphi)$$

that is, the moment of momentum also defines the particle state parity. When a particle spin equals 0, then its space parity is still determined as $(-1)^l$ ($l = j+s, j+s-1, \ldots \mid j-s \mid,$), where j is the total particle moment.

The theory invariance with respect to the space inversion demands invariability of quantities measured by experiment, for example:

$$\int \psi^\dagger(\mathbf{r}, t)\mathbf{p}\psi(\mathbf{r}, t)d\mathbf{r} = \int \psi'^\dagger(\mathbf{r}, t)\mathbf{p}'\psi'(\mathbf{r}, t)d\mathbf{r} =$$

$$= -\int \psi^\dagger(\mathbf{r}, t)P^\dagger\mathbf{p}P\psi(\mathbf{r}, t)d\mathbf{r} = \int \psi^\dagger(\mathbf{r}, t)\mathbf{p}P^\dagger P\psi(\mathbf{r}, t)d\mathbf{r}. \quad (5.7)$$

From Eq. (5.7) follows, that the parity operator must be unitary

$$P^\dagger P = PP^\dagger = I.$$

Up to now we spoke of the parity as a quantum number, characterizing a wavefunction of a particle in the coordinate space, that is, we considered only that part of the parity, which does not depend on particles type. In so doing we did not touch upon a subject, whether quantum system components possess any inner parity. However, if a number of particles can be changed, every particle must be ascribed an inner parity. In classical theory it is impossible, because there exists no notion concerning production and destruction of particles. For this reason the inner parity has no sense in the classical theory and, consequently, the parity conservation law has no classical analog, unlike the conservation laws of energy, momentum, and angular momentum.

Let us denote a wavefunction of a quantum system, consisting of particles $a, b\ldots$ by the expression $\psi_{ab\ldots}(\mathbf{r}_a, \mathbf{r}_a, \ldots)$. To act the parity operator is defined by the relation:

$$P\psi_{ab\ldots}(\mathbf{r}_a, \mathbf{r}_a, \ldots) = \epsilon_a\epsilon_b\ldots\psi_{ab\ldots}(-\mathbf{r}_a, -\mathbf{r}_a, \ldots),$$

where $\epsilon_a, \epsilon_b,\ldots$ are inner parities, chosen in such a way, so that P commutes with $\mathcal{H}$. Further on we shall speak of the resultant space parity of a particle, which is equal to the product of its inner and orbital (space) parities. System invariance with respect to the space inversion means, that if a system is in a state with a definite parity value, then only transitions to states with the same parity value are possible. It is obvious that the parity operator eigenvalues are multiplicative quantum numbers. Recall, that the majority of quantum numbers is not multiplicative. For example, the electric and baryon charges are additive quantum numbers. The question arises, how does one define the inner parity of object, whose inner structure is unknown and which may not possess any structure in general. One does not always manage to define inner parities of elementary particles in an unambiguous way. It forces us to adhere to a certain agreement, that is, to speak about relative inner parity. Let us explain what we mean by an example. Assume that we want to define the inner parity of the particle c by using a reaction:

$$a + b \rightarrow a + b + c, \quad (5.8)$$

which takes place with the parity conservation. Then we have:

$$\epsilon_a\epsilon_b(-1)^{l_i} = \epsilon_a\epsilon_b\epsilon_c(-1)^{l_f}, \quad (5.9)$$

where l_i (l_f) are the moment of momentum eigenvalues in the initial (final) state. From Eq. (5.9) follows:

$$\epsilon_c = (-1)^{l_f+l_i}.$$

In this example the weak point in the definition procedure of the inner parity of the particle c is very obvious: the inner parity can be unambiguously defined only for particles whose conserved quantum numbers (electric charge, baryon charge, strangeness, charm, etc.) equal to zero. Otherwise, we can speak only of relative parity. Thus, for example, if the particle c has the baryon charge, then only the following reaction is possible:

$$a + b \rightarrow a + b + c + \overline{d} \tag{5.10}$$

where d is baryon, and the parity conservation law defines only the relative parity:

$$\frac{\epsilon_c}{\epsilon_d} = (-1)^{l_f + l_i}.$$

Now we can define inner parities of all the baryons with respect to parities of some main baryons. Paying respect to the Sakata model [12] (it is assumed that all hadrons consist of p, n, and Λ) we take as main baryons p, n, Λ and let their parities be positive. The inner parity of a particle defines a kind of its wavefunction, namely, whether wavefunction is a true quantity (scalar, vector, etc.) or a pseudoquantity (pseudoscalar, pseudovector, etc.).

5.2   Time inversion

In the previous subsection we got acquainted with the space inversion transformation that had an obvious physical meaning. Really, it is logical to compare two physical systems with each other if they follow from each other by the mirror reflection operation ($\mathbf{r} \leftrightarrow -\mathbf{r}$). The operation of the time inversion $t \rightarrow -t$ has no so evident meaning in our world, where time always flows in one direction. Nevertheless, the time inversion is a very useful transformation and has a certain physical meaning, if one interprets it correctly. A simple task from the classical mechanics can serve as an example. Suppose, we have a system A and its Hamiltonian explicitly does not depend on time. The dynamics of A is defined by the Hamiltonian equations:

$$\dot{q}(t) = \frac{\partial}{\partial p(t)}\mathcal{H}(q(t), p(t)), \qquad \dot{p}(t) = -\frac{\partial}{\partial q(t)}\mathcal{H}(q(t), p(t)), \tag{5.11}$$

and specification of initial conditions for the generalized coordinates $q(t)$ and momenta $p(t)$:

$$q(t_i) = q^{(i)}, \qquad p(t_i) = p^{(i)}. \tag{5.12}$$

Solving (5.11) and using (5.12), we can determine the values $q(t)$ and $p(t)$ at any moment of time $t_f > t_i$ (t_f is the final moment of time)

$$q(t_f) = q^{(f)}, \qquad p(t_f) = p^{(f)}. \tag{5.13}$$

A trajectory on phase plane $A_{t_i \rightarrow t_f}$ corresponds to transition of the system A from the initial to the final state. Let us introduce new functions:

$$q'(t) = q(t_f - t), \qquad p'(t) = -p(t_f - t),$$

which satisfy the initial conditions:

$$q'(0) = q(t_f) = q^{(f)}, \qquad p'(0) = -p(t_f) = -p^{(f)}$$

and the motion equations:

$$\dot{q}'(t) = -\dot{q}(t_f - t) = -\frac{\partial}{\partial p(t_f - t)}\mathcal{H}(q(t_f - t), p(t_f - t)) = \frac{\partial}{\partial p'(t)}\mathcal{H}(q'(t), -p'(t)), \quad (5.14)$$

$$\dot{p}'(t) = \dot{p}(t_f - t) = -\frac{\partial}{\partial q(t_f - t)}\mathcal{H}(q(t_f - t), p(t_f - t)) = -\frac{\partial}{\partial q'(t)}\mathcal{H}(q'(t), -p'(t)). \quad (5.15)$$

At $t = t_f$ the functions $q'(t)$ and $p'(t)$ meet the following demands:

$$q'(t_f) = q(t_i) = q^{(i)}, \qquad p'(t_f) = -p(t_i) = -p^{(i)}, \qquad (5.16)$$

where we set $t_i = 0$. From the obtained relations follows, that the functions $q'(t)$ and $p'(t)$ characterize another mechanical system A^T, which passes from the initial state with $t = t_f$ to the final state with $t = t_i$, and the evolution of the system is described by the trajectory $A^T_{t_f \to t_i}$. Since under the time inversion the components of the generalized momentum reverse the sign $p(t) = (d/dt)mq \to p'(t) = -p(t)$, then boundary conditions of the system A^T are inverted in time with respect to boundary conditions of the system A. If one does not take into consideration irreversible processes (friction, for example), then we have

$$\mathcal{H}(q(t), -p(t)) = \mathcal{H}(q(t), p(t)), \qquad (5.17)$$

that is, both systems are described by the same equations. All this means, that the phase trajectory of the inverted system $A^T_{t_f \to t_i}$ follows from $A_{t_i \to t_f}$ by the replacement $p(t) \to -p(t)$ and the system A^T moves back in time. In other words, the Hamiltonian equations allow two solutions, which are connected with each other by the time inversion operation. It is also obvious, that the time inversion should be understood from the viewpoint of the motion conversion.

If one records some physical process with a video camera and plays it back, then one can observe a process, which is also possible. In some cases such a reversed-in-time process may appear to be of extremely low-probability, however, none of the classical physics laws would be violated during this process. Thus, for example, a reversed-in-time dive from a tower would be put into practice, if water molecules, moving suitably, gave back the diver, who had fallen into a pool, his initial energy and momentum, and, as a result, he could go up to the tower again. However, our life experience tells us that the probability of such events is extremely small. The principle of the entropy increase applied to macroscopic systems establishes that time flows in the direction of realization of the most probable processes, however, no law of classical physics prohibits the realization of any low-probability process. Since the principle of the entropy increase is only applicable to macrosystems (with large numbers of particles), but not to microworld events (processes, in which individual particles are participated), then we cannot use it to define time direction under the investigation of the microworld phenomena.

Now let us discuss the subject of the time inversion in quantum theory. All physical quantities are divided in two classes with respect to the time inversion. The first class consists of quantities that contain time in even powers and, consequently, are not changed under the time inversion (coordinates, total energy, kinetic energy, etc.). Quantities of the second class contain time in odd powers and reverse the sign at $t \to -t$ (momentum, total moment, spin, etc.). The transformation group of the time inversion is finite and contains two elements: the time inversion operator T and the identical transformation. The time reflection operation, applied to the system twice, returns it back to the initial state.

Let us discover the main properties of a wavefunction of a particle with arbitrary spin with respect to the time inversion. As we already know, for this purpose it is sufficient to consider the Schrödinger equation:

$$i\hbar\frac{\partial}{\partial t}\psi(\mathbf{r}, t) = \mathcal{H}\psi(\mathbf{r}, t).\tag{5.18}$$

The time inversion operation T yields the state $\psi(\mathbf{r}, -t)$, in which all physical quantities of the first class have the same values as in state $\psi(\mathbf{r}, t)$, and physical quantities of the second class have the opposite sign. Even if the Hamiltonian is invariant under the time inversion,

$$\mathcal{H} \overset{T}{\longrightarrow} \mathcal{H},$$

the corresponding equation in the system with inverse time, nevertheless, has the form:

$$-i\hbar\frac{\partial}{\partial t}\psi(\mathbf{r}, -t) = \mathcal{H}\psi(\mathbf{r}, -t),\tag{5.19}$$

that is, the motion equations for the functions $\psi(\mathbf{r}, t)$ and $\psi(\mathbf{r}, -t)$ are not identical.

Let us consider the equation complex conjugated with respect to Eq. (5.19)

$$i\hbar\frac{\partial}{\partial t}\psi^*(\mathbf{x}, -t) = \mathcal{H}^*\psi^*(\mathbf{x}, -t).\tag{5.20}$$

Since $\mathcal{H}$ is the Hermitian operator, then $\mathcal{H}$ and $\mathcal{H}^\dagger$ have the same eigenvalues (but, broadly speaking, different eigenfunctions). It means that there exists such an unitary operator V, that:

$$V\mathcal{H}^*V^{-1} = \mathcal{H}.$$

With regard to this relation, Eq. (5.20) takes the form:

$$i\hbar\frac{\partial}{\partial t}V\psi^*(\mathbf{x}, -t) = \mathcal{H}V\psi^*(\mathbf{x}, -t),\tag{5.21}$$

that is, $V\psi^*(\mathbf{x}, -t)$ is a solution inverted in time. Thus, an operator T', which transforms a wavefunction under the time inversion:

$$\psi(\mathbf{x}, -t) = T'\psi(\mathbf{x}, t),\tag{5.22}$$

is an antiunitary operator.

Let us recall the main properties of an antiunitary operator B:

$$B(c_1\psi_1(x) + c_2\psi_2(x)) = c_1^*B\psi_1(x) + c_2^*B\psi_2(x),\tag{5.23}$$

$$\int (B\psi_1(x))^\dagger(\psi_2(x))d\mathbf{x} = \int \psi_2(x)^\dagger\psi_1(x)d\mathbf{x}.\tag{5.24}$$

Since any antiunitary operator can be presented as a product of an unitary operator and complex conjugation operator K, then Eq. (5.22) acquires the form:

$$\psi(\mathbf{r}, -t) = T'K\psi(\mathbf{r}, t) = T\psi^*(\mathbf{r}, t),\tag{5.25}$$

where an explicit form of unitary operator T depends on representation wherein a wavefunction is specified. To replace a wavefunction by a complex conjugated one under the time inversion means, that the physical state of a system cannot be the eigenstate of the operator T. Due to this reason there is no corresponding quantum number. That is the main difference between the space and time inversion. A more detailed description of the time inversion transformation can be found in the book by Wigner [13].

Until 1964 it had been considered that in the microscopic scale the time direction was not measurable, that is, all laws of microworld were invariant under the time inversion. Observation of a decay $K_L^0 \to \pi^+\pi^-$ became a first indirect demonstration of the T-invariance violation [14] (we supply the long-living K^0-meson with the symbol L while its short-living fellow is denoted by K_S^0). The reaction:

$$K_L^0 \to \pi^0\pi^0,$$

which also confirms the existence of this effect, was registered later.

5.3   Charge conjugation

Besides the space and time inversions we can define the charge conjugation operation. This operation leaves invariable space-time particle characteristics (coordinates, momenta, spins, etc.), but for all that, changes every particle for antiparticle

$$C\psi(x) - \eta_c\psi^c(x), \tag{5.26}$$

where $\psi(x)$ ($\psi^c(x)$) is a wavefunction of a particle (antiparticle), η_c is a phase factor, for which $|\eta_c|^2$ is true, and C is a linear unitary transformation with properties:

$$C = C^\dagger = C^{-1}. \tag{5.27}$$

We emphasize, that a particle differs from an antiparticle not only in the sign of the electric charge (Q) and consequently, in the sign of the magnetic moment, but also in signs of other quantum numbers, such as the baryon charge (B), lepton number (L), strangeness (s), charm c, and so on. Therefore, the charge conjugation operator anticommutes with operators of corresponding "charges"

$$CQ = -QC, \qquad CB = -BC, \qquad CL = -LC, \qquad \ldots \tag{5.28}$$

Eqs. (5.28) are in conflict with the conditions of the simultaneous and precise eigenvalues measurement of the operators C, $Q, B, S, ..$:

$$[C, Q] = [C, B] = [C, L] = [C, S] = ... = 0. \tag{5.29}$$

Consequently, a charge parity (eigenvalue of the operator C) can be defined only for true neutral particles and systems, for example: π^0, γ, e^-e^+, $\mu^-\mu^+$... The charge parity is a multiplicative quantum number and it can take values ± 1. If an initial system has a definite C-parity, then from the invariance under the charge conjugation it follows that a final system must possess the same parity. Experimental data show that strong and electromagnetic interactions are symmetric with respect to the C-operation, while a weak interaction is not symmetric.

By way of exercise let us define the charge parities of some particles. We start with the photon. Since electric and magnetic fields are caused by static and moving charges, then:

$$\mathbf{E} \xrightarrow{\ C\ } -\mathbf{E}, \qquad \mathbf{B} \xrightarrow{\ C\ } -\mathbf{B}. \tag{5.30}$$

From Eq. (5.30) it is evident, that:

$$A_\mu \xrightarrow{\ C\ } -A_\mu, \tag{5.31}$$

that is, the charge parity of the photon is $\eta_\gamma = -1$. Recalling, that the π^0-meson decay:

$$\pi^0 \to 2\gamma$$

is motivated by an electromagnetic interaction, which conserves the C-parity, we obtain $\eta_{\pi^0} = 1$.

5.4   G-parity

So, we have learned, that eigenstates of the charge conjugation operator can be only states with the total electric charge being equal to zero. Taking into account the usefulness of selection rules with respect to the C-parity for neutral systems, we should like to have a conserved quantum number, somehow connected with the C-conjugation for charged states too. It appears to be possible for strong interactions because they conserve isotopic spin. We shall demonstrate this on the example of the isotopic triplet of the π-mesons.

Let us represent the triplet wavefunction in the form of a column matrix:

$$\begin{pmatrix} \psi_{\pi_1}(x) \\ \psi_{\pi_2}(x) \\ \psi_{\pi_3}(x) \end{pmatrix},$$

To define the neutral and charged pion fields we use the expressions:

$$\psi_{\pi^\pm}(x) = \frac{1}{\sqrt{2}} \left(\psi_{\pi_1}(x) \pm i\psi_{\pi_2}(x) \right), \qquad \psi_{\pi^0}(x) = \psi_{\pi_3}(x). \tag{5.32}$$

Fulfillment of the following relations is obvious:

$$C\psi_{\pi^0}(x) = \psi_{\pi^0}(x), \qquad C\psi_{\pi^\pm}(x) = \psi_{\pi^\mp}(x). \tag{5.33}$$

Consequently,

$$C \begin{pmatrix} \psi_{\pi_1}(x) \\ \psi_{\pi_2}(x) \\ \psi_{\pi_3}(x) \end{pmatrix} = \begin{pmatrix} \psi_{\pi_1}(x) \\ -\psi_{\pi_2}(x) \\ \psi_{\pi_3}(x) \end{pmatrix}. \tag{5.34}$$

Although the states $\psi_{\pi^\pm}$ are not eigenstates of the charge conjugation operator in themselves, and yet, it is possible to build linear combinations from them, which will be eigenstates of the operator C. However, such states do not possess a definite value of the electric charge, and this situation does not suit us at all. Earlier we were dealing with the isotopic spin operators T_i ($i = 1, 2, 3$) that were the rotation generators in the isospace. Looking at Eq. (5.34) we can clearly see the way of obtaining the eigenfunctions of such an operators combination that must contain the operator C. We should make a rotation in the isospace in such a way, that the first and third components of the wavefunction of the π-meson triplet reverse the signs. It can be reached by means of the operator of the rotation in the isotopic space about the second axis by the angle 180^0, that is, by operator $\exp(i\pi T_2)$:

$$\exp(i\pi T_2) \begin{pmatrix} \psi_{\pi_1}(x) \\ \psi_{\pi_2}(x) \\ \psi_{\pi_3}(x) \end{pmatrix} = \begin{pmatrix} -\psi_{\pi_1}(x) \\ \psi_{\pi_2}(x) \\ -\psi_{\pi_3}(x) \end{pmatrix}. \tag{5.35}$$

So, the required combination of the operators is $G = C \exp\left(i\pi T_2\right)$. From the relations:

$$G \begin{pmatrix} \psi_{\pi_1}(x) \\ \psi_{\pi_2}(x) \\ \psi_{\pi_3}(x) \end{pmatrix} = \begin{pmatrix} -\psi_{\pi_1}(x) \\ -\psi_{\pi_2}(x) \\ -\psi_{\pi_3}(x) \end{pmatrix}, \qquad G \begin{pmatrix} \psi_{\pi^+}(x) \\ \psi_{\pi^-}(x) \\ \psi_{\pi_0}(x) \end{pmatrix} = - \begin{pmatrix} \psi_{\pi^+}(x) \\ \psi_{\pi^-}(x) \\ \psi_{\pi_0}(x) \end{pmatrix}, \qquad (5.36)$$

it follows that all the π-mesons have the G-operator eigenvalues (G-parity) equal to -1. Therefore, the G-operator eigenvalues serve as a convenient generalization of the charge parity concept in the case with "charged" (having nonzero quantum "charges") particles.

The operation of the G-conjugation for an arbitrary state is defined as the charge conjugation followed by the rotation by the angle π about the second axis in the isotopic space. From this it follows that the Hamiltonian, being invariant under the charge conjugation and rotations in the isotopic space, is also invariant under the G-conjugation. Thus, G is the "good" quantum number only if the isospin is conserved, that is, the G-parity conservation law works only in the case of a strong interaction.

Since the charge parity is multiplicative and isotopic spin is additive, then the G-parity is a multiplicative quantum number. Thus, for example, a state, containing n of the π-mesons, has the G-parity equal to $(-1)^n$. For this reason transitions between odd and even numbers of the π-mesons under the action of strong interaction are forbidden. From the existence of the decays:

$$\rho^0 \rightarrow \pi^+ + \pi^-, \qquad \omega \rightarrow \pi^+ + \pi^- + \pi^0,$$

(here we are dealing with the ρ^0- and ω-meson with the mass 770 and 782 MeV/c^2) it follows immediately, that the G-parity of the ρ^0 meson equals $+1$, and the G parity of the ω-meson equals -1.

5.5 CPT theorem

CPT theorem concerns properties of the physical systems with respect to the product of the transformations C, P, and T, whose result is:

$$\mathbf{r} \rightarrow -\mathbf{r}, \qquad t \rightarrow -t, \qquad \text{particles} \rightarrow \text{antiparticles}.$$

(sometimes the CPT transformation is called a *strong reflection of the space-time*). This theorem had been proven before 1956. At that time there were no doubts about physical theories invariance with respect to every discrete operation taken separately. First it was established [15], that theories, invariant under the proper Lorentz transformations and the space inversions were also invariant under the time inversion or the charge conjugation. Then G. Luders managed to show [16], that the P invariant relativistic quantum theory of an arbitrary field was automatically invariant under the combined operation CT. In 1957 Pauli gave the general proof of the CPT theorem [17]. Another name of this theorem is the Pauli-Luders theorem. In the Lagrangian formulation of quantized field theory the contents of this theorem is as follows. *"The theory is invariant under the combined operation CPT if and only if the quantum theory of the given field meets the following demands: 1) locality; 2) invariance of the Lagrangian with respect to the proper Lorentz transformations; 3) hermicity of the Lagrangian; and 4) standard connection between the spin and statistics (particles with integer spin are subject to Bose-Einstein statistics, particles with half-integer spin are subject to Fermi-Dirac statistics)."* We underline, that the combined operator CPT is not proportional to the identity operator, that is, speech does not go about the identical transformation.

Violation of the P- and C-invariance, discovered in 1956, as well as the CP-asymmetry, discovered in 1964 hardly caused any changes in theoretical apparatus of quantized field theory, which could naturally involve these discoveries without violating any fundamental principles of theory. The probable experimental detection of the CPT-invariance violation will have tragic consequences for modern quantum field theory and will result in the radical revision of its main principles.

One consequence of the CPT-invariance is the fact that if the interaction is invariant with respect to one of the transformations C, P, or T it must be also invariant with respect to the product of two remaining operators. Analogously, if the interaction is not invariant with respect to one of the three transformations, it must not be invariant with respect to the product of two other ones. Really, already the first experiments on detection of the space parity violation in a weak interaction also demonstrated the violation of the charge symmetry. Thus, for example, at β^--decay:

$$X \to Y + e^- + \bar{\nu}_e, \tag{5.37}$$

where X (Y) is the mother (daughter) nucleus, the electrons have predominantly left-handed polarization, while in the charge conjugated decay the positrons have predominantly right-handed polarization.

For several years after the discovery of the charge and mirror symmetry there was a hope, that at least the CP-symmetry and, on the strength of the CPT-theorem, T-reversibility would remain the firm laws of the nature. However, as we already mentioned before, in 1964 the decay of the long-living neutral K_L^0-meson through the channel:

$$K_L^0 \to \pi^+ + \pi^-, \tag{5.38}$$

was discovered. It proved the absence of the T-symmetry and, consequently, the absence of the CP-invariance of the world we live in. Further detailed studies of (5.38) and other K_L^0-meson decays proved a violation of the CP- and T-invariance, but did not detect any hints on the ruin of the CPT-symmetry.

Unlike the C- and P-asymmetries, all the measured CP asymmetric effects are restricted only by the K_L^0-meson decay through the channels:

$$K_L^0 \to \pi^+ + l^- + \bar{\nu}_l, \qquad K_L^0 \to \pi^- + l^+ + \nu_l,$$

$$K_L^0 \to \pi^+ \pi^-, \qquad K_L^0 \to \pi^0 + \pi^0,$$

where $l = e^-, \mu^-, \tau^-$. As a measure of the CP-violation one may use the amplitudes ratio for the decays of the K_S^0- and K_S^0-meson (recall, that the K_S^0 decay goes with the CP-conservation):

$$\epsilon = \frac{A(K_L^0 \to \pi^+\pi^-)}{A(K_S^0 \to \pi^+\pi^-)} \approx 2.3 \times 10^{-3}. \tag{5.39}$$

The CP asymmetric effects are so little in other kaon decays as well. However, we have all reasons to expect that the effects of the CP-asymmetry turn out, to put it mildly, more significant during studies of the B^0- and $\overline{B^0}$-meson decays. Thus, the standard model predicts, that these effects can reach 0.9.

To find out the mechanism of the CP-violation it would be very important to measure the value of the neutron static (intrinsic) electric dipole moment (EDM) (not to be confused with a dynamic EDM, which arises on a particle in a strong electric field). Both the neutral and charged particles can possess the static EDM. The appearance of the static EDM of the neutron may be caused by the fact that the positive charge Q is distributed over the surface of one part of the neutron hemisphere, while the negative charge, equal to $-Q$, is spread over the other part of the neutron hemisphere. Then, by measuring a deformation, caused by such a charge distribution, we would actually measure the intrinsic EDM. Due to its electric neutrality, stability and a large size (compared to leptons) the neutron is a very convenient object to measure this moment. An interaction energy of a dipole with an external electric field has the form:

$$\mathcal{H}_{int} = (\mathbf{d}_0 \cdot \mathbf{E}) \propto (\mathbf{S} \cdot \mathbf{E}),$$

where d_0 is the intrinsic EDM value of the particle, $\mathbf{S}$ is the particle spin operator and $\mathbf{E}$ is the strength of the external electric field. Under the time inversion $\mathbf{S}$ reverses its sign, while $\mathbf{E}$ does not, that is, this interaction is not CP-invariant. Thus, the CP-invariance implies, that the intrinsic EDM of the neutron and other particles must be equal to zero. The latest experiments with ultracold neutrons gave the following result:

$$| d_n | < 0.97e \times 10^{-25} \text{ cm}$$

One of the significant consequences of the CPT-invariance is that particles and antiparticles must have absolutely identical masses, life-times, and their magnetic dipole moments must be identical in value, but opposite in sign. The CPT-theorem also predicts, that particles and antiparticles interact in the same way with a gravitation field. Up to now there is no experiment confirming violation of the CPT-invariance. The mass equality of particles and antiparticles has been most precisely measured for the K^0- and $\overline{K^0}$-mesons:

$$\frac{|m_{K^0} - m_{\overline{K^0}}|}{m_{K^0}} < 6 \times 10^{-19}.$$

The best accuracy in measuring the life-time (τ) for particles and antiparticles has been reached for muons:

$$\frac{\Gamma_{\mu^+ \to all} - \Gamma_{\mu^- \to all}}{\Gamma_{\mu^+ \to all}} < 3 \times 10^{-5}$$

(recall, that $\tau = \hbar/\Gamma$). The anomalous magnetic moments equality for the electron and positron has been checked with an accuracy of $\sim 10^{-12}$, while for μ^- and μ^+ —with an accuracy of $\sim 10^{-7}$.

In conclusion we address once more the invariance with respect to the charge conjugation. The C-invariance means that physical laws and equations, corresponding to them, are not changed if all particles are replaced by antiparticles. Another way of putting it is that the laws in the "world" and in the "antiworld" are equivalent. At this point the most experienced reader could make the remark: "Since all the matter in our galaxy consists of protons, neutrons, and electrons rather than of antiprotons, antineutrons, and positrons, our world is not C-invariant." Once again we stress, the C-invariance means, that particles and antiparticles interact in an identical way. From the C-invariance it does not in any way follow that in some region of the universe the equal amounts of particles and antiparticles are present. Really, in our local galactic cluster the amount of antimatter is overwhelmingly small: the antimatter fraction constitutes $< 10^{-4}$. To extrapolate this fact over the whole universe bears a name "baryon asymmetry of the Universe" (BAU). The quantity:

$$\delta = \frac{n_B - n_{\overline{B}}}{n_\gamma}, \tag{5.40}$$

where n_B, $n_{\overline{B}}$, n_γ are the density of baryons, antibaryons, and relic photons, is the quantitative measure of the BAU. According to the current data [18] its value is given by:

$$\delta \approx 6 \times 10^{-10}.$$

Explanation of the BAU origin and the value of δ is one of the key problems of modern cosmology and elementary particle physics. Needless to say, one may accept the point of view that the BAU is obliged by the role of an accident at the early stages of world evolution. Then, the value δ should be considered as an initial condition. There is nothing to contradict such an explanation, however, it is not solely possible.

It is more attractive to assume that at its birth moment the universe was still symmetric $n_B = n_{\overline{B}}$ while the asymmetry in the observed part of the universe appeared at some evolutionary stage. In this case two versions are possible: 1) the conservation law of the baryon number B is exact; 2) there are interactions leading to the nonconservation of B. In the first case for the BAU to appear either separation of matter and antimatter in macroscopic scales is needed (which is considered to be feasible with difficulty) or burial of antibaryons into black holes that could separate matter and antimatter when the CP-violation takes place. The second version is realized in the so-called great unification theories (GUTs). In such theories there exist superheavy particles, X-bosons, which are interaction carriers. Their decays are caused by the interaction to violate both the CP-invariance and the baryon number conservation law. For such particles the decays into the u-quarks

$$X \xrightarrow{r} u + u, \qquad \overline{X} \xrightarrow{\overline{r}} \overline{u} + \overline{u}, \tag{5.41}$$

and into the d-quark and a lepton

$$X \xrightarrow{1-r} \overline{d} + \overline{l}, \qquad \overline{X} \xrightarrow{1-\overline{r}} d + l, \tag{5.42}$$

prove to be allowed. In the case of the CP-violation we have $r \neq \overline{r}$. By this reason an excess of quarks over antiquarks (or baryons over antibaryons) appears under $r \geq \overline{r}$. However, that is possible only in the case of thermodynamic nonequilibrium, because in the equilibrium case to increase baryons production velocity leads to the growth of antibaryons production velocity. Thus, the demand, so that the X-decays are nonequilibrium, represents an obligatory condition. Otherwise, the CPT-symmetry conservation would ensure a system neutrality over all conserved charges, that is, in the thermodynamic equilibrium $B = 0$. Under expansion and cooling, according to the Big Bang theory, the universe passes the phase when the CP nonconserving interactions go out of equilibrium. Obviously, during a period that is equilibrium over interactions with the B nonconservation any initial value of B will be wiped out. Under exiting from this period the universe gains $B \neq 0$ at the sacrifice of the above-mentioned microprocesses. The GUT, based on the $SU(5)$-gauge group and supplemented by the Big Bang theory, leads to the value of δ in the interval from $\sim 10^{-6}$ to $\sim 10^{-12}$.

Problems

1.1. Examine four successive infinitesimal rotations:
(a) the rotation about the angle $\delta\varphi_1$ around the 1th axis;
(b) the rotation about the angle $\delta\varphi_2$ around the 2th axis;
(c) the rotation about the angle $-\delta\varphi_1$ around the 1th axis;
(d) the rotation about the angle $-\delta\varphi_2$ around the 2th axis.
Reveal that the sequence of these rotations is equivalent to the rotation of the second order about the angle $\delta\varphi_1 \times \delta\varphi_2$. Using this circumstance prove that the corresponding generators satisfy the relation

$$[J_1, J-2] = iJ_3.$$

1.2. Show that the product of two Levi-Chivita symbols is defined by the relation

$$\epsilon_{ijk}\epsilon_{i'j'k'} = \begin{pmatrix} \delta_{ii'} & \delta_{ij'} & \delta_{ik'} \\ \delta_{ji'} & \delta_{jj'} & \delta_{jk'} \\ \delta_{ki'} & \delta_{kj'} & \delta_{kk'} \end{pmatrix}.$$

1.3. Making use of the induction method prove the following expression for theconvolution of the Levi-Chivita symbols on several indices

$$\epsilon_{a_1\ldots a_l a_{l+1}\ldots a_n}\epsilon_{a_1\ldots a_l b_{l+1}\ldots b_n} = l! \begin{pmatrix} \delta_{a_{l+1}b_{l+1}} & \delta_{a_{l+1}b_{l+2}} & \cdots & \delta_{a_{l+1}b_n} \\ \cdots & \cdots & \cdots & \cdots \\ \cdots & \cdots & \cdots & \cdots \\ \delta_{a_n b_{l+1}} & \delta_{a_n b_{l+2}} & \cdots & \delta_{a_n b_n} \end{pmatrix}.$$

1.4. Let a G-group representation be irreducible. Show that in this case any linear operator U commuting with all the G-group representation operators is multiple to the unit one (Shur lemma). To put it differently, using the relations

$$T(g)U = UT(g), \qquad g \in G$$

prove the equality

$$U = \lambda I,$$

where λ is a constant. Take into account the fact that in a complex space any linear operator has one eigenvalue at least.

1.5. Find all possible states of the composite particle made of the particles with the spin $S = 1/2$ and $S = 3/2$.

1.6. Consider the particle to be described by the mixed spinor of the third rank $\psi_\gamma^{\alpha\beta}$. Find the irreducible representations corresponding to it.

1.7. Find the representation laws for the components of the symmetric $S_{\mu\nu}$ and antisymmetric $A_{\mu\nu}$ four-tensors under the homogeneous Lorentz group transformations.

1.8. Write down the homogeneous Lorentz group transformation formulae for the four-momentum p_μ when the relative velocity of motion of the reference frames under investigation has an arbitrary orientation.

1.9. Two particles having the masses $m_1 = m_2 = m$ and momenta p_1, p_2 come into collision. Determine the value of the Lorentz invariant $s = (p_1 + p_2)_\mu (p_1 + p_2)^\mu$ in the center-of-mass frame ($\mathbf{p}_1 + \mathbf{p}_2 = 0$). Find as well the first particle energy in the system in which the second particle is at rest (laboratory system).

1.10. A particle a collides with a rest particle b:

$$a + b \to \ldots$$

The particle masses (m_a, m_b) and the total energy of the a-particle (E_a) are given. Find:
(a) the laboratory frame reference velocity β_c with regard to the center-of-mass system;
(b) the Lorentz factor $\gamma_c = 1/\sqrt{1 - \beta_c^2}$;
(c) the total energy of the particles in the center-of-mass system.
Assuming the colliding particle masses to be equal, obtain the formula to relate the a-particle Lorentz factor with the Lorentz factor of the center-of-mass frame

$$\gamma_a = 2\gamma_c^2 - 1.$$

How does this formula transform for relativistic energies?

1.11. For second-order matrices, the matrices σ_μ and the unit matrix constitute a basis. Let us confront every real four-vector x_μ with the Hermitian matrix Y according to the following rule:

$$Y = x^\mu \sigma_\mu = \begin{pmatrix} x^0 + x^3 & x^1 - ix^2 \\ x^1 + ix^2 & x^0 - x^3 \end{pmatrix}.$$

Introduce the transformation

$$Y' = UYU^\dagger, \tag{I.1}$$

where U is unimodular matrix (det$U$=1), and find the relation between the matrix U and the homogeneous Lorentz group transformation matrix.

1.12. Find the explicit form of the transformation matrices U (see Eq. (P.1)) in two-dimensional space when:
(a) the U-matrix is Hermitian;
(b) the U-matrix is unitary.
Using the formula obtained in the previous problem

$$\Lambda^\mu_\nu(U) = \frac{1}{2}\mathrm{Sp}(\sigma^\mu U \sigma_\nu U^\dagger),$$

determine the explicit form of the corresponding Lorentz transformation for the four-vector x_μ.

1.13. The Poincare group representation is characterized by the total set of the mutually commuting operators. The Casimir operators $p_\mu p^\mu = m^2$ and $w = -W_\mu W^\mu$ (W_μ is the Pauli-Lubanski vector) always enter to this set. Along with them the momentum operator p_μ and the third spin projection operator S_3 to be defined by W_μ are also included in the total set. In the rest frame let the spin vector be equal to

$$\mathbf{S}_R = \frac{\mathbf{W}_R}{m},$$

and in an arbitrary reference frame represent a linear combination of the W_μ vector components with the coefficients depending on p_ν only.

Prove that the single axial-vector linear combination of the W_μ-operators, which turn to the vector $\mathbf{S}_R$ in the rest frame, is given by the expression

$$S_j = \frac{1}{m}\left(W_j - \frac{W_0 p_j}{m + p_0}\right) = \frac{1}{m}\left\{p_0 \mathcal{M}_j - [\mathbf{p} \times \mathcal{N}]_j - p_j \frac{(\mathbf{p} \cdot \mathcal{M})}{m + p_0}\right\}, \tag{I.2}$$

where $\mathcal{M}_{0j} = \mathcal{N}_j$, $\mathcal{M}_{12} = \mathcal{M}_3, \ldots$ and $\mathcal{M}_{\mu\nu}$ is the proper Lorentz group representation generator (the sum of the orbital and spin field moments). For proof one should take advantage of the following commutation relations

$$[\mathcal{M}_j, S_k] = i\epsilon_{jkl}S_l \qquad (I.3)$$

which are fulfilled owing to the fact that **S** is a three-vector.

1.14. Show that the relation (I.2) is nothing else than the spatial components of the Pauli-Lubanski vector to be transformed in the rest frame reference

$$S_j = \left(\Lambda_s^{-1}W\right)_j,$$

where Λ_s is the limited homogeneous Lorentz transformation, which works as follows

$$(\Lambda_s^{-1})_{\mu\nu}p^\nu = (m, 0, 0, 0).$$

1.15. There is a scalar field with a self-action $\lambda\Phi^4$. The theory Lagrangian is defined by the expression

$$\mathcal{L} = \frac{1}{2}[(\partial_\mu\Phi)(\partial^\mu\Phi) - m^2\Phi^2] - \frac{\lambda}{4!}\Phi^4.$$

Find the motion equations, the energy-momentum tensor and the angular tensor.

1.16. The equations system

$$\left.\begin{array}{l} \partial^\mu\psi_{\mu\nu\lambda} + m^2\psi_{\nu\lambda} + \partial_\nu\psi_\lambda - \partial_\lambda\psi_\nu = 0, \\ \psi_{\mu\nu\lambda} = \partial_\mu\psi_{\nu\lambda} - \partial_\nu\psi_{\mu\lambda} + \partial_\lambda\psi_{\mu\nu}, \\ \psi_\nu = \partial^\mu\psi_{\mu\nu}. \end{array}\right\} \qquad (I.4)$$

Find the Lagrangian to result in the equations.

1.17. Using the Lagrangian obtained in the previous problem write down the expressions for the energy-momentum tensor and the electric current.

1.18. The free Lagrangian of the charged scalar field is given by

$$\mathcal{L} = \frac{1}{2}[(\partial_\mu\Phi)^*(\partial^\mu\Phi) - m^2\Phi^*\Phi].$$

Having demanded the theory invariance with respect to the local gauge transformations build up the total Lagrangian describing the electromagnetic interaction of the charged scalar particles.

1.19. The Pauli principle for identical particles reads: *"a fermion system wavefunction is antisymmetric under rearrangement of any two particles while a boson system wavefunction is symmetric relative to the same rearrangement."* Making use of this principle find:
(a) the isotopic spin of the ρ-meson to decay through the channel

$$\rho \to \pi^+ + \pi^-$$

$(S_\rho = 1)$;
(b) the isotopic spin of the deuteron regarding it as the $^3S_1$ state of the (np) system.

1.20. Show that the recharge reaction of the slow π mesons under collision with the deuterium

$$\pi^- + d \to n + n + \pi^0$$

is forbidden when the orbital moment of the π^0 meson regarding the (nn) system and the orbital moment of the π^- meson regarding the deuterium are equal to zero. Under proof set the deuterium parity to be equal to $+1$.

1.21. Determine the positronium charge parity. Point out the more probable decay channels of the parapositronium ($^1S_0$) and the orthopositronium ($^3S_1$).

1.22. Using the G parity conservation law in the strong interaction establish whether the following reactions

$$p + \bar{p} \to \pi^+ + \pi^-,$$

$$p + \bar{p} \to \pi^0 + \pi^0$$

are possible.

References

[1] S. Tomonaga, Progr. Theor. Phys. **1**, 27 (1946).

[2] J. Schwinger, Phys. Rev. **75**, 651 (1949).

[3] H. Umezawa, *Quantum field theory*, (North-Holland PC, Amsterdam, 1956).

[4] F. Rohrlich, Phys. Rev. **80**, 666 (1950); P.T. Matthews, Phys. Rev. **76**, 1657 (1949).

[5] E. Stuckelberg, Helv. Phys. Acta, **11**, 225 (1938).

[6] O. M. Boyarkin, JETP, **48**, 13 (1978).

[7] A. Proca, J. phys. et rad. **7**, 347 (1936).

[8] O. M. Boyarkin, J. Phys. G: Nucl. Phys. **8**, 161 (1982).

[9] H. P. Stapp, Phys. Rev. **125**, 2139 (1962)

[10] E. P. Wigner, G. C. Wick, A. S. Wightman, Phys. Rev. **88**, 101 (1952).

[11] E. Noether, Nachr. Ges. Wiss. Gottingen, 171 (1918).

[12] S. Sakata, Progr. Theor. Phys. **16**, 686 (1956).

[13] E. P. Wigner, *Group Theory*, (New York, 1959).

[14] J. H. Christenson *et al.*, Phys. Rev. Lett. **13**, 138 (1964).

[15] L. C. Biedenharn, M. E. Rose, Phys. Rev. **83**, 459 (1951); H. A. Tolhoek, de Groot, Phys. Rev. **84**, 151 (1951).

[16] G. Luders, Zeit. Phys. **133**, 325 (1952).

[17] W. Pauli, in *Niels Bohr and the Development of Physics*, (Pergamon Press, London, 1957).

[18] D. N. Spergel *et al.*, Astrophys. J. Suppl. **170**, 377 (2007).

Part II

Birds's-Eye View on the Microworld

6

Fundamental interactions

Once, however, there was an eagle who became terribly bored with living all alone. The eagle began to brood. And the more he brooded, the more he began to think how wonderful it would be to live as the serf-owners of old lived. Why not gather a company of varlets and live like a lord? Let the crows bring him the latest gossip, the parrots turn somersaults, magpies cook porridge, horned owls, screech-owls and barn owls keep watch at night, and hawks, falcons and vultures supply him with food. And he, the eagle, would devote himself to bloody tyranny.

M.E. Saltykov Shchedrin, "The Eagle-Patron of Arts"

6.1 Species of interactions

The main purpose of physics is to explain all natural phenomena by means of a few simple fundamental principles. Since all matter consists of particles, then elementary particles physics must give the final answer. By this, one of the main subjects is the question about the nature of interactions. It appears that despite of the diversity of the world surrounding us, all interactions are reduced to four fundamental types, which differ greatly in intensity of their proceeding. Such a classification is not free from some conventionality due to the fact that the relative role of different interactions is changed while the energy of interacting particles grows. It means that the separation of interactions into classes, based on the comparison of processes intensity, can be made reliably only for not very high energies. Let us consider the types of fundamental interactions in order of decreasing their intensity.

Strong interaction. This is an overwhelming type of interaction in nuclear physics of high energies. Particles, which participate in strong, weak, and gravitation interactions, are called *hadrons*. Hadrons with half-integer spin are called *baryons*, while hadrons with integer spin are called *mesons*. In addition, the charged hadrons participate in electromagnetic interaction. Just the strong interaction causes couplings between protons and neutrons in the atomic nuclei and ensures an exclusive strength of these formations lying at the heart of matter stability in Earth conditions. The strong interaction is also responsible for the confinement of quarks inside hadrons. This interaction may be manifested not only as an ordinary attraction in the nucleus, but as a force, which causes instability of some elementary particles (particles to be decaying by means of the strong interaction are called *resonances*). Due to its great value, the strong interaction is a source of huge energy. In particular, the main part of heat in the Sun is produced by the strong interaction, when deuterium nuclei, made by the weak interaction, together with protons are synthesized into helium nuclei. The strong interaction is a short-range interaction with a radius of 10^{-13}

cm.

Only after the quark structure of hadrons had been discovered, the strong interaction theory ceased to resemble the plots of Russian popular tales. This theory, called *quantum chromodynamics* (QCD), resembles *quantum electrodynamics* (QED) in construction. However, the QED has a local gauge symmetry with respect to the group $U(1)_{em}$, while a local gauge symmetry of the QCD is associated with the $SU(2)_c$-group. Note, that both symmetries are internal ones, that is, they are connected with a system symmetry not in the ordinary space-time but in abstract spaces. The lower index c in the QCD symmetry group is caused by the fact, that quarks, beside ordinary quantum numbers, have three additional degrees of freedom, for which we use the conventional term "color" or color charge R (red), G (green), and B (blue). For both the QCD and QED internal symmetries are exact.

The subject of internal symmetry violation is of great importance in quantum field theory. There are two mechanisms of symmetry violations, namely, explicit and spontaneous. Under explicit violation the Lagrangian contains terms, which are not invariant with respect to a symmetry group. The value of these terms characterizes the degree of corresponding symmetry violation. Thus, for example, the Lagrangian of the strong interaction is variant under isotopic transformations, but the total Lagrangian also contains the electromagnetic and weak interactions, which explicitly violate the isotopic symmetry. For this reason the complete theory does not possess the exact isotopic invariance.

Under the spontaneous symmetry violation the Lagrangian possesses the invariance with respect to the transformations of the internal symmetry group; vacuum (vacuum is the state with the minimum energy), however, loses this invariance. Vacuum noninvariance reveals itself through the fact, that one or more components of a quantized field (as a rule these components correspond to scalar particles) acquire the nonzero vacuum averages $< 0|\varphi_i|0 >$ (or vacuum expectation values), which define various energy scales of the theory. Under the spontaneously violated local symmetry the corresponding gauge bosons, which are interaction carriers, prove to be massive particles, while under the exact symmetry these gauge bosons are massless. Thus, carriers of the strong interaction between quarks, which we call *gluons*, are massless particles.

In all observable hadrons the color charges of the quarks are compensated, that is, hadrons are colorless (white) formations. Hadron colorlessness can result either from mixing three main colors (true for baryons) or from the mixture of color and anticolor (true for mesons). In the strong interaction the color charges of the quarks play the same role as the electric charges of particles do in the electromagnetic interaction. The color charge is a source of the gluon field. As this takes place the gluon carries on itself both the color and anticolor charges, that is, its color composition is a product of color and anticolor. When the quark emits the gluon its color changes, depending on a gluon color. For instance, the red quark, emitting the red-antiblue gluon, turns blue. Analogously, the blue quark, absorbing the red-antiblue gluon, turns red, etc. A total of $3 \times 3 = 9$ "color-anticolor" combinations are possible. Among them there is one corresponding to the colorless state:

$$g_0 = R\overline{R} + B\overline{B} + G\overline{G}.$$

Since under its emitting or absorbing the quark state is not changed, then this combination cannot play the role of the gluon bearing interaction between quarks. Thus, only eight gluons are left. The gluons are electrically neutral, have zero-mass and spin equal to 1. All this makes them similar to the photons. But unlike the photons, the gluons have a "charge" of the field whose interaction they bear. The gluon can emit or absorb other gluons, changing its own color in so doing. That is, the gluons create the new gluon field around themselves. The photons are deprived of such a property, they have no electric charge and no new electric field is created around them. The electromagnetic field is the most intensive near the charge, which causes the field and further away it is permanently dispersed in space and weakened. Color-charged gluons produce around themselves new gluons, which produce new gluons and so on. The result is, that the gluon field is not decreasing, but increasing further away from the quark creating this field. In other words, effective color charges of the quarks and gluons are increasing as the distance grows. At the distances of the hadron size order ($\sim 10^{-13}$ cm) the color interaction becomes really strong. The perturbation theory, the main mathematical apparatus in microworld physics, is not applicable in this domain, so there are no reliable calculations. However, one could expect on qualitative grounds, that strengthening the interaction with distance must result in the fact that separating of isolated quarks at large distances becomes impossible. To put this another way, it brings to life imprisonment of the quarks in hadron prison. This phenomenon is called *confinement*.

The baryon consists of three differently colored quarks. The quarks are constantly exchanging the gluons and changing their own color. These changes, however, are not arbitrary. Mathematical apparatus of the QCD restricts severely this play of colors. At any moment of time the summarized color of three quarks must represent the sum $R + G + B$. Mesons consist of quark-antiquark pairs, every pair is colorless. Then, no matter what gluons the quark-antiquark pairs are exchanging the mesons also remain white formations. So, from the QCD standpoint, the strong interaction is nothing else but the tendency to maintain the $SU(3)_c$-symmetry, resulting in conservation of the white color of hadrons, while their components change their colors.

The strong interaction intensity is characterized by the so called QCD running coupling constant:

$$\alpha_s(q) = \frac{g_s^2}{4\pi},$$

where g_s is a gauge constant of the $SU(3)_c$-group. The term "running" reflects dependence of α_s on a distance or on a transferred momentum q. We are reminded, that to estimate a quantity order in the microworld one may use the Heisenberg uncertainty relation. Therefore, the transition to short (large) distances means transition to large (small) values of a transferred momentum $q \sim \hbar/r$.

Evolution of the running coupling constant of the QCD is governed by the equation:

$$\alpha_s(q) = \frac{12\pi}{(33 - 2n_f)\ln(q^2/\Lambda_{QCD}^2)}, \tag{6.1}$$

where n_f is a number of quark kinds (at given color) or a number of quark flavors, and Λ_{QCD} is a scale parameter of the QCD. Derivation of this equation is based on the use of the perturbation theory and the structure of the total Lagrangian describing the theory. At $q^2 \gg \Lambda_{QCD}^2$ the effective constant $\alpha_s(q)$ is small and, consequently, the perturbation theory describes the behavior of weakly interacting quarks and gluons successfully. At $q^2 \sim \Lambda_{QCD}^2$ it is impossible to use the perturbation theory while strongly interacting gluons and quarks start to form coupled systems, hadrons. Obviously, the parameter Λ_{QCD} defines the border between the world of quasifree quarks and gluons and the world of real hadrons. The value of Λ_{QCD} is not predicted by the theory. It is a free parameter that is determined from experiments. Nowadays, despite the joint efforts of experimentalists and theorists, the exact value of Λ_{QCD} remains unknown (its approximate value lays between 100 and 200 MeV). The formula (6.1) leads to the decrease of an effective interaction as a momentum grows and in the asymptotic ultraviolet limit the effective interaction tends to zero. Then the fields, participating in interaction, become free. The phenomenon of self-switching off interaction at short distances that is a reverse side of confinement, is called *asymptotic freedom*.

Since the hadrons are neutral with respect to the color, then at distances, bigger than the hadron size, there is no strong interaction between the hadrons at all. It is similar to the absence of electromagnetic forces between atoms at big distances, since they are electrically neutral. However, when two or more atoms approach at a distance, when their electron clouds are overlapping, the so-called Van der Waals forces, or chemical forces come into action. Their radius of action is of the atom size order. Molecular bond is caused by these forces. Its mechanism is based on the exchange of electrons between atoms, that is, the molecular bond is a complicated manifestation of the fundamental electromagnetic interaction between two spatially-distributed charged systems. Analogously, the hadron interaction also can be viewed as a complicated manifestation of the fundamental strong interaction between the color quarks, which becomes observable only under approaching the quark cores of the hadrons.

Electromagnetic interaction. The electromagnetic interaction keeps electrons within atoms and binds atoms in molecules and crystals. This interaction lies at the basis of nearly all the phenomena around us—chemical, physical, and biological. Elementary particles with the electric charge take part in the electromagnetic interaction. Neutral particles could also interact with the electromagnetic field due to multipole moments (dipole, quadrupole, anapole, etc.). However, only particles with a composite structure can have such moments. Thus, the electromagnetic interaction is not as universal as the gravitation one. It is easy to observe the electric and magnetic forces, acting between macroscopic bodies. These forces, just as the gravitation forces, are subjected to the inverse square law, that is, they are long-range forces. For instance the Earth magnetic field extends far into cosmic space, and the Sun magnetic field fills all the solar system.

Unification of the electric and magnetic forces into the classical theory of the electromagnetic field, made by Maxwell in the 1850s, is an example of the first unified field theory. The idea of unification can be illustrated by the example of the Lorentz force:

$$\mathbf{F} = q\{\mathbf{E} + [\frac{\mathbf{v}}{c} \times \mathbf{H}]\}. \tag{6.2}$$

Let us assume, that $|\mathbf{E}| \sim |\mathbf{H}|$. Then from Eq. (6.2) it follows, that at $|\mathbf{v}| \ll c$ the magnetic forces are very small compared to the electric ones and approach in their value to them only at $|\mathbf{v}| \to c$. Thus, the relative intensity of the forces is defined by the particle velocity, that is, there is the scale on which unification of the electric and magnetic fields takes place and this scale is determined by the light velocity. Since the energy is also growing at $|\mathbf{v}| \to c$, we can state, that unification occurs in the region of the ultrarelativistic energies of particles.

The quantum theory of the electromagnetic interaction of the electrons and positrons, called the QED, had been built up at the beginning of the 1950s. The QED is the most exact of all physical theories. Here the electromagnetic interaction is exhibited in its pure form. Unprecedented accuracy of calculations in the QED is caused by usage of apparatus of the perturbation theory on the small dimensionless parameter:

$$\alpha_{em}(q) = \frac{e^2}{4\pi\hbar c},$$

which is called a *fine structure constant*. $\alpha_{em}(q)$ is also a function of a distance or a transferred momentum. Its macroscopic value, which is defined at $q = m_e c$, equals to $1/137.0359895(61)$. The exact symmetry of the QED with respect to the local gauge group $U(1)_{em}$, whose gauge constant is equal to the electron charge, leads to the fact that the electromagnetic interaction carrier, the photon, has zero mass.

The QED perfectly describes not only the electrons, but electromagnetic properties of other charged leptons as well. Contrary to this, electromagnetic properties of the hadrons are not amenable to calculations because the hadrons are basically controlled by the strong interaction. We stress, that the QED is not only the first model of the quantum field theory (QFT), but it is also the simplest and the most extensively studied version of the QFT. Within the framework of the QED many fundamental concepts of the QFT were discovered and formulated. All this allows to build up more complicated quantum field theories in the image and similarity of the QED.

In the QED the phenomenon of vacuum polarization results in the screening of the electron charge by vacuum positrons. When polarizing the vacuum the electron attracts virtual positrons and repulses virtual electrons. As a result, the electron charge is partly screened, if one sees it from a large distance. If one penetrates deep inside of a cloud of virtual pairs, then screening would decrease and the effective electron charge would increase. In other words, contrary to the QCD running constant that is increasing with the distance, the QED running constant is decreasing as the distance grows. The calculations, made within the QED scope, define the evolution of $\alpha_{em}(q)$ by means of the equation:

$$\alpha_{em}(q) = \frac{3\pi\alpha_{em}(m_e c)}{3\pi - \alpha_{em}(m_e c)\ln\left[q^2/(4m_e^2 c^2)\right]}. \tag{6.3}$$

From Eq. (6.3) it follows that at $q \sim 80$ GeV/c the value of α_{em} is approximately equal to $\sim 1/128$.

Weak interaction. The weak interaction is destructive in its character because it is not able to create stable states of matter in the way, as, for example, the gravitation force maintains the existence of the solar system or the electromagnetic interaction ensures atom stability. In other words, the main destination of the weak interaction is to regulate a lifetime of inanimate matter. It is responsible for nuclear $\beta^{\pm}$-decays, for decays of particles not belonging to the resonance class (we call such particles "stable"). Only a few stable particles, for example, π^0-meson, η^0-meson, and Σ^0-hyperon decay due to the electromagnetic interaction.

If electromagnetic multipole moments of a neutrino are equal to zero, then all the processes with the neutrino participation are caused by the weak interaction only. The weak interaction is also responsible for nuclear and atomic processes going with parity violation. Particles, participating in the weak interaction and with the availability of the electric charge in the electromagnetic interaction but not participating in the strong ones, are called *leptons*.

In some cases the weak interaction also influences macroscopic objects. For example, it plays a key role in the Sun energy release, because deuterium nucleus production from two protons is caused by just this interaction:

$$p + p \to {}^2D + e^+ + \nu_e.$$

Neutrino emission in weak interaction processes defines stars evolution, especially at their final stages, initiates supernova explosions and pulsar production. If it were possible to switch off the weak interaction, then the matter around us would acquire quite another structure. It would contain all particles to decay due to the weak interaction (muons, $\pi^{\pm}$-mesons, K-mesons, etc.).

Intensity of the weak interaction is defined by a Fermi constant

$$G_F = 1.16639(1) \times 10^{-5}(\hbar^3 c^3) \text{ GeV}^{-2},$$

which is dimensional as we see. In the reference frame, where a particle rests the probability of the decay Γ due the weak interaction turns out to be proportional to $G_F^2 m^5$ (m is a mass of a decaying particle). In virtue of the Heisenberg uncertainty relation, an elementary particle lifetime τ is inversely proportional to Γ. For particles, decaying due to the weak interaction, the value of τ is quite large by microworld scales and lies in the interval 10^3–10^{-10} s. The lifetime of a particle decreases as the intensity of interaction, causing decay, grows. For particles, which instability is caused by the electromagnetic interaction, τ is of the order of $\sim 10^{-16}$ s, while for particles decaying because of the strong interaction, τ is of the order of 10^{-23}–10^{-24} s.

Notwithstanding the fact that the first process, caused by the weak interaction, the radioactive β^--decay of the nucleus, had been discovered by A. Becquerel in 1896, attempts of constructing the weak interaction theory was crowned with success only in the 1960s. For the construction of this theory Glashow, Salam, and Weinberg were awarded the Nobel prize in 1979. In this theory both the electromagnetic and weak interactions are the manifestations of one and the same interaction, which is called an *electroweak (EW) interaction*. The local symmetry with respect to transformations of the $SU(2)_{EW} \bigotimes U(1)_{EW}$-gauge group, possessing the two gauge coupling constants g and g', makes the base of the theory. In this case, there are two peculiarities, which make the EW interaction different from both the QED and QCD.

First, the local gauge symmetry of the EW interaction is spontaneously violated up to the local gauge symmetry of the QED:

$$SU(2)_{EW} \bigotimes U(1)_{EW} \rightarrow U(1)_{em}.$$

Second, from the very beginning the theory is not invariant with respect to the operation of the space inversion.

The nonviolated local symmetry $SU(2)_{EW} \bigotimes U(1)_{EW}$ demands the existence of four massless particles with spin 1, two of which are neutral and the remaining two are charged. It was known from experiments, that the action radius of the weak interaction R_W is extremely small $\sim 10^{-16}$ cm. Consequently, carriers of this interaction must have masses of the order of $\sim \hbar/(R_W c)$. To give the mass to the gauge bosons of the weak interaction, a doublet of massless scalar fields, Higgs bosons, consisting of neutral and charged components is introduced into the theory. In so doing, the neutral Higgs boson is not properly neutral. Due to the spontaneous symmetry violation (the nonzero vacuum average from a neutral component of the Higgs doublet is chosen) three of the gauge bosons of the group $SU(2)_{EW} \bigotimes U(1)_{EW}$ acquire masses, while the forth one remains massless. The massive gauge bosons, $W^\pm$ and Z, are identified with the gauge bosons of the weak interaction and the massless gauge boson, γ, is identified with the photon. Out of four massless scalar fields, one neutral field acquires mass and the remaining three leave physical sector, as if they were eaten by the gauge bosons while they are gaining their masses. From the massless vector field with two spin states and the massless scalar field the massive vector particle with three spin projections is produced, so, the number of degrees of freedom is conserved. The idea of generating the mass through interaction with the filled vacuum was stated for the first time by J. Schwinger [1]. In this work nucleons were the investigation subject. Schwinger's idea was transferred by P. Higgs on vector particles [2]. The mass production of the gauge field due to the spontaneous violation of the local symmetry is called a *Higgs mechanism*.

The neutral currents discovery (1973), that is, observing the processes going through the virtual Z-boson exchange, should be considered as the first confirmation of the Glashow-Weinberg-Salam (GWS) theory. The detection of the $W^\pm$- and Z-bosons at the proton-antiproton collider in CERN (1983) was the genuine triumph of the theory. Up-to-date values of the gauge boson masses are:

$$m_W = 80.41 \pm 0.10 \text{ GeV}/c^2, \qquad m_Z = 91.187 \pm 0.007 \text{ GeV}/c^2.$$

The predicted coincidence of the generations number of quarks and leptons (quark-lepton symmetry) also should be attributed to the success of the GWS theory.

We emphasize that, although the GWS theory unifies the weak and electromagnetic interactions, the gauge constants of the $SU(2)_{EW}$- and $U(1)_{EW}$-groups, g and g', are not connected with each other. These constants satisfy the relations:

$$\tan \theta_W = g'/g, \qquad g = e \sin \theta_W, \tag{6.4}$$

where there are no theoretical predictions for the angle θ_W (Weinberg angle) and it is defined from experiments only.

By now lots of data have been accumulated, which prove, that experiments fit the theory perfectly. However, the main problem in the EW interaction theory is not solved yet, namely, the mechanism of violating the original $SU(2)_{EW} \bigotimes U(1)_{EW}$-symmetry is not established. The most direct way to solve this problem is experimental searching for the Higgs boson. Since the theory does not predict its mass m_H, then the range of researching for it is rather wide.

The fact, that in the world surrounding us, we discriminate the electromagnetic and weak interactions, only means that their unification scale or the boundary of the spontaneous symmetry violation in the EW interaction theory lies on the higher energy scale $\sim m_W c^2 = 80.4$ GeV, that corresponds to the distances of the order of 10^{-16} cm.

To compare different interactions it is convenient to use dimensionless quantities. For this purpose we introduce a quantity α_2, which characterizes the intensity of the weak interaction according to the relations:

$$\alpha_2(q) = \frac{g^2}{4\pi}, \qquad \frac{g^2}{8m_W^2} = \frac{G_F}{\sqrt{2}(\hbar c)^3}. \tag{6.5}$$

Gravitation interaction. Although the gravitation interaction is the weakest of all, it possesses a cumulative effect. So, the gravitation interaction between two bodies is a cumulative sum of interactions between elementary masses that form these bodies. Since in the microworld the contribution of the gravitation interaction is very small compared to other interactions, it does not result in measurable effects on the subatomic level. However, on the macroscopic level the gravitation interaction is dominating: it keeps together parts of the terrestrial globe, unifies the Sun and planets into the solar system, connects stars in galaxies and controls the evolution of the whole universe.

Since the gravitation interaction was discovered in the first place, then just with its help the term "force" appeared in physics. The gravitation interaction is universal, because it is in operation between all bodies having the mass. It belongs to long-range interactions. Building the nonrelativistic gravitation theory was completed by I. Newton in 1687. According to this theory two mass points, having the masses m_1, m_2 and lying at a distance r, are attracted with the force, whose value and direction are given by the expression:

$$\mathbf{F} = G_N \frac{m_1 m_2 \mathbf{r}}{r^3},$$

where G_N is a Newton constant, $G_N = 6.67259(85) \times 10^{-8}$ cm^{3}g^{-1}s^{-2}. In this theory the force depends on the particles position at a given time only and so the gravitation interaction propagates instantly.

At an arbitrary mass distribution the gravitational force, operating on any point mass m_0 in the given spatial point, can be expressed as a product of m_0 on a vector $\mathbf{E}_g$ which is called the *gravitation field strength*. In the Newton theory the superposition principle is valid for the gravitation field. As this field is potential it is possible to introduce, by the usual fashion, the gravitation potential

$$\mathbf{E}_g = -\text{grad } \varphi_g.$$

The potential of a continuous distribution of a matter density $\rho(\mathbf{r})$ satisfies the Poisson equation

$$\Delta\varphi_g = -4\pi G_N \rho(\mathbf{r}). \qquad (6.6)$$

The Newton gravitation theory has allowed us to describe with great precision an extensive range of phenomena, including the motion of natural and artificial bodies in the solar system, the motion of celestial bodies in other systems: in binary stars, in stellar clusters, and in galaxies. On the basis of this theory the existence of the planet Neptune and the satellite of Sirius has been predicted. In modern astronomy the gravitational law of Newton is the foundation on the basis of which the motions, the structure of celestial bodies, their masses and evolution are calculated. The precise definition of the Earth's gravitational field allows us to define a mass distribution under Earth's surface and, hence, to directly solve the important applied problems.

As the Newton theory assumes the instantaneous propagation of the gravitation it cannot be made consistent with the special relativity theory, stating that the interaction propagation velocity cannot exceed c. It means, that this theory is inoperable when gravitational fields are so strong, that they accelerate bodies, moving in them, up to the velocities of the order of c. The velocity v, up to which the body falling freely from infinity ($v\,|_{t=0} \approx 0$) as far as some point with a gravitational potential $\varphi_g(\mathbf{r})$ has been accelerated, can be found from the relation

$$\frac{mv^2}{2} = m\varphi_g(\mathbf{r})$$

(we have put $\varphi(\infty) = 0$). Hence, the Newton theory of gravitation is applicable only in the case when

$$|\varphi_g| \ll c^2.$$

For the gravitational fields of usual celestial bodies this requirement is fulfilled. For example, on the Sun surface we have $|\varphi_g|/c^2 \approx 4 \times 10^{-6}$, and on the surface of the white dwarfs $|\varphi_g|$ is about 10^{-3}.

Besides the Newton theory is inapplicable under calculation of the particles motion even in a weak gravitational field with $|\varphi_g| \ll c^2$ if particles flying near massive bodies, had the velocity $v \sim c$ already far from them. Hence, it will lead to the improper answer under calculation of the light trajectory in a gravitational field. The Newton theory also is not used under investigation of the varying gravitational fields created by moving bodies (for example, binary stars) at distances $r > c\tau$, where τ is a period of revolution in a system of a binary star. Really, the Newton theory, based on the instantaneous propagation of the interaction, is unable to take into account the retardation effect that appears to be essential in this case.

The relativistic gravitation theory, that is, the general relativity theory (GRT) was built by Einstein in 1915. It changed drastically understanding gravitation in the classical, Newtonian, physics. In the Einstein theory gravitation is not a force, but a manifestation of the curvature of the space-time. The flat metric of Minkowski $g_{\mu\nu} = \mathrm{diag}(1, -1, -1, -1)$ in the space of the GRT is deformed into a metric

$$\eta_{\mu\nu}(x) = g_{\mu\nu} + h_{\mu\nu}(x),$$

where $h_{\mu\nu}(x)$ describes a massless spin-2 field whose source is a matter. The two postulates make the foundation of the GRT. The first one defines the form of the Lagrangian density $\mathcal{L}_g$, describing a propagation and a self-action of the gravitation field. On the basis of the second postulate, namely, an equivalence principle, the gravitation interaction is introduced by means of substitution $g_{\mu\nu} \to \eta_{\mu\nu}(x)$ into the Lagrangians of all the existing fields, that is, into $\mathcal{L}_{QCD} + \mathcal{L}_{EW}$. Variation of the total Lagrangian $\mathcal{L}_g + \mathcal{L}_{QCD} + \mathcal{L}_{EW}$ with respect to the gravitation potentials $\eta_{\mu\nu}(x)$ leads to the Einstein gravitation equations

$$R_{\mu\nu}(\eta) - \frac{1}{2}R(\eta)\eta_{\mu\nu}(x) = \frac{8\pi G_N}{c^4}T_{\mu\nu}(\eta), \qquad (6.7)$$

where $R_{\mu\nu}(\eta)$ is a Ricci tensor

$$R_{\mu\nu}(\eta) = \partial_\alpha\Gamma^\alpha_{\mu\nu} - \partial_\nu\Gamma^\alpha_{\mu\alpha} + \Gamma^\beta_{\alpha\beta}\Gamma^\alpha_{\mu\nu} - \Gamma^\beta_{\nu\alpha}\Gamma^\alpha_{\mu\beta},$$

$\Gamma^\lambda_{\mu\nu}$ are Christoffel symbols, which play the role of the gravitation field strength

$$\Gamma^\lambda_{\mu\nu} = \frac{1}{2}\eta^{\lambda\sigma}(\partial_\mu\eta_{\nu\sigma} + \partial_\nu\eta_{\mu\sigma} - \partial_\sigma\eta_{\mu\nu}) \qquad (6.8)$$

$R(\eta) = R_{\mu\nu}(\eta)\eta^{\mu\nu}$ and $T_{\mu\nu}$ is a symmetric energy-momentum tensor of matter.

Outwardly the Einstein equations (6.7) are similar to Eq. (6.6) for the Newtonian potential. In both cases the quantities characterizing the field stand in the left-hand side and the quantities characterizing matter that creates this field do in the right-hand side. However, between these equations there is a number of essential differences. Eq. (6.6) is linear and, consequently, satisfies the superposition principle. It allows computing a gravitational potential for any distribution of masses moving arbitrarily. The Newton gravitation field does not depend on the masses motion, therefore Eq. (6.6) in itself does not define immediately their motion. To describe the mass motion we must invoke the second Newton law. In the Einstein theory a pattern is quite different. As Eq. (6.7) are nonlinear, the superposition principle does not work any more. Further, in this theory it is impossible to set arbitrarily a right-hand side of Eq. (6.7) (that is, $T_{\mu\nu}$), which depends on a matter motion, and then to calculate the gravitational field $\eta_{\mu\nu}$. The solution of the Einstein equations leads both to the definition of the motion of matter generating the field and to the evaluation of the field itself. In so doing it is essential, that the gravitation field equations also contain the mass motion equations in the gravitation field. From the physical point of view it is equivalent to the fact, that in the Einstein theory the matter creates the space-time curvature which, in its turn, influences the motion of matter originating this curvature. In the GRT all particles move along extremal lines, called *geodesic curves*. In the flat space-time geodesic curves degenerate into straight lines. Notice, that in the ordinary field theory operating the flat space-time, the motion equations are also obtained by means of the extremum condition, however, this condition is imposed on the system action. The stronger is the gravitation field, the more appreciable is the curvature of the space-time. Thus, the nonrelativistic gravitation theory is not applicable, when the gravitation fields are very strong, as it occurs near collapsing objects like neutron stars or black holes. On the other hand, in weak fields one may be restricted by the calculation of small corrections to the Newton equations. The effects corresponding to these corrections, allow testing the GRT experimentally in the ordinary gravitational fields as well.

For today the experimental status of the basic statements of the Einstein theory is as follows. To check the principle of equivalence of the gravitational and inert masses is carried out with a precision of 10^{-12}. The theoretical formula for changing the light frequency (red shift) in the gravitational field that also is a consequence of the equivalence principle, is verified with a precision of 2×10^{-4}. The constancy with time of G_N being postulated by the theory was tested by radar observations of the motions both of planets (Mercury, Venus) and of spaceships, by measuring the Moon motion with the help of the laser, by observations of the motion of a neutron star, namely, pulsar PSR 1913+16, which forms part of a double star-shaped system. All the collection of the experimental data confirms the constancy of G_N with a precision of

$$\left| \frac{1}{G_N} \frac{dG_N}{dt} \right| < 10^{-11} \text{ years}^{-1}.$$

The GRT predicts bending the light ray when it is passing near the heavy mass. The analogous bending follows from the Newton theory as well, however, in the Einstein theory this effect is twice more. Numerous observations of this effect being done under passage of the light coming from the stars near the Sun (during the complete solar eclipse) have confirmed the GRT predictions with a precision up to $\sim 11\%$. The much more accuracy ($\sim 0.3\%$) has been already reached under the observation of the extraterrestrial point radiation sources.

The Einstein theory also predicts the slow rotation of the elliptic orbits of the planets spinning around the Sun. It should be emphasized that this rotation is not explained by the gravitational fields of other planets. The effect has the greatest magnitude for the Mercury orbit—$43''$ in a century. At present the verification precision of this prediction (precession of the Mercury perihelion) reaches 0.5%.

The GRT effects should be rather considerable when the stars are moving in a tight double system. With the greatest precision the motion of the pulsar PSR 1913+16 forming a part of the double star is explored. Here the orbit rotation due to the GRT effects attains 4.2% in one year, and for 14 observations years (1975–1989) has given $\sim 60^0$.

One further the GRT effect is the prediction, that the bodies, moving with variable acceleration, will radiate gravity waves. Despite numerous attempts one has not managed to register gravity waves as far. However, there are the serious grounds in support of their existence now already. For example, the observations of the pulsar PSR 1913+16 have confirmed an energy loss of the double system due to the radiation of the gravity waves. As a consequence of the effect, the period of the star circulation should decrease with time. The observations confirm the GRT prediction with the precision of 1%.

The GRT describes not only the large-scale phenomena in the universe, it also determines the evolution of the universe itself. And just here the final experimental checkout of the theory is being awaited. The first relativistic cosmological model, based on Eqs. (6.7), was built by Einstein in 1917. Proceeding from the considerations conventional for classical science, Einstein suggested that the universe, as a totality, should be eternal and invariable. However, Eqs. (6.7) were unable to describe the stationary universe. Because of this, Einstein has introduced the Λ-term, now known as the cosmological constant, in Eqs. (6.7)

$$R_{\mu\nu}(\eta) - \frac{1}{2} R(\eta) \eta_{\mu\nu}(x) = \frac{8\pi G_N}{c^4} T_{\mu\nu}(\eta) - \Lambda \eta_{\mu\nu}(x), \tag{6.9}$$

where $\Lambda > 0$. As a result, the last term in (6.9) describes the repulsive gravitational forces complementary to the attractive gravitational forces of the normal matter ($T_{\mu\nu}$). Nominally, the cosmological term is equivalent to the additional term of the energy-momentum tensor. Recalling the analogy between the Poisson equation for the gravitational potential in the Newtonian theory and the Einstein equation, the emergence of the similar term in the Newtonian gravitation theory were equivalent to the introduction of an additional force acting on the body from an object having a negative mass M_0

$$\mathbf{F_0} = G_N \frac{m M_0}{r^3} \mathbf{r}.$$

From the beginning, the function of the cosmological constant was to create or, what is more accurate, to describe antigravitation. Einstein assumed that in this way it is possible to balance the gravitation of the universe matter and ensure an immovability of matter distribution, that is, stationarity of the universe. Such a model cannot answer the question: "How and from where had the universe originated?" As regards this, the theory keeps a delicate silence. Nevertheless, no longer than 15 years later the astrophysical observations made the scientists give up the model of the stationary universe.

At the beginning of the 1920s A. Friedmann demonstrated that by the appropriate selection of a metric the equations of the GRT have nonstationary solutions with the cosmological term present as well. In Friedmann models based on the homogeneous and isotropic universe matter is considered as a continuous medium, uniformly filling the space and having specific values of the density ρ and pressure P at every instant of time. To analyze the motion of such a medium, the co-moving frame of reference is usually used, similar to the Lagrangian coordinates in the classical hydrodynamics. In this system, the matter is motionless, deformation of the matter being reflected by that of the reference system, and hence the problem is reduced to the description of the reference system deformation. The three-dimensional space of the co-moving frame of reference is referred to as a co-moving space. For a homogeneous and isotropic space the square of the four-dimensional interval ds may be represented in the form:

$$ds^2 = \eta_{\mu\nu}dx^\mu dx^\nu = c^2 dt^2 - a^2(t)\frac{dx^2 + dy^2 + dz^2}{1 + k(x^2 + y^2 + z^2)/4}, \tag{6.10}$$

where x, y, z are dimensionless space coordinates, $a(t)$ is a radius of a space curvature and $k = -1, 0, 1$. It should be noted that under selecting the metric we have already assumed that the universe is nonstationary. The space curvature is positive at $k = 1$ and negative at $k = -1$. Provided $k = 0$, the space is Eucledian (flat), and $a(t)$ has the meaning of a scale factor. Variation of $a(t)$ in time describes expansion or compression of the co-moving reference frame and hence the matter. The metric in (6.10) is known as the Friedmann–Robertson–Walker metric that forms a basis for modern cosmology. It is convenient to rewrite the expression (6.10) in the spherical coordinates:

$$ds^2 = c^2 dt^2 - a^2(t)\left[\frac{dr^2}{1 + kr^2} + r^2 d\theta^2 + r^2 \sin^2\theta d\phi^2\right], \tag{6.11}$$

that is, the nonzero components of the metric tensor $\eta_{\mu\nu}$ $(\mu, \nu = t, r, \theta, \phi)$ have the form

$$\eta_{tt} = 1, \qquad \eta_{rr} = -\frac{a^2(t)}{1 + kr^2}, \qquad \eta_{\theta\theta} = -a^2(t)r^2, \qquad \eta_{\phi\phi} = -a^2(t)r^2\sin^2\theta. \tag{6.12}$$

Using

$$\eta_{\mu\nu}\eta^{\nu\sigma} = \delta^\sigma_\mu,$$

we obtain

$$\eta^{tt} = 1, \qquad \eta^{rr} = -\frac{1 + kr^2}{a^2(t)}, \qquad \eta^{\theta\theta} = -a^{-2}(t)r^{-2},$$

$$\eta^{\phi\phi} = -a^{-2}(t)r^{-2}\sin^{-2}\theta. \tag{6.13}$$

To solve the problem about the deformation of the reference frame, it remains only to find the unknown function $a(t)$. The dynamics of the homogeneous and isotropic universe may be described similarly to a model for ideal liquid with the density $\rho(t)$ and pressure $P(t)$, averaged over all the galaxies, their clusters, and superclusters. Then, a hydrodynamic energy-momentum tensor for the matter is given by:

$$T_{\mu\nu} = P\eta_{\mu\nu} + (P + \rho c^2)U_\mu U_\nu, \tag{6.14}$$

where $U^t = 1$, $U^i = 0$. As seen from the calculations, for the metric (6.12) the following components of the Christoffel symbols are nonzero:

$$\Gamma^t_{ij} = \frac{a\dot{a}}{a^2}\eta_{ij}, \qquad \Gamma^i_{tj} = \frac{\dot{a}}{a^3}\delta^i_j, \qquad \Gamma^i_{jk} = \frac{1}{2}\eta^{il}\left(\frac{\partial\eta_{lj}}{\partial x^k} + \frac{\partial\eta_{lk}}{\partial x^j} - \frac{\partial\eta_{jk}}{\partial x^l}\right). \tag{6.15}$$

Taking into account Eq. (6.15) we obtain the following expression for the components of the Ricci tensor

$$R_{tt} = \frac{3\ddot{a}}{a}, \qquad R_{ti} = 0, \qquad R_{ij} = \frac{1}{a^2}\left(a\ddot{a} + 2\dot{a}^2 + 2k\right)\eta_{ij}. \tag{6.16}$$

The time components of the Einstein's equation give

$$\frac{\ddot{a}}{a} = -\frac{4\pi G_N}{3}\left(\rho + \frac{3P}{c^2}\right) + \frac{\Lambda c^2}{3} \tag{6.17}$$

while the purely space components of that lead to the relation

$$\frac{\ddot{a}}{a} + \frac{2\dot{a}^2}{a^2} + \frac{2k}{a^2} = 4\pi G_N\left(\rho - \frac{P}{c^2}\right) + \Lambda c^2. \tag{6.18}$$

Omitting $\ddot{a}$ from Eqs. (6.17), (6.18) we arrive at the first-order differential equation for $a(t)$

$$\frac{\dot{a}^2}{a^2} + \frac{k}{a^2} = \frac{8\pi G_N}{3}\rho + \frac{\Lambda c^2}{3}. \tag{6.19}$$

Eq. (6.17) describes changes in the expansion speed of the universe under the effect of gravitation. From this equation it follows that gravitation is due not only to the matter density but also to its pressure in the combination $\rho c^2 + 3P$, referred to as the effective gravitating energy of the matter ρ^G_{mat}. It is obvious that in this case the cosmological term will result in antigravitation since its effective gravitating energy is negative. To find the function $a(t)$ and determine a cosmological model by this means, it is necessary to know for some t the values of density $\rho(t_0) = \rho_0$ as well as the cosmological constant $\Lambda(t_0) = \Lambda_0$. Usually, instead of ρ_0 one uses the quantity $\Omega = \rho_0/\rho_c$, where $\rho_c = 3H^2/(8\pi G_N)$ is a critic matter density in the universe and H is a Hubble constant whose value is defined by experiments.

In whatever the scale the mass calculations for the universe should be performed, the deficiency of mass will be always revealed. Dynamically, a behavior of the galaxies themselves as well as (super)clusters is so as if they contain much more matter than is really available in their visible components, referred to as a luminous matter or baryon matter. The present-day value of the cosmic density (average over the whole observable world) for the baryon matter is determined by:

$$\Omega_B = \frac{\rho_B}{\rho_c} = 0.02 \pm 0.01. \tag{6.20}$$

Apart from the baryon matter in the universe there is a hidden mass (lately referred to as dark matter). Two special types of dark matter exist: a cold dark matter and a hot dark matter. The cold dark matter (CDM) is composed of nonrelativistic objects and its present density is given by:

$$\Omega_D = \frac{\rho_D}{\rho_c} = 0.3 \pm 0.1. \tag{6.21}$$

The CDM forms a vast invisible corona, or halo, around the stellar disk of the Milky Way. Similar dark halos seem to be present in all sufficiently massive isolated galaxies. The CDM is also contained in galactic clusters and superclusters. As with our galaxy, it makes up about 90% and sometimes more of the total mass for all these systems. There is no emission or absorption of electromagnetic waves by the CDM that manifests itself exclusively through the created gravitation. Unfortunately, the nature of the CDM has not been conclusively established up to the present. A wide variety of the possibilities is considered: from weakly interacting massive elementary particles to massive (exceeding a mass of the Sun) black holes, etc.

The third component of the cosmological medium is hot dark matter (HDM). The HDM comprises ultrarelativistic particles with masses equal to zero or of the order of $\sim$ eV. The density of this medium is determined by the expression:

$$\Omega_R = \frac{\rho_R}{\rho_c} = 0.8 \times 10^{-5}\alpha, \tag{6.22}$$

where the constant factor $1 < \alpha < 10$–30 includes the contribution of neutrinos, gravitons, and other possible ultrarelativistic particles.

In the final years of the twentieth century one more component of the cosmological medium, a cosmic vacuum, was discovered as a result of a series of experiments conducted by two big research collaborations of astronomers. As it turned out, a vacuum (a cosmic vacuum is often referred to as dark energy) predominates in the universe, with the energy density making it superior over all the "ordinary" forms of the cosmic matter taken together

$$\Omega_V = \frac{\rho_V}{\rho_c} = 0.73 \pm 0.01. \tag{6.23}$$

This means that 67% of the total matter of the world falls on the vacuum, 30%—on the cold dark matter, about 3%—on the baryon matter, and all the rest is connected with radiation.

The modern observations also indicate that cosmological expansion is proceeding with acceleration. On the other hand, acceleration may be exclusively due to antigravitation, whose origin is the cosmological term Λ in the Einstein equations. It is presently universally acknowledged that the cosmic vacuum is described by the cosmological constant. The vacuum density is related to the value of this constant by the following relation:

$$\rho_V = \frac{\Lambda}{8\pi G_N}.$$

From this point of view, we can relate the evolution of the universe to its genesis: expansion of matter stems from antigravitation of the cosmological vacuum, the matter per se appearing as a result of quantum fluctuations of the same vacuum.

Nevertheless, despite the impressive successes, the GRT has some unresolved problems. For example, the quantization of the GRT faces serious difficulties. So, it follows from the Einstein field equation, that the gravitation field theory does not belong to the class of renormalizable theories. Let us explain what we mean talking about a renormalization. As we know, the mathematical apparatus of quantum theory is mainly based on the usage of the perturbation theory series. In the four-dimensional field theories these series contain infinitely large quantities, which one must be removed in one way or another. Normally that is reached by means of the redefinition of finite number of physical parameters, such as the mass, the charge, etc. This procedure is called *renormalization* and the theories, in which it eliminates divergences, are called the *renormalizable theories*. For non-renormalizable theories there is no procedure to ensure convergence of the perturbation theory series. The presence of a dimensional interaction constant makes the ordinary renormalization procedure impossible. To eliminate divergences in the theory, we must summarize all the terms in the corresponding series of the perturbation theory. As a result, some divergences are canceled, and the remaining infinities are eliminated by the renormalization of the physical parameters of the theory. However, if the interaction constant is dimensional, then the terms in the perturbation theory series have the different dimensions and their summation has no sense.

In the GTR under expanding the metric tensor $\eta_{\mu\nu}$ in a power series near the flat space with the metric $g_{\mu\nu}$, the interaction constant κ appears

$$\eta_{\mu\nu} \approx g_{\mu\nu} + 2\kappa h_{\mu\nu}, \tag{6.24}$$

where κ is equal to $\sqrt{2G_N}$. Then in the first nonvanishing approximation the gravitation field equations have the form

$$\Box h_{\mu\nu} = G_N T_{\mu\nu}$$

and, as it is easy to see, in the stationary case they transfer to Newton's equations of the gravitation field. However, the interaction constant κ proves to be a dimensional quantity and, as a result, the ordinary theory of the renormalization does not work. Another difficulty connecting with the quantization of the GRT has the experimental nature. A particle, creating the gravitation field, a graviton, has not been yet discovered. The theory predicts for it zero mass, zero electric charge, and spin being equal to 2.

The Minkowski space considered in the special relativity theory (that is, in absence of gravitating bodies) has the high level of a symmetry that is described by the Poincare group. According to the relativity principle this group generates isomorphic sequences of events. In the space where the gravitation field presents, the symmetry disappears completely. Therefore, in such a space the relativity principle is not fulfilled. For this reason the title, the general relativity theory, belonging to Einstein, is inadequate and gradually disappears from literature, replaced with "the gravity theory."

Just as for other interactions, it is possible to introduce a nondimensional intensity of the gravitation interaction $\alpha_g(q)$. It is defined as follows:

$$\alpha_g(q) = \frac{G_N m^2}{4\pi\hbar c}. \tag{6.25}$$

The main difference in the above-enumerated interactions is the strength of their manifestation in nature. There are different ways to compare the interaction's intensity. One of them is based on values of corresponding energy effects. Thus, for example, the electromagnetic interaction can be characterized by the binding energy of an electron in the ground state of a hydrogen atom, $E_{em} \approx 10$ eV, while the energy effect of the strong interaction can be determined from the binding energy of nucleons in a nucleus, $E_s \approx 10$ MeV.

Another way is to compare the running coupling constants, which describe different interactions. However, since these quantities are energy functions, we must point out the energy value, at which the comparison takes place. One should remember, that the running constants of the groups $SU(2)_{EW}$ and $U(1)_{EW}$ (α_2 and $\alpha_1 = g'^2/4\pi$) cannot be identified with the running constants of the weak and electromagnetic interactions. The operation is legal only at energies much less than the energy, at which the spontaneous violation of the local symmetry of the electroweak interaction takes place. At the scale 1 GeV the running coupling constants of the strong, electromagnetic, and weak interactions are connected by the relation

$$\alpha_s : \alpha_{em} : \alpha_W \approx 1 : 10^{-2} : 10^{-6}.$$

As soon as the gravitation interaction is switched on, a confusing indefiniteness appears. What elementary particle should be taken as a standard? Now, the mass is the "charge" of the gravitation interaction, but the mass spectrum of elementary particles is continuous. So, for example, the ratio of the Coulomb and gravitation forces has the form

$$\frac{F_c}{F_g} \approx 10^{36} \tag{6.26}$$

for the protons and

$$\frac{F_c}{F_g} \approx 10^{43} \tag{6.27}$$

for the electrons. Using Eqs. (6.26) and (6.27) we arrive at two different intensity hierarchies

$$\left. \begin{array}{l} \alpha_s : \alpha_{em} : \alpha_W : \alpha_G \approx 1 : 10^{-2} : 10^{-6} : 10^{-38}, \\ \alpha_s : \alpha_{em} : \alpha_W : \alpha_G \approx 1 : 10^{-2} : 10^{-6} : 10^{-45}. \end{array} \right\} \tag{6.28}$$

6.2 On the path to a unified field theory

Grand Unification Theory. Extension of the electroweak theory to quarks and inclusion of the QCD resulted in unification of the strong and electroweak interactions. This created scheme was called the *standard model (SM)*. The picture of fundamental forces in this model is charmingly simple. The strong, weak, and electromagnetic interactions are caused by the existence of the local gauge symmetry group

$$SU(3)_c \bigotimes SU(2)_{EW} \bigotimes U(1)_{EW}$$

with its three gauge constants g_s, g, and g' and twelve gauge bosons being the carriers of the strong and electroweak interactions. At sufficiently small distances all these forces mainly resemble each other and lead to the potential of the Coulomb type $\sim g^2/r$. Scale of short distances for the strong interaction represents distances much smaller than the hadron size, that is, more small than 10^{-13} cm. For the electroweak interaction the scale of the small distances is the distance that is much smaller than the Compton wavelength of the $W^{\pm}$- and Z-bosons ($\lambda_c = \hbar/mc$), that is, more small than 10^{-16} cm. It is obvious, that at such short distances the existence of the masses on the gauge bosons is becoming inessential.

Since the SM gauge group is the production of three unbound sets of gauge transformations: the groups $SU(3)_c$, $SU(2)_{EW}$ and $U(1)_{EW}$, then the three gauge constants of these groups are not connected with each other. The gauge constants will be bound, if the SM gauge group proves to be embedded in a more wide group of gauge transformations G. Symbolically, it is written as follows:

$$G \ni SU(3)_c \bigotimes SU(2)_{EW} \bigotimes U(1)_{EW}.$$

As a result, all the interactions will be described by the unified gauge theory, the Grand Unification theory (GUT), with one gauge constant g_{GU} and at the same time all the other gauge constants are connected with g_{GU} in an unambiguous way, defined by the choice of the group G. The GUT symmetry must be spontaneously violated at supershort distances being many orders smaller than those, at which unification of the electromagnetic and weak interactions takes place. In other words, the strong interaction with the local $SU(3)_c$-symmetry, as well as the electroweak interaction with the local $SU(2)_{EW} \bigotimes U(1)_{EW}$-symmetry turn out to be the low energy fragments of the gauge interaction with the group G.

To estimate the distance scale, at which the Grand Unification takes place, one should turn to the equations defining evolution of the running constants of the strong and electroweak interactions. In so doing, it is necessary to represent these equations in such a form so that they determine the constants variation not as a function of the transferred momentum q, but as a function of variation of the mass scale μ. So, we suppose that the G-group exists. The charges of the $SU_c(3)$-, $SU_{EW}(2)$-, and $U_{EW}(1)$-subgroups are indicated by the symbols g_3, g_2, and g_1 respectively. When $\mu = M_{GU}$ these three constants are merged into the one constant of the Grand Unification, that is,

$$g_3(\mu) = g_2(\mu) = g_1(\mu) = q_{GU}, \qquad \text{at } \mu \geq M_{GU}.$$

Once $\mu < M_{GU}$, then the constants $g_i(\mu)$ are being separated and, at long last, shall pass into the phenomenological constants g, g', and g_s describing the observed interactions under energies less than $m_W c^2$.

Let us find the link between the interaction constants in the models with the groups $SU_c(3) \times SU_{EW}(2) \times U_{EW}(1)$ and G under $\mu \geq M_{GU}$. The Lagrangian determining the interaction between the quarks and gluons is connected with the $SU_c(3)$-group while the Lagrangian describing interaction between the fundamental fermions (the quarks and leptons) and the gauge bosons of the electroweak interaction is connected with the $SU_{EW}(2)$- and $U_{EW}(1)$-groups. Apart from the γ-matrices, the wavefunctions of bosons and fermions, these Lagrangians also contain generators and the interaction constants of the corresponding groups. When $G \supset SU(3)_c \bigotimes SU(2)_{EW} \bigotimes U(1)_{EW}$, then the generators of the $SU_c(3)$-, $SU_{EW}(2)$-, and $U_{EW}(1)$-subgroups must be the generators of the G-group too. It means that the normalization of all the generators must be the same. For non-Abelian groups the normalization is fixed by the nonlinear commutation relations between the generators defining their Lie algebra. For this reason, the normalization of the G-group generators is just the same as that of the $SU_c(3)$- and $SU_{EW}(2)$-group generators. From this follows

$$g_s(\mu) = g_3(\mu), \qquad g(\mu) = g_2(\mu). \tag{6.29}$$

Since there are no nonlinear constraints on the generator of the Abelian group $U_{EW}(1)$, then this generator should be multiplied by the coefficient c in order to ensure the right normalization. Thus, we have

$$g'(\mu) = cg_1(\mu), \tag{6.30}$$

where c is defined by the G-group structure. The relations (6.29), (6.30) are fulfilled in the limit of the G-group, that is, under $\mu \geq M_{GU}$. Now we should consider the region $\mu < M_{GU}$. The cause of changing the gauge coupling constants is the vacuum polarization, that is, it is stipulated by the processes of creation and consequent destruction of the virtual particles. If one neglects the contribution coming from the Higgs bosons then the coupling constant evolution in the $SU(3)_c$-, $SU(2)_{EW}$-, and $U(1)_{EW}$-gauge groups shall be described by the equations (see, for detail, Ref. [3]):

$$\frac{dg_i}{d(\ln \mu)} = -b_i g_i^3, \tag{6.31}$$

where

$$b_1 = -\frac{n_f}{24\pi^2}, \qquad b_2 = \frac{22 - 2n_f}{48\pi^2}, \qquad b_3 = \frac{33 - 2n_f}{48\pi^2}.$$

To integrate the relations (6.31) leads to the result

$$g_i^{-2}(\mu) = g_i^{-2}(\mu_0) + 2b_i \ln(\mu/\mu_o). \tag{6.32}$$

Thus, according to the theory, the dependence of $1/\alpha_i = 4\pi g_i^{-2}$ on $\ln M$ is linear, its slope value being defined by the polarization effect of the relevant vacuum. Discrepancy of the paths on which the three gauge coupling constants $1/\alpha_i(\mu)$ converge to $1/\alpha_{GU}$ is caused by discrepancy of the coefficients b_i in Eq. (6.32). It is easy to guess that the larger value of the slope of $1/\alpha_s$ compared to the slope of $1/\alpha_W$ is caused by the fact that the number of the gluons is larger than the number of the carriers of the weak interaction (the $W^{\pm}$- and Z-bosons) by a factor of two and, as a result, the gluons give the bigger antiscreening effect. In $1/\alpha$ the screening effect predominates (the tangent of the slope angle is negative by now) and for this reason the value of $1/\alpha$ drops with the growth of M.

On the unification scale $\mu = M_{GU}$ we have

$$g_{GU}^{-2} = g_i^{-2}(\mu_0) + 2b_i \ln(M_{GU}/\mu_o). \tag{6.33}$$

To find M_{GU} we should exclude n_f and g_{GU} from Eqs. (6.33). For this purpose we produce a linear combination

$$\frac{c^2}{g_1^2} + \frac{1}{g_2^2} - \frac{1+c^2}{g_3^2} = 2[c^2 b_1 + b_2 - (1+c^2)b_3] \ln \frac{M_{GU}}{\mu_0}, \tag{6.34}$$

where $g_i = g_i(\mu_0)$. With allowance made for Eq. (6.4) the left-hand side of Eq. (6.34) takes the form

$$\frac{1}{g'^2} + \frac{1}{g^2} - \frac{1+c^2}{g_3^2} = \frac{1}{e^2} - \frac{1+c^2}{g_3^2}. \tag{6.35}$$

To substitute (6.35) into (6.34) gives the desired equation for determining M_{GU}

$$\ln \frac{M_{GU}}{\mu_0} = \frac{48\pi^2}{22(1+3c^2)} \left[\frac{1}{e^2} - \frac{1+c^2}{g_3^2} \right] = \frac{6\pi}{11(1+3c^2)} \left[\frac{1}{\alpha} - \frac{1+c^2}{\alpha_s} \right], \tag{6.36}$$

where the absence of subscript "em" by α underlines the circumstance that the case in point is the fine structure constant not in the QED, but already in the more precise theory, namely, in the theory of the electroweak interaction. Setting the values of μ_0, $\alpha(\mu_0)$ and $\alpha_s(\mu_0)$ one may evaluate both the value M_{GU} at which the relation (6.36) is fulfilled and the value of the unified constant $\alpha_{GU}(M_{GU})$. Analogous calculations give one of the basic parameters of the GWS theory, $\sin \theta_W$,

$$\sin^2 \theta_W = \frac{1}{1+3c^2} \left[1 + 2c^2 \frac{\alpha}{\alpha_s} \right]. \tag{6.37}$$

Set $\mu_0 = m_W$ and take into account that $\alpha(m_W) \approx 1/128$ and $\alpha_s(m_W) \approx 0.1$. Further we choose the $SU(5)$-group as the G-group. That gives the value $5/3$ for the quantity c. Then for $\sin^2 \theta_W$ we obtain ~ 0.23 that is amazingly close to the experimental value. This suggests that to estimate the unification scale order one may use the $SU(5)$-group. Then Eq. (6.36) gives $\ln(M_{GU}/m_W) \approx 29$ and we obtain $M_{GU}/m_W \approx 4 \times 10^{12}$. It is not difficult to find $\alpha_{GU}(M_{GU})$, it proving close to 0.02. One may carry out the analogous calculations not concretizing the G-group, but under the condition that all the known fermions (or every fermion families individually) form the total representation of the G-group [4]. At $\mu_0 = m_W$ and with an allowance made for the contribution in vacuum polarization coming from the SM Higgs boson the dependence of the running constants α_s, α, and α_2 on μ is represented in Fig. 6.1. As it follows from Fig. 6.1, in this case the Grand Unification takes place at $M_{GU} \approx 2 \times 10^{14}$ GeV/c^2. The more detailed analysis leads to the result: $M_{GU} \approx 2 \times 10^{15} \Lambda_{QCD}$. The obtained values correspond to the distances $L_{GU} \approx \hbar c/E_{GU} \sim 10^{-28}$ cm. At distances, shorter than L_{GU}, the initial symmetry is

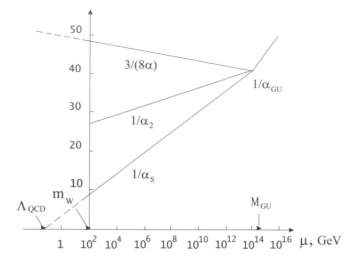

FIGURE 6.1

The μ-dependence of the running coupling constants α_s, α, and α_2.

restored and the interaction is described by the single constant α_{GU}, which evolutionary law is defined by the structure of the G-group.

At this point it is natural to pose a question: "whether it is possible to trust calculations, altogether ignoring the gravitation interaction." Gravitation effects are of the order of 1, when the masses of interacting particles have such values, that the potential gravitation energy is comparable to the particle rest energy

$$\frac{G_N M_P^2}{r} \approx M_P c^2. \tag{6.38}$$

If we set the distance between masses equal to the Compton wavelength, that is, to quantity $\hbar/M_P c$, then the condition (6.38) is realized at

$$M_P = \left(\frac{\hbar c}{G_N}\right)^{1/2} = 1.22 \times 10^{19} \text{ GeV}/c^2. \tag{6.39}$$

The obtained mass value is called the *Plank mass*. The time and the length, corresponding to it

$$t_P = \sqrt{\frac{\hbar G_N}{c^5}} \approx 5.4 \times 10^{-44} \text{ s}, \qquad L_P = \sqrt{\frac{\hbar G_N}{c^3}} \approx 1.6 \times 10^{-33} \text{ cm}, \tag{6.40}$$

are called the *Plank time and length*. Since $L_{GU} \gg L_P$, then on the Grand Unification scale we have the right to neglect gravitation effects. The choice of a GUT group G is defined by the demand: *it must contain $SU(3) \bigotimes SU(2)_{EW} \bigotimes U(1)_{EW}$ as a subgroup and has representations in which all the known quarks and leptons may be included.* Since the $SU(5)$-group is the minimal group satisfying this demand the first to be investigated was the $SU(5)$-model (Georgi-Glashow model [5]). The schema based on the $SO(10)$-group [6] is the direct generalization of the $SU(5)$-model. Probably, in laboratory conditions we shall never be able to produce energies, corresponding to the unification scale, consequently, the experimental check of the GUT is a very complicated task. Among the GUT consequences being available for observation we note the predictions of such effects as proton instability and neutron-antineutron oscillations (neutron transformation into antineutron in the vacuum and the reverse process). In the electroweak interaction the gauge constants g and g' are not bound together and their ratio $g/g' = \tan\theta_W$ is an experimentally obtained parameter. Opposite to that, the GUT allows us to calculate the Weinberg angle. The GUT models explain naturally electric charge quantization, manifested through the fact that quark charges are multiples of $e/3$, while lepton charges are equal either to e or to 0.

The GUTs have some cosmological consequences as well. According to an adopted point of view, our universe came into being approximately 2×10^{10} years ago as a result of the Big Bang and it is still expanding. This expansion is described by the GRT equations. The universe size changed from the value of the Plank length order to a contemporary value, which is of the order of 10^{28} cm. A substance, compressed in a volume of the order of L_P^3, began its evolution with the energy of the Plank order, that is, the early universe is a gigantic laboratory, where the GUT consequences could be checked. Within the GUT framework it is possible to get an explanation of the fact, that the matter at the moment is prevailing over the antimatter in the universe (baryon asymmetry). The value of the ratio of the baryons concentration n_B to the photons concentration n_γ in cosmic microwave background can be also obtained in the context of the GUT.

Along with the above-mentioned achievements, there are some weak points in the existing GUT. Let us enumerate some of them. The models have a rich variety of free parameters, which number exceeds the number of those in the SM. It is not possible to make any statement concerning the number of fermion generations within the framework of the models. The gravitation is excluded from the unification scheme. Serious difficulties are produced under explanation of difference by twelve orders of the distance scales at which the Grand Unification symmetry G and the electroweak interaction symmetry are broken (the hierarchies problem).

Supersymmetry. Is there a more "grandiose" unification, a unified field theory, which includes both the gravitation and the SM? Before we discuss the directions in which the construction of the theory is going, let us first get acquainted with one more type of symmetry— a supersymmetry (see, for review, [7]).

Up to now we have been considering the space-time and internal symmetries. Geometrical translations and rotations do not change the nature of a particle: the photon remains the photon after any space-time transformations. The internal symmetries can change the nature of a particle but not its spin value. So, under the action of the isotopic rotations the proton can turn into the neutron, but it cannot transform into the π^0-meson, for instance.

Unlike the above-mentioned symmetries, supersymmetry transformations can change not only the space-time coordinates of a particle and its nature, but its spin value as well. In other words, the supersymmetry implies the invariance of a physical system under fermion-boson transitions, that in its turn, allows us to call it the *Fermi-Bose symmetry*. The basis for supersymmetric theories is the extension of the space-time up to the superspace, which besides the normal space-time coordinates x^μ, also includes the spinor coordinates θ_α. The spinor variables θ_α anticommute with each other and commute with the space-time coordinates x^μ

$$\left.\begin{aligned} \{\theta_\alpha, \theta_\beta\} = 0 \qquad & \alpha, \beta = 1, 2, 3, 4, \\ [\theta_\alpha, x^\mu] = 0 \qquad & \mu = 0, 1, 2, 3. \end{aligned}\right\} \tag{6.41}$$

It follows from (6.41) that θ_α and x^μ can be viewed as odd and even generatrices of a Grassmann algebra, respectively. θ_α is usually called *Grassmann coordinates*. Anticommutation of θ_α is necessary to ensure the right connection between the spin and statistics. The desire to describe bosons and fermions on the same level became the reason of introducing θ_α. To simplify the matter, we may say, that the classical coordinates correspond to bosons, while the Grassmann coordinates θ_α correspond to fermions. The important consequence of the anticommutation of the Grassmann coordinates is their nilpotency

$$\theta_\alpha^2 = 0.$$

In the most general case, points in superspace are characterized by four even coordinates x^μ and $4N$ odd coordinates θ_α^j, where $j = 1, 2, .., N$ (N-extended supersymmetry). Let us restrict ourselves to the supersymmetry with $N = 1$. In the ordinary four-dimensional space-time there is the Poincare transformations group with 10 parameters, while in the superspace the expanded Poincare group with 14 parameters comes into action, where besides the rotations and translations in ordinary space the supertranslations

$$\left.\begin{array}{l} x'^\mu = x^\mu + \frac{i}{2}\bar\epsilon\gamma^\mu\theta, \\ \theta' = \theta + \epsilon, \end{array}\right\} \tag{6.42}$$

are added. In Eq. (6.42) θ is represented in the form of 4-columns consisting of θ_α, the spinor ϵ defined by 4 real parameters sets the supertranslation and $\bar\epsilon = \epsilon^\dagger\gamma_0$. Under the ordinary Poincare transformations θ and ϵ behave as ordinary spinors. Generators Q_α correspond to the supertranslations

$$Q_\alpha = \frac{\partial}{\partial\theta_\alpha} - \frac{i}{2}(\gamma^\mu\bar\theta)_\alpha\frac{\partial}{\partial x^\mu}. \tag{6.43}$$

Besides the ordinary relations of the Poincare algebra (1.3.46)–(1.3.48), the supersymmetry algebra also includes

$$[Q_\alpha, P_\mu] = 0, \tag{6.44}$$

$$[Q_\alpha, M_{\mu\lambda}] = \frac{1}{2}(\sigma_{\mu\lambda})_\alpha^\beta Q_\beta, \tag{6.45}$$

$$\{Q_\alpha, Q_\beta\} = -(\gamma^\mu C)_{\alpha\beta}P_\mu, \tag{6.46}$$

where $\sigma_{\mu\lambda} = i/2[\gamma_\mu, \gamma_\lambda]$, and C is a charge conjugation matrix. The first of these relations (6.44) means, that the supertranslation does not influence shifts in the classical space-time, the second one means, that Q_α is transformed under the Lorentz transformations as a ordinary spinor. Due to the anticommutation relation (6.46) internal degrees of freedom of a particle are connected with external space-time degrees of freedom. Both continuous transformations, carried by the generators P_μ and $M_{\mu\lambda}$, and the discrete ones, corresponding to the generators Q_α, turn out to be unified in superalgebra.

The procedure of theory construction is as follows. All the fields are substituted for a superfield $\Phi(x, \theta)$ and expansion in terms of the Grassmann variables θ_α is done. In so doing, it appears, that instead of an infinite series we obtain the finite number of expansion members of $\Phi(x, \theta)$. It is caused by properties of the Grassmann algebra, in which products can be composed of no more components than the number of algebra generatrices. As soon as two identical generatrices occur during expanding, by means of anticommutation relations (6.41) they can be placed together, and their square is equal to zero. At every member of the obtained series some tensor or spinor function appears according to whether the degree of θ is even or odd. The Lagrangian of the theory is defined in the form of invariant square expressions that consist of the superfield and first derivatives of the superfield with respect to both the classical and Grassmann coordinates. The superaction is defined as an integral of the Lagrangian over all variables. As a result of the integration over θ_α, the four-dimensional action is obtained, which is expressed only by means of four-dimensional spinor and boson functions, the expansion coefficients of $\Phi(x, \theta)$. Thus, the Grassmann variables are not included in the final result. Like fields in the Minkowski space, the superfields are classified according to eigenvalues of the Casimir operators, which are built out of supersymmetry group generators. Quantum numbers are the superspin and the superisospin, which generalize notions of the spin and isospin for ordinary fields. To single out irreducible representations from the superfields, either additional conditions are imposed (to eliminate the redundant superspins) or a gauge invariance demand is invoked. As we see, the prescriptions are the same as for ordinary fields.

Since every boson is associated with a supersymmetric fermion and vice versa, then the number of particles in the theory is doubled. The supersymmetric partners get their names either with the prefix "s" for scalar partners of normal fermions (for example, squark, selectron) or with the ending "ino" for fermionic partners of normal bosons (for example, photino, gravitino).

In supersymmetric theories divergences in perturbation theory series, corresponding to bosons and fermions have the opposite signs and mutually compensate each other. Thus, there is no need for the renormalization at all. In other words, the supersymmetry allows constructing finite, divergence-free theories. The inclusion of the supersymmetry in the SM led to come into being the minimal supersymmetric standard model. Discovery of the superpartners of the known fundamental particles will experimentally prove the existence of the supersymmetry in nature. By now, search of the superpartners have not given the positive results.

Supergravitation. In the 1960s some papers appeared in which the GTR was reformulated in the form of a gauge gravitation theory. For this purpose the symmetry group of the flat space-time, the 10-parameters Lorentz group, was chosen and its localization was carried out (that is, the transformation parameters became the coordinate functions). As a result, the gauge fields appeared, which were associated with the gravitation field of the GTR. The theory obtained was completely identical to the GTR and in fact, produced no new results. The development of the gauge interpretation of the gravitation was defined mainly by hopes to use it in the coming time for unification of the gravitation interaction with the other ones. The star hour of the gravitation gauge variant came in the 1970s when the supersymmetry theory had appeared. The theory, appearing as a result of the merging of two origins, the supersymmetry and the gauge principle, was called the *supergravitation* (see, for review, [8]). The supergravitation geometry is as simple and elegant as that of Einstein's GTR (the latter corresponds of the supergravitation with $N = 0$). The basis of supergravitation is a relativity principle, which reads, that "*the form of physical laws does not depend on the choice of a coordinate system in the superspace.*"

In the simple supergravitation ($N = 1$) the fourteen-parametric Lorentz group, extended by the supertranslation transformations is localized. In this variant of supergravitation the graviton and its superpartner gravitino (with spin 3/2) are the carriers of the gravitation field. In the simplest supergravitation extension ($N = 2$) a symmetry group with 18 parameters is localized, and consequently, there are more interaction carriers, they are: the graviton, two gravitinos, and a graviphoton. The $N = 2$ supergravitation represents the first supergravitation theory, which unifies the gravitation with the electromagnetism in principle. It becomes possible to unify particles with spin 2 and 1 due to the presence of the intermediate stage, the particle with the spin 3/2. In supergravitation the number of divergences is much less than the one in the ordinary gravitation theory. Many viable supersymmetric field theories contain supergravitation as an important component, which helps to spontaneously violate the supersymmetry. In such models the hierarchy problem finds its solution.

The 0(8)-supergravitation is the most extensive and promising supergravitation theory. It is probably nonrenormalizable. In addition to this confusing fact, the 0(8)-gauge group is too small to contain the minimal gauge group of the SM. In transition to groups, more extensive, than 0(8), particles with spin ≥ 3 appear. By now all the attempts to construct a consistent theory of interacting particles with the spin 3 and higher have been not successful.

It makes us suspect, that the 0(8)-gauge group is the limit of the supergravitation. Notice, that the SM group might appear within the frameworks of the $N = 8$ supergravitation. It turns out that on a mass shell the 0(8)-symmetry extends to the $SU(8)$-symmetry, which already contains the gauge group of the SM.

Kaluza-Klein five-dimensional world. To get acquainted with another direction in construction of an unified field theory, let us consider the Kaluza concept [9] later developed in the works by K. Klein [10]. The idea was to unify the GRT and the Maxwell electrodynamics on the basis of a hypothesis that our universe is the curved five-dimensional space-time. One of the coordinates denotes the time, while the other four are spatial coordinates. In such a space the five-dimensional interval square dS^2 is defined by the relation

$$dS^2 = g_{AB}(x)dx^A dx^B, \tag{6.47}$$

where $A, B=0,1,2,3,4$, and $g_{AB}(x)$ is a metric tensor with fifteen independent components. Then ten combinations $g_{AB}(x)$

$$g_{\mu\nu}(x) + \frac{g_{4\mu}(x)g_{4\nu}(x)}{g_{44}(x)}$$

are associated with ten components of the metric tensor of the GRT $\eta_{\mu\nu}(x)$. The following four combinations:

$$\frac{g_{4\nu}(x)}{\sqrt{-g_{44}(x)}}$$

are connected with four components of the electromagnetic potential $A_\mu(x)$. Let us explain why this operation is legal. Remember that the Christoffel symbols $\Gamma_{AC,B}$ are field strengths in the curved space-time. We set one of the indices C equal to 4, A and B equal to 0,1,2,3. Then, using (6.8), we obtain

$$\Gamma_{\mu 4,\nu} = \frac{1}{2}\left(\frac{\partial g_{\mu\nu}(x)}{\partial x^4} + \frac{\partial g_{\nu 4}(x)}{\partial x^\mu} - \frac{\partial g_{\mu 4}(x)}{\partial x^\nu}\right). \tag{6.48}$$

When one assumes, that the fifth coordinate is cyclic, then the expression (6.48) takes the form

$$\Gamma_{\mu 4,\nu} = \frac{1}{2}\left(\frac{\partial g_{\nu 4}(x)}{\partial x^\mu} - \frac{\partial g_{\mu 4}(x)}{\partial x^\nu}\right), \tag{6.49}$$

or, having set

$$F_{\mu\nu}(x) = \frac{c^2}{\sqrt{G_N}}\Gamma_{\mu 4,\nu}, \qquad A_\mu(x) = \frac{c^2}{2\sqrt{G_N}}g_{4\mu}(x), \tag{6.50}$$

for (6.49) we arrive at

$$F_{\mu\nu}(x) = \left(\frac{\partial A_\nu(x)}{\partial x^\mu} - \frac{\partial A_\mu(x)}{\partial x^\nu}\right), \tag{6.51}$$

that is, $g_{4\nu}(x)$ and $\Gamma_{\mu 4,\nu}$ can be really identified with the potential and the tensor of the electromagnetic field, respectively. The equations, governing system evolution, follow from the least-action principle

$$\delta(S_m + S_f) = 0, \tag{6.52}$$

where S_m and S_f are actions for the matter and the field, respectively. Variation of the first term in Eq. (6.52) results in five equations for geodesic lines, four of which coincide with the known four-dimensional equations for charged particles moving in the gravitation and electromagnetic fields

$$\frac{d^2x_\nu}{ds^2} = -\Gamma^{\mu\lambda}_\nu \frac{dx_\mu}{ds}\frac{dx_\lambda}{ds} + \frac{e}{m}F^\mu_\nu \frac{dx_\mu}{ds}, \tag{6.53}$$

and the fifth equation

$$\frac{dx_4}{ds} = -\frac{1}{2\sqrt{G_N}}\frac{q}{m} \tag{6.54}$$

shows, that while a body is moving in the gravitation and electromagnetic fields, its electric charge is conserved.

Varying only the potentials of the five-dimensional space-time, we obtain fifteen equations, which break down into the system out of ten of the ordinary four-dimensional GRT equations (6.7) and the system out of four of the Maxwell equations

$$\frac{\partial F^{\mu\nu}}{\partial x^{\nu}} = j^{\mu}, \qquad \frac{\partial F^{\mu\nu}}{\partial x^{\sigma}} + \frac{\partial F^{\nu\sigma}}{\partial x^{\mu}} + \frac{\partial F^{\sigma\mu}}{\partial x^{\nu}} = 0. \tag{6.55}$$

In so doing, the equation for the scalar component $g_{44}(x)$ again is out of use.

Despite obvious merits, the Kaluza-Klein theory leaves two questions to be completely open:

(i) What is the physical meaning of the fifth coordinate x_4 and why it is not observable?
(ii) Why all physical quantities are cyclic with respect to x_4?

The obvious answer could be that the manifestation region of the additional dimensionality is beyond the existing experimental technology. At the end of the twentieth such statements become the rule of good form and the mighty imagination of the theorist-physicists produces a great oasis of exotic phenomena in the energy scale close to the Plank energy. However, by 1938 only A. Einstein and P. Bergmann could come up with such an idea [11]. They suggested that the fifth coordinate can change from 0 to some value L, that is, the five-dimensional world is confined in a layer with thickness L. The assumption was also made, that any function $\Psi(x)$, related to physics, changes little along x_4 over a length of the layer, so that

$$L\frac{d\Psi(x)}{dx_4} \ll \Psi(x)$$

and in the average $\Psi(x)$ may be considered as a function only of the four-dimensional coordinates. Actually, instead of restricting x_4 values to quantity L, it is possible to assume, that the fifth coordinate varies within infinite limits, however, only functions, periodic in x_4 with the period L are under consideration. It means, that it is possible to glue together all the points, being distant from each other along x_4 on interval L, without any harm for generality done. As a result, we arrive at the five-dimensional space-time being closed by x_4. The world with such a property we shall call cyclic, closed, or compactified in the fifth coordinate. In such a theory there is no need for postulating the cyclic character of x_4, since the gauge principle of switching on electromagnetic interaction, which had been armed by us, resulted in the conclusion, that wavefunctions of charged particles have the form

$$\Psi(x) = \Psi(x_{\nu})\exp\left(\frac{iec}{2\sqrt{G_N}\hbar}x_4\right), \tag{6.56}$$

where $\Psi(x_{\nu})$ is an ordinary wavefunction in the four-dimensional space. The expression (6.56) describes the cyclic dependence of $\Psi(x)$ on x_4 with the period

$$L = \frac{4\pi\sqrt{G_N}\hbar}{ec} \approx 10^{-31} \text{ cm}. \tag{6.57}$$

Thus, the cyclic period or the world compactification scale in the fifth coordinate is infinitesimally small compared to the scale of phenomena, studied by contemporary physics.

Consequently, it is not surprising, that the fifth dimensionality has been skillfully hiding itself from experimentalists up to now. Of course, the extension of space-time dimensionalities can be generalized on larger dimensionalities as well.

Since the early universe was a substance compressed in the volume with the radius of the Plank length order, then it is easy to understand that the evolution of universe is completely defined by elementary particle physics. Consequently, the idea of the closed dimensionalities must find its place in the Big Bang theory, too. The universe is thought to have carried out the space-time compactification in additional dimensionalities at the early stages of evolution when its energy was rather high. At present, this hypothesis is the necessary attribute of the contemporary multidimension theories.

Superstring theory. The quantum field theory (QFT) and the GTR contain all human knowledge of the fundamental forces of nature. The QFT brilliantly explains all the microworld phenomena up to the distances 10^{-15} cm. The GTR, in its turn, is beyond any competition in describing both large-scale events in the universe, as well as the evolution history of the universe itself. The striking success of these theories is caused by the fact, that taken together, they explain both the behavior and structure of matter from subnuclear to cosmological scales. All attempts to unify these theories, however, have been always facing two main difficulties. The former deals with the appearance of divergences, while the latter demands to refuse one or more ideas about the nature of universe, which are highly respected nowadays. The QFT and the GTR are based on few postulates, which appear to be mandatory and natural. The postulates, the abandonment of which helps to unify two theories, are as follows: (i) space-time continuity; (ii) causality; (iii) theory unitarity (the absence of the states with the negative norm); (iv) interaction locality; and (v) point (structureless) particles. The first four items are rather serious. Against their background the abandonment of the fifth item looks like a child's prank.

Any local theory, operating with structureless objects contains divergences at energies, higher than the Plank energy E_P. It is caused, as we know, by the fact, that E_P is that very energy value, when the gravitation quantum theory is needed, while it is not renormalizable. To achieve renormalizability, another model can be introduced, in which at "low" energies (E_P) fundamental objects behave as point ones and their composite structure reveals itself only at energies, higher than E_P. If one adds to this hypothesis some latest inventions, concerning unification of all interactions (supersymmetry, Kaluza-Klein theory, etc.), then one more way is being opened to the unified field theory, called the *superstring theory* (the reader could find the superstring theory presentation in the book [12]).

String theories were used by hadron physics as early as the 1960s. Their creation was inspired by the fact, that at the distances of the order of 10^{-13} cm the gluon fields, binding the quarks, are concentrated in space not evenly but along the lines connecting the quarks. It resulted in the interpretation of the hadrons as nonlocal objects, one-dimensional strings, on the ends of which either the quarks for the baryons (fermion strings) or the quarks and antiquarks for the mesons (boson strings) are placed. Thus, the infinitely thin tube of the gluon field is modeled by a relativistic string. Relativistic string theories managed to explain many anomalies of the quark universe. Thus, for example, since the string energy is proportional to its length L, and the string mass square is $m^2 \sim L^2$, then the string angular moment of the string having the form of the linear segment is proportional to L^2. Really, experiments confirm linear dependence between the hadron spin J and the square of its mass $J \approx \alpha m^2$, where $\alpha \approx 4 \times 10^{-28} \mathrm{cm}^2/(\hbar c)^2$. The relativistic string, which binds the quark and antiquark, generates potential, linearly growing with distance, that is, it explains the quark confinement. The theory is built in such a way, that after the string breaks no free quarks appear, since at both newly formed string ends a pair "quark-antiquark" is produced.

However, the string relativistic theories have some very serious deficiencies, such as, for example, tachyon existence, massless hadrons with spin 2, etc. It also turned out, that consistent quantum theories could be only formulated in 26-dimensional space-time for bosonic strings and in 10-dimensional space-time for fermion strings.

The renaissance of interest in string theories took place after 1986 when M. Green and J. Schwarz proved [13], that the 10-dimensional gauge superstring theory, based on the internal symmetry group $SO(32)$ or $E_8 \otimes E_8$* (subscript indicates the group rank) can be used to unify all the interactions. Superstrings are one-dimensional in a spatial sense (two-dimensional, if the time is taken into account) objects with the typical length of the order of L_P. They are put in an n-dimensional ($n \geq 10$) space-time manifold. The observable space-time dimensionality is achieved by compactifying unnecessary dimensionalities at distances of the order of the Plank length. The theory contains a mechanism, ensuring a spontaneous compactification of additional dimensionalities. The initial symmetry is broken up to a symmetry group, involving the supergravitation and the supersymmetric GUT with fixed parameters and given particles content. If the gauge group $E_8 \otimes E_8$ is used then one E_8-group contains all the low energy physics, while the other E_8 manifests itself only in the gravitation interaction. Phenomenological properties of the superstring theory depend a lot on a compactification mechanism. As an example, let us pay attention to the following circumstance. Since the division into the normal space-time dimensionalities and compactified ones is not very strict, then it is possible, that some universes with nonconventional dimensionalities of the space-time exist.

The important difference between the superstring theory and the local field theory is that in the former theory the free superstring is characterized by an infinite number of supermultiplets, while in the latter one every field describes particles of only one kind. The superstrings have the same number of fermion and boson degrees of freedom. Superstring excitations (which are rotations, vibrations, or excitations of internal degrees of freedom) are associated with elementary particles observed. Particle mass scale is regulated by the superstring tension T with $\sqrt{T} \approx M_P c^2$. The number of states with masses, smaller than the Plank mass, is finite. It defines the number of elementary particles existing in nature. There is also a great number of excitations with masses, bigger than the Plank mass. The majority of these modes are unstable, however, there are also stable solutions with exotic characteristics (a magnetic charge, for example). It is remarkable, that in particle spectrum, which corresponds to superstring theory solutions, one massless state with the spin 2 appears, which is described by the GTR equations in a low energy limit, that is, it is the graviton.

Strings appear in two topologies: as open strings with free ends and as closed loops. Besides this, they can possess an internal orientation. Quantum numbers of the open strings are located at their ends, while in the closed loops quantum numbers are evenly spread along the string. The string interaction has the local character, despite the fact, that they are extended objects. When interacting the strings can scatter, produce new strings, and emit point particles as well.

It is evident that the choice of one or another model in the role of a candidate in an unified theory is the most principal question in the superstring theory. It is proved that all the known superstring theories are connected with each other by a duality transformation [14]. Discovery of the duality allows to reveal nonperturbative superstring properties that are connected with the existence of new stretched objects, D-brans.† The assumption was stated

---

*E_8 together with the groups G_2, E_4, E_6, and E_7 constitute the exceptional group class. The rank of the exceptional group is fixed, while the normal (regular) group can have any rank (for instance, $SU(2)$, $SU(3)$, $SU(5)$, and so on).

†The bran is a three-dimensional manifold with ordinary matter enclosed in a comprehensive multidimensional space. The D-brans include not only the matter fields but the gauge fields as well.

that all superstring theories are the special case of the unified fundamental theory, the *m*-theory, which lives in 11-dimensional space-time and describes 11-dimensional supergravity on rather small distances. The development of the superstring theory showed that it was a

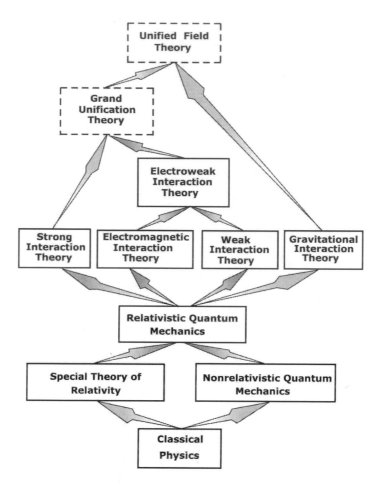

FIGURE 6.2

The building of twenty-first century physics.

fruitful generalization of the local field theory. However, nowadays the superstring theory still undergoes its development stage and has no experimental confirmations yet. Let us note that the experimental check of superstring models is very difficult due to many unknown parameters they contain. We would like to believe that the completed superstring theory would contain only two fundamental parameters: the tension and the superstring interaction constant.

In Fig. 6.2 the building of twenty-first century physics is pictured. Unfinished floors are designated by the dashed lines. Physics as a science originated from astronomy at the end of the sixteenth century. Two persons were at the cradle of physics, Tycho Brahe and Johannes Kepler. The former was the first experimentalist, whereas the latter was the first theorist. The majestic building of physics has been under construction for 400 years. And now, for its construction to be completed joint operations of already millions of physicists are required.

7

Atoms–Nuclei–Nucleons

The bullfinch was a sharp little fellow and a born misin-
-former. He received his education at a school for soldiers'
children and afterwards served in the army as a regimental
clerk. Having tackled the rules of punctuation, he began
to publish an uncensored periodical entitled "The Wood-
land Herald." Yet, try as he might, he somehow just
couldn't fall in with official requirements. Either he
touched on something forbidden, or he would omit a point
which not only could but should have been mentioned. And
naturally, he got rapped on the knuckles for this.
M.E. Saltykov Shchedrin, "The Eagle-Patron of Arts"

7.1   Atomism

The idea of fundamental particles has always symbolized the deepest aim of scientific
cognition–to express the complexity of nature through its simplest notions. Twenty-five
hundred years ago Greek philosophers laid the foundations of our understanding of the
nature of matter, when they made the first attempts to reduce the variety of the world to
the interaction of a few initial components, fundamental particles, or elements. In 4 **B.C.**
Phales suggested that water was the single primary element, of which all the existing world
is made up. Later Anaximen from Milet extended the list of primary elements to four,
such as soil, air, fire, and water. It is generally agreed that Democritus (460–370 **B.C.**) is
the creator of the atom idea, however, history also mentions his teacher Levkipus in this
connection.

A legend states that one time Democritus sat on the seashore with an apple in his hand.
His stream of thought was as follows: *"Suppose, I shall cut this apple in halves, and then*
every half I shall cut in halves again, so that there will be a quarter of an apple left, then
one eighth, one sixteenth, etc. The question is, if I continue to cut the apple to the end, will
the cut parts always possess properties of an apple? Or, maybe, at a certain moment the
remaining part of the apple will have no properties of an apple any more?" Upon serious
consideration the philosopher arrived at the conclusion that there was a limit for such a
division and he called the last indivisible part an "atom." His conclusions mentioned below,
are stated in his book *Great Diacosmos*.

"The beginning of the Universe is atoms and emptiness, other things exist only in our
imagination. There exists infinite number of worlds, and they all have their beginning and
end in time. And nothing appears from non-existence, nothing goes into non-existence.
Atoms are uncountable both in amount and in variety, whirling like the wind they rush in
the Universe and so all composite substances create: fire, water, air, soil. The point is that
these substances are compounds of some atoms. Atoms are unaffected by any influence, they

are unchangeable due to their hardness."

Democritus could not prove his statements, so he suggested his contemporaries trust his words. The majority of contemporaries did not believe him, and Aristotle was among them. Aristotle was the author of the opposite teaching. According to him, the process of the apple partition could be infinitely extended. Natural philosophy is not based on experiments and mathematics, it used the element of faith as the single criterion of truth. So it is not surprising, that both teachings for the ancients seemed to be equally reasonable and acceptable. It is difficult to tell, what has weighed down the weights bowl in favor of Aristotle's philosophy. Maybe, it was the gleam of military glory of Alexander the Great, whose teacher was Aristotle. Anyway, the teachings of Aristotle became dominant, while Democritus had been forgotten for many centuries.

In the 1640s Democritus's idea about atoms had been restored to life by the French philosopher Gassendet. When spring comes, all violets bloom at once. It was so with the atomic hypothesis. After twenty centuries of oblivion all the contemporary advanced scientists believed in the atomic theory, including the great Isaak Newton, whose credo was *"not to build any hypothesises."* One of the burning questions of atomism was, undoubtedly, the question whether the variety of bodies in nature means, according to Democritus, the same variety of atoms? If the answer is positive, then the atomic hypothesis does not bring us closer to an understanding the world. Luckily, the answer is negative. The variety of substances in nature is caused not by a variety of different types of atoms, but by the variety of different compounds of these atoms (nowadays, these compounds are called *molecules*). In 1808 D. Dalton, upon studying many chemical reactions, precisely formulated the notion of a chemical element: *"Chemical element is a substance, consisting of atoms of the same type."* It turned out, that there were not so many chemical elements. In 1869 D. Mendeleev managed to place all of them in one periodic table (at that time only 63 elements were discovered, while now their number reaches 120). The work by the botanist R. Brown (1827) may be considered as the first experimental proof in support of the atomic theory. He observed chaotic motions of flower pollen in water (Brownian motion). The discovery by Brown did not attract scientists's attention immediately and for a long time its nature remained unclear. Only seventy eight years later, the atomistic theory of Brownian motion was established in works by A. Einstein and M. Smoluchovski. In 1908 J. Perrin carried out a series of experiments to study Brownian motion. Not only did he prove experimentally the works by Einstein and Smoluchovski but he measured sizes and masses of atoms as well. The last and, probably, final proof of atomic matter structure was the work by E. Rutherford and Royds on measuring the number of α-particles in radium. By that time it had been known, that in minerals, containing α-radioactive elements (radium, thorium), helium is accumulated. The task was to determine the number of α-particles emitted by a sample and measure the volume of helium produced on the sample. In a second 13.6×10^{10} particles are emitted by one gram of radium. Having captured two electrons all these α-particles turn into helium atoms and occupy the volume of 5.32×10^{-9} cm^3. Consequently, 1 cm^3 contains $L = 2.56 \times 10^{19}$ atoms. Let us compare the value obtained with the Loschmidt number calculated as early as 1865 on the basis of the molecular-kinetic theory. One mole of helium (or any other gas) occupies a volume 2.241×10^{-2} cm^3/mole and contains 6.02×10^{23} atoms, that is, 1 cm^3 contains 2.68×10^{19} atoms. The coincidence is impressive. So, the existence of atoms got the final experimental proof. That has completed the ascent of physics on the first step of the "Quantum Stairway." A picture of the world on this step was very simple: *the matter in our universe consists of indivisible atoms of different types, which make all the elements of the Mendeleev periodic table.*

The variety of elements and the explicit systematization in the periodic table by Mendeleev inspire into us belief that far from being utopian, the result obtained is not final yet. The existence of the next step of the "Quantum Stairway" became obvious after a series of experiments, which may be called the "Roentgen" of atom according to Rutherford (1909–1911).

7.2   Rutherford model of the atom

The fact that all atoms contain electrons was the first important piece of information about the internal structure of an atom. The electrons have a negative electric charge, while atoms are electrically neutral. Consequently, every atom must contain enough of positively charged matter to compensate the negative charge of the electrons. In 1903 in the book *Electricity and Matter*, D. D. Thomson introduced a model of a radiating atom, which was satisfying to the totality of chemical and spectroscopic experiments existing at that time. Thomson's predecessor in the model was his famous namesake William Thomson (Lord Kelvin), who in 1902 suggested coming back to the atomic theory of F. Epinus (1759). The Epinus atom is a sphere, being uniformly charged with positive electricity, in the center of which a negative charged corpuscle is located. D. D. Thomson developed this model by assuming that the electrons rotate inside of a sphere, the number of the electrons and their orbit configuration depending on the atom's nature. The electrons are spaced as concentric rings (shell) and perform periodic motions, which cause the observed atom spectrum. Investigating the stability of the electron combination, Thomson gave the physical interpretation of valence. The atom model by Thomson was the first real attempt to explain the chemical properties of a substance and the periodic law by Mendeleev. His model, called "pudding with raisins" was warmly accepted by the scientific public. One should not think that the spirit of contradiction, being so characteristic for scientific creativity, has not led to the construction of some alternative atom models. So, for example, in 1901 A. Perren published the article titled "Nuclear planetary structure of the atom" in the scientific popular journal *Scientific Review*. Two years later the Japanese physicist Ch. Nagaoka (a disciple of Bolzman), suggested the model according to which an atom consists of a positive charged nucleus and of the electrons ring rotating around the nucleus (atom of Saturn type). First Nagaoka published his results in even less popular among physicists journal *Proceedings of Tokyo physical-mathematical society*. Then in 1904 he published the article on the same topic in the itPhilosophical Magazine. However, physicists paid no attention either to the Perren or Nagaoka models, although these models could be regarded as the predecessors of the atom nuclear model later suggested by Rutherford.

Nine more years passed before the atom model by Thomson was brought under serious experimental testing, results of which showed that the model had a single drawback, namely, it had nothing in common with physical reality. Creators of quantum theory liked to say: *"in order to check what is inside of a pudding with raisins one must simply put a finger in it."* In reality Rutherford used the method that was rather similar with the one described above. α-particles from radioactive sources were used as a probe, and thin foils of different substances with a typical thickness of about 10^4 atomic layers were used as targets. In Fig. 7.1, the scheme of these experiments done by G. Geiger and E. Marsden under the guidance of Rutherford is represented. An ampoule with a radioactive source (radium C, Po-214) was placed behind a lead screen with a small opening. α-particle beam was passing through a lead collimator and was directed at a target. After interacting with target atoms, α-particles got into a mobile screen of sulphureous zinc causing scintillations on it, which were registered by means of a microscope.

As we know from optics, to see an object, one must use the electromagnetic waves having

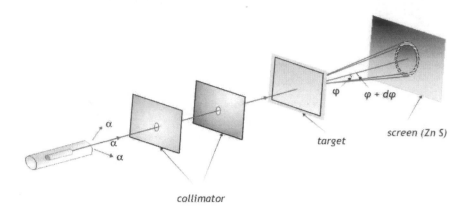

FIGURE 7.1

The scheme of Rutherford's experiments.

wavelengths of the same order as its size. The same rule takes place in the quantum world, only here we are dealing with de Broglie wavelength λ. Of course, in 1909 nobody knew about de Broglie waves, but nature once again proves to be favorable to man, and α-particles with energy 7.7 MeV (typical energy values in Rutherford experiments) had a wavelength of $\sim 5 \times 10^{-13}$ cm, which was quite enough to "see" a nucleus in an atom.

The Thomson atom model predicted that the main part of α-particles flies straight through a foil, while the remaining particles are subjected to the small deviation. In reality the behavior of most α-particles fitted in the frames of this model, but parallel with that some particles deviated on big angles. Let us show, quantitatively, that the Thomson model is not able to explain the results of these experiments. When, according to Thompson, one assumes that a positive charge Q is spread evenly all over the volume of a gold atom and does not take into account the influence of the electrons, then the electric field strength inside of an atom at a distance r from its center is given as:

$$E_+(r) = \frac{Qr}{4\pi R^3},$$

where R is an atom radius. The α-particle maximum deviation takes place, if it is slightly touching the atom surface where $E_+(r)$ reaches values of the order of 10^{13} V/cm. Thus, the maximum value of the force is determined by the expression:

$$F_{max} = \frac{Q|e|}{2\pi R^2}.$$

Since this force rapidly decreases with a distance, then for an approximate estimation, one can consider the interaction on the small interval L, which includes the distances before and after the contact between the α-particle and the atom. We accept $L \sim 2R$, and take the force value being maximum, that is, during the time interval $\Delta t = L/v_\alpha \approx 2R/v_\alpha$ the α-particle is subjected to F_{max}. In this case, the transverse momentum, transferred to the α-particle, is equal to

$$\Delta p_\alpha = F_{max}\Delta t \sim \frac{Q|e|}{\pi R v_\alpha}.$$

Consequently, the maximum deviation angle is as follows:

$$\theta_+ < \frac{Q|e|}{\pi R m_\alpha v_\alpha^2}. \tag{7.1}$$

Using the values 6.6×10^{-27} kg and 2×10^7 m/s for the mass and the velocity of the α-particle respectively, we obtain in the case of the gold atom $(Q = 79|e|)$

$$\theta_+ < 0.02^0.$$

Now let us take into account the influence of the atomic electrons on the α-particles motion. We assume that their initial velocity is equal to zero. Then the momentum transfer to an atomic electron is maximum under the head-on collision. From the conservation laws of the momentum and the energy (in a nonrelativistic case) it follows, that after a collision an electron has acquired the velocity

$$v_e = \frac{2m_\alpha}{m_\alpha + m_e} v_\alpha \approx 2v_\alpha,$$

where v_α is the initial velocity of the α-particle. Certainly, at such a collision the α-particle does not deviate. At a sliding collision the electron momentum change Δp would be already less than $2m_e v_\alpha$. To obtain a value of a maximum possible deviation θ_- under the scattering of the α-particle by the electron, we assume that the electron after a collision flies out at the right angle to the initial direction of the α-particle motion and has the momentum equal to $2m_e v_\alpha$. Then, the result follows

$$\theta_- \sim \frac{\Delta p_\alpha}{p_\alpha} < \frac{2m_e}{m_\alpha} \sim 0.02^0. \tag{7.2}$$

Notwithstanding the fact that the deviation of the incident α-particle caused by both the atomic electron and the positive charged sphere is as low as the order of 0.02^0, whether a series of such deviations could give rise to a big scattering angle. Let us suppose, that an average deviation about the angle $\bar\theta \sim 0.01^0$, caused either by the positive charged sphere or by the electron, occurs under a transition through one atomic layer. Using the statistical methods for obtaining the result of a sequence of random deviations, we find, that after passing through the $N = 10^4$ atomic layers, the total average deviation is equal to

$$\bar\theta_t = \bar\theta \sqrt{N} \sim 1^0. \tag{7.3}$$

Really, the experimentally measured average deviation represented about 1^0. However, some part of the α-particles were scattered through much bigger angles. For example, one of 8000 α-particles deviated through angle $\theta \geq 90^0$. The probability, that the α-particle is subjected to summarized scattering through the angle larger than θ under the average deviation $\bar\theta_t$ is

$$P(\geq \theta) = \exp\left(-\frac{\theta^2}{\bar\theta_t^2}\right). \tag{7.4}$$

For the Thomson atom model $P(\geq 90^0) = \exp(-8100) \sim 10^{-3500}$, that is, only one of 10^{3500} α-particles can be scattered through the angle $\geq 90^0$ that contradicts the experimental data.

To explain the α-particles scattering results Rutherford suggested the planetary atomic model, which the essence is as follows. Practically all the atom mass is concentrated in its nucleus that is located in the center and has the size of 10^{-13}–10^{-12} cm. The electrons are rotating around the nucleus at the distance of the order of 10^{-8} cm. Soon they found out, that the nucleus electric charge exactly equals the element number in the periodic table by Mendeleev. In the beginning of 1913 this idea was introduced by the Dutch physicist Vander Broek and some months later a Rutherford disciple G. Moseley produced its experimental proof. Moseley performed a set of experiments to measure the X-ray spectrum for various elements. It turned out that the X-ray wavelength systematically decreases while the atomic number Z in the periodical table increases. Moseley got the conclusion that this regularity is caused by increasing the atomic nucleus charge. The charge increases from atom to atom by one electronic unit and the number of such units coincides with the number of the element position in the Mendeleev table. Since the atom is electrically neutral, it means that the total number of the electrons in the atom is equal to Z as well.

Then, in the Rutherford atom at the nucleus surface the electric field strength is $> 10^{21}$ V/cm, which is almost eight orders higher than that at the atom surface. In Fig. 7.2 (R is the atom radius) the distribution of the electric field strength $E_+(r)$ in the atomic models by Thomson (Fig. 7.2(a)) and Rutherford (Fig. 7.2(b)) are showed for comparison. It is obvious, that the strong field in the planetary model can cause a big deviation and

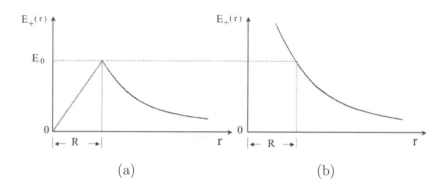

(a) (b)

FIGURE 7.2
The electric field strength in the Tomson atom model (a) and in the Rutherford atom model (b).

even scatter the α-particle backwards, when it is flying too close to the nucleus. However, the secret of the rapid development of physics lies in the fact that for confirming a physical hypothesis not only qualitative, but also quantitative, coincidences are needed. Let us prove that the Rutherford model correctly describes the scattering of α-particles by atoms.

To analyze both elastic and nonelastic collisions either the laboratory reference frame ("lab" frame) or the center-of-mass frame is used. The lab frame corresponds to the standard performance of experiments, namely, a beam of particle of type I strikes on a fixed target, built up from particles of type II. In the center-of-mass frame, however, equations, describing scattering processes, are much more simple, since the total system momentum equals zero $\mathbf{p}_1 + \mathbf{p}_2 = 0$. In the center-of-mass frame particle collision is reduced to a motion of one particle with a reduced mass

$$M = \frac{m_1 m_2}{m_1 + m_2}$$

in the field $U(\mathbf{r})$ of an immovable force center, located in a particles inertia center. In the lab frame scattering angles θ_1 and θ_2 (the second particle rested before collision) are connected with the scattering angle in the center-of-mass frame by the relations

$$\theta_2 = \frac{\pi - \chi}{2}, \qquad \tan \theta_1 = \frac{m_2 \sin \chi}{m_1 + m_2 \cos \chi}. \tag{7.5}$$

Notice that the center-of-mass frame can be practically realized under the performance of experiments with colliding beams.

In classical physics the collision of two particles is completely defined by their speeds and an impact parameter ρ. However, during real experiments we deal not with individual deviation of a particle but with the scattering of a beam, consisting of identical particles, striking on a scattering center at the same speed. Various particles in a beam have different impact parameters and, consequently, are scattered through different angles χ. Let dN be a number of particles, scattered in a unit of time through angles belonging to an interval χ and $\chi + d\chi$. Since dN is a function of a falling beam density, that is not convenient to characterize a scattering process. For this reason we use a quantity

$$d\sigma = \frac{dN}{n},$$

where n is a number of particles passing in the time unit through the unit of area of a beam cross section (we assume beam homogeneity over the whole section). In a given interval of angles only those particles are scattered, which fly with impact parameters enclosed in the interval between $\rho(\chi)$ and $\rho(\chi) + d\rho(\chi)$. The number of such particles is equal to the product of n and the area of the ring between circles with the radii $\rho(\chi)$ and $\rho(\chi) + d\rho(\chi)$, that is, $dN = 2\pi\rho(\chi)d\rho(\chi)n$. Thus, the effective cross section of scattering within the interval of the flat angles $d\chi$ (it is also called the *differential cross section*) is defined by the expression:

$$d\sigma = 2\pi\rho(\chi)\,|\,\frac{d\rho(\chi)}{d\chi}\,|\,d\chi. \tag{7.6}$$

Bearing in mind that the derivative $d\rho(\chi)/d\chi$ also could be negative, we used its absolute value only. Passing to the solid angle $d\Omega$, we obtain

$$d\sigma = \frac{\rho(\chi)}{\sin\chi}\,|\,\frac{d\rho(\chi)}{d\chi}\,|\,d\Omega. \tag{7.7}$$

To integrate the differential cross section over all values of the solid angle produces the total cross section σ. The cross section has the dimension of the area. it is the "useful" area of the interacting system, consisting of the incident particle and the target-particle. Thus, for the incident particle a target is like an area, and the hit in this area results in an interaction. The effective cross sections can be smaller or bigger than the geometrical cross sections of target particles, and can coincide with them as well. To obtain in the lab frame the scattering cross sections for the incident beam, one should express χ in terms of θ_1 and θ_2 by means of Eqs. (7.5).

In quantum theory the scattering problem is considered from other grounds since the conception of trajectories and impact parameters make no sense under the motion with a definite speed. Here the aim of the theory is to calculate the probability, that in the result of the collision the particles are scattered through one or another angle. Once again, we can introduce the conception of the effective cross section, which characterizes the transition probability of a system, consisting of two colliding particles, as a result of their elastic or nonelastic scattering to the definite final state.

The differential cross section of scattering within angles interval $d\Omega$ is equal to the ratio between the probability of such transitions $P_{i \to f}$ per the time unit to the incident particles flux j_0

$$d\sigma = \frac{W_{i \to f}}{j_0} d\Omega,$$

where $W_{i \to f} = P_{i \to f}/\Delta t$. Integration over the entire interval of the solid angle variation gives the total cross section

$$\sigma = \int_0^{4\pi} \frac{W_{i \to f}}{j_0} d\Omega.$$

In the CGS of units the square centimeter cm^2 is the unit of the effective cross section. This unit, however, is very large for the microworld and we use a unit having the order of a geometrical cross section of a nucleus 1 barn $= 10^{-26}$cm^2.

Let us calculate the differential cross section of the α-particles scattering in the Born approximation. In the center-of-mass frame the Born formula has the following form:

$$d\sigma = \frac{M^2}{4\pi^2\hbar^4} \mid \int U(\mathbf{r}) \exp\left(\frac{i}{\hbar}\mathbf{qr}\right)d\mathbf{r} \mid^2 d\Omega, \qquad (7.8)$$

where $d\Omega = 2\pi \sin\chi d\chi$, $U(\mathbf{r}) = Ze^2/2\pi r$, Z is an atomic number of a target, $\mathbf{q} = \mathbf{p} - \mathbf{p}'$, $\mathbf{p}$ and $\mathbf{p}'$ are the momenta of the particles before and after collision. Using the Poisson equation for the potential of a point charge, located in the beginning of the coordinate system

$$\Delta\left(\frac{e}{r}\right) = -4\pi e\delta(\mathbf{r}),$$

and the Fourier transformation for the delta function

$$\delta(\mathbf{r}) = \frac{1}{(2\pi)^3} \int \exp\left(i\mathbf{kr}\right)d\mathbf{k},$$

we obtain the following expression for the Fourier image of the potential energy

$$U(\mathbf{q}) = Ze^2 \int \frac{1}{2\pi r} \exp[\frac{i}{\hbar}(\mathbf{qr})]d\mathbf{r} = \frac{Ze^2}{4\pi^3} \int \frac{d\mathbf{k}}{\mathbf{k}^2} \int \exp\{i[\frac{(\mathbf{p}-\mathbf{p}')}{\hbar} + \mathbf{k}]\mathbf{r}\}d\mathbf{r}$$

$$= \frac{Ze^2(2\pi)^3}{4\pi^3} \int \frac{1}{\mathbf{k}^2}\delta(\frac{\mathbf{p}-\mathbf{p}'}{\hbar} + \mathbf{k})d\mathbf{k} = \frac{2Ze^2\hbar^2}{\mid \mathbf{p} - \mathbf{p}' \mid^2}.$$

Taking into account that

$$\mid \mathbf{p} - \mathbf{p}' \mid = 2p\sin\frac{\chi}{2},$$

where $\mid \mathbf{p} \mid = \mid \mathbf{p}' \mid = p$, we arrive at the result:

$$d\sigma = \left(\frac{Ze^2M}{4\pi p^2}\right)^2 \frac{d\Omega}{\sin^4(\chi/2)}. \qquad (7.9)$$

Notice some interesting peculiarities of the formula (7.9), which is called the *Rutherford formula*. When we do not take into consideration any relativistic effects, then solving the task exactly, we obtain the expression (7.9) as well (see, for example, [15]). The solution obtained does not depend on the potential energy sign, that is, the solution is the same for both attracting and repulsing force centers. Since the differential cross section does not contain the Plank constant, then one can analyze the Rutherford scattering by classical or quantum mechanical methods with the same success. The latter facts follow from the rule, which reads:"*if forces of an interaction between particles depend on a distance as r^n, then the cross section of scattering of such particles at each other is proportional to $\hbar^{4+2n}$.*"

If we try to integrate the expression (7.9) over the scattering angle, then we obtain infinity. It is caused by the long-range character of the electrostatic forces. For this reason the particles are scattered, no matter how far away from a scattering center they fly. To obtain a reasonable, that is, a finite result, we must take into consideration a screening effect of the electron shell. This is achieved under use of the potential

$$U(r) = \frac{Ze^2}{2\pi r} \exp\left(-\frac{r}{a}\right),$$

where a has the value of the order of the atom radius.

Now we must take the last step, namely, to compare the theoretical formula (7.9) with the experimental results. Let us turn to the lab frame and consider the cross section for the incident α-particles, taking into account that $m_N \gg m_\alpha$ (m_N is a nuclear mass). In this case $\theta \equiv \theta_1 \sim \chi$ and $M \sim m_\alpha$, so that

$$d\sigma = \left(\frac{Ze^2}{4\pi m_\alpha v_\alpha^2}\right)^2 \frac{d\Omega}{\sin^4 \frac{\theta}{2}}. \tag{7.10}$$

In Fig. 7.3 the results of calculations according to the formula (7.10) are plotted, which define the scattered particles number in relation to the angle θ at $Z = 79$. The experimental

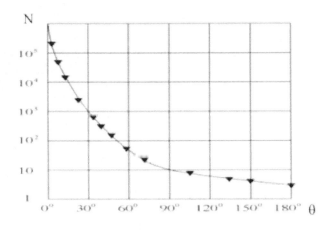

FIGURE 7.3

The θ angle dependence of the number of the scattered α-particles.

data obtained by Geiger and Marsden are displayed by small triangles. Excellent agreement between the theoretical and experimental results testifies that the internal structure of the atom indeed includes the hard and massive core, which is the nucleus.

It should be stressed that the formula (7.10) is valid under fulfillment the following conditions.

(i) The atom nucleus must have the mass exceeding the α-particle mass to such a degree that the recoil energy of the nucleus may be neglected safely.
(ii) The nucleus potential consists of the Coulomb part and the part that is responsible for nuclear attraction forces φ_N. For the α-particle not to be influenced by the nuclear forces the collision diameter d, defining the minimal distance on which the incoming particle can approach the scattering center, must greatly exceed the action radius of the potential φ_N. In this case the collision diameter is determined by the help of the energy conservation law

$$\frac{m_\alpha v_\alpha^2}{2} = \frac{Ze^2}{2\pi d}. \tag{7.11}$$

By a lucky chance both of these conditions has proved to be satisfied in experiments with gold targets. However, in later experiments with aluminum targets ($Z = 13$) deviations from the Rutherford formula were especially noticeable in the region of the large angles.

Using the Rutherford formula, we can approximately estimate the size of the atom nucleus. Let us assume, that the geometrical cross section of the nucleus is of the same order as the differential scattering cross-section of the α-particles deflected by the angle being greater than 90^0. Then for $Z = 79$ and $E_\alpha = 7.68$ MeV we have $d\sigma/d\Omega = 6.87 \times 10^{-28}$ m^2, which produces the wholly plausible result $R = 1.5 \times 10^{-14}$ m. If the core were absent in the atom center, that is, the Thomson model were true, then decreasing the α-particles deflected by big angles θ would take place along the shortest curve, which is a straight line (see Fig. 7.3). Thus, the excess of the differential cross section over the straight line crossing the horizontal axis provides direct confirmation of the fact that atoms are not indivisible elements of matter, but represent in themselves composite structures, consisting of the positive charged nucleus and the electrons.

Thus, according to Rutherford, the atom is similar to the solar system. The character of resemblance to the solar system represents as not only qualitative but also quantitative, which is very strange and has not found an exhaustive explanation till now. If one takes the ratio between the diameters of the Sun and the solar system, then this ratio proves to be approximately equal to one between the diameters of the nucleus and the atom. Further on, according to quantum theory, the electrons in the atom are not located at arbitrary distances from the nucleus. Their orbital radii are defined by the relation

$$r_n = \frac{4\pi\hbar^2}{me^2}n^2, \qquad n = 1, 2, 3...$$

It turns out, that the planets have analogous behavior, namely, distances between the planets and the Sun are not changed in a random manner but are subjected to a definite law. This fact was known to J. Kepler, but it was first mathematically formulated by D. Titius in 1772. Later on I. Bode made some corrections and the law was given the title the Titius-Bode law. If the distance from the Sun to Mercury is adopted as an 0.4 arbitrary unit, then the formula for the planet radii takes the form

$$R_n = 0.3n + 0.4,$$

where n = 0 (Mercury), 1 (Venus), 2 (Earth), 3 (Mars), 5 (Jupiter), 6 (Saturn), 7 (Uranus), 8 (Neptune), and 9 (Pluto). It is remarkable, that in this scheme a planet with the number n = 4, which should be spaced between Mars and Jupiter, is missing. Exactly in this place an asteroid belt is located. Astronomers suppose that this asteroid belt represents the fragments of the planet Phaeton that existed in former times.

As a result of the triumphal success of the Rutherford atom model the traditional picture of this model with its precise figure of the electron orbits becomes the universally recognized emblem of the past twentieth century. It has appeared in books covers, exhibitions and conferences booklets, and stamped papers of institutes and universities. Although nowadays we know that in reality there are no electron orbits in classical understanding, these figures have remained the deserved tribute of respect to the Rutherford atom model.

In the experiments described above Rutherford discovered not only the atomic nucleus he also established the main type of experiment in the microworld physics, which is particles scattering at each other. By the example of the Rutherford experiments we see that the scattering experiments are characterized by three time intervals. During the first interval the systems are prepared and brought into contact. During the second interval an interaction takes place. During the third interval the systems appeared in the result of an interaction (reaction products) moving towards measuring devices (detectors). Since during the first and the third intervals the systems are isolated from each other, then the systems investigations are reduced to the analysis of free motion equations. The description of the second interval demands deriving the solutions of the interacting systems equations and as a result this is one of the most important and fundamental physics problems.

7.3 Structure of the atomic nucleus

The discovery of the first isotopes in 1919 by F. Aston and establishing the integer numbers rule (all atomic or molecular masses are integer numbers within the limits of observation precision) set in turn the question about the composite structure of the nucleus. The long ago forgotten Proud hypothesis (1815), stating that hydrogen is the part of all atoms, became popular once again. Analogously to the term "protil," introduced by Proud, Rutherford suggests using the term "proton" for the hydrogen atomic nucleus (1920). As far back as World War I Rutherford started to investigate the collisions of the α-particles with the nuclei of such light elements as hydrogen, nitrogen, oxygen, and air. At the central collisions the nuclei emitted particles, causing scintillations at a screen made of sulphurous zinc. At that time a deep belief existed, that only nucleus components can be knocked out of the nucleus. Due to a lucky coincidence it really took place in the experiments on the α-particles scattering, because the α-particle energy E_α was not enough to produce even the lightest hadron π^0-meson ($E_\alpha < 135$ MeV). Based on studying the behavior of knocked-out particles in the magnetic field, in 1924 E. Rutherford and D. Chedvick presented in their paper the final proof that those particles were protons. They also suggested, that the α-particle was stuck in the nucleus. Using the Wilson camera, in 1925 P. Blackett made a photograph of the reaction, which showed that the α-particle is captured by the nucleus and its track is ended by the typical "fork," in which the short fat track belongs to the residual nucleus while the long thick track belongs to the knocked-out proton. The reaction, in which one was shown for the first time that the proton enters into the composition of the nucleus

$$\,_2^4He + \,_7^{14}N \rightarrow \,_8^{17}O + p,$$

was defined as an example of the first two-particle reaction.

Since nuclei masses are always larger than Z proton masses, then it is obvious, that some more particles must enter into the composition of the nucleus. It is natural to assume that the nucleus contains the electrons and that the protons exceeding the nuclear electrons cause the total positive nuclear charge. However, this hypothesis, despite its charming simplicity (no need to introduce new particles), ran into a sequence of serious contradictions. Let us enumerate them.

(i) The protons and the electrons are fermions having the spin $1/2$. Then, as follows from the rule of summing up the spin moments vectors, the nuclei with an even number of the protons n_p and the electrons n_e must have an integer spin, while the nuclei with odd n_p and n_e must have a half-integer spin. For example, for a deuteron (the nucleus of a deuterium atom) the spin should equal either $3/2$ or $1/2$ depending on the orientation of the electron and proton spins. In reality the observed deuteron spin is equal to 1, which is in no way reconciled with the hypothesis of the nuclear electrons.

(ii) If the electrons were present in the nucleus then the nuclear magnetic moment would be of the Bohr magneton order ($\mu_B = e\hbar/2m_e c$). However, as experiments exhibit, the nuclear magnetic moments are of the nuclear magneton order ($\mu_N = e\hbar/2m_p c$), which is 1000 times less than μ_B.

(iii) From the Heisenberg uncertainty relation

$$\Delta p \cdot \Delta x \geq \hbar/2$$

it follows that in the nucleus ($\Delta x \approx 10^{-14}$ m) Δp for the electron has of the order of $\hbar/\Delta x \approx 10^{-20}$ kg · m/sec. Since the momentum must be no less than this value, then for the energy of such an electron we obtain the value ~ 20 MeV. The maximum energy of the electrons emitted in the β-decay of various nuclides is in the interval 0.011–6.609 MeV that is much less than the energy of the electrons inside of the nucleus, if the electrons were inside of it.

In 1930 W. Bote and H. Becker obtained deeply penetrating radiation, which arises under the α-particles bombardment of light elements nuclei ${}^{9}_{4}Be$, ${}^{11}_{5}B$, and ${}^{19}_{9}F$. Since this radiation went through a lead layer several centimeters thick and was not deflected by the magnetic or electric fields, for several years unsuccessful attempts had been made to identify it with high energy photons. In February, 27, 1932 I. Chadwick published an article in the journal *Nature* [16], where he showed that the Bote-Becker penetrating radiation is a flux of neutral particles having masses close to the proton mass (nowadays these particles are called *neutrons, n*). After Chadwick's article, events developed at a striking quickness. In the same journal D. Ivanenko published a letter [17], in which he proposed the hypothesis of a proton-neutron nuclear structure. Soon the proton-neutron model was also introduced by W. Heisenberg in his article, published in the journal *Zeitschrift für Physik* (1932, June) [18]. Due to numerous similarities of the proton and the neutron (neutron spin equals 1/2 and $m_n \approx m_p$) a common name "nucleon" is used for both of them.

In the nucleus the nucleons form not such a hard lattice as the atoms in a crystal do, but rather a liquid structure in which the nucleons can move like molecules in a liquid. The successful explanation of an element's structure in the Mendeleev periodic table by means of only three types of particles, namely, the protons and the neutrons contained in the nucleus, which the electrons surround, provided all the reasons (at that time) to consider that e^-, p, and n are the structureless matter blocks, of which all the universe is built. Establishing this level of the matter structure was the third step of the "Quantum Stairway." The list of the "elementary" particles, known at that time, was rather short: electron, proton, neutron, photon, and positron (the electron antiparticle that was discovered in 1932 by K. Anderson in cosmic rays). So, the world used to be described by a fascinating simple scheme. One could think, that the third step, reached with such difficulty, was the last one on the way to the Creator. However, it was nothing else, but the illusion of an understanding that is like Kipling's cat who walks by herself and makes visits in accordance with her desire only.

8

From muon to gluon

The woodpecker was a humble scholar and lived a life of strict seclusion. He never saw anyone (many even thought that he drank as deep as all deep scholars), and sat on a pine-branch all day, pecking away. He'd pecked out a series of works on woodland history: "Goblin Genealogy," "Did Baba-Yaga the Witch Ever Wed?" "The Sex of a Sorceress and the Census," etc. But no matter how much he pecked, he couldn't find a publisher for his peckings. And so he, too, finally made up his mind to offer his services to the eagle as Courtyard Chronicler.
M.E. Saltykov Shchedrin, "The Eagle-Patron of Arts"

In 1936 muons $\mu^{\pm}$, discovered by K. Anderson and S. Niedermayer in cosmic rays, were added to the list of elementary particles. Before 1953, when the first accelerator was constructed (Brookhaven proton synchrotron with maximum energy 3 Gev), elementary particle investigation were intimately connected with cosmic ray investigation. In 1947 the group by S. Powell discovered π^+ and π^--mesons. The situation with elementary particles already became not entirely simple. The electrons and the nucleons are necessary to build atoms. The photons and the pions play the role of the carriers of the electromagnetic and nuclear forces, respectively. An electron antiparticle, a positron, can be viewed as a delicate hint (for the experienced mind) on the existence of antimatter searching, which should be continued until antinucleons are found. However, what does one do with muons, which do not find their place in this world scheme?

In the late 1940s–early 1950s a real demographic explosion had occurred in the elementary particle world. A whole zoo of new particles, called "strange" particles, had been discovered. A main peculiarity of those particles and the ones, discovered later, is that they are not the component of matter observed. They live for a very short time and decay into stable particles (protons, electrons, photons, and neutrinos). First particles from this group, K^+- and K^--mesons, Λ-hyperons were discovered in cosmic rays, the next ones—in accelerators. From the early 1950s accelerators become the main tool to investigate matter microparticles. Accelerators's energy is growing and the tendency for increasing the number of fundamental particles becomes more and more apparent. Nowadays the list of elementary particles has become tremendously large ~ 400. Properties of discovered elementary particles prove to be unusual in many respects. To describe them, characteristics taken from classical physics, such as electric charge, mass, momentum, angular moment, and magnetic moment, proved to be insufficient. It was necessary to introduce many new quantum numbers, having no classical analogs, which we call *internal quantum numbers*. The first reason for their introduction deals with additional degrees of freedom of elementary particles. The second reason is caused by striving to explain the nonobservation of some "acceptable" reactions*

---

* "Acceptable" reactions are reactions that are allowed by the conservation laws of energy, momentum, angular moment, and electric charge.

by means of the existence of an internal symmetry that leads to the conservation of a corresponding charge. The latter circumstance carries out the generous dispensation of the conserved charges to some groups of particles, according to the following principle: *"dynamic symmetry corresponds to a closed channel of an acceptable reaction."* It is possible, that a reader, being experienced by any sort of direct and inverse theorems about the existence and uniqueness of the solutions, will feel no deep satisfaction. As quieting reason, one may call attention to aesthetic attractiveness of the symmetric approach in all spheres of our life (it is hardly probable that the statue of Venus from Milos being deprived of symmetry could find a place within the walls of the Louvre).

Let us proceed to the classification of known particles. Elementary particles are divided into three categories:

(i) Hadrons, which participate in the strong, weak, and gravitation interactions. Being electrically charged they participate in electromagnetic interaction too.

(ii) Leptons, which do not participate in the strong interaction.

(iii) Field quanta, which carry the strong, electromagnetic, and weak interactions.

The hadrons are divided into baryons with a half-integer spin and mesons with an integer spin. Maximum values of the spin for hadrons are as follows: 6 for the mesons $a_6(2450)$ and $f_6(2510)$, and 11/2 for the baryons $N(2600)$ and $\Delta(2420)$ (in parenthesis the hadrons masses are given in MeV/c^2). Electron (e), muon (μ), tay-lepton (τ), and their neutrinos (ν_e, ν_μ, ν_τ), belong to the lepton class. They all have the spin 1/2. If one subtracts from the total number of discovered particles 12 interaction carriers (8 gluons, $W^\pm$, Z, and the photon) and 6 leptons, then the total number of the hadrons is obtained.

For the division of particles with the spin 1/2 into the leptons and the baryons to make sense, transitions between these particle kinds must be impossible. For example, the neutron must not decay into the electron-positron pair and the electron neutrino

$$n \to e^- + e^+ + \nu_e. \tag{8.1}$$

In reality, this decay has been never observed. Let us introduce two quantum numbers, namely, baryon B and lepton L charges connecting them with dynamic symmetries with respect to global* gauge transformations. For the baryons $B = 1$, for the antibaryons $B = -1$, for nonbaryons $B = 0$. In the case of the leptons, it is accepted to speak of not a lepton charge, but a lepton flavor. One discriminates the total lepton flavor L and individual lepton flavors L_e, L_μ and L_τ. For e^- (e^+)—$L_e = 1$ ($L_e = -1$), for μ^- (μ^+)— $L_\mu = 1$ ($L_\mu = -1$), for τ^- (τ^+)—$L_\tau = 1$ ($L_\tau = -1$) and

$$L = \sum_{i=e,\mu,\tau} L_i.$$

All nonlepton particles have $L = 0$. By now no reactions with violation of either the total or individual lepton flavors have been observed. However, there are no serious reasons in support of the lepton flavor conservation law. Consequently, many vital electroweak interaction theories predict the existence of processes, in which either the total or individual lepton flavor is not conserved. Scientists are intensively searching for reactions, which can help to establish upper limits on their cross sections. Some lepton decays going with the violation of L_i are given below:

$$\mu^- \to e^- + \gamma \quad (4.9 \times 10^{-11}), \tag{8.2}$$

---

*When the transformation parameters do not depend on coordinates the transformation is called global.

$$\mu^- \to e^- + e^+ + e^- \qquad (1.0 \times 10^{-12}), \tag{8.3}$$

$$\tau^- \to \mu^- + \gamma \qquad (3.0 \times 10^{-6}), \tag{8.4}$$

$$\tau^- \to e^- + \pi^0 \qquad (3.7 \times 10^{-6}), \tag{8.5}$$

where in brackets the upper bounds on their branchings* have been pointed. Comparison of theoretical expressions for partial decay width with experimental ones results in establishing the bounds on parameters of the theories in which these decays are allowed.

Despite the fact that the baryon charge conservation law ensures matter stability, its correctness is subjected to question, too. Within the framework of some GUTs B is not a conserved quantum number and that leads to proton instability. Some decay channels with the fixed upper limit on the proton lifetime with respect to the given channel are as follows:

$$p \to e^+ + \pi^0 \qquad (5.5 \times 10^{32} \text{ yrs}), \tag{8.6}$$

$$p \to e^+ + \nu_l + \nu_{l'} \qquad (1.1 \times 10^{31} \text{ yrs}), \tag{8.7}$$

$$p \to e^- + \mu^+ + \mu^+ \qquad (6 \times 10^{30} \text{ yrs}) \tag{8.8}$$

Since the age of our universe is as short as $\sim 10^9$ years, there are no special reasons for inconsolable grief over possible proton instability.

There is an important difference between the electric charge conservation and internal quantum numbers conservation. The electric charge is not a simple number, similar, for example, to the baryon charge, which is ascribed to various particles. The electric charge governs the system dynamics and is a source of the electromagnetic field in itself. The interaction between charged particles is carried out by means of the electromagnetic field whose quanta are nothing but the massless photons. Since with the help of the corresponding devices the electric field is easily measured, then it is possible to measure the electric charge of an object from a large distance, that is, without close contact with this object. Nothing similar takes place with the baryon charge. There exists no baryonic field connected with the baryon charge. For this reason it is impossible to measure the baryon charge (number) of an object at some distance away.

In a consistent description of interacting systems, the above-mentioned difference between the electric charge and "charges," not being the sources of physical fields, is in a different character of gauge transformations, responsible for the conservation laws. For a conserved quantity to be a field source, the theory must be invariant under local gauge transformations. Thus, the electric charge conservation corresponds to the invariance of the theory under the following transformations:

$$U_{em}(1) = \exp\left[ie\alpha(x)\right]. \tag{8.9}$$

Conserving the charges, not producing physical fields, are bound up with the invariance of the theory under global gauge transformations

$$U_N(1) = \exp\left[iN\alpha_N\right], \tag{8.10}$$

where $N = B, L_e, L_\mu, L_\tau, \ldots$

Let us proceed to the introduction of inexact internal quantum numbers, that is, numbers that are already conserved but not in all interactions. Above we mentioned the so-called strange particles that are produced by pairs (one or more) under colliding π-mesons with nucleons. Since the production of these particles was caused by a strong interaction the probability of their birth was large. However, they decayed into ordinary hadrons or leptons

---

*Branching is the ratio between the widths of the given decay channel (partial decay width) and the total decay width.

at the expense of only the weak interaction and as a result the probability of their decays was very small. The uncommon behavior of these particles once again reminded physicists of the famous phrase by F. Bacon: *"The perfect beauty without touch of strangeness is not available in the world."* When in 1954 at the Brookhaven Cosmotron these particles were obtained for the first time, among other processes the following one was observed:

$$\pi^- + p \to \Lambda + K^0.$$

The large value of its cross section indicated, that it is going on exclusively due to the strong interaction. On the other hand, the long lifetimes ($\sim 10^{-10}$ s) of the particles Λ and K^0 with respect to the decays

$$\Lambda \to p + \pi^-, \qquad K^0 \to \pi^+ + \pi^-$$

testified that these decays are caused by the weak interaction. For some reasons Λ and K^0 decays into lighter hadrons are forbidden due to the strong interaction and as a result, they live a long time. M. Gell-Mann and K. Nishijima independently from each other introduced a new additive quantum number, a strangeness s. They postulated its conservation in the strong and electromagnetic interactions and nonconservation in processes, caused by the weak interaction (for the weak interaction $|\Delta s| = 1$). For already known hadrons s takes the values -3, -2, -1, 0, 1. Further extension of the hadron sector demanded the introduction of such quantum numbers as a charm (c) and a beauty (b). They are also additive numbers and are conserved in the strong and electromagnetic interactions. In hadron decays due to the weak interaction they vary according to the rule

$$|\Delta c| = |\Delta b| = 1.$$

Using characteristics introduced by us, we can divide known baryons and mesons into the following families:

I. normal ($s = c = b = 0$) hadrons (nucleons, π-mesons);
II. strange ($c = b = 0, s \neq 0$) hadrons (Λ-, Σ-, Ξ- and Ω^--hyperons, K-mesons);
III. charmed ($s = b = 0, c \neq 0$) hadrons (Λ_c^+- and $\Sigma_c^{\pm,0}$- hyperons, D-mesons);
IV. beautiful ($s = c = 0, b \neq 0$) hadrons (Λ_b-baryons and B-mesons).

There are also mixtures of the last three families:
V. strange and charmed ($b = 0, s \neq 0, c \neq 0$) hadrons ($\Xi_c^+$- and Ω_c^0-baryons, D_s-mesons);
VI. strange and beautiful ($c = 0, s \neq 0, b \neq 0$) hadrons ($\Xi_b$-baryons, B_s^+-mesons);
VII. charmed and beautiful ($s = 0, c \neq 0, b \neq 0$) hadrons ($B_c$-mesons).

From all the plurality of the elementary particles only eleven are stable. They are: three neutrinos (ν_e, ν_μ, ν_τ), three antineutrinos ($\overline{\nu}_e, \overline{\nu}_\mu, \overline{\nu}_\tau$), the photon, the electron, the positron, the proton, and the antiproton. Other particles are unstable. Unstable particles can be divided in two classes: metastable particles and resonances. Metastable particles decay due to the weak or electromagnetic interactions, that is, they are tolerant to the decay caused by the strong interaction. Normally these particles are included to the class of stable particles. Particles-resonances decay predominantly due to the strong interaction (there may be the channels caused by the electromagnetic and weak interactions, but these channels are greatly suppressed). A typical resonance lifetime belongs to the interval 10^{-23}–10^{-24} (this time is necessary for a relativistic particle to cover the distance of the order of the hadron size $\sim 10^{-13}$ cm). Such short lifetimes do not enable one to register resonance traces in track detectors. Resonances are not observed in a free state, they reveal themselves while scattering in the form of quasi-stationary states of two or three strongly interacting particles. They possess such particle characteristics as the spin, the electric charge and they can be specified by internal quantum numbers, conserved in the strong interaction (isospin, parity, hypercharge, etc.). Resonances, however, have no definite mass value, unlike stable particles. They are described by a mass spectrum of a dispersion type, the maximum of this spectrum is called the itresonance mass m. The resonance mass spectrum width Γ supplies information about the probability of a resonance decay and must not exceed $mc^2/2$.

The nature of an unstable particle is the most transparent when one uses the concept of a quasi-stationary state. Considering an unstable particle f we can write down its total Hamiltonian H in the form:

$$\mathcal{H} = \mathcal{H}_f + \mathcal{H}_d,$$

where $\mathcal{H}_d$ is the part of the Hamiltonian, which is responsible for the decay. When neglecting $\mathcal{H}_d$, the particle becomes stable and, as this takes place, its states are eigenstates of the operator $\mathcal{H}_f$. In the case of the metastable particle, $\mathcal{H}_d$ contains the weak and electromagnetic interactions while in the case of the resonances it contains, in addition, the strong interaction. With $\mathcal{H}_d$, taken into account, the states of the particle f are quasi-stationary.

Let us find out the connection of the quasi-stationary state decay law with the function of energy distribution or, what is one and the same, with the mass spectrum of this state. Let $\Psi(\alpha, t = 0)$ be the initial state of a system. Here α denotes the plurality of variables, according to which the system states are classified. Now we expand $\Psi(\alpha, t = 0) = \Psi(\alpha, 0)$ in terms of the eigenfunctions of the energy operator $\psi(\alpha, E)$ (continuous spectrum)

$$\Psi(\alpha, 0) = \int a(E)\psi(\alpha, E)dE. \tag{8.11}$$

Then, at the time moment t, the state of a system is determined by the expression:

$$\Psi(\alpha, t) = \int \exp\left(\frac{iEt}{\hbar}\right)a(E)\psi(\alpha, E)dE. \tag{8.12}$$

From Eqs. (8.11) and (8.12) we obtain the probability of finding the system in the initial state after the time t

$$W(t) = \left|\int \Psi^\dagger(\alpha, 0)\Psi(\alpha, t)d\alpha\right|^2 = \left|\int \exp\left(\frac{iEt}{\hbar}\right)|a(E)|^2 dE\right|^2 =$$

$$= \left|\int \exp\left(\frac{iEt}{\hbar}\right)w(E)dE\right|^2, \tag{8.13}$$

where $w(E)dE = |a(E)|^2 dE$ is a function of the energy distribution for the initial $\Psi(\alpha, 0)$ state and, consequently, for the final $\Psi(\alpha, t)$) one. So, the decay probability of the state $\Psi(\alpha, 0)$ is only defined by the function of the energy distribution in this state.

In the case of a resting unstable particle, the energy distribution $dW(E) = w(E)dE$ is nothing else, but a particle mass spectrum. The states $\psi(\alpha, E)$ therewith include the decay product states as well. It becomes obvious from (8.13) that for a quasi-stationary state to decay it is necessary and sufficient that the energy distribution integral function $W(E)$ is continuous. Thus, the discrete mass spectrum is excluded. Really, the decay probability $[1 - W(t)]$ tends to unity under $t \to \infty$ only when $W(E)$ is a continuous function.

To obtain the radioactive decay law, familiar to us from nuclear physics

$$W(t) = W_0 \exp\left(-\frac{\Gamma t}{\hbar}\right), \tag{8.14}$$

where Γ is a total decay width of an unstable particle f, it is enough to assume, that the function $a(E)$ has the form:

$$a(E) = \frac{\Gamma/2}{E - E_0 + i\Gamma/2}, \tag{8.15}$$

where $E_0 = m_f c^2$. Indeed with the help of the residues theory we have

$$W(t) = | \int \frac{\Gamma/2}{E - E_0 + i\Gamma/2} \exp\left(\frac{iEt}{\hbar}\right) dE \ |^2 =$$

$$= | \ \frac{2\pi i \Gamma}{2} \exp\left(-\frac{\Gamma t}{2\hbar}\right) \exp\left(\frac{iE_0 t}{\hbar}\right) \ |^2 = (\pi \Gamma)^2 \exp\left(-\frac{\Gamma t}{\hbar}\right). \tag{8.16}$$

Thus, from (8.15) it follows, that the mass distribution of a resting unstable particle has the dispersion character

$$w(mc^2) dm = \frac{(\Gamma/2)^2}{(mc^2 - m_f c^2)^2 + \Gamma^2/4} dm. \tag{8.17}$$

Since unstable particles are characterized by the mass spectrum, then from the point of view of the Poincare group, they must be described by reducible representations. But the group theoretical definition of elementary character presupposes that the corresponding representation must be irreducible, so, from this point of view, unstable particles are not elementary particles. Before the quark structure of the hadrons was established, it was the reason to give concern. Moreover, the situation been has aggravated by the fact that in all of the known experiments, stable and unstable particles participate in interactions almost in the same way. Consequently, we have no reason to consider unstable particles to be less fundamental objects compared to stable particles. Moreover, if all hadrons (resonances and stable particles) are entered on a diagram, where the spin J and the mass square m^2 are plotted along the axes, then the states with the same B, I, and S are located on straight lines (Fig. 8.1) which are called *Regge trajectories*.* Regge [19], Chew and Frautschi [20] (see, for review, [21] as well) developed a strong interaction theory, in which hadrons are compared with poles of a scattering amplitude in a complex plane of spin values. These poles are called *Regge poles*. Regge trajectories are simply showing that the spin value is a function of the energy or, which is equivalent, of the mass $J(m^2)$. On every trajectory hadrons appear with an interval $\Delta J = 2$. Stable hadrons are the lowest located state on the Regge trajectories and that constitutes the only difference between them and resonances. Notice that just the striking linearity of the Regge trajectories detected at experiments gave impetus to creating the string models of hadrons. In these models the hadrons are considered as stretched objects, strings, the quantization of which results in the appearance of a particles sequence to be placed at linearly growing Regge trajectories.

Nowadays the difference between the physical and group theoretical concepts of the elementary character for resonances must be accepted with philosophical calmness since they are not structureless formations but made of the quarks.

If one neglects the interaction causing a decay, then formalism of relativistic quantum theory can be generalized to unstable states. Such an approximation is necessary to describe unstable particles scattering. As a rule, in scattering processes, particles in the initial and final states are not supposed to interact, that can be realized when they are separated at infinitely great distances from each other. In this case, one does not follow to forget that in reality unstable particles have been decayed long before asymptotical separation was achieved.

---

*The graph is called the Chew–Frautschi diagram.

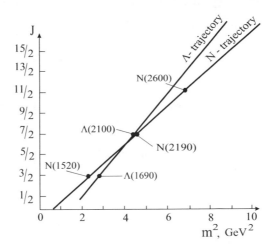

FIGURE 8.1

The Regge trajectories for baryon states with the negative parity.

Basic methods of resonances detection are based on the fact that resonances have the mass spectrum of the dispersion type. The first method deals with investigating the maxima in the total scattering cross section. To make it definite, let us assume, that we deal with the reactions:

$$a + b \rightarrow Y \rightarrow a + b, \tag{8.18}$$

$$d + f \rightarrow Y \rightarrow d + f, \tag{8.19}$$

$$a + b \rightarrow Y \rightarrow d + f, \tag{8.20}$$

where for Y both decay channels are possible

$$Y \rightarrow a + b, \qquad Y \rightarrow d + f. \tag{8.21}$$

Due to fulfillment of Eq. (8.21), these reactions are going through a s-channel.* The existence of such s-channel diagrams is a necessary condition for the resonance to be observed. This resonance peak, connected with Y, appears in all the above-mentioned reactions. If one presumes, that the reaction (8.19) is going on only by means of the production of the resonance-particle Y in a virtual state (the s-channel is only one of the reactions), then the total cross section of the elastic df-scattering as a function of the energy E near the resonance is defined by the Breit-Wigner formula[†] [22]:

$$\sigma(E) = \sigma_0 \frac{(\Gamma/2)^2}{(E - E_0)^2 + \Gamma^2/4}, \tag{8.22}$$

As we can see, this expression coincides with the masses distribution $w(E)$ to an accuracy of the kinematic factor. The energy E_0, corresponding to the cross section maximum

---

*A s-channel diagram is the Feynman diagram, where annihilation of initial particles takes place at one space-time point, while final particles production takes place at the other one.

[†]If other reaction channels are present, the possibility is not eliminated that even at resonance the contribution into the cross section coming from the s-channel diagram proves to be smaller (or the same order) as compared with one of the remaining diagrams. As a result the resonance peak in the cross section of the reaction in question is absent.

$\sigma(E) = \sigma_0$ being divided by c^2, defines what we agreed to consider the resonance mass. The maximum width informs us about the resonance decay probability. In this case the particles in the final state appear with retardation $\Delta t \sim \hbar/\Gamma$ as compared to scattering without the resonance production. The main drawback of this method is that it does not allow us to determine resonance quantum numbers completely.

The next method is the phase analysis method that is more universal because with its help, it is possible to define all the resonance characteristics (mass, width, spin, parity, isotopic spin, and so on). The method is based on measuring the differential cross section of elastic scattering $d\sigma = 2\pi|f(\theta, E)|^2 \sin\theta d\theta$ (θ is a scattering angle). If particles having the spin participate in scattering, then the scattering amplitude $f(\theta, E)$ is expanded as a series in the spherical functions $Y_m^l(\theta, \varphi)$. For spinless particles this expansion has the form:

$$f(\theta, E) = \sum_l \sqrt{4\pi(2l+1)} f_l(E) Y_l^0(\theta) = \sum_l (2l+1) f_l(E) \mathcal{P}_l(\theta), \tag{8.23}$$

where the coefficients $f_l(E)$ are partial waves of scattering with the moment l, which are determined from experimental data as complex functions of E. The resonance with the spin $J=l$ reveals itself as the Breit-Wigner contribution to $f_l(E)$. If the spins of two particles are equal to 0 and 1/2 respectively, then instead of Eq. (8.23) we have

$$f(\theta, E) = \sum_l \sqrt{4\pi(2l+1)}[\sqrt{\frac{l+1}{2l+1}} f_{l+\frac{1}{2}}(E) Y_{l+\frac{1}{2}}^{1/2}(\theta, \varphi) - \sqrt{\frac{l}{2l+1}} f_{l-\frac{1}{2}}(E) Y_{l-\frac{1}{2}}^{1/2}(\theta, \varphi)].$$
$$\tag{8.24}$$

In the case when there are three or more particles in the final state, the method of maxima in mass distributions is used to search the resonances. Let us consider the inelastic scattering reaction

$$a + b \rightarrow c_1 + ... + Y \rightarrow c_1 + ... + d + f, \tag{8.25}$$

where a resonance Y decays into stable particles d and f. During such scattering, the momentum and the energy of particles a and b are distributed between groups of particles $c_1 + ...$ and $d + f$ in the final state. If the resonance Y were the stable particle, then in its proper system (rest system) its energy, and the mass$\times c^2$ as well, would have the definite value. But the resonance is unstable and is specified by the distribution function $w(E)$, which in the particle rest system is directly connected with the mass spectrum of particle decay products. In other words, the total set of states in (8.24) consists of two-particle states $\psi_{df}(E)$ to have the same quantum numbers as does the Y-particle.* In distribution of the invariant mass square of the particles d and f

$$M_{df}^2 c^4 = (E_d + E_f)^2 - (\mathbf{p}_d + \mathbf{p}_f)^2 c^2, \tag{8.26}$$

the brightly expressed maximum will be observed. Thus, studying the distribution of the masses in complexes of particles belonging in the final states, one may obtain directly the distributions on the masses of unstable particles, which are the resonance states in such complexes. Of course, it does not necessarily mean that every peak in the mass distributions of the particles being the reaction products can be identified with the resonance, because kinematic peaks also occur, which are inherent in the given reaction only.† The difference between the real resonance and the ghost peak is that the energy, corresponding appearance of the real resonance in different experiments, is always the same, while the

---

*If the decay channel $Y \rightarrow d+f$ is not the only one, and there exists another one, for example $Y \rightarrow k+m+l$, then three-particle states $\psi_{kml}(E)$ will enter into Eq. (8.24) and so on.
†Such peaks are called *ghost ones*.

energy, connected with the ghost peak, is changed from one experiment to another. The distribution of the invariant mass square in the system $\Sigma^+\pi^-$, arising in the reaction

$$\pi^- + p \rightarrow \Sigma^+ + K^0 + \pi^- \tag{8.27}$$

at the momenta of the incident π^--meson from 2.2 to 2.4 GeV/c is shown in Fig. 8.2 (N is a number of events on $(50\ \text{MeV})^2$. Three peaks are distinguished against the background

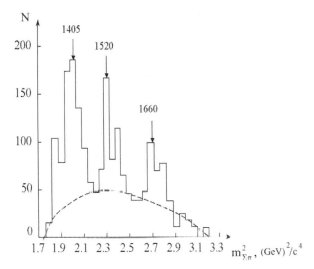

FIGURE 8.2

The distribution of the invariant mass square in the $\Sigma^+\pi^-$-system. The dashed line represents the expected distribution defined by the relativistic phase volume only

of uncorrelated events. All of them turn out to be the real resonances.

First resonances (Δ-resonances) were discovered in 1952 by E. Fermi under scattering of the π-mesons by protons

$$\pi + p \rightarrow \Delta \rightarrow \pi + p. \tag{8.28}$$

The total cross section of the pion-nucleon scattering versus the center-of-mass frame energy is displayed in Fig. 8.3. The solid line corresponds to σ_{π^+p}, while the dashed one describes σ_{π^-p}. As it follows from Fig. 8.3, at the π-mesons kinetic energy $T = 195$ MeV the resonance peak appears both in π^+p-scattering (Δ^{++}-resonance) and in π^-p-scattering (Δ^0-resonance). From the Gell-Mann-Nishijima formula

$$Q = S_3^{(is)} + \frac{B}{2}$$

it is evident that when the Δ^{++}- and Δ^0-resonance enter one and the same isotopic multiplet they have $S^{(is)} = 3/2$.

Further, by the example of the reaction (8.28), we show how to define such resonance characteristics as the mass and the isotopic spin from the experimental data. In the lab frame the proton rests, and consequently, the three-dimensional momentum and the energy of the Δ-resonance are defined by:

$$\mathbf{p}_\Delta = \mathbf{p}_\pi = \frac{1}{c}\sqrt{(T_\pi^r + m_\pi c^2)^2 - m_\pi^2 c^4}, \tag{8.29}$$

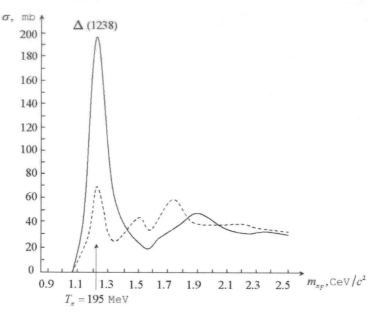

FIGURE 8.3

The total cross sections of the $\pi^+ p$- and $\pi^- p$- scattering.

$$E_\Delta = T_\pi^r + m_\pi c^2 + m_p c^2. \tag{8.30}$$

From Eqs. (8.29) and (8.30) it follows

$$m_\Delta = \frac{1}{c^2}\sqrt{E_\Delta^2 - \mathbf{p}_\Delta^2 c^2} = 1236 \text{ MeV}/c^2$$

(contemporary data lessen down the Δ mass value to 1232 MeV/c^2).

To find the isotopic spin value we use the isotopic invariance of the cross sections of the reactions (8.28). Since the isotopic spins of the nucleon and the π-meson are equal to 1/2 and 1 respectively, their sum could be equal either to 3/2 or 1/2. In order to do it we need the pion-nucleon states expansion in terms of the isotopic spin eigenstates. This expansion is given by the formulae (9.82), which we shall deduce later. Then, from the isotopic spin conservation law it follows that only the following scattering amplitudes exist:

$$(\Psi_{\pi^+ p}^f, \Psi_{\pi^+ p}^i) = (\Psi_{3/2,3/2}^f, \Psi_{3/2,3/2}^i) = A_{3/2}, \tag{8.31}$$

$$(\Psi_{\pi^- p}^f, \Psi_{\pi^- p}^i) = \frac{1}{3}A_{3/2} + \frac{2}{3}A_{1/2}, \tag{8.32}$$

$$(\Psi_{\pi^0 n}^f, \Psi_{\pi^- p}^i) = \frac{\sqrt{2}}{3}A_{3/2} - \frac{\sqrt{2}}{3}A_{1/2}. \tag{8.33}$$

The total cross sections of scattering in the region, where the multiple π-mesons production is insignificant, are given by:

$$\sigma_{\pi^+} = \rho \mid A_{3/2} \mid^2,$$

$$\sigma_{\pi^-} = \sigma_{\pi^- p \to \pi^- p} + \sigma_{\pi^- p \to \pi^0 n} = \rho \left(\frac{1}{3} \mid A_{3/2} \mid^2 + \frac{2}{3} \mid A_{1/2} \mid^2 \right), \tag{8.34}$$

where ρ is a kinematic factor (which is constant for all the three processes, if we neglect masses difference in isomultiplets). If we assume, that the isotopic spin of the Δ-resonance is equal to $3/2$, then the second term in Eq. (8.34) at $E = T_\pi^r$ goes to zero and we arrive at the relation:

$$(\sigma_{\pi^+}/\sigma_{\pi^-})|_{E=T_\pi^r} = 3,$$

which is in excellent accord with the experiment.

However, looking at this megapolis of the hadrons, we understand, that the above-mentioned division is nothing else, but a scheme of streets and squares that does not bring us closer to understanding the idea of the architect creating this elementary particles Babylon. Moreover, this picture is far from perfection and we do not experience reverential delight under a sight on it. Let us be reminded of the guiding thread, which brought us from the first step of the Quantum Stairway to the third one. The Mendeleev periodic table was built on the basis of an accidentally discovered periodic alteration of chemical properties of elements alongside with increasing their atomic masses (nucleus electric charge, to be exact). Sixty three years later after the creation of this table, with its help, we managed to construct hundreds of elements from only three fundamental (as it seemed then) particles p, n, and e^-. Let us concentrate our efforts to find a similar table for the hadrons.

9

Hadron families

*The bullfinch was the biggest success. By way of salu-
tation he read an article from the morning paper so
fluently that even the eagle thought he understood it.
Take life easy, the bullfinch read, and the eagle nodded:
jusso! All that he ever wanted, said the bullfinch,
was that his paper should sell well, and the eagle
repeated: jusso! The flunkey's life is better than his
master's, he chirped; the master has worlds of care,
the flunkey none–the good master looks after them.
Jusso! again said the eagle. They say that when he
had a conscience he had no trousers, but now that
it is dead he wears two pairs at once. And again the
eagle came in with his jusso!*
M.E. Saltykov Shchedrin, "The Eagle-Patron of Arts"

9.1   Yukawa hypothesis

In the language of quantum field theory, force field influence on a particle is interpreted as emission and absorption of the quanta of this field. For example, the electromagnetic interaction between two electrons appears due to photons exchange. Photon masslessness reveals itself in the long-range action of the Coulomb forces, that is, the Coulomb potential decreases with a distance as $1/r$. It is natural to expect that the massive carrier of an interaction will produce a short-range potential. In an analogy with the exchange nature of electromagnetic forces, in 1935 H. Yukawa [23] introduced a hypothesis about massive quantum exchange, between nucleons in a nucleus. The idea was, that the interaction between this massive quantum (π-meson) and nucleon N appears as a result of combination of virtual process

$$N \leftrightarrow N' + \pi. \tag{9.1}$$

In other words, the Hamiltonian, causing the interaction between nucleons in a nucleus H_{int} must be of the form:

$$H_{int} \sim \overline{N}(x)N(x)\pi(x). \tag{9.2}$$

Connections of this kind often occur in elementary particle physics and they are called *Yukawa couplings*.

Now we should more carefully consider the concept of virtual states. Physical laws describe only experimentally measurable quantities. In quantum theory the uncertainty relations for dynamical observables, represented by noncommuting operators, impose restrictions on a precision of simultaneous measurements of these quantities. Thus, the Heisenberg

uncertainty relations impose precision limits on simultaneous measurements of particle coordinates and momenta

$$\Delta p_i \Delta x_i \geq \frac{\hbar}{2}, \qquad (i = 1, 2, 3), \tag{9.3}$$

as well as its energy and time

$$\Delta E \Delta t \geq \frac{\hbar}{2}. \tag{9.4}$$

The uncertainty relation (9.4) can be interpreted by the way in which it is somewhat unusual
from the viewpoint of classical physics. In the microworld the processes, going on with the
violation of the energy conservation law on the value of ΔE are allowed, provided that this
violation lasts no longer than $\Delta t \sim \hbar/\Delta E$. In this case the violation cannot be registered
by any physical devices. Analogously, the momentum conservation law could be violated
on the value Δp in the region $\Delta x \sim \hbar/\Delta p$. We emphasize that the uncertainty relations
reflect the inner nature of the microworld and has nothing to do with the imperfection of
our measuring devices.

When considering interactions between particles by means of field quanta exchange, one
naturally thinks of ΔE to denote brought in or taken away quantum energy and Δt to
denote the exchange duration or the lifetime of this quantum. The particles states with
such lifetimes are called the *virtual states*. All the elementary particles can occur both
in real and virtual states. Inherent for real particles the connection between energy E,
momentum $\mathbf{p}$ and mass m

$$E^2 = \mathbf{p}^2 c^2 + m^2 c^4 \tag{9.5}$$

is violated in the case of the virtual particles due to the appearance of ΔE. Keeping in
mind the violation of this equality, they say the virtual particles lay beyond the mass shell.
So, according to the Yukawa hypothesis, the interaction between nucleons in the nucleus is
carried out by means of the virtual particle exchange. Emitting and absorbing the virtual
π-mesons, protons, and neutrons turn into each other. It is easy to guess that for $p \leftrightarrow n$
coupling to exist, π^+ and π^--mesons are necessary. Being based on the experimentally
determined principle of the nuclear forces charge independence

$$V_N = V_{pp} = V_{nn} = V_{np}, \tag{9.6}$$

where V_N is nucleons interaction potential, one might conclude that a neutral π^0-meson is
needed to describe $p \leftrightarrow p$ and $n \leftrightarrow n$ interactions. Of course, $p \leftrightarrow p$ and $n \leftrightarrow n$ interactions
can take place due to two charged π-mesons as well. Thus, for example, the interaction
between two protons takes place as follows. Both protons emit one π^+-meson each, in
so doing they turn into neutrons. Then, these neutrons absorb π^+-mesons and turn into
protons again. A corresponding mutual conversion chain for each proton has the form:

$$p \rightarrow \pi^+ + n, \qquad n + \pi^+ \rightarrow p, \tag{9.7}$$

and $p \leftrightarrow p$ interaction on its own is displayed by the reaction:

$$p + p \rightarrow n + \pi^+ + n + \pi^+ \rightarrow p + p. \tag{9.8}$$

It is quite obvious that emission of two π-mesons necessitates the larger value of ΔE and
consequently, lessens Δt. In a shorter lifetime π-mesons cover a shorter distance and that
reduces (approximately twice) the action radius of the nuclear forces between identical
nucleons, and, as a result, the condition (9.6) will be violated.

Let us estimate the π-meson mass by means of the uncertainty relation (9.4). We consider
the interaction between a proton and a neutron. The proton emits the π^+-meson and turns

into the neutron, while the initial neutron having absorbed the π^+-meson becomes the proton

$$p \to n + \pi^+, \qquad n + \pi^+ \to p. \tag{9.9}$$

The chain of the reactions (9.9) can be represented as follows:

$$p + n \xrightarrow{\pi^+} p + n. \tag{9.10}$$

If one neglects the proton recoil momentum, then the emission of the virtual π^+-meson leads to the energy violation by the value $\sim m_\pi c^2$ at the minimum. The virtual meson exists during the time $\Delta t \sim \hbar/\Delta E \sim \hbar/(m_\pi c^2)$. In Δt time the virtual Δt-meson, even if it moves at a maximum possible speed c, covers the distance, which defines the maximum value of the force field action radius

$$R \sim c\Delta t \sim \frac{\hbar c}{m_\pi c^2} \sim \frac{\hbar}{m_\pi c}. \tag{9.11}$$

Substituting in Eq. (9.11) the experimentally obtained value of the nuclear force action radius $R_n \sim 10^{-13}$, we obtain $m_\pi \approx 280 \, m_e$ (m_e is the electron mass).

In reality, Yukawa predicted, that the mass of the nuclear interaction carrier might be equal to $206 \, m_e$. That was caused by using the improper data concerning the value of R_n. Being guided by this number, experimentalists began to search and have found the particle with the mass $207 \, m_e$, which was called a *muon* (μ). However, the muons turned out to be weakly interacting particles. Consequently, they are not suitable for the role of the nuclear field quantum. As late as 1947, predicted by Yukawa π-mesons were discovered in cosmic rays.

The notion about the interaction law, caused by the π-meson exchange, is given by the the Yukawa potential, that is, the potential energy that is derived under the assumption that the interacting particles are immovable (particles masses are so large that one may precisely fix their positions and neglects their recoils under emitting and absorbing quanta). Since the π mesons have zero spin, their behavior is described by the Gordon-Klein equation. In a free case this equation can be obtained from the relativistic relation between the four-dimensional momentum and the mass (9.5) by substitution:

$$E \to i\hbar \frac{\partial}{\partial t}, \qquad \mathbf{p} \to -i\hbar \frac{\partial}{\partial \mathbf{r}},$$

which gives

$$\left(\frac{1}{c^2} \frac{\partial^2}{\partial t^2} - \Delta + \frac{m_\pi^2 c^2}{\hbar^2} \right) \psi(\mathbf{r}, t) = \left(\Box + k_0^2 \right) \psi(\mathbf{r}, t) = 0, \tag{9.12}$$

where $k_0 = m_\pi c/\hbar = \lambda_\pi^{-1}$ and λ_π is the Compton wavelength of the π-meson. Under switching on an interaction, a source density of the π-meson field must stand in the right-hand side of Eq. (9.12). Since nucleons, whose density is designated by $\rho(\mathbf{r}, t)$, provide such a source, then Eq. (9.12) takes the form:

$$\left(\Box + k_0^2 \right) \psi(\mathbf{r}, t) = \rho(\mathbf{r}, t). \tag{9.13}$$

In case of a static field (9.13) turns into

$$\left(\Delta - k_0^2 \right) \psi(\mathbf{r}) = -\rho(\mathbf{r}). \tag{9.14}$$

The Green functions method must be used to write down the solution of Eq. (9.14) without specifying its right-hand side form. The solution will look like

$$\psi(\mathbf{r}) = \int G(\mathbf{r} - \mathbf{r}')\rho(\mathbf{r}')d\mathbf{r}', \tag{9.15}$$

where $G(\mathbf{r})$ is the Green function, satisfying the equation:

$$\left(\Delta - k_0^2\right) G(\mathbf{r}) = -\delta(\mathbf{r}). \tag{9.16}$$

Inserting in Eq. (9.16) the Fourier expansions of the delta-function

$$\delta(\mathbf{r}) = \frac{1}{(2\pi)^3} \int \exp\left(i\mathbf{p}\cdot\mathbf{r}\right) d\mathbf{p} \tag{9.17}$$

and the Green function

$$G(\mathbf{r}) = \frac{1}{(2\pi)^3} \int G(\mathbf{p}) \exp\left(i\mathbf{p}\cdot\mathbf{r}\right) d\mathbf{p}, \tag{9.18}$$

we obtain the following expression for the Fourier image of the Green function:

$$G(\mathbf{p}) = \frac{1}{\mathbf{p}^2 + k_0^2}. \tag{9.19}$$

To calculate the Green function in the expression (9.18) it is convenient to turn to the spherical coordinate system with the axis z oriented along $\mathbf{p}$. Then we arrive at

$$G(\mathbf{r}) = \frac{1}{(2\pi)^3} \int_0^\infty \frac{p^2}{p^2 + k_0^2} dp \int_0^\pi \sin\theta \exp\left[ipr\cos\theta\right] d\theta \int_0^{2\pi} d\varphi =$$

$$= \frac{1}{2\pi^2} \int_0^\infty \frac{p\sin(pr)}{r(p^2 + k_0^2)} dp = \frac{1}{4\pi^2 i} \int_{-\infty}^\infty \frac{p\exp\left[ipr\right]}{r(p^2 + k_0^2)} dp, \tag{9.20}$$

where $p = |\mathbf{p}|$, $r = |\mathbf{r}|$. The remaining integral over p is most convenient to calculate by the method of the residues theory. The function under the integral sign is an analytical function of p, which exponentially decreases in the upper half-plane and has poles in points $\pm ik_0$. Consequently, the integral is equal to the residue in the point $p = +ik_0$ multiplied by $2\pi i$. Thus, the Green function is equal to

$$G(\mathbf{r}) = \frac{1}{4\pi|\mathbf{r}|} \exp\left[-k_0|\mathbf{r}|\right] \tag{9.21}$$

and the solution of Eq. (9.16) takes the form:

$$\psi(\mathbf{r}) = \frac{1}{4\pi} \int \frac{\exp\left[-k_0|\mathbf{r}-\mathbf{r}'|\right]}{|\mathbf{r}-\mathbf{r}'|} \rho(\mathbf{r}') d\mathbf{r}'. \tag{9.22}$$

Notice, in particular, that at $m_\pi = 0$ the expression (9.22) turns into a very familiar expression for the electromagnetic potential, created by given charges distribution $\rho(\mathbf{r}')$. As it is known the energy of the system under consideration is described by the Hamiltonian function H

$$H = \int \mathcal{H} d\mathbf{r},$$

where $\mathcal{H}$ is the Hamiltonian function density. Let us obtain $\mathcal{H}$ by using its connection with the Lagrangian $\mathcal{L}$. It is easy to check, that the Lagrange-Eulerian equations lead to Eq. (9.15) if $\mathcal{L}$ is defined by:

$$\mathcal{L} = \frac{1}{2}\left(\frac{\partial\psi(x)}{\partial x_\mu}\frac{\partial\psi(x)}{\partial x^\mu} - k_0^2\psi(x)\psi(x)\right) + \rho(x)\psi(x), \tag{9.23}$$

where we do not specify a π-meson field in question, since the theory is charge independent. Then, applying the usual rules, we obtain the following expression for H:

$$H = \frac{1}{2} \int [\left(\frac{\partial \psi(x)}{\partial x_0}\right)^2 + \left(\frac{\partial \psi(x)}{\partial x_k}\right)^2 + k_0^2 \psi(x)^2 - 2\rho(x)\psi(x)] d\mathbf{r}. \tag{9.24}$$

Passing to a stationary case, taking into account the three-dimensional Gauss theorem and using Eq. (9.22), we arrive at

$$H = \frac{1}{2} \int [\left(\frac{\partial \psi(\mathbf{r})}{\partial x_k}\right)^2 + k_0^2 \psi(\mathbf{r})^2 - 2\rho(\mathbf{r})\psi(\mathbf{r})] d\mathbf{r} = \frac{1}{2} \int \{\psi(\mathbf{r})[-\Delta + k_0^2]\psi(\mathbf{r}) - 2\rho(\mathbf{r})\psi(\mathbf{r})\} d\mathbf{r} =$$

$$= -\frac{1}{2} \int \psi(\mathbf{r})\rho(\mathbf{r}) d\mathbf{r} = -\frac{1}{8\pi} \int d\mathbf{r} \int \frac{\rho(\mathbf{r})\rho(\mathbf{r}')}{|\mathbf{r} - \mathbf{r}'|} \exp(-k_0|\mathbf{r} - \mathbf{r}'|) d\mathbf{r}'. \tag{9.25}$$

At $m_\pi = 0$ the expression obtained is nothing more nor less than the electrostatic interaction energy of two charge distributions. Then it follows from Eq. (9.25) that the interaction potential of two nucleons separated by the distance r is determined by the equation:

$$V_N(\mathbf{r}) = -\frac{C}{4\pi|\mathbf{r}|} \exp(-k_0|\mathbf{r}|), \tag{9.26}$$

where C is a constant connected with the nucleon "nuclear" charge. From the expression for $V_N(\mathbf{r})$ follows the same result as from the uncertainty relation: the nuclear forces have the finite action radius, approximately equal to the Compton wave length of the π-meson. Notice, that the negative sign in the expression for the Yukawa potential, testifying the attraction character of the nuclear forces, is caused by the choice of the sign in front of the nucleon field density function $\rho(\mathbf{r})$. One should emphasize, that the above-given discussion is classical in its essence. However, it is possible to show, that the same result will be obtained, when the π-mesons will be considered in the quantum mechanic way, while the nucleons will be described by the classical function of the source as before.

The process of emitting and absorbing the π-meson lasts no longer than 10^{-23} s. In all modern experiments such a process may be considered as an instant one. Roughly speaking, the protons spend one part of their life in a nucleus being the protons, while during the second part they are the neutrons. So, it is natural to consider the proton and the neutron as two different states of the same particle given the title *nucleon*. All nucleons consist of identical cores, surrounded by a cloud of virtual π-mesons. The only difference between p and n lies in the character of such a cloud. When two such clouds are approaching each other at the distance of the order of the π-meson Compton wave length, the π-meson exchange between clouds takes place.*. So the π-mesons are constantly scurrying between interacting nucleons. The situation partly reminds of the covalent coupling carried out by the electrons in the molecular ion H_2^+. Since, in this case, the electron can transfer from one to another proton, the exchange forces appear and they are added to the ordinary Coulomb forces.

Obviously, the π-meson clouds (fur coat), surrounding neutrons and protons, contribute to nucleon magnetic moments. Since such virtual conversion chains

$$p \to \pi^+ + n, \qquad n \to \pi^- + p$$

are possible for a proton and a neutron, then anomalous parts of magnetic moments of the particles, caused by the π-meson field, must be approximately equal in value and opposite in sign.

---

*At smaller distances exchanging the heavier particles (vector mesons ρ, φ, ω, etc.) begins to be more substantial.

Let us consider the process of originating the neutron magnetic moment due to its virtual dissociation into π^- and p. Since the π-meson has the zero-spin, it also has the zero intrinsic magnetic moment. Then only the π-meson with a nonzero orbital moment, for example, in p-state ($l = 1$), contributes to the neutron magnetic moment. For the angular moment conservation law in this virtual process to be fulfilled, the following demands must be met.

(i) The direction of the orbital moment of the virtual π-meson being equal to 1, must coincide with the neutron spin direction.

(ii) The virtual proton spin must be directed opposite to the neutron spin.

Since the π^--meson is negatively charged, then the neutron magnetic moment induced by the π^--meson, is negative as well. To estimate the neutron anomalous magnetic moment value, we must know how long the neutron exists in the dissociated state or, what amounts to the same thing, the transition probability into this state. Since $m_p \approx 6.72 m_\pi$, the magnetic moment of the system $p + \pi^-$ is in value of the order equal to $(-6.72 + 1)\mu_N$ ($\mu_N = e\hbar/(2m_p c)$). Then the neutron magnetic moment observed is

$$\mu_n = W_0 \mu_n^s - 5.72(1 - W_0)\mu_N, \tag{9.27}$$

where μ_n^s is the neutron intrinsic magnetic moment, that is, the quantity to be equal to zero, W_0 is the probability that the neutron will be found in the naked neutron state, $(1 - W_0)$ is the probability that the neutron will be found in the dissociated $p + \pi^-$-state. Experimental value of the neutron magnetic moment $\mu_n = -1.913 \ \mu_N$ is obtained if one sets $W_0 \approx 0.665$. Analogously, for the proton anomalous magnetic moment we have

$$\mu_p = W_0 \mu_N + 6.72(1 - W_0)\mu_N, \tag{9.28}$$

where we have taken into consideration that the intrinsic magnetic moment of the proton is equal to μ_N and assumed that the probabilities of the proton and neutron virtual dissociations are equal. Then the total proton magnetic moment is pretty close to its experimentally measured value $\mu_p = 2.793 \ \mu_N$.

The experimentally proved Yukawa theory of the nuclear forces is a milestone in the development of physics. It has finally strengthened the assurance that the quantum interpretation of the interaction as an exchange of virtual quanta is a correct one. Such an interaction interpretation underlies the foundation of modern physical theories. Taking into consideration virtual particles changes our concepts of the physical vacuum under transition from the classical to quantum theory. The example of the electrodynamics is very significant in this case. The electromagnetic field in the classical theory is defined by the values of the strengths of the electric and magnetic fields ($\mathbf{E}$ and $\mathbf{H}$) given in all points of space and in all moments of time. Under transition to the quantum electrodynamics in places of those strengths operators appear which, in particular, do not commute with operators, defining the number of photons in a given state. However, only physical quantities, to which commuting operators correspond, can simultaneously have definite values. If operators do not commute, then the more precisely the quantity corresponding to one of these operators is defined, the less information can be obtained for the second quantity. At the exact definition of $\mathbf{E}$ and/or $\mathbf{H}$ the number of photons is absolutely undefined. In the same way, if the number of photons is exactly defined, then field strengths are not defined.

In quantum theory we determine a vacuum as a state without real particles or as a state with the least energy. Then, since for the photon vacuum (electromagnetic vacuum) the number of particles is zero, that is, is exactly defined, the field strengths are not defined, and this fact, in particular, does not accept these strengths being equal to zero. Impossibility to simultaneously set both the field strengths and the photons number equal to zero make us consider the vacuum state in quantum theory not as the field absence, but as one of the possible field states having the definite properties that are displayed in real physical processes.

Virtual production and absorption of the photons should be viewed as the manifestation of the photon vacuum, or, in other words, as taking account of the photon vacuum effects. The concept of a vacuum being the lowest field energy state can be analogously introduced also for other particles. Considering interacting fields, the lowest energy state of all the systems could be called a *vacuum state*. If sufficient energy is supplied to a field in a vacuum state, then the field is exited, that is, the field quantum is produced. Thus, particle production can be described as a transition from an "unobserved" vacuum state to a real one. Without real particles and external fields a vacuum, as a rule, does not reveal itself through any phenomena due to its isotropy. The presence of real particles and/or external fields leads to the vacuum isotropy violation, since production of virtual particles and their subsequent absorption results in changes of the state of the real physical system. Virtual particle production and absorption of particles is limited by conservation laws, the electric charge conservation law among them. For this reason virtual production of a charged particle is impossible without charge change of real particles (if there are any). If real particle charges are invariable, then in virtual processes charged particles are created and destroyed in pairs only (*particle-antiparticle*). Thus, in the case of charged particles one can speak only of the *particles-antiparticles* vacuum: *electron-positron* vacuum, *proton-antiproton* vacuum, etc. From this it also follows, that since, for example, positrons and electrons can be produced only in pairs, one cannot speak of electrons as isolated and solitary types of matter, just as one impossibly draws a demarcation line between the electric and magnetic fields. The electron and positron fields make up the unified electron-positron field, and this circumstance remains imperceptible, provided the processes of pairs productions and pairs destructions may be neglected. Analogously, as in the case of the *photon-antiproton* vacuum, the *electron-positron* vacuum or any other *particles-antiparticles* vacuums will lead to observable effects, one of which is a change in the physical properties of particles. The above-mentioned effects of the charge screening and appearance of the nucleons anomalous magnetic moments can serve as an example.

So, in quantum theory every particle is enclosed by the fur coat consisting of clouds of virtual quanta, produced and subsequently absorbed by the particle. Quanta can belong to any field (electromagnetic, electron-positron, meson, etc.), with which the particle is interacting. The fur coat contains many layers with different densities. For example, since the meson interactions of nucleons are a hundred times more intensive than the electromagnetic ones, the meson fur coat of the proton should be several orders thicker than the electromagnetic one. The fur coat is not something like a solid state, since quanta, its components, are continuously produced and annihilated. One can say, that in the quantum theory a particle is suffering from striptease-mania, since one part of its lifetime it spent in the dressed state, while during the rest of the time it is naked.

9.2 Isotopic multiplets

Experimental confirmation of the Yukawa hypothesis has one more important consequence. So nuclear forces, acting between proton and neutron, proton and proton as well as between two neutrons, are practically the same (of course, particles are in the same states). In other words, the peculiarity of the interaction between nucleons is that switching off the proton charge liquidates the difference between proton and neutron. All this allowed us to consider a neutron and a proton as two states of one and the same particle, a nucleon. Different values of the isotopic spin operator projections correspond to the two possible states of the nucleon. Let us agreed to ascribe $I_3 = 1/2$ to the proton and $I_3 = -1/2$ to the neutron. Thus, a nucleon wavefunction can be presented in the form:

$$\Psi_N = N \bigotimes \psi(x; J), \tag{9.29}$$

where

$$N = \sum_{i=p,n} a_i \psi_i, \qquad \psi_p = \begin{pmatrix} 1 \\ 0 \end{pmatrix}, \qquad \psi_n = \begin{pmatrix} 0 \\ 1 \end{pmatrix}, \tag{9.30}$$

$|a_p|^2$ and $|a_n|^2$ define probability that the nucleon will be found in the proton and neutron states, respectively ($|a_p|^2 + |a_n|^2 = 1$) and $\psi(x; J)$ is a part of a wavefunction that includes coordinate and spin dependence. At this point we should recollect some facts, related to the ordinary spin. Although the spin describes particle behavior with respect to rotations in ordinary three-dimensional space, it cannot be connected with any particle spatial rotations. The spin can be related with indestructible particle rotation only in the internal space. Despite belonging to different kind of spaces, the spin and the orbital moment of momentum can be summarized, their sum being the total moment of momentum of a particle. A more convenient and quantum streamlined spin definition is simply indicating the existence on particles of the new (spin) degrees of freedom, which are the eigenvalues of the spin projection operator S_3, their number being equal to $2J + 1$. The isospin is also connected with particle behavior in the internal space, where the third axis is correlated with the electric charge. The number of the isospin degrees of freedom, that is, the number of possible values of $S_3^{(is)}$, is again equal to $2I + 1$. However, unlike the ordinary spin, the isospin does not contribute to quantities, which define the particle behavior in the ordinary space. It can be checked easily, that with our agreements concerning the eigenvalues of the operator $S_3^{(is)}$ for the nucleon, Gell-Mann-Nishijima formula

$$Q = I_3 + \frac{B}{2}, \tag{9.31}$$

where Q is a nucleon charge, expressed in units of $|e|$, takes place. From the aesthetic point of view it is attractive to explain the aforesaid by the existence of an isospin symmetry, which is the mathematical copy of the spin symmetry. It means that the isospin generators satisfy the same permutation relations as the operators of the ordinary angular moment, that is,

$$[I_k, I_m] = i\varepsilon_{kmn} I_n. \tag{9.32}$$

Thus, in the states space there is the $SU(2)$-group of special ($\det U = 1$) unitary ($U^\dagger U = I$) transformations, for which the states Φ and $\mathcal{M}\Phi$ ($\mathcal{M}$ is a transformation matrix of the $SU(2)$-group) describe one and the same phenomenon when only the strong interaction is taken into account. A nucleon is the most simple (spinor) representation of the rotations group in the isotropic space. In this case the generators of the $SU(2)$-group themselves form the group representation, that is, in the isospace the transformation matrix of a nucleon state at the rotations through the angle $\boldsymbol{\theta}$ is as follows:

$$U(\boldsymbol{\theta}) = \exp\left[i(\boldsymbol{\sigma} \cdot \boldsymbol{\theta})/2\right]. \tag{9.33}$$

Thus, the Pauli matrices, multiplied by 1/2, play roles of the representation generators $S_\alpha^{(is)}$ ($\alpha = 1, 2, 3$). Since

$$S_3^{(is)} \psi_p = \frac{1}{2} \begin{pmatrix} 1 & 0 \\ 0 & -1 \end{pmatrix} \begin{pmatrix} 1 \\ 0 \end{pmatrix} = \frac{1}{2} \begin{pmatrix} 1 \\ 0 \end{pmatrix} = \frac{1}{2} \psi_p, \tag{9.34}$$

$$S_3^{(is)}\psi_n = \frac{1}{2}\begin{pmatrix} 1 & 0 \\ 0 & -1 \end{pmatrix}\begin{pmatrix} 0 \\ 1 \end{pmatrix} = -\frac{1}{2}\begin{pmatrix} 0 \\ 1 \end{pmatrix} = -\frac{1}{2}\psi_n, \tag{9.35}$$

then the operator $S_3^{(is)}$ really is the isospin projection operator. From $S_1^{(is)}$, $S_2^{(is)}$ one may compose the operators

$$S_\pm^{(is)} = S_1^{(is)} \pm i S_2^{(is)} = \frac{1}{2}(\sigma_1 \pm i\sigma_2). \tag{9.36}$$

They influence the proton and neutron states in the following way:

$$S_+^{(is)}\psi_p = \begin{pmatrix} 0 & 1 \\ 0 & 0 \end{pmatrix}\begin{pmatrix} 1 \\ 0 \end{pmatrix} = 0, \qquad S_+^{(is)}\psi_n = \begin{pmatrix} 0 & 1 \\ 0 & 0 \end{pmatrix}\begin{pmatrix} 0 \\ 1 \end{pmatrix} = \psi_p, \tag{9.37}$$

$$S_-^{(is)}\psi_p = \begin{pmatrix} 0 & 0 \\ 1 & 0 \end{pmatrix}\begin{pmatrix} 1 \\ 0 \end{pmatrix} = \psi_n, \qquad S_-^{(is)}\psi_n = \begin{pmatrix} 0 & 0 \\ 1 & 0 \end{pmatrix}\begin{pmatrix} 0 \\ 1 \end{pmatrix} = 0. \tag{9.38}$$

From Eqs. (9.37) and (9.38) it is obvious, that the operator $S_+^{(is)}$ ($S_-^{(is)}$) raises (lowers) the isospin projection value by 1 for the nucleon states. In what follows we shall call $S_+^{(is)}$ and $S_-^{(is)}$ as the raising and lowering operators, for short. Their meaning can be understood without resorting to their obvious form (not to be related to concrete representation), but by using only the commutation relations

$$[S_+^{(is)}, S_-^{(is)}] = 2S_3^{(is)}, \qquad [S_3^{(is)}, S_\pm^{(is)}] = \pm S_\pm^{(is)}, \tag{9.39}$$

For this purpose it is enough to find the result of acting the operator $S_3^{(is)}$ on the state $S_\pm^{(is)}\Psi_{I,I_3}$,, where in the wavefunction Ψ_{I,I_3} for the sake of simplicity the spin and spatial variables are neglected. Eqs. (9.37) and (9.38) are the special cases of the relations

$$S_\pm^{(is)}\Psi_{I,I_3} = \sqrt{(I \mp I_3)(I \pm I_3 + 1)}\Psi_{I,I_3\pm 1}. \tag{9.40}$$

Since the obtained relations play an important role in obtaining the isospin or spin parts of wavefunctions for compound systems, we shall give their proof. We start from putting down the operator of the isotopic spin square as follows:

$$(\mathbf{S}^{(is)})^2 = S_-^{(is)} S_+^{(is)} + (S_3^{(is)})^2 + S_3^{(is)}. \tag{9.41}$$

Using definitions of the matrix elements of the operator A

$$(A)_{mn} = \int \Psi_{I,I_3=m}^\dagger A \Psi_{I,I_3=n} d\mathbf{r}, \tag{9.42}$$

we find the matrix elements for both parts of the operator equation (9.41). With allowance made for the diagonal form of the operators $(\mathbf{S}^{(is)})^2$, $S_3^{(is)}$ and taking into account the condition of the wavefunctions normalization, we obtain the result:

$$I(I + 1)\mathbf{I}_{mm} = (S_-^{(is)})_{mk}(S_+^{(is)})_{km} + (m^2 + m)\mathbf{I}_{mm}, \tag{9.43}$$

where $\mathbf{I}_{mm}$ are elements of the identity matrix. By means of the commutation relations it is possible to find nonzero matrix elements $(S_\pm^{(is)})_{km}$. Calculations for matrix elements of the commutators (9.39) between states with $I_3 = m$ and $I_3 = k$ leads to the following equations:

$$\left. \begin{array}{l} (S_3^{(is)} S_+^{(is)})_{mk} - (S_+^{(is)} S_3^{(is)})_{mk} = (S_+^{(is)})_{mk}, \\ (S_3^{(is)} S_-^{(is)})_{mk} - (S_-^{(is)} S_3^{(is)})_{mk} = -(S_-^{(is)})_{mk}, \end{array} \right\} \tag{9.44}$$

Since

$$(S_3^{(is)} S_\pm^{(is)})_{mk} = (S_3^{(is)})_{ml}(S_\pm^{(is)})_{lk} = m(S_\pm^{(is)})_{mk},$$

then the relations (9.44) take the form:

$$\left.\begin{array}{c} (m-k)(S_+^{(is)})_{mk} = (S_+^{(is)})_{mk}, \\ (m-k)(S_-^{(is)})_{mk} = -(S_-^{(is)})_{mk}. \end{array}\right\} \tag{9.45}$$

From Eqs. (9.45) it follows, that the matrix $(S_+^{(is)})_{mk}$ $((S_-^{(is)})_{mk})$ has nonzero elements only for transitions, corresponding to the increase (decrease) of the quantum number I_3 by unity $I_3 \to I_3 + 1$ $(I_3 \to I_3 - 1)$. Then the right-hand side of the expression (9.43) can be presented as follows:

$$(S_-^{(is)})_{m,m+1}(S_+^{(is)})_{m+1,m} + (m^2 + m)I_{mm} == |(S_-^{(is)})_{m,m+1}|^2 +$$

$$+(m^2 + m)\mathbf{I}_{mm} = |(S_+^{(is)})_{m+1,m}|^2 + (m^2 + m)\mathbf{I}_{mm}, \tag{9.46}$$

where we have taken into account

$$(S_+^{(is)})_{km} = (S_-^{(is)})_{mk}^*, \qquad (S_-^{(is)})_{km} = (S_+^{(is)})_{mk}^*$$

By means of (9.46) Eq. (9.43) leads to the result:

$$(S_+^{(is)})_{I_3+1,I_3} = \sqrt{I(I+1) - I_3(I_3+1)} = \sqrt{(I - I_3)(I + I_3 + 1)}, \tag{9.47}$$

$$(S_-^{(is)})_{I_3-1,I_3} = \sqrt{(I + I_3)(I - I_3 + 1)}, \tag{9.48}$$

which proves to be equivalent to the relations (9.40). Notice, that the proof of the relation (9.40) could be done with purely military straightforwardness. For this purpose one should build the "raising" and "lowering" operators from the orbital moment operators

$$L_\pm = L_x \pm iL_y, \qquad L_k = -i\hbar\varepsilon_{klm}x_l\partial_m,$$

and act by them on the eigenfunctions of the operators $\mathbf{L}^2$ and L_3, which are the spherical functions $Y_l^{m_l}(\theta, \varphi)$

$$\mathbf{L}^2 Y_l^{m_l}(\theta, \varphi) = l(l+1)\hbar^2 Y_l^{m_l}(\theta, \varphi), \qquad L_3 Y_l^{m_l}(\theta, \varphi) = m_l\hbar Y_l^{m_l}(\theta, \varphi).$$

As a result, we arrive at the required relation

$$L_\pm Y_l^{m_l}(\theta, \varphi) = \hbar\sqrt{(l \mp m_l)(l \pm m_l + 1)}Y_l^{m_l \pm 1}(\theta, \varphi). \tag{9.49}$$

For antinucleons the scheme of building the wavefunction must be changed. It is caused by the fact that the wavefunctions of the form $\overline{\psi}_p C$ and $\overline{\psi}_n C$ (C is the charge conjugation matrix) correspond to antinucleons. Thus, the relation

$$N' = U(\boldsymbol{\theta})N, \tag{9.50}$$

means, that

$$\overline{N}' = U^\dagger(\boldsymbol{\theta})\overline{N}. \tag{9.51}$$

Having used the obvious form of the transformation matrix $U(\boldsymbol{\theta})$, one can be easily persuaded, that

$$U^*(\boldsymbol{\theta}) = \exp\left[-i(\boldsymbol{\sigma}^* \cdot \boldsymbol{\theta})/2\right] = (i\sigma_2)\exp\left[i(\boldsymbol{\sigma} \cdot \boldsymbol{\theta})/2\right](-i\sigma_2) = (i\sigma_2)U(\boldsymbol{\theta})(-i\sigma_2). \tag{9.52}$$

Consequently, if we write down for the antinucleon

$$\xi = -i\sigma_2(\overline{N}C) = \begin{pmatrix} -\overline{\psi}_n C \\ \overline{\psi}_p C \end{pmatrix} = \begin{pmatrix} -\psi_{\overline{n}} \\ \psi_{\overline{p}} \end{pmatrix}, \tag{9.53}$$

then we get the transformation law

$$\xi' = U(\boldsymbol{\theta})\xi. \tag{9.54}$$

Since in the isospace, the transformation properties of ξ and N are the same, then working with ξ and not with the antinucleon doublet directly, we do not make any difference between particles and antiparticles. Thereby, there is no need to modify the theory of moments with an eye to extending this theory in the case of antiparticles.

Since nucleons are fermions, then, according to the Pauli principle, no more than one nucleon can present in one state, in this case the state now is characterized by one more additional number, the eigenvalue of the third component of the isospin operator. Consequently, a wavefunction of two nucleons must be completely antisymmetric with respect to the nucleons transposition.

Let us consider, as a case in point, a two-nucleon system, a deuteron, which is described by the production of two nucleon wavefunctions ψ_A and ψ_B. Since the isotopic spin is an additive quantum number, then the two-nucleon state can possess the isotopic spin being equal either to 1 or to 0. The isospin operator of a compound system with the wavefunction Ψ_{AB} is defined by the expression:

$$\mathbf{S}^{(is)} = \mathbf{S}_A^{(is)} + \mathbf{S}_B^{(is)}, \tag{9.55}$$

where the subscripts indicate, on which wavefunction of the subsystem the operator acts. It is convenient to represent the square of the operator $\mathbf{S}^{(is)}$ in the form:

$$(\mathbf{S}^{(is)})^2 = (\mathbf{S}_A^{(is)})^2 + (\mathbf{S}_B^{(is)})^2 + 2\mathbf{S}_A^{(is)} \cdot \mathbf{S}_B^{(is)} = (\mathbf{S}_A^{(is)})^2 + (\mathbf{S}_B^{(is)})^2 + (S_{A1}^{(is)} + iS_{A2}^{(is)})(S_{B1}^{(is)} -$$

$$-iS_{B2}^{(is)}) + (S_{A1}^{(is)} - iS_{A2}^{(is)})(S_{B1}^{(is)} + S_{B2}^{(is)}) + 2S_{A3}^{(is)}S_{B3}^{(is)} = (\mathbf{S}_A^{(is)})^2 + (\mathbf{S}_B^{(is)})^2 + S_{A+}^{(is)}S_{B-}^{(is)} +$$

$$+S_{A-}^{(is)}S_{B+}^{(is)} + 2S_{A3}^{(is)}S_{B3}^{(is)}. \tag{9.56}$$

Two-nucleon wavefunction in states with the definite values of I and I_3 is a linear combination of the states

$$\psi_p^{(A)}\psi_p^{(B)}, \qquad \psi_p^{(A)}\psi_n^{(B)}, \qquad \psi_n^{(A)}\psi_p^{(B)}, \qquad \psi_n^{(A)}\psi_n^{(B)}. \tag{9.57}$$

It is absolutely obvious, that the states

$$\psi_p^{(A)}\psi_p^{(B)} \qquad \text{and} \qquad \psi_n^{(A)}\psi_n^{(B)} \tag{9.58}$$

enter into the isotopic triplet ($I = 1$), and the values $I_3 = 1$ and $I_3 = -1$ are associated with them. A certain combination of the remaining states $\psi_p^{(A)}\psi_n^{(B)}$ and $\psi_n^{(A)}\psi_p^{(B)}$ must play the role of the third component of the triplet with $I_3 = 0$. Finally, the other orthogonal combination of the states $\psi_p^{(A)}\psi_n^{(B)}$ and $\psi_n^{(A)}\psi_p^{(B)}$ can only belong to the singlet state with $I = 0$ and $I_3 = 0$. To single out the triplet state with $I_3 = 0$, the state with $I_3 = 0$ must be acted upon by the lowering operator $S_-^{(is)} = S_{A-}^{(is)} + S_{B-}^{(is)}$, which, as we see, is symmetric with respect to the indices A and B. Since the state $\psi_p^{(A)}\psi_p^{(B)}$ is also symmetric with respect to these indices, then the emerging state has the form:

$$S_-^{(is)}\psi_p^{(A)}\psi_p^{(B)} = \text{const} \times \left(\psi_p^{(A)}\psi_n^{(B)} + \psi_n^{(A)}\psi_p^{(B)} \right), \tag{9.59}$$

where from the normalization condition it follows that const equals $1/\sqrt{2}$. Because the state with $I = I_3 = 0$ must be orthogonal to that defined by Eq. (9.59), then it is given by the expression:

$$\frac{1}{\sqrt{2}}\left(\psi_p^{(A)}\psi_n^{(B)} - \psi_n^{(A)}\psi_p^{(B)}\right). \tag{9.60}$$

Using the eigenvalue equations

$$\left.\begin{array}{l} (\mathbf{S}^{(is)})^2\Psi_{AB} = I(I+1)\Psi_{AB}, \\ (\mathbf{S}_A^{(is)})^2\psi_A = I_A(I_A+1)\psi_A, \\ (\mathbf{S}_B^{(is)})^2\psi_B = I_B(I_B+1)\psi_B, \end{array}\right\}, \tag{9.61}$$

and the expression for $(\mathbf{S}^{(is)})^2$ (9.56), one can easily check, that the symmetric states

$$\left.\begin{array}{ll} pp, & I_3 = 1 \\ \frac{1}{\sqrt{2}}\left(pn + np\right), & I_3 = 0 \\ nn, & I_3 = -1 \end{array}\right\} I = 1, \tag{9.62}$$

really form the isotriplet, and the antisymmetric state

$$\frac{1}{\sqrt{2}}\left(pn - np\right) \tag{9.63}$$

is the isosinglet (in the formulas (9.62) and (9.63) we switched over to more economic and obvious notations). The deuteron, which is the spatially symmetric two-nucleon state, must have such an isospin, that its total wavefunction is antisymmetric. Thus, the conclusion is, that the deuteron has an isospin equal to zero.

For a nucleon-antinucleon system $(N\overline{N})$ the isotriplet and the isosinglet states now have the form:

$$\left.\begin{array}{ll} -p\overline{n}, & I_3 = 1 \\ \frac{1}{\sqrt{2}}\left(p\overline{p} - n\overline{n}\right), & I_3 = 0 \\ n\overline{p}, & I_3 = -1 \end{array}\right\} I = 1, \tag{9.64}$$

$$\frac{1}{\sqrt{2}}\left(p\overline{p} + n\overline{n}\right). \qquad I = 0 \tag{9.65}$$

Isodoublets are also formed by some strange particles, for example, by K-mesons

$$\begin{pmatrix} K^+ \\ K^0 \end{pmatrix} \qquad (s = 1), \qquad \begin{pmatrix} \overline{K}^0 \\ K^- \end{pmatrix} \qquad (s = -1),$$

cascade Ξ-baryons

$$\begin{pmatrix} \Xi^0 \\ \Xi^- \end{pmatrix} \qquad (s = -2)$$

and so on. Since the masses of the strange particles, constituting the isodoublet are very close to each other

$$m_{K^\pm} = 493.577 \pm 0.016 \text{ MeV}/c^2 \qquad m_{K^0, \overline{K}^0} = 497.672 \pm 0.031 \text{ MeV}/c^2,$$

$$m_{\Xi^-} = 1321.32 \pm 0.13 \text{ MeV}/c^2 \qquad m_{\Xi^0} = 1314.9 \pm 0.6 \text{ MeV}/c^2,$$

then it is possible to speak of the confirmation of conserving the isospin under the strong interaction between strange particles. However, the appearance of a new quantum number, a strangeness, forces us to change the Gell-Mann-Nishijima formula (9.33) in the following way:

$$Q = I_3 + \frac{B+s}{2} = I_3 + \frac{Y}{2}, \qquad (9.66)$$

where Y is a particle hypercharge. The formula (9.66) can be viewed as nature's delicate hint at the structure of the gauge model of the electroweak interaction, which many years later will be destined to be discovered by Glashow, Weinberg, and Salam. Actually, the first term in the right-hand side of Eq. (9.66) is connected with the $SU(2)$-group, while the second one is associated with the $U(1)$-group. Both the isospin and hypercharge symmetries are not exact ones. However, the quantity in the left-hand side of Eq. (9.66), the electric charge, belongs among the exactly conserving quantities. So, if one suggests, that the Creator was using the same rules of the game to produce both the strong and electroweak interactions, then the electroweak interaction theory must contain the following elements:

(i) the weak isospin group $SU(2)_{EW}$ and the weak hypercharge group $U(1)_{EW}$;
(ii) the $SU(2)_{EW}$- and $U(1)_{EW}$-symmetries must be violated to the level of the $U(1)_{em}$-symmetry.

In Part II we shall learn that these very elements made the foundation of the Glashow-Weinberg-Salam model.

According to the Yukawa hypothesis, the nuclear forces between nucleons are caused by the $\pi^\pm$, π^0 meson exchanges (although, as we know at present, it is a very approximate statement). Consequently, the pion-nucleon interaction must be isotopically invariant as well. From the reaction

$$N \to N' + \pi,$$

which is the basis for the Yukawa hypothesis, follows that the π-meson must have the isospin equal either to $I = 1$ or $I = 0$. Since there are three π-mesons having the same spins, parities and almost equal masses, then it is natural to assume, that nature chose the possibility with $I = 1$. Thus, the π-mesons can be viewed as the different charge states of one and the same particle, whose wavefunction is transformed as a vector in the isospace, that is, has the following form:

$$\Psi_\pi = \Pi \bigotimes \Phi(x; J), \qquad (9.67)$$

where

$$\Pi = \sum_{i=1}^{3} b_i \psi'_i, \qquad \sum_{i=1}^{3} \mid b_i \mid^2 = 1, \qquad \psi'_{\pi+} = \frac{\alpha_+}{\sqrt{2}} \begin{pmatrix} 1 \\ i \\ 0 \end{pmatrix},$$

$$\psi'_{\pi^0} = \begin{pmatrix} 0 \\ 0 \\ 1 \end{pmatrix}, \qquad \psi'_{\pi-} = \frac{\alpha_-}{\sqrt{2}} \begin{pmatrix} 1 \\ -i \\ 0 \end{pmatrix},$$

and $\alpha_\pm$ are phase factors, which we choose later. Now the matrices

$$S_1^{(is)\prime} = \begin{pmatrix} 0 & 0 & 0 \\ 0 & 0 & -i \\ 0 & i & 0 \end{pmatrix}, \qquad S_2^{(is)\prime} = \begin{pmatrix} 0 & 0 & i \\ 0 & 0 & 0 \\ -i & 0 & 0 \end{pmatrix}, \qquad S_3^{(is)\prime} = \begin{pmatrix} 0 & -i & 0 \\ i & 0 & 0 \\ 0 & 0 & 0 \end{pmatrix}.$$

play the role of the representation generators. It is easy to check correctness of the relations:

$$S_3^{(is)\prime} \psi'_{\pi+} = \psi'_{\pi+}, \qquad S_3^{(is)\prime} \psi'_{\pi^0} = 0, \qquad S_3^{(is)\prime} \psi'_{\pi-} = -\psi'_{\pi-}.$$

Raising and lowering operators in this case are of the form:

$$
S_+^{(is)\prime} = \begin{pmatrix} 0 & 0 & -1 \\ 0 & 0 & -i \\ 1 & i & 0 \end{pmatrix}, \qquad
S_-^{(is)\prime} = \begin{pmatrix} 0 & 0 & 1 \\ 0 & 0 & -i \\ -1 & i & 0 \end{pmatrix}.
$$

Their action on the isotopic parts of the wavefunctions of the pion triplet is governed as follows:

$$
\left.\begin{aligned}
S_+^{(is)\prime}\psi_{\pi^0}' &= -\sqrt{2}(\alpha_+)^{-1}\psi_{\pi^+}', \qquad
S_+^{(is)\prime}\psi_{\pi^-}' = \sqrt{2}\alpha_-\psi_{\pi^0}', \\
S_-^{(is)\prime}\psi_{\pi^+}' &= -\sqrt{2}\alpha_+\psi_{\pi^0}', \qquad
S_-^{(is)\prime}\psi_{\pi^0}' = \sqrt{2}(\alpha_-)^{-1}\psi_{\pi^-}'.
\end{aligned}\right\}
\tag{9.68}
$$

The choice of phase factors in the form of

$$
\alpha_+ = -1, \qquad \alpha_- = 1
$$

makes the right-hand parts in Eq. (9.68) positive to turn the relations (9.68) in the particular case of the formula (9.40). However, it is often more convenient to work with the representation, in which the matrix $S_3^{(is)\prime}$ is diagonal. The transition to such a representation is carried out by the transformation:

$$
\psi_\pi = O\psi_\pi',
$$

where a matrix O has the form:

$$
O = \frac{1}{\sqrt{2}} \begin{pmatrix} 1 & -i & 0 \\ 0 & 0 & \sqrt{2} \\ 1 & i & 0 \end{pmatrix}.
$$

In the new representation the generators are defined according to

$$
S_1^{(is)} = \frac{1}{\sqrt{2}} \begin{pmatrix} 0 & -1 & 0 \\ -1 & 0 & 1 \\ 0 & 1 & 0 \end{pmatrix}, \qquad
S_2^{(is)} = \frac{1}{\sqrt{2}} \begin{pmatrix} 0 & i & 0 \\ -i & 0 & -i \\ 0 & i & 0 \end{pmatrix}, \qquad
S_3^{(is)} = \begin{pmatrix} 1 & 0 & 0 \\ 0 & 0 & 0 \\ 0 & 0 & -1 \end{pmatrix},
\tag{9.69}
$$

and the isospin parts of the π-meson wavefunctions have the form:

$$
\psi_{\pi^+} = \begin{pmatrix} 1 \\ 0 \\ 0 \end{pmatrix}, \qquad
\psi_{\pi^0} = \begin{pmatrix} 0 \\ 1 \\ 0 \end{pmatrix}, \qquad
\psi_{\pi^-} = \begin{pmatrix} 0 \\ 0 \\ 1 \end{pmatrix}.
\tag{9.70}
$$

Besides, there are isomultiplets formed of one particle only, so-called isosinglets. The isotopic parts of wavefunctions for such particles are isoscalars, which are invariant under the isotopic transformations. The Λ-hyperon can serve as an example.

To illustrate the power of the formula (9.40) under obtaining the spin or isospin parts of a wavefunction of composite systems, let us define the isotopic wavefunction of the pion-nucleon system (ΠN). According to the rule of the vector composition, the total isospin of the system can take values $I = 3/2$ or $I = 1/2$. Consequently, there are six states

$$
\left.\begin{aligned}
(3/2, 3/2), \quad (3/2, 1/2), \quad (3/2, -1/2), \quad (3/2, -3/2), \\
(1/2, 1/2), \quad (1/2, -1/2).
\end{aligned}\right\}
\tag{9.71}
$$

which must be expressed in terms of the π-meson and nucleon states. Obviously, the "highest" state is

$$
(3/2, 3/2) = \pi^+ p,
\tag{9.72}
$$

since it is the only one with $I_3 = 3/2$. Now let the operator $S_-^{(is)}$ act on both sides of Eq. (9.72). Then, according to (9.40), in the left-hand side we obtain

$$S_-^{(is)}(3/2, 3/2) = \sqrt{3}(3/2, 1/2). \tag{9.73}$$

In so doing, the right-hand side of Eq. (9.71) takes the form:

$$S_-^{(is)}(\pi^+ p) = S_{\pi-}^{(is)}(\pi^+ p) + S_{N-}^{(is)}(\pi^+ p) = (S_{\pi-}^{(is)}\pi^+ p) + (\pi^+ S_{N-}^{(is)} p) = \sqrt{2}(\pi^0 p) + (\pi^+ n). \tag{9.74}$$

Combining Eqs. (9.73) and (9.74) we get

$$(3/2, 1/2) = \sqrt{\frac{2}{3}}(\pi^0 p) + \sqrt{\frac{1}{3}}(\pi^+ n). \tag{9.75}$$

Action of the operator $S_-^{(is)}$ on Eq. (9.75) results in

$$S_-^{(is)}(3/2, 1/2) = 2(3/2, -1/2) \tag{9.76}$$

and

$$S_-^{(is)}[\sqrt{\frac{2}{3}}(\pi^0 p) + \sqrt{\frac{1}{3}}(\pi^+ n)] = \sqrt{\frac{2}{3}}[(S_\pi^{(is)}\pi^0 p) + (\pi^0 S_{N-}^{(is)} p)] + \sqrt{\frac{1}{3}}[(S_{\pi-}^{(is)}\pi^+ n) + (\pi^+ S_{N-}^{(is)} n)] =$$

$$= \sqrt{\frac{2}{3}}[\sqrt{2}(\pi^- p) + (\pi^0 n)] + \sqrt{\frac{1}{3}}[\sqrt{2}(\pi^0 n) + 0] = \frac{2}{\sqrt{3}}(\pi^- p) + 2\sqrt{\frac{2}{3}}(\pi^0 n). \tag{9.77}$$

Eqs. (9.76)–(9.77) produce

$$(3/2, -1/2) = \sqrt{\frac{1}{3}}(\pi^- p) + \sqrt{\frac{2}{3}}(\pi^0 n). \tag{9.78}$$

Now build-up of the state $(3/2, -3/2)$ is possible without using the operator $S_-^{(is)}$ because the only state with $S_3^{(is)} = -3/2$ is the combination:

$$(3/2, -3/2) = (\pi^- n). \tag{9.79}$$

The remaining two states with $(1/2, 1/2)$ and $(1/2, -1/2)$ can be easily obtained, according to their orthogonality to the states $(3/2, 1/2)$ and $(3/2, -1/2)$ respectively, that is, they have the form:

$$(1/2, 1/2) = \sqrt{\frac{2}{3}}(\pi^+ n) - \sqrt{\frac{1}{3}}(\pi^0 p), \tag{9.80}$$

$$(1/2, -1/2) = \sqrt{\frac{1}{3}}(\pi^0 n) - \sqrt{\frac{2}{3}}(\pi^- p). \tag{9.81}$$

Solving Eqs. (9.75), (9.78), (9.80) and (9.81), we obtain the final answer

$$\left.\begin{array}{l} (\pi^+ p) = (3/2, 3/2), \qquad (\pi^- n) = (3/2, -3/2), \\ (\pi^0 p) = \sqrt{\frac{2}{3}}(3/2, 1/2) - \sqrt{\frac{1}{3}}(1/2, 1/2) \\ (\pi^+ n) = \sqrt{\frac{1}{3}}(3/2, 1/2) + \sqrt{\frac{2}{3}}(1/2, 1/2), \\ (\pi^- p) = \sqrt{\frac{1}{3}}(3/2, -1/2) - \sqrt{\frac{2}{3}}(1/2, -1/2), \\ (\pi^0 n) = \sqrt{\frac{2}{3}}(3/2, -1/2) + \sqrt{\frac{1}{3}}(1/2, -1/2). \end{array}\right\} \tag{9.82}$$

Thus, we have learned the mechanism of defining the isospin (or any spinlike) part of the wavefunction for any compound system, that is, in this situation we have known the answer for questions "how", "where", ... and "from." Thereafter, it is high time to become aware of the existence of tabulated values of the numeric coefficients, which appear in the composition theory of moments (orbital, spinlike, total). By means of these coefficients the isospin function of the pion-nucleon system Φ_{I,I_3} is expressed in terms of the isofunctions of the π-meson and nucleon N as follows:

$$\Phi_{I,I_3} = \sum_{I_3'',I_3'} C_{I,I_3}^{I',I_3';I'',I_3''} \Pi_{I',I_3'} N_{I'',I_3''},$$

where $C_{I,I_3}^{I',I_3';I'',I_3''}$ are Clebsch-Gordan coefficients, which can be found in the *Review of Particle Physics* published every two years.

9.3   Unitary multiplets

By using the isospin symmetry we can reduce a number of independent elementary particles. For example, three π-mesons are simply three states of one and the same particle and so on. However, by the 1960s so many isotopic multiplets had been discovered, that the necessity of searching for a higher symmetry to unify particles into more densely populated families became natural. Since such a symmetry has thoroughly escaped observation up to now, it was obvious that somehow and somewhere this symmetry has passed a violation stage, that is, today it is approximate. As a violation criteria of this symmetry, once again they can take mass difference for particles, entering into new multiplets, which we are going to call *unitary multiplets*. Unitary multiplets must contain isomultiplets, that is, unitary symmetry group contains in itself a subgroup of the isotopic transformations. Thus, among the Casimir operators of an unitary symmetry group, the isotopic spin square is present. What other operators can apply for the role of invariants in a new group? If we agree once again to denote particle states with dots in some abstract space, then the dimensionality of a new space of states is defined by the number of invariants of an unitary symmetry group. From the physical point of view it is reasonable to demand, that "coordinates" of this space must be dependent (in the same way as space and time are interconnected in the special theory of relativity). Experiments, which had been carried out by that time, demonstrated, that the strangeness s and the isospin I are not independent. Let us explain what is meant by this. In the strong interaction s and I are the conserved quantities, and as a result, the following relation takes place:

$$\Delta s = \Delta I_3 = 0. \tag{9.83}$$

Behavior of the isospin and the strangeness is not so faultless as far as the weak interaction is concerned. Since the isospin and the strangeness are not defined for leptons, then it makes sense to analyze only semilepton and nonlepton weak interaction. In the former case the final state is formed of both leptons and hadrons, while in the latter case leptons are absent in the final state. Below we give some examples of typical semilepton decays and indicate the changes of the strangeness and the isotopic spin projection

$$n \to p + e^- + \bar{\nu}_e, \qquad \Delta s = 0, \qquad \Delta I_3 = 1 \tag{9.84}$$

$$\pi^+ \to \pi^0 + e^+ + \nu_e, \qquad \Delta s = 0, \qquad \Delta I_3 = -1 \tag{9.85}$$

$$\Lambda \to p + e^- + \overline{\nu}_e, \qquad \Delta s = 1, \qquad \Delta I_3 = 1/2, \tag{9.86}$$

$$K^+ \to \pi^0 + \mu^+ + \nu_\mu, \qquad \Delta s = -1, \qquad \Delta I_3 = -1/2. \tag{9.87}$$

For nonlepton decays the selection rules according to s and I_3 can be illustrated by the example of the reactions:

$$\Lambda \to p + \pi^-, \qquad \Delta s = 1, \qquad \Delta I_3 = -1/2, \tag{9.88}$$

$$K^+ \to \pi^+ + \pi^0, \qquad \Delta s = -1, \qquad \Delta I_3 = 1/2. \tag{9.89}$$

In the electromagnetic interaction the strangeness remains a good quantum number. Two typical decays

$$\pi^0 \to 2\gamma, \qquad \eta^0 \to 2\gamma, \tag{9.90}$$

demonstrate this circumstance

$$\Delta s = 0, \qquad \Delta I_3 = 0. \tag{9.91}$$

From the aforesaid it follows that in all the existing processes for a closed system the change of the strangeness entails the strictly defined change of the isotopic spin projection

$$\mid \Delta s \mid - 0 \qquad \longrightarrow \qquad \mid \Delta I_3 \mid - 1, 0, \tag{9.92}$$

$$\mid \Delta s \mid = 1 \qquad \longrightarrow \qquad \mid \Delta I_3 \mid = 1/2. \tag{9.93}$$

Thus, the choice of two quantum numbers, the isospin projection and the strangeness, for dependent coordinates of the unitary spin space is quite well grounded. In Fig. 9.1 we

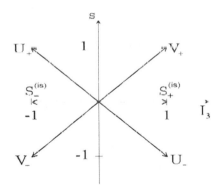

FIGURE 9.1
The unitary spin plane.

plotted the plane with the coordinates (s, I_3) and on it we shall specify the states of unitary multiplets. From the mathematical point of view, transitions between states must be carried out by means of operators, which we agree to denote by vectors. The raising and lowering isospin operators $S_+^{(is)}$ and $S_-^{(is)}$ allow crossing from one state to another inside of one and the same isomultiplet. Since acting $S_\pm^{(is)}$ on a state changes the values of I_3 by ± 1 and in the process $\Delta s = 0$, then the vectors with module 1, parallel to axis I_3, are correlated with these operators. It is obvious, that in order to ensure the transitions between isomultiplets, that is, the transitions with a strangeness change, the similar raising and lowering operators must exist. It is natural that these operators must be spinlike, that is, their algebra is determined by the same commutation relations as for ordinary spin algebra. The raising operator S_+^U is associated with the selection rules

$$\Delta s = 1, \qquad \Delta I_3 = -1/2, \tag{9.94}$$

while the lowering operator S_-^U describes the situation with

$$\Delta s = -1, \qquad \Delta I_3 = 1/2. \tag{9.95}$$

The operators S_+^U, S_-^U, and S_3^U are the group generators of the so-called U-spin, and for them the following commutation relations:

$$[S_+^U, S_-^U] = 2S_3^U \tag{9.96}$$

are valid. Obviously, without the introduction of one more type of generators it is impossible to obtain on the plane (s, I_3) the closed and symmetric (with respect to axis s) figures, which we are going to associate with unitary multiplets. The transitions with the selection rules

$$\left.\begin{array}{ll} \Delta s = 1, & \Delta I_3 = 1/2, \\ \Delta s = -1, & \Delta I_3 = -1/2, \end{array}\right\} \tag{9.97}$$

are carried out by the operators S_+^V and S_-^V. These operators alongside with S_3^V constitute a V-spin group, and for them the ordinary commutation relations for the moments are fulfilled

$$[S_+^V, S_-^V] = 2S_3^V. \tag{9.98}$$

Notice, that in the chosen scale the modules of the vectors, we correlate with the operators $S_\pm^U$ and $S_\pm^V$, are equal to $\sqrt{5}/2$, and tangents of the angles, to be formed by the operators S_+^U and S_+^V with the strangeness axis, are equal to $1/2$.

Nine generators $S_{1,2,3}^{(is)}$, $S_{1,2,3}^V$, and $S_{1,2,3}^U$ form the closed system and generate a group of the second rank $SU(3)$, an unitary spin group, which was proposed by Y. Neeman [24] and, regardless of him, M. Gell-Mann [25] to classify hadrons. In this scheme, just as in the Mendeleev periodic table, objects are placed in the order of increasing their masses and separation into families, unifying elementary particles with similar properties, is used. The proposed theory is very often called the *octal way*, since, according to its statements, the majority of particles is grouped in 8-plets (octets).

With regard to hadron classification over isotopic multiplets, practically all hadrons that were reliably established by the 1960s, could be unified in four families. These families are characterized by the baryon charge, the spin and the parity (J^p). Let us take the following notations. After a isomultiplet symbol the approximated value of average mass in MeV/c^2 will be given in parentheses, while the hypercharge value will be indicated under every multiplet. Then these families are as follows:

8 baryons with $1/2^+$ and $B = 1$:

$$\begin{array}{cccc} N(939) & \Lambda(1115) & \Sigma(1193) & \Xi(1317) \\ p, n & \Lambda^0 & \Sigma^+, \Sigma^0, \Sigma^- & \Xi^0, \Xi^- \\ +1 & 0 & 0 & -1 \end{array}, \tag{9.99}$$

8 pseudoscalar mesons and meson resonances with 0^-:

$$\begin{array}{cccc} \pi(137) & K(495) & \overline{K}(495) & \eta(548) \\ \pi^+, \pi^0, \pi^- & K^+, K^0 & \overline{K}^-, \overline{K}^0 & \eta \\ 0 & +1 & -1 & 0 \end{array}, \tag{9.100}$$

8 vector resonances with 1^-:

$$
\begin{array}{cccc}
\rho(770) & K^*(992) & \overline{K}^*(992) & \omega(782) \\
\rho^+,\rho^0,\rho^- & K^{*+},K^{*0} & \overline{K}^{*-},\overline{K}^{*0} & \omega^0 \\
0 & +1 & -1 & 0
\end{array}
\tag{9.101}
$$

9 baryon resonances with $\frac{3}{2}^+$ and $B=1$:

$$
\begin{array}{ccc}
\Delta(1232) & \Sigma^*(1385) & \Xi^*(1530) \\
\Delta^{++},\Delta^+,\Delta^0,\Delta^- & \Sigma^{*+},\Sigma^{*0},\Sigma^{*-} & \Xi^{*0},\Xi^{*-} \\
1 & 0 & -1
\end{array}
\tag{9.102}
$$

Subsequently, we shall find the explicit form of both the operators of the U-, V-spin and of the operators of the Y_U-, Y_V-hypercharge. We shall be also convinced, the commutation relations (9.96) and (9.98) are correct. Forestalling events we want to point out, that all of these operators are in fact definite combinations of the isospin and hypercharge operators. However, at the moment we are not interested in the explicit form of the operators $S_\pm^V$, $S_\pm^U$. Our task is, first of all, to be sure that they are able together with the operators $S_\pm^{(is)}$ to place all the known, by that time, hadrons in the superfamilies (9.99)–(9.102).

Let us start from the supermultiplet including the nucleons. For the sake of convenience on the abscissa we shall plot the hypercharge values and not the strangeness ones. In the plane (Y, I_3), which we are going to call the unitary spin plane, the point with the coordinates $Y = 1$, $I_3 = 1/2$ corresponds to the proton (Fig. 9.2). Let us assume, that p

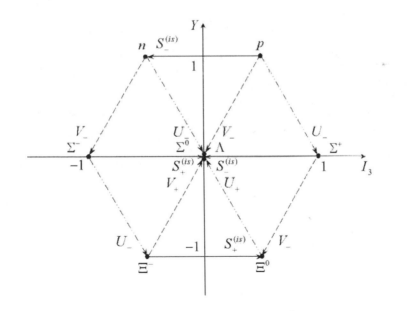

FIGURE 9.2

The baryon octet.

occupies the highest state, that is,

$$
S_+^{(is)}p = S_+^V p = S_+^U p = 0.
\tag{9.103}
$$

To obtain other particles, the lowering operators must act on the proton state. It gives

$$S_-^{(is)} p = n, \qquad (Y = 1, I_3 = -1/2), \tag{9.104}$$

$$S_-^V p = \Lambda, \Sigma^0 \qquad (Y = 0, I_3 = 0), \tag{9.105}$$

$$S_-^U p = \Sigma^+ \qquad (Y = 0, I_3 = 1). \tag{9.106}$$

Action of S_-^U on n brings forth the same result as the action of S_-^V on p, that is, it yields the state with $Y = 0 = I_3 = 0$ which we have already identified with the Λ- and Σ^0-hyperons. A new state with the quantum numbers $Y = 0$ and $I_3 = -1$, the Σ^--hyperon, can be obtained, if the operator S_-^V acts on n. It is important to remember that our final goal is to obtain a closed symmetrical figure. Consequently, the lowest state must have the quantum numbers with opposite signs compared to the highest state, that is, it has $Y = -1$, $I_3 = -1/2$. The Ξ^--hyperon is such a state and transition to it from the Σ^--state is carried out by the S_-^U-operator. Then, by means of the $S_+^{(is)}$-operator we arrive at the last state with $Y = -1$, $I_3 = 1/2$, which is the Ξ^0-hyperon. As one can see, the particle masses in the baryon octet are much more different from each other than those in isomultiplets. Thus, for example,

$$\frac{m(\Xi) - m(N)}{m(\Xi) + m(N)} \approx 17\%.$$

In other words, the unitary symmetry has been violated stronger than the isotopic one.

In the same way by means of the operators $S_\pm^{(is)}$, $S_\pm^U$ and $S_\pm^V$, the 0^+-mesons of (9.100) and the 1^--meson resonances of (9.101) can be grouped into octets. There are also unitary singlets, for example, the $\eta'(957)$-meson forms the 0^--singlet. Unlike mesons (where particles and antiparticles enter into one and the same families), antibaryons form individual families, which are the same as the baryon ones (9.102).

So, we introduced the new quantum number, the unitary spin, to be a generalization of the isospin, and involve both the isospin and the strangeness. Our world is made in such a way, that the strong interaction is approximately invariant with respect to the rotations in the unitary spin space. For this reason hadrons are grouped into the unitary multiplets. This is an axiom of the unitary symmetry theory. All particles of such superfamilies can be viewed simply as a set of the states of one and the same particle, which are degenerated in the electric charge and the hypercharge. According to concepts of the $SU(3)$-symmetry, baryons with spin $1/2$ must be unified in the unitary octet, while baryons with spin $3/2$ must be grouped into the unitary decuplet.

Nine baryon resonances with $3/2^+$ (9.102) might be placed in the decuplet with one vacant lower place (Fig. 9.3). From Fig. 9.3 it follows, that the difference of masses between neighboring isomultiplets is constant and it is approximately equal to 146 MeV/c^2. Thus, it is possible to predict the mass, the strangeness, and the electric charge of a missing member of the baryon decuplet $3/2^+$, which we are going to denote as Ω^-

$$m_\Omega = 1676 \text{ MeV}/c^2, \qquad s = -3.$$

The strangeness conservation law forbids the decay of this particle through the strong interaction. Actually, a decay channel with the least mass of strange particles

$$\Omega^- \to \Xi^0 + K^- \tag{9.107}$$

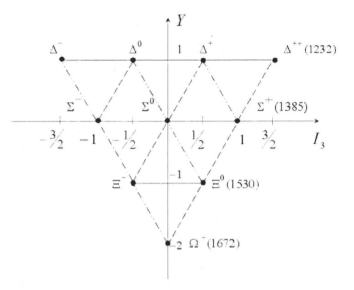

FIGURE 9.3

The baryon resonances decuplet.

turned out to be closed due to the energy arguments ($m_\Omega < m_{\Xi^0} + m_{K^-}$). The weak interaction with the following decay channels

$$\Omega \to \begin{cases} \Lambda + K^- \\ \Xi^- + \pi^0 \\ \Xi^0 + \pi^- \\ \Xi^- + \pi^+ + \pi \\ \Xi^0 + e^- + \overline{\nu}_e \end{cases} \tag{9.108}$$

can give the only chance for the Ω^--hyperon to maintain its status of an unstable particle. Then the calculations of the Ω^--hyperon lifetime gave the value of the order of 10^{-10} s. On the microworld scale it was a long-lived particle and its track had to be seen in a bubble chamber that had already existed by that time. In 1964 at the Brookhaven Accelerator the particle was discovered, which characteristics exactly coincided with the predicted ones. It was one of those miraculous enlightenments, which are so rare in the history of the human mind. Discovery of the Ω^--hyperon was similar to that of the planet Neptune by Leverrie— the discovery made on the pen's edge. It was regarded to be a brilliant proof of the hadrons classification according to the $SU(3)$-symmetry.

A certain isotopic multiplet is described by an isotopic wavefunction, being one of an irreducible representations of the $SU(2)$-group. In the same way a certain unitary multiplet can be described by a multi component unitary wavefunction, which is an irreducible representation of the $SU(3)$-group. Let us study in detail the $SU(3)$-group and in particular its Lie-algebra and its irreducible representations.

The number of linearly independent vectors, defined in the linear space in which the transformations matrix act, is called the *representation dimensionality*. In the case of an internal symmetry the number of particles in the corresponding multiplet is the representation dimensionality. The most simple representations, from which all the rest of the group representations can be built with the help of multiplication, are named by the fundamental ones. For the $SU(n)$-group those are n-component spinors. Thus, the triplet is the fundamental representation of the $SU(3)$-group. To be exact, there are two such representations: covariant and contravariant triplets, but we will discuss it later. Let us consider the transformations, which leave the three-component $SU(3)$-spinor invariant

$$\chi^k \longrightarrow \chi'^k = U_{kl}\chi^l, \tag{9.109}$$

where U are arbitrary and unimodular 3×3-matrices. Canonic representation of a matrix U has the form

$$U = \exp\left(\frac{i}{2}\alpha_a\lambda_a\right), \tag{9.110}$$

where $a = 1, 2, ..8$ and λ_a are 3×3-matrices satisfying the relations

$$\mathrm{Sp}\,(\lambda_a\lambda_b) = 2\delta_{ab}, \qquad \mathrm{Sp}\,\lambda_a = 0 \qquad \lambda_a^\dagger = \lambda_a. \tag{9.111}$$

We are reminded that the number of independent parameters and the number of generators of the $SU(n)$-group is equal to $n^2 - 1$. Here the matrices λ_a play the same role as the Pauli matrices do in the case of the $SU(2)$-symmetry, that is, $\lambda_a/2$ are the generators of the fundamental representation of the unitary spin. The standard writing of these matrices, introduced by Gell-Mann is as follows:

$$\lambda_1 = \begin{pmatrix} 0 & 1 & 0 \\ 1 & 0 & 0 \\ 0 & 0 & 0 \end{pmatrix}, \qquad \lambda_2 = \begin{pmatrix} 0 & -i & 0 \\ i & 0 & 0 \\ 0 & 0 & 0 \end{pmatrix}, \qquad \lambda_3 = \begin{pmatrix} 1 & 0 & 0 \\ 0 & -1 & 0 \\ 0 & 0 & 0 \end{pmatrix},$$

$$\lambda_4 = \begin{pmatrix} 0 & 0 & 1 \\ 0 & 0 & 0 \\ 1 & 0 & 0 \end{pmatrix}, \qquad \lambda_5 = \begin{pmatrix} 0 & 0 & -i \\ 0 & 0 & 0 \\ i & 0 & 0 \end{pmatrix}, \qquad \lambda_6 = \begin{pmatrix} 0 & 0 & 0 \\ 0 & 0 & 1 \\ 0 & 1 & 0 \end{pmatrix},$$

$$\lambda_7 = \begin{pmatrix} 0 & 0 & 0 \\ 0 & 0 & -i \\ 0 & i & 0 \end{pmatrix}, \qquad \lambda_8 = \frac{1}{\sqrt{3}}\begin{pmatrix} 1 & 0 & 0 \\ 0 & 1 & 0 \\ 0 & 0 & -2 \end{pmatrix}. \tag{9.112}$$

The choice of the generators in the form (9.112) is convenient, because the first three matrices $\lambda_{1,2,3}$ are the Pauli matrices (they form Lie algebra of the $SU(2)$-group). What this means is that the $SU(3)$-group contains the isotopic spin group as the subgroup. The matrices λ_3 and λ_8 commute with each other, that is, $SU(3)$ is really the group of the second rank. The permutation relations to characterize the group and to be satisfied by the matrices λ_a, resemble in form those for the matrices σ_l

$$[\lambda_a, \lambda_b] = 2if_{abc}\lambda_c. \tag{9.113}$$

Structural constants f_{abc} are real and antisymmetric with respect to all indices. They can be determined by means of the relations:

$$f_{abc} = \frac{1}{4i}\mathrm{Sp}([\lambda_a, \lambda_b]\lambda_c). \tag{9.114}$$

The components of f_{abc} being different from zero have the following values:

$$f_{123} = 1, \qquad f_{147} = f_{246} = f_{345} = f_{257} = -f_{156} = -f_{367} = \frac{1}{2}, \qquad f_{458} = f_{678} = \frac{\sqrt{3}}{2}. \tag{9.115}$$

For the matrices (9.112) antipermutation relations exist as well

$$\{\lambda_a, \lambda_b\} = 2d_{abc}\lambda_c + \frac{4}{3}\delta_{ab}, \tag{9.116}$$

where constants d_{abc} are completely symmetric with respect to transpositions of indices. By means of relations

$$d_{abc} = \frac{1}{4}\mathrm{Sp}(\{\lambda_a, \lambda_b\}\lambda_c), \qquad (9.117)$$

the nonzero components of d_{abc} can be obtained

$$\left.\begin{array}{l} d_{118} = d_{228} = d_{338} = -d_{888} = \frac{1}{\sqrt{3}}, \qquad d_{448} = d_{558} = \\ = d_{668} = d_{778} = -\frac{1}{2\sqrt{3}}, \qquad d_{146} = d_{157} = -d_{247} = \\ = d_{256} = d_{344} = d_{355} = -d_{366} = -d_{377} = \frac{1}{2} \end{array}\right\}. \qquad (9.118)$$

The matrices λ_a are usually called the *unitary spin operators*. As in the case of the normal spin, the sum of the matrix squares is proportional to an identity matrix

$$\lambda_a \lambda_a = \frac{16}{3}\begin{pmatrix} 1 & 0 & 0 \\ 0 & 1 & 0 \\ 0 & 0 & 1 \end{pmatrix} \qquad (9.119)$$

and that gives us the right to use the above-mentioned term. The λ_a-matrices can be also viewed as components of the eight-dimensional $SU(3)$-vector. The aforesaid is also valid for generators of any representation of the $SU(3)$-group. So, from generators of the unitary spin $S_a^{(un)}$ the operator of the unitary spin square can be obtained

$$\mathbf{S}^{(un)2} = S_a^{(un)} S_a^{(un)}, \qquad (9.120)$$

whose eigenvalues characterize the given representation. However, it is not the only Casimir operator in the $SU(3)$-group. If we introduce the $SU(3)$-vector

$$D_a = \frac{2}{3}d_{abc}S_b^{(un)}S_c^{(un)}, \qquad (9.121)$$

then it is easy to see that the quantity

$$F = S_a^{(un)}D_a, \qquad (9.122)$$

is the second Casimir operator. Although the D_a-vector is made of the $S_a^{(un)}$-generators, however, D_a and $S_a^{(un)}$ are linearly independent. It follows from the fact that after multiplying by $S_a^{(un)}$ they form two different Casimir operators of the $SU(3)$-group. These eight-dimensional $SU(3)$-vectors satisfy the commutation relations typical for moments theory:

$$[S_a^{(un)}, S_b^{(un)}] = if_{abc}S_c^{(un)}, \qquad (9.123)$$

$$[D_a, S_b^{(un)}] = if_{abc}D_c. \qquad (9.124)$$

Notice, that in the case of the $SU(2)$-group the well-known relation:

$$[p_k, M_l] = i\varepsilon_{klm}p_m, \qquad (9.125)$$

characterizing vector property of $\mathbf{p}$, is analog of (9.124). Representing $S_a^{(un)}$ and D_b in the form of 8×8-matrices

$$(S_a^{(un)})_{bc} = if_{abc}, \qquad (9.126)$$

$$(D_a)_{bc} = d_{abc}. \qquad (9.127)$$

one could be convinced of their linear independence. Equality (9.126) is proved by substitution of λ_a, λ_b, λ_c into the Jacobi identity

$$[[A, B], C] + [[B, C], A] + [[C, A], B] = 0, \tag{9.128}$$

by a consequent multiplication by every λ and by calculating the trace. Further, the validity of the choice of the matrix representation for the D_a-operator could be checked by the fulfillment of (9.124).

The connection of $S_a^{(un)}$ both with the isospin operators and with the U-, V-spin operators is given by:

$$\left.\begin{array}{ll} S_{\pm}^{(is)} = S_1^{(un)} \pm iS_2^{(un)}, & S_{\pm}^U = S_6^{(un)} \pm iS_7^{(un)}, \\ S_{\pm}^V = S_4^{(un)} \pm iS_5^{(un)}, & S_3^{(is)} = S_3^{(un)}, \quad Y = \frac{2}{\sqrt{3}}S_8^{(un)}. \end{array}\right\} \tag{9.129}$$

Since the hypercharge commutes with the third projection of the isospin, then the hypercharge operator can differ from $S_8^{(un)}$ only in a constant factor. If in Eq. (9.129) we set this factor equal to $2/\sqrt{3}$, then we arrive at correct expressions for hadron hypercharges. Taking into account (9.129), it is easy to check the validity of the following commutation relations:

$$[S_+^U, S_-^U] = \frac{3}{2}Y - S_3^{(is)} \equiv 2S_3^U, \tag{9.130}$$

$$[S_+^V, S_-^V] = \frac{3}{2}Y + S_3^{(is)} \equiv 2S_3^V. \tag{9.131}$$

The relations (9.130) and (9.131) prove, that the $SU(3)$-group besides the isospin subgroup $SU(2)$ really contains two more $SU(2)$-subgroups, the subgroups of the U- and V-spin. Since the charge operator

$$Q = S_3^{(un)} + \frac{1}{\sqrt{3}}S_8^{(un)}, \tag{9.132}$$

satisfies the commutation relations

$$[Q, S_{\pm}^U] = [Q, S_3^U] = 0, \tag{9.133}$$

then the operator $Y_U = \text{const} \times Q$ plays the role of the hypercharge operator for the U-spin. To obtain the correct values of hadron quantum numbers, the constant in definition of the U-hypercharge must be set to -1. Similar considerations concerning the operator Y_V lead to the result:

$$Y_V = S_3^{(is)} - \frac{1}{2}Y. \tag{9.134}$$

It is obvious, that classification of particles in the $SU(3)$-multiplet can be done according to both isomultiplets and multiplets of the U- and V-spin. In other words, states of particles in the unitary multiplet can be presented by points on the planes (Y_U, U_3) and (Y_V, V_3). The particles distribution of the baryon and meson octets according to the U- and V-spins is shown in Tables 9.1 and 9.2. Notice, that the U-multiplets contain particles with the same value of the electric charge.

The next step is to find an irreducible representations of the unitary spin group. A unitary scalar is the most simple irreducible representation. It describes particles forming unitary singlets. Thereupon the fundamental representation of the $SU(3)$-group, the unitary triplet, about which we discussed above, follows. Let us add some mathematical details to the aforesaid.

TABLE 9.1

The baryon octet.

U	Y_U	Baryons	Mesons
0	0	$-\frac{\sqrt{3}}{2}\Sigma^0 - \frac{1}{2}\Lambda^0$	$-\frac{\sqrt{3}}{2}\pi^0 - \frac{1}{2}\eta^0$
$\frac{1}{2}$	1	Σ^-, Ξ^-	π^-, K^-
$\frac{1}{2}$	-1	p, Σ^+	K^+, π^+
1	0	$n, -\frac{1}{2}\Sigma^0 + \frac{\sqrt{3}}{2}\Lambda^0, \Xi^0$	$K^0, -\frac{1}{2}\pi^0 + \frac{\sqrt{3}}{2}\eta^0, \overline{K}^0$

TABLE 9.2

The meson octet.

V	Y_V	Baryons	Mesons
0	0	$\frac{\sqrt{3}}{2}\Sigma^0 - \frac{1}{2}\Lambda^0$	$\frac{\sqrt{3}}{2}\pi^0 - \frac{1}{2}\eta^0$
$\frac{1}{2}$	1	Σ^+, Ξ^0	$\pi^+, \overline{K}^0$
$\frac{1}{2}$	-1	n, Σ^-	K^0, π^-
1	0	$p, \frac{1}{2}\Sigma^0 + \frac{\sqrt{3}}{2}\Lambda^0, \Xi^-$	$K^+, \frac{1}{2}\pi^0 + \frac{\sqrt{3}}{2}\eta^0, K^-$

In a space $\mathcal{E}_1$, transformed with respect to the fundamental representation, we introduce the orthonormalized basis vectors $\mathbf{e}_k$ $(k = 1, 2, 3)$, which can be chosen in the form:

$$\mathbf{e}_1 = \begin{pmatrix} 1 \\ 0 \\ 0 \end{pmatrix}, \qquad \mathbf{e}_2 = \begin{pmatrix} 0 \\ 1 \\ 0 \end{pmatrix}, \qquad \mathbf{e}_3 = \begin{pmatrix} 0 \\ 0 \\ 1 \end{pmatrix}. \tag{9.135}$$

For objects, defined in unitary spin spaces we are going to use the term "vector." The concrete nature of a vector is decoded in writing down its components. Thus, vectors in the space $\mathcal{E}_1$ are spinors of the first rank and arbitrary spinor $\boldsymbol{\Psi}$ is defined by the formula:

$$\boldsymbol{\Psi} = \Psi^k \mathbf{e}_k. \tag{9.136}$$

Under the $SU(3)$-transformations components of the spinor Ψ^k are transformed by the law:

$$\Psi'^k = U_{kl}\Psi^l. \tag{9.137}$$

Just as in the case of other groups, we continue to call spinors that are transformed by the fundamental representation *contravariant spinors of the first rank*. Thus, in the space $\mathcal{E}_1$ the contravariant spinors of the first rank are set.

Using the obvious form of the fundamental representation generators, it is easy to check the validity of the following relations:

$$\left. \begin{array}{ll} S_3^{(is)}\mathbf{e}_1 = \frac{1}{2}\mathbf{e}_1, & Y\mathbf{e}_1 = \frac{1}{3}\mathbf{e}_1, \\ S_3^{(is)}\mathbf{e}_2 = -\frac{1}{2}\mathbf{e}_2, & Y\mathbf{e}_2 = \frac{1}{3}\mathbf{e}_2, \\ S_3^{(is)}\mathbf{e}_3 = 0, & Y\mathbf{e}_3 = -\frac{2}{3}\mathbf{e}_3, \end{array} \right\}. \tag{9.138}$$

It is convenient to display the basis vectors of the space $\mathcal{E}_1$ on the unitary spin plane, as in Fig. 9.4. Clearly, according to our understandings, the unitary contravariant spinor χ with the components $\chi^1 = \chi^2 = \chi^3 = 1$ describes the unitary triplet of particles, in which the first two elements form the isodoublet $(I = 1/2)$ and the third one forms the isosinglet $(I=0)$. Let us agree to call it the fundamental triplet and display it as the triangle in the plane (I_3, Y).

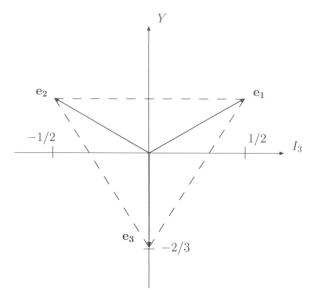

FIGURE 9.4
The basic vectors in the $\mathcal{E}_1$ space.

The space $\mathcal{E}_{\bar{1}}$, transforming with respect to the representation that is conjugate (contragradient) relative to the fundamental representation of the $SU(3)$-group, is also irreducible. In this space, just as in the previous case, arbitrary spinor $\mathbf{\Phi}$ can be presented in the form:

$$\mathbf{\Phi} = \Phi_k \mathbf{e}^k, \tag{9.139}$$

where the Φ_k-spinor components are transformed by means of the law:

$$\Phi'_k = U^*_{kl}\Phi_l = \Phi_l U^\dagger_{lk}. \tag{9.140}$$

The spinors, defined in the space $\mathcal{E}_{\bar{1}}$ shall be called the *covariant spinors of the first rank*. Thus, under transition from contravariant spinors to covariant ones, in the transformation matrix U the change

$$\lambda_a \rightarrow -\lambda^*_a \tag{9.141}$$

takes place. Inserting (9.141) into the expression for the operators $S_3^{(is)}$, Y and choosing the basis vectors of the space $\mathcal{E}_{\bar{1}}$ also in the form (9.135), we obtain the following eigenvalue equations:

$$\left.\begin{array}{ll} S_3^{(is)}\mathbf{e}^1 = -\frac{1}{2}\mathbf{e}^1, & Y\mathbf{e}^1 = -\frac{1}{3}\mathbf{e}^1, \\ S_3^{(is)}\mathbf{e}^2 = \frac{1}{2}\mathbf{e}^2, & Y\mathbf{e}^2 = -\frac{1}{3}\mathbf{e}^2, \\ S_3^{(is)}\mathbf{e}^3 = 0, & Y\mathbf{e}^3 = \frac{2}{3}\mathbf{e}^3. \end{array}\right\} \tag{9.142}$$

The basis vectors of the space $\mathcal{E}_{\bar{1}}$ on the unitary spin plane are displayed in Fig. 9.5. The components of the Hermitian conjugated covariant spinor $\Psi^\dagger_k$ are transformed as the components of the contravariant spinor Ψ^k. Since a Hermitian conjugated wavefunction is connected with antiparticles, then the covariant spinor $\boldsymbol{\eta}$ with the components $\eta_1 = \eta_2 = \eta_3 = 1$ describes the triplet of antiparticles. In what follows we are going to call it the *antitriplet* and depict it as the triangle on the unitary spin plane (Fig. 9.5).

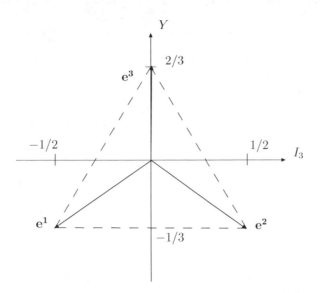

FIGURE 9.5
The basic vectors in the $\mathcal{E}_{\bar{1}}$ space.

In the case of the Isotopic spin group two doublets, transforming according to the rules:

$$\kappa' = U\kappa, \qquad \overline{\kappa}' = U^*\overline{\kappa}, \qquad (9.143)$$

are equivalent, since U and U^* or, which is equivalent, σ_i and $-\sigma_i^*$ are connected through the similarity transformation. This statement breaks down in the case of the matrices λ_a and $-\lambda_a^*$. For this reason the representations of the $SU(3)$-group are characterized by a number of the contravariant p and covariant q indices $D(p,q)$.

So, we got acquainted with the most simple representations: $D(0,0)$, $D(1,0)$, and $D(0,1)$. More complicated representations of the $SU(3)$-group are no longer irreducible. Consequently, a recipe is needed to divide these representations into direct sums of irreducible representations.

Let us first study some concrete examples, and then we shall try to summarize the results obtained. We start with the representation $U \bigotimes U$ in the space $\mathcal{E}_2$ with the basis $\mathbf{e}_{kl}$. The arbitrary contravariant spinor of the second rank with components

$$\Phi^{kl} = \begin{pmatrix} \Phi^{11} & \Phi^{21} & \Phi^{31} \\ \Phi^{12} & \Phi^{22} & \Phi^{32} \\ \Phi^{13} & \Phi^{23} & \Phi^{33} \end{pmatrix}$$

can be expanded into the symmetric part with six independent components

$$\Phi^{(s)kl} = \frac{1}{2}\left(\Phi^{kl} + \Phi^{lk}\right) \qquad (9.144)$$

and into the asymmetric part with three independent components

$$\Phi^{(a)kl} = \frac{1}{2}\left(\Phi^{kl} - \Phi^{lk}\right). \qquad (9.145)$$

Notice, that $\mathrm{Sp}(\Phi^{kl})$ is not a $SU(3)$-scalar. A scalar can be obtained by convolutions of covariant and contravariant indices. According to this, the space $\mathcal{E}_2$ expands into the two subspaces $\mathcal{E}_2^{(6)}$ and $\mathcal{E}_2^{(3)}$, whose bases are formed from six vectors

$$\mathbf{e}_{\{kl\}} = \mathbf{e}_{kl}^{(6)} = \begin{cases} \frac{1}{\sqrt{2}}\left(\mathbf{e}_{kl} + \mathbf{e}_{lk}\right), & k \neq l, \\ \mathbf{e}_{kl}, & k = l, \end{cases} \tag{9.146}$$

and three vectors

$$\mathbf{e}_{[kl]} = \mathbf{e}_{kl}^{(3)} = \frac{1}{\sqrt{2}}\left(\mathbf{e}_{kl} - \mathbf{e}_{lk}\right) \tag{9.147}$$

respectively. Using (9.146) and (9.147) we can write down any vector in the subspaces $\mathcal{E}_2^{(6)}$ and $\mathcal{E}_2^{(3)}$. For example, an arbitrary vector in $\mathcal{E}_2^{(6)}$ has the form:

$$\boldsymbol{\Phi}^{(6)} = \sum \Phi^{kl} \mathbf{e}_{kl}^{(6)} = \frac{1}{\sqrt{2}} \sum_{k>l} \Phi^{kl}\left(\mathbf{e}_{kl} + \mathbf{e}_{lk}\right) + \sum_k \Phi^{kk} \mathbf{e}_{kk}. \tag{9.148}$$

The vector $\boldsymbol{\Phi}$, belonging to the space $\mathcal{E}_2$, can be viewed as a direct product of two contravariant spinors of the first rank, or in other words, the components of the Φ^{kl}-spinor is the direct product of the components of the Φ^k- and Φ^l-spinors. The operation of antisymmetrization can be reduced to multiplying the starting spinor either by the quantities ε^{kmn}, ε_{kmn} or by their production. Thus, for example, since the relations:

$$V^p = \varepsilon^{pmn} \Phi_{mn}^{(a)} \qquad \Phi_{mn}^{(a)} = \frac{1}{2} \varepsilon_{mnp} V^p \tag{9.149}$$

are fulfilled, then the symmetric spinor of the second rank having the components $\Phi_{mn}^{(a)}$ is equivalent to the three-dimensional vector V_p. Consequently, from the outset we can deal not with $\Phi_{mn}^{(a)}$ but with the quantity $\varepsilon^{pmn} \Phi_{mn}$. Notice, that symmetrization of spinors of the highest ranks does not cause a transition from covariant to contravariant indices and vice versa. It is not true, however, in the case of spinors antisymmetrization (see, for example (9.149). Thus, we demonstrated that the product Φ^k and Φ^l is reducible and breaks down into the two irreducible representations, that is

$$\mathbf{3} \bigotimes \mathbf{3} = \mathbf{6} \bigoplus \bar{\mathbf{3}}.$$

Let us consider now the representation $U \bigotimes U^*$ in the space $\mathcal{E}_{1\bar{1}}$ with the basis $\mathbf{e}_l^k$. In this space vectors are mixed spinors of the second rank with components Φ_l^k

$$\Phi_l^k = \begin{pmatrix} \Phi_1^1 & \Phi_1^2 & \Phi_1^3 \\ \Phi_2^1 & \Phi_2^2 & \Phi_2^3 \\ \Phi_3^1 & \Phi_3^2 & \Phi_3^3 \end{pmatrix}. \tag{9.150}$$

The track of this spinor $\mathrm{Sp}(\Phi_l^k) = \Phi_1^1 + \Phi_2^2 + \Phi_3^3$ is a unitary scalar, which is not changed under the $SU(3)$-transformations. Subtracting the quantity $\delta_l^k \Phi_n^n/3$ from (9.150), we obtain the eight-component spinor

$$\Phi_l^k - \frac{1}{3}\delta_l^k \Phi_n^n =$$

$$= \begin{pmatrix} \frac{2}{3}\Phi_1^1 - \frac{1}{3}(\Phi_2^2 + \Phi_3^3) & \Phi_1^2 & \Phi_1^3 \\ \Phi_2^1 & \frac{2}{3}\Phi_2^2 - \frac{1}{3}(\Phi_1^1 + \Phi_3^3) & \Phi_2^3 \\ \Phi_3^1 & \Phi_3^2 & \frac{2}{3}\Phi_3^3 - \frac{1}{3}(\Phi_1^1 + \Phi_2^2) \end{pmatrix}. \tag{9.151}$$

which already is irreducible. According to this, the space $\mathcal{E}_{1\bar{1}}$ is decomposed into the two irreducible subspaces $\mathcal{E}_{1\bar{1}}^{(9)}$ and $\mathcal{E}_{1\bar{1}}^{(1)}$. The basis vector of the one-dimensional subspace $\mathcal{E}_{1\bar{1}}^{(1)}$ is given by:

$$\mathbf{e}_l^{(1)k} = \frac{1}{\sqrt{3}}\left(\mathbf{e}_1^1 + \mathbf{e}_2^2 + \mathbf{e}_3^3\right). \tag{9.152}$$

The basis vectors in the second subspace must be orthogonal to the vector (9.152). Obviously, six quantities $\mathbf{e}_l^k$ with $k \neq l$ meet this demand. The remaining two basis vectors have the form of linear combinations of the vectors $\mathbf{e}_1^1$, $\mathbf{e}_2^2$, and $\mathbf{e}_3^3$. They will be orthonormalized and orthogonal to the basis vector of the subspace $\mathcal{E}_{1\bar{1}}^{(1)}$, provided the following choice:

$$\frac{1}{\sqrt{2}} \left(\mathbf{e}_1^1 - \mathbf{e}_2^2 \right), \qquad \frac{1}{\sqrt{6}} \left(\mathbf{e}_1^1 + \mathbf{e}_2^2 - 2\mathbf{e}_3^3 \right)$$

is made. Thus, the basis vectors of the space $\mathcal{E}_{1\bar{1}}^{(9)}$ are as follows:

$$\mathbf{e}_2^1, \ \frac{1}{\sqrt{2}} \left(\mathbf{e}_1^1 - \mathbf{e}_2^2 \right), \ \mathbf{e}_1^2, \ \mathbf{e}_3^1, \ \mathbf{e}_3^2, \ \mathbf{e}_1^3, \ \mathbf{e}_2^3, \ \frac{1}{\sqrt{6}} \left(\mathbf{e}_1^1 + \mathbf{e}_2^2 - 2\mathbf{e}_3^3 \right). \tag{9.153}$$

Since the vector $\boldsymbol{\Phi}$, belonging to the space $\mathcal{E}_{1\bar{1}}$, is a direct product of the contravariant and covariant spinors of the first rank, then the representation to correspond to it breaks down into two irreducible ones

$$\mathbf{3} \bigotimes \bar{\mathbf{3}} = \mathbf{1} \bigoplus \mathbf{8}.$$

Thus, from the very beginning we managed to find among irreducible representations of the $SU(3)$-group the eight-dimensional representation, which can be used to describe the particle octets.

Let us determine the isospin content of the octet obtained, considering, that a mixed spinor with components Φ_l^k is composed of the product of the fundamental triplet and the fundamental antitriplet. If we denote an isospin multiplet by (I, Y), then a triplet and antitriplet are presented as $(1/2, 1/3) + (0, -2/3)$ and $(1/2, -1/3) + (0, 2/3)$, respectively. The product of the two isospin multiplets (I_1, Y_1) and (I_2, Y_2) contains multiplets with the hypercharge $Y = Y_1 + Y_2$ and the isospins, which according to the moment summation rule, take the values

$$I = \mid I_1 - I_2 \mid, \mid I_1 - I_2 \mid + 1, .., I_1 + I_2.$$

Thus, the isomultiplet multiplication rule has the form:

$$(I_1, Y_1)(I_2, Y_2) = (\mid I_1 - I_2 \mid, Y_1 + Y_2) + (\mid I_1 - I_2 \mid + 1, Y_1 + Y_2) + ...+$$

$$+ (I_1 + I_2, Y_1 + Y_2). \tag{9.154}$$

Using (9.154), we arrive at the relation

$$\mathbf{3} \bigotimes \bar{\mathbf{3}} = \mathbf{1} \bigoplus \mathbf{8} = (1/2, 1) \bigoplus 2(0, 0) \bigoplus (1, 0) \bigoplus (1/2, -1). \tag{9.155}$$

With the help of Eqs. (9.154) and (9.155) we finally obtain

$$\mathbf{8} = (1/2, 1) \bigoplus (0, 0) \bigoplus (1, 0) \bigoplus (1/2, -1), \tag{9.156}$$

that is, the octet consists of the two isodoublets with $Y = 1$ and $Y = -1$, the isotriplet with $Y = 0$, and the isosinglet with $Y = 0$.

For symmetrization and antisymmetrization of the covariant or contravariant spinors of the rank $n \geq 2$ it is convenient to use projection operators. Quantities to symmetrize and antisymmetrize spinors on two indices can be built from the Kronecker symbols

$$\mathcal{S}_{mn}^{kl} = \frac{1}{2} \left(\delta_m^k \delta_n^l + \delta_n^k \delta_m^l \right) \tag{9.157}$$

$$\mathcal{A}_{mn}^{kl} = \frac{1}{2} \left(\delta_m^k \delta_n^l - \delta_m^l \delta_n^k \right). \tag{9.158}$$

It is easy to see that the quantities $\mathcal{A}_{mn}^{kl}$ and $\mathcal{S}_{mn}^{kl}$ form the set of the projection operators. Actually they meet all the demands, imposed on these operators

$$\mathcal{A}^2 = \mathcal{A}, \qquad \mathcal{S}^2 = \mathcal{S}, \qquad \mathcal{A} + \mathcal{S} = I, \qquad \mathcal{A}\mathcal{S} = 0, \tag{9.159}$$

where for the sake of simplicity we neglected spinor indices. Antisymmetrization can be reduced to multiplication of the initial spinor by the quantities ε^{kmn}, ε_{kmn} or by their product. Thus, for example, an antisymmetric spinor of the second rank with components $\Phi_{mn}^{(a)}$

$$\Phi_{mn}^{(a)} = \mathcal{A}_{mn}^{kl} \Phi_{kl}$$

is equivalent to a three-dimensional vector V_p, since the following relations:

$$V^p = \varepsilon^{pmn} \Phi_{mn}^{(a)} \qquad \Phi_{mn}^{(a)} = \frac{1}{2} \varepsilon_{mnp} V^p \tag{9.160}$$

take place. Consequently, from the very beginning we can deal not with $\Phi_{mn}^{(a)}$, but with the quantity $\varepsilon^{pmn} \Phi_{mn}$. Analogously, we could use the quantity $\varepsilon^{pmn} \Phi_{pmn}$ instead of the spinor $\Phi_{\tilde{m}\tilde{n}\tilde{p}}$, where the tilde above the indices indicates the spinor antisymmetry with respect to transpositions according to these indices.

Let us use the methods of symmetrization and antisymmetrization over spinor indices to obtain irreducible representations of the space $\mathcal{E}_3$, which is transformed by means of the representation $U \otimes U \otimes U$. We consider an arbitrary contravariant spinor of the third rank Φ^{pkl} (27 components). First, with the help of the symmetrizing projector we single out from it all the components, symmetric in all three indices. Performing this operation by stages with every pair of indices, we obtain

$$\Phi^{\dot{p}\dot{k}l} = \mathcal{S}_{mn}^{pk} \Phi^{mnl}, \tag{9.161}$$

$$\Phi^{p\dot{k}\dot{l}} = \mathcal{S}_{mn}^{kl} \Phi^{pmn}, \tag{9.162}$$

$$\Phi^{\dot{p}k\dot{l}} = \mathcal{S}_{mn}^{pl} \Phi^{mkn}, \tag{9.163}$$

where the dot over the indices denotes the spinor symmetry with respect to transposition on those indices. Summarizing Eqs. (9.161)–(9.163), we get the spinor to be symmetric in all the three indices:

$$\Phi^{\dot{p}\dot{k}\dot{l}} = \frac{1}{6} \left(\Phi^{pkl} + \Phi^{plk} + \Phi^{lpk} + \Phi^{lkp} + \Phi^{klp} + \Phi^{kpl} \right). \tag{9.164}$$

There are ten independent components of this spinor: three with $p = k = l$, six with only two coinciding indices, and one with three different p, k, l. The same sequence of operations with the antisymmetrizing projector $\mathcal{A}$ produces a completely antisymmetric spinor:

$$\Phi^{\tilde{p}\tilde{k}\tilde{l}} = \frac{1}{6} \left(\Phi^{pkl} - \Phi^{plk} + \Phi^{lpk} - \Phi^{lkp} + \Phi^{klp} - \Phi^{kpl} \right). \tag{9.165}$$

The spinor $\Phi^{\tilde{p}\tilde{k}\tilde{l}}$ has only one independent component, since the spinor being antisymmetric in all the three indices is equivalent to a scalar V

$$V = \frac{1}{6} \varepsilon_{pkl} \Phi^{pkl}.$$

In the initial spinor let us single out components to be symmetric in p, k and antisymmetric in k, l

$$\Phi^{\dot{p}\dot{k}l} = \mathcal{S}_{mn}^{pk} \Phi^{mnl}, \tag{9.166}$$

$$\Phi^{p\bar{k}\bar{l}} = \mathcal{A}^{kl}_{mn}\Phi^{pmn}. \tag{9.167}$$

However, not all components in Eqs. (9.166), (9.167) are independent. Some of them are already taken into consideration at definition of the spinors, which are symmetric and antisymmetric in all the three indices (for example, the first two terms in Eq. (9.164), the first and the last terms in Eq. (9.165)). Summing up (9.166) and (9.167) and subtracting from the obtained result the parts found before, we arrive at:

$$\Phi^{p\dot{k}\dot{l}} = \frac{1}{3}\left(\Phi^{pkl} + \Phi^{kpl} - \Phi^{plk} - \Phi^{lpk}\right). \tag{9.168}$$

Continuing manipulations with the projectors (9.157) and (9.158), we obtain the expression for the spinor, which is antisymmetric with respect to p, k and symmetric with respect to k, l indices:

$$\Phi^{\bar{p}\dot{k}\dot{l}} = \frac{1}{3}\left(\Phi^{pkl} + \Phi^{plk} - \Phi^{kpl} - \Phi^{klp}\right). \tag{9.169}$$

It is clear that the number of independent components q of the $\Phi^{p\dot{k}\dot{l}}$- and $\Phi^{\bar{p}\dot{k}\dot{l}}$-spinors coincide and it is consequently equal to:

$$q = \frac{27 - 10 - 1}{2} = 8.$$

Thus, the spinor Φ^{pkl} falls into the following irreducible representations:

$$\mathbf{3}\bigotimes\mathbf{3}\bigotimes\mathbf{3} = \mathbf{1}\bigoplus\mathbf{8}\bigoplus\mathbf{8}\bigoplus\mathbf{10}. \tag{9.170}$$

Let us determine (I, Y)-content of the representations in (9.170). According to (9.154) we have

$$\mathbf{3}\bigotimes\mathbf{3}\bigotimes\mathbf{3} = (3/2, 1)\bigoplus 2(1/2, 1)\bigoplus 3(1, 0)\bigoplus 3(0, 0)\bigoplus$$
$$\bigoplus 3(1/2, -1)\bigoplus(0, -2). \tag{9.171}$$

Taking into consideration the isotopic structure of the octet (9.156) obtained before, we define the (I, Y)-content of the decuplet from Eq. (9.171)

$$\mathbf{10} = (3/2, 1)\bigoplus(1, 0)\bigoplus(1/2, -1)\bigoplus(0, 2). \tag{9.172}$$

Now we demonstrate how to fulfill splitting on irreducible representations in the case of mixed spinors. Recall, that in the $SU(3)$-group there are three invariants, namely, δ^k_l, ε_{ijk}, and ε^{ijk}. The Kronecker symbol invariance follows from the unitarity of the $SU(3)$-transformations:

$$\delta'^k_l = U^m_l U^{*k}_n \delta^n_m = \delta^k_l. \tag{9.173}$$

The antisymmetric quantities ε_{ijk} and ε^{ijk} are invariant due to the unimodularity of the $SU(3)$-group:

$$\varepsilon'_{ijk} = U^m_i U^n_j U^p_k \varepsilon_{mnp} = (\det U)\varepsilon_{ijk} = \varepsilon_{ijk}. \tag{9.174}$$

Thus, the projectors, which symmetrize $\mathcal{S}^{pk}_{mn}$ and antisymmetrize $\mathcal{A}^{kl}_{mn}$, where

$$\mathcal{A}^{kl}_{mn} = \frac{1}{2}\varepsilon^{klp}\varepsilon_{mnp}, \tag{9.175}$$

are built of the $SU(3)$-group invariants. Besides these operations, there exist the operation of convolution in the upper as well as in the low indices and the operation of a track vanishing for a mixed spinor. They are both carried out by means of the invariant δ^m_n.

The mixed spinor for which the above-mentioned operations can not be fulfilled is already irreducible.

By the example of the spinor Φ_{pq}^{mn} we illustrate how to obtain irreducible spinors by means of the $SU(3)$-group invariants.

(i) Multiplicating of Φ_{pq}^{mn} by δ_m^p and δ_n^q produces a scalar:

$$\Psi = \Phi_{pq}^{pq}.$$

(ii) Multiplicating of Φ_{pq}^{mn} by δ_m^p and vanishing of the track results in the eight-components quantity:

$$\Psi_q^n = \Phi_{pq}^{pn} - \frac{1}{3}\delta_q^n(\Phi_{pm}^{pm}).$$

Analogously, multiplicating of Φ_{pq}^{mn} by δ_n^q and vanishing of the track leads to:

$$\Psi_p^m = \Phi_{pq}^{mq} - \frac{1}{3}\delta_p^m(\Phi_{nq}^{nq}).$$

(iii) Multiplication of Φ_{pq}^{mn} by ε_{mnl} and symmetrization with respect to the indices lpq gives ten-components quantity:

$$\Psi_{l\dot{p}\dot{q}} = S(lpq)\varepsilon_{mnl}\Phi_{pq}^{mn},$$

where $S(lpq)$ symbolizes symmetrization with respect to the indices lpq, that, as we know, is reduced to the multiplication of the initial spinor by the Kronecker symbols combinations. Analogously, the multiplication of Φ_{pq}^{mn} by ε^{pql} and symmetrization with respect to indices lmn produces another irreducible spinor:

$$\Psi^{l\dot{m}\dot{n}} = S(lmn)\varepsilon^{pql}\Phi_{pq}^{mn}.$$

(iv) Symmetrization with respect to the upper and low indices and vanishing of the track results in the 27-components quantity:

$$\Phi_{\dot{p}\dot{q}}^{\dot{m}\dot{n}} - \frac{1}{5}\left(\delta_p^m\Phi_{\dot{s}\dot{q}}^{\dot{s}\dot{n}} + \delta_q^m\Phi_{\dot{p}\dot{s}}^{\dot{s}\dot{n}} + \delta_p^n\Phi_{\dot{s}\dot{q}}^{\dot{m}\dot{s}} + \delta_q^n\Phi_{\dot{p}\dot{s}}^{\dot{m}\dot{s}}\right) + \frac{1}{20}\left(\delta_p^m\delta_q^n + \delta_q^n\delta_p^m\right)\Phi_{\dot{r}\dot{s}}^{\dot{r}\dot{s}}.$$

If one takes into consideration that the spinor Φ_{pq}^{mn} can be presented as the product of the spinors Φ_p^m and Φ_q^n then all the above-mentioned splitting into irreducible spinors can be symbolically expressed as follows:

$$8 \bigotimes 8 = 1 \bigoplus 8 \bigoplus 8 \bigoplus 10 \bigoplus \overline{10} \bigoplus 27.$$

Acquired experience allows us to come to generalizations, namely, to formulate the irreducibility condition for the highest dimension representations (representations, not belonging to: $D(0,0)$, $D(1,0)$, and $D(0,1)$). A mixed spinor is irreducible, if being multiplied by all the three invariants of the $SU(3)$-group—δ_l^k, ε_{ijk}, ε^{ijk}—it produces zero. We are reminded that we replace antisymmetrization with multiplication by ε_{ijk}, ε^{ijk} or by their product. If one takes into account that the product of a symmetric spinor on the completely antisymmetric quantity ε_{ijk} or ε^{ijk} is equal to zero, then the aforesaid means, that the components of an irreducible spinor must be separately symmetric with respect to the upper and low indices and its track must turn to zero.

Let us connect the number of covariant and contravariant indices of an irreducible representation with the number of particles in a multiplet n, and in so doing we determine the form of a function $n(p,q)$. Obviously, $n(p,q)$ is also a number of independent components of a corresponding irreducible $SU(3)$-spinor $\Psi_{l_1\cdots l_q}^{k_1\cdots k_p}$. Sometimes we use the term "spinor" to denote vector components in the unitary spin space, only if it causes no misunderstandings. Consider a contravariant spinor $\Psi^{k_1 \cdots k_p}$ symmetrized with respect to all indices (consequently, irreducible), or representation $D(p,0)$. Find the quantity $n(p,0)$. By p_1, p_2, p_3 we denote the number of ones, twos, and threes among the indices p. Since $p_1 + p_2 + p_3 = p$, then with the given p_2 the number p_1 can run the $p - p_2 + 1$-values from 0 to $p - p_2$. Therefore, we have

$$n(p,0) = \sum_{p_2=0}^{p} (p - p_2 + 1) = \frac{1}{2}(p+1)(p+2). \tag{9.176}$$

Analogously, the number of components of a symmetrized covariant spinor of the rank q is equal to

$$n(0,q) = \frac{1}{2}(q+1)(q+2). \tag{9.177}$$

Then the number of components of the spinor $\Psi^{k_1 \cdots k_p}_{l_1 \cdots l_q}$ be given by:

$$n(p,0)n(0,q) = \frac{1}{4}(q+1)(q+2)(p+1)(p+2), \tag{9.178}$$

if all the tracks:

$$\Psi^{k_1 \ldots m \ldots k_p}_{l_1 \ldots m \ldots l_q}$$

would be arbitrary. With allowance made for Eq. (9.178) the number of such tracks or conditions:

$$\Psi^{k_1 \ldots m \ldots k_p}_{l_1 \ldots m \ldots l_q} = 0 \tag{9.179}$$

is equal to the number of components of a mixed spinor with $q - 1$ contravariant indices and $p - 1$ covariant ones

$$n(q-1,0)n(0,p-1) = \frac{1}{4}q(q+1)p(p+1). \tag{9.180}$$

Thus, the multiplicity of the representation $D(p,q)$ is defined by the expression:

$$n(p,q) = n(p,0)n(0,q) - n(q-1,0)n(0,p-1) = \frac{1}{2}(p+1)(q+1)(p+$$

$$+q+2). \tag{9.181}$$

Now we can establish the multiplicity of any spinor. For example, using the relation (9.170), we obtain for the spinor Φ^{pmq}:

$$D(1,0) \bigotimes D(1,0) \bigotimes D(1,0) = D(0,0) \bigoplus 2D(1,1) \bigoplus D(3,0), \tag{9.182}$$

where it has been allowed for that the transition from contravariant indices to covariant indices has taken place during antisymmetrization.

The next step is to determine the explicit form of wavefunctions of unitary multiplets. Let us consider a space transformed under the representation to be the product of the fundamental representations

$$\underbrace{U \bigotimes \cdots \bigotimes U}_{p \ times} \bigotimes \underbrace{U^* \cdots \bigotimes U^*}_{q \ times}. \tag{9.183}$$

We introduce the orthonormalized basis $\mathbf{e}_{k_1\cdots k_p}^{l_1\cdots l_q}$. Then an arbitrary vector in this space is defined by the formula:

$$\mathbf{\Psi} = \Psi_{l_1\cdots l_q}^{k_1\cdots k_p}\mathbf{e}_{k_1\cdots k_p}^{l_1\cdots l_q}. \tag{9.184}$$

The components of the vector $\mathbf{\Psi}$ are transformed according to the law

$$\Psi_{l_1\cdots l_q}^{\prime k_1\cdots k_p} = U_{k_1 m_1}\cdots U_{k_p m_p}U_{l_1 n_1}^{*}\cdots U_{l_q n_q}^{*}\Psi_{n_1\cdots n_q}^{m_1\cdots m_p}. \tag{9.185}$$

The action of the representation generators on the basis vectors is determined by the formula:

$$S_a^{(un)}\mathbf{e}_{k_1\cdots k_p}^{l_1\cdots l_q} = \sum_{r=1}^{q}(\lambda_a)_{l_r l_r'}\,\mathbf{e}_{k_1\cdots k_p}^{l_1\cdots l_{r-1}l_r' l_{r+1}\cdots l_q} - \sum_{r=1}^{p}(\lambda_a^T)_{k_r k_r'}\,\mathbf{e}_{k_1\cdots k_{r-1}k_r' k_{r+1}\cdots k_p}^{l_1\cdots l_q}, \tag{9.186}$$

where we took into consideration the explicit form of the generators for the contravariant and covariant spinors of the first rank. Let us denote the numbers of the covariant (contravariant) indices, equal to one, two and three by q_1, q_2 and q_3 (p_1, p_2 and p_3), respectively. Then from Eq. (9.186) and the particular form of the λ_a-matrices one can see, that the basis vector with the given numbers of indices, equal to one, two, and three correspond to the following eigenvalues of the $SU(3)$-group invariants

$$S_3^{(is)}\mathbf{e}_{k_1\cdots k_p}^{l_1\cdots l_q} = \{\frac{1}{2}[q_1 - p_1] - \frac{1}{2}[q_2 - p_2]\}\mathbf{e}_{k_1\cdots k_p}^{l_1\cdots l_q}, \tag{9.187}$$

$$Y\mathbf{e}_{k_1\cdots k_p}^{l_1\cdots l_q} = \{\frac{1}{2\sqrt{3}}[q_1 - p_1] + \frac{1}{2\sqrt{3}}[q_2 - p_2] - \frac{1}{\sqrt{3}}[q_3 - p_3]\}\mathbf{e}_{k_1\cdots k_p}^{l_1\cdots l_q}. \tag{9.188}$$

As before, we shall consider the eigenvalues of the operators $S_3^{(is)}$ and Y to be components of two-dimensional vectors on the unitary spin plane, that is, the end of the vector $\mathbf{e}_{k_1\cdots k_p}^{l_1\cdots l_q}$ corresponds to the definite state of the $SU(3)$-multiplet.

As an example, we examine the irreducible representation $D(1,1)$, according to which the mixed spinor of the second rank ψ_l^k with the zero-track is transformed. As we already know, the vectors

$$\mathbf{f}^a = (\psi_l^k)^a\mathbf{e}_k^l, \qquad a = 1, 2, \ldots 8, \tag{9.189}$$

making up the basis in the space $D(1,1)$ can be chosen in the form:

$$\left.\begin{array}{llll} \mathbf{f}^1 = \mathbf{e}_2^1, & \mathbf{f}^2 = \frac{1}{\sqrt{2}}\left(\mathbf{e}_1^1 - \mathbf{e}_2^2\right), & \mathbf{f}^3 = \mathbf{e}_1^2, & \mathbf{f}^4 = \mathbf{e}_3^1, \\ \mathbf{f}^5 = \mathbf{e}_3^2, & \mathbf{f}^6 = \mathbf{e}_1^3, & \mathbf{f}^7 = \mathbf{e}_2^3, & \mathbf{f}^8 = \frac{1}{\sqrt{6}}\left(\mathbf{e}_1^1 + \mathbf{e}_2^2 - 2\mathbf{e}_3^3\right), \end{array}\right\} \tag{9.190}$$

Then the nonzero components of the vectors $(\psi_l^k)^a$ of the basis $\mathbf{f}^a$ have the following values

$$\left.\begin{array}{llll} \left(\psi_1^2\right)^1 = 1; & \left(\psi_1^1\right)^2 = \frac{1}{\sqrt{2}}, & \left(\psi_2^2\right)^2 = -\frac{1}{\sqrt{2}}; \\ \left(\psi_2^1\right)^3 = 1; & \left(\psi_1^3\right)^4 = 1; & \left(\psi_2^3\right)^5 = 1; & \left(\psi_3^1\right)^6 = 1; \\ \left(\psi_3^2\right)^7 = 1; & \left(\psi_1^1\right)^8 = \left(\psi_2^2\right)^8 = \frac{1}{\sqrt{6}}, & \left(\psi_3^3\right)^8 = -\frac{2}{\sqrt{6}}. \end{array}\right\} \tag{9.191}$$

Using Eqs. (9.187) and (9.188), it is easy to demonstrate that the vectors $\mathbf{f}^a$ describe the following isomultiplets of the baryon $SU(3)$-octet:

$$\left.\begin{array}{lll} I = 1, Y = 0: \Sigma^+ = \mathbf{f}^1, & \Sigma^0 = \mathbf{f}^2, & \Sigma^- = \mathbf{f}^3; \\ I = \frac{1}{2}, Y = 1: p = \mathbf{f}^4, & n = \mathbf{f}^5; \\ I = \frac{1}{2}, Y = -1: \Xi^- = \mathbf{f}^6, & \Xi^0 = \mathbf{f}^7; \\ I = 0, Y = 0: \Lambda^0 = \mathbf{f}^8. \end{array}\right\} \tag{9.192}$$

Now we introduce the quantity

$$\mathbf{B}_l^k = \left(\psi_l^k\right)^a \mathbf{f}^a,$$

whose matrix components represent the wavefunction of the baryon octet

$$(B_l^k) = \begin{pmatrix} \frac{1}{\sqrt{2}}\Sigma^0 + \frac{1}{\sqrt{6}}\Lambda^0 & \Sigma^+ & p \\ \Sigma^- & -\frac{1}{\sqrt{2}}\Sigma^0 + \frac{1}{\sqrt{6}}\Lambda^0 & n \\ \Xi^- & \Xi^0 & -\frac{2}{\sqrt{6}}\Lambda^0 \end{pmatrix}. \tag{9.193}$$

To obtain a normalized wavefunction, corresponding to the definite particle of the octet, in Eq. (9.193) its symbol should be replaced by 1 and all the matrix elements B_l^k unrelated to this particle must be set equal to zero.

Meson octets have a similar form. Thus, for example, the wavefunctions of the pseudoscalar and vector meson octets are defined by the expressions:

$$(P_l^k) = \begin{pmatrix} \frac{1}{\sqrt{2}}\pi^0 + \frac{1}{\sqrt{6}}\eta^0 & \pi^+ & K^+ \\ \pi^- & -\frac{1}{\sqrt{2}}\pi^0 + \frac{1}{\sqrt{6}}\eta^0 & K^0 \\ K^- & \overline{K^0} & -\frac{2}{\sqrt{6}}\eta^0 \end{pmatrix} \tag{9.194}$$

$$(V_l^k) = \begin{pmatrix} \frac{1}{\sqrt{2}}\rho^0 + \frac{1}{\sqrt{6}}\omega^0 & \rho^+ & K^{*+} \\ \rho^- & -\frac{1}{\sqrt{2}}\rho^0 + \frac{1}{\sqrt{6}}\omega^0 & K^{*0} \\ K^{*-} & \overline{K^{*0}} & -\frac{2}{\sqrt{6}}\omega^0 \end{pmatrix}, \tag{9.195}$$

respectively.

It is obvious that any matrix with the zero-track could be expanded in terms of the total system of the linearly independent matrices λ_a

$$\Phi = \lambda_a \Phi_a,$$

where the coefficients Φ_a present the $SU(3)$-vector components. This, in its turn, exhibits the wavefunctions of the baryon or meson octets in the form of the eight-dimensional vectors, as for example:

$$B_a = \frac{1}{2}\mathrm{Sp}(B\lambda_a). \tag{9.196}$$

Notice, that the P- and V-matrices are Hermitian ones, that is, they transfer to themselves under the transposition and replacement of particles by antiparticles. Obviously, the baryon matrices do not have such a property, since baryons and antibaryons form independent multiplets. The computations, similar to the above, produce the following expression for the antibaryons octet matrix

$$(\overline{B}_l^k) = \begin{pmatrix} \frac{1}{\sqrt{2}}\overline{\Sigma^0} + \frac{1}{\sqrt{6}}\overline{\Lambda^0} & \Sigma^- & \overline{\Xi}^- \\ \Sigma^+ & -\frac{1}{\sqrt{2}}\overline{\Sigma^0} + \frac{1}{\sqrt{6}}\overline{\Lambda^0} & \overline{\Xi}^0 \\ \overline{p} & \overline{n} & -\frac{2}{\sqrt{6}}\overline{\Lambda^0} \end{pmatrix}. \tag{9.197}$$

With expressions for unitary wavefunctions near at hand, one can use the Lagrangian formalism to describe the behavior of the baryon and meson superfamilies. However, there is one "but" connected with particle masses in an unitary multiplet. The free Lagrangian of the baryon octet has the form:

$$\mathcal{L} = \frac{i}{2}\left[\overline{B}_l^k(x)\gamma_\mu\partial^\mu B_k^l(x) - \partial^\mu\overline{B}_l^k(x)\gamma_\mu B_k^l(x)\right] - m_0\overline{B}_l^k(x)B_l^k(x), \tag{9.198}$$

where we assumed, that all the baryons possess the same mass m_0. However, in reality the $SU(3)$-symmetry has been violated and the particle masses in the multiplet differ from each other. Consequently, this factor should be necessarily taken into consideration under accomplishing the exact calculations.

The world is made in such a way, that the weaker the interaction, the less symmetric it is. The stronger interaction behaves as though it does not observe slight violations of definite conservation laws. As a result, such an interaction conserves the given physical quantity and consequently, it is more symmetric. The total interaction between hadrons can be presented as a sum of a hypothetical super strong interaction (with the $SU(3)$-symmetry group), the strong interaction (which violates the unitary symmetry, but conserves the isotopic one), and last the electromagnetic and weak interactions (which violate the isotopic invariance). The division of the strong interaction into the super strong and the normal strong interaction, is, of course, rather conditional. In fact, there is no super strong interaction at all. There are only high energy regions, where mass differences of particles in multiplets are insignificant. When the strong, electromagnetic and weak interactions are switched off, the exact $SU(3)$-symmetry takes place and all the particles in the unitary multiplet are degenerated in the mass (m_0 is a degeneracy mass). Switching on the strong interaction makes the mass operator in the free Lagrangians dependent on the isospin and the hypercharge

$$(\Psi_\alpha^{(n)}, \mathcal{M}\Psi_\alpha^{(n)}) = (\Psi_\alpha^{(n)}, \mathcal{M}_0\Psi_\alpha^{(n)})+$$

$$+(\Psi_\alpha^{(n)}, \Delta\mathcal{M}\Psi_\alpha^{(n)}) \equiv M = m_0 \cdot 1^{(n)} + \Delta m^{(n)}(I, Y), \tag{9.199}$$

where $\Psi_\alpha^{(n)}$ denotes the unitary wavefunction written in the form of the column matrix, the indices n and α characterize the representation and the particle state in a multiplet, respectively.

Let us consider the concrete case, the baryon octet. From the multiplication rules for the representations:

$$\mathbf{8} \bigotimes \mathbf{8} = \mathbf{1} \bigoplus \mathbf{8} \bigoplus \mathbf{8} \bigoplus \mathbf{10} \bigoplus \overline{\mathbf{10}} \bigoplus \mathbf{27}. \tag{9.200}$$

it follows that the masses of the octet members can be expressed in terms of 6 constants that are connected with contributions coming from the following multiplets: one from a singlet, two from octets, two from decouplets, and one from a $\mathbf{27}$-representation. It is known from experiments that states with fixed Y, belonging to the given isomultiplet, are still degenerate on a mass. For this reason, in expansion of the mass matrix in terms of irreducible representations, only operators with $Y = T = 0$ appear. Since the $\mathbf{10}$- and $\overline{\mathbf{10}}$-representations contain no members with $Y = T = 0$, then they do not contribute to M. Further, following Gell-Mann [25] we assume (and every assumption is good if it brings the theory in accordance with experiments) that the value of contribution to the mass operator decreases while the representation dimension grows. In other words in (9.199) the term $m_0 \cdot 1^{(n)}$ should be interpreted as the contribution from the singlet representation (zero-order approximation on the strong interaction) while that $\Delta m^{(n)}(I, Y)$ should be interpreted as the contribution from the octet representation (first-order approximation on the strong interaction). Since the operator $\Delta\mathcal{M}$ enters into the total set of operators, defining a system state, then it must commute with operators of the hypercharge and the isospin. It is possible, provided it is either the third or eighth component of the $SU(3)$-vector. In the unitary spin group there are two independent vectors $S_a^{(un)}$ and D_a. Consequently, Eq. (9.199) can be presented as follows:

$$(\Psi_\alpha^{(9)}, \mathcal{M}\Psi_\alpha^{(9)}) = m_0 \cdot 1^{(9)} + c_1(\Psi_\alpha^{(9)}, S_8^{(un)}\Psi_\alpha^{(9)}) + c_2(\Psi_\alpha^{(9)}, D_8\Psi_\alpha^{(9)}), \tag{9.201}$$

where c_1, c_2 are real constants and for $\Delta\mathcal{M}$ we stopped our choice on the eighth component of the $SU(3)$-vector. Employing Eqs. (9.118), (9.121), and (9.129), it is easy to obtain

$$D_8 = \frac{1}{3\sqrt{3}}\left(3S_k^{(is)2} - \frac{3}{4}Y^2 - S_a^{(un)2}\right). \tag{9.202}$$

Inserting (9.202) into (9.201), we arrive at Gell-Mann-Okubo mass formula [25,26]

$$(\Psi_\alpha^{(9)}, \mathcal{M}\Psi_\alpha^{(9)}) = m_0 + AY + B[I(I+1) - \frac{1}{4}Y^2], \tag{9.203}$$

which defines particle masses in all unitary multiplets according to the values of real constants m_0, A, and B. Notice, that the formula (9.203) is valid when particles are described by the first-order equations. The first-order equations formalism is employed both for particles with the spin $1/2$ and for particles with the spin 0 and 1. However, as a rule, the second-order equations are most commonly used for bosons. In this case, the Gell-Mann-Okubo formula acquires the form:

$$M = \mu_0^2 + A'Y + B'[I(I+1) - \frac{1}{4}Y^2].$$

There are four different masses in the baryon octet: m_N, m_Ξ, m_Σ, and m_Λ. Writing Eq. (9.203) for these masses and excluding the parameters m_0, B and C, we obtain the formula

$$m_\Sigma + 3m_\Lambda = 2(m_N + m_\Xi), \tag{9.204}$$

which is fulfilled with an accuracy of 0.5%.

Switching on the electromagnetic interaction initiates further masses splitting in a unitary multiplet. To define an electromagnetic addition to particle masses, the division of unitary superfamilies into the U-spin multiplets proves useful to us. At this point it is appropriate to recollect the first work in the teaching strategy, which so magnificently covered the very essence of this problem. It was written in the twelfth century by canon Chue (St. Victoria Abbey, Paris). Unlike a modern "opus magnum" devoted to the teaching strategy, the work contained only one sentence: *study everything, afterwards you shall see—there is nothing unnecessary*. Since the U-spin multiplets group particles with the equal value of the electric charge, one can assume that the electromagnetic interaction conserves the U-spin, that is, it acts identically on all members of a U-multiplet. Let us denote mass corrections to charged and neutral particles by the symbols $\delta_\pm$ and δ_0, respectively, and use the Table 9.1, which defines the baryon octet division into the U-multiplets. Switching on the electromagnetic interaction modifies the masses of the members of the two U-doublets

$$m_p = m_N^{(0)} + \delta_+, \qquad m_{\Sigma^+} = m_\Sigma^{(0)} + \delta_+, \tag{9.205}$$

$$m_{\Sigma^-} = m_\Sigma^{(0)} + \delta_-, \qquad m_{\Xi^-} = m_\Xi^{(0)} + \delta_-, \tag{9.206}$$

where the upper index 0 denotes the particle mass in the isomultiplet before the electromagnetic interaction is switched on. Analogously, for the member of the U-spin triplet we get

$$m_n = m_N^{(0)} + \delta_0, \qquad m_{\Sigma^0} = m_\Sigma^{(0)} + \delta_0. \tag{9.207}$$

Excluding $\delta_\pm$, δ_0, and $m_{N,\Sigma,\Xi}^{(0)}$ from Eqs. (9.205)–(9.207), we arrive at Coleman-Glashow formula [27]:

$$m_{\Xi^-} - m_{\Xi^0} = m_{\Sigma^-} - m_{\Sigma^+} + m_p - m_n, \tag{9.208}$$

which is fulfilled with an accuracy of $\sim 1.4\%$.

Now we proceed to the build-up of the $SU(3)$ invariant interaction Lagrangian for the baryon octet. Let us restrict ourselves to the case of a baryon-meson interaction. Of course, it must be a generalization of the isotopically invariant interaction between nucleons and π-mesons:

$$\mathcal{L}^{(is)} = ig_{N\pi}\overline{N}_\alpha(\sigma_k)_{\alpha\beta}\gamma_5 N_\beta \pi_k, \tag{9.209}$$

where $\alpha, \beta = 1, 2$, $k = 1, 2, 3$ and $\gamma_5 \equiv \gamma^5 = i\gamma^0\gamma^1\gamma^2\gamma^3$. The obvious analog of (9.209) is the expression:

$$ig_f\overline{B}_a(S_c^{(un)})_{ab}\gamma_5 B_b P_c. \tag{9.210}$$

However, contrary to the $SU(2)$-group where there was only one independent $SU(2)$-vector, the $SU(3)$-group has two independent $SU(3)$-vectors, $S_a^{(un)}$ and D_a. Thus, the $SU(3)$ invariant interaction Lagrangian has the form:

$$\mathcal{L}^{(un)} = \mathcal{L}_1^{(un)} + \mathcal{L}_2^{(un)} = ig_f\overline{B}_a(S_c^{(un)})_{ab}\gamma_5 B_b P_c + ig_d\overline{B}_a(D_c)_{ab}\gamma_5 B_b P_c =$$

$$= g_f f_{cab}\overline{B}_a\gamma_5 B_b P_c + ig_d d_{cab}\overline{B}_a\gamma_5 B_b P_c = -\frac{i}{4}\text{Sp}([\overline{B}\gamma_5, B]P) + \frac{i}{4}\text{Sp}(\{\overline{B}\gamma_5, B\}P). \tag{9.211}$$

In the next to last line of Eq. (9.211) we have used matrix representations of the operators $S_a^{(un)}$ and D_a.

Everything, we have yet known about the $SU(3)$-symmetry and about symmetry in general, concerns only kinematic aspects of symmetry. However the real power of symmetry reveals itself under investigation of physical systems evolution. For example, symmetry allows us to obtain relationships between the cross sections of different reactions without using the motion equations. It will be very instructive to show how to find such relationships by the example of the $SU(3)$-symmetry.

Suppose, we study the reaction:

$$1 + 2 \rightarrow 3 + 4. \tag{9.212}$$

We shall denote the wavefunction of the initial (final) state by Φ_i (Φ_f). Since particles are free before and after interaction, then:

$$\Phi_i = \Phi_1(\mathbf{p}_1, J_1)\Phi_2(\mathbf{p}_2, J_2), \qquad \Phi_f = \Phi_3(\mathbf{p}_3, J_3)\Phi_4(\mathbf{p}_4, J_4),$$

where the wavefunction $\Phi_n(\mathbf{p}_n, J_n)$ ($n = 1, 2, 3, 4$) depends on the momentum and the spin of nth-particle. An amplitude of a transition from the initial to the final state is given by the expression:

$$M_{i\rightarrow f} = \Phi_4^\dagger\Phi_3^\dagger S\Phi_1\Phi_2, \tag{9.213}$$

where the operator S derived the name the S-matrix is defined by the form of the interaction Lagrangian for particles 1, 2, 3, and 4. In the case when we deal not with particles scattering, but with the scattering of $SU(3)$-multiplets, the wavefunctions and the S-matrix contain dependence on the unitary spin as well. Since the spin matrices and the unitary spin ones act in different spaces, they commute with each other. It leads to the fact that the spatial and unitary parts are separated. Thus, the wavefunction of the l-multiplet is the production of the spatial part $\Phi_l(\mathbf{p}, J)$, which is common for all multiplet particles, and the unitary part $\Phi_l^{(un)}$. Then, the amplitude of the reaction (9.212) can be presented in the following form:

$$M_{i\rightarrow f} = \sum_j\sum_i R_j\Gamma_i T_{ji}, \tag{9.214}$$

where Γ_i are $SU(3)$-invariant combinations, R_j are relativistically invariant spin combinations and T_{ji} are relativistically invariant functions on the momenta of the initial and final particles. To define the set Γ_i, all the possible independent unitary invariants ($SU(3)$-scalars) must be composed out of the unitary wavefunctions of particles, entering into the reaction. The number of independent $SU(3)$-scalars might be reduced in case we demand that the amplitude of the reaction (9.212) is invariant under other transformations (P, C, T, etc.).

Let us come back to the case of baryons and consider elastic scattering of the pseudoscalar meson octet by the baryon octet:

$$P + B \to P' + B', \tag{9.215}$$

where for the sake of convenience the final states of the meson and baryon octets are marked with the primes. To define the number of independent unitary scalars, corresponding to the reaction (9.215) we use the fact that the $\mathbf{8} \otimes \mathbf{8}$-products of the wavefunctions, both of the initial state and of the final one, are separated into irreducible representations according to the formula (9.198). Since the unitary symmetry allows only transitions between equal representations,* then from (9.200) it follows that in the amplitude of the reaction (9.215) the matrix elements of only the following eight transitions:

$$\left. \begin{array}{cccc} \mathbf{1} \to \mathbf{1}, & \mathbf{8} \to \mathbf{8}, & \mathbf{8'} \to \mathbf{8}, & \mathbf{8} \to \mathbf{8'}, \\ \mathbf{8'} \to \mathbf{8'}, & \mathbf{10} \to \mathbf{10}, & \mathbf{10} \to \mathbf{10}, & \mathbf{27} \to \mathbf{27}. \end{array} \right\} \tag{9.216}$$

are not equal to zero. In other words, the amplitude of the process (9.215) contains eight independent unitary scalars Γ_i, which must be built from products:

$$\overline{B}'^{k}_{l} \overline{P}'^{m}_{n} B^{i}_{p} P^{r}_{j}$$

by means of summation over all the possible pairs of the upper and down indices. Summation over one pair of indices for two matrices means multiplication of these matrices:

$$B^{k}_{l} P^{m}_{k} - (BP)^{m}_{l} \, ,$$

while summation over the second pair of indices gives the track of this product:

$$(BP)^{m}_{m} = \mathrm{Sp}\,(BP) \, .$$

From products of four unitary wavefunctions eight independent invariants, differing from each other by summation order, can be constructed

$$\left. \begin{array}{ll} \Gamma_1 = \mathrm{Sp}(\overline{B}'B)\mathrm{Sp}(\overline{P}'P), & \Gamma_2 = \mathrm{Sp}(\overline{B}'\overline{P}')\mathrm{Sp}(BP), \\ \Gamma_3 = \mathrm{Sp}(\overline{B}'P)\mathrm{Sp}(B\overline{P}'), & \Gamma_4 = \mathrm{Sp}(\overline{B}'B\overline{P}'P), \\ \Gamma_5 = \mathrm{Sp}(\overline{B}'BP\overline{P}'), & \Gamma_6 = \mathrm{Sp}(B\overline{B}'\overline{P}'P), \\ \Gamma_7 = \mathrm{Sp}(B\overline{B}'P\overline{P}'), & \Gamma_8 = \mathrm{Sp}(\overline{B}'\overline{P}'BP) - \mathrm{Sp}(\overline{B}'PB\overline{P}'). \end{array} \right\} \tag{9.217}$$

The number of independent quantities Γ_i decreases, if the theory invariance under the time inversion is taken into account. The process (9.215) contains the same baryon and meson octets in the initial and final states, that is, $B' = B$ and $P' = P$. Consequently, at the T-inversion it turns into itself. Since at the time inversion the initial particles become final ones and vice versa, then the unitary wavefunctions are transformed according to the law:

$$B^{k}_{l} \to \overline{B}'^{l}_{k}, \qquad \overline{B}'^{k}_{l} \to B^{l}_{k}, \qquad P^{k}_{l} \to \overline{P}'^{l}_{k}, \qquad \overline{P}'^{k}_{l} \to P^{l}_{k}. \tag{9.218}$$

---

*This is caused by the orthogonality of representations.

Let us find out, how the Γ_i quantities behave under the T-inversion. We start with the last combination in Eq. (9.217). Since:

$$\mathrm{Sp}(\overline{B}PBP) = \overline{B}_i^k \overline{P}_k^m B_m^j P_j^i \to B_k^i P_m^k \overline{B}_j^m \overline{P}_i^j = \overline{B}_j^m P_m^k B_k^i \overline{P}_i^j = \mathrm{Sp}(\overline{B}PB\overline{P})$$

and

$$\mathrm{Sp}(\overline{B}PB\overline{P}) \to \mathrm{Sp}(\overline{B}PBP),$$

takes place, then under the time inversion Γ_8 reverses its sign and, as a result, proves to be forbidden by the theory T-invariance. It can be shown in a similar way, that all the remaining combinations in (9.217) are T-invariants. Now we can proceed to establishing connections between the cross sections both of elastic and of nonelastic scattering of particles entering into the baryon and meson octets. Consider the processes:

$$\pi^+ + p \to \pi^+ + p, \tag{9.219}$$

$$\pi^- + p \to \pi^0 + n, \tag{9.220}$$

$$\pi^+ + p \to K^+ + \Sigma^+, \tag{9.221}$$

$$K^+ + p \to K^+ + p, \tag{9.222}$$

$$K^- + p \to K^- + p, \tag{9.223}$$

$$\pi^- + p \to \pi^- + p, \tag{9.224}$$

$$K^- + p \to \pi^- + \Sigma^+. \tag{9.225}$$

From the explicit form of the wavefunctions of the baryon and meson octets (see Eqs. (9.193)–(9.195)) it follows, that for the process (9.219) $B_1^3 = \overline{B}_3^1 = P_1^2 = \overline{P}_2^1 = 1$, and all the other components are equal to zero. As a result, from seven independent unitary scalars Γ_i only two are different from zero

$$\Gamma_1 = \Gamma_7 = 1.$$

Consequently, the amplitude of the process (9.219) is defined by the expression:

$$M(\pi^+ p \to \pi^+ p) = \sum_j (T_{1j} + T_{7j}) R_j. \tag{9.226}$$

In an analogous way the expressions for the amplitudes of the processes (9.220)–(9.225) can be obtained. Excluding quantities $T_{ij} R_j$ from them we arrive at the following relations between amplitudes:

$$M(\pi^+ p \to \pi^+ p) - M(\pi^- p \to \pi^- p) = \sqrt{2} M(\pi^- p \to \pi^0 n), \tag{9.227}$$

$$M(K^- p \to K^- p) - M(\pi^- p \to \pi^- p) = M(K^- p \to \pi^- \Sigma^+), \tag{9.228}$$

$$M(K^+ p \to K^+ p) - M(\pi^+ p \to \pi^+ p) = M(\pi^+ p \to K^+ \Sigma^+). \tag{9.229}$$

Eq. (9.227) could be obtained even from the isotopic invariance while Eqs. (9.228) and (9.229) are consequences of only the unitary symmetry. Passing in (9.228)–(9.229) from the amplitudes to the cross sections, we obtain the so-called triangle inequalities:

$$\sqrt{\sigma(K^- p \to \pi^- \Sigma^+)} \geq | \sqrt{\sigma(K^- p \to K^- p)} - \sqrt{\sigma(\pi^- p \to \pi^- p)} |, \tag{9.230}$$

$$\sqrt{\sigma(\pi^+ p \to K^+ \Sigma^+)} \geq | \sqrt{\sigma(\pi^+ p \to \pi^+ p)} - \sqrt{\sigma(K^+ p \to K^+ p)} | . \tag{9.231}$$

It is also possible to demonstrate, that

$$
\left.
\begin{aligned}
M(K^-p \to \pi^+\Sigma^-) &= M(K^-p \to K^0\Xi^0), \\
M(\pi^-p \to K^+\Sigma^-) &= M(\overline{K^0}p \to K^+\Xi^0).
\end{aligned}
\right\}
\tag{9.232}
$$

takes place. This, in its turn, leads to the equalities between the cross sections:

$$
\left.
\begin{aligned}
\sigma(K^-p \to \pi^+\Sigma^-) &= \sigma(K^-p \to K^0\Xi^0), \\
\sigma(\pi^-p \to K^+\Sigma^-) &= \sigma(\overline{K^0}p \to K^+\Xi^0).
\end{aligned}
\right\}
\tag{9.233}
$$

However, the above-mentioned considerations apply only to the idealized case of the exact unitary symmetry. In reality, the $SU(3)$-symmetry is very approximate, as there is sizeable mass splitting in the $SU(3)$-multiplets. Because of this, to compare relations obtained with experiments, symmetry violation must be taken into account. In the first approximation one could allow for only mass splitting, that is, substitutes physical masses into the Lagrangians, but considers the interaction Lagrangians to be $SU(3)$ invariant as before. This procedure leads to a quite good accord with the experiment. The next stage in taking into account unitary symmetry violation, is to introduce corrections to the interaction Lagrangians, which brings in the appearance of additional parameters in the theory. Due to a large number of such parameters, calculations become ineffective.

The wavefunctions of unitary multiplets are eigenfunctions of the Casimir operators $S_a^{(un)2}$ and $S_a^{(un)}D_a$. Thus, every irreducible representation can be marked with the eigenvalues of these operators. Labeling of the $SU(3)$ multiplets with only the eigenvalues of the operator of the unitary spin square is generally accepted. From Eq. (9.186) it is easy to obtain

$$
S_a^{(un)2}D(p,q) = gD(p,q) = [p + q + \frac{1}{3}(p^2 + pq + q^2)]D(p,q).
\tag{9.234}
$$

Then, for the unitary multiplets of the lowest dimensions we have

$$
g =
\begin{cases}
0 & \text{for} & \mathbf{1}, \\
\frac{4}{3} & \text{for} & \mathbf{3}, \mathbf{3}, \\
3 & \text{for} & \mathbf{8}, \\
6 & \text{for} & \mathbf{10}.
\end{cases}
\tag{9.235}
$$

The classification of hadrons according to unitary multiplets resembles one of chemical elements in the Mendeleev periodic table. Like the Mendeleev table, the hadrons classification unostentatiously points up (but again for the experienced mind) to a composite structure of hadrons. In this sense the $SU(3)$-group of the isospin and the hypercharge has served its historical mission. It has prepared all the conditions for the next step up the Quantum Stairway, going out on the quark-lepton level of matter structure. However unlike the Moor,* who had to disappear after his task was realized, the $SU(3)$-symmetry, as we shall see later, settles down firmly in physics of the strong interaction. True, it must slightly change its role.

---

*Here a personage of the *Othello* tragedy is born in mind.

10

Quark "atoms"

It was now the nightingale's turn. But he came a cropper at the very first note. He sang of a servant's delight at God having sent him a master, of noble eagles that never grudged servants money for vodka. In a word, the nightingale did his best to sing in the servile vein, but try as he would, the "art" that lived in him refused to be curbed. The bird himself was servile from beak to tail (he even picked up a second-hand necktie somewhere and curled his hair) but his "art" just could not be kept within the bounds of servility and broke through again and again.

M.E. Saltykov Shchedrin, "The Eagle-Patron of Arts"

10.1 Hypothesis of fundamental triplets

In 1964 M. Gell-Mann [29] and G. Zweig [30] independently from each other hypothesized that all hadrons are built of three particles of the unitary triplet. Wishing to emphasize the unusual properties of new blocks of matter, Gell-Mann called them "quarks." The term was borrowed from *Finnegans Wake* by James Joyce. If one compares this novel with *War and Peace* by L. Tolstoy, then the first thing that comes to mind is that *Finnegans Wake* was written in at least 26-dimensional space-time, which never experienced the joy of compactification. During the act a protagonist Humphrey Chimpden Earwicher is constantly changing his appearance. He is reincarnated at one moment into Mark, the King of Cornwell, at another into his sons, Sham and Shaun, and so on. Earwicher's children (he has a daughter as well) are far from being simple and they also can be transformed into their father. There is an episode in the novel when the protagonist being reincarnated into King Mark, sends his nephew, the knight Tristan, by a wedding boat to bring the king's bride Isolde. As expected, during their travels Tristan and Isolde happen to have been struck down by Cupid's arrow practically on the spot. The seagulls circling above the ship are completely informed about the events taking place on the ship, as demonstrated by their song starting with the words: "Three quarks for Mister Mark." If one distracts from the remaining part of the song, then the above-mentioned phrase can be unambiguously viewed as a prediction of the fundamental triplet. Setting the imagination free, one can assume that Mister Earwicher with his transformations chain reproduces the hadrons spectrum and his children are nothing else but three quarks. However, the whole content of the bird's opus suggests that the phrase "three quarks" may be treated as evidence that the old king was deceived triply. Time will show whether that is the excited indication on the analogous fate of the quark hypothesis, the development of which physicists have been devoted to already for half a century.

Earlier we demonstrated how to build singlets, octets, and decouplets from the unitary triplets with precisely the same isotopic structure as experimentally observed unitary multiplets of hadrons. It appeared that the hypothesis of the fundamental triplets was obvious even for philosophers who are constantly loitering around the building of modern physics and whose basic activities consist in abusing the terminology specially invented by just them. However, such models were constantly neglected due to the inertia of thinking because they demanded fragmentation of the electric and baryon charges of particles entering into the fundamental triplet.

Originally a quark family included only three particles and three corresponding antiparticles. Two fundamental triplets, which we denote by 3 and $\bar{3}$, had the following form:

$$q = \begin{pmatrix} q_1 \\ q_2 \\ q_3 \end{pmatrix} = \begin{pmatrix} u \\ d \\ s \end{pmatrix}, \qquad \bar{q} = (\bar{q}^1, \bar{q}^2, \bar{q}^3) = (\bar{u}, \bar{d}, \bar{s}), \qquad (10.1)$$

where symbols u, d, and s ("up," "down", and "strange") are used for components of the quark triplet q_α ($\alpha = 1, 2, 3$). It is also possible to say, that a quark dwells in three aroma (flavor) states and attribute the flavor meaning to the index α. u- and d-quark form the isotopic doublet ($I_3^u = 1/2$ and $I_3^d = -1/2$), while s-quark forms the isotopic singlet. Only the s-quark has a nonzero strangeness ($s = -1$). Since we are going to construct particles with arbitrary spin values, then the quarks must have spin $1/2$. We also ascribe to all the three quarks the baryon charge $1/3$. Then it becomes obvious, that in the unitary spin plane the quark and antiquark triplets are denoted by the same triangles as the fundamental triplet and antitriplet of the $SU(3)$-group, respectively (see Figs. 10.1). In its turn, from the

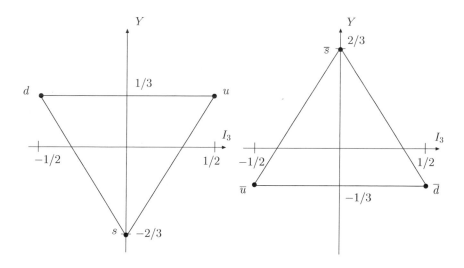

FIGURE 10.1
The quark and antiquark triplets.

Gell-Mann-Nishijima formula it follows that the quark charges in units of $|e|$ are fractional quantities

$$Q_u = 2/3, \qquad Q_d = Q_s = -1/3.$$

Further, we should define the quark contents of hadrons. It is evident, that mesons must contain an even number of quarks, while baryons must contain an odd number of quarks.

Let us assume that all mesons are built from quark-antiquark pairs

$$M_i^k = \bar{q}_i q^k,$$

and all baryons are built from three quarks

$$B^{ikl} = q^i q^k q^l.$$

Now hadrons made of the quarks should be placed into the corresponding unitary multiplets. In our forthcoming design activity we are going to use the composition law of the unitary spins as our basic instrument. This law must be a generalization of the composition rule for the ordinary spins (2.23). Let us formulate this law in such a way that it become applicable to any spin, whether it is the ordinary, isotopic, unitary spin, and so on. Thus, to obtain all the possible spin states of a coupled two-particle system, it is necessary:

(i) to superpose the center of a spin diagram of the first particle on every state of a spin diagram of the second particle;
(ii) to mark the states obtained;
(iii) to single out diagrams with the equal spin value out of all states.

Let us begin the hadrons construction process with a meson sector. Making the center of diagram $\bar{3}$ coincident with every out of three states of the diagram 3, we obtain nine possible states of the $\bar{3}3$ system (Fig. 10.2). These states are separated into the $SU(3)$-octet and

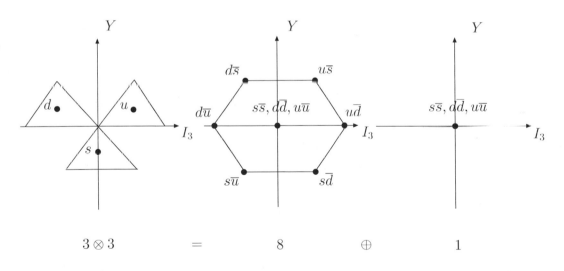

$$3 \otimes 3 \qquad = \qquad 8 \qquad \oplus \qquad 1$$

FIGURE 10.2
The meson multiplets.

the $SU(3)$-singlet. Three of them, namely, I, II, III being linear combinations $u\bar{u}$, $d\bar{d}$, and $s\bar{s}$ falls in one and the same point of the diagram with $I_3 = Y = 0$. Consequently, they demand our great attention. The unitary singlet I is nothing else but

$$I = \text{const} \left(\bar{u}, \bar{d}, \bar{s} \right) \begin{pmatrix} u \\ d \\ s \end{pmatrix},$$

where const is defined from the normalization condition. It gives

$$I = \frac{1}{\sqrt{3}} \left(u\bar{u} + d\bar{d} + s\bar{s} \right). \tag{10.2}$$

Two remaining states belong to the octet. We assign the state II to the isospin triplet, whose contents could be reconstructed by the replacement $p \to u$ and $n \to d$ in the formulas (9.62). So,

$$(d\bar{u}, II, -u\bar{d}),$$

where

$$II = \frac{1}{\sqrt{2}} \left(u\bar{u} - d\bar{d} \right).$$

Next, we find that the isospin singlet, the state III, being orthogonal to I and II is defined by the expression:

$$III = \frac{1}{\sqrt{6}} \left(u\bar{u} + d\bar{d} - 2s\bar{s} \right).$$

The spin J of a composite meson is defined by

$$\mathbf{J} = \mathbf{S} + \mathbf{L},$$

where $\mathbf{S}$ is the spin of a $(q\bar{q})$-system being equal 1 or 0, $\mathbf{L}$ is a relative orbital moment of quarks. Since q and $\bar{q}$ have opposite internal parities, the meson parity is determined by the expression

$$P = -(-1)^l,$$

where l is an orbital quantum number and we have taken into account that under the space inversion $\theta \to \pi - \theta$, $\varphi \to \varphi + \pi$ the wavefunction angular part $Y_l^{m_l}(\theta, \varphi)$ of a $q\bar{q}$-pair acquires the factor $(-1)^l$. It should also take into consideration that meson multiplets possess the definite charge parity (the $q\bar{q}$-system is neutral). To find a proper value of the charge conjugation operator C we must carry out the replacement $q \leftrightarrow \bar{q}$, interchange the quarks, and exchange their spin directions. This operation results in

$$C = -(-1)^{s+1}(-1)^l = (-1)^{s+l},$$

where s is a spin quantum number and the common minus sign is connected with the permutation of fermions, while the factor $(-1)^{s+1}$ is associated with the symmetry properties of the $q\bar{q}$-pair spin states (see Eqs. (9.62) (9.63)). In such a way under $l = 0, 1$ the three quarks model predicts the existence of meson octets and singlets with the following quantum numbers

$$l = 0, \qquad j = 0, 1 \to \left. \begin{matrix} 0^{-+} \\ 1^{--} \end{matrix} \right\}$$

$$l = 1, \qquad j = 0, 1, 2 \to \left. \begin{matrix} 1^{+-} \\ 2^{++} \\ 1^{++} \\ 0^{++} \end{matrix} \right\}$$

where the symbol J^{PC} is used for marking the states. The states with $l = 0$ can be viewed as the basic ones, while the states with $l > 0$ can be viewed as the orbital excitations. It should be stressed that the quark filling of meson nonets with different values of J^{PC} is the same. Below we give the meson nonet with $J^{PC} = 0^{-+}$ consisting of the octet

$$\left.\begin{array}{lll} K^0 = d\bar{s}, \qquad K^+ = u\bar{s}, \qquad \pi^+ = u\bar{d}, \qquad \overline{K}^0 = s\bar{d}, \\[2mm] K^- = s\bar{u}, \qquad \pi^- = d\bar{u}, \qquad \pi^0 = \dfrac{1}{\sqrt{2}}\left(u\bar{u} - d\bar{d}\right), \\[2mm] \eta = \dfrac{1}{\sqrt{6}}\left(u\bar{u} + d\bar{d} - 2s\bar{s}\right). \end{array}\right\} \tag{10.3}$$

and the singlet

$$\eta' = \sqrt{\frac{1}{3}}\left(u\bar{u} + d\bar{d} + s\bar{s}\right). \tag{10.4}$$

So, all the discovered mesons can be placed into the $SU(3)$-multiplets $q\bar{q}$. Up to 1971, when the first reliable data confirming the existence of quarks into hadrons were obtained, such tests have been the main argument in favor of the quark hypothesis.

We are coming now to the investigation of the quark structure of baryons. First, we combine two quarks. We plot the already familiar to us result

$$3 \bigotimes 3 = 6 \bigoplus \bar{3}$$

in Fig. 10.3. We remind that the sextet 6 is symmetric with respect to transposition of two

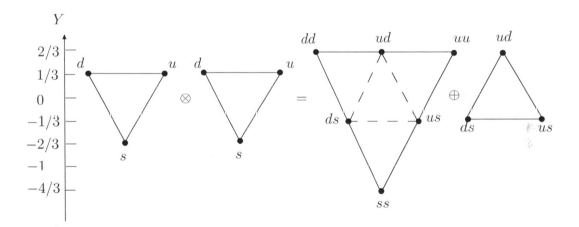

FIGURE 10.3
The graphical interpretation of the formula $3 \bigotimes 3 = 6 \bigoplus \bar{3}$.

quarks while the triplet $\bar{3}$ is antisymmetric. In Fig. 10.3 we specify only the quark filling in every point of the diagram. For the wavefunction of particles to be exactly defined we must take into account the symmetry properties of multiplets. So, the state ud belonging to the sextet is described by:

$$\psi_s(ud) = \frac{1}{\sqrt{2}}\left(ud + du\right),$$

whereas the wavefunction of the analogous state in the triplet has the form:

$$\psi_t(ud) = \frac{1}{\sqrt{2}}\left(ud - du\right).$$

Let us add one more quark. The final result of decomposition

$$3 \otimes 3 \otimes 3 = (6 \oplus \bar{3}) \otimes 3 = (6 \otimes 3) \oplus (\bar{3} \otimes 3) = 10 \oplus 8 \oplus 8 \oplus 1, \qquad (10.5)$$

is displayed in Fig. 10.4, where the octet following the decuplet appears under summarizing the unitary spins of the sextet and the triplet, while the second octet comes from $\bar{3} \otimes 3$.

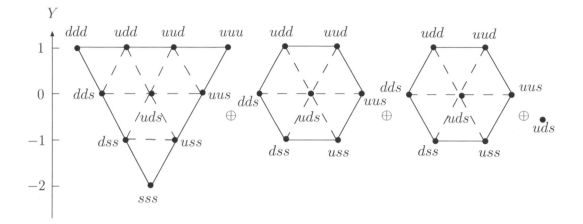

FIGURE 10.4

The graphical interpretation of the formula $3 \otimes 3 \otimes 3 = 10 \oplus 8 \oplus 8 \oplus 1$.

Eqs. (9.164), (9.165), (9.168), and (9.169) prove to be useful under finding the wavefunctions of the multiplets corresponding to the division (10.5). These equations could be presented in the following form:

$$B^{ikl}(\mathbf{10}) = \frac{1}{6} \left(q^i q^k q^l + q^i q^l q^k + q^l q^i q^k + q^l q^k q^i + q^k q^l q^i + q^k q^i q^l \right), \qquad (10.6)$$

$$B^{ikl}(\mathbf{1}) = \frac{1}{6} \left(q^i q^k q^l - q^i q^l q^k + q^l q^i q^k - q^l q^k q^i + q^k q^l q^i - q^k q^i q^l \right), \qquad (10.7)$$

$$B_{i[kl]}(\mathbf{8}) = \frac{1}{3} \left(q^i q^k q^l + q^k q^i q^l - q^i q^l q^k - q^l q^i q^k \right), \qquad (10.8)$$

$$B_{[ik]l}(\mathbf{8}) = \frac{1}{3} \left(q^i q^k q^l + q^i q^l q^k - q^k q^i q^l - q^k q^l q^i \right), \qquad (10.9)$$

which is more convenient for practical use. For the multiplets with mixed symmetry we are constrained by the indication of only antisymmetry on indices and it was done with the help of the square brackets. The choice of the quantities $B^{[ik]l}(\mathbf{8})$ and $B^{i[kl]}(\mathbf{8})$ is not unique, that is, they are ambiguous. Moreover, one may be convinced by the simple check that $B^{[ik]l}(\mathbf{8})$ and $B^{i[kl]}(\mathbf{8})$ are not orthogonal to each other. It is evident that this circumstance could become a source of trouble at a later time. Let us correct this deficiency. With the help of Jacobi identities:

$$B^{i[kl]}(\mathbf{8}) + B^{l[ik]}(\mathbf{8}) + B^{k[li]}(\mathbf{8}) = 0, \qquad (10.10)$$

$$B^{[ik]l}(\mathbf{8}) + B^{[li]k}(\mathbf{8}) + B^{[kl]i}(\mathbf{8}) = 0, \qquad (10.11)$$

one may build the mutually orthogonal representations:

$$B^{ikl}(8) = \frac{2}{3\sqrt{3}} \left(q^i q^k q^l + q^i q^l q^k + q^k q^i q^l + q^k q^l q^i - 2q^l q^i q^k - 2q^l q^k q^i \right), \tag{10.12}$$

$$B'^{ikl}(8) = \frac{1}{3} \left(q^i q^k q^l - q^i q^l q^k + q^k q^i q^l - q^k q^l q^i \right). \tag{10.13}$$

Notice, that the constant quantities standing in front of the parenthesis are not the normalizing coefficients. They are no more than the coefficients of splitting $B^{\alpha\beta\gamma}$ into irreducible tensors, that is,

$$B^{ikl} = \frac{1}{6} \left(B^{ikl}(10) + B^{ikl}(1) \right) + \frac{2}{3\sqrt{3}} B^{ikl}(8) + \frac{1}{3} B'^{ikl}(8).$$

To put this another way, to obtain the unitary spin wavefunctions from the quantities (10.6), (10.7), (10.12), and (10.13) one should additionally allow for the normalization condition. As an example, we use Eqs. (10.7), (10.12), and (10.13) in order to get the wavefunctions of baryons with the quark filling udd ($i = 1$, $k = l = 2$). These functions have the form:

$$\psi(udd)(10) = \frac{1}{\sqrt{3}} \left(udd + dud + ddu \right), \tag{10.14}$$

for the decuplet,

$$\psi(udd)(8) = \frac{1}{\sqrt{6}} \left(2udd - dud - ddu \right), \tag{10.15}$$

for the first octet, and

$$\psi'(udd)(8) = \frac{1}{\sqrt{2}} (dud - ddu), \tag{10.16}$$

for the second octet.

We denote the relative angular moment of two quarks by $\mathbf{L}_1$ and the angular moment of the third quark as related to the masses center of the first two quarks by $\mathbf{L}_2$. Then the total angular moment of three quarks is determined by the expression:

$$\mathbf{L} = \mathbf{L}_1 + \mathbf{L}_2.$$

Assuming the quark parities to be positive we find that for the low-laying baryon states ($\mathbf{L} = \mathbf{L}_1 = \mathbf{L}_2 = 0$) the parity is equal to $+1$. According to the vector summation rules the resulting spin of the three quarks may equal either $3/2$ or $1/2$. Thus, the low-laying baryons multiplets are characterized by the values $3/2^+$ and $1/2^+$. Baryons having the lowest mass values are precisely placed into the decuplet $3/2^+$ and the octet $1/2^+$. Heavier baryons must have $l = 1$, that is, either $l_1 = 1$ and $l_2 = 0$ or $l_1 = 0$ and $l_2 = 1$ (in both cases their parity is negative -1). And really, existing baryon resonances are finely placed into the following multiplets: singlets and decuplets with $\frac{1}{2}^-$ and $\frac{3}{2}^-$, octets with $\frac{1}{2}^-$, $\frac{3}{2}^-$ and $\frac{5}{2}^-$.

10.2 X-ray photography of nucleons

Today, building the quark model of hadrons could be viewed as some kind of analog of establishing the nucleon structure of atomic nuclei. However, initially the quark model was considered the only formal scheme that was very convenient to systematize hadrons. At the end of the 1960s physicists had obtained, at their disposal, new possibilities for investigating the structure of hadrons. The created sources of high-energy electrons probed distances up to 10^{-15} cm, that is, on two orders smaller than the hadron size. To use electrons is convenient for two reasons. First, they are structureless particles and, second, they do not participate in a strong interaction. Since the electromagnetic interaction of point particles has been studied thoroughly, the theoretical analysis of the experimental results is greatly facilitated. At that time it was already known that hadrons have specific structures that are described by electromagnetic formfactors. By formfactors we understand the function characterizing the space distribution of the electric charge and multipole moments inside hadrons (further, for the sake of simplicity, we shall talk about the magnetic dipole moment only). Thus, carrying out the "Roentgen" of a proton with the help of scattered electrons one may investigate the proton structure more carefully and, by doing so, one gets down to the experimental checkout of the quark hypothesis.

The first stage in similar kinds of investigations consists of the analysis of elastic scattering. Consequently, we should start with the reaction

$$e^- + p \to \gamma^*, Z^* \to e^- + p, \tag{10.17}$$

where the intermediate step in (10.17) means that interaction is performed by exchanges of the virtual photon and Z-boson. For the sake of simplicity, we shall take into account the photon exchange only and shall be constrained by the second order of the perturbation theory. So, our task is to define a differential cross section of the process (10.17).

To calculate cross sections of elementary particle processes in quantum field theory a method of Feynman diagrams [30] exists. The successive presentation of this method will be given in Part IV. Here we set forth some preliminary information to understand the calculation technique.

Feynman diagrams, or Feynman graphs, outwardly resemble displays of trajectories of all particles taking part in an interaction. The classical description, however, is not applicable in this case, so that interpretation of graphs as classical trajectories is incorrect, one can speak only of outer similarity. By the example of nonrelativistic quantum mechanics we became aware of the fact, that only in the case of the most simple potentials we managed to obtain exact solutions for quantum systems. For this reason, the perturbation theory method is the main mathematical method in the quantum theory. The essence of the perturbation theory method is most transparent in the Green function formalism. So, if a system wavefunction in an initial state $\psi(\mathbf{r}_i, t_i)$ is known, then its value at an arbitrary moment of time $t = t_f$ in the point with coordinates $\mathbf{r}_f$ is given by the expression:

$$\psi(\mathbf{r}_f, t_f) = \int G(\mathbf{r}_f - \mathbf{r}_i, t_f - t_i)\psi(\mathbf{r}_i, t_i)d^4x_i. \tag{10.18}$$

The Green function in nonrelativistic quantum mechanics is represented by the series in the perturbation $H_{int}(\mathbf{r}, t)$

$$G(\mathbf{r}_f - \mathbf{r}_i, t_f - t_i) = G^{(0)}(\mathbf{r}_f - \mathbf{r}_i, t_f - t_i) -$$

$$-\frac{i}{\hbar}\int G^{(0)}(\mathbf{r}_f - \mathbf{r}_1, t_f - t_1)H_{int}(\mathbf{r}_1, t_1)G^{(0)}(\mathbf{r}_1 - \mathbf{r}_i, t_1 - t_i)d^4x_1 +$$

$$+\left(-\frac{i}{\hbar}\right)^2 \int d^4x_1 \int G^{(0)}(\mathbf{r}_f - \mathbf{r}_1, t_f - t_1)H_{int}(\mathbf{r}_1, t_1)G^{(0)}(\mathbf{r}_1 - \mathbf{r}_2, t_1 - t_2)H_{int}(\mathbf{r}_2, t_2) \times$$

$$\times G^{(0)}(\mathbf{r}_2 - \mathbf{r}_i, t_2 - t_i)d^4x_2 + ..., \tag{10.19}$$

where $G^{(0)}(\mathbf{r}, t)$ is the Green function in zero approximation. The expression (10.19) brings us to the idea that the evolution of a quantum system can be plotted as a chain of transitions from the initial state to the final one through the totality of points (vertices) corresponding to interaction acts. In so doing the number of vertices corresponds to the perturbation theory order. These are nonrelativistic prerequisites for appearance of the Feynman diagram method. This method is also unambiguously connected with the step-by-step inclusion of every order of the perturbation theory. The diagrams topology is completely defined by the form of the interaction Lagrangian. By the aspect of the Feynman diagram, it is immediately possible to write down the corresponding expression for a scattering amplitude, that is, there is a set of rules, the so-called Feynman rules connecting any diagram element with particle characteristics. To depict a free particle one or another line is introduced (which is, of course, only a graphic symbol of particles propagation), while lines knot (vertex) corresponds to particles interaction. External lines describe real initial and final particles and internal ones do virtual particles. Since particles in initial and final states are considered to be free (it is guaranteed by experiment conditions), then real particles correspond to wavefunctions, which are solutions of free equations of corresponding fields. Virtual particles, describing interaction, are put in correspondence of the Green functions of the equations, which they would satisfy, if they were real. These functions will be referred to the propagators. A coupling constant, characterizing the given interaction, must be in every diagram vertex. In the case of the QED it equals $\sqrt{e^2/(4\pi\hbar c)}$. Moreover, in every vertex the momentum conservation law must be taken into consideration (interaction, taking place at a vertex, can take place at any space point, so that $\Delta x_i = \infty$, and it means, that a momentum is precisely defined). In every vertex of "charge" (electric, baryon, lepton, strange, and so on) conservation laws that are valid for the given interaction, must be fulfilled. The Feynman diagrams can be plotted in the coordinate and momentum spaces, in the latter case the four-dimensional momentum of a corresponding particle is ascribed to every line. Let us display bosons with a wavy line, fermions with a solid line. Sometimes a symbol of a particle will be also mentioned near every line. We agree to direct time axes from the left to the right in all the diagrams. In fermion lines and ones corresponding to charged bosons we put arrows that denote a particle propagation direction. Arrows, directed to the time axis denote particles, while antiparticles are indicated by arrows being opposite to the time axis direction, that is, in the Feynman diagram formalism antiparticles are moving backwards in time. The easiest way to understand that is to address the holes theory that was proposed by Dirac to overcome difficulties connected with the appearance of negative energies. In this theory the states with the positive energy and whose dependence on time has the form $\exp(-iEt)$, are identified with the electrons, while the negative energy states with the wavefunctions $\psi \sim \exp(iEt)$ are identified with the positrons. It is obvious that the change of the time direction leads to changing the sign in the exponent $\exp(iEt)$, that is, to solutions with the positive energy.

In the QED the interaction Hamiltonian is given by the expression:

$$\mathcal{H}_{int} = e\overline{\psi}_l(x)\gamma_\mu\psi_l(x)A^\mu(x), \tag{10.20}$$

($l = e, \mu, \tau$), that is, $\mathcal{H}int$ has the trilinear structure. It means that in every vertex three lines are encountered, namely, two fermion (electron and/or positron) and one photon lines. As this takes place, the electric charge conservation law demands direction invariability of the fermion line over its length.

The Feynman diagram corresponding to the process (10.17) is given in Fig. 10.5. The circle in the vertex, describing the interaction of the virtual photon with the proton, stresses the circumstance that the proton, unlike the electron, has an internal structure. When the proton were the point particle with the charge $|e|$ and the magnetic moment $e\hbar/2m_p c$

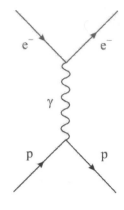

FIGURE 10.5

The Feynman diagram for the process $e^- p \to \gamma^* \to e^- p$.

predicted by the Dirac theory, then the cross section of the elastic electron-proton scattering would follow from the cross section of the process

$$e^- + \mu^+ \to e^- + \mu^+, \tag{10.21}$$

under changing the muon mass by the proton one. It proves to be convenient to calculate the cross section of the reaction (10.21) and then, using the obtained result, define the cross section of the elastic electron-proton scattering taking into consideration the proton composite structure.

Before proceeding to researching the reaction (10.21), let us agree about the choice of measurement units. The reasonably chosen units comprise a convenient tool under the description of a definite phenomena region. The fundamental constants of the quantum field theory (QFT) $\hbar$ and c enter into the majority of theory equations. One may "unload" the QFT formulae when one sets these constants to be equal to 1, that is, one chooses the action quantum $\hbar$ as an action unit and does the light velocity c as a velocity unit. The system of units with $\hbar = c = 1$ derives the name of the natural system of units (NSU). The description of the NSU is given in Appendix 1. Further, unless otherwise specified, we shall use the NSU over the course of the book. It is self-evident that when comparing between theory and experiment we should pass to one of an ordinary system of units in formulae obtained.

According to the formulae to be obtained in Section 18, in the momentum representation the differential cross section of the process (10.21) is given by the expression

$$d\sigma = |\mathcal{M}_{if}|^2 \frac{d^3 p_f^{(e)} d^3 p_f^{(\mu)}}{(2\pi)^2 |\mathbf{v}^{(e)}|} \delta^{(4)}(p_i^{(\mu)} + p_i^{(e)} - p_f^{(\mu)} - p_f^{(e)}), \tag{10.22}$$

where j_0 is an initial density of the flux of particles participating in the reaction,[*] $p_i^{(e)}$ and $p_i^{(e)}$ ($p_f^{(\mu)}$ and $p_f^{(\mu)}$) are four-dimensional momenta of initial and final electrons (muons), and $\mathcal{M}_{i \to f}$ is a matrix element defined by the Feynman rules

$$\mathcal{M}_{i \to f} = \left(\overline{\psi}(p_f^{(e)}) \sqrt{\frac{m_e}{E_f^{(e)}}} \right) (-ie\gamma^\nu) \left(\psi(p_i^{(e)}) \sqrt{\frac{m_e}{E_i^{(e)}}} \right) \left[\frac{-ig_{\nu\tau}}{(p_f^{(e)} - p_i^{(e)})^2} \right] \times$$

---

[*]In the lab frame (the initial muon is in rest) $j_0 = |\mathbf{v}^{(e)}|$ ($\mathbf{v}^{(e)}$ is an initial electron velocity).

$$\times \left(\overline{\psi}(p_i^{(\mu)}) \sqrt{\frac{m_\mu}{E_i^{(\mu)}}} \right) (-ie\gamma^\tau) \left(\psi(p_f^{(\mu)}) \sqrt{\frac{m_\mu}{E_f^{(\mu)}}} \right). \qquad (10.23)$$

Let us elucidate the origin of the factors belonging to $\mathcal{M}_{i \to f}$. The factors $\overline{\psi}(p_f)\sqrt{m/E_f}$ and $\psi(p_i)\sqrt{m/E_i}$ are associated with the leptons in the final and initial states, respectively. The factors $-ie\gamma_\nu$ and $-ie\gamma_\tau$ correspond to two vertices in which the electromagnetic interaction of the point-like leptons occurs; ν is the polarization index of the photon created in the electron vertex, τ is the polarization index of the photon annihilated in the muon vertex. The factor standing in the square bracket describes the virtual photon propagation between the electron and muon vertices. The order of writing the spinor matrices in (10.23) is determined by the direction of the diagram detour, which is opposite to the fermion lines direction, that is, the diagram detour is performed from the final fermion state to the initial one (for antifermions, on the contrary).

When we are not interested by the particles polarization in the initial and final states, then in $\mid \mathcal{M}_{if} \mid^2$ one should fulfill averaging over the initial polarizations and summing over the final ones. It means that we must pass to the quantity

$$\frac{1}{4} \sum_{i,f} |\mathcal{M}_{if}|^2 = g_0 \sum_{i,f} \mid [\overline{\psi}_\alpha(p_f^{(e)})(\gamma^\nu)_{\alpha\beta}\psi_\beta(p_i^{(e)})][\overline{\psi}_m(p_f^{(\mu)})(\gamma_\nu)_{mn}\psi_n(p_i^{(\mu)})] \mid^2, \qquad (10.24)$$

where

$$g_0 = \frac{c^4 m_e^2 m_\mu^2}{4q^4 E_i^{(\mu)} E_i^{(e)} E_f^{(\mu)} E_f^{(e)}},$$

$q_\nu = (p_i^{(e)} - p_f^{(e)})_\nu = (p_f^{(\mu)} - p_i^{(\mu)})_\nu$ and the factor $1/4$ appears at the cost of averaging over polarizations of the initial electron and muon. Since in Eq. (10.24) the matrix indices, which define the multiplication order, are represented in explicit form, then the factors, entering into this expression, may be interchanged arbitrarily. Grouping the electron and muon parties separately, we obtain

$$\frac{1}{4} \sum_{i,f} |\mathcal{M}_{if}|^2 - g_0 \sum_{i,f} [\overline{\psi}_\alpha(p_f^{(e)})(\gamma^\nu)_{\alpha\beta}\psi_\beta(p_i^{(e)})\overline{\psi}_{\alpha'}(p_i^{(e)})(\gamma^\tau)_{\alpha'\beta'}\psi_{\beta'}(p_f^{(e)})] \times$$

$$\times [\overline{\psi}_m(p_f^{(\mu)})(\gamma_\nu)_{mn}\psi_n(p_i^{(\mu)})\overline{\psi}_{m'}(p_i^{(\mu)})(\gamma_\tau)_{m'n'}\psi_{n'}(p_f^{(\mu)})] =$$

$$= g_0 \sum_{i,f} [\psi_{\beta'}(p_f^{(e)})\overline{\psi}_\alpha(p_f^{(e)})(\gamma^\nu)_{\alpha\beta}\psi_\beta(p_i^{(e)})\overline{\psi}_{\alpha'}(p_i^{(e)})(\gamma^\tau)_{\alpha'\beta'}] \times$$

$$\times [\psi_{n'}(p_f^{(\mu)})\overline{\psi}_m(p_f^{(\mu)})(\gamma_\nu)_{mn}\psi_n(p_i^{(\mu)})\overline{\psi}_{m'}(p_i^{(\mu)})(\gamma_\tau)_{m'n'}]. \qquad (10.25)$$

To summarize over the spin lepton states (the factor $1/2$ converts such summarizing into averaging) is fulfilled with the help of the relations

$$\sum \psi_\alpha^\epsilon(p)\overline{\psi}_\beta^\epsilon(p) = \left(\frac{\epsilon\hat{p} + m}{2m} \right)_{\alpha\beta}, \qquad (10.26)$$

where $\hat{p} = \gamma_\nu p^\nu$, $\epsilon = 1$ for particles and $\epsilon = -1$ for antiparticles (the derivation of (10.26) will be done in Part III). Thus, Eq. (10.25) may be put down as:

$$\frac{1}{4} \sum_{i,f} |\mathcal{M}_{if}|^2 = g_0 L^{\nu\tau} M_{\nu\tau}, \qquad (10.27)$$

where we shall call $L^{\nu\tau}$ and $M_{\nu\tau}$ as the electron and muon tensor, respectively. For the electron tensor we have

$$L^{\nu\tau} = \sum_{i,f} \psi_{\beta'}(p_f^{(e)})\overline{\psi}_\alpha(p_f^{(e)})(\gamma^\nu)_{\alpha\beta}\psi_\beta(p_i^{(e)})\overline{\psi}_{\alpha'}(p_i^{(e)})(\gamma^\tau)_{\alpha'\beta'} =$$

$$= \left(\frac{\hat{p}_f^{(e)} + m_e}{2m_e}\right)_{\beta'\alpha} (\gamma^\nu)_{\alpha\beta} \left(\frac{\hat{p}_i^{(e)} + m_e}{2m_e}\right)_{\beta\alpha'} (\gamma^\tau)_{\alpha'\beta'} =$$

$$= \frac{1}{4m_e^2}\mathrm{Sp}[(\hat{p}_f^{(e)} + m_e)\gamma^\nu(\hat{p}_i^{(e)} + m_e)\gamma^\tau] = \frac{1}{4m_e^2}\mathrm{Sp}[\gamma^\nu(\hat{p}_i^{(e)} + m_e)\gamma^\tau(\hat{p}_f^{(e)} + m_e)], \quad (10.28)$$

where the symbol "Sp" means the operation of taking the matrices track and the cyclicity property

$$\mathrm{Sp}[ABCD] = \mathrm{Sp}[DABC] = \mathrm{Sp}[CDAB] = \mathrm{Sp}[BCDA]$$

has been allowed for. As far as the muon tensor is concerned, the analogous operations give

$$M_{\nu\tau} \to L_{\nu\tau}(p^{(e)} \to -p^{(\mu)}, m_e \to m_\mu). \quad (10.29)$$

The track in the expressions (10.28) and (10.29) could be found by the application of the formulae

$$\mathrm{Sp}(\gamma^\mu\gamma^\nu) = 4g^{\mu\nu}, \qquad \mathrm{Sp}(\gamma^\mu\gamma^\nu\gamma^\lambda\gamma^\sigma) = 4(g^{\mu\nu}g^{\lambda\sigma} + g^{\mu\sigma}g^{\nu\lambda} - g^{\mu\lambda}g^{\nu\sigma}) \quad (10.30)$$

and by considering the fact, that the track of the odd number of the γ-matrices is equal to zero. The calculations give

$$L^{\nu\tau} = [g^{\nu\tau}(m_e^2 - p_i^{(e)}p_f^{(e)}) + p_i^{(e)\nu}p_f^{(e)\tau} + p_i^{(e)\tau}p_f^{(e)\nu}]. \quad (10.31)$$

Multiplying the lepton tensors and neglecting the electron mass, we obtain for (10.27) the following expression in the lab frame:

$$\frac{1}{4}\sum_{i,f}|\mathcal{M}_{if}|^2 = \frac{e^4}{2q^4 E_i^{(\mu)} E_i^{(e)} E_f^{(\mu)} E_f^{(e)}} \left[m_\mu^2(q^2 + 4E_i^{(e)} E_f^{(e)}) - \right.$$

$$\left. -q^2 m_\mu(E_i^{(e)} - E_f^{(e)})\right] = \frac{2e^4 m_\mu^2}{q^4 E_i^{(\mu)} E_f^{(\mu)}} \left(\cos^2\frac{\theta}{2} - \frac{q^2}{2m_\mu^2}\sin^2\frac{\theta}{2}\right), \quad (10.32)$$

where θ is the angle between the momenta of the impacting and outgoing electrons.

Under the deduction of (10.32) we have made use of the relations

$$q^2 = (p_i^{(e)} - p_f^{(e)})^2 \approx -2(p_f^{(e)}p_i^{(e)}) \approx$$

$$\approx -2E_f^{(e)} E_i^{(e)}(1 - \cos\theta) = -4E_f^{(e)} E_i^{(e)}\sin^2\frac{\theta}{2} = -\frac{4E_i^{(e)2}\sin^2\frac{\theta}{2}}{1 + \frac{2E_i^{(e)}}{m_\mu}\sin^2\frac{\theta}{2}}, \quad (10.33)$$

$$q^2 = -2(qp_i^{(\mu)}) = -2(E_i^{(e)} - E_f^{(e)})m_\mu. \quad (10.34)$$

The latter is the consequence of the operation of taking the square of the identity $q + p_i^{(\mu)} = p_f^{(\mu)}$.

Let us substitute (10.32) into the expression for the differential cross section (10.22). The presence of the delta functions presents an opportunity to make the integration over the three-dimensional momenta of the final particles. Taking into consideration the delta function property

$$\delta(x^2 - a^2) = \frac{1}{2a}[\delta(x - a) + \delta(x + a)], \tag{10.35}$$

and accounting that the muon is on the mass shell and its energy is positive, we can rewrite the integral over $\mathbf{p}_f^{(\mu)}$ for the cross section part, depending on $\mathbf{p}_f^{(\mu)}$, in the form:

$$I_{p_f^{(\mu)}} = \int \frac{d^3 p_f^{(\mu)}}{2E_f^{(\mu)}} \delta^{(4)}(p_i^{(\mu)} + q - p_f^{(\mu)}) = \int d^3 p_f^{(\mu)} dE_f^{(\mu)} \delta^{(4)}(p_i^{(\mu)} + q - p_f^{(\mu)}) \theta(E_f^{(\mu)}) \delta(p_f^{(\mu)2} - m_\mu^2), \tag{10.36}$$

where

$$\theta(x) = \begin{cases} 1, & \text{at } x > 0, \\ 0, & \text{at } x < 0. \end{cases}$$

Further calculations of $I_{p_f^{(\mu)}}$ are trivial and the final result is given by the expression:

$$I_{p_f^{(\mu)}} = \delta\left((p_i^{(\mu)} + q)^2 - m_\mu^2\right) = \delta\left(2p_i^{(\mu)}q + q^2\right) = \frac{1}{2m_\mu}\delta\left(E_i^{(e)} - E_f^{(e)} + \frac{q^2}{2m_\mu}\right) =$$

$$- \frac{1}{2m_\mu}\delta\left(E_i^{(e)} \quad E_f^{(e)} - \frac{2E_i^{(e)}E_f^{(e)}}{m_\mu}\sin^2\frac{\theta}{2}\right) = \frac{1}{2bm_\mu}\delta\left(E_f^{(e)} - \frac{E_i^{(e)}}{b}\right), \tag{10.37}$$

where

$$b = 1 + \frac{2E_i^{(e)}}{m_\mu}\sin^2\frac{\theta}{2}. \tag{10.38}$$

So, our cross section acquires the form

$$\frac{d\sigma}{d^3 p_f^{(e)}} = \frac{e^4}{q^4 (2\pi)^2 b}\left(\cos^2\frac{\theta}{2} \quad \frac{q^2}{2m_\mu^2}\sin^2\frac{\theta}{2}\right)\delta\left(E_f^{(e)} - \frac{E_i^{(e)}}{b}\right). \tag{10.39}$$

Further, the passage to the spherical coordinate system with the vector $\mathbf{p}_i^{(e)}$ directing along the axis z allows us to present $d^3 p_f^{(e)}$ in the form:

$$d^3 p_f^{(e)} = | \mathbf{p}_f^{(e)} |^2 d | \mathbf{p}_f^{(e)} |^2 d\varphi d(\cos\theta^{(e)}) - (E_f^{(e)})^2 dE_f^{(e)} d\Omega.$$

Using the expression obtained, one may carry out integrating the expression (10.39) without any trouble and obtains the Mott formula [15]:

$$\frac{d\sigma}{d\Omega} = \left(\frac{\alpha}{2E_i^{(e)}\sin^2(\theta/2)}\right)^2 \frac{E_f^{(e)}}{E_i^{(e)}}\left[\cos^2\frac{\theta}{2} - \frac{q^2}{2m_\mu^2}\sin^2\frac{\theta}{2}\right]. \tag{10.40}$$

We would do well to discuss the case of the elastic scattering of the electron by a spinless (scalar) particle h, that is, to consider the reaction

$$e^- + h \rightarrow e^- + h.$$

The electromagnetic interaction Hamiltonian of the scalar particle described by the wavefunction $\varphi(x)$ has the form

$$\mathcal{H}_{int}^{S=0} = ie[\varphi^*(x)\partial_\mu\varphi(x) - \partial_\mu\varphi^*(x)\varphi(x)]A^\mu(x) + e^2\varphi^*(x)\varphi(x)A_\mu(x)A^\mu(x).$$

In the momentum representation the vertex corresponding to the interaction between the photon with the polarization ν and the scalar particle is determined by

$$e(p_\nu + p'_\nu), \tag{10.41}$$

where p_ν (p'_ν) is the four-dimensional momentum of the h-particle entering (coming out of) the vertex. Then in order to define the cross section in the expression (10.27) one would make the replacement:

$$M_{\nu\tau} \to (p_i + p_f)_\nu (p_i + p_f)_\tau,$$

which would lead to the following value of the cross section:

$$\left(\frac{d\sigma}{d\Omega}\right)_0 = \left(\frac{\alpha}{2E_i^{(e)} \sin^2(\theta/2)}\right)^2 \frac{E_f^{(e)}}{E_i^{(e)}} \cos^2 \frac{\theta}{2}. \tag{10.42}$$

Thus, the factor $-ie\gamma_\mu$ corresponds to the Feynman diagram vertex describing the electromagnetic interaction of the point particles with the spin 1/2. For particles having the composite structure, such as the proton, the form of the vertex has a more complicated view and this form is not known on the whole. However, one may define the form of the proton electromagnetic vertex (Fig. 10.6) from conditions that are natural enough for the quantum theory. We bear in mind the following conditions: the relativistic covariance and the four-dimensional current conservation law or, what is the same, gradient invariance. The

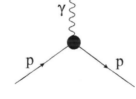

FIGURE 10.6
The electromagnetic proton vertex.

number of elements from which one may construct the Lorentz covariant vector operator of vertex is not quite large. They are:
four-vectors

$$p_{\pm\nu} = (p_f \pm p_i)_\nu,$$

two four-dimensional tensors

$$\varepsilon_{\mu\nu\lambda\sigma}, \qquad g^{\mu\nu},$$

and the following Dirac matrix combinations

vectors	pseudovectors
γ_μ	$\gamma_5\gamma_\mu$
tensors	pseudotensors
$\sigma_{\mu\nu}$	$\gamma_5\sigma_{\mu\nu}$
$\gamma_\mu\gamma_\nu\gamma_\lambda$	$\gamma_5\gamma_\mu\gamma_\nu\gamma_\lambda$

where

$$\sigma_{\mu\nu} = \frac{i}{2}[\gamma_\mu, \gamma_\nu], \qquad \gamma_5 = i\gamma^0\gamma^1\gamma^2\gamma^3$$

(all the other Dirac matrix products may be reduced to one of the listed above fourteen combinations by application of the anticommutation relations $\{\gamma_\mu, \gamma_\nu\} = 2g_{\mu\nu}$). From these quantities one can build twelve independent four-dimensional vectors $v_{i\nu}$:

1. $p_{+\nu}$	**2.** $p_{-\nu}$	**3.** γ_ν
4. $\sigma_{\nu\mu}p^{+\mu}$	**5.** $\sigma_{\nu\mu}p^{-\mu}$	**6.** $\hat{p}_+p_{-\nu}$
7. $\hat{p}_-p_{+\nu}$	**8.** $\hat{p}_+p_{+\nu}$	**9.** $\hat{p}_-p_{-\nu}$
10. $\gamma_5\varepsilon^{\nu\mu\lambda\tau}\gamma_\mu p_{+\lambda}p_{-\tau}$	**11.** $\hat{p}_+\hat{p}_-p_{+\nu}$	**12.** $\hat{p}_+\hat{p}_-p_{-\nu}$

(multiplying every vector by γ_5, one may also introduce twelve pseudovectors that would prove to be claimed when the parity violation effects are being taken into account, that is, when the weak interaction is switched on). We insert the same amount of the analytical functions F_i of invariant variable $q^2 = p^2_{-\nu}$ and represent the proton electromagnetic vertex in the form:

$$-i|e|\Lambda_\nu = -i|e|\sum_{i=1}^{12} F_i(q^2)v_{i\nu}. \tag{10.43}$$

From Fig. 10.5 it is obvious that the vertex operator, entering into the reaction amplitude, is confined between the wavefunctions of the free proton. Consequently, we are interested in the quantity

$$J_\nu = |e|\overline{\psi}(p_f)\Lambda_\nu\psi(p_i), \tag{10.44}$$

which represents the four-dimensional vector of the electromagnetic transition current for the proton. Then, using the free Dirac equation for the spinors $\psi(p)$ and $\overline{\psi}(p)$

$$(\hat{p} - m_p)\psi(p) = 0, \qquad \overline{\psi}(p)(\hat{p} - m_p) - 0, \tag{10.45}$$

one may reduce the number of the four-dimensional vectors $v_{i\nu}$ at the cost of redefining arbitrary functions $F_i(q^2)$. Really, to apply Eqs. (10.45) reduces **6,8** to **1,2** and converts **7,9** into zero. Further, in consequence of the relations

$$\overline{\psi}(p_f)\hat{p}_+\hat{p}_-\psi(p_i) = \overline{\psi}(p_f)[-2m_p^2 + 2(p_ip_f)]\psi(p_i),$$

11,12 and **4,5** are also transformed to the kind of **1,2**. The inclusion of the identity

$$\varepsilon^{\nu\mu\lambda\tau}\gamma_5\gamma_\mu = \frac{1}{3}(\sigma^{\lambda\tau}\gamma^\nu + \sigma^{\tau\nu}\gamma^\lambda + \sigma^{\nu\lambda}\gamma^\tau) \tag{10.46}$$

and the Dirac equation reduces **10** to **1** and **2**. All this allows us to represent the transition current in the form:

$$|e|\overline{\psi}(p_f)\Lambda_\nu\psi(p_i) = |e|\overline{\psi}(p_f)[F_1(q^2)p_{+\nu} + F_2(q^2)p_{-\nu} + F_3(q^2)\gamma_\nu]\psi(p_i). \tag{10.47}$$

Since the current J_ν is conserved, then multiplying the expression (10.47) by $p_{-\nu}$ gives

$$\overline{\psi}(p_f)F_2(q^2)q^2\psi(p_i) = 0.$$

This relation, in its turn, means that $F_2(q^2)$ is equal to zero for all real values of $q^2 \neq 0$. Then, from the analyticity condition of the formfactors it follows

$$F_2(q^2) = 0$$

for any q^2.

Carrying out the Hermitian conjugation of the space components of the expression (10.47), we obtain

$$\{\overline{\psi}(p_f)[F_1(q^2)\mathbf{p}_+ + F_3(q^2)\boldsymbol{\gamma}]\psi(p_i)\}^\dagger = \{\psi^\dagger(p_f)\gamma_0[F_1(q^2)\mathbf{p}_+ + F_3(q^2)\boldsymbol{\gamma}]\psi(p_i)\}^\dagger =$$

$$= \psi^\dagger(p_i)[F_1^*(q^2)\mathbf{p}_+ - F_3^*(q^2)\boldsymbol{\gamma}]\gamma_0\psi(p_f) = \overline{\psi}(p_i)[F_1^*(q^2)\mathbf{p}_+ + F_3^*(q^2)\boldsymbol{\gamma}]\psi(p_f), \qquad (10.48)$$

where we have chosen the γ-matrices representation with anti-Hermitian γ-matrices and taken into consideration that the matrices $\boldsymbol{\gamma}$ and γ_0 anticommute with each other. When $p_i = p_f$ the transition current is nothing more nor less than the electromagnetic current J_ν. Since the three-dimensional electromagnetic current is Hermitian ($\mathbf{J}^\dagger = \mathbf{J}$), then comparing the left-hand and right-hand sides of Eq. (10.48), we have drawn the conclusion:

$$F_1(q^2) = F_1^*(q^2), \qquad F_3(q^2) = F_3^*(q^2), \qquad (10.49)$$

that is, F_1 and F_3 are real quantities. Further, the relation

$$p_\nu\psi(p) = (i\sigma_{\nu\mu}p^\mu + m\gamma_\nu)\psi(p) \qquad (10.50)$$

appears to be useful. To prove it one should multiply the equation

$$\gamma_\mu p^\mu \psi(p) = m\psi(p)$$

by γ_ν on the left and transform the left-hand side by the following manner

$$\gamma_\nu\gamma_\mu p^\mu \psi(p) = \frac{1}{2}([\gamma_\nu, \gamma_\mu] + \{\gamma_\nu, \gamma_\mu\})p^\mu\psi(p) = (-i\sigma_{\nu\mu} + g_{\nu\mu})p^\mu\psi(p).$$

It is convenient to represent the quantities $F_1(q^2)$ and $F_3(q^2)$ as follows:

$$F_1(q^2) = -\frac{a^{(p)}\mathcal{F}_2(q^2)}{2m_p}, \qquad F_3(q^2) = \mathcal{F}_1(q^2) + a^{(p)}\mathcal{F}_2(q^2), \qquad (10.51)$$

where $a^{(p)}$ is the value of the proton anomalous magnetic moment expressed in terms of the nuclear magneton units. Then, with an allowance made for Eq. (10.50), the expression for the transition current takes the form:

$$J_\nu = |e|\overline{\psi}(p_f)[F_1(q^2)p_{+\nu} + F_3(q^2)\gamma_\nu]\psi(p_i) = |e|\overline{\psi}(p_f)\{[\mathcal{F}_1(q^2) + a^{(p)}\mathcal{F}_2(q^2)]\gamma_\nu -$$

$$- \frac{a^{(p)}\mathcal{F}_2(q^2)}{2m_p}p_{+\nu}\}\psi(p_i) = |e|\overline{\psi}(p_f)\left(\mathcal{F}_1(q^2)\gamma_\nu + i\frac{a^{(p)}}{2m_p}\mathcal{F}_2(q^2)\sigma_{\nu\mu}q^\mu\right)\psi(p_i) =$$

$$= |e|\overline{\psi}(p_f)\Lambda_\nu\psi(p_i), \qquad (10.52)$$

where

$$\Lambda_\nu = \left(\mathcal{F}_1(q^2)\gamma_\nu + i\frac{a^{(p)}}{2m_p}\mathcal{F}_2(q^2)\sigma_{\nu\mu}q^\mu\right). \qquad (10.53)$$

So, the proton electromagnetic vertex $-i|e|\Lambda_\nu$ has been represented in terms of two formfactors $\mathcal{F}_{1,2}$ that hold all information concerning the proton structure. From Eq. (10.53) the sense of the transition $F_i(q^2) \to \mathcal{F}_i(q^2)$ has become clear as well. Now, the electromagnetic vertex operator consists of two terms where the former describes the Dirac-type interaction and the latter does the Pauli-type interaction. Under $\mathcal{F}_i(q^2) \to 1$ the quantity Λ_ν transfers into the vertex operator of the point particle having the spin 1/2 and the magnetic moment $(1 + a^{(p)})\mu_N$, as demands the correspondence principle.

When $q^2 \to 0$, then gamma-raying of the proton is performed by long-wavelength photons that do not "see" the internal proton structure. In this case we simply observe the point particle. For this reason the nucleon formfactors must be chosen in such a way that the following conditions are fulfilled:

$$\left.\begin{array}{lll} \mathcal{F}_1(0) = 1, & \mathcal{F}_2(0) = 1 & \text{for proton,} \\ \mathcal{F}_1(0) = 0, & \mathcal{F}_2(0) = 1 & \text{for neutron.} \end{array}\right\}.$$

When calculating the differential cross section of the elastic electron-proton scattering, we shall make use of the expression (10.53) as the proton electromagnetic vertex. Then, the proton tensor is defined by:

$$\mathcal{P}^{\lambda\sigma} = \frac{1}{4}\text{Sp}[\Lambda^\lambda(\hat{p}_f + m_p)\Lambda^\sigma(\hat{p}_i + m_p)]. \tag{10.54}$$

Multiplying (10.31) by (10.54), we work out the final result, namely, the Rosenbluth formula [31]:

$$\frac{d\sigma}{d\Omega} = \left(\frac{d\sigma}{d\Omega}\right)_0 \left\{\mathcal{F}_1^2(q^2) - \frac{a^{(p)2}q^2}{4m_p^2}\mathcal{F}_2^2(q^2) - \frac{q^2}{2m_p^2}\left[\mathcal{F}_1(q^2) + a^{(p)}\mathcal{F}_2(q^2)\right]^2 \tan^2\frac{\theta}{2}\right\}. \tag{10.55}$$

The factor, standing in the braces of Eq. (10.55), describes the manner in which the scattering process is changed because of the proton structure. When the proton were the point particle, like the muon, then at any q^2 one would have

$$a^{(p)} = 0 \qquad \text{and} \qquad \mathcal{F}_1(q^2) = 1$$

and the expression (10.55) would coincide with the Mott formula. As one should expect, deviations from the Mott formula are larger in the region of small wavelengths of the virtual photon, that is, in the region of the big values of q^2. For example, when the electric charge distribution inside the proton is described by the exponential law, then we obtain at $q^2 \to \infty$

$$\frac{d\sigma}{d\Omega} - \left(\frac{d\sigma}{d\Omega}\right)_0 q^{-8}.$$

In order to measure the formfactors experimentally it is convenient to redefine them in such a way that the interference term $\mathcal{F}_1(q^2)\mathcal{F}_2(q^2)$ will be absent in the cross section (10.55). With this object in mind we introduce the electric and magnetic nucleon formfactors [32] (they are called *Sachs formfactors* as well):

$$G_E^N(q^2) = \mathcal{F}_1^N + \frac{a^{(N)}q^2}{4m_p^2}\mathcal{F}_2^N, \qquad G_M^N(q^2) = \mathcal{F}_1^N + a^{(N)}\mathcal{F}_2^N, \tag{10.56}$$

where $N = n, p$. Then, the formula (10.55) takes the form:

$$\frac{d\sigma}{d\Omega} = \left(\frac{d\sigma}{d\Omega}\right)_0 \left[\frac{(G_E^p)^2 - q^2(G_M^p)^2/(2m_p)^2}{1 - q^2/(2m_p)^2} - \frac{q^2}{2m_p^2}(G_M^p)^2 \tan^2\frac{\theta}{2}\right]. \tag{10.57}$$

From definition of $G_E^N(q^2)$ and $G_M^N(q^2)$ follows

$$\left.\begin{array}{ll} G_E^p(0) = 1, & G_M^p(0) = 1 + a^{(p)} = 2.79\mu_N, \\ G_E^n(0) = 0, & G_M^p(0) = a^{(n)} = -1.91\mu_N. \end{array}\right\} \tag{10.58}$$

This suggests that $G_E^N(q^2)$ and $G_M^N(q^2)$ must be related with distributions of the charge and magnetic moment of a nucleon.

The experimental method of determining the formfactors is simple in its description at least. Let us fix q^2 and plot the quantity

$$f(\tan^2 \frac{\theta}{2}) = \frac{d\sigma}{d\Omega} \left(\frac{d\sigma}{d\Omega}\right)_0^{-1}$$

along the ordinate, whereas the values of $\tan^2(\theta/2)$ are along the abscissa. Then the function f is represented as the straight line (Rosenbluth straight line) whose slope is

$$-\frac{q^2 [G_M^p(q^2)]^2}{(\sqrt{2}m_p)^2}$$

and whose ordinate in the point

$$\tan^2 \frac{\theta}{2} = -\frac{1}{2[1 - q^2/(2m_p)^2]}$$

equals

$$(G_E^p)^2 [1 - q^2/(2m_p)^2]^{-1}.$$

In other words, the straight line slope gives the value of $(G_M^p)^2$ while the ordinate does the value of $(G_E^p)^2$. Repeating this procedure under the different values of q^2 one may define the formfactors as the functions of q^2. True enough, since we are dealing with the formfactor squares, then the confused ambiguity, concerning the formfactor signs, is left. However, it will disappear if we take into account the formfactor values at $q^2 = 0$ (the sole exception is provided by $G_E^n(q^2)$ because $G_E^n(0) = 0$).

Note some more peculiarities of measuring the proton formfactors. With an allowance made for the value of q^2, which is given by Eq. (10.36), we obtain for $(d\sigma/d\Omega)_0$

$$\left(\frac{d\sigma}{d\Omega}\right)_0 = -\frac{\alpha^2}{q^2} \left(\frac{E_f^{(e)}}{E_i^{(e)}}\right)^2 \tan^{-2} \frac{\theta}{2}.$$

Then, at the electron backward scattering ($\theta = 180^0$) the differential cross section (10.55) takes the form:

$$\frac{d\sigma}{d\Omega} = \frac{\alpha^2}{q^2} \left(\frac{E_f^{(e)}}{E_i^{(e)}}\right)^2 \tan^{-2} \frac{\theta}{2} \left\{\frac{q^2}{2m_p^2} \tan^2 \frac{\theta}{2} [G_M^p(q^2)]^2\right\} = \left(\frac{\alpha E_f^{(e)}}{\sqrt{2}m_p E_i^{(e)}}\right)^2 [G_M^p(q^2)]^2,$$

which defines $G_M^p(q^2)$ directly. At great values of q^2 the contribution coming from $G_M^p(q^2)$ is essentially suppressed so far as the factor $[1 - q^2/(2m_p)^2]^{-1}$ stands in front of $[(G_E^p(q^2)]^2$ while $-q^2/(\sqrt{2}m_p)^2$ stands in front of $[G_M^p(q^2)]^2$.

The neutron formfactors are found from the data of scattering off electrons by the deuteron. For example, under the analysis of the reaction

$$e^- + d \to n + p + e^-$$

one should subtract the contribution coming from the electron-proton scattering and make the small correction on the nucleon coupling. The obtained data are in accordance with the assumption

$$G_E^n(q^2) \approx 0, \qquad G_M^n(q^2) \approx a^{(n)} G_E^p(q^2).$$

In Fig. 10.7 we display the proton formfactor dependence on the square of the transferred momentum. For point particles the formfactor is a constant. The formfactor dependence

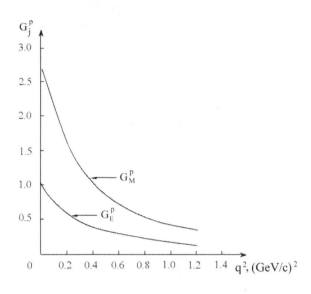

FIGURE 10.7

The transferred momentum square dependence of the proton formfactors.

on q^2, which is observed in experiments means the nucleons have a specific structure. The experimental data (see, for review, [33]) are agreed with the so-called scale relations, the scaling law

$$G_E^p(q^2) = \frac{G_M^p(q^2)}{2.79} = \frac{G_M^n(q^2)}{-1.91} = G(q^2), \qquad (10.59)$$

where in the region $q^2 \leq 0.5$ (GeV)^2 the unified formfactor $G(q^2)$ is well described by the empiric dipole formula [34]:

$$G(q^2) = \left(1 \quad q^2/m_0^2\right)^{-2}, \qquad (10.60)$$

$m_0 = 0.71$ GeV. At $q^2 \geq 0.5$ GeV^2 deviations of $\leq 20\%$ from the dipole formula are observed. Notice, all the experimental data concerning the elastic ep-scattering may be described if one assumes that the scaling law is valid and the unified formfactor has the form of the sum of two poles

$$G(q^2) = \frac{b}{1 - q^2/m_1^2} + \frac{1 - b}{1 - q^2/m_2^2}, \qquad (10.61)$$

where

$$b = -0.33, \qquad m_1 = 1.31 \text{ GeV}, \qquad m_2 = 0.64 \text{ GeV}.$$

If the proton were the static target, then the Fourier images of the formfactors would be interpreted as a distribution of the electric charge and the magnetic moment. Unfortunately, a proton recoil spoils all the state of affairs. However, one could find a Lorentz frame of reference in which the physical meaning of the formfactors becomes more transparent. The Breit reference system (the "brick wall" reference system):

$$\mathbf{p}_i = -\mathbf{p}_f.$$

has such a property. In this reference frame the proton behaves like a ball jumping aside a brick wall. As in the Breit reference frame the four-dimensional vector q_ν is remaining spacelike as before

$$q_\nu^2 = -\mathbf{q}^2 < 0,$$

this reference frame really exists and, consequently, could be connected with the lab frame by the Lorentz transformation. Using (10.52) we can find the components of the proton transition current in the Breit reference frame

$$p_i^\mu = (E, \mathbf{p}), \qquad p_f^\mu = (E, -\mathbf{p}).$$

If one denotes the eigenvalues of the proton helicity operator in the initial and final states by $\alpha_{i,f}$ and takes into account the fact that at $\mathbf{p} \to -\mathbf{p}$ the helicity changes signs, the normalization condition:

$$\psi^{(\alpha)\dagger}(p)\psi^{(\alpha')}(p) = \frac{E}{m_p}\delta_{\alpha\alpha'}, \qquad \overline{\psi}^{(\alpha)}(p)\psi^{(\alpha')}(p) = \delta_{\alpha\alpha'},$$

leads to the result:

$$\psi^{(-\alpha_i)\dagger}(p_f)\psi^{(\alpha_i)}(p_i) = \frac{E}{m_p}, \qquad \overline{\psi}^{(-\alpha_i)}(p_f)\psi^{(\alpha_i)}(p_i) = 1.$$

Then the time component of the current is different from zero if and only if $\alpha_f = -\alpha_i$ and its connection with the formfactors has the following form:

$$J_0 \equiv \rho = |e|\overline{\psi}(p_f)\left\{[\mathcal{F}_1^p(q^2) + a^{(p)}\mathcal{F}_2^p(q^2)]\gamma_0 - \frac{E}{m_p}a^{(p)}\mathcal{F}_2^p(q^2)\right\}\psi(p_i) = |e|\left[\mathcal{F}_1^p(q^2)+\right.$$

$$\left.+a^{(p)}\left(1 - \frac{E^2}{m_p^2}\right)\mathcal{F}_2^p(q^2)\right] = |e|G_E^p(q^2),$$

that is,

$$G_E^p(q^2) = \frac{\rho}{|e|}. \tag{10.62}$$

Analogous manipulations with the space component of the current lead to the relation:

$$G_M^p(q^2) = \frac{\mathbf{J}}{|e|\overline{\psi}(p_f)\boldsymbol{\gamma}\psi(p_i)}. \tag{10.63}$$

From (10.62) it is obvious that we achieved the goal to be thought–in the "brick wall" reference frame the formfactor $G_E^p(q^2)$ could be interpreted as the Fourier image of the static distribution of the electric charge

$$G_E^p(q^2) = \int \rho(\mathbf{r}) \exp(i\mathbf{q} \cdot \mathbf{r})d\mathbf{r}, \tag{10.64}$$

(for a pointlike charge $\rho(\mathbf{r}) = \delta(\mathbf{r})$ and the formfactor is simply equal to 1). Notice, that such a interpretation has not been assigned a literal meaning. Since every value $\mathbf{q}$ corresponds to its reference frame, then the function $\rho(\mathbf{r})$ does not refer to the definite reference frame. However, in the nonrelativistic limit $\mathbf{q}^2 \ll m_p^2$ changing of the proton energy during scattering may be neglected. Then the Breit reference frame coincides with the proton rest reference frame and the function $\rho(\mathbf{r})$ gains the real meaning of the space charge distribution.

Based on the experimental data by measuring the formfactors, it is easy to determine the space size and the charge distribution character of the proton. At small values of $|\mathbf{q}|$ the exponent in Eq. (10.64) could be expanded into the series:

$$G_E^p(q^2) \approx \int \rho(\mathbf{r})\left(1 + i(\mathbf{q} \cdot \mathbf{r}) - \frac{(\mathbf{q} \cdot \mathbf{r})^2}{2} + ...\right)d\mathbf{r}.$$

Assuming the charge distribution to be spherically symmetric $(\rho(\mathbf{r}) = \rho(r))$ and directing the z axis along the vector q, we obtain

$$G_E^p(q^2) \approx 1 - \frac{1}{2} \int \rho(r)(\mathbf{q} \cdot \mathbf{r})^2 d\mathbf{r} \approx 1 - \frac{\mathbf{q}^2}{6} \int \rho(r) r^2 (4\pi r^2 dr) \approx 1 - \frac{\mathbf{q}^2}{6} < r^2 >,$$

where $< r^2 >$ is an average value of the proton radius square and we have taken into account that $\int \rho(r) d\mathbf{r} = 1$. Measurements of the nucleon formfactors lead to the conclusion that the average radius of the proton and the neutron has the order of 0.8 Fermi. Roughly the same radius is found from the experimental data for the magnetic moment distribution.

To establish the electromagnetic structure of the proton we should check what forms of the charge distribution lead to the dipole formula (10.60), which works well in the region of small values of q^2. It turns out that the positive result is provided by the exponential distribution

$$\rho(\mathbf{r}) = \rho(r) = \frac{m_0^3}{8\pi} \exp(-m_0 r), \tag{10.65}$$

where m_0 is the mass of a particle carrying an interaction between nucleons. Really, substituting (10.65) into (10.64) and choosing the spherical coordinate system with the axis z along the vector $\mathbf{q}$, we arrive at the result (θ is the azimuthal angle!):

$$G_E^p(q^2) = \frac{m_0^3}{8\pi} \int_0^\infty dr \int_0^{2\pi} d\varphi \int_0^\pi d\theta r^2 \sin\theta \exp(i \mid \mathbf{q} \mid r \cos\theta \quad m_0 r) - \left(1 - q^2/m_0^2\right)^{-2}. \tag{10.66}$$

So, the distribution of the charge and the magnetic moment for the proton is described by the sufficiently simple function, the exponent. Since the quantity $\rho(r)$ tends to the constant under $r \to 0$, then it is obvious that a nucleon does not have any hard core, that is, there are no the congestion of charges in the center, as it was in the case of an atom. From this, it does not follow in any way that structural elements are generally absent inside a nucleon. On the contrary, the distribution nonhomogeneity of the charge and the magnetic moment testifies to presence of such objects. In order to "see" and investigate the properties of blocks that constitute a nucleon one needs to increase the resolution of our devices. This means in the quantum language that we should further decrease the de Broglier wavelength of the virtual photon, that is, the value of q^2 should be increased.

Without doubt the energy growth of the incoming electrons leads to changing the character of the collisions with nucleons. So, at definite values of q^2 the proton may be excited into one of the nucleons resonances ($\Delta(1238)$, $N^*(1520)$, $N^*(1688)$, and the like) and emits the π^0-meson under returning to the final state (quasi-elastic scattering). On further increasing the values of q^2, multiple hadrons generation, a proton break-up into numerous fragments (deep inelastic scattering), will take place. In such a situation, identification of the final proton state is out of the question. In Fig. 10.8 we represent the Feynman diagram for this case. Here, P is an initial four-dimensional momentum of the proton and P_f is a four-dimensional momentum of a final hadron state. It appears, that now we already have two independent variables, as distinguished from the elastic scattering. To make sure of this we introduce the quantity

$$\nu = \frac{(q \cdot P)}{m_p}, \tag{10.67}$$

which simply is an electron energy loss in the proton rest system, that is,

$$\nu = E_i^{(e)} - E_f^{(e)}.$$

Let us see what role plays the quantity ν in the electron-proton collisions. The mass of a final state is defined according to

$$M_f^2 = P_f^2 = (q + P)^2 = m_p^2 + q^2 + 2m_p\nu. \tag{10.68}$$

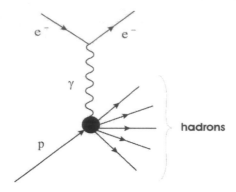

FIGURE 10.8

Deep inelastic collisions of electrons with hadrons.

From Eq. (10.68) we find the values of q^2 for the elastic scattering

$$q^2 = -2m_p\nu, \tag{10.69}$$

for exciting a nucleon resonance with the mass m_N

$$q^2 = -2m_p\nu - m_p^2 + m_N^2, \tag{10.70}$$

and for the deep inelastic scattering (the proton converts into a continuous spectrum through the resonances region)

$$q^2 = -2m_p\nu - m_p^2 + M_f^2, \tag{10.71}$$

where M_f^2 forms a continuous spectrum. We select the scattered electron in the final state and do not concretize the hadrons system on which the proton is decayed. Such reactions are called *inclusive ones*, since they include all that is possible into the hadrons system. Then the measured cross section is the cross sections sum including different final states of hadrons. From (10.71) follows, since M_f^2 may take any values, then the quantities q^2 and ν are already independent variables in the case of the deep inelastic scattering. Therefore, the additional kinematic degree of freedom appears under the inelastic electron-proton scattering. One is clearly expecting, thanks to this circumstance, the inelastic scattering investigation to give more information than the elastic processes investigation. In the case of the inelastic scattering the electron-proton interaction dynamics will already be defined by the formfactors $W_{1,2}(\nu, q^2)$ (since they determine the proton structure they are called the *structure functions*). Their precise form can be established only within the sequential theory of the strong interaction, that is, the QCD. However, in order to find $W^{\lambda\sigma}$ we again call to the relativistic and gauge invariance for help. Let us go in this way.

There are two vectors P_λ and q_λ in our disposal. As we are going to parameterize the cross section, which has already been summed up and averaged over spins, the matrices γ_λ are not included in consideration. The set of independent tensors that are built from the vectors P_λ and q_λ are as follows:

$$P^\lambda P^\sigma, \qquad q^\lambda q^\sigma, \qquad P^\lambda q^\sigma, \qquad q^\lambda P^\sigma. \tag{10.72}$$

Based on the fact that the metric tensor $g^{\lambda\sigma}$ can also participate in our constructions, the most general expression for the hadron tensor takes the form:

$$W^{\lambda\sigma} = a_1 g^{\lambda\sigma} + a_2 P^\lambda P^\sigma + a_3 q^\lambda q^\sigma + a_4 P^\lambda q^\sigma + a_5 q^\lambda P^\sigma, \tag{10.73}$$

where a_i is a function of the scalars q^2 and ν. From the current conservation law follows

$$q_\lambda W^{\lambda\sigma} = q_\sigma W^{\lambda\sigma} = 0. \tag{10.74}$$

Then, multiplying (10.73) by q_λ (q_σ) and equaling the coefficients at q^σ and P^σ (q^λ and P^λ), we arrive at four equations:

$$a_1 + a_3 q^2 + a_4 (Pq) = 0, \tag{10.75}$$

$$a_2 (Pq) + a_5 q^2 = 0, \tag{10.76}$$

$$a_1 + a_3 q^2 + a_5 (Pq) = 0, \tag{10.77}$$

$$a_2 (Pq) + a_4 q^2 = 0. \tag{10.78}$$

Subtracting (10.76) from (10.78), we obtain $a_4 = a_5$. This, in its turn, means that Eqs. (10.75) and (10.77) coincide, that is, there are only two equations to define four quantities. We choose a_1 and a_2 as the independent quantities. From Eqs. (10.77) and (10.78) we find

$$a_4 = -\frac{a_2 (Pq)}{q^2}, \qquad a_3 = -\frac{a_1}{q^2} + \left[\frac{(Pq)}{q^2}\right]^2 a_2. \tag{10.79}$$

Using the relations (10.79) and introducing the designations

$$a_1 = -W_1(\nu, q^2), \qquad a_2 = \frac{W_2(\nu, q^2)}{m_p^2},$$

we deduce for the hadron tensor $W^{\lambda\sigma}$ the final expression:

$$W^{\lambda\sigma} = \left(-g^{\lambda\sigma} + \frac{q^\lambda q^\sigma}{q^2}\right) W_1(\nu, q^2) + \frac{1}{m_N^2}\left[P^\lambda - \frac{(Pq)q^\lambda}{q^2}\right]\left[P^\sigma - \frac{(Pq)q^\sigma}{q^2}\right] W_2(\nu, q^2), \tag{10.80}$$

where $W_{1,2}(\nu, q^2)$ are functions defining the nucleon structure. Multiplying (10.80) with the electron tensor $L_{\lambda\sigma}$ (Eg. (10.31)) and neglecting the electron mass we obtain for the doubly differential cross section of the inclusive ep-scattering in the lab frame the following expression:

$$\frac{d^2\sigma}{d\Omega dE_f^{(e)}} = \sigma_0\left[2W_1(\nu, q^2)\tan^2\frac{\theta}{2} + W_2(\nu, q^2)\right], \tag{10.81}$$

where

$$\sigma_0 = \left(\frac{\alpha}{2E_i^{(e)}\sin^2(\theta/2)}\right)^2 \cos^2\frac{\theta}{2}. \tag{10.82}$$

From (10.81) it is evident that the functions $W_{1,2}$ have the dimensionality of length. The comparison of (10.82) with (10.42) makes obvious the fact that σ_0 is nothing else but the cross section of the elastic scattering off electrons by the point spinless particle having an infinitely large mass.

The structure functions $W_{1,2}(\nu, q^2)$ represent the inelastic analog of the formfactors of the elastic scattering. The technique of their experimental determination is also simple. When one fixes the variables q^2, ν and plots the values of

$$f\left(\tan^2\frac{\theta}{2}\right) = \frac{d^2\sigma}{d\Omega dE_f^{(e)}}\sigma_0^{-1}$$

along the ordinate whereas the values of $\tan^2(\theta/2)$ along the abscissa, then $f(\tan^2(\theta/2))$ is displayed as a straight line. Its slope equals $2W_1(\nu, q^2)$ and its ordinate equals $W_2(\nu, q^2)$ at the point $\tan^2(\theta/2) = 0$.

The functions $W_{1,2}(\nu, q^2)$ could be determined under investigating the process for different scattering angles as well. Obviously, at small angles $W_2(\nu, q^2)$ dominates but when the angles are close to 180^0, $W_1(\nu, q^2)$ will prevail. In Fig. 10.9 we represent $W_2(\nu, q^2)$ as a function of ν for different values of q^2 at $\theta = 6^0$. At small values of ν the peaks of $W_2(\nu, q^2)$

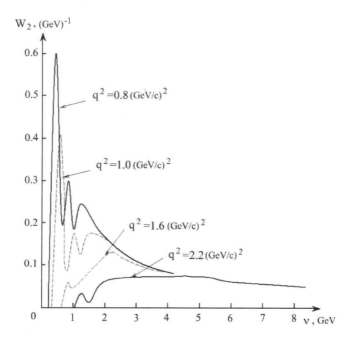

FIGURE 10.9
The $W_2(\nu, q^2)$-dependence of ν for different values of q^2.

are associated with the elastic formfactor and the resonances excitation. At $\nu \geq 3$ (GeV)^2 the final states hit on the continuous spectrum region and the curve is flattened. When $\nu \geq 4$ (GeV)^2 all the experimental points of Fig. 10.9 lay on the same curve for any values of q^2. Let us try to understand all the ensuing consequences.

We define the nondimensional variable

$$x = -\frac{q^2}{2m_p\nu}, \qquad 0 \leq x \leq 1, \tag{10.83}$$

which is connected with the invariant mass of a hadron system M_f by the relation:

$$M_f^2 = m_p^2 - q^2\left(\frac{1}{x} - 1\right).$$

It is clear that the case $x = 1$ corresponds to the elastic eN-scattering. One logically passes to the dimensionless structure function as well

$$\nu W_2(\nu, q^2) = G_2(x, q^2/m_p^2). \tag{10.84}$$

At $\nu \geq 4$ GeV^2 the quantity $G_2(x, q^2/m_p^2)$ becomes the function of the variable x only. In other words, when one fixes x, then the plot of $G_2(x, q^2/m_p^2)$ versus x is represented by the straight line being parallel to the ordinate. Experiments show that the other structure function has the analogous behavior

$$m_p W_1(\nu, q^2) = G_1(x, q^2/m_p^2), \tag{10.85}$$

where we have again passed to the dimensionless variable. Thus, in the deep inelastic region

$$| q^2 | \gg m_N^2, \qquad \nu \gg m_N, \tag{10.86}$$

the structure functions $G_{1,2}(x, q^2/m_p^2)$ become independent on any scale or are scale invariant.

Recall, that the scale invariance, the scaling, is the invariance of physical theory with respect to space-time transformations

$$\mathbf{x} \to \rho\mathbf{x}, \qquad t \to \rho t, \tag{10.87}$$

where $\rho > 0$ is a numerical parameter of transformation. In quantum theory the transformations (10.87) are supplemented by ones

$$\mathbf{p} \to \rho\mathbf{p}, \qquad E \to \rho E. \tag{10.88}$$

Physical quantities are changed in accordance with their dimensionalities under the scale transformations. So, the vector potential of the electromagnetic field and the current are transformed by the law

$$\mathbf{A} \to \rho^{-1}\mathbf{A}, \qquad \mathbf{j} \to \rho^{-3}\mathbf{j}.$$

It is evident that dimensionless quantities are scale invariants. Particle masses also fall into this category. When masses or other dimensional quantities, which are not changed under the scale transformations, do not enter into motion equations or boundary conditions, then the corresponding theories are scale invariant. Free Lagrangians of the photon and gluon fields possess the scale invariance. Clearly, in the real world where the gravitational, weak, electromagnetic, and strong interactions are switched on, the scaling does not take place. In the first place the absence of the scale invariance is caused by the fact that for physical particles the relation

$$E^2 = m^2 + \mathbf{p}^2$$

must be fulfilled. It is obvious that this relation is not invariant with respect to the transformations (10.88). On the other hand, there are no reasons that would obstruct exhibiting the scaling in nature. As we saw, the scale invariance of the dimensionless structure functions, taking place in the deep inelastic ep-scattering, is one of such examples.

Even before the first experiments for studying the inclusive electron-proton scattering, J. Bjorken [35] predicted the scaling of the functions $G_{1,2}(x, q^2/m_p^2)$ in the deep inelastic region, that is

$$\left. \begin{array}{l} G_1(x, q^2/m_p^2) \to F_1(x), \\ G_2(x, q^2/m_p^2) \to F_2(x), \end{array} \right\} \tag{10.89}$$

when

$$| q^2 | \to \infty, \qquad \nu \to \infty$$

and x is fixed. For this reason, the phenomena of the scale invariance of the structure functions is named by Bjorken scaling. Later experiments, fulfilled on electrons, muons, and neutrinos beams, displayed that the Bjorken scaling is not exact (true, violation is insignificant and may be considered as a correction to the basic effect). We shall not go into causes of scaling violation since that is beyond the framework of the book (successive explanation could be obtained within the QCD only). At a given stage the most important thing for us is to understand the conclusions that follow from the scale invariance.

The behavior of the functions $G_{1,2}(x, q^2/m_p^2)$ is greatly distinguished from the corresponding behavior of the elastic formfactors $G_{E,M}(q^2)$. Whereas $G_{E,M}(q^2)$ sharply fall down with the increase of $\mid q^2 \mid$, at $\mid q^2 \mid \to \infty$ the functions $G_{E,M}(q^2)$ do not depend on q^2 at all. This looks like the proton would not have the electromagnetic formfactors in the deep inelastic scattering region. In other words, in this case the proton behaves as a point particle. However, the proton is not the point particle and its size, defined from the elastic ep-scattering, is far from small (~ 1 Fermi). The only reasonable explanation of the experiments on the ep-scattering resides in the fact that the electric charge inside the proton is concentrated in several points, that is, in particles entering into the composition of the proton. All this may be formulated in a somewhat different way. The presence of the scaling means that such dimensional parameters as the proton mass m_p and the corresponding length $\sim 1/m_p$ do not play any significant dynamical role in the deep inelastic scattering processes, that is, there is no distances scale in this case for the proton. We shall explain the aforesaid in the example of an atom. In an atom, aside from its own size, there is one more distances scale, the size of the atom nuclear. Thanks to the uncertainty relation, the scale of distances is inversely proportional to that of energies. The existence of the two scales in an atom leads to the fact that processes, taking place with it at low and high energies, are distinguished from each other in a radical way. When the distances are bigger than the nuclear size R_N, the system is approximately described by the Coulomb potential. However, when the distances have the order of R_N or are smaller than R_N, the Yukawa potential works. In that case when there is only one scale, the qualitative change of the processes character is not in progress under the energy increase. To put it otherwise, for the proton in particular, and for hadrons in general, the second scale is equal to infinity in the case of the inclusive scattering. This means, if a hadron really represents the composite particle, then particles, entering the hadron, should have the negligibly small size, that is, they should be pointlike or, what is the same, structureless. At collisions with these hadron blocks the electrons can be often scattered through large angles just as α-particles were scattered in the Rutherford experiments when they were finding their way into an atom's nuclei.

10.3   Parton model

To prove the scale invariance of the structure functions $G_{1,2}(x, q^2/m_p^2)$ Bjorken used the hadron current algebra. One naturally expected the appearance of the model that could explain the scaling from the point of view of the hadron structure. With this in mind, in 1969 R. Feynman [36] suggested the parton model, the brief content of which we shall give below (the reader could find the detailed description of the model in the book [37]).

An extended nucleon (and any hadron as well) does not represent a formation that is smeared by the continuous way in the space, but it consists of weakly confined point particles (partons). The scattering of the high energy electron occurs on partons incoherently, that is, the electron is elastically scattered on one of partons, not affecting others. Every nucleon participates in the inelastic reaction by only one parton (active parton), which transfers the fraction x_i (i is an index of parton kind) of the hadron four-dimensional momentum P_μ.

Thus, for the ith parton we have

$$\mathbf{p}_i = x_i \mathbf{P}, \qquad E_i = x_i E, \tag{10.90}$$

When one denotes the partons number density, the parton distribution function, by $f_i(x_i)$, then $f_i(x_i)dx_i$ defines the number of partons with the four-dimensional momentum $x_i P_\mu$ in the range from x_i to $x_i + dx_i$. However, under fulfillment of (10.90) the parton mass proves to be a variable quantity that seems strange at least. The situation is clearing up in the reference frame where the time component of the vector q_μ is equal to zero (such a system really exists since $q^2 < 0$). To find it we consider the reference frame K, which moves as related to the lab frame with the velocity v being parallel to the vector $\mathbf{q}$. Using the Lorentz transformation, we obtain

$$q_0 = \frac{q_0' - v \mid \mathbf{q}' \mid}{\sqrt{1 - v^2}}, \tag{10.91}$$

where we have supplied the quantities in the lab frame by the prime. When one chooses the velocity v to be equal to

$$v = \frac{q_0'}{\mid \mathbf{q}' \mid}, \tag{10.92}$$

then from Eq. (10.91) follows that we have achieved our goal—$q_0 = 0$. It is clear that in the reference frame K the nucleon momentum is given by:

$$\mid \mathbf{P} \mid = \frac{m_N v}{\sqrt{1 - v^2}}. \tag{10.93}$$

Using (10.92) and taking into consideration $q_0' = \nu$, we arrive at

$$\mid \mathbf{P} \mid = \frac{m_N \nu}{\sqrt{Q^2}} \tag{10.94}$$

and

$$\mathbf{P}^2 = \frac{m_N^2 \nu^2}{Q^2} = \frac{m_N \nu x}{2}, \tag{10.95}$$

where we have passed to the positive quantity $Q^2 = -q^2$ for the reasons of convenience. In the deep inelastic region the relations (10.86) take place, therefore

$$\mid \mathbf{P} \mid \gg m_N \qquad \mathbf{P} \to \infty. \tag{10.96}$$

Thus, in the Lorentz system where $q_0 = 0$ and the relation (10.96) is fulfilled (the system of infinity momentum (SIM)), one may escape the question about a variable mass of a parton, if one assumes both m_i and m_N being equal to zero. The notion of partons makes sense only in the reference frame where a nucleon moves with the relativistic velocity. This circumstance is the reflection of the already known fact that only the high energy virtual photon ($\mid q^2 \mid \geq 1$—2 GeV^2) may discern a parton. In the SIM the transverse component (as related to the direction of the nucleon motion) of the parton momentum appears to be negligibly small. Really, in the rest system of a nucleon the mean square longitudinal and transverse parton momenta are equal to each other. It is clear, that in the SIM, whose velocity as related to the lab frame is close to the light velocity ($v = \nu/\sqrt{\nu^2 - q^2} \approx 1$) the longitudinal parton momentum is much bigger than the transverse parton momentum. The transition to the SIM has one more advantage, namely, it sheds a light on a parton behavior in a nucleon. In this system the interaction acts frequency of partons with each other decreases, by virtue of the relativistic delay of the time. Thus, in the short time interval between interactions of the virtual photon with the parton, the parton behaves as a nearly free particle. Scattered partons (active partons) and residues of the initial hadrons, which did not take part in interactions (the set of passive partons), turn into final hadrons thanks to the strong interaction. The final hadrons produce two jets, one in the direction of a scattered parton, and the other in the direction of an initial nucleon. A hadron jet represents the set of hadrons having small (the order of 300 GeV) transverse momenta relative to the motion of the parent particle. The jet's existence already on its own serves as evidence of the weakness of hadron matter interaction on the small distances. Indeed, if the hadron matter produced what amounts to dense high excited cluster, then the isotropic configuration with a large value of an average transverse momentum (the order of an collision energy) would be natural for outgoing secondary particles.

Now we show how the parton model explains the scaling phenomenon. Incoherence (probabilities are summarized rather than scattering amplitudes) of the lepton-parton scattering means that every parton with the corresponding distribution function $f_i(x_i)$ gives independent contribution into the lepton-parton scattering cross section, that is,

$$d\sigma = \sum_i d\sigma_i \sim L_{\lambda\sigma} W^{\lambda\sigma}.$$

The hadron tensor $W^{\lambda\sigma}$ could be represented in the following form:

$$W^{\lambda\sigma} = \sum_i \int_0^1 dx_i \frac{f_i(x_i)}{2m_N p_{i0}} W_{(i)}^{\lambda\sigma}, \tag{10.97}$$

where $W_{(i)}^{\lambda\sigma}$ is a parton tensor, and the factor $(2m_N p_{i0})^{-1}$ was introduced for reasons of convenience. Assume that partons have the spin equal to $1/2$. As we are not interested in a parton polarization we should carry out an average over the initial spin parton states and a summation over the final ones. Thus, to find the parton tensor one may use the corresponding expression for the pointlike fermion (see, Eq. (10.31)) and presents (10.97) in the form:

$$W^{\lambda\sigma} = \frac{1}{m_N} \sum_i Q_i^2 \int_0^1 dx_i f_i(x_i) \int d^4p_i' \theta(p_{i0}')\delta(p_i'^2 - m_i^2)\delta^{(4)}(p_i + q - p_i')\frac{1}{4} \times$$

$$\times \mathrm{Sp}[\gamma^\lambda(\hat{p}_i' + m_i)\gamma^\sigma(\hat{p}_i + m_i)], \tag{10.98}$$

where p_i and p_i' are the ith-parton momentum in the initial and final states, respectively, and Q_i is an electric charge of ith-parton. In the expression (10.98) the first delta function $\delta(p_i'^2 - m_i^2)$ ensures locating a parton on the mass shell while the second one $\delta^{(4)}(p_i + q - p_i')$ guarantees fulfillment of the four-dimensional momentum conservation law under the interaction between partons and virtual photons. Having taken the track and having carried out the integration over four-dimensional momenta of final partons, we obtain the following expression for the hadron tensor:

$$W^{\lambda\sigma} = \sum_i Q_i^2 \int_0^1 dx_i f_i(x_i)\frac{1}{2m_N^2\nu}\delta(x_i - x)[2p_i^\lambda p_i^\sigma + q^\lambda p_i^\sigma + p_i^\lambda q^\sigma - g^{\lambda\sigma}(p_i q)], \tag{10.99}$$

where we have neglected a parton mass and taken into account that

$$\delta(q^2 + 2p_i q) = \frac{1}{2m_N}\delta\left(\frac{q^2}{2m_N} + \frac{x_i Pq}{m_N}\right) = \frac{1}{2m_N\nu}\delta\left(x_i - x\right).$$

The comparison of (10.99) with (10.80) allows us to determine the contribution coming from the ith-parton to the structural nucleon functions $W_{1,2}$:

$$W_{1,2} = \sum_i \int_0^1 dx_i f_i(x_i) W_{1,2}^{(i)},$$

where

$$W_1^{(i)} = \frac{1}{2m_N} Q_i^2 x_i \delta(x_i - x), \qquad W_2^{(i)} = \frac{Q_i^2 x_i^2}{\nu} \delta(x_i - x). \tag{10.100}$$

Thus, due to the momentum conservation law expressed by the function $\delta(x_i - x)$, the scaling variable x proves to coincide with a parton fraction of a nucleon momentum. The relations (10.100) express the extremely instructive fact—a virtual photon can be absorbed by ith-parton only at the right value of the variable $x_i = -q^2/(2m_N\nu)$. Substituting (10.100) into (10.97) and considering the definitions (10.84) and (10.85), we get the Bjorken scaling for the structural nucleon functions:

$$\left. \begin{aligned} m_N W_1 &= \mathcal{C}_1(x, q^2/m_p^2) = \frac{1}{2} x \sum_i Q_i^2 f_i(x), \\ \nu W_2 &= \mathcal{C}_2(x, q^2/m_p^2) = x^2 \sum_i Q_i^2 f_i(x). \end{aligned} \right\} \tag{10.101}$$

From (10.101) it follows Callan-Cross formula [38]:

$$\mathcal{C}_2(x, q^2/m_p^2) = 2x \mathcal{C}_1(x, q^2/m_p^2). \tag{10.102}$$

The relations (10.101) establish a link between the structural nucleon functions and the distribution functions of partons interacting with a photon.

So, the parton model proves to be extremely fruitful. First and foremost, with its help one managed to prove the scaling behavior of the structure functions $G_{1,2}(x, q^2/m_p^2)$. It gets worse and worse as it goes on. This model gives an opportunity to establish the spin of partons participating in the electromagnetic interaction. Really, the scaling relations (10.101) takes place only at the parton spin being equal to 1/2. For example, if the parton has zero spin, the replacement

$$\frac{1}{4} \mathrm{Sp}[\gamma^\lambda(\hat{p}_i' + m_i)\gamma^\sigma(\hat{p}_i + m_i)] \rightarrow (p_i' + p_i)^\lambda (p_i' + p_i)^\sigma$$

would be done in the expression for the parton tensor (10.98). Then, as is easy to see, the function F_2 remains unchanged while $F_1 = 0$.

Evidently, it would be rather attractive to identify the partons with the quarks. From the quantum laws, the quarks in hadrons can exist both in real and in virtual states. We agree to call the real quarks by the valent quarks.* So, three valent quarks enter baryons whereas mesons consist of valent quark-antiquark pairs. Just the valent quarks define additive quantum numbers of hadrons (electric charge, strange, baryon charge, and so on). Thanks to the uncertainty relation, quark-antiquark pairs could be supplemented to the valent quarks in a short time. One naturally calls the quarks forming the sea of virtual quark-antiquark pairs by the sea quarks. For the belief to the quark-parton model to be strengthened, the electric charges of the quarks should be defined. Measuring the nonelastic formfactors and using the relations (10.101) one may determine the quark charges. Generally speaking, this is not so simple because a nucleon consists of a large number of partons interacting with

---

*Such quarks are also named by the block or constituent quarks.

each other. As a consequence of detailed measurements and comparing the experiments on various hadron targets they managed to select the contributions coming from different kinds of partons and define the electric charges of partons. It appeared that they coincide with the electric charges of the quarks, namely, are equal to 2/3 and −1/3.

To investigate the electron-nucleon scattering gives the opportunity to define the nucleon momentum fractions, which are transferred by quarks and antiquarks, when the nucleon moves with the big velocity. The quark and antiquark contributions to the nucleon momentum could be expressed in terms of the quark and antiquark

$$\sum_i \int x[f_i(x) + \overline{f}_i(x)]dx = 1 - \epsilon, \tag{10.103}$$

where we have taken into account the contributions coming both from valent quarks and from the sea ones. If one assumes that the quarks and the antiquarks are the sole pretenders on the partons role, then ϵ must take the value 0. By the end of the 1960s of the twentieth century, the experiments on probing the nucleons by means of the electrons were fulfilled at SLAC (Stanford Linear Accelerator Center) and some years later (in 1973) the similar raying of nucleons, with the help of the neutrino beams, was carried out at CERN (Conseil Europeen pour la Recherche Nucleaire). These experiments gave

$$\epsilon \approx 0.5.$$

This result means that nearly 50% of the nucleon momentum is transferred by partons that do not take part in both the electromagnetic and weak interactions. In the QCD only the gluons, which are the carriers of the strong interaction between the quarks, may pretend to the role of such particles.

As discussed earlier, the distribution functions of the valent and sea quarks in a nucleon may be measured in the processes of the deep inelastic ep- or μp-scattering. However, the best processes for the experimental definition of quark distributions are the processes of the inclusive scattering of neutrino or antineutrino from protons. This is caused by the fact that the neutrino interacts only with the d- and $\overline{u}$-quarks while the antineutrino does with the $\overline{d}$- and u-quarks. The quark distributions are described by sufficiently simple functions that once again confirms the truth every physicist must know: *"The Creator is not ill-intentioned."* For example, the distribution functions of the valent quarks in the proton have the form:

$$x f_{u_V}(x) \approx 2.04\sqrt{x}(1 - x)^{2.5}, \qquad f_{d_V}(x) \approx 0.57(1 - x).$$

At present we discuss the question about the quarks masses. The particle mass could be exactly determined by its energy only for the free particle. Since the free quarks were not discovered up till now, then one assigns the precise meaning to their masses with difficulty. The quark mass problem is very similar to the problem concerning the electron mass in solid state physics. The electron, when it is moving in a solid, behaves as a particle with the "effective" mass m_{ef}, which is significantly distinguished from its true mass m. Moreover, m_{ef} may depend on the motion features, because in the reality the masses difference $\Delta m = m_{ef} - m$ is caused by the interaction of the electron with objects surrounding it. In this sense the masses of all the quarks are "effective," because they are defined in the processes in which the quarks are interacting with other particles. For this reason the quark mass values, found under analyzing the energy levels location of the bound states (quarkoniums.*), may appreciably differ from those values that correspond to

---

*Bound state to consist of heavy quarks and a corresponding antiquark is named the quarkonium.

the weak decay of the quarks. They distinguish the current and constituent (block) quark masses. Since, in the quantum field theory interactions are formulated on the language of currents and potentials, it is natural to call the quarks entering into Lagrangians by the current quarks. Thus, the current masses concern to naked quarks and they do not take into account the contributions coming from their gluons and quark-antiquarks fur coats. In the QCD the current quark mass depends on the momentum transferred to the quark and is decreased with the growth of this momentum. So, at the scale ~ 2 GeV the current quark masses are confined into the intervals

$$1.5 \text{ MeV} < m_u < 3.3 \text{ MeV}, \qquad 3.5 \text{ MeV} < m_d < 6.0 \text{ MeV}, \qquad m_s = 104^{+26}_{-34} \text{ MeV}.$$
(10.104)

Thanks to the fur coats contribution, the block masses exceed the corresponding current masses on 300 GeV approximately. From (10.104) it becomes clear that the success of the $SU(2)$-symmetry is basically caused by a closeness of the masses of the u- and d-quarks. The symmetry with respect to the $SU(3)$-group, which includes the more heavier s-quark, is already violated much stronger. However, by the irony of fate, just the approximate $SU(3)$-symmetry found the exact symmetry status in the same quark walk of life, provided the condition that it is being used to describe interactions between quarks in themselves.

10.4 Quantum chromodynamics. The first acquaintance

So, late in the 1960s, the hypothetical quarks have been acquiring the status of objects the reality of which, while indirectly, manifests in experiments. However, the quark model has a lot of unresolved problems, as before. One of these problems is connected with the quarks statistics. As an example, we consider the Ω^--hyperon entering into the $3/2^+$ baryon decuplet. Its total wavefunction is the product of three wavefunctions which express the dependence on the space-time variables, the ordinary and unitary spins, that is,

$$\Psi_\Omega = \Psi(\mathbf{r})\Phi(\mathbf{S})\Theta(\mathbf{S}^{(un)}), \tag{10.105}$$

where the quark filling of the Ω^--hyperon may be schematically represented in the following way

$$\Theta(\mathbf{S}^{(un)}) - | s \uparrow s \uparrow s \uparrow >, \tag{10.106}$$

(the arrow on the quark symbol defines the direction of the spin projection). It is clear that $\Theta(\mathbf{S}^{(un)})$ is completely symmetric with respect to any transposition of the s-quarks. In the case of parallelism of three spins $\Phi(\mathbf{S})$ is also symmetric. Since the symmetry character of the space part of the wavefunction is determined by the factor $(-1)^l$ (for the $3/2^+$-decuplet, $l = 0$), then $\Psi(\mathbf{r})$ proves to be symmetric under the three quarks transposition as well. Thus, the total wavefunction of three s-quarks system appears symmetric. However, the quarks have the half-integer spin, consequently, they obey to the Fermi-Dirac statistics and the Pauli principle is valid for them. As a result, the total wavefunction Ψ_Ω must be antisymmetric. To prove the Pauli theorem about the connection of the spin with the statistics is based on such fundamental concepts of the quantum theory as microcausality and locality. The refusal from this theorem would be tantamount to the volley of Aurora cruiser that has announced the beginning of the October revolution in Russia in 1917. But the evolutions way always was more preferable than the mysterious ways of revolutions. The painless solution of the conflict with the statistics proves to be possible by introducing the new discrete variable [39], called a *color* [40], which is assigned to all the quarks independently of the flavor. By the example of the Ω^--hyperon, it is clear that the minimum number of the new variable values should be equal to 3. We are restricted by three values for the color degree of freedom and shall designate them as R (red), G (green), and B (blue). To introduce the color allows putting three quarks into one and the same quantum mechanical state inside a hadron. For the Pauli principle to be fulfilled, the baryon wavefunctions must be antisymmetrized on the color variables. So, the Ω^--hyperon wavefunction part, connected with the unitary spin, has the form:

$$\Theta(\mathbf{S}^{(un)}) = \frac{1}{\sqrt{6}} \sum_{i,j,k=1}^{3} \varepsilon_{ijk} s_i s_j s_k, \qquad (10.107)$$

where the coefficient $1/\sqrt{6}$ is related with the normalization, and not entirely convenient indices R, G, B are simply replaced by 1,2,3.

When we are talking about the kinematic aspects of the quark systems, then three internal degrees of freedom, the colors, are conveniently considered as the eigenvalues of the color spin operator (Fig. 10.10). It should be stressed that by definition the color spin operator has nonzero values for quarks and gluons only. If the hypothesis of the quarks confinement

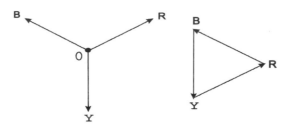

FIGURE 10.10
The color spin operators.

is valid then the color spin belongs to the number of the hidden degrees of freedom. This, in its turn, means that all the particles, which are being observed in the free state, must have the color spin equal to zero. As it follows from Fig. 10.10 there are two ways to obtain a colorless (or white) hadron. The first way is to mix three different colors (or anticolors) while in the second way every color is mixed with the corresponding anticolor. Then, for any baryons, whose wavefunction with consideration both for flavor and for color will be written in the form

$$B_{\alpha\beta\gamma} = \frac{1}{\sqrt{6}} \sum_{i,j,k=1}^{3} \varepsilon_{ijk} q_{\alpha i} q_{\beta j} q_{\gamma k}, \qquad (10.108)$$

the color spins are mutually compensated and the total baryon color equals zero. Analogously, mesons prove to be colorless because they consist of a mixture of color and corresponding anticolor in equal proportion

$$M_{\beta}^{\alpha} = \frac{1}{\sqrt{3}} \sum_{i=1}^{3} \overline{q}^{\alpha i} q_{\beta i}. \qquad (10.109)$$

Let us express the hadron colorless on the strict mathematical language. Assume that a quark changes its color, to say, it goes from one color state to another. Moreover, we shall consider that this new state is the linear combination of all the possible old states (for the sake of simplicity, we omit the flavor indices of the quarks)

$$q'_i = \sum_{i=1}^{3} U_{ij} q_j. \tag{10.110}$$

Now we require hadrons to have one and the same view both in the old and new variables, that is, hadrons should be the invariants of the transformation (10.110)

$$\sum_{i=1}^{3} q'_i \bar{q}'_i = \sum_{i=1}^{3} q_i \bar{q}_i, \tag{10.111}$$

$$\sum_{i,j,k=1}^{3} \varepsilon_{ijk} q'_i q'_j q'_k = \sum_{i,j,k=1}^{3} \varepsilon_{ijk} q_i q_j q_k. \tag{10.112}$$

As the antiquarks $\bar{q}$ are transformed by the complex conjugated matrices U^*, then from (10.111) follows

$$\sum_{i=1}^{3} U_{ij} U^*_{im} = \delta_{jm},$$

that is, the transformation (10.111) is performed by the unitary matrices. In its turn Eq. (10.112) leads to the condition:

$$\det U = 1 \tag{10.113}$$

which means the transformation (10.111) is special. So, the transformations in question constitute the special unitary group in three-dimensional color space, the $SU(3)$-group, that is, the explicit form of (10.110) is as follows:

$$q'_j = [\exp{(i\alpha_a T_a)}]_{jk} q_k, \tag{10.114}$$

where $a = 1, 2, ...8$, α_a are real parameters and $T_a = \lambda_a/2$ are group generators. Further, in order to distinguish this group from the group on flavors we shall be talking about it as the color group $SU(3)_c$. The observed hadrons are invariant with respect to the transformations of the $SU(3)_c$-group or, what is the same, represent the color singlets. The color spin introduction liquidates not only the conflict with the statistics, it also forbids the existence of bound qq-states (of the type RB, GR, and so on) on the strength of the postulate: *only the colorless hadrons are observable*. Thus, the color scheme explains the exceptional role of the quark combinations qqq, $\overline{qqq}$, and $q\bar{q}$ in nature.

Now we address the dynamical aspects of the color hypothesis. It is evident that when we are considering the quarks dynamics the quark interactions should be taken into account. In this case, the role of the color degree of freedom is most of all similar to the role of the electric or gravitational charges and, let them forgive us for this terminological liberty, we shall be speaking of the color charge rather than the color spin.

The quantum field theory describing interactions between quarks, the QCD, is built by the analogy with the quantum electrodynamics (QED). Recall that the electromagnetic field appears as the compensated field that ensures the charged fields invariance with respect to the local one-parametric Abelian group $U(1)_{em}$. In 1954 C. N. Yang and R. L. Mills [41] investigated the local generalization of the non-Abelian groups by an example of three-parametric group $SU(2)$. As a result they came to recognize that in this case the local gauge invariance of the theory already demands introducing a three-parametric compensated field which is interpreted as a field of strongly interacting vector bosons. An obvious generalization of this fact lies in a statement: *in the case of n-parametric local gauge group the theory invariance demands introducing n-parametric compensated field.* However the Yang-Mills theory faced two troubles there and then. First, these vector bosons, similar to the photon, would have zero mass and lead to long-range forces. But in this case, such particles should be already observed. Second, due to the big value of the interaction coupling constant the perturbation theory does not work, which does not allow making any conclusions from the theory at that time. In this connection the Yang-Mills theory has attracted purely academic interest first and it was practically impossible to make out the prototype of the future theory of the strong interaction in its outlines. There were not such notions as quark and gluon. They appeared a decade later and, in the beginning, they were not connected with a mathematical apparatus of non-Abelian theories in any way. Still ten years were necessary in order for the synthesis of these two ideas to led to the QCD formulation.

The QCD is based on developing the formalism stated in Section 4.5. But now, in place of the $U(1)_{em}$-gauge group, we are dealing with the phase transformations group of the color quark fields, the $SU(3)_c$-group. We shall consider for simplicity that the quarks have one flavor only (one flavor approximation). Then the free quarks Lagrangian is given by the expression:

$$\mathcal{L}_0 = \frac{i}{2}[\overline{q_k}(x)\gamma_\mu\partial^\mu q_k(x) - \partial^\mu\overline{q_k}(x)\gamma_\mu q_k(x)] - m\overline{q_k}(x)q_k(x), \qquad (10.115)$$

where $k = R, B, G$. Let us investigate the consequences of the invariance of $\mathcal{L}_0$ with respect to the local gauge transformations of the non-Abelian group $SU(3)_c$

$$q'_k(x) = U_{kj}(x)q_j(x) = \{\exp[-ig_s\alpha_a(x)T_a]\}_{kj}\, q_j(x), \qquad (10.116)$$

where $T_a = \lambda_a/2$ and λ_a are the Gell-Mann matrices (9.112). So, our task is to ensure the local $SU(3)_c$ invariance of the Lagrangian $\mathcal{L}_0$. Pass in (10.116) to infinitesimal transformations

$$q'_k(x) = [\delta_{kj} + i\alpha_a(x)(T_a)_{kj}]q_j(x). \qquad (10.117)$$

In this case the derivative of the quark field function is transformed by a law

$$\partial_\mu q'_k(x) = [\delta_{kj} + i\alpha_a(x)(T_a)_{kj}]\partial_\mu q_j(x) + i(T_a)_{kj}q_j(x)\partial_\mu\alpha_a(x) \qquad (10.118)$$

and violates the invariance of $\mathcal{L}_0$. To rescue the situation we introduce eight gauge fields $G^a_\mu(x)$ and build covariant derivatives

$$D^\mu_{kj}(x) = \delta_{kj}\partial^\mu + ig_s(T_a)_{kj}G_{a\mu}(x), \qquad (10.119)$$

where g_s is a gauge constant of the $SU(3)_c$ group. Further, we replace the ordinary derivatives with the covariant ones in $\mathcal{L}_0$

$$\mathcal{L} = \frac{i}{2}[\overline{q_k}(x)\gamma_\mu D^\mu_{kj}(x)q_j(x) - D^{\mu\dagger}_{kj}\overline{q_k}(x)\gamma_\mu q_j(x)] - m\overline{q_k}(x)q_k(x) =$$

---

*Gauge fields which are introduced to ensure the local non-Abelian gauge invariance are now called the *Yang-Mills fields* and the equations they satisfy in the free case

$$\partial^\nu G_{a\mu\nu} + g_s\varepsilon_{abc}G^\nu_b G_{c\mu\nu} = 0,$$

where ε_{abc} are structural constants of the local gauge group under consideration, are given the title Yang-Mills equations.

$$= \frac{i}{2}[\overline{q_k}(x)\gamma_\mu\partial^\mu q_k(x) - \partial^\mu\overline{q_k}(x)\gamma_\mu q_k(x)] - m\overline{q_k}(x)q_k(x) - g_s[\overline{q_k}(x)\gamma_\mu(T_a)_{\overline{k}j}q_j(x)]G_a^\mu(x).$$
$$(10.120)$$

The invariance of the Lagrangian will be provided under a condition:

$$\overline{q_k}'(x)D_{\overline{k}j}^{\mu'}(x)q_j'(x) = \overline{q_k}(x)D_{\overline{k}j}^\mu(x)q_j(x),$$

what, in its turn, gives

$$D_{\overline{k}j}^{\mu'}(x) = U_{ki}(x)D_{il}^\mu(x)U_{jl}^\dagger(x). \tag{10.121}$$

By analogy with the QED we demand that the transformation law of the gauge fields $G_\mu^a(x)$ has the form:

$$G'_{a\mu}(x) = G_{a\mu}(x) - \frac{1}{g_s}\partial_\mu\alpha_a(x). \tag{10.122}$$

However, in this case, the last quantity in (10.120) is not the invariant with respect to the $SU(3)_c$ transformations. Really, taking into account the algebra of the λ_a-matrices we obtain

$$[\overline{q_k}(x)\gamma_\mu(T_a)_{\overline{k}j}q_j(x)]' = [\overline{q_k}(x)\gamma_\mu(T_a)_{\overline{k}j}q_j(x)] + i\alpha_b(x)\overline{q_k}(x)\gamma_\mu(T_aT_b - $$

$$-T_bT_a)_{\overline{k}j}q_j(x) = [\overline{q_k}(x)\gamma_\mu(T_a)_{\overline{k}j}q_j(x)] - f_{abc}\alpha_b(x)(\overline{q_k}(x)\gamma_\mu(T_c)_{\overline{k}j}q_j(x)),$$

From the obtained expression follows that the gauge invariance of the Lagrangian (10.120) will be restored if we replace the transformation law (10.122) with

$$G'_{a\mu}(x) = G_{a\mu}(x) - \frac{1}{g_s}\partial_\mu\alpha_a(x) \quad f_{abc}\alpha_b(x)G_{c\mu}(x). \tag{10.123}$$

Now we should supplement the Lagrangian $\mathcal{L}$ by that of the free gauge bosons $\mathcal{L}_G$. Thanks to presence of the last term in Eq. (10.123), the field tensor $G_{a\mu}(x)$ has a more complicated form than its analog in the QED. It is not difficult to show that the gauge invariance will be ensured by the following choice of $\mathcal{L}_G$

$$\mathcal{L}_G = -\frac{1}{4}G_{\mu\nu}^a(x)G_a^{\mu\nu}(x), \tag{10.124}$$

where

$$G_a^{\mu\nu}(x) = \partial^\mu G_a^\nu(x) - \partial^\nu G_a^\mu(x) - g_s f_{abc}G_b^\mu(x)G_c^\nu(x). \tag{10.125}$$

Thus, using only the requirement of the Lagrangian invariance with respect to the local gauge group $SU(3)_c$, we get the total Lagrangian of the QCD

$$\mathcal{L}_{QCD} = \frac{i}{2}[\overline{q_k}(x)\gamma_\mu\partial^\mu q_k(x) - \partial^\mu\overline{q_k}(x)\gamma_\mu q_k(x)] - m\overline{q_k}(x)q_k(x) - $$

$$-g_s[\overline{q_k}(x)\gamma_\mu G_{\overline{k}j}^\mu(x)q_j(x)] - \frac{1}{4}G_{\mu\nu}(x)G^{\mu\nu}(x) \tag{10.126}$$

where

$$G_{\overline{k}j}^\mu(x) = \frac{1}{2}G_a^\mu(x)(\lambda_a)_{kj}, \qquad G^{\mu\nu}(x) = \partial^\mu G^\nu(x) - \partial^\nu G^\mu(x) - ig_s[G^\mu(x), G^\nu(x)],$$

$G_{\overline{k}j}^\mu(x)$ is a four-dimensional gluon field potential in a point x (every component of $G_{\overline{k}j}^\mu$ represents a 3×3 Hermitian matrix in the color space). The number of quark field phases, we can change by arbitrary ways, is equal to eight. Consequently, to compensate all changing of the phases we need eight gluons as well. Since the introduction of the gluon mass term leads to the violation of the local gauge invariance, the carriers of the strong interaction, the gluons, are massless. From (10.126) follows that the wavefunctions of the gluon field possess two color indices j and $\overline{k}$, that is, the two color gluons bear the color and anticolor charge. It is clear, that the combination deprived with the color charge $g_0 = R\overline{R} + B\overline{B} + G\overline{G}$ is a color singlet. Exchange of this singlet changes by no means the color state of the quark, and consequently, g_0 cannot pretend on the role of particle bearing the color interaction between the quarks. In order to establish the wavefunction color parts of eight gluons left we appeal to the original version of the quark theory, the fundamental triplets hypothesis. Let us set up a correspondence between three quark flavors and three quark colors. Then the gluons by their own structure are analogous to the mesons. Therefore, to determine the color parts of the gluon wavefunctions we can use the quark filling of the meson octet. Having fulfilled the replacements

$$u \to R, \qquad d \to G, \qquad s \to B,$$

in Eq. (10.3), we arrive at the following expressions for the color parts of the gluon wavefunctions:

$$\left.\begin{array}{l} g_1 = G\overline{B}, \qquad g_2 = R\overline{B}, \qquad g_3 = R\overline{G}, \qquad g_4 = B\overline{G}, \\[2mm] g_5 = B\overline{R}, \qquad g_6 = G\overline{R}, \qquad g_7 = \frac{1}{\sqrt{2}}\left(R\overline{R} - G\overline{G}\right), \\[2mm] \qquad\quad g_8 = \frac{1}{\sqrt{6}}\left(R\overline{R} + G\overline{G} - 2B\overline{B}\right). \end{array}\right\} \qquad (10.127)$$

It is obvious from the form of $\mathcal{L}_{QCD}$ that in the QCD the kinetic energy of the gluons is not already purely kinetic, since it contains interactions between the gluons ($\sim g_s G_\mu^a G_\nu^b G_\lambda^c$ and $\sim g_s^2 G_\mu^a G_\nu^b G_\lambda^c G_\sigma^d$). Thus, in the QCD the Feynman diagrams include vertices in which only the gluons are met. In other words, the gluons possess nonlinear self-interaction (self-interaction is the typical feature of non-Abelian gauge theories) and this circumstance is caused by the fact that they have the color charge. The inclusion of the gluons contribution into the vacuum polarization explains the quarks behavior specificity at large transfers of the momentum. With penetrating in the gluon fur coat, which surrounds the quark, the quark color charge is being decreased. This means that in the limit of infinitely small distances separating the quarks, the color interaction between them is switched off and they behave very similar to free particles (asymptotic freedom). In the case of the deep inelastic scattering off electrons by protons the quarks, which exhibit themselves as partons, are in the protons just in the same condition.

To build the successive quantum theory of quark-gluon interactions we need to use the so-called interaction representation in which quarks and gluons are described by the free equations of motion. It is apparent that in the QCD such an operation is not absolutely lawful because quarks and gluons are not observed in the free states. Quarks in hadrons might be considered as free particles at small distances and only in this case to use the QCD, based on the perturbation theory methods, (perturbative QCD) is lawful. With the growth of the distances between quarks their effective coupling constant increases and, as a result, the perturbative QCD does not cease to work. By now the quarks confinement has not received the final understanding within the QCD, that is, it has remained the only hypothesis confirmed by experiments.

We will briefly discuss some models that explain the confinement of quarks and gluons inside hadrons. So, one may account for the confinement because the hadrons in the color states are much heavier than those in the colorless states and, for this reason, the latter are not observed in up-to-date experiments. The analogy with atoms helps to understand this idea. Let the neutral (nonionized) atoms correspond to the white hadrons while the charged ions correspond to color states of the hadrons. It is clear that the ions have a larger energy and are going to come back in the neutral atom state. Such a tendency is explained by the fact that the electromagnetic interaction, which could be approximately described by the Coulomb potential $V_c \sim \alpha_{em}/r$ in the atom case, acts between the electric charges.

One naturally assumes that there also exists color interaction to hinder the quark flying out of the hadron. This idea has found embodiment in the descriptive bag model. The simplest variant, the MIT bag model [42] (Massachusetts Institute of Technology), is based on the assumption: *a hadron represents the bag with the sharp borders that hinders all color objects fly out.* The hadron system is described by the Lagrangian function

$$L = \int d\mathbf{r}[\mathcal{L}_{QCD} - f(\mathbf{r})], \tag{10.128}$$

where $f(\mathbf{r})$ defines the walls pressure and, consequently, makes a provision for the confinement of both quarks and gluons. If the quarks become widely separated, then the gluon fields, propagating between the quarks, are stretched into straight lines and the bag takes the form of the tube. In the case of interaction of quarks with antiquarks the picture looks the most simple. When one forgets about the nonlinearity for the time being, then the system $q\bar{q}$ would be similar to the electric dipole whose distribution of field lines in the space is displayed in Fig. 10.11(a). The presence of an interaction between gluons results

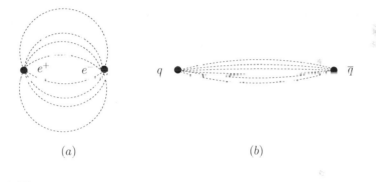

(a)   (b)

FIGURE 10.11
The field lines distribution in the $q\bar{q}$-system in (a) and (b).

in compressing these lines into the color field tube, the gluon guide (Fig. 10.11(b)). The interaction structure is such that the tube cross section area S is kept constant under the growth of the distance between the members of the pair $q\bar{q}$ (r). But the field line numbers depend on the source charges only. Consequently, the field strength in the gluon guide is being kept constant and the field energy increases proportionally to the tube volume Sr. Since the cross section area S is fixed the energy of the system $q\bar{q}$ is linearly increasing with the growth of r. It means that any process, in which a finite amount of the energy is transferred to this system, is not able to separate the quark from the antiquark. In reality the flux tube cannot be stretched infinitely since the production of the quark-antiquark pair from the vacuum has become energetically more profitable, which corresponds to the transition of the one-hadron state into the two-hadron state. In the first approximation the effective potential of interaction between q and $\bar{q}$ has the form:

$$V^c(r) = -\frac{4}{3}\frac{\alpha_s}{r} + Ar + V_0, \tag{10.129}$$

where A and V_0 are positive constant quantities, that is, it looks like a whirlpool (Fig. 10.12). The first Coulomb-like term $\sim 1/r$ accords with the one-gluon exchange (the analog of the one-photon exchange in the electrodynamics), and the linear term $\sim r$ providing

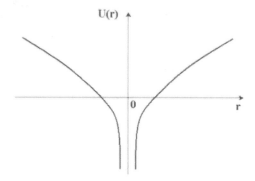

FIGURE 10.12

The effective potential of interaction between q and $\bar{q}$.

the quark confinement inside the hadron is explained by the contribution of multigluons exchanges. In the next approximation the relativistic corrections, the spin-spin and spin-orbital interactions, are taken into account in the expression for the potential $V^c(r)$. As we see, with growing the distance to the source $V^c(r)$ is increased. Just the same reasons can be applied to any composite color singlet. If one tries to disjoin some color part of this system (say, a qq-pair in a baryon) from others, then its energy will linearly be increased resulting in the confinement of color components.

One more sufficiently perspective attempt of the confinement explanation is the so-called Wilson lattice theory [43]. In lattice theories one is assumed that the space and time do not make up a continuum but represent the points discrete set that resemble the crystal lattice (most commonly, the cubic one). The quarks are placed in the lattice sites and the field lines of the gluons field connect the lattice arbitrary site with its nearest neighbors only. Further, it is assumed that the energy of the interaction between quarks or between quarks and antiquarks is proportional to a length of a string to connect them. One may draw an uncounted set of strings between two points of a such a space-time. In the lattice theory the quantum mechanics average is fulfilled on these strings. Integrals appearing in the process may be analytically calculated in the region of so-called strong coupling when the lattice step is much bigger than the typical scale of the quantum fluctuations of the gluon fields ($\sim 10^{-13}$ cm). To pass to the continuous space-time is realized by the way of decreasing the lattice step. Since in this case the lattice sites are merged with each other, then the problem of calculating the large multiplicity integrals appears. The problem is usually resolved with the Monte Carlo method. As this method could be applied to the finite multiplicity integrals the lattices with the finite number of the sites along every one of four axes are only considered. In the long run one may show that under certain conditions the quarks confinement taking place in the strong coupling region is persisting under decreasing the lattice step as well. Detailed description of the lattice QCD could be found in the book by Creutz [44].

Since the gluons bear the color they cannot exist in free states as well. If the QCD is true, then one should expect the existence of hadrons containing the gluons only. The most simple bound gluon state is two gluons forming the color singlet. Of course one may build the color singlets from three and more gluons. Hadrons of these kinds are called *gluonium* or *glueball*. Different glueballs may be discriminated by the spin and the mass only. From the theoretical point of view the gluonium identification seems to be very difficult because it is impossible to point to the decays or other properties of the gluonium that would certainly discriminate the gluonium from the quarkonium having the same quantum numbers. On the other hand, calculations show that the most intensive gluonium formation must occur in those reactions and decays in which the gluons rather than the quarks are produced at small distances. Examples are found in decays of heavy mesons ψ and Υ. However, up to now the gluonium has not been discovered. The numerical calculations on a computer being done within the QCD allow us to obtain the definite predictions concerning the masses of the most light glueballs. In doing so the typical masses scale proves to be of the order of 1.5 GeV.

One such confirmation of the existence of the quarks and the gluons is the process of detecting the hadron jets. To understand it we recall how elementary particles are observed in the Wilson chamber. The track left by the particle is not a manifestation of the particle itself at all. It is the result of the interaction between the particle and matter filling the chamber. This interaction leads to the production of great numbers of ions along the particle trajectory. It is evident that the interaction distorts the elementary particle motion. For high energy particles (only such particles have left the track) such distortions (for example, the track jitter in a transverse plane) are negligibly small and we could state the trajectory of the particle itself as being observed directly. Analogously, the multiparticle cluster of fast hadrons with small transverse momenta, the hadron jet, is the track for a parton that is outgoing from the deep inelastic scattering region. The jet is not only the visualization way of the "free" quark or gluon but the form of their existence as well. Here a vacuum, a structure of which is set by the QCD, plays the role of an environment filling the track chamber. In this vacuum, just the same as in the QED vacuum, small-scale fluctuations (SCF) are present thanks to the asymptotic freedom phenomenon. Apart from the SCF, gluon field long-wavelength fluctuations (LWF) caused by nonlinear character of the QCD exist. At this time, the notions of "small" or "large" are determined from the viewpoint of a parton, that is, the LWF are realized on the distances of the quark universe radius (~ 1 Fermi). The LWF corresponds to the really strong interquarks interaction. They play the role of "matter" with which a high-energy parton interacts. The average transverse momentum $< p_T >$ of creating hadrons just conforms to the LWF scale: $< p_T > \approx 300$ MeV ≈ 1 Fermi^{-1}. Thus the hadron jet created by the quark or the gluon may be identified with the quark itself or gluon itself in just the same sense as the chain of drops in the Wilson chamber is considered by the particle trajectory.

In 1975, based upon the results of analyzing the e^+e^--annihilation into hadrons at the electron-positron storage ring SPEAR, the Stanford research group announced the discovery of quark jets. It appeared that when the jet energy grows the average angle of the jet spread decreases, that is, hadrons are increasingly gathered round the direction of flying away q and $\bar{q}$. At the jet energy of the order of 18 GeV the hadrons that constitute the jet occupy only 2% of the total solid angle. Investigations have also shown that the correspondence $quark \leftrightarrow jet$ has a universal character. This means the composition of hadrons in jets (relationships between p, n, π, K, and so on) and hadrons distribution over momenta do not depend on what concrete reaction the given flavor quark, the jet primogenitor, is produced.

The first indirect manifestations of gluon jets (DESY, PETRA, 1979) were connected with researching the decays of the Υ-meson* into hadrons. According to the QCD the three gluons must be produced at annihilation of the $b\bar{b}$-pair, that is, the decays in question must have the three-jets nature. True enough, in the case of the Υ-mesons it is difficult to observe three gluon jets directly because the gluons energy is too small as yet. For this reason two

---

*This meson represents the bound state of two quarks b and $\bar{b}$ to be dealt within the next subsection.

indirect methods are used. The essence of the former is as follows. Usually, at large energies in the result of the electron-positron annihilation two quark jets in the opposite directions are created. If one increases the energy E_{cm} to an extent that the Υ-meson begins to be born, then the situation will be changed. In the final state instead of the quark-antiquark pair three gluons will be created, their momenta being allocated now on the whole space. As a consequence, at $E_{cm} \geq m_{\Upsilon}$ the two-jets structure of the created hadrons, which existed at $E_{cm} < m_{\Upsilon}$, should disappear. Just that effect was registered at DESY.

The second method of the indirect observing of the gluon jets was based on kinematics of the events. At the electron-positron pair annihilation the Υ-meson is created in the rest. Consequently, the total momentum of the created gluons must be equal to zero as well. But three three-dimensional vectors, the sum of which equals zero, must lay in the same plane. With good precision this should also be fulfilled for the particle momenta of the final hadron state. Of course this plane is varied from one decay to another. The analysis of the Υ-decays showed the final hadron momenta do lay in the same plane.

In the same year, a little bit later, under increasing the PETRA energy up to 20 GeV the bremsstrahlung of the gluon by the quark was observed in the following process:

$$e^+ + e^- \to q + \overline{q} + g.$$

The additional hadrons created by the bremsstrahlung gluon should lead to the azimuthal asymmetric thickening of one of the quark jets. With the energy increasing and the enhancement of the statistics set, one managed to select the gluon from the basic jet and measure the distributions both over energy and over the angle of flying out of the gluon jet. Measurements have shown the new jet behaves similarly to the bremsstrahlung photon in the reaction:

$$e^+ + e^- \to \mu^+ + \mu^- + \gamma,$$

that is, just as the particle with the spin 1 should behave.

Once one has managed to "see" the quark, the next task was to count the quarks number (with allowance made for the color and the flavor). For this purpose the process of the e^+e^--annihilation into hadrons appeared to be the most suitable

$$e^+ + e^- \to q + \overline{q} \to 1 \text{ jet} + 2 \text{ jet}. \tag{10.130}$$

The quark and the antiquark, which are produced at the second stage of the reaction, cannot exist in the free states since they are the color objects. They extract the color quark-antiquark pair suitable to them from a vacuum and recombine with this pair into colorless hadrons that fly apart as two jets directed oppositely. At sufficiently high energies the contributions coming from created quark-antiquark pairs with different colors and flavors are incoherent, that is, the total cross section of the reaction (10.130) is presented in the form of the cross sections sum over quarks of all the colors and flavors. The cross section obtained may be compared with that of the e^+e^- annihilation into the other point particle, muons

$$e^+ + e^- \to \mu^+ + \mu^-. \tag{10.131}$$

Both the quarks and charged leptons are described by the Dirac equation. However, mass and charge entering into this equation are quite different for quarks and leptons. In the case of high energies one may neglect their masses to get the following expression for the ratio of the cross sections of the reactions (10.130) and (10.131):

$$R = \frac{\sigma_{e^+e^- \to hadrons}}{\sigma_{e^+e^- \to \mu^+\mu^-}} = \frac{\sum_i \sigma_{e^+e^- \to q_i \overline{q_i}}}{\sigma_{e^+e^- \to \mu^+\mu^-}} = \sum_i \left(\frac{e_{q_i}}{e}\right)^2. \tag{10.132}$$

If the quarks did not have the color degree of freedom, this ratio would be[*]

$$R = \left(\frac{e_u}{e}\right)^2 + \left(\frac{e_d}{e}\right)^2 + \left(\frac{e_s}{e}\right)^2 = \frac{4}{9} + \frac{1}{9} + \frac{1}{9} = \frac{2}{3}. \qquad (10.133)$$

The experiment gave three times as much value. But such trebling must appear with the regard for the color degrees of freedom. Really, every quark-antiquark pair of the given flavor may be created in three different color states.

The origin of the ordinary strong interaction between real colorless hadrons (for example, between nucleons in the atom nuclear) is explained by approximately the same way as the origin of the chemical couplings of the electromagnetic nature between electrically neutral atoms and molecules in matter, that is, by means of interaction of the charged structural components. In particular, the π-mesons exchange between nucleons[†] may be connected with the quark-antiquark pair production inside a nucleon and a subsequent conversion of these pairs into the π-mesons that can fly out of the nucleons because they are colorless objects.

10.5   Heavy quarks and flavor symmetries

Thus, we have managed to build all the hadrons, which were known by 1975, with the help of the u_i, d_i, and s_i-quarks. And the understanding illusion visited us once again: *all matter in the universe consists of combinations of nine quarks and four leptons (electron, electron neutrino, muon, and muon neutrino)*. It seemed that the problem of matter structure was close to the completion. It only remained for us to make more precise the properties of the fields describing the weak and strong interactions. True enough one "but" was again, namely, the so-called quark-lepton symmetry.

Stable matter consists of the electrons, the u and d-quarks. We have every reason to believe that the electron neutrino is a stable particle as well. We call this totality by the first generation of the quarks and leptons. Further, we introduce a new quantum number S^W with the same algebra as the ordinary spin has and consider the representation with the weight $1/2$. Let us place the first generation particles into the weak isospin doublets

$$\begin{pmatrix} \nu_e \\ e^- \end{pmatrix}, \qquad \begin{pmatrix} u \\ d \end{pmatrix}. \qquad (10.134)$$

Notice, the charges difference of neutrinos and charged leptons are equal to those of the *up* and *down* quarks while the algebraic sum of the charges of quarks (with allowance made for trebling over the color) and leptons is equal to zero. There exists the second lepton generation too

$$\begin{pmatrix} \nu_\mu \\ \mu^- \end{pmatrix}, \qquad (10.135)$$

whereas the second quark generation is kept unfilled

$$\begin{pmatrix} ? \\ s \end{pmatrix}. \qquad (10.136)$$

---

[*]Running ahead we stress, it is valid only when energies are smaller than the creation threshold of hadrons, which consist of more heavier quarks, that is $E_{e^-e^+} < 3$ GeV.

[†]Such a treatment of nuclear forces hold good when the distance between nucleons exceeds 8×10^{-14} cm.

If one assumes that the quark-lepton symmetry is the immovable law of nature, then the existence of the quarks of a new flavor is needed.

In the autumn of 1974 in Brookhaven National Laboratory the group under the supervision of C. Ting began investigating the process of the electron-positron pair production at the collisions of the protons with the helium target (pp-collisions) in the mass region between 2 and 4 GeV. It was discovered that majority of the e^-e^+-pairs has the mass being approximately equal to 3.1 GeV, that is, the pronounced maximum takes place in the cross section of the process

$$p + He \rightarrow e^+ + e^- + X, \tag{10.137}$$

where we have designated the undetected particles plurality by the symbol X. If this maximum corresponds to the true resonance, then it should be present in the cross sections of some other reactions. And really, at about the same time the group working at the SPEAR facility (B. Richter as a supervisor) discovered this resonance under investigating the processes

$$e^+ + e^- \rightarrow \text{hadrons}, \qquad e^+ + e^- \rightarrow e^+ + e^-, \qquad e^+ + e^- \rightarrow \mu^+ + \mu^-. \tag{10.138}$$

Both groups simultaneously reported about the discovery of the new particle with the mass ~ 3.1 GeV. Since it was highly difficult to decide who bore the palm, then the new particle was called the double name J/ψ (the symbol J means Ting's name in Chinese and the name ψ was proposed by the SPEAR Collaboration). The name J/ψ has been saved not only as a tribute of respect to its ground breakers but also due to a play on words, $J/\psi=$ gi/psi=gipsy, which has been so amusing to the romantic soul of a physicist.

The particle J/ψ has the spin 1, the negative parity and it is long-lived by the microworld standards, that is, it has the anomalously small decay width $\Gamma \approx 70$ keV whereas ordinary resonances have $\Gamma = 100$—200 MeV. To clarify, the true nature of the new particle was being proceeded by about three years. Among the working hypotheses that have appeared, there was even the hypothesis: *J/ψ is not a hadron but the long-awaited neutral intermediate boson, one of carriers of the weak interaction.* With the passage of time larger and larger arguments arise what begin to turn the scale in favor of the hypothesis: *J/ψ consists of a quark and a antiquark, $c\bar{c}$, with a new flavor.* From the simplest estimation $m_c \simeq m_{J/\psi}/2 = 1.55$ GeV it follows at once that the c-quarks should be much more heavier than the u, d, and s-quarks. To match the theory and experiments one should make an assumption that the c-quark is the carrier of the new quantum number, the charm, $c = 1$ for c-quark and $c = -1$ for $\bar{c}$. Then J/ψ represents the particle with the hidden charm and, on its structure, is very similar to φ-meson which is constituted of the strange quark and antiquark (the particle with the hidden strange). This resemblance has served as a key to understand the small decay width of the discovered particle. In Figs.10.13(a) and 10.13(b) the quark diagrams corresponding to the φ-meson decay are displayed. The distinction from Feynman diagrams resides in the fact that here the quarks, when $t = \pm\infty$, are not free particles because hadrons hold them captive as before. Besides, in these diagrams the strong interaction between quarks are not usually displayed. Just the same as in the case of the Feynman diagrams the arrow directed backwards the time* corresponds to the antiparticle. In Fig. 10.13(a) the s-quarks entering into the φ-meson composition get over to the K-mesons composition. In the second case (Fig. 10.13(b)) the strange quarks are annihilated and instead of them the pairs of the u- and d-quarks appear. The experiments show the φ-meson predominantly decays through the channel $K + \overline{K}$ (the relative probability, the branching, $Br \approx 84\%$) and very reluctantly decays through the channel $\pi^+ + \pi^- + \pi^0$ ($Br \approx 15\%$). At this example

---

*On the diagrams the time flies from left to right.

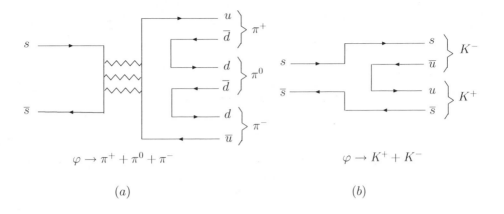

FIGURE 10.13
The quark diagrams (a) and (b) of the φ-meson decay.

we see how to work the approximate semi-phenomenological rule by Okubo Zweig–Iizuka (OZI) [45], which assumes the systematization of relative amplitudes of reactions of hadrons interaction depending on a topology of quark diagrams to display these processes. The largest degree of suppression is present in the diagrams where the quarks and antiquarks lines going out of one and the same hadron are connected with each other and represent the block that is not related with the remaining part of the diagram. In this case the quark-antiquark pair belonging to one and the same hadron disappears. An alternative to such a process is the process where the same quark and antiquark pass into the different hadrons of the final state. All this shows once again the behavior uncommonness of the quarks. In processes of inclusive scattering the created quark-antiquark pairs behaved in such a way as though they beforehand deduced in what groups on two (mesons) or three (baryons) they should be unified. The OZI rule may be also understood as the manifestation of the specific quark "thinking," namely, the quarks choose the possible variant of the future hadron prison already at their creating. The OZI rule, as applied to the J/ψ-particle, means J/ψ would predominantly decay through the particles containing the c-quark and the light u-and d-quarks (Fig. 10.14(b)). These particles having the explicit charm are called the *D-meson*.

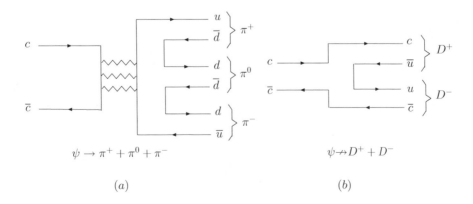

FIGURE 10.14
The quark diagrams (a) and (b) of the J/ψ-particle decay.

However, by a lucky chance, the decay

$$J/\psi \to D^+ + D^- \tag{10.139}$$

proved to be forbidden energetically ($2m_D > m_{J/\psi}$). The decays into the π- and K-mesons were allowed energetically but they were connected with the annihilation of the c-quarks and, as a result, were greatly suppressed by virtue of the OZI rule (Fig. 10.14(a)). Thus, the small value of the J/ψ-meson decay width served as the indication of the existence of the new kind of quark c.

About a week later after the J/ψ-meson discovery, at SPEAR the narrow resonance ψ' placed at slightly more high energy was detected. With further increasing the energy one had found some resonances both with the spin 1 and with the spin 0 in the neighborhood of 4 GeV. If one displays the mass spectrum of these particles (we call them the ψ-*particles*) graphically, then this spectrum will resemble the atom spectral lines picture (Fig. 10.15). According to contemporary concepts, this mass spectrum can be related to the energy levels

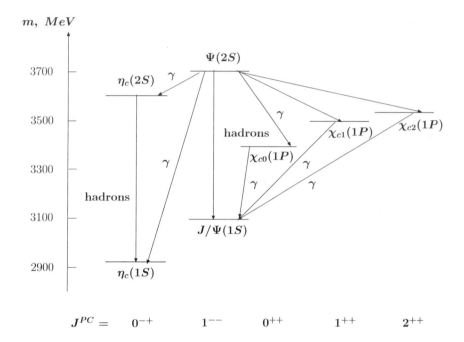

FIGURE 10.15
The charmonium spectrum.

spectrum of the system $c\bar{c}$. By analogy with the well-known positronium the bound system $c\bar{c}$ is called the *charmonium*. However, unlike the positronium where the binding forces have the electromagnetic nature, the charmonium owes its existence to the strong interaction. The particles denoted by a prime represent the radial excitations of the underlying states. To classify the charmonium levels we make use of the spectroscopic designations $^{2S+1}L_J$. Of course, this is a nonrelativistic classification but the nonrelativistic approach is lawful because of the c-quark large mass (by now experiments give the current c-quark mass value in the interval from 1.15 to 1.35 GeV).

From the data concerning only ψ-particles it was impossible to make the final conclusion in favor of the existence of the c-quark. The hypothesis about the charm quark is ultimately confirmed only after the discovery of hadrons with the explicit charm (1976): mesons

$$D^+ = c\bar{d}, \qquad D^0 = c\bar{u},$$

$$D^- = d\bar{c}, \qquad \overline{D}^0 = u\bar{c},$$

$$F^+ = c\bar{s}, \qquad F^- = s\bar{c},$$

and baryons

$$\Lambda^+ = cdu, \qquad \Sigma_c^+ = udc, \qquad \Sigma_c^{++} = uuc, \dots$$

These discoveries furnished the genuine triumph of the quark theory of hadrons structure.

Since the mass of the discovered fourth quark was much bigger than the masses of the u-, d-, and s-quarks it was already hard to say of the existence of the $SU(4)$-symmetry (flavor symmetry) in the hadron world. Notice, the $SU(2)$-symmetry violation is $\sim 1\%$ while the $SU(2)$-symmetry violation is $10 \div 20\%$. According to a contemporary point view, a violation of a $SU(n_f)$-symmetry is caused by the quark masses difference; namely, the more apparent this difference the stronger the violation of the corresponding $SU(n_f)$-symmetry.

However, flavor symmetries are being used as before to classify hadrons. Now, to graphically picture the $SU(4)$-group representations one should draw on a three dimensional coordinate system. Let us agree to plot the charm value on the axis z, the isospin projection value on the axis y, and the strange value on the axis x. In Figs. 10.16, and 10.17 the $SU(4)$ 16-plet for pseudoscalar mesons and the $SU(4)$ 20-plet baryons are shown, respectively. The shaded areas correspond to the $SU(3)$-multiplet with the equal value of the charm. Soon

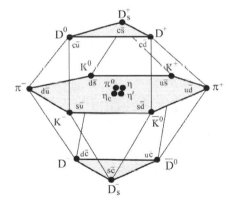

FIGURE 10.16

The 16-plet for the pseudoscalar mesons made of the u, d, s, and c quarks.

after the J/ψ-meson discovery in 1975 one more charged lepton, a τ-lepton, was detected in the independent experiments series. The direct confirmation of the existence of the ν_τ-neutrino will have taken place much later (2001). However, even in 1975, the majority of physicists based upon the idea of the unity of nature, was considering that the τ-lepton also has the neutrino satellite. Thus, the quark-lepton symmetry that had been rebuilt with the c-quark discovery was violated again. The new lepton pair should correspond a new quark pair. The first confirmation of this hypothesis happened in 1977. In the Fermi laboratory (FERMILAB) at the proton accelerator with the energy up to 400 GeV, a new particle with the mass 9.45 GeV and $J^P = 1^-$ was discovered. It was called Υ-*particle (upsilon)*.

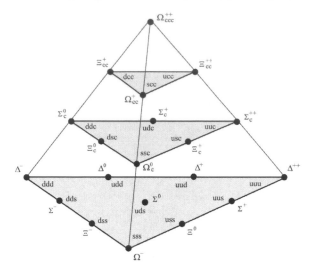

FIGURE 10.17

The 20-plet of baryons made of the u, d, s, and c quarks.

Subsequently, in DESY (Hamburg) and at Cornell University (USA) at investigation of the electron-positron collisions the existence of the Υ-particle was confirmed and its excited states (Υ', Υ'' and so on) were detected. Since the properties of the Υ-particle are very similar to those of the J/ψ-meson then one assumed that it is the bound state of the quark-antiquark pair of the fifth kind, $\Upsilon = b\bar{b}$. As the particles belonging to the Υ family (we call it *upsilonium*) have masses around 10 GeV, then the mass of the quarks entering into them must be by no means smaller than 5 GeV (according to the contemporary data the mass of the current b-quark lays in the interval from 4 to 4.4 GeV). The investigations of the Υ-meson properties have shown the b-quark has $I = 0$ and $Q = -1/3$. Physicists were going on the already well-paved road for the small width of the Υ-meson decay to be explained. The new quantum number, the beauty $b = 1$, which is conserved in the strong interaction only, was assigned to the b-quark. Then the OZI rule suppressed decays into hadrons that do not contain b-quarks and nature, for its part, in order not to lead the researchers into temptation, gave orders in such a way that the decays into a pair of mesons with the explicit beauty proves to be forbidden energetically (the most light beauty mesons $B^+ = u\bar{b}$ and $B^0 = d\bar{b}$ have the mass 5270 MeV). As we know the quantity

$$R = \frac{\sigma_{e^+e^- \to hadrons}}{\sigma_{e^+e^- \to \mu^+\mu^-}}$$

is a sensitive tool to define the quark flavors number. Between thresholds of a $q_i\bar{q}_i$-pairs production the quantity R is constant and when the next i threshold had been achieved it was abruptly increased on the value $3Q_i^2$. In Fig. 10.18 we display the experimentally measured values of R^{exp} as a function of energy in the center-of-mass frame. The resonant levels of the charmonium and upsilonium are put on the stepped monotonous behavior. Upon subtracting the resonant contributions R^{exp} is well matched with the experimental value, which is the powerful argument in favor of the b-quark existence.

To date, several tens of mesons and several baryons with the explicit beauty are known. Contrary to them the upsilonium represents the state with the hidden beauty.

Thus, in the late eighties physicists have at their disposal the two quarkoniums, the charmonium and upsilonium. In the QCD the quarkonium role is similar to the hydrogen

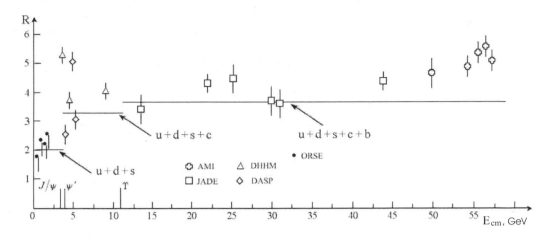

FIGURE 10.18

The experimental values of R^{exp} as a function of the electron-positron beam energy.

role in Atomic Physics. The quarkoniums serve as probes of the strong interaction at small distances. In particular quarkonium decays are the gluons source, that is, here gluon properties and specificity of gluons conversion into hadrons could be investigated.

After the b-quark discovery the hunting season is opening on its partner on the weak isospin doublet, which was called the t-quark. Its name originates from the first letter of the word *top*, though one may consider that the origin is the word *truth*. At first, the indirect evidences for the existence of the sixth quark were obtained. Among those, the results of investigating the process

$$e^+ + e^- \rightarrow b + b \qquad (10.140)$$

at the CERN electron-positron collider called LEP (Large Electron Positron) were the most convincing. To study the asymmetry in the scattering of the b-quark confirmed the b-quark does represent the member of the $SU(2)$ doublet of the weak isospin, as the electroweak interaction model by Glashow, Weinberg, and Salam (GWS) [46] predicted. Recall, by that time the GWS theory had gone through the basic stage of its experimental checkout. Moreover, the detailed measurements of the properties of the $W^{\pm}$- and Z-bosons that were fulfilled at LEP, SLC (Stanford Linear Collider, e^+e^--collider), CERN $Sp\bar{p}S$ ($p\bar{p}$-collider), and FERMILAB Tevatron ($p\bar{p}$-collider) gave the unquestionable proofs in favor of the existence of the heavy quark connected with the b-quark by means of the electroweak interaction. In April 1994 the first experiments, which directly point the t-quark existence, appeared. They were carried out at FERMILAB Tevatron, which represents the six-kilometer storage ring where the protons and antiprotons having the energy 1.8 TeV in the center-of-mass frame are rotated in opposite directions. The $p\bar{p}$ collisions occur in two points of the ring where two detectors system, CDF and D⊘ detectors, are placed. The collider luminosity* was being constantly increased and in 1994 it reached the value $L = 10^{31}$ cm$^{-2} \cdot$ s^{-1}, which allowed one to detect the t-quark. At looking for the t-quarks one is guided by the fact

---

*The luminosity L is the collisions number in a second on the unit section, that is, L, being multiplied by the cross section of the given process σ that expressed in cm^2, gives the number of the corresponding events in a second.

that the cross sections both of their productions and of their consequent decays depend on the t-quark mass value. The first to raise the low bound on m_t to ~ 175 GeV was CDF Collaboration. Already in 1994 they observed twelve events with $m_t \sim 175$ GeV that correspond to the selection criterions for the t-quark. But the expected background constituted roughly six events and, although the interpretation of the obtained results as the t-quark observation offered the most probable, the statistical providing of the result was insufficient to recognize it as the t-quark discovery. And only in February 1995, in one and the same day, both Collaborations, CDF, and D0 sent off the reports about the t-quark detection to the press.

According to the QCD, at energy in the center-of-mass frame $E = 1.8$ TeV, the $t\bar{t}$ pair production in $p\bar{p}$ collisions mainly occurs at the cost of the subprocesses:

$$q + \bar{q} \to t + \bar{t}, \qquad g + g \to t + \bar{t}, \qquad (10.141)$$

where we have designated the gluons by the symbol g. We choose the axis z along the proton beam direction and shall study such variables of the final products of a reaction as the transverse momentum relative to z axis p_T and connected with it the transverse energy E_T. The standard model (SM) predicts that the t-quark decays through the channel $t \to W^+ b$ just about always. Distribution over the transverse momentum for one of the decay products has the peculiarity that enables one not only to detect such decays but to define the mass of the unstable particle as well. This method based on merely kinematic arguments became very popular after it had been used under discovering the W-boson in $p\bar{p}$ collisions (in that case, the electrons and the positrons with very high transverse momenta, $p_T \sim 40$ GeV, were detected). At sizeable excess of the decaying particle mass m_i over the decay product masses, the events concentration maximum is observed close to the transverse momentum value of one of the decay products k_T to be approximately equal to $m_i/2$. However, this method of the t-quark identification did not work since the W-boson was the unstable particle with a very small lifetime while the b-quark served as a source of producing the hadron jets. The W-boson decays into $q\bar{q}'$-pairs ($u\bar{d}$ or $\bar{s}$) with the probability 2/3 and decays into one of three lepton families ($\bar{l}\nu_l$) with the probability 1/3. Thus, there are 6 high-energy point fermions that could be either charged and neutral leptons or quarks giving rise to the production of the hadron jets. It was very difficult to select the decays with the τ-leptons from the hadron background and, for this reason, they were not taken into consideration. High background (signal-to-noise correlation was smaller than 10^{-4}) made an inclusion in analysis of final states with six hadron jets impossible. As a result the following channels decay of the $t\bar{t}$-pair were available to observe

$$t\bar{t} \to e^+ \nu_e b e^- \bar{\nu}_e \bar{b}, \qquad (1/81), \qquad (10.142)$$

$$t\bar{t} \to \mu^+ \nu_\mu b \mu^- \bar{\nu}_\mu \bar{b}, \qquad (1/81), \qquad (10.143)$$

$$t\bar{t} \to e^\pm \nu_e b \mu^\mp \bar{\nu}_\mu \bar{b}, \qquad (2/81), \qquad (10.144)$$

$$t\bar{t} \to e^\pm \nu_e b q \bar{q}' \bar{b}, \qquad (12/81), \qquad (10.145)$$

$$t\bar{t} \to \mu^\pm \nu_\mu b q \bar{q}' \bar{b}, \qquad (12/81), \qquad (10.146)$$

where the numbers in the parentheses denote the branching predicted by the SM. First three dilepton channels prove to be the most clear but they have very small statistics. Last two channels, *lepton+hadron jets*, have high statistics (close to 30% on the total decay width $t\bar{t}$) but suffer from extremely large background. At $p\bar{p}$ collisions the total cross section of the $t\bar{t}$-pair production depends not only on the parton distributions inside the proton and the antiproton but on the t-quark mass as well. So, at $m_t = 100, 155$ GeV it has the values $\sim 10^2$ and 10 pb, respectively. It should be stressed that these values are very small by the strong interaction standards. When $E = 1.8$ TeV the $p\bar{p}$ collisions also lead to the production of hadron jets and lepton pairs in the final state at the cost of the production of the W- and Z-gauge bosons in virtual states. It has been just these events that constitute the main background for the $t\bar{t}$-pair production. To compare, we give the values of the competed cross sections:

(i) for the W-production—$\sigma_W \sim 20$ nb;
(ii) for the Z-production—$\sigma_Z \sim 2$ nb;
(iii) for the WW-production—$\sigma_{WW} \sim 10$ pb;
(iv) for the WZ-production—$\sigma_{WZ} \sim 5$ pb.

Let us consider the criterions used by CDF and D0 Collaborations to select the signal from the background. The heavy quark pair production ($m_t = 173.8 \pm 5.2$ GeV) and its consequent decay generates the final states with more large average energy than the background events. To investigate the distribution of the final hadrons and leptons over the transverse momenta proves to be useful in this case as well. High-energy electrons, muons, and hadrons were recorded with the help of different detectors and were easily distinguishable from each other. As for the neutrino is concerned the disbalance of the total transverse energy E_T or the missing transverse momentum p_T testify to its existence in the final state. It is evident that the quantity

$$H_T = \sum_{i=1}^{N_r} E_T^i,$$

where the summation is realized over all hadron jets and basic electron clusters by which are meant a leptons plurality, including even if one electron, is very useful for the kinematic analysis. At the investigation of the final states *lepton+hadron jets,* two methods were used by D0 Collaboration to select the signal from the background. The former based upon the kinematic analysis (KA) resided in the demand $H_T > 200$ GeV and in the presence of at least four hadron jets with $E_T > 15$ GeV. The second method was connected with the b-quark identification (B tagging-out) through the decays

$$b \to \mu^- + \bar{\nu}_\mu + X, \qquad b \to \mu^+ + \nu_\mu + X. \tag{10.147}$$

Studying the data on *lepton+hadron jets* the CDF Collaboration already used other methods. One of them was also connected with the b-quarks and received the name: "tagging-out of the second vertex (TSV)." The tracks of charged products of the b-quarks decay are detected in the drift chamber. Only the tracks with $p_T > 1.5$ GeV (decays of the b-quarks to be produced from the $t\bar{t}$-pairs and the W-bosons are their sources) were of interest. The b-quark decay point in the silicon stripped detector was rebuilt by the extrapolations method. It allowed selecting the signal from a background by means of a comparison of a different events intensity. The second method received the name "tagging-out of soft leptons (TSL)," was based on detecting the low-energy ($p_T \sim 2$ GeV) muons and electrons near hadron jets.

In Table 10.1 we give the expected number of the background events and the number of the events connected with the t-quark production in the observation results by CDF and D0 Collaborations. As it follows from Table 10.1, in all channels both Collaborations observed the sizeable excess of the signal over a background that undoubtedly testifies to the $t\bar{t}$-pairs production in the $p\bar{p}$ collisions.

To account for the experimental data one should assign to the t-quark a new quantum number t (the top or the truth) equal to 1 which along with the strange, the charm, and the

TABLE 10.1
The observation results for the t-quark production at
CDF and D0.

Sampling	Background	Signal
dileptons (CDF)	1.3 ± 0.3	6
dileptons (D0	0.65 ± 0.15	3
leptons+jets (D0 KA)	0.93 ± 0.5	8
leptons+jets (D0 B-tagging-out)	1.21 ± 0.26	6
leptons+jets (CDF TSV)	6.7 ± 2.1	27
leptons+jets (CDF TSL)	15.4 ± 2.3	23

beauty is conserved in the strong interaction but is not conserved in the electromagnetic and weak interactions.

Below in Table 10.2. we give the additive quantum numbers of all the six quarks. It was

TABLE 10.2
The quantum numbers of the quarks.

q	Q	J	I	I_3	B	s	c	b	t
u	2/3	1/2	1/2	1/2	1/3	0	0	0	0
d	-1/3	1/2	1/2	-1/2	1/3	0	0	0	0
c	2/3	1/2	0	0	1/3	0	1	0	0
s	-1/3	1/2	0	0	1/3	-1	0	0	0
t	2/3	1/2	0	0	1/3	0	0	0	1
b	-1/3	1/2	0	0	1/3	0	0	-1	0

natural to wait for the existence of a quarkonium consisting of the t-quark and t-antiquark ($\bar{t}$), the *toponium*. However, by now the toponium has not been discovered. It is not inconceivable that nature presented us only two kinds of the simplest "quark" atoms with the great numbers of energy levels. Recall, that in Atomic Physics the hydrogen atom was such a present. But in Nuclear Physics, mildly speaking, nature already was a little bit miserly and constrained itself by the deuteron only which has got none of the excited levels.

Introducing the b- and t-quarks extends the flavor symmetry of the strong interaction up to the $SU(6)$-group. However, thanks to the sharp gradation of the quark masses this symmetry is strongly violated. On the other hand, since up to the present hadrons containing the constituent t-quark have not been discovered, then for practical calculations we could successfully use the $SU(5)$ flavor symmetry whose violation degree is much less ($m_b \ll m_t$). The flavor symmetry plays a double role, it not only defines the classification of hadrons on various multiplets but it also establishes a series of dynamical relations between amplitudes of different processes of hadrons interaction. In the low energy region ($E \leq m_0$, where m_0 is an average mass value in a multiplet) strongly violated flavor symmetry possesses a weak predictive force and is unlikely applicable for practical use. However, at high energies ($E \gg m_0$) it becomes the useful tool of investigation.

Among unsolved problems of quark mechanics the problem of existing quarks in the free state appears to be principal. There is nothing to prevent us from conjecturing that the negative results on the quarks production are caused only by the insufficient energy of contemporary accelerators to create the quarks in a free state. This, in turn, means that at the hadrons production the giant quark mass is eaten by such huge binding energies. And only the fact of the free quarks existence can prove this statement.

Looking for the free quarks is being carried on two directions. The former is a geophysical-chemical approach, which is based on the assumption: the stable quarks occur in matter surrounding us. In this case the negative charged quarks will be captured by nuclei and form either quark atoms or quark ions with the noninteger charge. It is evident that the created compounds should have specific physical and chemical properties. Thus, the experiments of this approach are aimed for discovering the characteristic manifestation of the fractionally charged particles existence: the lowered ionization constituting 1/9 or 4/9 of the integer charged particle ionization; an unusual value of e/m in mass-spectroscopic experiments; an anomalous behavior of a levitating matter grain in an electrostatic field; a nonstandard position of spectral lines in quark atoms, and so on. The quarks were being sought in terrestrial matter, in lunar soil, in meteorites. To look for quark atoms in solar matter with the help of spectroscopic methods was also carried out. It is natural that the most reliable constraints on the free quarks existence have been obtained during searching for the quarks in the stable matter of the Earth. Different variants of experiments result in that the upper limit values of a possible quarks concentration in matter lay in the interval from 5×10^{-15} to $5 \times 10^{-28} \frac{\text{quarks}}{\text{nucleon}}$.

The second direction includes the attempts of detecting the quarks in cosmic rays and accelerators directly. In experiments on accelerators the free quarks are not discovered up to the masses 250 GeV in $p\bar{p}$ collisions (CDF, 1992) and 84 GeV in $e^- e^+$ collisions (LEP, DELPHI, 1997) at the production cross sections higher than 10 and 1 pb, respectively. Experiments on detecting the quarks in collisions of cosmic rays with particles in the upper atmosphere layers, which were fulfilled in a wide interval of energies and, consequently, did not have severe constraints on the mass of created quarks, have not also brought to success and have set the constraint on the quarks flux from the cosmos: $< 2.1 \times 10^{-15} \frac{\text{quarks}}{\text{cm}^2 \cdot \text{steradian} \cdot \text{s}}$ (KAM2, 1991).

Meanwhile, if the free quarks existence is not forbidden in principle, then they would be created at early stages of the universe evolution when the temperature was very high, say $kT > 2m_q$. At such a temperature the quarks are in the state of the thermodynamic equilibrium with other fundamental particles (number of created quarks is equal to that of annihilating quarks). At $kT \sim m_q$ the equilibrium is violated: reactions of the quarks production have been switched off and the quarks start to burn away. This burning away takes place at the cost of reactions of the kind

$$q + \bar{q} \to \text{mesons}, \qquad q + \bar{q} \to \bar{q} + \text{baryon}. \qquad (10.148)$$

Since the reactions (10.145) are exothermic, then their cross sections tend to constant values whose sum is denoted by σ_0. If one assumes that the cross section of the quarks destruction has the typical nuclear scale, say $\sigma_0 \sim m_\pi^{-2}$, one may show the quark-to-proton-concentration ratio in the present universe must be

$$\frac{n_q}{n_p} \sim 10^{-12}.$$

This number is greater than the gold abundance on the Earth but the quark Klondike never has been opened to date. One would assume the quarks are unstable particles and all this relic quark sea has had time to disappear by now. However, on the strength of the electric charge conservation law, at least one of the quarks must be stable and should live till the present day. So, either the free quarks are really absent in nature or their production cross section has as minimum the atom scale rather than the nuclear one.

It should be noted that some experimental groups make reports concerning the free quark observations every now and then. So, the Stanford University group investigated a behavior of a niobium ball $\sim 10^{-4}$g that levitated in the nonuniform magnetic field. They observed the cases when the electric charge of the ball was equal to $\pm e/3$. As this takes place, the corresponding quarks concentration has the order of $\approx 10^{-20}$ $\frac{quarks}{nucleon}$ [47]. However, these data are not confirmed by other investigations yet. And the result is included in a category of the reliable one, if and only if, it is independently obtained by several different groups using different experimental methods.

Physics development shows that nature gives answers to correctly posed questions only. There is no sense in the question that tormented ancient scholastics: "How many angels could be place into a sword tip?" The question, physicists tortured themselves at the very outset of Quantum Theory, appeared to be senseless: "What is the electron—the particle or the wave?" Maybe, when we are trying to detect the quarks in a free state we are in the analogous situation? It is not expected that the quarks represent the specific kind of quasi-particles, field quanta, which describe collective oscillations of corresponding freedom degrees of a hadron. We faced such formations in other regions of physics and before. Among these are: the magnon, the quantum of the spin oscillations in magneto-ordered systems; the plasmon, the quantum of the charge density oscillations in conductive mediums; and the phonon, the quantum of the elastic oscillations of the atoms or molecules in crystal lattice. At switching off interaction similar particles are pulled down into compound parts and stop their existence. For example, the phonon decays and turns into a plurality of independent motions of particles, which constitute a crystal. Then the quarks have the sense only as dynamical essences inside hadrons in just the same way as the phonons that cannot exist outside a crystal. However, be it as it may, looking for the free quarks are continued. The problems of their detection at LHC (Large Hadron Collider) and accelerators of the next generation, NLC (Next Liner Collider), FMC (First Muon Collider), and so on, are intensively discussed.

11

Passing glance on the theory of electroweak interaction

The point is that the owl and the hawk who had taken joint leadership of education made a gross blunder. They took it into their heads to teach grammar to the eagle himself. But a still greater blunder was that like all tutors, neither one nor the other gave their pupil a breather. The owl dogged his footsteps, shouting out bbb, kkk, zzz, while the hawk unceasingly tried to impress on him that without the four rules of arithmetic one cannot divide one's booty.
M.E. Saltykov Shchedrin, "The Eagle-Patron of Arts"

11.1   Spontaneous symmetry breaking

The QCD is not the sole descendant of the Yang-Mills theory. The electroweak interaction theory was also built on the basis of the non-Abelian gauge theory. However, the way of its production was already more complex. The point is that gauge fields are long-range by its nature. This fact immediately leads to the zero mass of the interaction carriers as it takes place both in the QCD and in the QED. The weak interaction exists only at very small distances and the weak interaction carriers, the $W^{\pm}$- and Z-bosons, must have a huge mass. So, it was necessary to combine two incompatible things, namely, the local gauge invariance and the nonzero mass of the $W^{\pm}$- and Z-bosons. It appeared that one needs to use the idea of a spontaneous symmetry breaking to solve this problem.

An evolution of any physical system is defined by two factors: (i) a form of the Lagrangian (or the Hamiltonian), and (ii) initial conditions. As an example, we consider a smooth surface in the form of a peaked hat resting on a horizontal ground (Fig. 11.1). We place a ball on the hat top and shall consider that the gravitational force is the sole force acting on the system. It is clear that the system possesses the explicit symmetry with respect to rotations about a horizontal axis passing through the hat center. However, the system is not stable. Really, if we move the ball out of position, that is, change the initial condition, it will be rolled down and the system symmetry will be violated. Having stopped on the fixed place of the hat brim, the ball sets the selected direction from the central axis. The system has found a stability at the cost of the symmetry violation. Since the state with the violated symmetry has more low energy, the ball prefers to roll on the hat brim. In the stable configuration the initial rotational symmetry of the gravitational force exists as before but now it exists in the hidden form. The observed system state does not reflect the symmetry of the interaction that is present in the system.

In quantum field theory two analogous conditions define a character of a symmetry manifestation: (i) a form of the Lagrangian (or the Hamiltonian); and (ii) a vacuum form. In

FIGURE 11.1

The example of the spontaneous symmetry violation.

the QCD and the QED both the Lagrangian and the vacuum state were invariant under the symmetry transformations. In such cases we say the symmetry of the theory has not been violated. In 1960 Y. Nambu and J. Goldstone [48] showed that there exist theories whose Lagrangian is invariant with respect to symmetry transformations while a vacuum is not invariant. In this case we use the term "*spontaneous symmetry breaking.*" Since we do not want spontaneously to break the Lorentz invariance, spontaneous symmetry breaking can be realized only in the presence of scalar fields (fundamental or composite). With other words the vacuum is filled with scalars (with zero four-dimensional momenta) carrying the quantum number of the broken symmetry.

We investigate the spontaneous breaking of the local gauge symmetry by the example of the $U(1)$-group. Consider the world that consists of charged scalar particles only and is defined by the Lagrangian

$$\mathcal{L} \equiv T - V(\varphi) = [\partial_\mu \varphi(x)]^* [\partial^\mu \varphi(x)] - \mu^2 \varphi^*(x)\varphi(x) - \lambda[\varphi^*(x)\varphi(x)]^2 \qquad (11.1)$$

with $\lambda > 0$. When $\mu^2 > 0$, then the Lagrangian will describe a self-interacting (according to the law $[\varphi^*(x)\varphi(x)]^2$) scalar field with the mass μ^2. In this case the value $\varphi(x) = 0$ corresponds to the vacuum (the minimum of $V(\varphi)$. In other words, the average over the vacuum of $\varphi(x)$ turns into zero ($< 0|\varphi(x)|0 >= 0$). However, we wish to study the case with $\mu^2 < 0$. If one introduces the field functions according to the relation

$$\varphi(x) = \frac{\varphi_1(x) + i\varphi_2(x)}{\sqrt{2}},$$

the Lagrangian (11.1) takes the form:

$$\mathcal{L} = \frac{1}{2}\{[\partial_\mu \varphi_1(x)]^2 + [\partial_\mu \varphi_2(x)]^2 - \mu^2[\varphi_1^2(x) + \varphi_2^2(x)] - \lambda[\varphi_1^2 + \varphi_2^2(x)]^2/2\}. \qquad (11.2)$$

From this writing it is evident that the minima of the potential $V(\varphi)$ lie on the circle of the radius v in the plane φ_1, φ_2 (see, Fig. 11.2)

$$\varphi_1^2 + \varphi_2^2 = v^2, \qquad v^2 = -\frac{\mu^2}{\lambda}. \qquad (11.3)$$

This means the constant scalar field, the so-called scalar field vacuum condensate, exists in the vacuum. This quantity, that is, the energy shift of the ground state, cannot be measured

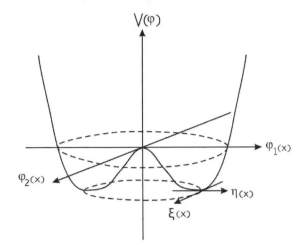

FIGURE 11.2
The potential of the complex scalar field.

directly since the quantity under test is the difference between the given energy and the vacuum energy,

In order to make the Lagrangian (11.2) be invariant with respect to the local gauge transformations of the

$$\varphi'(x) = \exp[i\alpha(x)]\varphi(x),$$

we introduce the covariant derivative

$$D_\mu(x) = \partial_\mu + ieA_\mu(x), \tag{11.4}$$

where $A_\mu(x)$ is transformed by the law

$$A'_\mu(x) = A_\mu(x) - \frac{1}{e}\partial_\mu\alpha(x). \tag{11.5}$$

Having supplemented the obtained Lagrangian with the free gauge bosons Lagrangian we arrive at

$$\mathcal{L} = [\partial_\mu - ieA_\mu(x)]\varphi^*(x)[\partial^\mu + ieA^\mu(x)]\varphi(x) - \mu^2\varphi^*(x)\varphi(x) - \lambda[\varphi^*(x)\varphi(x)]^2 -$$
$$-\frac{1}{4}F_{\mu\nu}(x)F^{\mu\nu}(x). \tag{11.6}$$

Within elementary particle physics in the majority of instances we cannot obtain exact solutions. We most commonly have to use the perturbation theory series expansion and calculate fluctuations close to a minimum energy. If we try to carry out the expansion in a neighborhood of the unstable point $\varphi = 0$, the perturbation theory series will not converge. The correct activity method is to carry out the expansion in a neighborhood of the minimum of the potential $V(\varphi)$, that is, in a neighborhood of the stable vacuum. For the minimum we choose the point $\varphi_1 = v, \varphi_2 = 0$. Notice, the vacuum is not already invariant with respect to the $U(1)$-group , that is, the symmetry appears to be spontaneously broken. To expand $\mathcal{L}$ in a neighborhood of the vacuum we introduce real fields $\eta(x)$ and $\xi(x)$, which describe quantum fluctuations around this minimum

$$\varphi(x) = \frac{\eta(x) + i\xi(x) + v}{\sqrt{2}}. \tag{11.7}$$

To substitute (11.7) into the Lagrangian (11.6) leads to the expression

$$\mathcal{L}' = \frac{1}{2}\{[\partial_\mu\xi(x)]^2 + [\partial_\mu\eta(x)]^2\} - v^2\lambda\eta^2(x) + \frac{1}{2}e^2v^2 A_\mu(x)A^\mu(x) - \frac{1}{4}F_{\mu\nu}(x)F^{\mu\nu}(x) + \mathcal{L}_{int},$$
(11.8)

where $\mathcal{L}_{int}$ denotes the terms describing interactions of $\eta(x)$, $\xi(x)$, and $A_\mu(x)$ fields. From the Lagrangian (11.8) follows

$$m_\xi = 0, \qquad m_\eta = \sqrt{2\lambda v^2}, \qquad m_A = ev. \tag{11.9}$$

In summary, we have attained the goal to be sought. Our gauge bosons have found the mass. However, as this takes place, the other problem connected with the occurrence of a massless scalar particle has appeared. Such particles are called *Goldstone bosons*. Let us gain an understanding of the situation. Having given the mass to the field $A_\mu(x)$, we thereby increased the number of polarization degrees of freedom from 2 to 3, because now the field $A_\mu(x)$ can have the longitudinal polarization too. But the simple shift of the field variables that is given by Eq. (11.7) cannot create new degrees of freedom in any way. It is obvious that not all the fields entering into $\mathcal{L}'$ correspond to the physical particles. It is beyond doubt that just the Goldstone boson arose suspicion. Let us show that it does not really belong to the physical sector. Since the theory is gauge invariant, we can carry out any gauge transformation (fix the gauge*) and the physical contents of the theory is kept invariable. The approximate equality

$$\varphi(x) = \frac{\eta(x) + i\xi(x) + v}{\sqrt{2}} \approx \frac{\eta(x) + v}{\sqrt{2}} \exp\left[\frac{i\xi(x)}{v}\right] \tag{11.10}$$

which is true to lowest order in $\xi(x)$ can suggest the required transformation form. It is clear that we should introduce new fields of the kind

$$\varphi'(x) = \exp\left[-\frac{i\xi(x)}{v}\right]\varphi(x) = \frac{\eta(x) + v}{\sqrt{2}},$$

$$A'_\mu(x) = A_\mu(x) + \frac{1}{ev}\partial_\mu\xi(x).$$

Further, taking into account

$$D_\mu\varphi(x) = \exp\left[\frac{i\xi(x)}{v}\right]\left[\partial_\mu\varphi'(x) + ieA'_\mu(x)\varphi'(x)\right] = \exp\left[\frac{i\xi(x)}{v}\right]\frac{1}{\sqrt{2}}\{\partial_\mu\eta(x) + ieA'_\mu(x)[\eta(x) + v]\},$$

and

$$|D_\mu\varphi(x)|^2 = \frac{1}{2}|\partial_\mu\eta(x) + ieA'_\mu(x)[\eta(x) + v]|^2, \qquad F_{\mu\nu} = \partial_\mu A'_\nu(x) - \partial_\nu A'_\mu(x),$$

we rewrite the Lagrangian (11.6) in the following form:

$$\mathcal{L}'' = \frac{1}{2}|\partial_\mu\eta(x) + ieA'_\mu(x)[\eta(x) + v]|^2 - \frac{\mu^2}{2}[\eta(x) + v]^2 - \frac{\lambda}{4}[\eta(x) + v]^4 - \frac{1}{4}F_{\mu\nu}(x)F^{\mu\nu}(x) = \mathcal{L}_0 + \mathcal{L}_{int},$$
(11.11)

where

$$\mathcal{L}_0 = \frac{1}{2}[\partial_\mu\eta(x)]^2 - \frac{\mu^2}{2}\eta^2(x) - \frac{1}{4}[\partial_\mu A'_\nu - \partial_\nu A'_\mu]^2 + \frac{1}{2}(ev)^2 A'_\mu(x)A^{\mu\prime}(x),$$

---

*The gauge that excludes the Goldstone boson is called the *unitary one*.

$$\mathcal{L}_{int} = \frac{1}{2}e^2 A'_\mu(x) A^{\mu\prime}(x) \eta(x)[\eta(x) + 2v] - \lambda v^2 \eta^3(x) - \frac{\lambda}{4}\eta^4(x).$$

To sum up, the field $\xi(x)$ disappears from the Lagrangian, that is, the seeming additional degree of freedom connected with the Goldstone boson proved spurious in practical situations. Obviously, this degree of freedom corresponds to the liberty of the choice of the gauge transformation. The Lagrangian describes a mere two interacting massive particles, the vector gauge boson $A_\mu(x)$ with the mass $M = ev$ and the scalar particle $\eta(x)$ with the mass $m = \sqrt{2}\mu$ (the Higgs boson). Having absorbed the Goldstone boson the massless gauge field became massive. The number of degrees of freedom has been conserved since the massive vector particle with three spin projections appeared from the massless vector field, having two spin states, and from the massless scalar particle. In field theory this phenomenon was discovered in 1964 and named the *Higgs mechanism*.

11.2   Glashow–Weinberg–Salam theory

The symmetry $SU(2)_L \times U(1)_Y$ was first proposed by S. Glashow (1961) and then it was expanded to include the massive vector bosons (W and Z) by S. Weinberg and A. Salam (1967–1968). At building the electroweak interaction theory by Glashow, Weinberg, and Salam (GWS) [46] one may single out four basic stages. It should be stressed that the analogous division takes place for any version of an electroweak theory as well.

First, one should choose a gauge group G in such a way that it includes all the necessary vector particles. The gauge symmetry $SU(2)_L \times U(1)_Y$ * is the base of the GWS theory. Here $SU(2)_L$ represents a group of the weak isospin while $U(1)_Y$ represents a group of the weak hypercharge (the term "weak" is used in order to stress the difference from the corresponding characteristics of the strong interaction). Thus, the weak isospin $\mathbf{S}^{\mathbf{W}}$ and the weak hypercharge Y^W (more precisely, $Y^W/2$) are generators of the gauge transformations of $SU(2)_L$ and $U(1)_Y$, respectively. The weak hypercharge is assigned to every field so that the analog of the formula by Gell-Mann-Nishijima is fulfilled

$$Q = S_3^W + \frac{Y^W}{2}. \tag{11.12}$$

The unbroken local $SU(2)_L \times U(1)_Y$-symmetry demands the existence of four massless vector bosons. Three of them W^1, W^2, W^3 represent gauge bosons of the non-Abelian group $SU(2)_L$ and their interaction is characterized by the gauge constant g. B describes the gauge field of the Abelian group $U(1)_Y$ and its interaction is determined by the gauge constant g'.

In the second stage one should choose the representation of the symmetry group for the matter particles (leptons and quarks). Since the theory will describe weak processes which, as it is well known, do not conserve the parity, then the theory must be explicitly mirror-asymmetric from the beginning. This asymmetry is realized as follows. The left-hand components of the fermions

$$\psi_L(x) = \frac{1}{2}(1 - \gamma_5)\psi(x)$$

---

*We used the designation $SU(2)_{EW} \times U(1)_{EW}$ for this group before.

form the weak isospin doublets with respect to the $SU(2)_L$-group

$$\begin{pmatrix} \nu_{eL} \\ e_L^- \end{pmatrix}, \quad \begin{pmatrix} \nu_{\mu L} \\ \mu_L^- \end{pmatrix}, \quad \begin{pmatrix} \nu_{\tau L} \\ \tau_L^- \end{pmatrix}, \qquad S^W = \frac{1}{2}, \ Y^W = -1 \tag{11.13}$$

$$\begin{pmatrix} u_L \\ d_L \end{pmatrix}, \begin{pmatrix} c_L \\ s_L \end{pmatrix}, \begin{pmatrix} t_L \\ b_L \end{pmatrix}, \qquad S^W = \frac{1}{2}, \ Y^W = \frac{1}{3}, \tag{11.14}$$

while the right-hand components of all fermions excepting the neutrinos

$$\psi_R(x) = \frac{1}{2}(1 + \gamma_5)\psi(x)$$

represent the weak isospin singlets

$$e_R^-, \ \mu_R^-, \ \tau_R^-, \qquad S^W = 0, \ Y^W = -2,$$

$$u_R, \ c_R, \ t_R, \qquad S^W = 0, \ Y^W = \frac{4}{3}, \tag{11.15}$$

$$d_R, \ s_R, \ b_R, \qquad S^W = 0, \ Y^W = -\frac{2}{3}.$$

The absence of the neutrino singlets in the GWS theory was connected with the fact that the neutrino was considered the massless particles by the time of this theory production.

At the $SU(2)_L \times U(1)_Y$ global transformations the transformation law of the left-hand and right-hand components of the field $\psi(x)$ has the form:

$$\psi_L'(x) = \exp[i(\boldsymbol{\alpha} \cdot \mathbf{S}^W + \frac{\beta Y^W}{2})]\psi_L(x), \qquad \psi_R'(x) = \exp[\frac{i\beta Y^W}{2}]\psi_R(x). \tag{11.16}$$

We begin our consideration with the leptons. The lepton sector of the GWS theory is described by the Lagrangian

$$\mathcal{L}_l = i \sum_{l=e,\mu,\tau} [\overline{\psi}_{lL}(x)\gamma^\mu\partial_\mu\psi_{lL}(x) + \overline{\psi}_{lR}(x)\gamma^\mu\partial_\mu\psi_{lR}(x)]. \tag{11.17}$$

To introduce the mass terms into the Lagrangian (11.17) directly

$$m_l[\overline{\psi}_{lL}(x)\psi_{lR}(x) + \overline{\psi}_{lR}(x)\psi_{lL}(x)]$$

violates the gauge invariance. This makes us use the mechanism of the mass generation at the cost of the spontaneous symmetry breaking not only for the weak interaction carriers but for leptons as well. For this purpose the Lagrangian of the Yukawa type that describe interactions between leptons and Higgs fields is used

$$\mathcal{L}_Y = - \sum_{l=e,\mu,\tau} f_l[\overline{\psi}_{lL}(x)\psi_{lR}(x)\varphi(x) + \varphi^\dagger(x)\overline{\psi}_{lR}(x)\psi_{lL}(x)]. \tag{11.18}$$

In order to make the neutrino massive it is sufficient to introduce the neutrino singlets in the theory. Notice, there is no theoretical principle that fixes the Yukawa constants f_l and they unfortunately remain arbitrary parameters of the theory.

In the third stage one needs to localize the gauge group in question, that is, carry out the replacement

$$\boldsymbol{\alpha} \to \boldsymbol{\alpha}(x), \qquad \beta \to \beta(x).$$

This, as it is known, demands the transition to the covariant derivatives and an introduction of the free gauge bosons Lagrangian. In the case of the $SU(2)_L \times U(1)_Y$-gauge group the covariant derivatives for the fields entering into the Lagrangian have the form:

$$D_\mu = \partial_\mu - ig\mathbf{S}^W \cdot \mathbf{W}_\mu - ig'\frac{Y^W}{2}B_\mu. \tag{11.19}$$

Then, recalling that at $S^W = 1/2$ the matrices $\sigma_k/2$ are the generators of the $SU(2)$ transformations, we obtain for the fields $\psi_L(x)$ and $\psi_R(x)$

$$D_\mu\psi_{lL}(x) = \left[\partial_\mu - \frac{ig}{2}\boldsymbol{\sigma} \cdot \mathbf{W}_\mu(x) + \frac{ig'}{2}B_\mu(x)\right]\psi_{lL}(x), \tag{11.20}$$

$$D_\mu\psi_{lR}(x) = [\partial_\mu + ig'B_\mu(x)]\psi_{lR}(x). \tag{11.21}$$

The free gauge bosons Lagrangian is determined by the expression

$$\mathcal{L}_0 = -\frac{1}{4}W_{a\mu\nu}(x)W_a^{\mu\nu}(x) - \frac{1}{4}B_{\mu\nu}(x)B^{\mu\nu}(x),$$

where $W_{\mu\nu}^a(x)$ is a tensor of a non-Abelian field

$$W_{a\mu\nu}(x) = \partial_\mu W_{a\nu}(x) \quad \partial_\nu W_{a\mu}(x) + g\varepsilon_{abc}W_{b\nu}(x)W_{c\mu}(x), \qquad a, b, c - 1, 2, 3,$$

$B_{\mu\nu}(x)$ is a tensor of an Abelian field

$$B_{\mu\nu}(x) = \partial_\mu B_\nu(x) - \partial_\nu B_\mu(x).$$

To pass to the covariant derivative leads to the appearance of two basic interactions in the total Lagrangian: interaction of the weak currents isotriplet $\mathbf{J}^\mu(x)$ with three vector bosons $\mathbf{W}^\mu(x)$

$$g\mathbf{J}_l^\mu(x) \cdot \mathbf{W}_\mu(x) = g\overline{\psi}_{lL}(x)\gamma^\mu \mathbf{S}^W \cdot \mathbf{W}_\mu(x)\psi_{lL}(x), \tag{11.22}$$

and interaction of the weak hypercharge current $j_l^{Y\mu}(x)$ with fourth vector boson $B_\mu(x)$

$$\frac{g'}{2}j_l^{Y\mu}(x)B_\mu(x) = g'\overline{\psi}_l(x)\gamma^\mu\frac{Y^W}{2}\psi_l(x)B_\mu(x), \tag{11.23}$$

where

$$\psi_l(x) = \psi_{lL}(x) + \psi_{lR}(x).$$

In the fourth stage we must give the mass both to the weak interaction carriers and to the leptons. For this purpose we shall make use of the mechanism of the spontaneous symmetry breaking on the chain

$$SU(2)_L \times U(1)_Y \to U(1)_{em}.$$

The supplementary two-component complex scalar Higgs field $\varphi(x)$ (four degrees of freedom) is introduced

$$\varphi(x) = \frac{1}{\sqrt{2}}\begin{pmatrix} i\Phi_1(x) + \Phi_2(x) \\ H(x) - i\Phi_3(x) \end{pmatrix} = \begin{pmatrix} \varphi^+(x) \\ \varphi^0(x) \end{pmatrix}, \qquad S^W = 1/2, \ Y^W = 1, \tag{11.24}$$

where $\text{Im } \Phi_i(x) = 0$ and $\text{Im } H(x) = 0$. The Lagrangian describing the Higgs fields doublet, in addition to the kinetic energy, contains the potential energy of self-interaction as well

$$\mathcal{L}_H = |D_\mu\varphi(x)|^2 - V(\varphi), \tag{11.25}$$

where

$$V(\varphi) = \mu^2 \varphi^\dagger(x)\varphi(x) + \lambda[\varphi^\dagger(x)\varphi(x)]^2 \tag{11.26}$$

with $\lambda > 0$ and $\mu^2 < 0$. The spontaneous symmetry breaking is realized by the shift of the neutral Higgs field component on the real constant $v = \sqrt{-\mu^2/\lambda}$

$$\varphi(x) = \frac{1}{\sqrt{2}} \begin{pmatrix} i\Phi_1(x) + \Phi_2(x) \\ H(x) - i\Phi_3(x) + v \end{pmatrix} = \varphi'(x) + \xi_0, \tag{11.27}$$

where

$$\xi_0 = \frac{1}{\sqrt{2}} \begin{pmatrix} 0 \\ v \end{pmatrix}$$

and $< 0|\varphi'(x)|0 >= 0$. Then the parametrization of fluctuations close to ξ_0 in the lowest order in powers of Φ takes the form:

$$\varphi(x) = \frac{1}{\sqrt{2}} \exp[i\boldsymbol{\sigma} \cdot \boldsymbol{\Phi}(x)/v] \begin{pmatrix} 0 \\ v + H(x) \end{pmatrix}. \tag{11.28}$$

Notice, at any choice of $\varphi(x)$ the symmetry violation inevitably results in the appearance of the mass on the corresponding gauge bosons. But when the invariance, both of the Lagrangian and of the vacuum with respect to some gauge transformation subgroup, is conserved, then the gauge bosons connected with these subgroups are kept massless. Under the choice

$$< 0|\varphi(x)|0 >= \frac{1}{\sqrt{2}} \begin{pmatrix} 0 \\ v \end{pmatrix}$$

with $S^W = 1/2$, $S_3^W = -1/2$ and $Y^W = 1$ both the $SU(2)_L$- and $U(1)_Y$-gauge symmetries are violated. Since the generators of the groups $U(1)_{em}$, $SU(2)_L$ and $U(1)_Y$ satisfy the relation (11.12), then

$$Q\xi_0 = 0, \tag{11.29}$$

or

$$\xi_0' = \exp[i\alpha(x)Q]\xi_0 = \xi_0. \tag{11.30}$$

Thus, both the final Lagrangian and the vacuum are invariant with respect to the $U(1)_{em}$-group transformations what ensures the zero-mass photon.

As a result of the shift in $|D_\mu\varphi(x)|^2$ the terms which are bilinear on the components of W_μ^a and B_μ appear. They give the contribution to the mass matrix of the gauge bosons

$$\left| [-\frac{ig}{2}\boldsymbol{\sigma} \cdot \mathbf{W}_\mu(x) - \frac{ig'}{2}B_\mu(x)]\varphi(x) \right|^2 =$$

$$= \frac{1}{8} \left| \begin{pmatrix} gW_\mu^3(x) + g'B_\mu(x) & g[W_\mu^1(x) - iW_\mu^2(x)] \\ g[W_\mu^1(x) + iW_\mu^2(x)] & -gW_\mu^3(x) + g'B_\mu(x) \end{pmatrix} \begin{pmatrix} 0 \\ v \end{pmatrix} \right|^2 =$$

$$= \frac{g^2v^2}{8} \left\{ [W_\mu^1(x)]^2 + [W_\mu^2(x)]^2 \right\} + \frac{v^2}{8} \left[gW_\mu^3(x) - g'B_\mu(x) \right]^2. \tag{11.31}$$

As it follows from Eq. (11.31) the fields $W_\mu^3(x)$ and $B_\mu(x)$ prove to be mixed. This is no surprise, since they have the identical quantum numbers. For diagonalization of the last term in (11.31) we pass to a new basis

$$\begin{pmatrix} Z_\mu \\ A_\mu \end{pmatrix} = \frac{1}{\sqrt{g^2 + g'^2}} \begin{pmatrix} g & -g' \\ g' & g \end{pmatrix} \begin{pmatrix} W_\mu^3 \\ B_\mu \end{pmatrix} = \begin{pmatrix} \cos\theta_W & -\sin\theta_W \\ \sin\theta_W & \cos\theta_W \end{pmatrix} \begin{pmatrix} W_\mu^3 \\ B_\mu \end{pmatrix}, \tag{11.32}$$

where

$$\tan\,\theta_W = \frac{g'}{g}.$$

In the new basis the term $v^2[gW_\mu^3(x) - g'B_\mu(x)]^2/8$ takes the form:

$$\frac{1}{2}m_Z^2 Z_\mu(x)Z^\mu(x),$$

where

$$m_Z = \frac{v\sqrt{g^2 + g'^2}}{2}. \tag{11.33}$$

Further crossing to complex self-conjugate fields

$$W_\mu^\pm = \frac{W_\mu^1 \mp iW_\mu^2}{\sqrt{2}}, \qquad W_\mu \equiv W_\mu^-,\ W_\mu^* \equiv W_\mu^+ \tag{11.34}$$

and taking into account (11.33), we rewrite Eq. (11.31) in the form:

$$m_W^2 W_\mu^*(x)W^\mu(x) + \frac{1}{2}m_Z^2 Z_\mu(x)Z^\mu(x), \tag{11.35}$$

where

$$m_W = \frac{gv}{2}. \tag{11.36}$$

Thus, the weak interaction carriers have acquired the mass whereas the photon is kept massless. The massless gauge bosons had two polarization states. After they had acquired the mass the number of their polarization states increased by one. They borrow these three additional degrees of freedom from the Higgs bosons. However, the Higgs field had four degrees of freedom. What destiny has the last Higgs component? It appeared that the remaining Higgs boson becomes massive and passes into the physical particles sector (the physical Higgs boson).

Substituting (11.28) into (11.27), one may be convinced that the field $H(x)$ has the mass $m_H^2 = 2\lambda v^2$ while the fields $\Phi_i(x)$ have been kept massless, that is, they represent the Goldstone fields. To eliminate the Goldstone bosons we use the parametrization of the field $\varphi(x)$ in the form (11.29). We carry out the $SU(2)_L$-gauge transformation in the total Lagrangian written in terms of $\varphi(x)$

$$\varphi'(x) = \mathcal{U}(x)\varphi(x) = \frac{1}{\sqrt{2}}\begin{pmatrix} 0 \\ v + H(x) \end{pmatrix},$$

$$\frac{\sigma_k}{2}W_{k\mu}'(x) = \mathcal{U}(x)\left(\frac{i}{g}\partial_\mu + \frac{\sigma_k}{2}W_{k\mu}(x)\right)\mathcal{U}^{-1}(x),$$

$$\psi_L'(x) = \mathcal{U}(x)\psi_L(x), \qquad B_\mu'(x) = B_\mu(x), \qquad \psi_R'(x) = \psi_R(x)$$

where

$$\mathcal{U}(x) = \exp[-i\boldsymbol{\sigma}\cdot\boldsymbol{\Phi}(x)/v].$$

The transformation law for $W_k^\mu(x)$ follows from the Lagrangian invariance with respect to $\mathcal{U}(x)$, what is ensured by the condition

$$\mathcal{U}^{-1}(x)D_\mu'(x)\mathcal{U}(x)\psi_L(x) = D_\mu(x)\psi_L(x).$$

It is not difficult to show the fields $\Phi_1(x), \Phi_2(x)$, and $\Phi_3(x)$ are not already contained in the final Lagrangian. To sum up, three Goldstone bosons are eliminated by the gauge transformation from the theory (to be gauged) while the liberated three degrees of freedom cross into the transverse components of the $W^{\pm}$- and Z-bosons, which became massive.

The physical Higgs boson is not isolated from the rest of the part of the model. It interacts both with the leptons and with the gauge bosons. The corresponding Lagrangian is given by the expression:

$$\mathcal{L}_H = -\sum_l f_l \bar{l}(x) l(x) H(x) + \frac{g^2}{4} \left[W^*_\mu(x) W^\mu(x) + \frac{1}{2\cos^2\theta_W} Z_\mu(x) Z^\mu(x) \right] \left[H^2(x) + 2vH(x) \right].$$

(11.37)

For the massive neutrinos, as stated before, we should introduce the neutrino singlets in the theory. This leads to the appearance of the following term in $\mathcal{L}_H$:

$$-\sum_l f_{\nu_l} \bar{\nu}_l(x) \nu_l(x) H(x).$$

The leptons also get the masses in the result of the shift of the field φ on the constant (11.28)

$$m_l = \frac{f_l v}{\sqrt{2}}, \qquad m_{\nu_l} = \frac{f_{\nu_l} v}{\sqrt{2}}.$$

(11.38)

As it follows from Eqs. (11.37) and (11.38) the coupling constants describing the interaction of the Higgs bosons with the W- and Z-bosons are proportional to gm_W and gm_Z respectively, that is, they are much larger than the coupling constants determining the interaction between the Higgs bosons and the fermions.

Let us consider the sum of the terms (11.22) and (11.23) that describe the interaction between the leptons and the gauge bosons of the $SU(2)_L \times U(1)_Y$-group. Taking into account the explicit form of the Pauli matrices, we may present an interaction of the weak currents isotriplet $\mathbf{J}_\mu(x)$ with three $\mathbf{W}^\mu(x)$-bosons in the form:

$$g\mathbf{J}_l^\mu(x) \cdot \mathbf{W}_\mu(x) = \frac{g}{2}(\bar{\nu}_{lL}(x), \bar{l}_L(x))\gamma^\mu \begin{pmatrix} 0 & W^1_\mu(x) - iW^2_\mu(x) \\ W^1_\mu(x) + iW^2_\mu(x) & 0 \end{pmatrix} \begin{pmatrix} \nu_{lL}(x) \\ l(x) \end{pmatrix} +$$

$$+ \frac{g}{2}(\bar{\nu}_{lL}(x), \bar{l}_L(x))\gamma^\mu \begin{pmatrix} W^3_\mu(x) & 0 \\ 0 & -W^3_\mu(x) \end{pmatrix} \begin{pmatrix} \nu_{lL}(x) \\ l_L(x) \end{pmatrix} =$$

$$= \frac{g}{\sqrt{2}} \left[\bar{\nu}_{lL}(x)\gamma^\mu l_L(x) W^*_\mu(x) + \bar{l}_L(x)\gamma^\mu \nu_{lL}(x) W_\mu(x) \right] + g J_l^{3\mu}(x) W^3_\mu(x), \quad (11.39)$$

where

$$g J_l^{3\mu}(x) W^3_\mu(x) = \frac{g}{2}[\bar{\nu}_{lL}(x)\gamma_\mu \nu_{lL}(x) W^3_\mu(x) - \bar{l}_L(x)\gamma_\mu l_L(x) W^3_\mu(x)]. \quad (11.40)$$

Interaction of the weak hypercharge current $j_l^{Y\mu}(x)$ with the fourth vector boson $B_\mu(x)$, in its turn, may be written as:

$$\frac{g'}{2} j_l^{Y\mu}(x) B_\mu(x) = -\frac{g'}{2}\bar{\psi}_{lL}(x)\gamma^\mu \psi_{lL}(x) B_\mu(x) - g'\bar{\psi}_{lR}(x)\gamma^\mu \psi_{lR}(x) B_\mu(x) =$$

$$= -\frac{g'}{2}[\bar{\nu}_L(x)\gamma^\mu \nu_{lL}(x) + \bar{l}_L(x)\gamma^\mu l_L(x) + 2\bar{l}_R(x)\gamma^\mu l_R(x)] B_\mu(x). \quad (11.41)$$

Now we should unify the last term in (11.39) with (11.41). It is evident at a glance that the choice of the basis in the form (11.32) provides the absence of the electromagnetic interaction of the neutrino. Taking into consideration Eq. (11.32), we get without trouble

$$g J_l^{3\mu}(x) W^3_\mu(x) + \frac{g'}{2} j_l^{Y\mu}(x) B_\mu(x) = -\frac{g'g}{\sqrt{g^2 + g'^2}} \left[\bar{l}_L(x)\gamma^\mu l_L(x) + \bar{l}_R(x)\gamma^\mu l_R(x) \right] A_\mu(x) +$$

$$+ \left[\frac{g^2 + g'^2}{2\sqrt{g^2 + g'^2}} \bar{\nu}_{lL}(x)\gamma^\mu \nu_{lL}(x) - \frac{g^2 - g'^2}{2\sqrt{g^2 + g'^2}} \bar{l}_L(x)\gamma^\mu l_L(x) + \right.$$

$$\left. + \frac{g'^2}{\sqrt{g^2 + g'^2}} \bar{l}_R(x)\gamma^\mu l_R(x) \right] Z_\mu(x). \tag{11.42}$$

Since Eq. (11.12) takes place, then the following relation:

$$(j_\mu)_{em} = J_\mu^3 + \frac{1}{2} j_\mu^Y$$

must be fulfilled, that is, the electromagnetic interaction $-j_{em}^\mu(x)A_\mu(x)$ has to be included into interactions (11.22) and (11.23). Then it becomes clear that the first term in (11.42) describes the electromagnetic interaction of the charged leptons and the multiplier $gg'/\sqrt{g^2 + g'^2}$ is nothing more nor less than the electric charge

$$|e| = \frac{gg'}{\sqrt{g^2 + g'^2}}. \tag{11.43}$$

Taking into consideration this circumstance and allowing for Eqs. (11.39), (11.42), we obtain the final expression for the interaction Lagrangian of the gauge bosons with the leptons

$$\mathcal{L}_G = \frac{g}{2\sqrt{2}} \left[\bar{\nu}_l(x)\gamma^\mu(1 - \gamma_5)l(x)W_\mu^*(x) + \bar{l}(x)\gamma^\mu(1 - \gamma_5)\nu_l(x)W_\mu(x) \right] - e\bar{l}(x)\gamma^\mu l(x)A_\mu(x) +$$

$$+ \frac{g}{4\cos\theta_W} \left\{ \bar{\nu}_l(x)\gamma^\mu[1 - \gamma_5]\nu_l(x) + \bar{l}(x)\gamma^\mu[4\sin^2\theta_W - 1 + \gamma_5]l(x) \right\} Z_\mu(x). \tag{11.44}$$

Since all the terms in the Lagrangian (11.44) represent the quantities of the type *current×potential*, then the weak interaction caused by the exchanges of the W- and Z-bosons is commonly called an *interaction of the charged and neutral currents*, respectively.

To develop the electroweak interactions theory for quarks is performed in full analogy with the above-mentioned scheme for the leptons. There is the following correspondence between the quarks and the leptons:

$$\begin{pmatrix} \nu_{eL} \\ e_L^- \end{pmatrix} \leftrightarrow \begin{pmatrix} u_L^\alpha \\ d_L^\alpha \end{pmatrix}, \qquad \begin{pmatrix} \nu_{\mu L} \\ \mu_L^- \end{pmatrix} \leftrightarrow \begin{pmatrix} c_L^\alpha \\ s_L^\alpha \end{pmatrix}, \qquad \begin{pmatrix} \nu_{\tau L} \\ \tau_L^- \end{pmatrix} \leftrightarrow \begin{pmatrix} t_L^\alpha \\ b_L^\alpha \end{pmatrix}, \tag{11.45}$$

$$\left. \begin{matrix} \nu_{eR} \leftrightarrow u_R^\alpha, & e_R^- \leftrightarrow d_R^\alpha, & \nu_{\mu R} \leftrightarrow c_R^\alpha, \\ \mu_R^- \leftrightarrow s_R^\alpha, & \nu_{\tau R} \leftrightarrow t_R^\alpha, & \tau_R^- \leftrightarrow b_R^\alpha, \end{matrix} \right\} \tag{11.46}$$

where α is a color quark index.* But what actually happens is that, instead of the d-, s- and b-quarks, their linear combinations enter into singlets and doublets of the weak isospin. We shall designate them by symbols d', s', and b'. They are connected with the unprimed quarks by the relations:

$$q'^d = \begin{pmatrix} d' \\ s' \\ b' \end{pmatrix} = \mathcal{M}^{CKM} q^d = \mathcal{M}^{CKM} \begin{pmatrix} d \\ s \\ b \end{pmatrix} =$$

$$= \begin{pmatrix} c_{12}c_{13} & s_{12}c_{13} & s_{13}e^{-i\delta_{13}} \\ -s_{12}c_{23} - c_{12}s_{23}s_{13}e^{i\delta_{13}} & c_{12}c_{23} - s_{12}s_{23}s_{13}e^{i\delta_{13}} & s_{23}c_{13} \\ s_{12}s_{23} - c_{12}c_{23}s_{13}e^{i\delta_{13}} & -c_{12}s_{23} - s_{12}c_{23}s_{13}e^{i\delta_{13}} & c_{23}c_{13} \end{pmatrix} \begin{pmatrix} d \\ s \\ b \end{pmatrix}, \tag{11.47}$$

---

*As the electroweak interaction does not change the quark flavor, further the flavor index will be neglected.

where $\mathcal{M}^{CKM}$ is the Cabibbo–Kobayashi–Maskawa matrix (CKM), $c_{ij} = \cos\theta_{ij}^{CKM}$, $s_{ij} = \sin\theta_{ij}^{CKM}$, i, j are generation indices to equal $1, 2, 3$, θ_{ij}^{CKM} are mixing angles and a phase multiplier $e^{\pm i\delta_{13}}$ describes the CP parity violation. So in Eqs. (11.45) and (11.46) one should make the replacement:

$$q_i^d \to q_i'^d = \mathcal{M}_{ij}^{CKM} q_j^d.$$

The CKM matrix is unitary. Its elements could be determined from the weak decays of hadrons and from the experiments on the deep inelastic scattering of the neutrino by hadrons. Since the matrix $\mathcal{M}^{CKM}$ is the product of three noncommuting matrices of the rotations in three planes $[x_1 x_2]$, $[x_1 x_3]$, and $[x_2 x_3]$ of the abstract space, then the parametrization (11.47) is not unique.*

In place of Eqs. (11.22) and (11.23) we have for the quarks

$$g\mathbf{J}_q^\mu(x)\cdot\mathbf{W}_\mu(x) + \frac{g'}{2} j_q^{Y\,\mu}(x) B_\mu(x) = g\overline{\psi}_{qL}(x)\gamma^\mu \mathbf{S}^W\cdot\mathbf{W}_\mu(x)\psi_{qL}(x) + g'\overline{\psi}_q(x)\gamma^\mu \frac{Y^W}{2}\psi_q(x) B_\mu(x).$$
(11.48)

Passing in (11.48) from the gauge basis (W^1, W^2, W^3, B) to the mass eigenstates one (W, W^*, Z, γ) with the help of formulae (11.32), (11.34), we arrive at the following expression for the Lagrangian describing the interaction between quarks and gauge bosons:

$$\mathcal{L}_q = \frac{g}{\sqrt{2}} \left[\overline{q}_{iL}^u(x)\gamma^\mu \mathcal{M}_{ik}^{CKM} q_{kL}^d W_\mu^*(x) + \overline{q}_{kL}^d(x)\gamma^\mu \mathcal{M}_{ki}^{CKM*} q_{iL}^u W_\mu(x) \right] +$$

$$+ \frac{g}{4\cos\theta_W} \left[\overline{q}_i^u(x)\gamma^\mu \left(1 - \frac{8}{3}\sin^2\theta_W + \gamma_5\right) q_i^u(x) - \overline{q}_i^d(x)\gamma^\mu \left(1 - \frac{4}{3}\sin^2\theta_W + \right.\right.$$

$$+ \gamma_5\Big) q_i^d(x) \Big] Z_\mu(x) + \frac{e}{3} \left[2\overline{q}_i^u(x)\gamma^\mu q_i^u(x) - \overline{q}_i^d(x)\gamma^\mu q_i^d(x) \right] A_\mu(x). \qquad (11.49)$$

We define the Yukawa Lagrangian for the quarks by analogy with the lepton case. Then, after spontaneous symmetry breaking we get the Lagrangian determining the interaction of quarks with the physical Higgs boson in the form:

$$\mathcal{L} = -\sum_i f_{q_i} \overline{q}_i(x) q_i(x) H(x),$$

where the Yukawa constant f_{q_i} determines the mass of the q_i-quark

$$m_{q_i} = \frac{f_{q_i} v}{\sqrt{2}}.$$

The Fermi theory was a predecessor of the GWS theory. This theory most advantageously managed with the description of the weak interaction in the low energy region. Then, according to the correspondence principle, at low energies the GWS theory must reproduce the results of the former theory. We take advantage this circumstance to define the linkage between the parameters of the new and old theories. It may be beneficial to recall some essential points of the Fermi theory.

By the early 1930s it was established that the beta-radioactivity of all nuclei is caused by a pair of fundamental reactions where the proton and neutron are interconverted

$$n \to p + e^- + \overline{\nu}_e, \qquad (11.50)$$

---

*We have used the parametrization accepted in Review of Particle Physics.

$$p \to n + e^+ + \nu_e. \tag{11.51}$$

The electron and the antineutrino or the positron and the neutrino, which appear as a result of the beta-decay, are created because they do not exist in the radioactive nucleus. This phenomenon is analogous to the process of the photon emitting by the electron when it transits with the one orbit on the other orbit located close to the nucleus. E. Fermi used this analogy and already known mathematical apparatus of the quantum theory of the electromagnetic interaction in his weak interaction theory proposed in 1934. The interaction Lagrangian by Fermi has the form of the product *current×current*

$$\mathcal{L}_F = \frac{G_F}{\sqrt{2}} j_\mu^\dagger(x) j^\mu(x). \tag{11.52}$$

The weak current entering into (11.52) is built out of the wavefunctions of the particles which are pairwise unified: neutron—proton, electron—electron antineutrino and so on. So, for the process (11.50) the one current is nucleon and it transfers the neutron to the proton. The other current is lepton and it creates the pair, the electron and the electron antineutrino. These currents belong to the class of the charged currents since they change the electric charge of the particles to interact. In the both currents the charge is increased by $|e|$: the positive charged proton arises from the neutral neutron while the electron antineutrino does from the electron. The interaction (11.52) gave the name *four-fermion contact interaction* *. The structure of the charged weak currents has been finally established in the middle (19)50s. Analysis of experiments being done by that time led to the conclusion that the weak current represents the difference of the vector V and the axial vector A ($V - A$ structure). Thus the lepton part of the current is defined by the expression:

$$j_l^\sigma(x) = \bar{e}(x)\gamma^\sigma(1 - \gamma_5)\nu_e(x) + \bar{\mu}(x)\gamma^\sigma(1 - \gamma_5)\nu_\mu(x) + \bar{\tau}(x)\gamma^\sigma(1 - \gamma_5)\nu_\tau(x). \tag{11.53}$$

After determination of the composite structure of nucleons the quarks occupy the nucleons place in the nucleon part of the charged weak current

$$j_q^\sigma(x) = \bar{d'}(x)\gamma^\sigma(1 - \gamma_5)u(x) + \bar{s'}(x)\gamma^\sigma(1 - \gamma_5)c(x) + \bar{b'}(x)\gamma^\sigma(1 - \gamma_5)t(x). \tag{11.54}$$

If one is constrained by the first order of the perturbation theory in the weak interaction constant G_F under calculations, then the Fermi theory gives a fine accordance with an experiment. However the corrections of the high orders in G_F represent the integrals which become infinite at the large energies, that is, physically meaningless. Consequently the Fermi theory must be suitably reconstructed. The refusal of interaction locality is the most evident way. In other words, the analogy between the quantum electrodynamics and the weak interaction theory should be deeper, namely, the weak interaction is also carried by gauge bosons.

To connect the GWS theory parameters with the Fermi constant G_F we consider the muon decay through the channel

$$\mu^- \to e^- + \bar{\nu}_e + \nu_\mu. \tag{11.55}$$

The corresponding diagrams in the momentum representation for the Fermi theory and the GWS theory are displayed in Figs.10.3 and 10.4. In the GWS theory the amplitude of the decay (11.55) is given by the expression:

$$\mathcal{A}_{WSG} = \frac{g^2}{8} \sqrt{\frac{m_\mu m_{\nu_\mu} m_e m_{\nu_e}}{E_\mu E_e E_{\nu_e} E_{\nu_\mu}}} \bar{\nu}_\mu(p_{\nu_\mu}) \gamma^\sigma (1 - \gamma_5) \mu(p_\mu) \left(\frac{g_{\sigma\beta} - q_\sigma q_\beta / m_W^2}{m_W^2 - q^2} \right) \times$$

---

*Since four fermion wavefunctions enter into the Lagrangian we call the interaction by four-fermion. As the interaction takes place in one and the same point x we name it by contact.

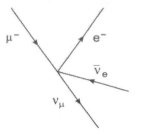

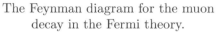

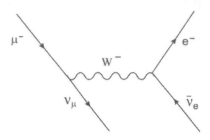

FIGURE 11.3

The Feynman diagram for the muon
decay in the Fermi theory.

FIGURE 11.4

The Feynman diagram for the
muon decay in the GWS theory.

$$\times \overline{e}(p_e)\gamma^\beta(1 - \gamma_5)\nu_e(p_{\nu_e}), \tag{11.56}$$

where $q = p_\mu - p_{\nu_\mu}$ and the expression standing in the parenthesis describes the propagation of the virtual W^- boson, that is, is the propagator of this particle. In the Fermi theory the decay amplitude has the form:

$$\mathcal{A}_F = \frac{G_F}{\sqrt{2}}\sqrt{\frac{m_\mu m_{\nu_\mu} m_e m_{\nu_e}}{E_\mu E_e E_{\nu_e} E_{\nu_\mu}}}\overline{\nu}_\mu(p_{\nu_\mu})\gamma^\sigma(1 - \gamma_5)\mu(p_\mu)\overline{e}(p_e)\gamma_\sigma(1 - \gamma_5)\nu_e(p_{\nu_e}). \tag{11.57}$$

Under small q the expressions (11.56) and (11.57) must coincide. As in this case the W-boson propagator is reduced to $g_{\sigma\beta}/m_W^2$, the required linkage is given by:

$$\frac{g^2}{8m_W^2} = \frac{G_F}{\sqrt{2}}. \tag{11.58}$$

The constant g characterizes emitting and absorbing the $W^\pm$-bosons, much as e defines emitting and absorbing the photons. From (11.43) follows that $e > g$ and, therefore, the weak interaction is in essence stronger than the electromagnetic one. However, as it was shown, the weak process amplitudes are proportional to g^2/m_W^2 at low energies.*. So, owing to that the W-bosons are very heavy, the weak interaction processes appear to be much orders of magnitude weaker than the electromagnetic processes.

Not only does the GWS theory unify the electromagnetic and weak interactions, but it also predicts the existence of new phenomena in the weak interaction physics, the neutral currents. In 1973 the first reactions caused by the neutral currents were observed

$$\nu_\mu + p \to \nu_\mu + p + \pi^+ + \pi^-. \tag{11.59}$$

Information confirming the neutral currents existence also follows from experiments on observing the parity violation in atom physics. The interaction constant of the neutral currents proves to be approximately the same as that of the charged currents.

The GWS theory predicts the linkage between the masses of the W- and Z-bosons as well. From the relations (11.33) and (11.36) follows

$$m_Z = \frac{m_W\sqrt{g^2 + g'^2}}{g} = \frac{m_W}{\cos\theta_W}. \tag{11.60}$$

---

*Recall, that the separation into the electromagnetic and weak interactions has a sense only at energies < 100 GeV.

Using Eqs. (11.43) and (11.58), we obtain

$$m_W = \sqrt{\frac{\pi \alpha_{em}}{G_F \sqrt{2}}} \sin^{-1} \theta_W. \tag{11.61}$$

Having done three independent experiments one may define the constants G_F, $\sin \theta_W$, α_{em}, the knowledge of which determines not only the masses of the W- and Z-bosons but the vacuum average v (vacuum expectation value) of the Higgs field as well

$$v = \frac{m_W \sin \theta_W}{\sqrt{\pi \alpha_{em}}}.$$

The Weinberg angle value may be found out of different experiments concerning nuclear physics, physics of the weak interactions at low energies, and high energy physics. By 1983 the results of $\sin^2 \theta_W$ determination have come to be matched. The averaged value was given by:

$$\sin^2 \theta_W \approx 0.23. \tag{11.62}$$

Then substituting the values

$$\alpha_{em} \approx \frac{1}{137}, \qquad G_F = 1.17 \times 10^{-5} \text{ GeV}^{-2},$$

into Eqs. (11.60), (11.61) and using the value of $\sin^2 \theta_W$, we find

$$m_W \approx 80 \text{ GeV}, \qquad m_Z \approx 91 \text{ GeV}. \tag{11.63}$$

It is evident that the discovery of the W- and Z-bosons would be the deciding step on the way of the GWS theory checkout. For these purposes the proton-antiproton collider was built in CERN. It began operation in summer 1981. The direct production of the W-boson with the subsequent decay through the electron and electron antineutrino

$$u + d \to W^{-*} \to e^- + \overline{\nu}_e \tag{11.64}$$

is displayed in Fig. 11.5. The cross section of the reaction (11.64) is a function of the

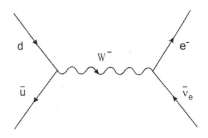

FIGURE 11.5
The Feynman diagram for the process $u + \overline{d} \to W^{-*} \to e^- + \overline{\nu}_e$.

colliding quarks energy and as soon as the energy in the center-of-mass frame is approaching to m_W, the W-boson is exhibiting as a resonance. In the resonance region the cross section has the sharp maximum of the height and the width of which are predicted by the GWS theory. At the resonance the cross section value can be calculated by application of the Breit-Wigner formula (8.22).

It is beyond doubt that to directly observe the quark collisions is impossible, since the quarks in the free states are unavailable for us. The proton-antiproton collisions are just the most changing. A monochromatic proton beam may be considered as the quark beam with the wide distribution over momenta, that is, when P is a proton momentum then $x_i P$ is a momentum of an ith-quark ($0 \leq x_i \leq 1$). An antiproton beam looks analogous. In the process of the W production the quark picks for itself the antiquark with suitable momentum. In the center-of-mass frame the energy of the impacts of the quark with the antiquark $s_{q\bar{q}}$ is connected with $s_{p\bar{p}}$ by the relation:

$$s_{q\bar{q}} = s_{p\bar{p}} x_q x_{\bar{q}}.$$

The distribution functions of the quarks both in the proton and in the antiproton are such that in order to provide the right correlation between the proton quark and the antiproton antiquark the following condition:

$$x_q \approx x_{\bar{q}} \geq 0.25 \tag{11.65}$$

must be fulfilled. Thus, there is one wide region of optimal energies for the $p\bar{p}$ impacts at the given W-boson mass. For $m_W = 80$ GeV it is given by:

$$400 \leq \sqrt{s_{p\bar{p}}} \leq 600 \text{ GeV}.$$

In actual fact the following processes are investigated at the CERN $p\bar{p}$-collider:

$$p + \bar{p} \rightarrow W^{\pm} + X, \tag{11.66}$$

where X is arbitrary hadrons plurality. To detect the $W^{\pm}$-bosons was carried out through the lepton decays

$$W^+ \rightarrow e^+ + \nu_e, \qquad W^- \rightarrow e^- + \bar{\nu}_e. \tag{11.67}$$

Such processes are represented by diagrams that include the elements both of the quarks diagrams and of the Feynman ones. So the diagram pictured in Fig. 11.6 corresponds to the process

$$p + \bar{p} \rightarrow W^{-*} \rightarrow e^- + \bar{\nu}_e + X. \tag{11.68}$$

To obtain the cross section of the reaction (11.68) one should integrate the cross section of the reaction (11.64) at the resonance over the distributions of the quarks in the proton and the antiproton.

Since the W-boson mass is large, then charged leptons $l^{\pm}$, appearing under the W boson decay, have a large transverse momentum. Thus, the detection of $l^{\pm}$ has not caused any troubles. The neutrino recording was based on purely kinematic reasons. For this purpose the special detector was used. Its sensitivity relative to all charged or neutral particles that were created in the impact process, was uniform in the whole volume over the whole solid angle. Since interactions are observed in the center-of-mass frame any appreciable disbalance of a momentum signals the presence of one or more numbers of noninteracting particles (presumably, neutrinos). Calorimeters are ideally suited for the role of such detectors because their energy registration efficiency may be done to be sufficiently homogeneous for different particles to hit in them. Notice, in the reference frame where the W-boson rests the momentum carried away by the neutrino p_ν is equal to $m_W/2$, that is, it is very large.

In January 1983 two independent Collaborations UA1 and UA2 working at the CERN $p\bar{p}$-collider presented the first results on detecting the W-bosons in the reactions (11.66), (11.67).

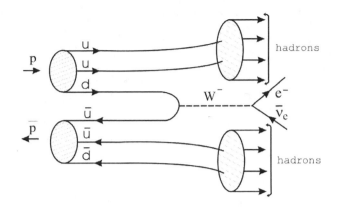

FIGURE 11.6

The Feynman diagram for the process $p + \bar{p} \to W^{-*} \to e^- + \bar{\nu}_e + X$.

In June 1983 the group UA1 had reported the observation of the first five cases of the creation and the decay of the Z-bosons. In August of the same year the group UA2 had detected the eight analogous events. The Z-bosons are created in the reaction

$$p + \bar{p} \to Z + X \tag{11.69}$$

and are detected through the decays

$$Z \to e^- + e^+, \qquad Z \to e^- + e^+ + \gamma, \qquad Z \to \mu^- + \mu^+. \tag{11.70}$$

These experiments have led not only to the discovery of the W and Z-bosons, they have also shown that their properties are exactly described by the GWS theory. It should be noticed that the mass values (11.63) are approximate rather than precise. To obtain the precise values of the gauge boson masses we must take into consideration the interaction of particles with the vacuum or, what is the same, incorporate the higher orders of the perturbation theory (radiative corrections) under calculating the cross sections. To influence the radiative corrections (RC) on the values of α_{em} and $\sin^2 \theta_W$ is especially significant. The calculations showed that the inclusion of the RC changes the gauge bosons mass values in the formulae (11.63) on 5% approximately.

12

Fundamental particles of the Standard Model

*Once more the blackbirds chanted "The Fruits of Science Feed the
Young." But all realized that the "Golden Age" was drawing to a close.
The sole prospect now was the descending darkness of ignorance, accom-
panied by its inevitable companions—dissension and strife.*
M.E. Saltykov Shchedrin, "The Eagle-Patron of Arts"

The evolution of our notions about the matter structure, the four steps on the Quantum
Stairway, may be schematically represented by Fig. 12.1. Every stage of this Stairway

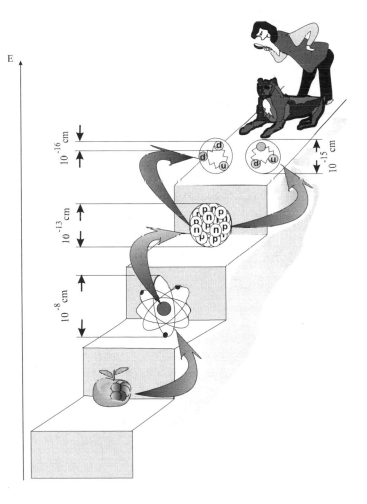

FIGURE 12.1
The Quantum Stairway.

presents a separate region of physics and, consequently, the phenomena range described by it is characterized first and foremost by particle velocities and a region size that is available for particle motions. An object having an intrinsic structure from the viewpoint of the higher stage is considered as a structureless one in all phenomena at any underlying stage. So an atom is supposed to be a point particle in Classical Physics, a nucleus—in Atomic Physics, a nucleon—in Nuclear Physics.

Before when we talked about the quark-lepton symmetry we were pointing the way of its storing, placing the quarks and the leptons into the weak isospin doublets. In this case the down quarks proved to be subjected to mixing. The absence of similar mixing in the lepton sector would shake not only our belief in the quark-lepton symmetry, but it would also make us refuse the belief in the unity of nature. The analogous phenomenon proves to has taken place in the lepton sector as well. But we have learned about it much later, in 2001. The experiments showed that the electron, muon, and tau-lepton neutrinos are not the states with definite mass value but they represent the mixtures of the physical states ν_1, ν_2, and ν_3. The corresponding mixing matrix in the neutrino sector $\mathcal{M}^{NM}$ has the same form as the CKM matrix, that is,

$$\mathcal{M}^{NM} = \mathcal{M}^{CKM}(\theta_{ij}^{CKM} \to \theta_{ij}^{NM}), \qquad (12.1)$$

where θ_{ij}^{NM} are neutrinos mixing angles. Moreover, the mixing angles in the lepton sector are connected with those in the quark sector by the relations:

$$\theta_{12}^{NM} + \theta_{12}^{CKM} = \frac{\pi}{4}, \qquad \theta_{23}^{CKM} + \theta_{23}^{NM} = \frac{\pi}{4}, \qquad \theta_{13}^{NM} \sim \theta_{13}^{CKM}.$$

Let us look aside from the existence of the right-handed weak isospin singlets for the time being and represent the quarks and the leptons by means of three generations of the $SU(2)_L$-doublets

$$\begin{pmatrix} \nu_{eL} \\ e_L^- \end{pmatrix}, \qquad \begin{pmatrix} u_L \\ d_L' \end{pmatrix}^\alpha \qquad (12.2)$$

$$\begin{pmatrix} \nu_{\mu L} \\ \mu_L^- \end{pmatrix}, \qquad \begin{pmatrix} c_L \\ s_L' \end{pmatrix}^\alpha, \qquad (12.3)$$

$$\begin{pmatrix} \nu_{\tau L} \\ \tau_L^- \end{pmatrix}, \qquad \begin{pmatrix} t_L \\ b_L' \end{pmatrix}^\alpha. \qquad (12.4)$$

The processes of annihilating the electrons-positrons into hadrons have set the upper bound on the quarks size. By now this bound is $\leq 10^{-16}$ cm. Such a limitation exists for leptons as well. Within the accuracy of the modern experiment the particles entering into both groups are considered as point particles, that is, at present we have all the reasons to believe that these particles are fundamental.

It is convenient to present the fundamental particles of the SM in the three-dimensional coordinate system $\mathcal{M}$, in which the operator eigenvalues of the weak isospin projection on the third axis S_3^W are plotted along the z axis while the plane xy is used for the definition of the eigenvalues of the color spin operator $S^{(col)}$. Draw a regular-shaped triangular prism in $\mathcal{M}$ (Fig. 12.2). We take into account the quark-lepton symmetry and start the process of placing the particles with the first fermion generation. The upper (lower) components of the weak isospin doublets will be put in the upper (lower) base of the prism. Three color states of the u-quark are placed in the vertices of the upper base (the points corresponding to $S_3^W = 1/2$) and three color states of the d'-quark are placed in the vertices of the low base ($S_3^W = -1/2$). The colorless leptons will be housed in the straight line $x = y = 0$: the electron neutrinos in the center of the upper base while the electrons in the center

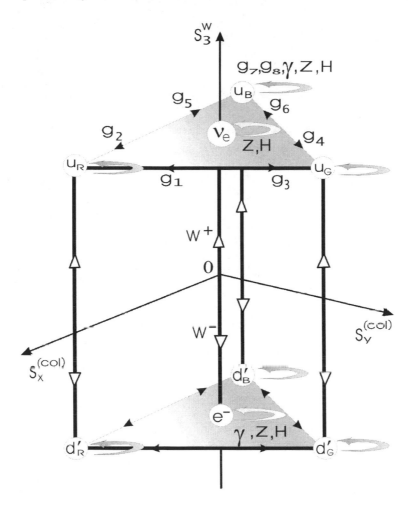

FIGURE 12.2
The fundamental particles of the SM.

of the lower base. Two analogous prisms will correspond to the fundamental fermions of the second and third generations. The quarks and the leptons form the first group of the fundamental particles, the matter particles group. The next group, the interaction carriers group, constitute four gauge bosons of the electroweak interaction W^-, W^+, Z, γ, eight gluons of the QCD g_i and yet-undiscovered carrier of the gravitational interaction, the graviton G. We call this group the *gauge bosons group*. Let us be distracted from the graviton existence and display the remaining interaction carriers in the form of the arrows, meaning the result of interactions between the bosons and the fermions, at the same figure. Emitting or absorbing the charged $W^\pm$-bosons leads to the transitions between the vertices of the fermion triangles with $\Delta S_3^W = \pm 1$, that is, it produces the motion along the z axis. From eight existing gluons two

$$g_7 = \frac{1}{\sqrt{2}} \left(R\overline{R} - G\overline{G} \right), \qquad g_8 = \frac{1}{\sqrt{6}} \left(R\overline{R} + G\overline{G} - 2B\overline{B} \right)$$

are responsible for the processes without a change of the quarks color whereas six

$$g_1 = G\overline{B}, \qquad g_2 = R\overline{B}, \qquad g_3 = R\overline{G}, \qquad g_4 = B\overline{G}, \qquad g_5 = B\overline{R}, \qquad g_6 = G\overline{R}$$

—for the processes changing the quarks color. The clockwise motions and counterclockwise motions along the perimeters of the equilateral fermion triangles correspond to the gluons to change the color.

We are coming now to displaying the gauge bosons, which do not effect the change of the electric charge, of the weak isospin, of the color and, consequently, do not alter the fermion position in the $\mathcal{M}$ space. Such particles are the photon, the neutral gauge boson, and two remaining gluons. Apart from the above-mentioned fundamental particles, there is one more particle that has the common features with both groups but still stands aside from them, that is, it actually forms the third group of the fundamental particles. This is the Higgs boson having the zero-spin and electric charge to be equal to zero. It does not enter into the matter composition and its interaction with all the fundamental fermions is not attached to the definite class. However, according to the SM all the fundamental particles acquired their masses thanks to the Higgs boson. It may be an echo of the spontaneous symmetry breaking, which happened (if any) in the epoch of the early Universe.* Under emitting and absorbing the Higgs boson the fermion does not alter its position in the $\mathcal{M}$ space as well. Let us agree the bosons, whose interactions with the matter particles result in

$$\Delta Q = \Delta S_3^W = \Delta S^{(col)} = 0,$$

will be displayed by the arrows that begin and end on the fermions.

The first generation plays a particular role. In effect, all we see around us in nature consists of the first generation fermions. All the members of the second and third generations are unstable, the exception is probably provided by the neutrinos. They appear only in accelerators and in phenomena produced by the cosmic rays. The fermions of the second and third generations played the important role in the early universe, in the first instant of the Big Bang. In particular, the neutrino flavors number has defined the quantitative ratio between hydrogen and helium in the universe. The second and third generations made an impact on the mass values of the first generation particles.

We pay attention to the trick nature has used choosing the mass values of the first generation particles. Let us consider the simplest but the most important atom of our universe, the hydrogen atom. The stability of the atom is caused by the fact that the reaction

$$p + e^- \to n + \nu_e \tag{12.5}$$

is energetically forbidden because the masses of the electron and the proton, which constitute it, satisfy an inequality

$$m_e < \Delta m, \tag{12.6}$$

where $\Delta m = m_n - m_p \approx 1.3$ MeV (we have neglected the neutrino mass). It is clear that the fragile equilibrium expressed by Eq. (12.6) may be violated even by an insignificant change of m_u, m_d, and m_e. Consequences of the hydrogen instability would be catastrophic. If the hydrogen that represents the main fuel for the stars of universe would be absent, then the

---

*In November 2000 the reports about the Higgs boson observation with the mass equal to 115 GeV came from two groups L3 and ALEPH (LEP) that investigated the reaction

$$e^+ + e^- \to Z + H.$$

However, two other Collaborations working at LEP, OPAL, and DELPHI did not confirm the results of their colleagues. In 2001 LEP ceased the operation, leaving the Higgs boson history unwritten.

ordinary stars did not exist and the universe took a totally another form. Then in order to make our world stable, say in store, why not increase the value of Δm in the relation (12.6). However, the other problem connected with the deuterium is waiting for us in this way. Its nucleus, the deuteron, possesses the most small binding energy $E_c \approx 2.24$ MeV. A guarantor of the deuteron stability is the fact that in it the decay of the neutron through the channel

$$n \to p + e^- + \bar{\nu}_e \tag{12.7}$$

is energetically unprofitable. In this case, the energy conservation law demands

$$m_p + m_n - E_c = m_p + m_p + m_e + T, \tag{12.8}$$

where T is a kinetic energy of the decaying particles. From the positivity of T it follows that the decay is forbidden under condition

$$E_c + m_e > \Delta m. \tag{12.9}$$

Thus, if we made Δm too large and violated (12.9), then the deuterium would be unstable that led to its complete lack in nature. However, the deuterium production is the first step in the chain of nuclear transformations tracing from hydrogen to more heavy elements that were not in the early universe. In the case of deuterium lack, the routine way of producing the elements that are heavier than the hydrogen would become impossible and the universe would acquire an entirely different appearance.

The inequality (12.9) demands Δm must be sufficiently small. However, Δm is nothing else than masses difference of particles entering the same isomultiplet. If Δm is compared with masses difference of particles belonging to other isomultiplets, it proves to be minimal. For example,

$$m_{\pi^\pm} - m_{\pi^0} = 4.6 \text{ MeV}, \qquad \text{and} \qquad m_{\Xi^-} - m_{\Xi^0} = 6.48 \text{ MeV}.$$

A smallness of Δm is supplied with the fact that the masses of the u and d quarks are closely related.

The small ratio of the electron and the proton masses is the cause of such an important phenomenon as the exact localization of the nucleus in the electrons cloud that, in its turn, defines a molecules architecture. Otherwise the stable configurations, characterizing the life itself and much in the world surrounding us, would not exist. All this looks as though nature was finely tuning the masses of the first generation fermions.

Nature often resorts to a similar fine-tuning of physics laws. Here the example with such composite systems as stars will be indicative. Their existence is caused by all four interactions. The gravitational interaction tends to compress a star. As the star consistence increases, the temperature in the star center grows until a thermonuclear reaction, burning hydrogen into helium, is begun. The strong and weak interactions govern this reaction. The appearance of new energy source slows down the compression process since radiation exerts pressure on the star's outward layers. At that, the main effect is caused by electromagnetic radiation. Finally, the release rate of the thermonuclear energy rises to an extent that the electromagnetic radiation balances the gravitational forces action in all volume of a star's matter. At these conditions the system structure critically depends on the interactions magnitude and thereby on numerical values of fundamental constants. Equilibrium between the gravitational and electromagnetic interactions within stars proves to be kept with striking accuracy. Calculations fulfilled by B. Carter [49] show that changing any interaction only on 10^{-40} of its value would result in catastrophe for stars of the Sun's type.

Let us return to Fig. 12.2. Generally speaking, the prism presented in it displays only the weak isospin left-handed doublets. To represent the right-handed fermion singlets we need one more figure in the $\mathcal{M}$ space. Having combined the upper and lower bases of the prism in the point $S^W = 0$ we obtain the sought figure. In so doing the u and d quarks of the identical color fall into the same vertex of the obtained triangle. Now transitions between the right-handed quarks of different (transitions associated with the $W^\pm$-bosons) flavors are absent. As for the right-handed neutrino, its position on the figure depends on a way that is used to impart the mass to this particle. In the most simple case, when a right-handed neutrino $SU(2)_L$-singlet is introduced into the theory, the right-handed neutrino falls into the same point as for the right-handed electron. Inasmuch as the symmetry between the right-handed singlets plane and the prism is absent, the presented picture loses some portion of its attraction.

To rectify the situation one should remove the asymmetry between the left and the right, that is, apply to models possessing the mirror symmetry (P-invariance) before the spontaneous breaking symmetry. The SM gauge group extension on the factor $SU(2)$ is the primary choice. The model with the gauge group $SU(2)_L \times SU(2)_R \times U(1)_{B-L}$ was proposed in [50] and later was generalized for the case of arbitrary orientation of the $SU(2)_R$-generator in the group space [51]. In this model all the fundamental fermions enter into the theory in the symmetric way, that is, they constitute the left-handed and right-handed doublets

$$\begin{pmatrix} \nu_{eL} \\ e_L^- \end{pmatrix}, \qquad \begin{pmatrix} N_{eR} \\ e_R^- \end{pmatrix}, \qquad \begin{pmatrix} u_L \\ d_L' \end{pmatrix}^\alpha, \qquad \begin{pmatrix} u_R \\ d_R' \end{pmatrix}^\alpha, \tag{12.10}$$

$$\begin{pmatrix} \nu_{\mu L} \\ \mu_L^- \end{pmatrix}, \qquad \begin{pmatrix} N_{\mu R} \\ \mu_R^- \end{pmatrix}, \qquad \begin{pmatrix} c_L \\ s_L' \end{pmatrix}^\alpha, \qquad \begin{pmatrix} c_R \\ s_R' \end{pmatrix}^\alpha, \tag{12.11}$$

$$\begin{pmatrix} \nu_{\tau L} \\ \tau_L^- \end{pmatrix}, \qquad \begin{pmatrix} N_{\tau R} \\ \tau_R^- \end{pmatrix}, \qquad \begin{pmatrix} t_L \\ b_L' \end{pmatrix}^\alpha, \qquad \begin{pmatrix} t_R \\ b_R' \end{pmatrix}^\alpha \tag{12.12}$$

Apart from three additional right-handed neutrinos two additional charged $W_R^\pm$-bosons and one neutral Z'-boson are present in the theory. Then, in the $\mathcal{M}$-space two rectilinear triangular prisms correspond to every fermion generation. The prism associated with the left-handed doublets is analogous to that displayed in Fig. 12.2. In the prism associated with the right-handed doublets the motion along z-axis is generated by the $W_R^\pm$-bosons while transitions not to affect the fermions position are caused by the Z'-boson. In the simplest version of this model [52], after a spontaneous symmetry violation we are left with 8 physical Higgs bosons : two singly charged scalars (a motion along prism edges correlates with them) and six neutral scalars (arrows to begin and finish on fermions may again be used for these particles).

At a sight in Fig. 12.2 it is difficult to get rid of the temptation to declare the quarks and the leptons, which belong to the base triangles, by the quark-lepton octet of some symmetry group. The fermions of the second and third generations would also constitute the analogous octets. Having stood on this point of view, we reduce all the plurality of the fundamental matter particles to three superparticles, each of them could be in eight states. However, such a simple and elegant scheme has its own exotics. The term a "multiplet" has a sense, if and only if, the transitions between all its states are allowed. But in our case only the *quark↔quark* and *lepton↔lepton* transitions exist whereas the *lepton↔quark* transitions are forbidden. This circumstance is a consequence of a lack of interconnections between the electroweak and QCD fields within the SM. But if we choose the symmetry group, which allow us to place all the fermions of each generation into a fundamental octet, and subsequently gauge this symmetry and spontaneously break it, then we obtain frighteningly huge numbers, both of the Higgs bosons and of the gauge bosons. It is clear that such a scheme of the universe hardly has the right to life, since, as Aristotle said: *"Nature always realizes the best of possibilities."*

The $SU(5)$-group is a minimum group of the GUT that includes the SM group as a subgroup. In this model one cannot manage to place all the known fundamental fermions of each generation into one representation. However, it could be done with the help of two representations, the quintet and decuplet representations of the $SU(5)$-group. The quintet for the first generation has the form:

$$(\overline{d}_R', \overline{d}_G', \overline{d}_B', e^-, \nu_e),$$

while the corresponding decuplet is represented by an antisymmetric matrix

$$\frac{1}{\sqrt{2}}\begin{pmatrix} 0 & \overline{u}_B & -\overline{u}_G & -u_R & -d_R' \\ -\overline{u}_B & 0 & \overline{u}_R & -u_G & -d_G' \\ \overline{u}_G & -\overline{u}_R & 0 & -u_B & -d_B' \\ u_R & u_G & u_B & 0 & -e^+ \\ d_R' & d_G' & d_B' & e^+ & 0 \end{pmatrix}.$$

In so doing all the fermion fields are considered by the left-hand chiral fields, that is, the wavefunctions of all the fields are multiplied by the quantity $(1 - \gamma_5)/2$. Evidently, if one displays the fermion multiplets of the $SU(5)$-group in the coordinate system $\mathcal{M}$, then the obtained picture will not yet hold the same aesthetic appeal as it was in the case of Fig. 12.2. As the $SU(5)$-group has 24 generators, then the corresponding gauge transformation is achieved by 24 gauge bosons. Twelve of them are the gauge bosons of the SM. The remaining bosons, the $X_i^\pm$ and $Y_i^\pm$ bosons ($i = 1, 2, 3$), have the masses $\sim M_{GU}$ and the charges $\pm 4e/3$, $\pm e/3$. There is also a great deal of physical Higgs bosons. For example, this number equals 16 in the version proposed by Georgi and Glashow.

Under increasing the GUT group dimensionality, the number of both of the physical Higgs bosons and of the gauge bosons grow. So, in the $SO(10)$-group the number of the gauge bosons reaches 45.

In these examples we see one of possibilities—the achieved stage on the Quantum Stairway is the last stage but the list of the fundamental particles may be wider. Increasing the list may occur both through the physical Higgs bosons and through the gauge bosons. There are also no reasons to believe that the number of fermion generations may not be greater than three. The other possibility remains to be open, namely, that all the fundamental particles, or some of them at least, for example, the quarks and the leptons actually represent an objects consisting of subparticles, preons. Then some new forces responsible for unifying the preons into the quarks and the leptons must exist. In this case the strong and electroweak forces appear to be no more fundamental than the chemical or nuclear forces and the attempt of building the GUT from the elementary quarks and leptons will be doomed to failure. Are the preons the last undivided matter blocks or is climbing to the next stage of the Quantum Stairway waiting for us? It is clear only that as early as the first century bfB.C. Lucretius said: *"But without doubt, the well-known limit of breaking in pieces has been set."*

13

Technical equipment of particle physics

In less than a month not a trace remained of the recent "Golden Age." To vindicate themselves the falcon and the vulture joined hands temporarily and put all the blame on education. Science, they said, was, of course, beneficial, but only if applied at the proper time. After all, our grandsires did without it, and so could we. And to prove that all the harm really lay in education they began to discover conspiracies, invariably such as involved at least a prayer-book or an elementary school reader. An interminable nightmare of searches, investigations and trials followed

M.E. Saltykov Shchedrin, "The Eagle-Patron of Arts"

13.1   Accelerators

From the time of the atomic nucleus discovery by Rutherford it has been clear that a study of the nuclei structure requires the production of accelerated beams. Natural sources of accelerated particles, radioactive elements, exhibit very low intensities, restricted energies, and are absolutely uncontrollable. To produce high-energy particles, the developments of special accelerating facilities were being started. Presently, giant accelerators symbolize modern elementary particle physics. Experiments involving elementary particles are impossible without the use of accelerators, and the progress of physics is inconceivable without such experiments. During the twentieth century typical accelerator energies were varying from a few eV to several TeV. In other words, the attainable energies were growing exponentially, being doubled every 2.5 years. It is obvious that such a rapid energy growth cannot be expected in the twenty-first century.

Modern accelerator facilities include three basic blocks:

(i) an accelerator that effects the kinetic energy increase, formation, and ejection of high-intensity particle beams;
(ii) a detector that comprises a system for registration of the interaction processes and analysis of the reaction products; besides, a particular section of the detector may be used as a target;
(iii) equipment for input/output, storage and processing of the experimental information; a unit for automatic control of the whole accelerating system.

According to the acceleration principle, facilities may be classed as synchrotron and linear accelerators; depending on the collision method, they may be subdivided into the stationary-target and colliding-beam machines. Both stable, for example, electron, proton, neutrino, and unstable particles, for example, muons, may be accelerated in the process.

The first accelerator designed by P. Van de Graaff in 1931 was of the high-voltage linear type. It represented a combination of the high-voltage source (generator) and acceleration vacuum tube, where the charged particles, moving between the generator poles, acquired an energy corresponding to the voltage at the poles. High-voltage accelerators are intended to accelerate light (electrons) as well as heavy particles (protons, ions). Their merits are as follows: continuous operation, high energy stability of the accelerated particles, and small energy spread in the beam ($\Delta E \leq 0.01\%$). High-voltage accelerators with energies up to 10–20 MeV are still in current use for preliminary acceleration in large accelerators.

As the attainment of high potential difference (for example, 10^6 V) presents technical difficulties, it is expedient to cause the accelerated particles to rotate in the magnetic field, progressively accelerating them by low-voltage pulses applied to the accelerating electrodes at a certain instance of time and at a frequency equal to the particle rotation frequency. In this way acceleration is imparted only to the particles entrapped into the accelerating gap at proper times. Because of this, only parts of the beam are accelerated (particle clusters) rather than the whole beam. This principle was used for the creation of the first cyclic accelerator (cyclotron) constructed by E. Lawrence in 1931. In a cyclotron the charged particles are accelerated from zero to maximum energy under the effect of an alternating electric field with the constant period T_a. Curving of the orbits is provided by a constant magnetic field directed perpendicular to the orbit plane. The rotation period of a particle is determined by the expression:

$$T^f = \frac{2\pi mc}{eB\sqrt{1 - v^2/c^2}} = \frac{2\pi E}{ecB}, \tag{13.1}$$

where B is a magnetic field induction, v is a particle motion velocity on an orbit, and E is a total particle energy. It should be noted that Eq. (13.1) is written in the CGS. In this section, for the sake of obviousness, it is convenient to use an ordinary system of units. For nonrelativistic velocities $E \approx mc^2$ the period is constant. Provided in this case T^f is a multiple of T_a, a prolonged resonance may be observed between the particle rotation in the magnetic field and variations in the accelerating voltage. Accelerated particles are moving along the spiral orbits with an ever-growing radius

$$R = \frac{mcv}{eB}. \tag{13.2}$$

And acceleration takes place until the particle motion is in resonance with the accelerating field. When relativistic velocities are attained by the accelerated particles, the total particle energy begins to grow, causing the resonance disturbance, and hence acceleration of the particles is terminated.

Independently of one another, V. I. Vecsler (USSR, 1944) and E. McMillan (USA, 1945) have proposed the phase stability principle, which ensures the resonance condition at any relativistic velocity. By Eq. (13.1), the relationship between the total particle energy and accelerating field frequency ω_a should be as follows:

$$E = \frac{ecBq}{\omega_a}, \tag{13.3}$$

where $q = 1, 2, 3 \ldots$. According to the stable-phase mechanism, the particle energy automatically takes the value close to the resonance one, with a relatively slow variation in time of the accelerating electric field frequency and magnetic field induction. The finding of the stable phase principle has resulted in the advent of new types of accelerators. As follows from (13.3), an increase in the equilibrium (resonance-associated) energy of a particle requires a decrease of the accelerating field frequency (synchrocyclotron) or increased induction of the twisting magnetic field (synchrotron), or else variation of both the indicated frequency and induction (synchrophasotron) or, finally, an increase in the acceleration multiplicity, that is, in the value q (microtron).

The stable phase ensures stability of the particle motion in the azimuthal direction (in the direction of a particle orbit). The transverse (orbit-perpendicular) motion stability, or focusing, is realized by an adequate selection of the radius-variation of the magnetic field. To achieve stability in both transverse directions, it is necessary to attain a minor radial decrease of the magnetic field.

Synchrotrons are used for the formation of a beam of relativistic electrons. Synchrophasotrons serve as accelerators of heavy particles (protons, ions). In accelerators of both types the radius of the equilibrium orbit is constant. This circumstance allows one to make the magnetic system in the form of the ring. The principal limitations of ring accelerators are as follows: (i) accelerated particles in the twisting magnetic field sustain the energy loss due to magneto-bremsstrahlung (synchrotron radiation) whose power is given by the formula:

$$\mathcal{P} = \frac{C_\gamma c E^4}{R^2}, \tag{13.4}$$

where

$$C_\gamma = \frac{2e^2}{3(mc^2)^4};$$

and (ii) bulky magnetic system.

To increase the energy of an accelerated particle considering these limitations, it is required to enlarge the ring radius of the accelerator. Indeed, with growing R losses by radiation drop, the magnetic field value necessary for retention of the particle on the orbit is decreased. So, proton synchrophasotron with a maximum energy of 500 GeV, constructed in the FERMILAB in 1972, was 2 km in diameter.

Proton and ion linear accelerators are based on the same principle as cyclic: a particle moving in the resonance falls within the accelerating voltage phase in every gap. Nevertheless, the particle motion proceeds along a straight line, and the gaps along this line are arranged at certain intervals in order that the particle transit time from gap to gap be equal to the period of the accelerating electric field T_a or be a multiple of this period. Here the phase stability is also necessary for matching of the transit time between the gaps with T_a and for focusing in the transverse direction.

Linear electron accelerators are considerably different from the proton ones. Considering that the velocity of relativistic electrons is practically constant

$$v = \frac{pc^2}{E} = \frac{pc^2}{\sqrt{p^2c^2 + m_e^2c^4}} \approx c,$$

synchronism is ensured because the accelerating electromagnetic wave propagates at a velocity of light thus excluding the necessity for the phase-stability mechanism. As

$$\frac{dp_\perp}{dt} = 0, \qquad \text{then} \qquad \frac{mv_\perp}{\sqrt{1 - v^2/c^2}} = \text{const},$$

and the transverse motion velocities $v_\perp$ are rapidly falling with an increase in v. Consequently, there is no need in focusing too. The transverse Coulomb repulsion of electrons in the beam is insignificant due to a nearly absolute compensation, owing to the magnetic attraction of the currents. The energy losses by synchrotron radiation in a linear accelerator are also insignificant. To transit to high energies, however, calls for increasing of an accelerator's length. For example, the linear electron accelerator at the Stanford Linear Accelerator Center (SLAC) constructed in 1966 and having a maximum energy of 25 GeV was 3 km in length.

For the most part, the primary electron source in accelerators is represented by the so-called electron gun including a thermionic cathode and electron-optical system. A source of protons and weakly-ionized heavy ions is plasma, from where they are pulled by the external electric field. Positrons, antiprotons, and greatly-charged ions are generated due to interactions between the primary electron, proton, or ion beam and matter. Vacuum within the volume, where the particle motion takes place, in all accelerators is of the order of 10^{-5}–10^{-7} mm Hg to lower particle scattering from the residual gas.

The concept of colliding-beam accelerators put forward by D. Kernst in 1956 has led to revolution in the technology. Its realization enables one to attain a critical energy increase for the colliding particles and hence to investigate the matter structure at still closer distances.

With a stationary target the kinetic energy of an incoming particle (shell) is only partly transferred to the reaction energy; some part of the kinetic energy is spent for the target recoil energy. In case the target is more massive than a shell the recoil energy is low, otherwise the collision efficiency decreases drastically. Because of a relativistic increase of mass, the energy loss by recoil is growing with the particle velocity approaching the velocity of light. A character of interaction is determined by the particle energy in the center-of-mass frame rather than in the lab frame as the major part of the energy in the lab frame is transformed into the kinetic energy of the reaction products.

In the present-day accelerators the colliding beams are not strictly opposite, intersecting at a small angle. During processing the results of such experiments all kinematic characteristics are transformed to the center-of-mass frame for subsequent analysis. On collision of two particles with arbitrary momenta $\mathbf{p}_a$ and $\mathbf{p}_b$ the transition to the center-of-mass frame is realized by the Lorentz transformation at the appropriate velocity

$$\mathbf{v} = \frac{(\mathbf{p}_a + \mathbf{p}_b)c^2}{E_a + E_b}. \tag{13.5}$$

Actually, directing the axis x towards the sum of momenta $\mathbf{p}_a + \mathbf{p}_b = \mathbf{p}$ and using the Lorentz transformation for the four-dimensional momentum, in a new coordinate system (marked with asterisk) we get:

$$p_x^* = \frac{p_x - (E_a + E_b)v/c^2}{\sqrt{1 - v^2/c^2}}, \qquad p_y^* = p_y, \qquad p_z^* = p_z. \tag{13.6}$$

Setting the vector $\mathbf{p}^*$ to zero, we can find the center-of-mass frame velocity relative to the reference frame, where

$$\mathbf{p} \neq 0, \qquad v = \frac{\mathbf{p}c^2}{E_a + E_b},$$

to coincide with Eq. (13.5). The center-of-mass frame velocity with respect to the lab frame (b particle rests) may be determined by the expression:

$$v = \frac{\mathbf{p}_a c^2}{E_a + m_b c^2}. \tag{13.7}$$

With the use of (13.7) it is possible to relate the particle energies and momenta for both systems. For instance, in case of the particle a we have:

$$\mathbf{p}_a^* = \frac{\mathbf{p}_a - E_a \mathbf{v}/c^2}{\sqrt{1 - v^2/c^2}} = \frac{m_b \mathbf{p}_a}{\sqrt{m_a^2 + m_b^2 + 2E_a m_b/c^2}}, \tag{13.8}$$

$$E_a^* = \frac{m_a^2 c^2 + m_b E_a}{\sqrt{m_a^2 + m_b^2 + 2E_a m_b/c^2}}. \tag{13.9}$$

The relation (13.9) enables one to gain some insight into the energy advantage of accelerators of the collider type. For simplicity, we consider the collision of identical particles. Passing to kinetic energy in the formula (13.9) we find:

$$T = 4T^* + \frac{2(T^*)^2}{m_b c^2}. \tag{13.10}$$

From (13.10) it follows that to produce the energy $T^* = 200$ GeV one can realize the variant of a stationary-target electron accelerator at an energy of $T \approx 1.6 \times 10^8$ GeV. As regards the energy, the merits of colliders are obvious. At the same time, one should never forget that cheese may be free-of-charge in a mousetrap only. The principal drawback of colliders is the low frequency of the reactions. This peculiarity is easily comprehended when we compare shooting at a large stationary target and shooting at the bullets that fly towards each other. The efficiency of colliders may be improved using a higher particles density of the beam. This is attained by the use of storage rings, where the accelerated particles are stored throughout many acceleration cycles. Moreover, the focusing system may be constructed so that maximum beam compression can be provided at the point of collision, contributing to the enhanced beam density and higher probability of the reaction. We introduce the quantity known as accelerator luminosity that defines the number of events in a unit time at the unit interaction cross section (1 cm^2). For stationary-target accelerators the luminosity L is equal to

$$L = n_0 l N, \tag{13.11}$$

where n_0 is particles density in the target, l is a target thickness along the beam, and N is particles flux outgoing from the accelerator. In the case of colliding beams the accelerator luminosity is given by the expression:

$$L = \frac{N_1 N_2 l f}{l_c S}. \tag{13.12}$$

Here N_1 and N_2 is the total particles number in the beams, l — cluster extent, l_c—collision length ($l_c > l$), f—rotation frequency of the particles in the accelerator in terms of s^{-1}, S—cross section area of a larger cluster measured in cm^2. Luminosity dimensionality is cm$^{-2} \cdot$ s^{-1}. The luminosity multiplied by the process cross section σ in cm^2 gives the relevant number of events per second

$$R = \sigma L. \tag{13.13}$$

Modern high-energy accelerators are colliders. Also, these accelerators may be operated in the stationary-target mode. All of them represent synchrotrons, with the exception of the linear electron-positron collider at SLC (Stanford). This collider obtains energy up to 100 GeV at the luminosity $L = 2.5 \times 10^{30}$ cm^{-2}s^{-1}. In linear accelerators there is a single region for interaction of colliding particles, and therefore its cross section area S should be exceptionally small with a diameter of the order of $\sim 10^{-4}$ cm. At SLC the beam of accelerated electrons (positrons) is divided into hyperdense clusters (4×10^{10} particles) 0.1 cm in length, with a cross section width of 1.5×10^{-4} cm and height of 0.5×10^{-4} cm. The SLC is about 1.5 km in length.

The operation of the most powerful cyclic electron-positron collider LEP located at Geneva was terminated in 2001. Its maximum energies were in excess of 200 GeV. To decrease synchrotron radiation, the perimeter of LEP was increased to 26.66 km. Its luminosity amounted to $L = 5 \times 10^{31}$ cm^{-2}s^{-1}. In this collider the clusters of accelerated electrons and positrons were denser by the order of magnitude (6×10^{11} particles), whereas their space dimensions were much greater: extent of 1 cm, height—8×10^{-4} cm and width—2×10^{-2} cm. The rotation period was 22×10^{-6} s. During acceleration time of 550 s the particles covered a distance of the ring approximately 2.5×10^7 times, however, keeping their orbits to an accuracy of 2×10^{-3} cm. The energy of LEP is a limit for circular electron-positron colliders. Further energy enhancement for e^-e^+-machines is possible only in case of linear facilities. Such colliders are already at the stage of conception [53]; their putting into service is expected in 10–12 years. The typical energy for such a linear accelerator (so-called linac) will reach 500–1000 GeV. The energy of linear colliders cannot be increased without limit. This is associated with the material structure of the accelerator. In order for a particle to be accelerated to energies of the order of 1000 TeV or higher on the typical distances of order of 100 km, it is required to create an accelerating gradient of the electric field in the region of 10^8 V/cm. Unfortunately, such strong fields will break away electrons from atoms, altering the structure of any material. An effort to realize acceleration using a field of this strength will result in the destruction of the accelerator. It is hoped that the exit from this deadlock may be found with the help of nanotechnologies. There is reason to believe that advances in nanotechnology will enable the creation of microscopic accelerator cells with the requisite accelerating gradient. In this case, these cells will have such property that after their disruption they could be regenerated in a short time.

It is clear that for the effective acceleration the gain in energy of a particle per cycle should be higher than the total radiation loss $\mathcal{P}$ determined by the formula (13.4). Since $\mathcal{P} \sim m^{-4}$, in case of protons the attainable acceleration energies will be much higher. Presently, the highest energy (2 TeV) is provided by the FERMILAB proton-antiproton collider. Its performance is as follows: luminosity—2.1×10^{32} cm^{-2}s^{-1}, rotation period—3.8×10^{-6} s, total acceleration time—10 s, proton-antiproton cluster extent—38 cm, radius of $p(\bar{p})$-beam—34×10^{-4} cm (29×10^{-4} cm). The ring length of this accelerator is equal to 6.9 km.

A proton LHC (Large Hadron Collider, 2009) has been constructed on the basis of LEP to reach fantastic energies 14 TeV by today's standards. However, this is not the limit as in principle the circular proton machines are capable of providing the energies from 100 to 1000 TeV. Because of this, the creation of another proton supercollider is technically possible. At the present time this idea is put forward for discussion. A tentative name for this machine is VLHC (Very Large Hadron Collider). Its contemplated running into operation should not be expected earlier than in 20-30 years.

The basis of the scientific program of LHC is search experiments. In the first place it is an experimental looking for Higgs bosons and superpartners of ordinary particles. Also, it is planned to realize a search for preons and heavy gauge bosons $W^{\pm\prime}$ and Z', additionally to the SM gauge bosons. Another problem is associated with a search for the formation of quark-gluon plasma using the potentialities of LHC. The detection of such processes is a real challenge for researchers as the cross sections of the expected processes are extremely small. Moreover, some of the background events possess the intensities by milliards higher than that of an event under study. Reliability of the obtained results will be ensured by the simultaneous use of two different detectors operated by different research teams. Two all-purpose detectors ATLAS and CMS are constructed to facilitate the solution of the principal tasks of LHC. It should be noted that the detectors are constructed on the basis of dissimilar conceptions. Their magnetic systems, the general construction, detecting devices are considerably different. It is obvious that the coincidence of the results obtained at ATLAS and CMS should be indicative of their maximum reliability.

A new ALICE detector that is also created at the present time is intended for investigation of collisions with ultrarelativistic energies: nucleus-nucleus (Pb-Pb, Ca-Ca) as well as proton-proton and proton-nucleus. Since these collisions exhibit a hyperhigh energy density (5.5 TeV per each pair of colliding nucleons), the occurrence of quark deconfinement and the formation of quark-gluon plasma may be expected.

Electron and proton beams may be ejected from accelerators and directed to the external targets, both hydrogen and nuclear. The produced charged $\pi^{\pm}$-mesons, K-mesons, or $\bar{p}$ may be focused into the secondary beams for their further use during the experiments. This process may continue to produce muon and neutrino beams. Since the decay of the π^- results in two weakly interacting particles μ^- and $\bar{\nu}_\mu$, after its passage through the absorber the π-meson beam is turned to the muon beam with the neutrino contaminant. At a sufficiently large thickness of the absorbing material muons disappear, leaving a pure muon antineutrinos beam. The K^+- and π^+-mesons (the K^-- and π^--mesons) resultant from the proton bombardment of a beryllium oxide target are used as the neutrinos (antineutrinos) source. Basically, the beam comprises muon neutrinos and antineutrinos, the electron component being strongly suppressed. The energy of neutrinos within the beam is uniformly distributed from zero to $E_{max} = r_{K,\pi} p_\mu c$, where p_μ is a muon momentum and

$$ r_{K,\pi} = \frac{m^2_{K,\pi} - m^2_\mu}{m^2_{K,\pi}}, $$

($r_K = 0.954$, $r_\pi - 0.427$). The Earth shield serves as a muon absorber. An absorber 1 km in length provides absorption of muons with the energy up to 200 GeV. Increasing the maximum energy of muons necessitates further growth of the absorber length. As the neutrino is electrically neutral, focusing of the neutrino beam is accomplished indirectly. In Fig. 13.1 the scheme of the K^-- or π^--meson decays into the muon and the muon antineutrino is displayed. The angle ϕ the muon antineutrino trajectory makes with the

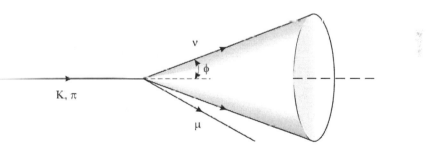

FIGURE 13.1

The scheme of producing the neutrino beams.

original meson trajectory is determined uniquely by the muon antineutrino energy and the muon momentum:

$$ \phi = \frac{m_{K,\pi} c}{p_\mu} \sqrt{\frac{r_{K,\pi} p_\mu c}{E_\nu} - 1}. \tag{13.14} $$

Thus, it is evident that changing the muon beam momentum with the help of quadrupoles and band-pass magnets one can gain focusing of the neutrino beam. Muon storage rings are used as neutrino fabrics, where electron neutrinos and muon antineutrinos (or electron antineutrinos and muon neutrinos) are produced approximately in equal proportions already within the beam. The production of focused high-energy neutrino beams is required mainly for so-called "long-baseline" oscillation experiments, in the process of which a neutrino beam created by the accelerator penetrates the Earth's thickness and is detected by an underground detector. The principal objective of such experiments is a study of neutrino oscillations (transitions $\nu_i \leftrightarrow \nu_{j \neq i}$, where $i, j = e, \mu, \tau$).

Compton scattering of high-energy electrons from laser photons makes it possible to produce γ-beams with the energy amounting to 80% of that for the primary electrons (this procedure was first realized at SLAC in 1963). This opens up possibilities for the creation of $\gamma\gamma$-colliders with the luminosity to $10\% L_e$, where L_e is the luminosity of a parent electron collider (e^+e^- or e^-e^-). Besides, a γ-source may be provided by the classical photon bremsstrahlung of e^- or e^+ beam. The next generation electron-positron colliders, for example, NLC (Next Linear Collider) with a maximum energy of 500 GeV, are so designed that they can be operated both in the $\gamma\gamma$- and $e^{\pm}\gamma$-modes. In this way combined colliders may be constructed in addition to the available electron and proton colliders. Presently, this class of accelerators is represented by the DESY synchrotron (Hamburg), version HERA (e^-p-collider), that is used to study collisions of electrons (with the energy 30 GeV) and protons (with the energy 820 GeV).

Probably, our reader is of the opinion that only stable particles may be the candidates for the function of collider-accelerated particles as the limited lifetime of unstable particles will preclude their acceleration to very high energies. However, this is a mere delusion, so to speak, a tribute to the nonrelativistic style of thinking. Just recall the time deceleration phenomenon associated with the moving clock. In the lab frame, an unstable particle covers to its decay the distance that is much greater than may be derived from nonrelativistic considerations by means of simple multiplication of its velocity by its proper lifetime. Precisely this principle is the basis for the construction of muon colliders (MC). The greatest difficulties during the construction of the MC are associated with the fact that over the lifetime of a muon, that is equal to $\tau_\mu = 2.2$ ms in the proper reference frame, muon beams should be stored, cooled, accelerated, and brought into interaction with each other. In the lab frame the muon lifetime is increased due to the relativistic factor by the value

$$\kappa_0 = \frac{1}{\sqrt{1 - v^2/c^2}} = \frac{m_\mu c^2}{m_\mu c^2 \sqrt{1 - v^2/c^2}} = \frac{E_\mu}{m_\mu c^2}.$$

Then the intensity of muon decay along the beam trajectory may be written in the form:

$$\frac{dN}{dl} = -\frac{N}{L_\mu \kappa_0}, \tag{13.15}$$

where $L_\mu = c\tau_\mu \approx 660$ m and it is assumed that $v \approx$. In the absence of the acceleration mode this leads to the ordinary exponential form of the beam loss:

$$N = N_0 \exp\left(-\frac{l}{L_\mu \kappa_0}\right). \tag{13.16}$$

In the acceleration section κ is not a constant

$$\kappa = \kappa_0 + \kappa' l = \kappa_0 + \frac{eV_g}{m_\mu c^2} l, \tag{13.17}$$

where eV_g is a acceleration gradient. Substituting expression (13.17) for κ_0 into (13.15), we obtain

$$N(l) = N_0 \left(\frac{\kappa_0}{\kappa_0 + \kappa' l}\right)^{1/(L_\mu \kappa')},$$

to give the relation

$$\frac{N(l)}{N_0} = \left(\frac{E_i}{E_f}\right)^{1/(L_\mu \kappa')}, \tag{13.18}$$

where E_i (E_f) is an initial (final) muon energy. From (13.18) it follows that the decay losses in the muon beam are minimal for

$$L_\mu \kappa' \gg 1. \tag{13.19}$$

Substitution of the numerical values into (13.19) demonstrates that the normal functioning of a muon collider requires the existence of the acceleration gradient

$$eV_g \gg 0.16 \text{ MeV/m}$$

throughout the whole muon system.

It is well known that the creation of $e^- e^+$ colliders with multiTeV energy is restricted by two factors: (i) increasing losses by synchrotron radiation; and (ii) drastic increase in the material costs as two linear accelerators will be required to avoid considerable synchrotron radiation in the storage rings.

The bremsstrahlung of muons is negligible. They may be accelerated and stored in the rings, whose radius is considerably less as opposed to hadron colliders with comparable energies. Unlike hadron colliders, where the background appears at the point of particles interaction and comes from the accelerator as well, the background of the MC may be found in the detectors only. Also, the MC exhibits high monochromaticity. The roof-mean-square deviation R from the Gaussian energy distribution in the beam falls within the interval from 0.04% to 0.08%. Owing to cooling of the muon beam, R may be decreased down to 0.01%. Thus, the energy resolution of the beam in the MC is much higher than that in $e^- e^+$ colliders. Another advantage of the MC is its fast rearrangement for operation in the $\mu^- \mu^-$- or $\mu^+ \mu^+$-mode. Since the construction of the MC includes special storage rings to provide optimization of the luminosity for some energy, the MC is an ideal instrument for investigation of resonances with an extremely small decay width (e.g., the Higgs boson of the SM). A quantity determining the energy spread in the beam $\sigma_{\sqrt{s}}$ is an essentially important characteristic for the collider. In the case of the MC this quantity is given by the expression:

$$\sigma_{\sqrt{s}} = (7 \text{ MeV}) \left(\frac{R}{0.01\%}\right) \left(\frac{\sqrt{s}}{100 \text{ GeV}}\right). \tag{13.20}$$

Note that the detection and examination of a particular particle in the s-channel may be successful provided $\sigma_{\sqrt{s}}$ is of the same order as the total decay width of this particle.

At present two projects are investigated for the construction of the MC. The first, First Muon Collider, allows for building a MC having the center-of-mass frame energy $\sqrt{s} = 0.5$ TeV and luminosity $L \sim 10^{33}$ cm^{-2}s^{-1}. The second, Next Muon Collider, is a collider of much higher power with the following characteristics: $\sqrt{s} \sim 4$ TeV, $L \sim 10^{35}$ cm^{-2}s^{-1}.

13.2 Detectors

For progress in experimental particle physics the development of accelerators is an essential but insufficient requirement. Another prerequisite is the development of experimental methods, the creation of more and more sophisticated detectors, computerization, and other modern technologies. In order to propose an experiment for the testing of a particular hypothesis or explaining experimental results, a physicist-theorist must necessarily know the advanced elementary particle detection techniques. Below is a short summary of the most extensively employed types of detectors.

The functioning of a detector is based on ionization or excitation of the atoms within the detector material by the accelerated charged particles. A charged particle leaves a track of the ionized or excited atoms that enables one to judge the form of a particle path. Uncharged particles (photons, neutrons, neutrinos, etc.) do not ionize the material, exhibiting themselves by the secondary charged particles, produced as a result of interactions between the uncharged particles and detector material. For instance, photons are detected by the produced electron-positron pairs and Compton electrons of recoil; neutrinos—by the generation of the μ^+-mesons upon their collision with protons and electrons, by the formation of recoil electrons or due to the inverse β-decay within the detector material. Passing of the particles through the material may be also accompanied by the formation of free electrons, ions, positrons subsequently annihilating with electrons to photons, and also by different reactions, thermal phenomena, and the like. Because of this, the particles may be detected by electric pulses appearing at the detector's exit, by photoemulsion blackening or else by changes in the structure of solid material in the detector. Electric signals of the particles are generally so weak that an additional system is required for the signal amplification. The most important characteristics of a detector are as follows: efficiency, that is, detection probability for a particle finding its way into the active volume of a detector; detector memory time, that is, the time period when the changes within the detector volume due to transit of the detected particle are retained; dead time of a detector, that is, the time of returning to the initial sensitivity after the regular action; energy, spatial and temporal resolution, that is, the energy, position, and time determination accuracy for a particle detected. According to the information obtained about the particles, all detectors may be conventionally subdivided into two classes. The first class comprises detectors signaling about the particles by short electrical pulses. This class, in turn, may be further subdivided into two subclasses: (a) detectors of single particles; and (b) detectors of electromagnetic and nuclear cascades. The second class is represented by track registration detectors providing direct observation of particle tracks.

1 a. Single Particle Detectors

Ionization Chamber (IC). IC is the simplest form of such a detector. It comprises two electrodes, and the interelectrode space is filled with gaseous, liquid, or solid material. Under the effect of moving charged particles, the material within the chamber is ionized to produce free electrons and ions. Migrating between the electrodes, electrons and ions create a current pulse in the external circuit of the chamber. IC may be used to detect both the particle fluxes (measurement of the average output current) and single particles (measurement of output pulses). Actually, IC may be used for the detection of all particle types, although by the appropriate selection of the detector material and electric field one is enabled to adjust the chamber for the detection of a certain particle type.

Proportional Multiwire Chambers (PMC). PMC represent a modern prototype of proportional counters, gas-discharge detectors, where the electric signal amplitude at the output is proportional to the energy spent by the particle on gas ionization. PMC consist of numerous parallel, small-diameter ($\sim 2 \times 10^{-3}$ cm) anodic wires fixed between two flat cathodes, solid or wire-type but with wires of greater diameter. Each of the anodic wires is functioning as an independent detector. Under the effect of an electric field, the primary electrons formed by the particle entering in the detector are moving to the anode to get into a high-strength field, where they are greatly accelerated causing the secondary gas ionization, that is, electron avalanches are observed. The spatial resolution of PMC is minor: $\geq 7 \times 10^{-2}$ cm. However, short dead time $\geq 3 \times 10^{-8}$ s make them most widely used position detectors.

Drift Chambers (DC). DC are used as position detectors. These gas-discharge devices involve wire electrodes, where the particle positions are determined by the drift time of the electron in a homogeneous and constant electric field, from the place of their origination to anodic wires. The field around the anode is inhomogeneous, resulting in electron acceleration and hence in the formation of electron avalanches. The spatial resolution of DC achieves 10^{-6} cm. As the dead time (electron drift time) of DC is long ($\sim 10^{-6}$ s), these chambers cannot function in intensive loading conditions.

High-precision Position Detectors (HPD). HPD are used to reconstruct the particle positions and paths at the vertex of the event under study or in its neighborhood. A typical task for HPD is a search for the "second vertex" resultant from the decay of the short-lived (with a lifetime of 10^{-12}–10^{-13} s) particle that was produced at the first vertex.

Most popular are microstrip semiconductor detectors structured as follows: strips of a conducting material are deposited as electrodes on one of the surfaces of a silicon monocrystal, whereas the other surface is metallized. The voltage applied to these electrodes makes up a few Volts. An ionizing particle in transit through the crystal is forming the electron-hole pairs migrating towards the electrodes to create the current pulses. The spatial resolution d of microstrip detectors is determined by the strip width and interstrip gaps, reaching 10^{-7} cm (higher resolution is demonstrated only by nuclear photoemulsions $d \sim 10^{-8}$ cm). The temporal resolution of these detectors is of the order of 10^{-8} s.

Scintillation Counter (SC). SC comprises a scintillating material (special liquids, plastics, crystals, noble gases), where a charged particle effects both ionization and excitation of the atoms and molecules forming this scintillator. Recovering to ground state, these atoms and molecules are emitting photons incident on the cathode of the photomultiplier (PM) to knock off photoelectrons from the cathode. Owing to these electrons, a pulse with amplitude proportional to the energy transferred by the particle to the scintillator is formed at the anode of the PM. The accuracy of the measured particle energy provided by the SC is within 10%. Since under the effect of charged particles the majority of scintillators reveal the characteristic luminescence time of about 2×10^{-8} s, the transit time of a particle through the counter may be determined with high accuracy. The detection efficiency for charged particles is close to 100%. Neutrons may be detected (by recoil protons) with the use of hydrogen-containing scintillators, and photons—with the use of higher-density scintillators like iodide sodium. Detection of neutrinos requires the use of combined scintillators. So, in their experiment (1953) F. Reines and C. L. Cowan have used the hydrogen-containing scintillator with an addition of a cadmium salt for the detection of electron antineutrino in the reaction

$$p + \overline{\nu}_e \rightarrow n + e^+. \tag{13.21}$$

The first scintillation flare was caused by annihilation of the positron with the electron of scintillator, while the second flare occurring in $(5 - 10) \times 10^{-6}$ s was due to the cadmium atom returning into the ground state after absorption of a neutron.

Cherenkov Counters (CC). The operation principle of CC is based on the detection of Cherenkov radiation. This radiation is generated when charged particles are moving in a transparent medium at the speed v exceeding the speed of light in the medium

$$v \geq \frac{c}{n}, \tag{13.22}$$

where n is a refractive index of this medium. Light is emitted only in a forward direction along the motion path of a particle, forming a cone with an axis in the direction of $\boldsymbol{v}$, and the cone angle, emission angle, is determined by the relation

$$\cos\theta = \frac{c}{vn}. \tag{13.23}$$

This phenomenon is similar to the acoustic cone of the airplane flying at a supersonic speed. A light flare following the particle motion in the medium is detected by the PM. The counter is used for the detection of relativistic particles as well as estimation of their charge, speed, and motion direction (to within 10^{-4}). Measuring the particle momentum by deflection in a magnetic field, one is able to measure its speed with the help of the CC and hence to determine a mass of this particle. Relation (13.22) forms the basis for the operation of threshold or integrating CC that can detect all particles having the speeds above threshold $v > v_t = c/n$. And the counter provides radiation measurements over the whole range of angles from 0 to $\theta_{max} = \arccos(/vn)$. The operation of angular or differential CC is based on relation (13.23). These counters detect particles at the speeds from v_0 to $v_0 + \Delta v$. Emission of the particles parallel to the optical axis of the counter is collected only in a narrow range of angles, from θ_0 to $\theta_0 + \Delta\theta$. The CC comprise a radiation-generating medium, collecting optics that directs this radiation to the photocathode of PM, PM and recording system. A new type of gas CC, RICH (Ring Image Cherenkov detector), has been proposed recently. This detector provides detection and imaging of the particles by their Cherenkov radiation. The energy region, where the CC offer mass separation for the particles, has an upper limit as the difference in the speeds of particles distinguished by their masses is decreased with growing energy. For instance, the speeds separation of the π- and K-mesons by threshold gas CC is possible up to the energies amounting to several dozens of GeV, whereas by differential gas CC with a compensation of radiation dispersion—up to several hundred GeV.

Transition Radiation Detectors (TRD). The operation of these detectors is based on the formation of an electromagnetic field by the moving charged particle. The electromagnetic field of this particle changes its configuration when the particle goes from a medium with the dielectric permeability ϵ_1 to that with ϵ_2. This process is accompanied by the emission of transient radiation. The distinguishing feature of this radiation is the fact that its properties are determined by the relativistic factor

$$\gamma = \frac{1}{\sqrt{1 - v^2/c^2}} = \frac{E}{mc^2}.$$

In this case the radiation intensity is proportional to the particle energy, and radiation is concentrated within a cone of angle $\theta = \gamma^{-1}$. In layered structures, where a particle crosses the interface repeatedly, the intensity of transient radiation may be resonantly amplified providing a means for the detection and identification of ultrarelativistic particles with $\gamma > 10^3$. The Lorentz factor $\gamma \sim m^{-1}$ and intensity of transient radiation at constant energy E will be greater for particles with lower mass. This allows for mass separation over the energy range hardly accessible for gas CC. TRD consist of a layered medium, usually comprising a multitude of foils from lightweight materials (Li, Al), which are stretched perpendicular to the particle direction, and a radiation detector. In some TRD the radiator may be represented by ordinary porous materials like foamed plastics. The radiation detection is most commonly done by proportional wire chambers filled with heavy gases.

1 b. Electromagnetic and Hadron Cascade Detectors

The operation principle of these detectors, also referred to as Total-Absorption Detectors (TAD), is based on total absorption of the cascades created by the detected particles within the detector material. TAD enable detection of the integrated Cherenkov radiation for all the particles forming the electron-photon shower (total-absorption CC) or integrated energy spent by all the particles for ionization (calorimeters). In electromagnetic cascades this energy is practically equal to the energy of the primary electron or photon. In hadron cascades ionization requires the major part of the energy possessed by the primary particle but some part of its energy (up to 20–30%) is spent for nuclear disintegration, then carried away by neutrinos formed as a result of particle decays, and is not detected by calorimeters.

Total-Absorption Cherenkov Counters. These detectors provide detection of photons and electrons with estimation of their energy. The blocks of lead glass serve as radiators. Their size must be sufficient for absorption of the main part of the shower produced by the primary particle. Cherenkov radiation is detected by the PM.

Calorimeters (C). C are intended to measure the energy of particles, both charged and neutral, beginning from 10^2 GeV and higher. Interacting with the nuclei within the material of C, a high-energy particle produces a cascade of the high-energy secondary particles, in turn, interacting with the material to generate new particles. Because of this, electron-nuclear showers occurring in the active volume of the device rapidly moves in a direction of the primary particle, and its energy is spent for ionization of the material. Provided a layer of the material in C is sufficiently large, all shower particles are left in the material, and the number of created ions is proportional to the energy of the primary particle. Then these ions are collected at the calorimeter electrodes, and their total charge is measured to determine the primary particle energy with an accuracy of 10–15%. The use of C makes it possible to locate the shower origination and determine its spatial development. The simplest C are constructed as "sandwiches" consisting of alternated layers of heavy material and ionization detectors. In electromagnetic C such a sandwich comprises thin layers of lead and scintillator. Its total thickness may reach a few dozen centimeters.

Hadron cascades are slowly developing in the majority of materials, penetrating deeper than the electromagnetic ones. Because of this, hadron C possess a much greater thickness, up to several meters, with thicker layers of the material (commonly iron) and scintillator, that may be replaced by other ionization measuring detectors.

2. Track Detectors

These detectors enable observation of particle tracks. Track detectors, being subjected to magnetic fields, make it possible to determine a sign of the electric charge for the particles and to measure their momenta by the path curvature to a high degree of accuracy. The first track detector was constructed by Ch. Wilson in 1912 and received the name the Wilson chamber. Its operation is based on condensation of supersaturated vapor and the formation of visible little drops of liquid at the ions originating along the path of a fast charged particle. The device comprises a closed vessel, having windows intended for the track observation, filled with gas and saturated vapors of some liquid substance, for example, methyl alcohol. On rapid adiabatic rarefying this gas is cooled, whereas the vapor becomes supersaturated. After photographing the tracks, the chamber regains its initial state due to fast gas compression, causing evaporation of the droplets at the ions with the formation of saturated vapor, and the former ions influenced by the electric field are collected from the tracks at the electrodes.

Bubble Chamber (BC). BC is one of the main types of track devices in high-energy physics. It comprises a large container several meters in diameter filled with a transparent superheated liquid. Its boiling is delayed owing to the high pressure that is 5–20 times higher than the atmospheric pressure. The abrupt decrease of the pressure results in superheating of the liquid, and in case an ionizing particle is passing through the chamber an additional heating leads to drastic boiling of the liquid in a narrow channel along the particle path. Its trajectory is marked by a chain of vapor bubbles. These bubbles are allowed to grow for a period of 10 ms, afterwards they are photographed by stereoscopic cameras. Subsequently, the starting pressure is applied to the liquid causing collapse of the bubbles, and the chamber is ready for operation again. In the BC the most common is liquid hydrogen. Deuterium, propane, and such heavy liquids as xenon and Freon are more rarely used. The latter are of special interest, especially for the detection of neutrino interactions. BC is usually placed in a strong magnetic field, and by the track curvature one can measure the particle momenta to a high accuracy. The spatial resolution of BC comes to 10^{-2} cm. The main advantage of BC is the possibility to use its working liquid both as a target for the incoming particles and as a detector for the reactions proceeding upon particle collisions with electrons and nuclei of the liquid. This advantage is especially marked in studies of complex processes involving large numbers of particles. The principal disadvantage of BC is its "uncontrollability": it is impossible to realize its response on a signal of fast detectors that previously select the required events. The response of BC with a period of ~ 1 s is synchronized with points in time of fast beam ejection from the accelerator.

Spark Chambers (SC). SC are the controlled gas discharge detectors including a series of parallel metallic plates placed into the container filled with inert gas. These plates, alternately, are connected to the high-voltage source or have an earth connection. Provided an ionizing particle crosses the working volume of SC, by command of the monitor counters a short high-voltage pulse (10–20 kV/cm) is applied. Originating at the points of particle pass, spark discharges parallel to the electric field are photographed or located by the magnetostriction method.

Streamer Chambers (StC). StC are the counter-controlled gas discharge detectors, where discharges are formed exclusively along the particle tracks. These chambers contain two flat parallel electrodes positioned at a distance measuring several dozens of centimeters. To the electrodes a very short ($\leq 10^{-8}$ s) high-voltage (10–50 kV/cm) pulse is applied. In these conditions the discharges, originating at an ionizing particle passage, are terminated and take the form of short ($\sim 10^{-1}$ cm) luminous channels (streamers) aligned with the field. Their photographs are made to obtain the track images. As contrasted to SC, StC are isotropic, that is, they are capable of reproducing the tracks of every spatial orientation and can measure the particles ionization.

As a rule, all the available particle detectors are combined detector systems (CDS) featuring a series of detectors integrated in one detecting unit. CDS represent the major element of the modern accelerator. Their size measures dozens of meters, mass amounts to $\sim 10^4$ t, and the number of information channels may be as great as 10^6. The personnel required for their operation runs into hundreds of people, whereas the construction expenditures comprise a significant part of the total cost for the whole accelerating complex. The majority of CDS are similar in structure, though the choice, amount, dimensions, and arrangement of elements are dependent on the specific task at hand. Most typical elements are as follows: target, vertex detector surrounding the target that indicates the reaction products and determines their escape direction; position detectors localizing trajectories of primary and secondary particles; spectrometric detectors measuring the momenta of secondary particles or their energy; and identifiers of secondary particles. Large-scale CDS are given proper names such as ATLAS, ALICE, DELPHI, etc. Now the particle fluxes passing through CDS are as great as 10^8 s^{-1}. Unfortunately, the difficulties in processing the measurement results in the case of numerous information channels and high detection rates generally prevent a real-time analysis. Considering this situation, information is recorded and processed on completing the experiment.

13.3 Neutrino Telescopes

Among the fundamental particles, the neutrino holds a special place since it plays an important role in large-scale events of the universe. The finding that the neutrino has a mass makes this particle a worthy candidate for the role of a particle constituting the hot dark matter. This, in turn, enables one to evaluate the average matter density in the Universe, age of the Universe, and its further fate. The problems associated with detection of cosmic neutrinos are the subject matter of the neutrino astrophysics. The neutrino astrophysics may be considered as a compound part of the elementary particle physics. And this is related not only with the fact that both physics divisions are concerned with the universe structure. The other important aspect is that solution of the problems concerning the generation and detection of the neutrinos depends upon the character and intensity of interaction between the elementary particles. Because of this, it is obvious that neutrino telescopes (NT), being basic instruments of the neutrino astrophysics, are useful for studies of particle physics too.

Depending on the detection technique, all available NT, may be subdivided into two classes: NT operating in the continuous counting mode and NT operating in the discrete counting mode. The first class includes NT using the radiochemical methods. The second class involves NT intended for real-time detection of the particles, the production of which is initiated by the interaction between the neutrinos and the working medium of NT.

The operation of radiochemical NT is based on investigating the process of the inverse β-decay due to the incoming neutrino

$$\nu_e + X \to e^- + Y, \tag{13.24}$$

where X are nuclei of the elements determining the initial composition of NT detector. As a rule, nuclei of Y originating in the detector are radioactive, their half-life $T_{1/2}$ determining the duration of an active measurement stage $t_a \sim (2\ \ 3)T_{1/2}$. The formation rate of daughter nuclei is given by the expression:

$$R = N \int \Phi(E)\sigma(E)dE, \tag{13.25}$$

where $\Phi(E)$ is the neutrino flux incident on the detector, N are numbers of detector atoms, and σ is the cross section of the process (13.24). At the incident neutrino flux 10^{10} cm^{-2}·s^{-1} (approximately amounting to the flux of solar neutrinos incident on the Earth) and σ of the order of 10^{-45} cm^2 the provision of a single useful event a day necessitates about 10^{30} atoms. Consequently, a mass of this detector should be in the region of several kilotons. Chemical analysis of the detector material takes place in time t_a, and the nuclei number Y is indicative of the capture rate for the neutrinos. The advantage of radiochemical NT is the possibility for varying the reaction energy threshold with changes in the detector material. This makes them indispensable in studies of low-energy neutrino fluxes. At the same time, radiochemical NT feature an inability to measure such neutrino characteristics as hitting time for the detector, energy, and trajectory direction. The latter is rather discouraging as we have no chances to distinguish between, for example, the solar neutrino and the neutrino produced by the terrestrial source.

As an example of a radiochemical NT, we consider the Homestake facility that was the first to study the fluxes of the solar neutrinos (1967–2001). The neutrinos were detected with the use of the chlorine-argon method, that is, operation of this NT was based on the chemical reaction

$$^{37}Cl + \nu_e \to ^{37}Ar + e^-. \tag{13.26}$$

Isolation of the useful event was realized using the decay of

$$^{37}Ar \rightarrow {}^{37}Cl + e^- + \overline{\nu}_e, \tag{13.27}$$

whose half-life is 35 days. This facility, representing a vessel with a capacity of 390 l filled with 610 t of perchloroethylene (C_2Cl_4), was located in the gold-bearing mine (Homestake, South Dakota, USA) at a depth of 1 480 m. As argon atoms are produced in the form of a volatile compound, they were isolated approximately once a month. The obtained argon was subjected to multistage processing. At the final stage, special small-size proportional chambers were filled with this argon. Then the chambers were shielded with lowest-activity lead to provide observation of decays (13.27).

The operation of the second-type NT may be based, for example, on the detection of elastic scattering

$$\nu_l + e^- \rightarrow \nu_l + e^-, \tag{13.28}$$

where $l = e, \mu, \tau$. The spectrum of recoil electrons is given by the following expression:

$$\frac{d\sigma}{dT} = \sigma_0 \left[g_1^2 + g_2^2 \left(1 - \frac{T}{E_\nu} \right)^2 - g_1 g_2 \frac{m_e T}{E_\nu^2} \right]. \tag{13.29}$$

Here T is a kinetic energy of recoil electrons, $\sigma_0 = 8.8 \times 10^{-45}$ cm^2,

$$g_1 = \sin^2 \theta_W \pm \frac{1}{2}, \qquad g_2 = \sin^2 \theta_W,$$

where "+" sign is associated with the ν_e-scattering, while "-" sign—with the ν_μ- and ν_τ-scattering. Because of this, in the second case the scattering cross section is approximately one-sixth of that in the first case making it possible to distinguish between the neutrino kinds. The cross section of the ν_e-scattering integrated with respect to the energy is simple in form

$$\sigma(\nu_e e) = 9 \times 10^{-44} \frac{E_\nu}{10 \text{ MeV}} \text{ cm}^2. \tag{13.30}$$

The angular distribution of recoil electrons is characterized by a sharp maximum in the forward direction (relative to the incoming neutrino) whose width is $\Delta\theta \sim \sqrt{m_\nu/E_\nu}$. This enables one to determine the direction of a neutrino source by the trajectories of scattered electrons. Another advantage of such a facility is the possibility for the detection of individual acts when the neutrino is hitting the detector, permitting measurements of time and energy for the neutrinos (real-time operation). A limitation of NT based on recoil electrons resides in the presence of a considerable background as the detected electrons may be produced during the processes of elastic scattering involving any neutral particles. However, in the case of the high-energy neutrinos the background elimination is greatly facilitated.

Also, the operation of the second-type NT may be based on the detection of particles produced in reactions with the high-energy neutrino, for example,

$$\nu_l(\overline{\nu}_l) + N \rightarrow l(\overline{l}) + N', \qquad \nu_l(\overline{\nu}_l) + N \rightarrow l(\overline{l}) + N' + X, \tag{13.31}$$

$$\nu_l(\overline{\nu}_l) + {}^{16}O \rightarrow l(\overline{l})^{16}O + \pi^+(\pi^-), \tag{13.32}$$

where $N = p, n$ and X denotes hadron collections. Since the secondary high-energy electrons and hadrons are responsible for the nuclear-electromagnetic shower and the main free path of these electrons within the detector material is very short, they are practically indistinguishable from hadrons. Compared to electrons, the mean free path of high-energy muons is very long as their energy losses for bremsstrahlung, formation of the $e^- e^+$-pairs and nuclear interactions are small. Of special importance is the fact that muons are moving practically in the same direction as the neutrinos producing them. The average angle between the ν_μ- and μ-trajectories expressed in degrees is determined by the expression

$$< \theta > \approx 2.6 \times \sqrt{\frac{100}{E_\mu(\text{GeV})}}.$$

In the active volume of NT, bremsstrahlung photons, the $e^- e^+$-pairs and hadrons originate along the muon trajectory to initiate the nuclear-electromagnetic showers. With $E_\mu \approx 100$ TeV particular minor showers are overlapping, and the whole muon trajectory is glowing due to the Cherenkov radiation of these showers. Based on the direction and intensity of the Cherenkov radiation, one can determine the trajectory and energy for muon.

Hadrons generated in the reactions described by (13.31) and (13.32) also initiate nuclear-electromagnetic showers, the direction and energy of which is determined by the Cherenkov radiation. The detection of showers may be performed using the acoustic method as well [51]. In this case the detected signal is represented by the pressure pulse in the active volume of NT conditioned by the drastic heating of a narrow channel within the shower due to ionization energy losses of the electrons. For instance, in water an acoustic signal is propagating in the form of a thin disk, with a thickness of about the shower length $s \sim 5$ m and characteristic radius $R \sim 1$ km. By this method the detecting element is represented by hydrophones detecting signals perpendicular to the shower axis.

In NT based on detection of the secondary muons the effective volume of the detector is considerably greater than the physical volume owing to the detection of muons generated within a thick layer of the material surrounding the detector. When the neutrinos are detected with the use of a detection mechanism on the basis of hadron showers, however, this is not the case. Short lengths of hadron showers enable their detection by the Cherenkov radiation only within the physical volume of NT.

As a working material (detector) for NT of the second type one can use water or arctic ice. Arctic ice represents a sterile medium with lower concentration of radioactive elements than in sea or lake water. The use of arctic ice as a detector contributes considerably to the sensitivity of NT. For instance, an NT positioned at a depth of about 1 km makes it possible to separate the background muon (atmospheric) flux that is 100 times greater compared to the limiting flux for the deep-sea NT DUMAND (Deep Underwater Muon and Neutrino Detector), which was positioned in sea water at a depth of 4.5 km. NT AMANDA located at the South Pole is intended for studies of high-energy neutrino fluxes. Deep-water NT on muons, BAIKAL NT-200 (the Baikal Lake), is an example too.

Among NT of the second type one may name the Super-Kamiokande facility (Japan, Kamioka) constructed jointly with the U.S. specialists (1996). This NT is located in the mine with a shielding depth of 2700 m in water equivalent; absorption of particle fluxes by the rock is equivalent to a water thickness of 2700 m. The principal element of this facility is a water Cherenkov detector in the shape of a cylinder, 39 m in diameter, and 41 m in height, that contains 50000 t of water and provides ring imaging of the detected particles. The detector is optically subdivided into the internal (working) volume scanned by 11200 photoelectronic multipliers (PM), and also the outer (shielding) volume containing 2200 PM and operating in the anticoincidence mode. This NT can investigate the fluxes of both the solar and atmospheric neutrinos.

The main source of the solar neutrinos is a series of thermonuclear fusion reactions at the central part of the Sun, resultant in hydrogen-to-helium transformation without catalysts, the hydrogen cycle. This chain may be represented as a multistage process

$$4p \rightarrow {}^4He + 2\nu_e + 2e^+ + 26.73 \text{ MeV} - E_\nu, \tag{13.33}$$

where E_ν is an energy carried away by the electron neutrino, its average value being $\sim$ 0.6 MeV ($E_{max} < 18.8$ MeV). In Super-Kamiokande the detection of the solar neutrino is realized using the neutrino elastic scattering reaction from electrons with the energy threshold 5.5 MeV.

Cosmic rays interacting with atomic nuclei initiate in the atmosphere surrounding the Earth the production of pions, kaons and muons, the decay channels of which involve electron and muon neutrinos as well as antineutrinos

$$\left.\begin{array}{l}\pi^\pm \to \mu^\pm + \nu_\mu(\overline{\nu}_\mu) \to e^\pm + \nu_e(\overline{\nu}_e) + \nu_\mu(\overline{\nu}_\mu), \\ K^\pm \to \mu^\pm + \nu_\mu(\overline{\nu}_\mu) \to e^\pm + \nu_e(\overline{\nu}_e) + \nu_\mu(\overline{\nu}_\mu).\end{array}\right\} \tag{13.34}$$

The neutrino flux is formed in the region of 10–20 km altitudes above sea level, its energy varying from 100 MeV to 1 000 GeV. Since the dominant interaction type for the neutrinos with such high energies is interaction with the target nuclei, in case of Super-Kamiokande the detection of the atmospheric neutrinos is performed using reactions (13.31) and (13.32).

The second-type NT, Sudbury Neutrino Observatory (SNO), came into use in Canada in May 1999. At this facility the detector is represented by 1000 t of heavy water (D_2O) enabling investigation of the solar neutrino with the help of the following processes:

$$\nu_e + d \to p + p + e^-, \tag{13.35}$$

$$\nu_l + e^- \to \nu_l + e^-, \tag{13.36}$$

$$\nu_l + d \to \nu_l + n + p, \tag{13.37}$$

Reaction (13.35) is sensitive to ν_e-neutrino, whereas reactions (13.36) and (13.37) are sensitive to the neutrinos of all three kinds. For reactions (13.35) and (13.36) the energy threshold equals 5 MeV, and that for reaction (13.37) is 2.225 MeV.

The multipurpose NT named KamLAND (Japan, Kamioka) using 1000 t of an ultra-pure liquid scintillator as a detector came into operation in Spring 2001. Although the solar neutrino may be detected by KamLAND, its main function is to observe the oscillations in the total neutrino flux coming from ten reactors localized in the region at 80–350 km from the detector.

The flux of the solar neutrinos originating in reaction

$$^7Be + e^- \to {}^7Li + \nu_e$$

(so-called beryllium neutrinos) is especially sensitive to the neutrino characteristics. Real-time measurements of this monoenergetic ($E_\nu = 0.86$ MeV) flux are the principal objectives of NT named BOREXINO (Gran Sasso, Italy) that is based on recoil electrons with the threshold 250 keV and came to operation early in 2002. Recoil electrons caused by $\nu_l e$-scattering (the cross sections for ν_μ and ν_τ are smaller than for ν_e) produce light flare in the bulk of the liquid scintillator, that is detected by the PM. Nylon sphere contains 300 t of ultra pure pseudocumene, and 100 t of pseudocumene contained in the central region comprise the effective (sensitive) volume. The nylon sphere is, in turn, surrounded by pseudocumene filling the corrosion-proof steel sphere 13.7 m in diameter that contains optical elements surrounding the nylon sphere. The whole construction is submerged into the reservoir with purified water having a mass of 2500 t. The experimental threshold is set as 0.25 MeV because the energy spectrum of recoil electrons is continuous up to 0.66 MeV. At these low energies the control of natural radioactivity caused by radioactive isotopes being present everywhere is the greatest problem. By the present time extensive research has been conducted with the aim to select materials and realize their purification to extremely high levels of radioactive purity. Simultaneously, the measuring techniques for ultra low radioactivity levels have been developed. The attained results are quite impressive: $10^{-16} - -10^{-17}$ (gram of contaminant per gram of material) for ^{232}Th and ^{238}U.

Problems

2.1. A μ^--meson has been captured by an Al^{27} atom. Define the orbit radius of the μ^--meson and its residence time inside the nucleus.

2.2. When one tries to integrate the expression (7.10) over the scattering angle then infinity will be obtained. The reason consists in the long-range character of the electrostatic forces. The particles are scattered, no matter how far away from a force center they fly. To get a finite result one should allow for a screening effect of the electron shell. This is attained under using the potential

$$U(r) = \frac{Ze^2}{2\pi r} \exp\left(-\frac{r}{a}\right),$$

where a has the value of the order of the atom radius. Using this potential, find the total cross section of the α-particles scattering.

2.3. Consider the reactions mentioned below

1) $\Sigma^- + p \to \Lambda^0 + n,$ 2) $\Sigma^+ + p \to K^+ + p,$ 3) $p \to n + e^+ + \bar{\nu}_e,$

4) $\Omega^- \to n + \pi^+,$ 5) $\Xi^0 \to \Lambda^0 + \overline{K}^0,$ 6) $\pi^- + p \to \Sigma^- + K^+,$

7) $\pi^- + p \to K^- + \Sigma^+,$ 8) $\eta \to \pi^+ + \pi^- + \gamma,$ 9) $\mu^+ \to e^+ + e^- + e^+.$

Point out which of them are forbidden by conservation laws. Name these laws.

2.4. Examine the resonance production processes

$$\pi^+ + p \to \Delta^{++} \to \pi^+ + p,$$

$$\pi^- + p \to \Delta^0 \to \pi^- + p.$$

Thinking Δ^{++} and Δ^0 to enter into one and the same isotopic multiplet and taking into consideration the experimental results

$$\left.\frac{\sigma_{\pi^+ p}}{\sigma_{\pi^- p}}\right|_{E=T_\pi^r} = 3,$$

where T_π^r is the kinetic energy of the π-mesons at which the Δ-resonances appear, determine the Δ-resonances isotopic spin.

2.5. Protons being bombarded by K^--mesons with the kinetic energy $T_K = 790$ MeV, the following reaction

$$K^- + p \to (\pi^- \Lambda) \to \Lambda + \pi^+ + \pi^-$$

is observed. At the same time the emergent π^+-mesons have the kinetic energy $T_\pi^* = 300$ MeV. Find the mass of the $(\pi^- \Lambda)$-resonance and its decay energy.

2.6. Protons are bombarded by π^--mesons with the momentum 1.14 GeV, with the reaction events

$$\pi^- + p \to n + 2\gamma$$

being registered. Two peaks in distribution on the angle of divergence of two γ-quants in the center-of-mass frame are observed. The former is located near the angle 25^0 while the latter—near the angle 100^0. Find for each of these peaks the masses of corresponding unstable particles.

2.7. Find the angular distribution of secondary particles in two-body decay in flight $X \to 1 + 2$ when in the rest frame of the X-particle this distribution is isotropic.

2.8. For the decay

$$X \to 1 + 2$$

find the distribution of secondary particles on the angle of their divergence in the lab frame if in the rest frame of the X-particle at $m_1 = m_2$ the decay has an isotropic character.

2.9. Two particles with the isospins $I = 1/2$ and $I = 1$ make up a coupled state. Using the properties of the lowering and rasing isospin operators $S_{\pm}^{(is)} = S_1^{(is)} \pm S_2^{(is)}$ build isospin wavefunctions for the following states: $\Psi(3/2, 1/2), \Psi(3/2, -1/2), \Psi(1/2, 1/2)$ where $\Psi(I, I_3)$ is a wavefunction of a system with the total isospin I and the projection I_3.

2.10. Taking into account the $f_1(1420)$-resonance spin to be equal to zero, define isotopic relations between strong decay probabilities of the $f_1(1420)$-resonance through the channels:

$$f \to \overline{K}^0 K^+ + \pi^-, \qquad f \to K^0 K^- + \pi^+,$$

$$f \to K^+ K^- + \pi^0, \qquad f \to K^0 \overline{K}^0 + \pi^0.$$

Thinking the C-parity of $f(1420)$ to equal $+1$, point out the basic peculiarities of detecting the third decay channel.

2.11. Allowing for the isospin conversation law in the strong interaction, find the ratio of the total cross sections for the following reactions:

$$p + p \to d + \pi^+, \qquad n + p \to d + \pi^0.$$

2.12. Ascertain on what isospin channels the following reactions:

$$\pi^- + p \to K^0 + \Sigma^0, \qquad \pi^- + p \to K^+ + \Sigma^-,$$

$$\pi^+ + p \to K^+ + \Sigma^+$$

could be proceeded. Assuming the channel with the definite isospin to dominate, find the ratio of the total cross sections for these processes.

2.13. Define an isospin change and a third isospin projection change in the following weak decays

$$K^+ \to \pi^+ + \pi^0, \qquad \overline{K}^0 \to \pi^+ + \pi^-.$$

2.14. Let us consider a single π-meson production processes to be observed under neutrino scattering by nucleons:

$$\nu_\mu + p \to \pi^+ + p + \mu^-, \tag{II.1}$$

$$\nu_\mu + n \to \pi^0 + p + \mu^-, \qquad \nu_\mu + n \to \pi^+ + n + \mu^-. \tag{II.2}$$

If we assume the first pion-nucleon $\Delta(1232)$-resonance contribution to dominate, then a $\pi - N$-system will have the fixed isospin $I = 3/2$. The strangeness not being changed in the weak processes ($\Delta s = 0$), if that is the case the hadron state isospin is varied by 1 ($\Delta I = 1$). Show, then the cross section of the reaction (II.1) is three times bigger than that of the reaction (II.2). Assuming $\Delta I = 2$, find the ratio of these reactions as well. [NOTE: If one assumes a certain fictitious particle with the isospin equal ΔI, a spurion, to be added to the left-hand side of the equations describing the decay processes, then the above-mentioned decays may be considered as the processes going with the isospin conservation].

2.15. Proceeding from the rule $\Delta I = 1/2$, prove the correctness of the following relations between three-pion decay probabilities of charged and neutral kaons:

$$\Gamma(K_L \to 3\pi^0) = \frac{3}{2}\Gamma(K_L \to \pi^+\pi^-\pi^0), \qquad \Gamma(K^+ \to \pi^+\pi^-\pi^+) = 4\Gamma(K_L \to \pi^+\pi^0\pi^0).$$

NOTE: In obtaining the given relations, one has assumed all three π-mesons to be in the S relative state. Then any pair of π-mesons must be in a symmetric isospin state, that is, either in the state with $I = 0$, or in the state with $I = 2$, or in the mixed state with $I = 0$ and $I = 2$. The third π-meson with $I = 1$ must be arranged with that pair so as in the case of the K^+-decays the resultant three-pion state has $I = 1$ and $I_3 = 1$ while in the case of the K^--decays it has $I = 1$ and $I_3 = 0$].

2.16. Taking into account the parities of K^0- and $\overline{K}^0$-mesons to be negative, build from them two states with a definite CP-parity.

2.17. Under K^0-meson decays an uncommon phenomenon is observed. One half of the K^0-mesons decays during the short time τ_S through the channels:

$$K^0 \to \pi^- + \pi^+, \qquad K^0 \to \pi^0 + \pi^0,$$

while the latter half of the K^0-mesons does during the time $\tau_L = 651\tau_S = 5.77 \times 10^{-8}$ s through the channels:

$$K^0 \to \pi^0 + \pi^- + \pi^+, \qquad K^0 \to \pi^0 + \pi^0 + \pi^0.$$

Being based on the CP-parity conservation law, explain this phenomenon.

2.18. (Regeneration of K_1^0-mesons) Consider the K^0-mesons beam moving in a vacuum during the time much bigger than the K_1^0 lifetime. Assume this beam hits on a target consisting of heavy nuclei with the mass M that is much more than the incoming particles energy. The number of nuclei per the target length unit is equal to N. Find the K_1^0-mesons production probability on coming out of the target. Consider contributions of different target nuclei to be added coherently.

2.19. Being based both on the U-spin conservation law and on the energy reasons, determine which of the reactions mentioned below

$$\Delta^- \to \Sigma^- + K^0, \qquad \Sigma^{*-} \to \Sigma^- + \pi^0,$$

$$\Xi^{*-} \to \Xi^- + \eta, \qquad \Xi^{*-} \to \Xi^- + \pi^0$$

are allowed and find the relations between allowed reaction amplitudes.

2.20. If one considers a photon as a scalar relative to the U-spin, then which of the following decays are allowed in the $SU(3)$-symmetry scheme

$$\Xi^{*-} \to \Xi^- + \gamma, \qquad \Sigma^{*-} \to \Sigma^- + \gamma,$$

$$\Delta^+ \to p + \gamma, \qquad \Sigma^{*+} \to \Sigma^+ + \gamma.$$

2.21. Allowing for the quark filling of a meson octet 0^{-+} and setting $m_u = m_d =$, show the validity of the following mass relation

$$2(m_K + m_{\overline{K}}) = 3m_\eta + m_\pi.$$

2.22. Taking into account the quark filling of a baryon decuplet $3/2^+$ and setting $m_u = m_d =$, obtain the equal intervals rule

$$m_\Omega - m_\Xi = m_\Xi - m_\Sigma = m_\Sigma - m_\Delta.$$

2.23. Taking into account the quark filling of a baryon octet $1/2^+$ and setting $m_u = m_d =$, prove the validity of the relation

$$2(m_N + m_\Xi) = 3m_\Lambda + m_\Sigma.$$

2.24. What amount of gluons could appear under annihilation of neutral mesons 0^{-+} and 1^{--}?

2.25. Making use of the parton model, find the ratio of the total cross sections of πN- and NN-interactions and compare it with the experimental value. Think the quark and antiquark total cross sections to be connected with the relations

$$\sigma_{qq} = \sigma_{q\bar{q}} = \sigma_{\bar{q}\bar{q}}.$$

2.26. Taking into consideration the quark filling of vector ρ-, ω- and ϕ-mesons

$$\rho = \frac{1}{\sqrt{2}}\left(u\bar{u} - d\bar{d}\right),$$

$$\omega = \frac{1}{\sqrt{2}}\left(u\bar{u} + d\bar{d}\right), \qquad \phi = s\bar{s},$$

compute the probabilities ratio of their decays through the e^-e^+-pair. Under computations use the Feynman diagram for the subprocess

$$q + \bar{q} \to \gamma^* \to e^- + e^+.$$

Compare results obtained with the experimental data.

2.27. Find a quark wavefunction of a proton with the up spin.

2.28. Compute proton and neutron magnetic moments in the quark model.

2.29. Consider a electron-positron bound state, a positronium. Analyze possible decay channels of the positronium S-state with the spin 0 (parapositronium) and 1 (orthopositronium).

2.30. Using spectroscopic designations $(n^{2S+1}L_j)$ propose a classification of low energy levels for charmonium and bottonium. Find quantum characteristics J^{PC} of these states and establish a correspondence of charmonium and bottonium to experimentally observable states of J/Ψ- and Υ-families.

2.31. A collision of a high-energy electron with a proton is described by an energy ν, transmitted to a proton in the proton rest frame, and a quantity $q^2 = p^2 - \nu^2$, where p is a momentum transmitted to the proton. Neglecting the electron mass, show that the following relations:
$$q^2 = 2EE'(1 - \cos\theta), \qquad \nu = |E' - E|,$$

where E and E' is the electron energy before and after collision and θ is a scattering angle, are fulfilled.

2.32. Using the previous problem result consider the interaction of a relativistic electron with a proton in the center-of-mass frame. Consider that a momentum of each particle is equal to p and the quantities ν and q are known. Find the energy transmitted to the proton in the center-of-mass frame. Neglect the electron mass and think that $p \gg M$, where M is the proton mass. Estimate the value of the electron interaction time with the proton.

2.33. An electron flies at a proton having a momentum p in the center-of-mass frame. Calculate the effective mass of a quark entering into the proton at the quark interaction with the electron. Base calculations on the following assumptions. Before collision the quark possessed the transverse momentum $p_\perp$ to be perpendicular to the proton center-of-mass direction. Under flying out the quark gets the longitudinal proton momentum xp, where $0 < x < 1$. By definition

$$m_{eff} = p\Delta E,$$

where ΔE is an energy spent on the free quark production. Making calculations, assume

$$p^2 x^2 \gg m_q^2 + p_\perp^2, \qquad p^2(1-x)^2 \gg M + p_\perp^2,$$

where M is the proton mass.

2.34. Using the parton model, show at what conditions one may consider a quark to be free under its interaction with a high-energy electron. Think the effective quark mass to be known.

2.35. Determine the energy distributions of muons and neutrinos resultant the decay

$$\pi^+ \to \mu^+ + \nu_\mu$$

in the reference frame where π^+-meson is moving with a momentum p. Take into account that in the π^+-meson rest frame the angle distribution of muons (neutrinos) is isotropic.

2.36. What distance does a beam $\mu, \pi^+, \pi^0, K^+, n$ pass in a vacuum before its intensity is halved? Take on the neutron energy to be equal to 14 MeV and the rest of particles energy—1 GeV.

2.37. Proton and electron circular accelerators have the equal radius 350 m. Estimate the maximum allowable energy (neglecting synchrotron emission losses) electrons and protons could get under $H = 20$ kGs. What are synchrotron emission losses at one rotation under maximum allowable energy of particles?

2.38. Provide a comparative characteristic of proton accelerators: (i) accelerator with colliding beams (proton and antiproton currents $i_p = i_{\bar{p}} = 2.8$ mA; clusters rotation frequency $f = 10^4$ Hz; the number of clusters on an orbit $N = 1$; a beam size $r_x = r_y = 0.02$ mm; momenta $p_p = p_{\bar{p}} = 32$ GeV/c); and (ii) accelerator with a fixed target (a momentum $p_p = 400$ GeV/c, a proton beam intensity $I = 10^{12}$ /, a target—Be, $\rho = 1.85$ g/cm^3, a beam target size—10 cm).

Compare: carrying out experiments time; an energy available for new particles production; possibilities for generating new particle beams; and possibilities for detecting collision products.

2.39. Evaluate the speed of a set of statistics in a process of the Z-boson resonance production at $e^- e^+$-collider under the following conditions: electron and positron currents are $i_e = i_{\bar{e}} = 3$ mA, clusters rotation frequency is $f = 10^4$ Hz, the number of clusters to be simultaneously found on an orbit equals $N = 1$, and a beam cross section is $S = 0.04$ mm^2, $m_Z = 91.2$ GeV/c^2. In order to solve the problem use the Breit-Wigner formula

$$\sigma = \frac{g\pi}{p^2} \frac{\Gamma(Z \to e^+ e^-)\Gamma(Z \to \mu^+ \mu^-)}{(E - m_z)^2 + \Gamma_{tot}^2/4},$$

where

$$g = \frac{2J_Z + 1}{(2J_{e^-} + 1)(2J_{e^+} + 1)}, \qquad E = E_{e^-} + E_{e^+},$$

$J_{Z,e\bar{e}}$ are spins of Z, e^-, e^+, E_{e^-} (E_{e^+}) is the electron (positron) energy, $\Gamma(Z \to e^+ e^-)$ and $\Gamma(Z \to \mu^+ \mu^-)$ are the partial widths of decays, Γ_{tot} is the Z-boson total decay width being equal to 2.49 GeV and $\Gamma(Z \to e^+ e^-) = \Gamma(Z \to \mu^+ \mu^-) = 0.0336\,\Gamma_{tot}$.

2.40. At colliding $e^- e^+$-beam rings the reaction

$$e^- + e^+ \to \pi^+ + \pi^-$$

is investigated. It is ascertained that the total cross section takes on a value 1.8×10^{-30} cm^2 near the resonance $e^- + e^+ \rightarrow \rho \rightarrow \pi^+ + \pi^-$. Estimate the amount of such events for one hour when the ring radius is $r = 10$ m, the electron and positron beam currents are $i = 10$ mA, the cross section area of the beams is $S = 0.1$ cm^2 and at every rotation the electron and positron clusters are collided twice ($N = 2$).

References

[1] J. Schwinger, Ann. Phys. **2**, 407 (1957).

[2] P. Higgs, Phys. Lett. **B12**, 132 (1964).

[3] P. Langacker, Phys. Repts. **72**, 185 (1981).

[4] H. Georgi, H. Quinnn, S. Weinberg, Phys. Rev. Lett. **33**, 451 (174).

[5] H. Georgi, and S. L. Glashow, Phys. Rev. Lett. **32**, 438 (1974).

[6] H. Georgi, *Particles and Fields*, (AIP, New York, 1975); H. Fritzcsh and P. Minkowski, Ann. Phys. **93**, 193 (1975).

[7] J. Wess and J. Bagger, *Supersymmetry and Supergravity*, (Princeton University Press, Princeton, New Jersey, 1983).

[8] *Supergravity'81*, Proc. the 1st School on Supergravity, 22 April-6 May, 1981, Trieste, Italy (ed. by S. Ferrara and J. G. Taylor).

[9] Th. Kaluza, Zitzungsber. Preuss. Acad. Wiss. Berlin, Math. Phys. **K1**, 966 (1921).

[10] K. Klein, Zeit. Phys. **37**, 895 (1926); Nature, **118**, 516 (1926).

[11] A. Einstein, P. Bergmann, Annals Math. **39**, 683 (1938).

[12] M. Green, J. Schwartz and E. Witten, *Superstring theory*, (Cambridge University Press, Cambridge, 1987).

[13] M. Green and J. Schwarz, Phys. Lett. **B151**, 21 (1985).

[14] J. Polchinski, Phys. Rev., Lett. **75**, 4274 (1995); E. Witten, Nucl. Phys., **B443**, 85 (1995).

[15] N. F. Mott, H.S. W. Massay, *The Theory of Atomic Collisions*, (Clarendon, 1965).

[16] J. Chadwick, Nature **129**, 312 (1932).

[17] D. Ivanenko, Nature **129**, 798 (1932).

[18] W. Heisenberg, Zeit. Phys. **77**, 1 (1932).

[19] T. Regge, Nuovo Cimento **14**, 957 (1959).

[20] G. Chew, S. Frautschi, Phys. Rev. Lett. **5**, 12 (1960).

[21] P. D. Collins and E. J. Squires, *Regge Poles in Particle Physics*, (Springer-Verlag, New York, 1968).

[22] G. Breit, E. P. Wigner, Phys. Rev. **49**, 519 (1936); 642 (1936).

[23] H. Yukawa, Proc. Phys. Math. Soc. Japan, **17**, 48 (1935).

[24] Y. Neeman, Nucl. Phys. **26**, 222 (1961).

[25] M. Gell-Mann, Phys. Rev. **125**, 1067 (1962).

[26] S. Okubo, Progr. Theor. Phys. **27**, 949 (1962).

[27] S. Coleman, S. L. Glashow, Phys. Rev. Lett. **6**, 423 (1961).

[28] M. Gell-Mann, Phys. Lett. **8**, 214 (1964).

[29] G. Zweig, CERN preprints 8182/Th. 401 and 8419/Th. 412, 1964.

[30] R. P. Feynman, Phys. Rev. **76**, 769 (1949).

[31] M. N. Rosenbluth, Phys. Rev. **79**, 615 (1950).

[32] R. G. Sachs, Phys. Rev. **126**, 2256 (1962).

[33] R. E. Taylor, in *Proc. 1975 Int. Symp. on Lepton and Photon Interactions at High Energies*, (ed. W.T.Kirk), (Stanford, California, 1975).

[34] J. R. Dunning *at al.* Phys. Rev. **141**, 1286 (1966).

[35] J. D. Bjorken, Phys. Rev. **179**, 1547 (1969).

[36] R. P. Feynman, Phys. Rev. Lett. **23**, 1415 (1969).

[37] R. P. Feynman, *Photon-hadron interaction*, (Reading, Massachusets, Benjamin, 1972).

[38] C. G. Jr. Callan, D. I. Cross, Phys. Rev. Lett. **22**, 156 (1969).

[39] O. W. Greenberg, Phys. Rev. Lett. bf13, 598 (1964); M. Han, Y. Nambu, Phys. Rev. **B139**, 1006 (1965).

[40] H. Fritzsch, M. Gell-Mann, H. Leutwyler, Phys. Lett. **B47**, 365 (1973).

[41] C. N. Yang, R. L. Mills, Phys. Rev. **96**, 191 (1954).

[42] A. Chodos, R. L. Jaffe, K. Johnson, C. B. Thorn, V. E. Weiskopf, Phys. Rev. **D9**, 3471 (1974); W. Bardeen, M. S. Chanowitz, S. D. Drell, M. Weinstein, T. M. Yan, Phys. Rev. **D11**, 1094 (1974).

[43] K. G. Wilson, Phys. Rev. **D10**, 2445 (1974).

[44] M. Creutz, Quarks, *Gluons and Lattices*, (University Press, Cambridge, 1983).

[45] S. Okubo, Phys. Lett. **5**, 165 (1963); G. Zweig, CERN preprint Th/401, 1964; J. Iizuka, K. Okada, O. Shito, Prog. Theor. Phys. **35**, 1061 (1965).

[46] S. L. Glashow, Nucl. Phys. **22**, 579 (1961); S. Weinberg, Phys. Rev. Lett. **19**, 1264 (1967); A. Salam, in *Elementary particle physics*, (ed. N.Svartholm), (Almqvist and Wilsell, Stockholm, 1968).

[47] G. S. Larue *et al.*, Phys. Rev. Lett. **38**, 1011 (1977); **42**, 142 (1979).

[48] Y. Nambu, Phys. Rev. Lett. **4**, 380 (1960); J. Goldstone, Nuovo Cimento **9**, 154 (1961).

[49] B. Carter, in *Proceedings of the International Astronomical Union Symposium 63: Confrontation of Cosmological Theories with Observational Data*, (ed. M. S. Longair), (D. Reidel, Boston, 1974), p. 291.

[50] R. N. Mohapatra, J. C. Pati, Phys. Rev. **D11**, 566 (1975); R. N. Mohapatra, G. Senjanovic, Phys. Rev. **D23**, 165 (1981).

[51] O. M. Boyarkin, Phys. Rev. **D50**, 2247 (1994).

[52] G. Senjanovic, Nucl. Phys. **B153**, 334 (1979).

[53] R. W. Assmann *et al.*, "A3-TeV e^+e^- linear collider based on CLIC technology," SLAC-REPRINT-2000-096.

Part III

Quantum Idyll—Free Fields

14

Scalar field

> Bruin the First had a long career behind him;
> he could build a den and root up trees; hence, with certain
> allowances, he could pass for an expert engineer. His greatest
> merit, however, was his ardent desire to figure in the annals
> of History. For this reason he preferred the glories of
> bloodshed to everything else world. And so, whatever he
> spoke of, be it commerce, industry or science, he always
> ended the same: "Blood, sirs, blood, that's what we need".
> M.E. Saltykov Shchedrin, "Bears in Government"

14.1 Klein-Gordon equation

Interest in a free-field theory is by no means purely academic. Before and after the particle collision in the process of scattering (known as the key experiment in the microworld) the distance between these particles is so great that their interaction is practically zero. Because of this, we can consider the particles at the initial and final states to be free. To describe such a physical situation, an adequate mathematical operation—the representation of interaction—may be used for going from equations of interacting fields to the free-field equations. Of course, we have to do with a certain idealization; there is no possibility to exclude interactions all together, while interactions between the scattered particles are excluded. In other words, exact equations for particles at the initial and final states have to include the terms determining the interactions with the environment.

Let us consider the simplest example of a relativistically invariant wave equation

$$(\Box + m^2)\psi(x) = 0, \tag{14.1}$$

proposed by E. Schrödinger in 1926 [1] along with his famous non-relativistic equation. Later this equation was considered by Gordon [2], Klein [3], Fock [4], and Kudar [5]. Now one is customary to call Eq. (14.1) a Klein-Gordon equation. From the beginning, there was a tendency to use Eq. (14.1) for the description of a relativistic electron. But later with a deeper insight into the problem it was accepted that the equation is associated with spinless particles and should be related to scalar and pseudoscalar mesons ($\pi^{\pm}, \pi^0, K^{\pm}, K^0, ...$). After the discovery of a compound hadron structure, a role of the Klein-Gordon equation was reduced to the description of Higgs bosons. It should be noted that in the majority of SM extensions, the Higgs sector includes neutral as well as charged scalar and pseudoscalar bosons. Owing to its remarkable simplicity, the Klein-Gordon equation still remains a proving ground for verification of the principal statements of the quantum field ideology.

The wavefunction of a spinless particle $\psi(x)$ is a single-component quantity that in the process of inhomogeneous Lorentz transformations

$$x' = \Lambda x + a$$

is transformed as

$$\psi'(x') = \psi(x) \tag{14.2}$$

or

$$\psi'(x) = \psi(\Lambda^{-1}(x - a)). \tag{14.3}$$

Provided the wavefunction is of positive (negative) parity, Eq. (14.1) describes a scalar (pseudoscalar) particle. For a truly neutral particle the wavefunction is real, being complex otherwise. The generalities of the second quantization are most fully reflected for a field that includes both particles and antiparticles. In this case Eq. (14.1) may be derived from the Lagrangian density

$$\mathcal{L}(x) = \partial_\mu \psi^*(x) \partial^\mu \psi(x) - m^2 \psi^*(x)\psi(x). \tag{14.4}$$

Using a common procedure, we find the following dynamic variables:

$$T_{\mu\nu}(x) = \partial_\mu \psi^*(x)\partial_\nu \psi(x) + \partial_\nu \psi^*(x)\partial_\mu \psi(x) - g_{\mu\nu}\mathcal{L}(x), \tag{14.5}$$

the energy-momentum tensor,

$$M^\lambda_{\mu\nu}(x) = \partial^\lambda \psi^*(x)[x_\nu \partial_\mu \psi(x) - x_\mu \partial_\nu \psi(x)] + \partial^\lambda \psi(x)[x_\nu \partial_\mu \psi^*(x) -$$

$$-x_\mu \partial_\nu \psi^*(x)] + \mathcal{L}(x)(x_\mu \delta^\lambda_\nu - x_\nu \delta^\lambda_\mu) \tag{14.6}$$

the angular momentum tensor, and ($\alpha = e, s, c, b, ..$)

$$j^{(\alpha)}_\mu(x) = (\rho(x), \mathbf{j}(x)) = i\alpha[\psi^*(x)\partial_\mu \psi(x) - \partial_\mu \psi^*(x)\psi(x)] \tag{14.7}$$

the "current" four-dimensional vector ($\alpha = e, s, c, b, ..$).

For neutral particles the four-dimensional vector of the electromagnetic current ($\alpha = e$) becomes zero. But in the case when particles are not truly neutral, the "current" associated with a quantum number distinguishing particle and antiparticle is nonzero. For instance, with the use of Eq. (14.1) for the description of K^0-meson the current related to the quantum number called the *strangeness*, $j^{(s)}_\mu(x)$, is different from zero.

According to de Broglie, a free particle is described by a plane monochromatic wave (de Broglie wave). In consequence, the solution to the Klein-Gordon equation is given by:

$$\phi_{\mathbf{k}}(x) = A \exp(-ikx). \tag{14.8}$$

Solutions of the plane-wave type will be always understood as a limiting case for the corresponding solutions of the wave-packet type, meeting the orthonormality requirements. For the solutions of (14.8) orthonormalization is as follows:

$$i \int d^3r \left\{ \phi^*_{\mathbf{k}}(x)[\partial^0 \phi_{\mathbf{k}'}(x)] - [\partial^0 \phi^*_{\mathbf{k}}(x)]\phi_{\mathbf{k}'}(x) \right\} \equiv$$

$$\equiv i \int d^3r[\phi^*_{\mathbf{k}}(x) \overleftrightarrow{\partial^0} \phi_{\mathbf{k}'}(x)] = 2\omega_{\mathbf{k}}\delta(\mathbf{k}' - \mathbf{k}), \tag{14.9}$$

$$i \int d^3r[\phi^*_{\mathbf{k}}(x) \overleftrightarrow{\partial^0} \phi^*_{\mathbf{k}'}(x)] = i \int d^3r[\phi_{\mathbf{k}}(x) \overleftrightarrow{\partial^0} \phi_{\mathbf{k}'}(x)] = 0. \tag{14.10}$$

It should be noted that the description of a free particle by a plane wave (a complete set includes the momentum and energy operators) is not unique. For a complete set one may take the operators of energy, angular momentum square, and angular momentum projection onto the chosen direction in space. In the latter case the eigenfunction of these operators and hence the solution for the free Klein-Gordon equation will be given by the quantity:

$$\phi_{lm_z}(r,\theta,\varphi) = A\frac{\exp(-i\omega t + ikr)}{r}Y_l^{|m_z|}(\theta,\varphi), \qquad (14.11)$$

where $Y_l^{|m_z|}(\theta,\varphi)$ is the spherical function.

Substituting (14.8) into (14.1), we obtain:

$$E = \omega = \pm\sqrt{\mathbf{k}^2 + m^2}.$$

And the appearance of the negative-energy solutions is like a bolt from the blue. Negative values of the energy in the free case are physically meaningless. Within the scope of classical physics we could ignore their appearance altogether. Indeed, in the classical physics the energy of a particle may vary only continuously. Because of this, a particle with a positive energy at the initial moment of time has no chance of going to the negative-energy state, due to the $2mc^2$ interval of "unobservable energies" separating the positive-energy and negative-energy zones. At the same time, in a quantum theory it is natural to deal precisely with discrete transitions from one level to the other. And, in particular, the transitions between the states with negative and positive energies are not forbidden. In this way the solutions with negative energy could not be simply neglected without violation of the basic principles inherent in the quantum theory. For example, if we would discard a partial solution associated with the negative energy and retain only the positive-energy solution, then the obtained system of functions will be incomplete.

There is one more difficulty connected with the negative probabilities. Let us derive a relativistic equation of continuity. To this end, we write an equation for the complex-conjugate wave function:

$$(\Box + m^2)\psi^*(x) = 0. \qquad (14.12)$$

Multiplying (14.1) by $\psi^*(x)$ on the left, multiplying (14.12) by $\psi(x)$ on the right, and then subtracting one equation from the other, we have:

$$\partial^\mu j_\mu(x) = 0,$$

where the probability current is determined by the following expression:

$$j_\mu(x) = \psi^*(x)\partial_\mu\psi(x) - \psi(x)\partial_\mu\psi^*(x).$$

It is obvious that the probability density

$$j_0(x) \equiv \rho(x) = \psi^*(x)\partial_t\psi(x) - \psi(x)\partial_t\psi^*(x)$$

is not always positive. Since Eq. (14.1) involves the second derivative with respect to time, in a particular instant of time $\psi(x)$ and $\partial_t\psi(x)$ may be given both independently and arbitrarily, that may yield negative values of $\rho(x)$. Besides, it is clear that if $\rho(x)$ will be greater than zero for the states with a positive energy, then the solutions with negative energies result in a negative probability density. Thus, there is no way for interpretation of the Klein-Gordon equation as a single-particle equation with a wavefunction $\psi(x)$, that is, we have to turn down the ideology of a nonrelativistic (primary quantized) theory. The difficulties, resultant from the appearance of the physically-meaningless solutions with negative energies, are due to "illegality" of considering the problem of a single-particle motion at relativistic energies. At such energies we have to deal with the particle-antiparticle pair production (and their annihilation), and this problem concerns a variable number of the particles, that is, we have to solve the problem using methods of a quantum field theory.

14.2   Quantization

In 1934 the Klein-Gordon equation was reanimated by Pauli and Weisskopf who have used the second quantization method. By this method, the field functions should be considered as the operators governed by certain commutation relations rather than as *c*-numbers. These operators influence the state vector Φ that is describing the state of a field as a particular quantum-mechanical system with a variable number of the particles. In nonrelativistic quantum mechanics (NQM) the wavefunction of a system may have been specified in any representation, no matter coordinate, momentum, or of some other type. In a second quantized theory the state vector is determined in the space of the particle numbers (quanta of a given field) or, simply speaking, in the occupation numbers space (Fock space).

Recall how the states of physical systems are described in NQM. As is known, two methods may be used. In the first method, the operators relating to the physical quantities are assumed to be independent of time, whereas the state wavefunction is considered variable in time, its variation being described by the Schrödinger equation. The time dependence of the average values of the physical quantities appears only through the time dependence of the state wavefunction according to the formulae

$$\overline{f}(t) = \int \Psi^*(q,t)\hat{f}\Psi(q,t)dq.$$

This representation was named the Schrödinger representation. However, the quantum mechanical apparatus may be equivalently formulated in some other way, carrying over the time dependence from the wavefunctions to the operators (Heisenberg representation). The state vector and operators of the both representations are connected by the relations:

$$\Psi^S(q,t) = \hat{S}\Psi^H(q,0), \qquad \hat{f}^S = \hat{S}\hat{f}^H\hat{S}^{-1},$$

where

$$\hat{S} = \exp\left(-\frac{i}{\hbar}\hat{H}t\right).$$

Formally, equations of motion for the operators in Heisenberg's representation are similar in form to the classical mechanics equations for the corresponding quantities. On account of this reason, the procedure of second quantization is conveniently considered in Heisenberg's representation.

Let the field described by the function $\psi(x)$ be confined within a cube of volume $V = L^3$, where L is the cube edge length. We expand the operator field functions in a complete set of solutions for the Klein-Gordon equation:

$$\psi(x) = \frac{1}{\sqrt{V}} \sum_{\mathbf{k}} \frac{1}{\sqrt{2\omega}} [a_{\mathbf{k}}^{(+)} \exp[i(\mathbf{kr}) - i\omega t] + a_{\mathbf{k}}^{(-)} \exp[i(\mathbf{kr}) + i\omega t]], \qquad (14.13)$$

$$\psi^*(x) = \frac{1}{\sqrt{V}} \sum_{\mathbf{k}} \frac{1}{\sqrt{2\omega}} [a_{\mathbf{k}}^{(+)*} \exp[-i(\mathbf{kr}) + i\omega t] + a_{\mathbf{k}}^{(-)*} \exp[-i(\mathbf{kr}) - i\omega t]], \qquad (14.14)$$

where the quantities $a_{\mathbf{k}}^{(+)}$ and $a_{\mathbf{k}}^{(-)}$ are operators,[*] and $\psi^{(+)}(x)$ ($\psi^{(-)}(x)$) represent a sum of the solutions with a positive (negative) frequency, satisfying the orthonormalization conditions:

$$i \int d^3r [\psi_{\mathbf{k}}^{(q)*}(x) \overleftrightarrow{\partial^0} \psi_{\mathbf{k}'}^{(q')}(x)] = 2\omega_{\mathbf{k}} \delta_{qq'} \delta_{\mathbf{k}'\mathbf{k}}. \qquad (14.15)$$

---

[*]Hereinafter, the cap over operators that is usual for the NQM will be omitted.

Now we find the permutation relations for the operator field functions. As we still believe in the unity of nature, a change over to quantum field theory should be a natural continuation of that thread, which has led us to the NQM from classical physics. Recall that dynamic variables in the NQM have acquired the status of operators, and their commutation relations were found with the use of quantum Poisson brackets:

$$[a, b]' = \sum_r \left(\frac{\partial a}{\partial q_r} \frac{\partial b}{\partial p_r} - \frac{\partial a}{\partial p_r} \frac{\partial b}{\partial q_r} \right), \qquad (14.16)$$

where a and b are arbitrary operators. The Poisson brackets in turn were simply related to the commutator

$$[a, b]' = -i[a, b]. \qquad (14.17)$$

Specifically, for the coordinate and momentum operators the relations were as follows:

$$[q_r(t), k_s(t)] = i\delta_{rs}, \qquad (14.18)$$

where we have written down the time argument for the operators to emphasize that they were determined at the same instant of time. In quantum field theory, the wavefunctions are playing the role of the coordinates, and the canonically conjugate momenta are determined with the help of the Lagrangian just the same way as in the classical physics

$$\pi(x) - \frac{\partial \mathcal{L}}{\partial(\partial^0 \psi(x))} = \partial_0 \psi^*(x), \qquad \pi^*(x) = \frac{\partial \mathcal{L}}{\partial(\partial^0 \psi^*(x))} = \partial_0 \psi(x). \qquad (14.19)$$

Quite logically it may be assumed that the commutative law of the operator field functions in its form is similar to (14.18). It should be remembered, however, that the field functions are dependent on the spin, leading to different statistics for fermions and bosons. Without proof we accept the Pauli theorem that relates the transformational properties of a field and its quantization method: *"The fields describing the integer-spin particles are quantized with the use of commutators (Bose-Einstein quantization); whereas the fields describing the half-integer-spin particles are quantized using anticommutators (Fermi-Dirac quantization)."* Note that for bosons as well as fermions the permutation relations of the field functions are c-numbers.

In case of a spinless particle, as a field analogue of (14.18), we have the commutation relations:

$$[\psi(\mathbf{r}', t), \pi(\mathbf{r}, t)] - [\psi(\mathbf{r}', t), \partial_0 \psi^*(\mathbf{r}, t)] = i\delta(\mathbf{r}' - \mathbf{r}), \qquad (14.20)$$

$$[\psi^*(\mathbf{r}', t), \pi^*(\mathbf{r}, t)] - [\psi^*(\mathbf{r}', t), \partial_0 \psi(\mathbf{r}, t)] = i\delta(\mathbf{r}' - \mathbf{r}). \qquad (14.21)$$

All other commutators from the pairs of "coordinate" and "momentum" operators are zero. The above commutation relations were determined at equal times only. Under incoincident moments of time their determination necessitates the knowledge of solutions for the motion equations. Using (14.20), we may find the commutators of the operators $a_{\mathbf{k}}^{(+)}$ and $a_{\mathbf{k}}^{(-)}$. Substitute the expansions (14.13) and (14.14) into (14.20)

$$[\psi(\mathbf{r}, t), \partial_0 \psi^*(\mathbf{r}', t)] = \frac{i}{V} \sum_{\mathbf{k}\mathbf{k}'} \frac{\omega'}{\sqrt{4\omega\omega'}} \left([a_{\mathbf{k}}^{(+)}, a_{\mathbf{k}'}^{(+)*}] \times \right.$$

$$\left. \times \exp\{i[(\mathbf{kr}) - (\mathbf{k}'\mathbf{r}') - (\omega - \omega')t]\} - [a_{\mathbf{k}}^{(-)}, a_{\mathbf{k}'}^{(-)*}] \exp i\{[(\mathbf{kr}) - (\mathbf{k}'\mathbf{r}') + (\omega - \omega')t]\} \right), \qquad (14.22)$$

where we have took into account that $a_{\mathbf{k}}^{(+)}$ and $a_{\mathbf{k}}^{(-)}$ are describing the independent degrees of freedom of a field and hence must commute

$$[a_{\mathbf{k}'}^{(+)*}, a_{\mathbf{k}}^{(-)*}] = [a_{\mathbf{k}'}^{(+)}, a_{\mathbf{k}}^{(-)}] = 0. \qquad (14.23)$$

It is easily seen that to obtain the delta function on the right-hand side of (14.22), the following quantization rules should be applied:

$$[a_{\mathbf{k}}^{(+)}, a_{\mathbf{k'}}^{(+)*}] = -[a_{\mathbf{k}}^{(-)}, a_{\mathbf{k'}}^{(-)*}] = \delta_{\mathbf{kk'}}. \tag{14.24}$$

Actually, in this case the relation (14.22) takes the form:

$$[\psi(\mathbf{r}, t), \partial_0 \psi^*(\mathbf{r'}, t)] = \frac{i}{V} \sum_{\mathbf{k}} \exp\{i[(\mathbf{kr}) - (\mathbf{kr'})]\}. \tag{14.25}$$

To exchange integration with summation in the right-hand side of (14.25), we should remember that the condition of vanishing the field function and its first space derivative at the boundaries of a cube with the edge L results in the following relations:

$$k_i L = 2\pi n_i, \qquad i = x, y, z.$$

Then the total number ΔN of $\mathbf{k}$-vectors over the interval $d^3 k$ is equal to

$$\Delta N = \frac{V d^3 k}{(2\pi)^3}.$$

Provided the volume V is large, and the functions under the summation sign are slowly varying, a change from summation to integration is realized according to

$$\frac{1}{V} \sum_{\mathbf{k}} \to \frac{1}{(2\pi)^3} \int d^3 k. \tag{14.26}$$

By substitution of (14.26) into the right-hand side of (14.25) with subsequent comparison with (14.20) we are convinced that the commutation relations (14.24) are valid.

The above-stated quantization method is referred to as a *canonical quantization*. With this formalism the time and space coordinates are not equal in rights. This circumstance casts some doubt on the relativistic invariance of a second-quantized theory based on this approach. But in actual fact the doubts are unfounded, and by the canonical field quantization we obtain the expressions possessing the relativistic invariance. At the same time, the canonical formalism may be "shaped" into the covariant form without difficulties. And as the reader could guess, this may be performed using the same method that has provided a relativistic covariance for the evolution equation of the state vector in Chapter 1. We revert to this problem at the end of Section 14.5.

14.3   Production and destruction operators

The next step of the second quantization procedure for a field should be establishing a physical meaning for the operators $a_{\mathbf{k}}^{(+)}$ and $a_{\mathbf{k}}^{(-)}$. We introduce the quantities $N_{\mathbf{k}}^+ = a_{\mathbf{k}}^{(+)*} a_{\mathbf{k}}^{(+)}$ and $N_{\mathbf{k}}^- = a_{\mathbf{k}}^{(-)*} a_{\mathbf{k}}^{(-)}$. An equation for the eigenvalues of the operator $N_{\mathbf{k}}^+$ is of the form:

$$a_{\mathbf{k}}^{(+)*} a_{\mathbf{k}}^{(+)} \Phi_{n_{\mathbf{k}}^+} = n_{\mathbf{k}}^+ \Phi_{n_{\mathbf{k}}^+}. \tag{14.27}$$

In Eq. (14.27) $\Phi_{n_{\mathbf{k}}^+}$ is the eigenfunction meeting the normalization condition:

$$\int \Phi_{n_{\mathbf{k}}^+}^* \Phi_{n_{\mathbf{k}}^+} d\tau \equiv (\Phi_{n_{\mathbf{k}}^+}, \Phi_{n_{\mathbf{k}}^+}) = 1, \tag{14.28}$$

where integration is performed over all the variables on which $\Phi_{n_{\mathbf{k}}^+}$ is dependent. From Eq. (14.27) it follows that:

$$n_{\mathbf{k}}^+ = (\Phi_{n_{\mathbf{k}}^+}, a_{\mathbf{k}}^{(+)*} a_{\mathbf{k}}^{(+)} \Phi_{n_{\mathbf{k}}^+}) = (a_{\mathbf{k}}^{(+)} \Phi_{n_{\mathbf{k}}^+}, a_{\mathbf{k}}^{(+)} \Phi_{n_{\mathbf{k}}^+}) \geq 0. \tag{14.29}$$

Let us define the function $\Phi_{n_{\mathbf{k}}^+}^{(-1)} = a_{\mathbf{k}}^{(+)} \Phi_{n_{\mathbf{k}}^+}$. For $n_{\mathbf{k}}^+ > 0$ this function is nonzero as

$$a_{\mathbf{k}}^{(+)*} \Phi_{n_{\mathbf{k}}^+}^{(-1)} = a_{\mathbf{k}}^{(+)*} a_{\mathbf{k}}^{(+)} \Phi_{n_{\mathbf{k}}^+} = n_{\mathbf{k}}^+ \Phi_{n_{\mathbf{k}}^+} \neq 0. \tag{14.30}$$

It is seen that it is the eigenfunction of the operator $N_{\mathbf{k}}^+$:

$$N_{\mathbf{k}}^+ \Phi_{n_{\mathbf{k}}^+}^{(-1)} = a_{\mathbf{k}}^{(+)*} a_{\mathbf{k}}^{(+)} a_{\mathbf{k}}^{(+)} \Phi_{n_{\mathbf{k}}^+} = (a_{\mathbf{k}}^{(+)} a_{\mathbf{k}}^{(+)*} - 1) a_{\mathbf{k}}^{(+)} \Phi_{n_{\mathbf{k}}^+} =$$

$$= (n_{\mathbf{k}}^+ - 1) \Phi_{n_{\mathbf{k}}^+}^{(-1)}. \tag{14.31}$$

The relation (14.31) makes it possible to write

$$a_{\mathbf{k}}^{(+)} \Phi_{n_{\mathbf{k}}^+} = \alpha \Phi_{n_{\mathbf{k}}^+ - 1}, \tag{14.32}$$

where α is a constant whose form will be stated later. In similar way, we can show that the operator $N_{\mathbf{k}}^+$ also has $n_{\mathbf{k}}^+ - 2, n_{\mathbf{k}}^+ - 3, ...$ as its eigenvalues. But it is obvious that such a series should be broken because negative eigenvalues $n_{\mathbf{k}}^+$ are impossible. This is possible only in the case when $n_{\mathbf{k}}^+$ represents a series of positive integers, $n_{\mathbf{k}}^+ = 0, 1, 2, ...$ Clearly, these numbers may be as great as you like. For demonstration we consider the following functions:

$$\Phi_{n_{\mathbf{k}}^+}^{(1)} = a_{\mathbf{k}}^{(+)*} \Phi_{n_{\mathbf{k}}^+}, \qquad \Phi_{n_{\mathbf{k}}^+}^{(2)} = a_{\mathbf{k}}^{(+)*} a_{\mathbf{k}}^{(+)*} \Phi_{n_{\mathbf{k}}^+}, \qquad \Phi_{n_{\mathbf{k}}^+}^{(3)} = a_{\mathbf{k}}^{(+)*} a_{\mathbf{k}}^{(+)*} a_{\mathbf{k}}^{(+)*} \Phi_{n_{\mathbf{k}}^+}, ...$$

For $\Phi_{n_{\mathbf{k}}^+}^{(1)}$ we have an equation for the eigenvalue:

$$N_{\mathbf{k}}^+ \Phi_{n_{\mathbf{k}}^+}^{(1)} = a_{\mathbf{k}}^{(+)*} a_{\mathbf{k}}^{(+)} a_{\mathbf{k}}^{(+)*} \Phi_{n_{\mathbf{k}}^+} = a_{\mathbf{k}}^{(+)*} (1 + a_{\mathbf{k}}^{(+)*} a_{\mathbf{k}}^{(+)}) \Phi_{n_{\mathbf{k}}^+} =$$

$$= (n_{\mathbf{k}}^+ + 1) \Phi_{n_{\mathbf{k}}^+}^{(1)}. \tag{14.33}$$

Then it is readily seen that the eigenvalues of the operator $N_{\mathbf{k}}^+$ are represented by a numerical series $n_{\mathbf{k}}^+ + 1, n_{\mathbf{k}}^+ + 2, n_{\mathbf{k}}^+ + 3.....$ Also, from (14.33) it follows that

$$a_{\mathbf{k}}^{(+)*} \Phi_{n_{\mathbf{k}}^+} = \alpha' \Phi_{n_{\mathbf{k}}^+ + 1}, \qquad \alpha' = \text{const.} \tag{14.34}$$

The properties of the functions $\Phi_{n_{\mathbf{k}}^+}^{(-1)}$ and $\Phi_{n_{\mathbf{k}}^+}^{(1)}$ enable one to find the matrix elements for the operators $a_{\mathbf{k}}^{(+)}$ and $a_{\mathbf{k}}^{(+)*}$:

$$\left. \begin{array}{l} < n_{\mathbf{k}}^{+\prime} \mid a_{\mathbf{k}}^{(+)} \mid n_{\mathbf{k}}^+ > \equiv (\Phi_{n_{\mathbf{k}}^+}^{(-1)}, a_{\mathbf{k}}^{(+)} \Phi_{n_{\mathbf{k}}^+}), \\ < n_{\mathbf{k}}^{+\prime\prime} \mid a_{\mathbf{k}}^{(+)*} \mid n_{\mathbf{k}}^+ > \equiv (\Phi_{n_{\mathbf{k}}^+}^{(1)}, a_{\mathbf{k}}^{(+)*} \Phi_{n_{\mathbf{k}}^+}). \end{array} \right\} \tag{14.35}$$

Clearly, these matrix elements are nonzero only in the case when $n_{\mathbf{k}}^{+\prime} = n_{\mathbf{k}}^+ - 1$ and $n_{\mathbf{k}}^{+\prime\prime} = n_{\mathbf{k}}^+ + 1$. From (14.35) it follows immediately that:

$$< n_{\mathbf{k}}^+ - 1 \mid a_{\mathbf{k}}^{(+)} \mid n_{\mathbf{k}}^+ > = \sqrt{n_{\mathbf{k}}^+}, \tag{14.36}$$

$$< n_{\mathbf{k}}^+ + 1 \mid a_{\mathbf{k}}^{(+)*} \mid n_{\mathbf{k}}^+ >= \sqrt{n_{\mathbf{k}}^+ + 1}, \tag{14.37}$$

where we ignore the phase term that is unity in absolute value. Now we can find the constants α and α', that leads to:

$$a_{\mathbf{k}}^{(+)} \Phi_{n_{\mathbf{k}}^+} = \sqrt{n_{\mathbf{k}}^+} \Phi_{n_{\mathbf{k}}^+ - 1}, \qquad a_{\mathbf{k}}^{(+)*} \Phi_{n_{\mathbf{k}}^+} = \sqrt{n_{\mathbf{k}}^+ + 1} \Phi_{n_{\mathbf{k}}^+ + 1}. \tag{14.38}$$

From Eqs. (14.38) the physical meaning of the operators $a_{\mathbf{k}}^{(+)*}$ and $a_{\mathbf{k}}^{(+)}$ becomes quite transparent. Let the field under consideration contain $n_{\mathbf{k}}^+$ positive-energy particles with the momentum $\mathbf{k}$. If $\Phi_{n_{\mathbf{k}}^+}$ denotes the field state vector, then, according to (14.38), the operation of $a_{\mathbf{k}}^{(+)*}$ reduce to increasing the quanta number $n_{\mathbf{k}}^+$ by 1, whereas the operation $a_{\mathbf{k}}^{(+)}$—to decreasing the quanta number $n_{\mathbf{k}}^+$ by 1. In this way $a_{\mathbf{k}}^{(+)*}$ represents the production operator of a particle with the momentum $\mathbf{k}$ and the positive energy ω, and $a_{\mathbf{k}}^{(+)}$ represents the annihilation operator of the same particle. It should be emphasized that in the quantum theory, as distinct from the classical one, a particle is not understood as something localized in space. Here it means nothing else but field excitations. But these excitations are discrete objects satisfying the proper relativistic relations between the energy and momentum. Therefore, they are termed as particles.

Let us return back to expressions (14.13) and (14.14). The form of the terms involved in the first sums of both expressions enables one to infer that annihilation of the particles having the positive energy is associated with the time factor $\exp(-i\omega t)$, and their production—with $\exp(i\omega t)$. Then the way to reconstruction of the remaining terms becomes quite obvious. To make their form meeting the norms of a quantum behavior, it is necessary to interpret $a_{\mathbf{k}}^{(-)}$ as a production operator and $a_{\mathbf{k}}^{(-)*}$—as an annihilation operator for some other particles, which, as it is found out later, are antiparticles. Consequently, $a_{\mathbf{k}}^{(-)}$ is replaced by the antiparticle production operator $b_{-\mathbf{k}}^*$, whereas $a_{\mathbf{k}}^{(-)*}$—by the antiparticle annihilation operator $b_{-\mathbf{k}}$. Also, for simplicity, we omit the frequency index of the operators $a_{\mathbf{k}}^{(+)}$, $a_{\mathbf{k}}^{(+)*}$. Replacing the notation for the summation variable $\mathbf{k}$ by $-\mathbf{k}$ (in so doing the exponential factor takes the relativistically-covariant form) in the second sums of expressions (14.13) and (14.14), we have the following expansion of the field functions:

$$\psi(x) = \psi^{(+)}(x) + \psi^{(-)}(x) = \frac{1}{\sqrt{V}} \sum_{\mathbf{k}} \frac{1}{\sqrt{2\omega}} [a_{\mathbf{k}} \exp[i(\mathbf{kr}) - i\omega t] + b_{\mathbf{k}}^* \exp[-i(\mathbf{kr}) + i\omega t]], \tag{14.39}$$

$$\psi^*(x) = \psi^{(+)*}(x) + \psi^{(-)*}(x) = \frac{1}{\sqrt{V}} \sum_{\mathbf{k}} \frac{1}{\sqrt{2\omega}} [a_{\mathbf{k}}^* \exp[-i(\mathbf{kr}) + i\omega t] + b_{\mathbf{k}} \exp[i(\mathbf{kr}) - i\omega t]], \tag{14.40}$$

where the operators $a_{\mathbf{k}}$ and $b_{\mathbf{k}}$ satisfy the commutation relations, this time symmetric

$$[a_{\mathbf{k}}, a_{\mathbf{k}'}^*] = \delta_{\mathbf{kk}'}, \tag{14.41}$$

$$[b_{\mathbf{k}}, b_{\mathbf{k}'}^*] = \delta_{\mathbf{kk}'}, \tag{14.42}$$

Thus, all the operators $a_{\mathbf{k}}$, $b_{\mathbf{k}}$ become multiplied by the exponents having the "proper" time dependence ($\sim \exp(-i\omega t)$), that is, all the solutions obtained have the "correct" energy sign. As a result, in "charged" fields we come to the understanding that two kinds of the particles (particles per se and antiparticles) are possible, which are acting jointly and on equal footing. It is evident that the sign changes in the exponent $\exp(i\omega t)$ (going to the solutions with the positive energy) may be attained by varying the direction in the course of time. In other words, a solution with negative energy provides the description for a particle moving backwards in time or, equivalently, an antiparticle with the positive energy, whose motion in time is in the forward direction. Note that from the mathematical viewpoint, the relations (14.39) and (14.40) represent nothing else but the Fourier series expansion of the $\psi(x)$- and $\psi^*(x)$-functions. Possibly, an enlightened reader has already guessed that expressions (14.39) and (14.40) may be derived on the basis of the relativistic covariance requirement. The Lorentz transformations, the rotation of a four-dimensional coordinate system, do not take the axis t beyond the boundaries of the corresponding light-cone cavity, which is a manifestation of the existence of the limit signals propagation velocity. But from the pure mathematical viewpoint, as a rotation we can also consider the four-inversion (PT-transformation):

$$\mathbf{r} \to -\mathbf{r}, \qquad t \to -t, \tag{14.43}$$

because the determinant of this transformation equals $+1$. However, considering that in this case the time axis is shifted from one cavity of the light cone into the other, such a transformation is physically unrealizable. Then any expression invariant under the Lorentz transformations should be invariant under the four-inversion. For the operator of a scalar field this implies:

$$\psi(\mathbf{r}, t) = \psi(-\mathbf{r}, -t), \tag{14.44}$$

and that is only possible when a charge conjugation is performed in $\psi(-\mathbf{r}, -t)$, meaning interchangeability of particles and antiparticles:

$$a_{\mathbf{k}} \to b_{\mathbf{k}}^*, \qquad b_{\mathbf{k}} \to a_{\mathbf{k}}^*. \tag{14.45}$$

A set of (14.43) and (14.45) is nothing else but CPT-transformation. To obtain expansion of the field functions for a free scalar field in terms of the solution of the Klein-Gordon equations in the form of (14.39) and (14.40), it is enough to demand these expansions be meeting the CPT-invariance requirement.

14.4 Wave-corpuscle dualism

The commutation conditions imposed on the operators $a(\mathbf{k})$ and $b(\mathbf{k})$ have to result in the adequate corpuscular pattern of a scalar field. This implies that the dynamic field invariants (energy, momentum, angular momentum square, angular momentum projection, electric charge, etc.) should be equal to a sum of the dynamic invariants for separate particles, quanta of the scalar field. Provided these requirements are met, we will have a direct support for the validity of the derived commutation relations.

We assume that Eq. (14.1) describes a field of the charged particles and then restrict ourselves to finding the energy, momentum, and electric charge of the field. From (14.5) the energy is determined as:

$$E = \int T^{00} d^3x = \int d^3x \left[\sum_\mu \partial_\mu \psi^*(x) \partial_\mu \psi(x) + m^2 \psi^*(x) \psi(x) \right] =$$

$$= \frac{1}{V} \sum_{\mathbf{k}\mathbf{k}'} \int d^3x \left\{ \frac{1}{\sqrt{4\omega\omega'}} [\partial_\mu \psi^{(+)*}(k) \partial_\mu \psi^{(+)}(k) + \partial_\mu \psi^{(+)*}(k) \partial_\mu \psi^{(-)}(k) + \right.$$

$$+ \partial_\mu \psi^{(-)*}(k) \partial_\mu \psi^{(+)}(k) + \partial_\mu \psi^{(-)*}(k) \partial_\mu \psi^{(-)}(k) + m^2 [\psi^{(+)*}(k) \psi^{(+)}(k) +$$

$$+ \psi^{(+)*}(p) \psi^{(-)}(k) + \psi^{(-)*}(k) \psi^{(+)}(k) + \psi^{(-)*}(k) \psi^{(-)}(k)] \},$$

where we have introduced the designations

$$\left. \begin{array}{l} \psi^{(+)}(k) = a_{\mathbf{k}} \exp\left[i(\mathbf{kr}) - i\omega t\right] = \psi^{(+)}(\mathbf{k}) \exp\left(-i\omega t\right), \\ \psi^{(-)}(k) = b_{\mathbf{k}} \exp\left[-i(\mathbf{kr}) + i\omega t\right] = \psi^{(-)}(\mathbf{k}) \exp\left(i\omega t\right). \end{array} \right\} \tag{14.46}$$

It may be easily shown that the terms, which contain the products of the functions $\psi^{(\pm)}(k)$ and $\psi^{(\pm)*}(k)$ having the same sign of frequency, make no contributions into the dynamic invariant E. For example,

$$\frac{1}{V} \sum_{\mathbf{k}\mathbf{k}'} \int \frac{d^3 x}{\sqrt{4\omega\omega'}} [\partial_\mu \psi^{(+)*}(k)\partial_\mu \psi^{(-)}(k') + m^2 \psi^{(+)*}(k)\psi^{(-)}(k')] =$$

$$= \frac{1}{V} \sum_{\mathbf{k}\mathbf{k}'} \frac{1}{\sqrt{4\omega\omega'}} \exp\left[i(\omega + \omega')t\right](m^2 - k_\mu k'_\mu)a_{\mathbf{k}}^* b_{\mathbf{k}'} \times$$

$$\times \int d^3 x \exp\left[-i(\mathbf{k}' + \mathbf{k})\mathbf{r}\right] = \frac{1}{V} \sum_{\mathbf{k}\mathbf{k}'} \frac{1}{\sqrt{4\omega\omega'}} \exp\left[i(\omega + \omega')t\right](m^2 - k_\mu k'_\mu) \times$$

$$\times a_{\mathbf{k}}^* b_{\mathbf{k}'} (2\pi)^3 \delta(\mathbf{k} + \mathbf{k}') = \sum_{\mathbf{k}} \frac{1}{2\omega} \exp\left(2i\omega t\right)(m^2 - \omega^2 + \mathbf{k}^2)a_{\mathbf{k}}^* b_{-\mathbf{k}} = 0. \tag{14.47}$$

In the derivation of (14.47) we have allowed for the relation:

$$\delta_{\mathbf{k}\mathbf{k}'} = \frac{(2\pi)^3}{V} \delta(\mathbf{k} - \mathbf{k}'), \tag{14.48}$$

the validity of that may be easily demonstrated using a definition of the δ-function and assuming $\mathbf{k} = \mathbf{k}'$. Finally, for the field energy we have:

$$E = \sum_{\mathbf{k}} \omega_{\mathbf{k}} [a_{\mathbf{k}}^* a_{\mathbf{k}} + b_{\mathbf{k}} b_{\mathbf{k}}^*]. \tag{14.49}$$

By the substitution of (14.13), (14.14) into (14.5) and (14.7) we can determine the momentum and electric charge of the field by the following relations:

$$\left. \begin{array}{l} \mathbf{P} = \sum_{\mathbf{k}} \mathbf{k}[a_{\mathbf{k}}^* a_{\mathbf{k}} + b_{\mathbf{k}} b_{\mathbf{k}}^*], \\ Q = \sum_{\mathbf{k}} q[a_{\mathbf{k}}^* a_{\mathbf{k}} - b_{\mathbf{k}} b_{\mathbf{k}}^*], \end{array} \right\} \tag{14.50}$$

where q is an electric charge of particles. Note that Q is not positively defined.

With the use of the commutation relations we can rewrite the expression for the field energy (14.49) as:

$$E = \sum_{\mathbf{k}} \omega_{\mathbf{k}} [a_{\mathbf{k}}^* a_{\mathbf{k}} + b_{\mathbf{k}}^* b_{\mathbf{k}} + 1] = \sum_{\mathbf{k}} \omega_{\mathbf{k}} [(N_{\mathbf{k}}^a + \frac{1}{2}) + (N_{\mathbf{k}}^b + \frac{1}{2})]. \tag{14.51}$$

Two very important conclusions may be drawn from (14.51). First, the operators $N_{\mathbf{k}}^a = a_{\mathbf{k}}^* a_{\mathbf{k}}$ and $N_{\mathbf{k}}^b = b_{\mathbf{k}}^* b_{\mathbf{k}}$ are associated with the particles and antiparticles number, respectively. Second, in (14.51) we recognize a well-known expression from the NQM for the energy of a two-dimensional harmonic oscillator, which enables us to treat the field under consideration as a set of the infinite number of harmonic oscillators. Also, expression (14.51) reveals that the field energy has the lowest value when the particle and antiparticle numbers equal zero. Such a lowest-energy state of the field is called the *vacuum state*. Adding a constant to E, if required, we can shift the reference point for the energy so that a minimum value of the energy can be zero. Now we assume that there is a single zero-energy state $\Phi_0 \equiv| 0 >$, the vacuum state. It should satisfy the conditions:

$$D_i\Phi_0 = 0, \qquad a_{\mathbf{k}}\Phi_0 = 0, \qquad b_{\mathbf{k}}\Phi_0 = 0, \qquad (14.52)$$

where D_i is a set of the dynamic field invariants. It is supposed that a vector of the vacuum state is normalized to unity:

$$(\Phi_0, \Phi_0) = 1.$$

An arbitrary normalized state, involving $n_{\mathbf{k}_1}^+$ particles with the momentum $\mathbf{k}_1$, $n_{\mathbf{k}_1'}^-$ antiparticles with the momentum $\mathbf{k}_1'$, $n_{\mathbf{k}_2}^+$ particles with the momentum $\mathbf{k}_2$, $n_{\mathbf{k}_2'}^-$ antiparticles with the momentum $\mathbf{k}_2'$ and so on, may be written in the following form:

$$\Phi_{n_{\mathbf{k}_1}^+ n_{\mathbf{k}_1'}^- \ldots} \equiv| n_{\mathbf{k}_1}^+, n_{\mathbf{k}_2}^+, \ldots; n_{\mathbf{k}_1'}^-, n_{\mathbf{k}_2'}^- \ldots >=$$

$$= \frac{1}{\sqrt{n_{\mathbf{k}_1}^+ n_{\mathbf{k}_2}^+ \ldots n_{\mathbf{k}_1'}^- n_{\mathbf{k}_2'}^-}} [a_{\mathbf{k}_1}^*]^{n_{\mathbf{k}_1}^+} [a_{\mathbf{k}_2}^*]^{n_{\mathbf{k}_2}^+} [b_{\mathbf{k}_1'}^*]^{n_{\mathbf{k}_1'}^-} [b_{\mathbf{k}_2'}^*]^{n_{\mathbf{k}_2'}^-} \ldots | 0 > . \qquad (14.53)$$

Considering the commutation relations, the momentum and charge of this field may be represented as:

$$\mathbf{P} = \sum_{\mathbf{k}} \mathbf{k}[(N_{\mathbf{k}}^a + \frac{1}{2}) + (N_{\mathbf{k}}^b + \frac{1}{2})], \qquad (14.54)$$

$$Q = \sum_{\mathbf{k}} q[(N_{\mathbf{k}}^a - (N_{\mathbf{k}}^b - 1)]. \qquad (14.55)$$

As seen from (14.55), the electric charges of particles and antiparticles are opposite in sign. But if in the quantization process we used anticommutators rather than commutators, the energy and momentum would not be positively defined, whereas the charges of particles and antiparticles would not have the opposite signs.

The vacuum average of E:

$$(\Phi_0, E\Phi_0) = \sum_{\mathbf{k}} \omega_{\mathbf{k}}$$

proves to be equal to infinity. Similar infinities are encountered when a vacuum average is taken for the other dynamics invariants as well. These difficulties may be obviated in two ways: to subtract an infinite constant (for the energy, for example, $\sum_{\mathbf{k}} \omega_{\mathbf{k}}$ is such a constant) or to use the freedom of changing the operators order in the expressions for the dynamic invariants. Taking the latter approach, we write the quantities expressed in terms of the free-field operators in the form of normal products (N-products). This means an expansion in terms of the production and annihilation operators with their subsequent reordering: all annihilation operators should be to the right of all the production ones. Since separation of the field function $\psi(x)$ into positive and negative frequencies (expansion in terms of the production and annihilation operators) is relativistically invariant ($k_\mu x^\mu \equiv kx =$ inv), the definition of the normal product is relativistically invariant too. The normal product of the operators will be denoted either by the symbol N or by the colons that are placed on each side of the expression containing these operators. Conventionally, all the dynamic variables quadratically dependent on the operators with identical arguments (for example, Lagrangian, momentum energy tensor, electromagnetic current, etc.) by definition are written in the form of the normal product. In this case the Lagrangian density for a scalar field is of the following form:

$$\mathcal{L}(x) =: \partial_\mu \psi^*(x)\partial^\mu \psi(x) - m^2\psi^*(x)\psi(x) : . \qquad (14.56)$$

Then a four-dimensional vector of the field momentum and charge contain no infinite expressions at all:

$$P_\mu = \sum_{\mathbf{k}} k_\mu [N_{\mathbf{k}}^a + N_{\mathbf{k}}^b],$$

$$Q = \sum_{\mathbf{k}} q [N_{\mathbf{k}}^a - N_{\mathbf{k}}^b].$$

Thus, the dynamic field invariants are actually expressed as sums of the dynamic invariants for the field quanta, that is, the corpuscular properties of a scalar field are adequately described by the quantization conditions (14.41) and (14.42).

For better understanding of the second quantization procedure we proceed from the established equivalence between a wave field and a set of harmonic oscillators. All the results associated with the second quantization procedure—commutation relations (14.41) and (14.42), matrix elements of the production (14.36) and annihilation (14.37) operators, expression for the field energy (14.51), and the like—may be derived from the problem for a harmonic oscillator. For this purpose one should specify the Hamiltonian function:

$$H = \frac{1}{2}(p_j^2 + \omega^2 q_j^2), \qquad j = a, b,$$

where ω is a frequency, p_j and q_j are momenta and coordinates of the oscillator satisfying the commutation relations

$$[q_a, p_b] = i\delta_{ab},$$

and, for simplicity, the mass of the oscillator is made equal to unity. And then with the use of the production and annihilation operators we construct for oscillators their coordinate and momentum operators as follows:

$$a_{\mathbf{k}} = \frac{1}{\sqrt{2\omega}}(p_a + i\omega q_a), \qquad a_{\mathbf{k}}^* = \frac{1}{\sqrt{2\omega}}(p_a - i\omega q_a),$$

$$b_{\mathbf{k}} = \frac{1}{\sqrt{2\omega}}(p_b + i\omega q_b), \qquad b_{\mathbf{k}}^* = \frac{1}{\sqrt{2\omega}}(p_b - i\omega q_b).$$

Next, it is necessary to represent the oscillator in the excited state ($N_{\mathbf{k}}^{a,b} \geq 1$) as a collection of excitation quanta, each having the energy $\omega_{\mathbf{k}}$. The equidistant energy levels result in a very important property: impossibility to distinguish between the excitation quanta for each of the oscillators, of course without reference to the values of other quantum numbers. Note that the particles are associated with the excitation quanta of the oscillator and not with the oscillator itself. In other words, the appearance of the particles is due to the quantization procedure, or more precisely due to quantization of the oscillator energy. The wavefunctions of the oscillator, the eigenfunctions of the Hamiltonian operator, may be derived from each other with the use of the operators $a_{\mathbf{k}}^*$ and $b_{\mathbf{k}}^*$

$$a_{\mathbf{k}}^* \Phi_{N_{\mathbf{k}}^a} = \sqrt{1 + N_{\mathbf{k}}^a}\,\Phi_{N_{\mathbf{k}}^a+1}, \qquad b_{\mathbf{k}}^* \Phi_{N_{\mathbf{k}}^b} = \sqrt{1 + N_{\mathbf{k}}^b}\,\Phi_{N_{\mathbf{k}}^b+1}.$$

Thus, the state vector of a system is dependent on the numbers determining a degree of the quantum states occupation with the field particles, that is, it turns to be given in the occupation numbers space.

14.5 Commutation relations. Properties of commutator functions

Let us find the commutation relations between the field functions at arbitrary times. As it has been mentioned previously, this is only possible when we know the solutions for equations of motion. In the case of a free-field they are given by expressions (14.39) and (14.40). It should be noted that, if we were interested in the commutation relations for a particle located, for example, in a homogeneous magnetic field, expansion of the field function would involve the solutions for the Klein-Gordon equation in a magnetic field (Laguerre polynomials) rather than de Broglie plane waves. With the use of (14.39) and (14.40) we obtain:

$$[\psi(x), \psi^*(x')] = \frac{1}{V} \sum_{\mathbf{kk'}} \frac{1}{\sqrt{4\omega\omega'}} \{[a_\mathbf{k}, a^*_\mathbf{k'}] \exp(ik'x' - ikx) + [b^*_\mathbf{k}, b_\mathbf{k'}] \exp(-ik'x' + ikx)\} =$$

$$= \frac{1}{V} \sum_\mathbf{k} \frac{1}{2\omega}(\exp[-ik(x - x')] - \exp[ik(x - x')]) = -i\Delta_0(x - x'; m^2), \tag{14.57}$$

where

$$\Delta_0(x; m^2) \equiv \Delta_0(\mathbf{r}, t; m^2) = \frac{1}{(2\pi)^3} \int \exp(-i\mathbf{kr})\frac{\sin\omega t}{\omega}d^3k. \tag{14.58}$$

As seen, the function $\Delta_0(x; m^2)$ (called *Pauli-Jordan function*) is a relativistic invariant. To illustrate, we represent it as:

$$\Delta_0(x; m^2) = \frac{i}{(2\pi)^3} \int \exp(-ikx)\varepsilon(\omega)\delta(k^2 - m^2)d^4k, \tag{14.59}$$

where $d^4k = dk_1dk_2dk_3d\omega$, $\varepsilon(\omega) = \theta(\omega) - \theta(-\omega)$ and

$$\theta(x) = \begin{cases} 1, & \text{when} \quad x > 0, \\ 0, & \text{when} \quad x < 0. \end{cases}$$

It is obvious that all factors under the integral sign, except for $\varepsilon(\omega)$, are relativistically invariant. The presence of the delta function $\delta(k^2 - m^2)$ in (14.59) tell us that integration is carried out on the surface $k^2 - m^2 = 0$. Therefore, the frequency sign (ω)in this case is relativistically invariant. This suggests a relativistic invariance of the function $\varepsilon(\omega)$.

From the viewpoint of mathematics, the Pauli-Jordan function is of the same class as the δ-function, that is, the class of the improper or generalized functions. As distinct from the normal functions, the generalized ones are determined by setting the integration rules for their products with sufficiently regular functions rather than by setting a correspondence between the values of the function and its argument.

It is useful to divide the Pauli-Jordan function into the frequency components:

$$\Delta_0(x; m^2) = \Delta_0^+(x; m^2) + \Delta_0^-(x; m^2), \tag{14.60}$$

where

$$\Delta_0^\pm(x; m^2) = \pm\frac{i}{(2\pi)^3} \int \exp(-ikx)\theta(\pm\omega)\delta(k^2 - m^2)d^4k. \tag{14.61}$$

They may be used to express the nonzero commutators associated with the negative and positive frequency components of the field function:

$$[\psi^{(+)}(x), \psi^{(+)*}(x)] = -i\Delta_0^+(x; m^2),$$

$$[\psi^{(-)}(x), \psi^{(-)*}(x)] = -i\Delta_0^-(x; m^2).$$

Later it will be demonstrated that $\Delta_0^{\pm}(x; m^2)$ may be used to express not only the Pauli-Jordan function but other commutator functions as well. Because of this, we begin our study of the above-mentioned class of functions with the calculations of the integrals (14.61). Carrying out the integration with respect to k_0 and angular variables in space $\mathbf{k}$, we represent these expressions in the form

$$\Delta_0^+(x; m^2) = \frac{1}{4\pi r} \frac{\partial}{\partial r} F(x), \qquad \Delta_0^-(x; m^2) = \frac{1}{4\pi r} \frac{\partial}{\partial r} F^*(x), \qquad (14.62)$$

where

$$F(x) = \frac{i}{2\pi} \int_{-\infty}^{\infty} \frac{\exp[i(\omega t + kr)]}{\omega} dk, \qquad k = |\mathbf{k}|, \qquad r = |\mathbf{r}|.$$

By this means the problem of finding $\Delta_0^{\pm}(x; m^2)$ is reduced to calculation of the function $F(x)$. Having done the replacement of the variable

$$k = m \sinh y, \qquad \omega = m \cosh y,$$

we for $F(x)$ obtain

$$F(x) = \frac{i}{2\pi} \int_{-\infty}^{\infty} \exp[im(t \cosh y + r \sinh y)] dy. \qquad (14.63)$$

For four cases possible: 1) $t > 0$, $t > r$; 2) $t > 0$, $t < r$; 3) $t < 0$, $|t| > r$; 4) $t < 0$, $|t| < r$, we use the following substitutions:

1) $\quad t = \sqrt{\lambda} \cosh y_0, \quad r = \sqrt{\lambda} \sinh y_0,$ 2) $\quad t = \sqrt{-\lambda} \sinh y_0, \quad r = \sqrt{-\lambda} \cosh y_0,$
3) $\quad t = -\sqrt{\lambda} \cosh y_0, \quad r = \sqrt{\lambda} \sinh y_0,$ 4) $\quad t = -\sqrt{-\lambda} \sinh y_0, \quad r = \sqrt{-\lambda} \cosh y_0,$

where $\lambda = x_\mu^2$. Then taking into consideration the integral representation of cylindrical functions [6], for four different time domains we get:

1) $\quad \dfrac{i}{2\pi} \displaystyle\int_{-\infty}^{\infty} \exp[im\sqrt{\lambda} \, \cosh(y + y_0)] dy = -\frac{1}{2} J_0(m\sqrt{\lambda}) - \frac{i}{2} N_0(m\sqrt{\lambda}),$

2) $\quad \dfrac{i}{2\pi} \displaystyle\int_{-\infty}^{\infty} \exp[im\sqrt{-\lambda} \sinh(y + y_0)] dy = \frac{i}{\pi} K_0(m\sqrt{-\lambda}),$

3) $\quad \dfrac{i}{2\pi} \displaystyle\int_{-\infty}^{\infty} \exp[-im\sqrt{\lambda} \cosh(y - y_0)] dy = \frac{1}{2} J_0(m\sqrt{\lambda}) - \frac{i}{2} N_0(m\sqrt{\lambda}),$

4) $\quad \dfrac{i}{2\pi} \displaystyle\int_{-\infty}^{\infty} \exp[-im\sqrt{-\lambda} \sinh(y - y_0)] dy = \frac{i}{\pi} K_0(m\sqrt{-\lambda}),$

or

$$F(x) = F(t, \lambda) = \begin{cases} \frac{1}{2i} N_0(m\sqrt{\lambda}) - \frac{1}{2}\varepsilon(t) J_0(m\sqrt{\lambda}) & \text{when } \lambda > 0, \\ \frac{i}{\pi} K_0(m\sqrt{-\lambda}), & \text{when} \qquad \lambda < 0, \end{cases} \qquad (14.64)$$

where $J_0(z)$ is a Bessel function of the zero order, $N_0(z)$ is a Neumann function of the zero order, $K_0(z)$ is a Hankel function of an imaginary argument of the zero order. At $z \approx 0$ these cylindrical functions are represented by the series:

$$\left. \begin{aligned} J_0(z) &= 1 - \left(\tfrac{z}{2}\right)^2 + O(z^4), \\[6pt] N_0(z) &= \tfrac{2}{\pi}\left[1 - \left(\tfrac{z}{2}\right)^2\right] \ln \tfrac{z}{2} + \tfrac{2}{\pi}\mathbf{C} + O(z^2), \\[6pt] K_0(z) &= -\left[1 + \left(\tfrac{z}{2}\right)^2\right] \ln \tfrac{z}{2} - \mathbf{C} + O(z^2) \end{aligned} \right\} \qquad (14.65)$$

(**C** is the Euler constant to be equal to 0.577215...), and in the region of great values of z they are described by the asymptotic formulae:

$$\left.\begin{array}{l} J_0(z) = -\frac{2}{\sqrt{\pi z}}\sin z, \\[2mm] N_0(z) = \frac{2}{\sqrt{\pi z}}\sin z, \\[2mm] K_0(z) = \sqrt{\frac{2}{\pi z}}e^{-z}. \end{array}\right\} . \tag{14.66}$$

In (14.62) we proceed from the differentiation with respect to r to that with respect to λ and take into consideration a jump of the function $F(t,\lambda)$ at the point $\lambda = 0$. Then, having regard to the following relations for the cylindrical functions

$$Z_1(x) = -\frac{\partial}{\partial x}Z_0(x), \tag{14.67}$$

where $Z = J, N, K$, we derive the required expressions:

$$\Delta_0^{\pm}(x; m^2) = \frac{\varepsilon(t)\delta(\lambda)}{4\pi} - \frac{m\theta(\lambda)}{8\pi\sqrt{\lambda}}\left[\varepsilon(t)J_1(m\sqrt{\lambda}) \pm iN_1(m\sqrt{\lambda})\right] \mp \frac{im\theta(-\lambda)}{4\pi^2\sqrt{-\lambda}}K_1(m\sqrt{-\lambda}). \tag{14.68}$$

It should be noted that all singularities of the functions $\Delta_0^{\pm}(x; m^2)$ are positioned at the light cone only ($\lambda - 0$), as at the space and time infinities these functions, in accordance with (14.66) and (14.67), are decreased as $|\lambda|^{-3/4}\exp(-m\sqrt{|\lambda|})$ and $|\lambda|^{-3/4}$, respectively.

Under investigating a regularity of different commutator functions we need the expansions of $\Delta_0^{\pm}(x; m^2)$ in a light con neighborhood. Using Eqs. (14.65) and (14.67), we arrive at

$$\Delta_0^{\pm}(x; m^2) = \frac{1}{4\pi}\varepsilon(t)\delta(\lambda) \pm \frac{i}{4\pi^2\lambda} \mp \frac{im^2}{8\pi^2}\ln\frac{m|\lambda|^{1/2}}{2} - \frac{m^2}{16\pi}\varepsilon(t)\theta(\lambda) + O(\sqrt{|\lambda|}\ln|\lambda|). \tag{14.69}$$

For the expressions (14.69) there are four singularity types at the light cone: the pole λ^{-1}, $\ln|\lambda|$, $\delta(\lambda)$ and the jump $\theta(\lambda)$. The term $\varepsilon(t)$ creates no additional singularities beyond the light cone because, due to factors $\delta(\lambda)$ and $\theta(\lambda)$ which stand before it, jumps of $\varepsilon(t)$ appear at the origin only, that is, again at the light cone. So the functions $\Delta_0^{\pm}(x; m^2)$ are singular.

Using (14.60) and (14.68), we can represent the Pauli-Jordan function in the form, exhibiting both its covariant properties and its singularities

$$\Delta_0(x; m^2) = \frac{1}{2\pi}\varepsilon(t)\delta(\lambda) - \frac{m}{4\pi\sqrt{\lambda}}\theta(\lambda)\varepsilon(t)I_1(m\sqrt{\lambda}). \tag{14.70}$$

It follows from (14.70) that this function is zero beyond the light cone ($\lambda < 0$). Because of this, all (anti)commutators of the field operators, the arguments of which are separated by a space-like interval, go to zero.

Making use of (14.69), we get that in the neighborhood of the light cone the commutator function $\Delta_0(x; m^2)$ is determined by

$$\Delta_0(x; m^2) = \frac{1}{2\pi}\varepsilon(t)\delta(\lambda) - \frac{m^2}{8\pi}\varepsilon(t)\theta(\lambda) + O(\lambda). \tag{14.71}$$

According to (14.71), the function $\Delta_0(x; m^2)$ has at the light cone a δ-shaped singularity and a jump of $m^2\varepsilon(t)/8\pi$ under the transition from the space-like domain to the time-like one.

14.6　Microcausality

The equality of the $\Delta_0(\mathbf{r}, t; m^2)$-function to zero over the whole domain exterior to the light cone may be demonstrated from more common considerations. Really, recall that $\Delta_0(\mathbf{r}, t; m^2)$-function is relativistically invariant and that each world point of the space-like domain, with the use of the Lorentz transformations, may be reduced to a point with $t = 0$. And since we have:

$$\Delta_0(\mathbf{r}, t; m^2)|_{t=0} = 0,$$

then for the Pauli-Jordan function the following is true:

$$\Delta_0(\mathbf{r}, t; m^2) = 0, \qquad |t| < |\mathbf{r}|. \tag{14.72}$$

As a result, commutators of the operator field functions disappear beyond the light cone. This property of the commutators (and anticommutators in the case of Fermi fields) was called *microcausality* or *local commutativity*. As is known, the commutation relations allow the following physical interpretation. Provided the operators A and B associated with some physical quantities a and b do not commute with each other and meet the commutation relations

$$[A, B] = iC,$$

where C in the most general case is an operator, then a system has no states, for which a and b may have exact values simultaneously. Errors of measuring the quantities a and b are in the intervals Δa and Δb obeying the uncertainty relations:

$$\Delta a \cdot \Delta b \geq \frac{1}{2} <C>.$$

In this case from the commutation relations (14.57) it follows that

$$\Delta\psi(x)\Delta\psi^*(x') = \begin{cases} \geq \Delta_0(x - x'; m^2) & \text{for time-like interval,} \\ 0 & \text{for space-like interval.} \end{cases} \tag{14.73}$$

It seems that microcausality, reflecting the independence of physical events separated by the space-like interval, represents a certain modernization of the causality principle (future has no influence on the past). However, in the quantum field theory the notion of a point event is devoid of a direct physical meaning. This is due to the Heisenberg uncertainty relations setting the lower limit to the extent and duration of any act associated with the fields interaction, measuring the dynamic invariants of a field, etc. To illustrate, the coordinate of a particle at rest may be accurately fixed merely within its Compton wavelength. Local fields represent mathematical idealization, and only the fields weight-averaged over a small space-time domain are physically meaningful. In other words, the microcausality conditions should be considered as extrapolation of a physical causality condition to the domain of short distances and small time intervals. Verification of the quantum electrodynamics has provided support for the microcausality conditions up to the distances on the order of 10^{-16} cm and time intervals on the order of 10^{-26} s. Let us return back to the relation (14.73). Since $\Delta_0(x - x'; m^2)$-function has a δ-like singularity at the light cone $(x - x')^2 = 0$ and the δ-function is physically meaningful only after integration with respect to the domain containing the δ-like singularity, then, to clarify a physical meaning of (14.73), we should take the field values not at the discrete world points but averaged over the space-time domains Ω and Ω'. Now instead of (14.73), we have

$$\Delta\{\psi(x)\}_\Omega \Delta\{\psi^*(x')\}_{\Omega'} = \begin{cases} \geq \{\Delta_0(x - x'; m^2)\}_{\Omega\Omega'} & \text{for time-like interval,} \\ 0 & \text{for space-like interval,} \end{cases} \tag{14.74}$$

where $\{\}_\Omega$ ($\{\}_{\Omega'}$) denotes the operation of averaging over the domain Ω (Ω'). So, when the operators representing a measured value at different points of the Minkowsky space x and x' commute with each other, the associated quantity may be measured at these points so that one measurement cannot influence the other. This is due to the fact that in this case:

$$| \mathbf{r} - \mathbf{r}' | > | t - t' |,$$

and so the field perturbation caused by measurement procedure in one domain, even with the light-speed propagation, there is no possibility to reach the other domain. If Ω and Ω' are coupled by the light signals, at least partially, one measurement act may influence the results of the other. No doubts that in this place a natural question on the reader appears: *"What observables may be associated with the operators $\psi(x)$ and $\psi^*(x')$?"* We know that in the NQM the wavefunction itself has no physical meaning, as distinct from its squared modulus. Because of this, it would be strange to expect that in the quantum field theory the wavefunction, even in the operator form, acquires a status of the observed quantity. Of course, this is not the case. In the quantum field theory the observables (charge, energy, etc.) are in essence bilinear combinations of the field functions, with the electromagnetic field strengths making a nice exception. However, bilinear combinations meet the causality principle only at a condition that for the field operators the microcausality principle is fulfilled.

14.7   Relativistically covariant scheme of the canonical quantization

As follows from the integral representation of the function $\Delta_0(x; m^2)$ (14.58), it is a singular solution for the Klein-Gordon equation

$$(\Box + m^2)\Delta_0(x; m^2) = 0, \tag{14.75}$$

satisfying the initial conditions:

$$\Delta_0(\mathbf{r}, t; m^2)|_{t=0} = 0, \qquad \frac{\partial}{\partial t}\Delta_0(\mathbf{r}, t; m^2)|_{t=0} = \delta^{(3)}(\mathbf{r}). \tag{14.76}$$

By definition, this function is an even function of the space coordinates and an odd time function:

$$\Delta_0(-\mathbf{r}, t; m^2) = \Delta_0(\mathbf{r}, t; m^2), \qquad \Delta_0(\mathbf{r}, -t; m^2) = -\Delta_0(\mathbf{r}, t; m^2).$$

Also, it is easy to see that its positive- and negative-frequency components are the solutions for the Klein-Gordon equation:

$$(\Box + m^2)\Delta_0^+(x; m^2) = 0, \qquad (\Box + m^2)\Delta_0^-(x; m^2) = 0. \tag{14.77}$$

So, an explicit form of the quantities to characterize the field demonstrates that, despite a selected role of time in the canonical quantization procedure, the resultant theory is relativistically invariant. Based on this reasoning, the existence of a canonical quantization procedure possessing an explicit relativistic invariance at all the intermediate stages is apparent [7]. We describe its principal points.

Consider the space-like hypersurface σ, whose normal vector is n^μ. Any four-dimensional vector may be resolved into the longitudinal and transverse components with respect to the vector n^μ:

$$a^\mu = a^{\mu T} + n^\mu a_n, \qquad (14.78)$$

where

$$a^{\mu T} = (\delta^\mu_\nu - n^\mu n_\nu)a^\nu, \qquad a_n = n_\nu a^\nu.$$

All scalar products in this representation are of the form:

$$a^\mu b_\mu = a_n b_n - a^{\mu T} b^T_\mu.$$

The "momentum" $\pi(x)$ canonically conjugate to the "coordinate" of the field $\psi(x)$ may be determined in the relativistically invariant form

$$\pi(x) = \frac{\partial \mathcal{L}}{\partial \left(\frac{\partial \psi(x)}{\partial x_n} \right)}. \qquad (14.79)$$

From the Lagrangian (14.4) we can find

$$\pi(x) = \partial_n \psi^\dagger(x) = \partial_\mu \psi^*(x) n^\mu = \pi^c_\mu(x) n^\mu. \qquad (14.80)$$

Next, we suppose that the commutation relations between the field coordinates and momenta are given as follows:

$$\int_\sigma [\psi(x), \pi^c_\mu(x')] d\sigma^\mu(x') = i, \qquad (14.81)$$

$$\int_\sigma [\psi^\dagger(x), \pi^{c\dagger}_\mu(x')] d\sigma^\mu(x') = i, \qquad (14.82)$$

where x and x' belong to one and the same space-like hypersurface σ. Substituting the expression for $\pi^c_\mu(x')$ into (14.81), we have:

$$\int_\sigma [\psi(x), \frac{\partial \psi^*(x')}{\partial x'^\mu}] d\sigma^\mu(x') = i. \qquad (14.83)$$

The relation of (14.83) is nothing else but a covariant generalization of the ordinary canonical commutation relation:

$$[\psi(t, \mathbf{r}), \frac{\partial \psi^*(t, \mathbf{r}')}{\partial t}] = i\delta^{(3)}(\mathbf{r}' - \mathbf{r}), \qquad (14.84)$$

having been integrated with respect to the three-dimensional volume dV' including the point $\mathbf{r}$. To prove this statement, we show that the left-hand side of the expression (14.83) is independent on the choice of the space-like hypersurface σ, that is, its derivative $\delta/\delta\sigma(x')$ goes to zero. Taking into consideration that $\psi^*(x')$ obeys the Klein-Gordon equation, we actually obtain:

$$\frac{\delta}{\delta\sigma(x')} \int_\sigma [\psi(x), \frac{\partial \psi^*(x'')}{\partial x''_\mu}] d\sigma_\mu(x'') = [\psi(x), \square' \psi^*(x')] = m^2 [\psi(x), \psi^*(x')] = 0. \qquad (14.85)$$

Then for hypersurface σ in the left-hand side of (14.83) we can select the hyperplane $t' = t =$const thus proving our statement.

When $\mathbf{r}' \neq \mathbf{r}$, the right-hand side of the expression (14.83) goes to zero. In this case the corresponding covariant generalization of (14.83), similar to other field commutators to be equal to zero, represents simply the relation:

$$[\psi(x'), \psi^*(x) = 0, \qquad (x'_\mu - x_\mu)^2 < 0, \qquad (14.86)$$

that means commutation of the field variables associated with two different points of the space-like hypersurface.

Let us show that the function $\Delta_0(x; m^2)$ makes it possible to solve a Cauchy problem for the Klein-Gordon equation. It means that using this function, we can find a Klein-Gordon equation solution at the arbitrary space-like hypersurface σ in terms of the values of the solution and its derivative at any earlier space-like hypersurface σ_0:

$$\psi(x) = \int_{\sigma_0} d\sigma^\mu(x') \left[\frac{\partial \Delta_0(x - x'; m^2)}{\partial x'^\mu} \psi(x') - \Delta_0(x - x'; m^2) \frac{\partial \psi(x')}{\partial x'^\mu} \right]. \tag{14.87}$$

To demonstrate the validity of Eq.(14.87) we write its right-hand side in the form:

$$I_\sigma = \int_{\sigma_0} d\sigma^\mu F_\mu$$

and act upon that by the operator $\delta/\delta\sigma$. Since both $\psi(x')$, and $\Delta_0(x - x'; m^2)$ are the solutions for the Klein-Gordon equations, we can derive:

$$\frac{\delta}{\delta\sigma(x)} I_\sigma = \lim_{\Omega \to 0} \frac{\int_\sigma d\sigma^\mu F_\mu(x) - \int_{\sigma_0} d\sigma^\mu F_\mu(x)}{\Omega(x)} = \frac{\partial F_\mu(x)}{\partial x_\mu} = 0. \tag{14.88}$$

As follows from (14.88), I_σ is independent on a choice of the space-like hypersurface σ. Then nothing prevents us from choosing as such the hyperplane $t = $ const again. Owing to the properties of $\Delta_0(x; m^2)$, from (14.87) we obtain the identity $\psi(x) = \psi(x)$, which was to be proved.

Now finding of the commutation relations at an arbitrary instant of time becomes fairly trivial. To find the value of $[\psi(x), \psi^\dagger(x')]$, one should express the operator $\psi(x)$ in terms of the field variables at the space-like hypersurface passing through the point x', and, next, makes use of the well-known commutation relations at this surface. By application of Eqs. (14.87) and (14.83) we get

$$[\psi(x), \psi^\dagger(x')] = \int_\sigma d\sigma^\mu(x'') \left\{ \frac{\partial \Delta_0(x - x''; m^2)}{\partial x''^\mu} [\psi(x''), \psi^\dagger(x')] - \right.$$

$$\left. - \Delta_0(x - x''; m^2) [\frac{\partial \psi(x'')}{\partial x''^\mu}, \psi^\dagger(x')] \right\}. \tag{14.89}$$

Since $x' \in \sigma$, $x'' \in \sigma_0$ and the integral in the right-hand side is independent on a choice of σ, then, passing to the hyperplane $t=$const, we obtain the same commutation relations as in the ordinary canonical formalism:

$$[\psi(x), \psi^\dagger(x')] = -i\Delta_0(x - x'; m^2). \tag{14.90}$$

14.8   Green function of the scalar field

For a scalar field the Green function is a solution of the equation:

$$(\Box + m^2)G(x) = \delta(x). \tag{14.91}$$

Performing an integral Fourier expansion of $G(x)$ and using an integral representation of the delta-function, with the help of Eq. (14.91) we obtain the following formal expression for the Green function:

$$G(x) = \frac{1}{(2\pi)^4} \int G(k) \exp(-ikx) d^4k, \qquad (14.92)$$

where

$$G(k) = \frac{1}{m^2 - k^2}.$$

This expression is indefinite because no rules for detour of the poles $(k_0)_{1,2} = \pm\sqrt{\mathbf{k}^2 + m^2}$ are set. Indefiniteness reveals the fact that a complete solution for equation (14.91) is given as a sum of a partial solution for the inhomogeneous equation and solutions of the corresponding homogeneous equation, $\Delta_0^+(x)$ and $\Delta_0^-(x)$, taken with arbitrary factors. We shall demonstrate that, setting the rules for detour of the poles or, which is equivalent, imposing the boundary conditions for $G(x)$, we uniquely determine these factors.

Let us consider a retarded Green function, for that the following initial condition is true:

$$\Delta^{ret}(x) = 0, \qquad t < 0. \qquad (14.93)$$

In order to represent $\Delta^{ret}(x)$ in the form close to (14.92), we note that on multiplication of this function by $\exp(-\epsilon t)$, because of (14.93), it assumes no additional properties:

$$\Delta^{ret}(x) \exp(-\epsilon t) = G_\epsilon, \qquad (14.94)$$

and may be represented as a limit:

$$\Delta^{ret}(x) = \lim_{\epsilon \to +0} G_\epsilon. \qquad (14.95)$$

In accordance with the definition of (14.94), the function G_ϵ satisfies the equation:

$$\left[\left(\frac{\partial}{\partial t} + \epsilon \right)^2 - \Delta + m^2 \right] G_\epsilon = \delta(x)$$

taking in the momentum representation in the limit $\epsilon \to +0$ the following form:

$$\frac{1}{m^2 - (k_0 + i\epsilon)^2 + \mathbf{k}^2} \to \frac{1}{m^2 - k^2 - 2i\epsilon k_0}.$$

And according to (14.94), the retarded Green function may be represented as follows:

$$\Delta^{ret}(x) = \frac{1}{(2\pi)^4} \int \frac{d^4k}{m^2 - k^2 - 2i\epsilon k_0} \exp(-ikx). \qquad (14.96)$$

It is obvious that expression (14.96) meets the condition of (14.93). To make sure that this is true, it suffices to integrate with respect to the variable k_0 using a theory of residues. An infinitesimal addition $2i\epsilon k_0$, in the denominator of (14.96) indicates that detour of both poles $(k_0)_{1,2} = \pm\sqrt{\mathbf{k}^2 + m^2} = \pm\omega$ in a complex plane of the variable k_0 should be from above, that is, integration follows the contour L_C (Fig. 14.1). In accordance with Jordan's lemma, the contour L may be always closed by semicircles $L_\smile$ and $L_\frown$ of radius R (Fig.14.2) so that in the limit $R \to \infty$ the additional contour integrals along $L_\smile$ and $L_\frown$ turn to zero. To this end, it is sufficient to require the fulfillment of the condition:

$$\exp(-izt)|_{R \to \infty} \to 0, \qquad (14.97)$$

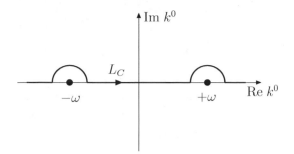

FIGURE 14.1

The integration contour for the function Δ^{ret}.

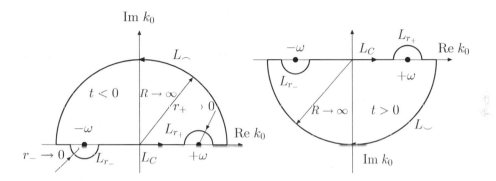

FIGURE 14.2

The closed integration contours in the complex plane of k_0.

where $z = \operatorname{Re} k_0 + i\operatorname{Im} k_0$, under $R \to \infty$. In this case from (14.96) it follows that for $t < 0$ the integration should be over the upper half-plane with a counterclockwise detour of the closed contour, and for $t > 0$—clockwise over the lower half-plane. At $t < 0$ there are no poles within the contour of integration, and we obtain (14.93). On the other hand, using a theory of residues, at $t > 0$ we find:

$$\Delta^{ret}(x) = \frac{1}{(2\pi)^3 i} \int d^3k \left[\frac{\exp\left(ik_0 t\right) - \exp\left(-ik_0 t\right)}{2k_0}\right]_{k_0=\sqrt{\mathbf{k}^2+m^2}} \exp\left(-i\mathbf{k}\mathbf{r}\right) = \Delta_0(x).$$

(14.98)

Thus

$$\Delta^{ret}(x) = \theta(t)\Delta_0(x).$$

(14.99)

Performing similar calculations for the advanced Green function

$$\Delta^{adv}(x) = 0, \qquad \text{at} \qquad t > 0$$

(14.100)

we get

$$\Delta^{adv}(x) = \frac{1}{(2\pi)^4} \int \frac{d^4k}{m^2 - k^2 + 2i\epsilon k_0} \exp\left(-ikx\right) = -\theta(-t)\Delta_0(x).$$

(14.101)

14.9   Causal Green function

Causal link of the production and annihilation processes at various space-time points is established by the causal Green function. Since in a theory of interacting fields this function gives a description for the virtual particles propagation, it is often called the *propagator*. The process of a scalar particle production at the point x and its subsequent annihilation at the point x' $(t < t')$ is associated with the matrix element:

$$\Phi_1^*(x')\Phi_1(x) = \Phi_0^*\psi^{(-)}(x')\psi^{(-)*}(x)\Phi_0 = -i\Delta_0^+(x - x'; m^2). \tag{14.102}$$

But when a particle is produced at the point x' and annihilated at the point x, instead of (14.102) we obtain:

$$\Phi_1^*(x)\Phi_1(x') = \Phi_0^*\psi^{(-)}(x)\psi^{(-)*}(x')\Phi_0 = i\Delta_0^-(x - x'; m^2). \tag{14.103}$$

To characterize both processes, we introduce the causal function in accordance with the following relation:

$$\Delta^c(x) = \theta(t)\Delta_0^-(x; m^2) - \theta(-t)\Delta_0^+(x; m^2). \tag{14.104}$$

As seen from (14.104), the causal function represents a sum of the partial solution for inhomogeneous equation (14.91) and general solution for equation (14.1) because:

$$\Delta^c(x) = \Delta^{ret}(x) - \Delta_0^+(x; m^2). \tag{14.105}$$

Thus, it satisfies all the requirements for the general solution of the equation:

$$(\Box + m^2)\Delta^c(x) = \delta(x). \tag{14.106}$$

Let us find an expression for the causal function in the momentum representation. For this purpose we rewrite the expression for $\Delta^{ret}(x)$ in the following form:

$$\Delta^{ret}(x) = \frac{i}{(2\pi)^4} \lim_{\epsilon \to +0} \int d^4k \exp\left(-ikx\right)\left[-\int_{-\infty}^0 d\beta\theta(-k_0)\times\right.$$

$$\left.\times \exp\left[-i\beta(m^2 - k^2 - 2i\epsilon k_0)\right] + \int_0^\infty d\beta\theta(k_0)\exp\left[-i\beta(m^2 - k^2 - 2i\epsilon k_0)\right]\right] =$$

$$= \frac{1}{(2\pi)^4}\left[P\int\frac{d^4k}{m^2 - k^2}\exp\left(-ikx\right) + i\pi\int d^4k\delta(m^2 - k^2)\exp\left(-ikx\right)\right], \tag{14.107}$$

where the symbol P denotes that the principal-value integral is taken. The principal value exists for the improper integral under consideration because: (i) the function $f(k) = 1/(m^2 - k^2)$ is continuous over the interval $(-\infty, \infty)$, going to zero only at the points $(k_0)_{1,2} = \pm\sqrt{\mathbf{k}^2 + m^2}$; (ii) the first-order and second-order derivatives, $(\partial f(k)/\partial k_0)$ and $(\partial^2 f(k)/\partial k_0^2)$, at singular points are nonzero. In this way the integral:

$$\int\frac{d^4k}{m^2 - k^2}\exp\left(-ikx\right)$$

is divergent, but there are limits for both singular points. For instance, for the point $(k_0)_2$ it is of the form:

$$\lim_{\eta \to +0}\left[\int_{-\infty}^{k_0^{(2)}-\eta}\frac{dk_0}{m^2 - k^2}\exp\left(-ikx\right) + \int_{k_0^{(2)}+\eta}^{k_0^{(1)}-\eta}\frac{dk_0}{m^2 - k^2}\exp\left(-ikx\right)\right].$$

Using (14.61), (14.105) and taking into account

$$P\frac{1}{a} + i\pi\delta(a) = \frac{1}{(a - i\epsilon)},$$

we get

$$\Delta^c(x) = \frac{1}{(2\pi)^4} \int \frac{d^4 k}{m^2 - k^2 - i\epsilon} \exp(-ikx). \tag{14.108}$$

To study a behavior of the causality function, we establish its explicit form. This operation presents no difficulties as we have already obtained the associated expressions for $\Delta_0^{\pm}(x; m^2)$. With allowance made for (14.68) and (14.104), we find:

$$\Delta^c(x) = \frac{1}{4\pi}\delta(\lambda) - \frac{m}{8\pi\sqrt{\lambda}}\theta(\lambda)\left[J_1(m\sqrt{\lambda}) - iN_1(m\sqrt{\lambda})\right] +$$

$$+ \frac{im}{4\pi^2\sqrt{-\lambda}}\theta(-\lambda)K_1(m\sqrt{-\lambda}). \tag{14.109}$$

So, the causal function is other than zero both within and beyond the light cone.

With the help of (14.69), we can obtain expansion of $\Delta^c(x)$ in the neighborhood of the light cone:

$$\Delta^c(x) = \frac{1}{4\pi}\delta(\lambda) + \frac{1}{4\pi^2 i\lambda} - \frac{m^2}{16\pi}\theta(\lambda) + \frac{im^2}{8\pi^2}\ln\frac{m|\lambda|^{1/2}}{2} + O(\sqrt{|\lambda|}\ln|\lambda|). \tag{14.110}$$

As follows from (14.110), at the light cone the causal function has all four singularities possible for commutator functions: $\delta(\lambda)$, pole λ^{-1}, jump $\theta(\lambda)$ and $\ln|\lambda|$.

14.10   Chronological product. Convolution of operators

Except for the normal product of the operators, an important role is played by the chronological product, the T-product. The factors of a T-product have chronological order, namely: the factor on the right is associated with a lower value of time, and that on the left—with the greater one. In the case of a scalar field this means that:

$$T(\psi(x)\psi^*(x')) = \begin{cases} \psi(x)\psi^*(x'), & t > t' \\ \psi^*(x')\psi(x), & t < t'. \end{cases} \tag{14.111}$$

Let us show that definition of the T-product is relativistically invariant. Indeed, if x and x' are separated by a time-like interval, this statement is apparent because in this case the notion of earlier and later time is absolute in character. But if x and x' are separated by a space-like interval, and a sign of $t - t'$ is varying in the process of the Lorentz transformation, the operators $\psi(x)$ and $\psi^*(x')$ commute with each other. Because of this, in all frames of reference they may be positioned in the order set by the chronological product. Since bosons are quantized by the commutation relations, if required, reordering of the factors both in the normal and chronological products leaves the sign unaltered. As may be seen later, for fermions, where quantization involves anticommutators, this is not the case.

Consider the difference between chronological and normal products of the field functions $\psi(x)$ and $\psi^*(x')$. This difference, referred to as convolution or pairing of the operators, is denoted by the superscript s_i, where i takes on values 1,2,3... and its value is the same for each pair of the operators,

$$T(\psi(x)\psi^*(x')) - N(\psi(x)\psi^*(x')) \equiv \psi^{s_1}(x)\psi^{*s_1}(x'). \tag{14.112}$$

It is easily demonstrated with direct calculations that, both for $t > t'$ and for $t < t'$, convolution of the operators $\psi^{s_1}(x)\psi^{*s_1}(x')$ is given by one and the same expression:

$$\psi^{s_1}(x)\psi^{*s_1}(x') = \frac{1}{V}\sum_{\mathbf{k}}\frac{1}{2\omega}[\exp[i\mathbf{k}(\mathbf{r} - \mathbf{r}') - i\omega\mid t - t'\mid]], \tag{14.113}$$

that contains no operators, that is, is a c-number. Going from the sum to the integral, we obtain

$$\psi^{s_1}(x)\psi^{*s_1}(x') = \Delta'(x - x'), \tag{14.114}$$

where

$$\Delta'(x) = \frac{1}{2(2\pi)^3}\int\theta(k_0)\exp(i\mathbf{k}\mathbf{r} - ik_0\mid t\mid)\frac{d^3k}{k_0} \tag{14.115}$$

and $k_0 \equiv \omega$. Further, it is essential to know a Fourier-component of the $\Delta'(x)$ function

$$\Delta'(x) = \frac{1}{(2\pi)^4}\int d^4k\Delta'(k)\exp(-ikx). \tag{14.116}$$

It may be found by the formula:

$$\frac{\exp(-ik_0\mid t\mid)}{\omega} = \frac{1}{\pi i}\int_{L_I}dk_0\frac{\exp(-ik_0t)}{m^2 - k^2}, \tag{14.117}$$

with integration over the complex plane k_0 along the contour L_I shown in Fig. 14.3. The

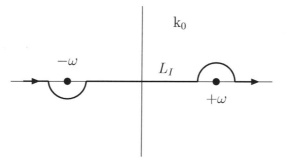

FIGURE 14.3
The integration contour L_I.

validity of (14.117) is apparent as for $t > 0$ the integral is determined by the residue at the point $(k_0)_1 = \omega$ and for $t < 0$—by the residue at the point $(k_0)_2 = -\omega$. In Eq. (14.117) from integration along the contour L_I we may proceed to integration along the real axis, with shifting of the ω-pole to the lower half-plane and $-\omega$-pole—to the upper half-plane (Fig. 14.4). To this end, we replace $m^2 - k^2$ by $m^2 - k^2 - i\epsilon$ in the denominator of the

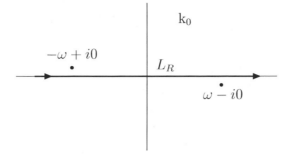

FIGURE 14.4

Crossing to integration along the real axis.

integrand, where ϵ is an infinitesimal positive number. Substituting (14.117) into (14.115) and comparing with (14.116), we derive the following expression for $\Delta'(k)$:

$$\Delta'(x) = \frac{-i}{(2\pi)^4} \int \frac{d^4k}{m^2 - k^2 - i\epsilon} \exp(-ikx), \qquad (14.118)$$

that is a clear demonstration of the coincident (within a constant factor) of operators convolution and causal Green function. As follows from the definition of the operators convolution (14.112), it is related to the vacuum average of the T-product for these operators

$$\psi^s(x)\psi^{*s}(x') = <T(\psi(x)\psi^*(x'))>_0 = -i\Delta^c(x - x'). \qquad (14.119)$$

Also, it is obvious that convolution of the two commuting operators becomes zero:

$$\psi^{s_1}(x)\psi^{s_1}(x') = \psi^{*s_1}(x)\psi^{*s_1}(x') = 0 \qquad (14.120)$$

So far on the expansion of the field functions into the solutions for free equations of motion we have used the discrete momentum representation. It turned to be very useful for consideration of the field from the viewpoint of the wave-particle dualism. It was helpful for making clear a field model based on a set of an infinite number of harmonic oscillators. At the same time, in practice a ordinary (continuous) momentum representation is more convenient for the calculations. This becomes particularly apparent under a changeover from the coordinate representation to the momentum representation in matrix elements associated with the Feynman diagrams. In what follows we use the continuous momentum representation. A changeover from the discrete representation to the continuous one is as follows. First, in the expressions (14.13) and (14.14) we must replace summation with integration that is achieved by the use of (14.26). Next, we go from the operators with respect to the discrete variable $a_{\mathbf{k}}^{(\pm)}$ to new operators $a^{(\pm)}(\mathbf{k})$. Recall that in the case of a discrete spectrum we have selected the normalization corresponding to a single particle within the volume $V = 1$. Retention of the previous normalization requires the fulfillment of the condition:

$$\sum_{\mathbf{k}} |a_{\mathbf{k}}^{(\pm)}|^2 = \int d^3k |a^{(\pm)}(\mathbf{k})|^2,$$

whence the desired relation is found:

$$\frac{\sqrt{V}}{(2\pi)^{3/2}} a_{\mathbf{k}}^{(\pm)} \to a^{(\pm)}(\mathbf{k}).$$

Because of this, modification involves also the commutation relations for the production and annihilation operators. Here we need replacement of the Kronecker symbols by the delta-functions. As a result, we have:

$$\psi(x) = \psi^{(+)}(x) + \psi^{(-)}(x) = \frac{1}{(2\pi)^{3/2}} \int \frac{d^3k}{\sqrt{2k_0}} [a(\mathbf{k})e^{-ikx} + b^*(\mathbf{k})e^{ikx}], \qquad (14.121)$$

$$\psi^*(x) = \psi^{(+)*}(x) + \psi^{(-)*}(x) = \frac{1}{(2\pi)^{3/2}} \int \frac{d^3k}{\sqrt{2k_0}} [a^*(\mathbf{k})e^{ikx} + b(\mathbf{k})e^{-ikx}] \qquad (14.122)$$

and

$$[a(\mathbf{k}), a^*(\mathbf{k}')] = [b(\mathbf{k}), b^*(\mathbf{k}')] = \delta(\mathbf{k} - \mathbf{k}'). \qquad (14.123)$$

With the use of the orthonormality conditions the expansions (14.121) and (14.122) may be inverted, giving the following equations:

$$a(\mathbf{k}) = \frac{i}{(2\pi)^{3/2}} \int \frac{d^3x}{\sqrt{2k_0}} \left(e^{ikx} \overset{\leftrightarrow}{\partial_0} \psi(x) \right), \qquad (14.124)$$

$$b(\mathbf{k}) = \frac{i}{(2\pi)^{3/2}} \int \frac{d^3x}{\sqrt{2k_0}} \left(e^{ikx} \overset{\leftrightarrow}{\partial_0} \psi^*(x) \right). \qquad (14.125)$$

The equations for $a^*(\mathbf{k})$ and $b^*(\mathbf{k})$ follow from (14.124) and (14.125) under complex conjugation. Using the obtained relations, we can rewrite the system state vector (14.53) in terms of the operator field functions in the configuration space as follows:

$$\Phi_{n_{\mathbf{k}_1}^+ n_{\mathbf{k}_1'}^- \dots} = \int \dots \int f(x_1, \dots, x_n) \psi^{\alpha_1}(x_1) \dots \psi^{\alpha_n}(x_n) d^4x_1 \dots d^4x_n \Phi_0, \qquad (14.126)$$

where $\alpha_i = \pm$, $\psi^+(x) \equiv \psi(x)$, $\psi^-(x) \equiv \psi^*(x)$ and the weight function $f(x_1, \dots, x_n)$ has a meaning of the quantum-mechanical wavefunction for a particles system in the configuration space. It should be noted that by definition of the vacuum state, the operator field functions in (14.126) contain only the parts associated with the particles and antiparticles production. Definition of the state vector in the form of (14.126) is referred to as the Fock state-vector representation.

For a truly neutral scalar field the above second quantization formalism has no appreciable changes. The Lagrangian density in this case is given by:

$$\mathcal{L}(x) =: \frac{1}{2}\partial_\mu\psi(x)\partial^\mu\psi(x) - \frac{m^2}{2}\psi^*(x)\psi(x) : . \qquad (14.127)$$

Expansion of the field function with respect to the solutions for the Gordon-Klein equation involves the particles operators only:

$$\psi(x) = \frac{1}{(2\pi)^{3/2}} \int \frac{d^3k}{\sqrt{2k_0}} [a(\mathbf{k}) \exp(-ikx) + a^*(\mathbf{k}) \exp(ikx)]. \qquad (14.128)$$

The principal difference from the case of a complex scalar field is the fact that all "currents" of the real scalar field are zero.

14.11   First-order equation for scalar particles

The Klein-Gordon equation is inferred from the relativistic relations between the energy, momentum, and mass. It means that all the components of the operator field functions describing the particles with an arbitrary spin have to obey this equation, or else, as they often say, must obey the Klein-Gordon condition. A relativistic wave equation for an arbitrary spin field may be represented as:

$$\Lambda_{\alpha\beta}(\partial)Q_\beta(x) = 0, \tag{14.129}$$

where $\beta = 1, 2, ...r$ $(r > 2J + 1)$,

$$\Lambda_{\alpha\beta}(\partial) = (i\Gamma^\mu \partial_\mu - m)_{\alpha\beta},$$

Γ^μ are r-rank square matrices, the properties of which are determined by the transformation laws of the $Q_\beta(x)$-function and by the covariance conditions of equation (14.129). The existence of a "conjugate" field function is needed for the Lagrangian density to be built. This function is determined by:

$$\overline{Q}_\beta(x) \equiv Q_\alpha^* \eta_{\alpha\beta}$$

where η is a nonsingular matrix. An equation for $\overline{Q}_\beta(x)$ is of the form:

$$\overline{\Lambda}_{\alpha\beta}(\partial)\overline{Q}_\alpha(x) = 0, \tag{14.130}$$

where $\overline{\Lambda}(\partial) \equiv \eta\Lambda(\partial)$. The Klein-Gordon condition restricts Eqs. (14.129) and (14.130) as regards the existence of such a $d_{\tau\alpha}(\partial)$ operator that would meet the relations:

$$d_{\tau\alpha}(\partial)\Lambda_{\alpha\beta}(\partial) = (\Box + m^2)\delta_{\tau\beta}, \tag{14.131}$$

$$\overline{\Lambda}_{\alpha\beta}(\partial)\overline{d}_{\beta\tau}(\partial) = (\Box + m^2)\delta_{\tau\beta}, \tag{14.132}$$

$(\overline{d}(\partial) \equiv d(\partial)\eta^{-1})$, and would be dependent on the operators ∂_μ

$$d(\partial) \equiv [d_{\tau\alpha}(\partial)] = \alpha + \alpha_\mu \partial^\mu + ... + \alpha_{\mu_1....\mu_n}\partial^{\mu_1}....\partial^{\mu_n} + ..., \tag{14.133}$$

where the coefficients $\alpha, \alpha_\mu....$ are matrices of the r-order. Note that in the most general case the quantities $Q_\beta(x)$ belong to the reducible representation.

Substitution of (14.133) into (14.131) leads to the following recurrent relations:

$$\left. \begin{array}{ll} \alpha - -mI, \quad i\alpha\Gamma_\mu - m\alpha_\mu = 0, \quad i(\alpha_\nu\Gamma_\mu + \alpha_\mu\Gamma_\nu) - 2m\alpha_{\mu\nu} = 2g_{\mu\nu}, \\ \sum^{(P)}(i\Gamma_{\mu_1}\alpha_{\mu_2,...\mu_l} - m\alpha_{\mu_1,...\mu_l}) = 0 \quad\quad l > 2, \end{array} \right\} \tag{14.134}$$

where the symbol $\sum^{(P)}$ denotes summation over all the terms resultant from various rearrangement of the indices.

To have an explicit form of the $d_{\alpha\beta}(\partial)$-operator, we need to know a maximum order of the involved derivatives (n_{max}). We prove a theorem that states that n_{max} is given by the relation [8]:

$$n_{max} = 2J, \quad\quad m \neq 0, \tag{14.135}$$

where J is a maximum spin value of different fields described by the field function Q_α.

In the process of the Lorentz transformation the operator $d_{\alpha\beta}(\partial)$ is transformed in the same way as $Q_\alpha \bigotimes Q_\beta$. Since the quantities Q_α describe the fields with a maximum spin J, they form an invariant space for the representation of a three-dimensional rotations group that is decomposed into irreducible representations $D_J, D_{J-1},....$ Because of this, the representation $Q_\alpha \bigotimes Q_\beta$ may be decomposed into $D_{2J}, D_{2J-1},...$ Similarly, $\partial_{\mu_1}...\partial_{\mu_l}$ are associated with a family of the irreducible representations $D_l, D_{l-1},..$ until some of the operators ∂_μ begin to combine into the scalar operator $\Box$. So, in $d(\partial)$ the term $\alpha_{\mu_1....\mu_l}\partial^{\mu_1}....\partial^{\mu_l}$ may appear only in the following form:

$$\alpha_{\mu_1....\mu_l}\partial^{\mu_1}....\partial^{\mu_l} = \alpha'_{\mu_1....\mu_{2J}}\partial^{\mu_1}....\partial^{\mu_{2J}}(\Box)^{\frac{l-2J}{2}}, \tag{14.136}$$

where $l - 2J > 0$. It is evident that the relation (14.136) goes to zero for odd $l - 2J$. Then from the last line of the recurrent formulae (14.134) we obtain that the quantity $\alpha_{\mu_1 \ldots \mu_l}$ is also zero for even positive values of $l - 2J$, that is,

$$\alpha_{\mu_1 \ldots \mu_l} = 0 \qquad l > 2J, \qquad (14.137)$$

which was to be proved.

Let us perform the procedure of transformation from the second-order equation to the first-order one (linearization) for a spinless particles field. We introduce the following quantities:

$$\psi_\mu(x) = \frac{i}{m} \partial_\mu \psi(x).$$

Then Eq. (14.1) is replaced by the system:

$$\left. \begin{array}{l} i\partial_\mu \psi(x) - m\psi_\mu(x) = 0, \\ i\partial_\mu \psi^\mu(x) - m\psi(x) = 0, \end{array} \right\} \qquad (14.138)$$

that may be represented in the form of the first-order matrix equation:

$$(i\beta^\mu \partial_\mu - m)\Psi(x) = 0, \qquad (14.139)$$

where

$$\Psi(x) = \begin{pmatrix} \psi_0(x) \\ \psi_1(x) \\ \psi_2(x) \\ \psi_3(x) \\ \psi(x) \end{pmatrix}$$

and the β-matrices have the following form:

$$\left. \begin{array}{c} \beta_0 = \begin{pmatrix} 0 & 0 & 0 & 0 & 1 \\ 0 & 0 & 0 & 0 & 0 \\ 0 & 0 & 0 & 0 & 0 \\ 0 & 0 & 0 & 0 & 0 \\ 1 & 0 & 0 & 0 & 0 \end{pmatrix}, \quad \beta_1 = \begin{pmatrix} 0 & 0 & 0 & 0 & 0 \\ 0 & 0 & 0 & 0 & -1 \\ 0 & 0 & 0 & 0 & 0 \\ 0 & 0 & 0 & 0 & 0 \\ 0 & 1 & 0 & 0 & 0 \end{pmatrix}, \\[3em] \beta_2 = \begin{pmatrix} 0 & 0 & 0 & 0 & 0 \\ 0 & 0 & 0 & 0 & 0 \\ 0 & 0 & 0 & 0 & -1 \\ 0 & 0 & 0 & 0 & 0 \\ 0 & 0 & 1 & 0 & 0 \end{pmatrix}, \quad \beta_3 = \begin{pmatrix} 0 & 0 & 0 & 0 & 0 \\ 0 & 0 & 0 & 0 & 0 \\ 0 & 0 & 0 & 0 & 0 \\ 0 & 0 & 0 & 0 & -1 \\ 0 & 0 & 0 & 1 & 0 \end{pmatrix}. \end{array} \right\} \qquad (14.140)$$

Eq. (14.139) was proposed in [9] and called the *Duffin-Kemmer equation*.

Since the resultant representation includes the representation $(0,0)$, according to which the scalar $\psi(x)$ is transformed, and the representation $(1/2, 1/2)$ defining the transformation law of the vector $\psi_\mu(x)$, then:

$$n_{max} = 2.$$

Solving the system of Eqs. (14.134) for $n_{max} = 2$, we obtain both the expression for $d(\partial)$:

$$d(\partial) = \frac{1}{2m}(\beta^\mu \beta^\nu + \beta^\nu \beta^\mu)\partial_\mu \partial_\nu - i\beta^\mu \partial_\mu - \frac{1}{m}(\Box + m^2), \qquad (14.141)$$

and the permutation relations determining the Duffin-Kemmer matrix algebra:

$$\beta^\mu \beta^\nu \beta^\lambda + \beta^\lambda \beta^\nu \beta^\mu = g^{\mu\nu} \beta^\lambda + g^{\lambda\nu} \beta^\mu. \qquad (14.142)$$

It is easily seen that the determinant of each β_μ-matrix is zero, that is, there are no matrices inverse to β_μ. This may be easily demonstrated without recourse to the explicit form of matrices β_μ, proceeding from their algebra only. Actually, from (14.142) it follows that:

$$\left. \begin{array}{ll} \beta_\mu \beta_\nu \beta_\mu = 0 & (\mu \neq \nu), \\ \beta_\mu^3 = \beta_\mu, & \text{(no summation on } \mu). \end{array} \right\} \tag{14.143}$$

As a consequence, in case the matrices inverse to β_μ were existing, the second relation (14.143) would give $\beta_\mu^2 = I$, and hence from the first relation it would follow that $\beta_\mu \beta_\nu = 0$ ($\mu \neq \nu$) and $\beta_\mu = 0$.

It is clear that the Duffin-Kemmer equation and relations of (14.142) are invariant with respect to the transformation:

$$\Psi'(x) = U\Psi(x), \qquad \beta'_\mu = U\beta_\mu U^{-1}, \tag{14.144}$$

where U is an arbitrary unitary matrix. Thus, the physical results obtained by the Duffin-Kemmer theory should be expressed in terms of the quantities to be invariant with respect to the transformations (14.144). For the expressions involving β-matrices this means that they may contain only such combinations of β-matrices, which may be reduced to tracks. Indeed, for the product of an arbitrary number of β-matrices the only possible quantity being invariant under the transformation (14.144) is a track as:

$$\text{Sp} \{\beta_{\mu_1} \beta_{\mu_2}, ..., \beta_{\mu_k}\} = \text{Sp} \{U^{-1} U \beta_{\mu_1} \beta_{\mu_2}, ..., \beta_{\mu_k}\} = \text{Sp} \{U \beta_{\mu_1} \beta_{\mu_2} ..... \beta_{\mu_k} U^{-1}\} =$$

$$= \text{Sp} \{U \beta_{\mu_1} U^{-1} U \beta_{\mu_2} U^{-1} ..... U \beta_{\mu_k} U^{-1}\} = \text{Sp} \{\beta'_{\mu_1} \beta'_{\mu_2} ..... \beta'_{\mu_k}\} = \text{inv},$$

where we have used:

$$\text{Sp}\,(ABC) = \text{Sp}\,(CAB) = \text{Sp}\,(BCA). \tag{14.145}$$

To calculate the tracks of β-matrices, we use the following procedure. Introduce the projective operators:

$$P = \frac{1}{3}\left(\beta^\mu \beta_\mu - 1\right), \qquad \overline{P} = \frac{1}{3}\left(4 - \beta^\mu \beta_\mu\right),$$

separating in the five-dimensional space of the $\Psi(x)$-functions one-dimensional space of the $\psi(x)$-scalars and four-dimensional space of the $\psi_\mu(x)$-vectors, respectively. Using (14.142), we make sure that the following relations:

$$\left. \begin{array}{ll} P\beta_\nu = \beta_\nu \overline{P}, & \overline{P}\beta_\nu \beta_\sigma = \beta_\nu \beta_\sigma \overline{P}, \\ P\beta_\nu \beta_\sigma = \beta_\nu \beta_\sigma P, & P\beta_\nu P = \overline{P}\beta_\nu \overline{P} = 0. \end{array} \right\} \tag{14.146}$$

are valid. Now with the help of (14.146) we may find the formula of a track for the product of any number n of β-matrices, So, for even n, after simple transformations, we have:

$$\text{Sp}\{\beta_{\mu_1} \beta_{\mu_2} ..... \beta_{\mu_n}\} = \text{Sp}\{\beta_{\mu_1} \beta_{\mu_2} ..... \beta_{\mu_n} P^2\} + \text{Sp}\{\beta_{\mu_1} \beta_{\mu_2} ..... \beta_{\mu_n} \overline{P}^2\} =$$

$$= \text{Sp}\{P\beta_{\mu_1} \beta_{\mu_2} ..... \beta_{\mu_n} P\} + \text{Sp}\{\overline{P}\beta_{\mu_1} \beta_{\mu_2} ..... \beta_{\mu_n} \overline{P}\} =$$

$$= \text{Sp}\{\overline{P}\beta_{\mu_2} \beta_{\mu_3} ..... \beta_{\mu_n} \beta_{\mu_1} P\} + \text{Sp}\{\overline{P}\beta_{\mu_1} \beta_{\mu_2} ..... \beta_{\mu_n} \overline{P}\} =$$

$$= \text{Sp}\{(\overline{P}\beta_{\mu_2} \beta_{\mu_3} \overline{P}) ..... (\overline{P}\beta_{\mu_n} \beta_{\mu_1} \overline{P})\} + \text{Sp}\{(\overline{P}\beta_{\mu_1} \beta_{\mu_2} \overline{P}) ..... (\overline{P}\beta_{\mu_{n-1}} \beta_{\mu_n} \overline{P})\}. \tag{14.147}$$

Taking into account that for the five-row β-matrices the following is satisfied:

$$\overline{P}\beta_\mu \beta_\sigma \overline{P} = g_{\mu\sigma},$$

finally we have:

$$\text{Sp}\{\beta_\mu\beta_\nu\beta_\lambda\beta_\kappa.....\beta_\rho\beta_\sigma\beta_\tau\} = g_{\mu\nu}g_{\lambda\kappa}....g_{\sigma\tau} + g_{\nu\lambda}...g_{\rho\sigma}g_{\tau\mu}. \tag{14.148}$$

In a similar way we may show that a track for an odd number of β-matrices equals zero.

As is seen in the first-order equation, the number of components of the spinless-particle wavefunction Q_α is found to be redundant. From Section 3.6 it follows that for the reduction in the number of independent components down to $2J + 1$ the field functions should meet the additional conditions contained in the equation of motion themselves. As follows from (14.143), with the use of $1 - \beta_\mu^2$ we can eliminate some of the β-matrices in Eq. (14.139). It is clear that the matrix β_0 is the best candidate for the elimination, since in this case we arrive at the relation containing no time derivative. Multiplication of Eq. (14.139) on the left by the matrix $1 - \beta_0^2$ results in the following:

$$(1 - \beta_0^2)\Psi(x) = \frac{i}{m}\beta_k\partial_k\beta_0^2\Psi(x). \tag{14.149}$$

Because $1 - \beta_0^2 = \text{diag}(0, 1, 1, 1, 0)$ and $\beta_0^2 = \text{diag}(1, 0, 0, 0, 1)$, expression (14.149) makes it possible to express three components of the field function $(1 - \beta_0^2)\Psi(x)$ in terms of the two components $\beta_0^2\Psi(x)$, that could be thought to be dynamically independent. These two components, however, prove to be equal in the rest frame of a particle, that is apparent from the additional condition of (14.149) written in the momentum representation. Then their relationship may be established in any Lorentz system, that is, the number of independent components is actually equal to 1.

Let us derive an equation for the "conjugate" field function $\overline{\Psi}(x)$. Performing Hermitian conjugation from (14.139), allowing for the properties of β-matrices, we get:

$$i\partial_\mu\Psi^\dagger(x)\beta^{\mu\dagger} + m\Psi^\dagger(x) = i\left(\partial_0\Psi^\dagger(x)\beta^0 + \partial_k\Psi^\dagger(x)\beta_k\right) + m\Psi^\dagger(x) = 0. \tag{14.150}$$

The obtained equation is not of the correct form as it is deliberately relativistically non-invariant. To eliminate this deficiency, we introduce the matrix $2\beta_0^2 - 1 = \eta_0$ with the apparent properties:

$$[\beta_0, \eta_0] = 0, \qquad \{\beta_k, \eta_0\} = 0, \tag{14.151}$$

and define

$$\overline{\Psi}(x) = \Psi^\dagger(x)\eta_0.$$

Taking into consideration Eq. (14.151), we get an equation for the conjugate field function:

$$i\frac{\partial\overline{\Psi}(x)}{\partial x^\mu}\beta^\mu + m\overline{\Psi}(x) = 0, \tag{14.152}$$

which has the correct form due to the manipulations undertaken. Inclusion of $\overline{\Psi}(x)$ allows for definition of the Lagrangian:

$$\mathcal{L} =: \frac{1}{2}\overline{\Psi}(x)(i\beta_\mu\partial^\mu - m)\Psi(x) - \frac{1}{2}(i\partial^\mu\overline{\Psi}(x)\beta_\mu + m\overline{\Psi}(x))\Psi(x) : \tag{14.153}$$

and construction of the dynamic variables on its basis. Of particular importance is a feature of the Lagrangians leading to the first-order wave equations: when the wave equations are satisfied, these Lagrangians are going to zero.

We consider a relativistic covariance of the Duffin-Kemmer equation. The physics expressed with the help of any relativistic equation should be independent of the selected Lorentz frame of reference. Because of this, an adequate description of physical phenomena necessitates that the equation itself be similarly invariant relative to the selection of a frame of reference. Let us demonstrate that the Duffin-Kemmer equation is in a form invariant with respect to the Poincare transformations:

$$x' = \Lambda x + a,$$

on condition that the transformed field function is determined by:

$$\Psi'(x') = S(\Lambda)\Psi(x) = S(\Lambda)\Psi[\Lambda^{-1}(x'-a)], \qquad (14.154)$$

and the matrix Λ meets special requirements, the finding of which is our principal objective. Invariance of this equation means that in a new frame of reference the field function $\Psi'(x')$ obeys the equation:

$$(i\beta^\mu \partial'_\mu - m)\Psi'(x') = 0. \qquad (14.155)$$

It should be noted that the Lorentz transformations do not "touch upon" the matrices β^μ. Using the relations (14.154) and

$$\frac{\partial}{\partial x^\mu} = \frac{\partial x'^\nu}{\partial x^\mu} \frac{\partial}{\partial x'^\nu} = \Lambda^\nu_\mu \partial'_\nu, \qquad (14.156)$$

Eq. (14.139) may be rewritten as:

$$i\Lambda^\nu_\mu \beta^\mu \partial'_\nu [S^{-1}(\Lambda)\Psi'(x')] - mS^{-1}(\Lambda)\Psi'(x') - 0. \qquad (14.157)$$

After the multiplication of Eq. (14.157) by $S(\Lambda)$ on the left we have:

$$iS(\Lambda)\Lambda^\nu_\mu \beta^\mu S^{-1}(\Lambda)\partial'_\nu \Psi'(x') - m\Psi'(x') = 0. \qquad (14.158)$$

Consequently, the Duffin-Kemmer equation will be relativistically covariant when the transformation matrix for the field function (14.154) satisfies the condition:

$$\Lambda^\nu_\mu \beta^\mu - S^{-1}(\Lambda)\beta^\nu S(\Lambda). \qquad (14.159)$$

With the use of the Poincare transformation property

$$\Lambda^{\nu\mu} \Lambda_{\sigma\mu} = \delta^\nu_\sigma,$$

we can easily verify that the matrices:

$$\beta'^\nu = \Lambda^\nu_\mu \beta^\mu,$$

obey the same permutation relations as the matrices $\beta\mu$.

By a system of four Eqs. (14.159) we can determine the transformation matrix for the field functions of the Duffin-Kemmer equation in the case of the Poincare transformations. To illustrate, we perform the procedure for the case of space inversion, that is, in the case when the matrix Λ is of the following form:

$$\Lambda = \begin{pmatrix} 1 & 0 & 0 & 0 \\ 0 & -1 & 0 & 0 \\ 0 & 0 & -1 & 0 \\ 0 & 0 & 0 & -1 \end{pmatrix}. \qquad (14.160)$$

Multiplying the equality (14.159) on the left by $S(\Lambda)$, we bring it into the form:

$$\beta_\nu S(\Lambda) = S(\Lambda)\Lambda^\mu_\nu \beta_\mu. \qquad (14.161)$$

Now with allowance made for the explicit form of the Λ matrix we get:

$$\left.\begin{array}{ll}\beta_0 S(\Lambda) = S(\Lambda)\beta_0, & \beta_1 S(\Lambda) = -S(\Lambda)\beta_1, \\ \beta_2 S(\Lambda) = -S(\Lambda)\beta_2, & \beta_3 S(\Lambda) = -S(\Lambda)\beta_3. \end{array}\right\} \tag{14.162}$$

The relations (14.162) are satisfied if $S(\Lambda)$ is selected as follows: $S(\Lambda) = \eta_0$.

It is possible that the reader is inspired with the illusion that under the transition from the second- to the first-order equation the number of roots of the associated characteristic equation is reduced to one, with elimination of the unwanted negative root. But this is not the case, since *"Nothing originates from nothing."* Indeed, the substitution of the de Broglie wave:

$$\Psi(x) = \varphi(k)\exp(-ikx),$$

into Eq. (14.139) gives

$$(\hat{k} - m)\varphi(k) = 0, \tag{14.163}$$

where $\hat{k} = \beta^\mu k_\mu$. By Eq. (14.142), for any vector a_μ the following relation is satisfied:

$$\hat{a}(\hat{a}^2 - a^2) = 0, \tag{14.164}$$

then from (14.163) and (14.164) follows:

$$k^2 - m^2 = 0,$$

As before, there are two solutions with opposite signs of the energy.

Proceeding from the same ideology as in the case of the second-order equation, we can perform a series expansion of the operator field functions in terms of the solutions for the free equation:

$$\Psi(x) = \frac{1}{(2\pi)^{3/2}} \int \frac{d^3k}{\sqrt{2k_0}}[a(\mathbf{k})\varphi^{(+)}(k)\exp(-ikx) + b^*(\mathbf{k})\varphi^{(-)}(k)\exp(ikx)], \tag{14.165}$$

$$\overline{\Psi}(x) = \frac{1}{(2\pi)^{3/2}} \int \frac{d^3k}{\sqrt{2k_0}}[a^*(\mathbf{k})\overline{\varphi}^{(+)}(k)\exp(ikx) + b(\mathbf{k})\overline{\varphi}^{(-)}(k)\exp(-ikx)], \tag{14.166}$$

where the particles (antiparticles) production $a^*(\mathbf{k})$ $(b^*(\mathbf{k}))$ and annihilation $a(\mathbf{k})$ $(b(\mathbf{k}))$ operators satisfy the commutation relations (14.123), and $\varphi^{(\pm)}(k)$ obey the normalization conditions:

$$\overline{\varphi}^{(\pm)}(k)\beta_0\varphi^{(\pm)}(k) = \pm 2k_0. \tag{14.167}$$

It is easy to derive nonzero commutation relations for the operator field functions:

$$[\Psi_\alpha(x), \overline{\Psi}_\beta(x')] = -id_{\alpha\beta}(\partial)\Delta_0(x - x'; m^2) \tag{14.168}$$

and the expression for the causal Green function:

$$D_{\alpha\beta}^c(x) = d_{\alpha\beta}(\partial)\Delta_0^c(x) = \frac{1}{(2\pi)^4} \int d^4k \frac{d_{\alpha\beta}(k)}{m^2 - k^2 - i\epsilon}\exp(-ikx). \tag{14.169}$$

The fact that the operator $d_{\alpha\beta}(\partial)$ is so frequently used in our expressions indicates its exclusive role in the Duffin-Kemmer theory. Note that this operator is also very important for the particles with a nonzero spin value. Point to one more activity sphere of the operator $d_{\alpha\beta}(\partial)$. It is connected to computing of the bilinear combinations of the form:

$$\mathcal{P}^{(\pm)}(k) = \varphi^{(\pm)}(k)\overline{\varphi}^{(\pm)}(k), \tag{14.170}$$

which is quite often found in calculations of the scattering cross sections. As $\mathcal{P}^{(+)}(k)$ and $\mathcal{P}^{(-)}(k)$ represent solutions of the equations:

$$(\hat{k} - m)\mathcal{P}^{(+)}(k) = 0, \tag{14.171}$$

$$(\hat{k} + m)\mathcal{P}^{(-)}(k) = 0, \tag{14.172}$$

they should be coincident, within a numerical factor, with the operators $d(k)$ and $\bar{d}(k)$:

$$\mathcal{P}^{(+)}(k) = c_+ d(k), \qquad \mathcal{P}^{(-)}(k) = c_- \bar{d}(k). \tag{14.173}$$

The factors c_+ and c_- may be determined from the normalization conditions (14.167) as follows:

$$\overline{\varphi}_\alpha^{(+)}(k)(\beta_0)_{\alpha\tau}\varphi_\tau^{(+)}(k) = \mathrm{Sp}\{\overline{\varphi}_\alpha^{(+)}(k)(\beta_0)_{\alpha\tau}\varphi_\tau^{(+)}(k)\} =$$
$$= \mathrm{Sp}\{(\beta_0)_{\alpha\tau}\varphi_\tau^{(+)}(k)\overline{\varphi}_\alpha^{(+)}(k)\} = \mathrm{Sp}\{(\beta_0)\mathcal{P}^{(+)}(k)\} = 2k_0, \tag{14.174}$$

$$\overline{\varphi}^{(-)}(k)\beta_0\varphi^{(-)}(k) = \mathrm{Sp}(\beta_0\mathcal{P}^{(-)}(k)) = -2k_0. \tag{14.175}$$

Using the track calculation rules for the Duffin-Kemmer matrices, with the help of (14.174) and (14.175) we get

$$c_+ = c_- = \frac{1}{2}.$$

Considering that the particles are at the mass surface, we finally arrive at:

$$\varphi^{(+)}(k)\overline{\varphi}^{(+)}(k) = \frac{\hat{k}(\hat{k} + m)}{2m}, \qquad \varphi^{(-)}(k)\overline{\varphi}^{(-)}(k) = \frac{\hat{k}(\hat{k} - m)}{2m}. \tag{14.176}$$

With the use of (14.153), (14.165), and (14.166), we can calculate the energy, momentum, charge, and other dynamic invariants of a spinless particle. In so doing we obtain the same expressions as in the case of the Klein-Gordon equation. Because of this, it may be inferred that the descriptions of a free scalar field both in the Duffin-Kemmer and Klein Gordon form are equivalent. Also, this equivalence is retained on inclusion of the interactions as both formalisms result in the same values of the quantities measured in the process of experiments.

Thus, even though the equation order was lowered, the Duffin-Kemmer formalism has not given a simpler pattern of a scalar field compared to the Klein-Gordon formalism. However, our time was not wasted. It was demonstrated that the characteristics of the arbitrary-spin particles are expressed in terms of the corresponding quantities of a scalar field. Moreover, this makes it possible: (i) to simplify, using the formalism of the first-order equations, a derivation procedure for the commutation relations, causal Green function, and other quantities associated with the arbitrary-spin particles; and (ii) to formulate the relativistic correspondence principle that is very useful in practical calculations: *scattering cross sections of spin particles σ_s include the cross sections of the scalar particles σ_0, that is,*

$$\lim_{s \to 0} \sigma_s = \sigma_0.$$

15

Particles with spin 1/2

*And sure enough, the forest subjects hadn't winked
twice, when Bruin the First, the new governor, arrived. He
reached his residence early in the morning of St. Michael's Day
and immediately resolved: tomorrow would see bloodshed. What
made him adopt this decision is unknown: for as a matter of
fact he was not really spiteful—just a brute, that's all.*
M.E. Saltykov Shchedrin, "Bears in Government"

15.1 Dirac equation

As it was shown, the single-particle Klein-Gordon equation suffers from two disadvantages:
(i) the probability density is not positively defined; and (ii) the existence of the solutions
with a negative value of the energy. Trying to overcome these problems, in 1928 P. Dirac
proposed an equation that was later named after him. The Dirac equation is of special
importance because it allows for the description of such fundamental particles of matter as
leptons and quarks.

At least now one easily understands the train of reasoning that led Dirac to his equation
[10]. To avoid the appearance of negative probabilities, it is essential that an expression for
the probability density be free from the time derivatives. Consequently, a wave equation
should involve the time derivatives of the first order, and not higher. Allowing for that a
relativistic covariance requires complete symmetry with respect to all the space and time
coordinates, the equation should be of the first order for the space derivatives as well. If, in
so doing, one takes on arms the principle of superposition, then the conclusion follows—the
wavefunction should satisfy the first-order linear differential equation for all four coordinates. Then with due regard for the Klein-Gordon equation, we lead to the problem that
is already known

$$(\Box + m^2)\psi_r = d_{rs}(\partial)\,(i\gamma_\mu\partial^\mu - m)_{s\kappa}\,\psi_\kappa = 0. \tag{15.1}$$

Its solution necessitates specification of the order for the operator $d_{rs}(\partial)$. It is natural
that in this situation unity has the greatest attractiveness and Dirac's choosing fell upon
it. Then it becomes clear that the operator $d_{rs}(\partial)$ should be looked for in the form

$$d_{rs}(\partial) = -\,(i\gamma_\mu\partial^\mu + m)_{rs}\,.$$

Since we have:

$$(i\gamma_\mu\partial^\mu + m)\,(-i\gamma_\nu\partial^\nu + m) = \gamma_\mu\gamma_\nu\partial^\mu\partial^\nu + m^2 =$$

$$= \frac{1}{2}(\gamma_\mu\gamma_\nu + \gamma_\nu\gamma_\mu)\partial^\mu\partial^\nu + m^2,$$

then Eq. (15.1) is satisfied when the matrices γ_μ prove to be meeting the anticommutation relations:

$$\{\gamma_\mu, \gamma_\nu\} = 2g_{\mu\nu}. \tag{15.2}$$

A set of hypercomplex numbers, for which the relation (15.2) is true, forms the Clifford algebra that exists for any space supplied by the metric $g_{\mu\nu}$. If the space dimension is n, the dimension of algebra (that is, the number of independent elements involved in the expansion of any $n \times n$-matrix) is 2^n. Note, that considering γ_μ-matrices, there is no need to assume that they possess a certain property with respect to the Hermitian conjugation [11].

By Dirac it was supposed that a relativistic free electron is described by the equation:

$$(i\gamma_\mu \partial^\mu - m)\psi(x) = 0. \tag{15.3}$$

It seems that the electron may be equally well described by the following equation:

$$(i\gamma_\mu \partial^\mu + m)\psi(x) = 0.$$

However, a physical meaning of the solutions and the conclusions made on their basis are the same for both equations. This is conditioned by the fact that the relation of Clifford's algebra (15.2) is invariant in respect of the transformation:

$$\gamma_\mu \to -\gamma_\mu.$$

The condition (15.2) defines the matrices γ_μ ambiguously. When the wavefunction ψ is subjected by the transformation:

$$\psi' = U\psi, \tag{15.4}$$

and for the matrices γ^μ a transformation of similarity is performed:

$$\gamma'_\mu = U\gamma_\mu U^{-1}, \tag{15.5}$$

where U is an arbitrary unitary matrix, the form of both the Dirac equation and relation (15.2) remains unaltered. Therefore, the γ-matrices are determined to within a unitary transformation, and a particular representation of these matrices may be chosen by different ways. At the same time, the opposite is true [12] (the fundamental theorem for the γ-matrix): *if two sets of matrices γ_μ and γ'_μ satisfy the relations (15.2), then there exists the matrix U relating them by the relation (15.5).*

Let us define a rank of the matrices γ_μ. To this end, we find a determinant of the product $\gamma_\mu\gamma_\nu$ for $\mu \neq \nu$. Considering that:

$$\gamma_\mu\gamma_\nu = -I\gamma_\nu\gamma_\mu,$$

where I is the unit matrix, we get:

$$\det(\gamma_\mu\gamma_\nu) = \det(\gamma_\mu)\det(\gamma_\nu) = \det(-I)\det(\gamma_\nu)\det(\gamma_\mu). \tag{15.6}$$

Since the determinants represent ordinary c-numbers, then from Eq. (15.6) follows

$$\det(-I) = 1,$$

or

$$(-1)^n = 1.$$

It is obvious that the value $n = 2$ is excluded because there are no linearly independent two-row matrices satisfying the condition (15.2). Linearly independent two-row matrices are represented by three Pauli matrices $\sigma_{1,2,3}$ and the unit matrix that naturally commutes (and not anticommutes) with all $\sigma_{1,2,3}$. In case $n = 4$ there is a possibility to construct matrices with the desired properties. By a straightforward checking procedure we can see that the following two choices of the gamma-matrices, namely the Dirac-Pauli representation:

$$\gamma^k = \begin{pmatrix} 0 & \sigma_k \\ -\sigma_k & 0 \end{pmatrix}, \qquad \gamma^0 = \begin{pmatrix} I & 0 \\ 0 & -I \end{pmatrix}, \tag{15.7}$$

and Weyl representation (chiral representation):

$$\gamma^k = \begin{pmatrix} 0 & -\sigma_k \\ \sigma_k & 0 \end{pmatrix}, \qquad \gamma^0 = \begin{pmatrix} 0 & I \\ I & 0 \end{pmatrix}, \tag{15.8}$$

satisfy all the above-mentioned requirements, being coupled with each other by the unitary transformation. In both representations γ_0 is the Hermitian matrix, whereas the matrices γ_k are anti-Hermitian.

It should be noted that the matrix $\gamma_5 \equiv \gamma^5 \equiv i\gamma^0\gamma^1\gamma^2\gamma^3$ is diagonal in the representation (15.8):

$$\gamma_5 = \begin{pmatrix} I & 0 \\ 0 & -I \end{pmatrix}, \tag{15.9}$$

while in the representation (15.7) it has the form:

$$\gamma_5 = \begin{pmatrix} 0 & I \\ I & 0 \end{pmatrix}. \tag{15.10}$$

From the γ-matrices we may form 16 linearly independent elements:

$$\Gamma_j = I, \gamma_\mu, \sigma_{\mu\nu}, \varepsilon_{\mu\nu\lambda\sigma}\gamma^\nu\gamma^\lambda\gamma^o, \gamma_5, \tag{15.11}$$

where

$$\sigma_{\mu\nu} = \frac{i}{2}(\gamma_\mu\gamma_\nu - \gamma_\nu\gamma_\mu).$$

Using the same arguments as in the case of the Duffin-Kemmer equation, we can obtain an equation for the "conjugate" wavefunction:

$$i\partial^\mu\overline{\psi}(x)\gamma_\mu + m\overline{\psi}(x) - 0, \tag{15.12}$$

where $\overline{\psi}(x) = \psi^\dagger(x)\gamma^0$. Equations of motion follow from the Lagrangian density:

$$\mathcal{L} = \frac{1}{2}\overline{\psi}(x)(i\gamma_\mu\partial^\mu - m)\psi(x) - \frac{1}{2}(i\partial^\mu\overline{\psi}(x)\gamma_\mu + m\overline{\psi}(x))\psi(x), \tag{15.13}$$

with independent variations in terms of $\overline{\psi}(x)$ and $\psi(x)$. It should be noted that the Lagrangian becomes zero when the equations of motion hold.

Multiplying Eq. (15.12) on the right by $\psi(x)$ whereas Eq. (15.3) on the left by $\overline{\psi}(x)$ and adding the obtained equations, we get the continuity equation:

$$\partial^\mu j_\mu(x) = 0, \tag{15.14}$$

where $j_\mu(x)$, the four-dimensional vector of the current density, is given by the following expression:

$$j_\mu(x) = \overline{\psi}(x)\gamma_\mu\psi(x). \tag{15.15}$$

As:

$$j_0(x) \equiv \rho(x) = \overline{\psi}(x)\gamma_0\psi(x) = \sum_{i=0}^{3} \mid \psi_i \mid^2, \qquad (15.16)$$

the probability density turns to be positively defined, and the single-particle Dirac equation overcomes one of the barriers that the Klein-Gordon equation could not leave behind. But, as far as the appearance of negative energies is concerned, this trouble has remained common for both equations.

We rewrite the Dirac equation (15.3) in the Schrödinger-like form:

$$i\partial_t\psi(x) = H'\psi(x), \qquad (15.17)$$

where

$$H' = i(\boldsymbol{\alpha} \cdot \boldsymbol{\nabla}) + m\gamma_0, \qquad \alpha_k = \gamma_0\gamma_k.$$

The plane de Broglie wave

$$\psi(x) = \psi(p)\exp\left(-ipx\right) \qquad (15.18)$$

should be the solution of Eq. (15.3) in a free case. This is true on condition that the bispinor $\psi(p)$ obeys the equation:

$$H\psi(p) = p_0\psi(p), \qquad (15.19)$$

with the Hamiltonian function:

$$H = (\boldsymbol{\alpha} \cdot \mathbf{p}) + m\gamma_0. \qquad (15.20)$$

From (15.19) it follows that for a Dirac particle at rest the equation in eigenvalues is fulfilled:

$$m\gamma_0\psi(p) = p_0\psi(p).$$

With the use of a diagonal representation for the γ_0-matrix (15.7) we come to the conclusion that we have two solutions with the energy equal to m and two solutions with the energy equal to $-m$. To obtain a result in an arbitrary reference frame, we have to use the solution existence condition for a system of homogeneous linear equations (15.19), that is, the determinant composed of the coefficients under unknown functions $\psi(p)$ should become zero. Calculating the determinant, we get:

$$p_0^2 = \mathbf{p}^2 + m^2,$$

or

$$p_0 = \pm E_p,$$

where $E_p = \sqrt{\mathbf{p}^2 + m^2}$. Thus, in the Dirac equation the solutions with the negative energy are also present.

At the same time, the achievements of the Dirac equation are so impressive (it gave the correct values for: spin, spin magnetic-dipole moment for the electron, fine-structure of energy levels for hydrogen, etc.) that it was impossible to bury it in oblivion. The only possibility was to reconstruct its ideology. In 1930 Dirac has put forward a so-called theory of holes [13]. According to this theory, the vacuum is not actually emptiness, as we perceive it, being rather a system, where all the negative energy levels have two electrons with differently directed spin projections—similar to the situation in dielectrics with their valence zone completely filled by the electrons. Since electrons are governed by the Pauli principle, none of them is able to make a series of transitions from a positive level to negative-energy levels, falling to the values of $E = -\infty$ and donating infinitely high energy in the form of electromagnetic radiation. Following Dirac's example, the vacuum becomes a complex system, in a sense including a part of the universe—electrons occupying all the levels with negative energy. Dirac's vacuum is no longer a complete emptiness of Democritus; it becomes a prototype of the vacuum of the modern quantum field theory.

What happens when an electron from a negative-energy level is excited to go to the positive-energy one? A vacant place, referred to as a hole, with the energy $-E_p$ should be immediately filled by the electron having the positive energy E_p and emitting the energy $2E_p$ in the process. Then a hole is considered as the absence of the electron or presence of the particle with the positive charge $+|e|$. Considering that normally all the negative-energy levels are filled, a hole is associated with a decrease of the negative or increase of the positive energy. So, a hole behaves like a positively charged particle with the positive energy. If such a particle is found, it may be filled by the positive-energy electron. Such a transition should represent the process when positively- and negatively-charged particles mutually destroy each other. At first Dirac posited that a positively charged "electron" (antielectron) was represented by the proton. As atoms include both electrons and protons, this statement could be verified without difficulty. It suffices to calculate a probability of the transition:
$$electron + hole \rightarrow vacuum,$$
derive the lifetime of the normal matter t_m, and compare the results obtained with the experiment. The value of t_m calculated by Oppenheimer [14] was on the order of 10^{-10} s, that, to put it mildly, is at variance with the stability of our world. Fortunately, soon all these problems were solved due to the experimental results. C. Andersen [15] has reported about the discovery of a particle in cosmic rays, possessing the same mechanical characteristics (spin, mass, etc.) as an electron but having the electromagnetic characteristics equal in the value and opposite in sign. This particle is actually an antielectron and later it has been called the positron. The prediction of the positron, and antiparticles in general, was a historic triumph for the Dirac theory. Symmetry between the matter and antimatter was for the first time accepted in physics to become one of the major ideas in a theory of elementary particles.

Presenting the multiple-particle formalism, the holes theory leads to significantly complicated calculations in practice. To illustrate, a wavefunction describing at least one electron should take into account all the filled states with the negative energy. Though the holes theory is close in spirit to the quantum field theory we still have to refuse it. And this is caused not only by the elaborate computations. For example, in the case of bosons the holes theory is useless because the reasoning of Dirac was based on the Pauli principle. On the other hand, even for fermions, the concept of the unobserved infinite charged sea is a bit mystical. Because of this, the use of the second quantization to interpret the wave equations of the quantum theory is physically more reasonable.

15.2 Calculating γ-matrix tracks

As the Dirac equation is invariant under the transformations (15.5) and (15.6), the dynamic observables should be also invariant with respect to these transformations. In other words all the products of γ-matrices in the dynamic observables should be reduced to the tracks (spurs), which are the only invariants relative to the transformation of similarity (15.6).

Let us find the formulae to calculate the spurs of the γ-matrices. First, we demonstrate that the spur of the odd number of these matrices is zero. For simplicity, we perform this with the help of the matrix γ_5. Using (14.145) and

$$\{\gamma_\mu, \gamma_5\} = 0, \qquad \gamma_5^2 = 1,$$

we get:

$$\mathrm{Sp}\,\{\gamma_{\mu_1}\gamma_{\mu_2}.....\gamma_{\mu_n}\} = \mathrm{Sp}\,\{\gamma_5\gamma_5\gamma_{\mu_1}\gamma_{\mu_2}.....\gamma_{\mu_n}\} = \mathrm{Sp}\,\{\gamma_5\gamma_{\mu_1}\gamma_{\mu_2}.....\gamma_{\mu_n}\gamma_5\} =$$

$$= (-1)^n\mathrm{Sp}\,\{\gamma_{\mu_1}\gamma_{\mu_2}.....\gamma_{\mu_n}\} = 0. \tag{15.21}$$

If this product contains an even number of matrices that is equal to n, then the permutation relations enable us to reduce the spur of this product to a sum of the product spurs containing $n-2$ matrices only. For instance, for $n=2$ we have:

$$\mathrm{Sp}\,(\gamma_\mu\gamma_\nu) = \mathrm{Sp}\,(-\gamma_\mu\gamma_\nu + 2g_{\mu\nu}).$$

whence:

$$\mathrm{Sp}\,(\gamma_\mu\gamma_\nu) = 4g_{\mu\nu}. \tag{15.22}$$

Next, we find $\mathrm{Sp}\,(\gamma_\mu\gamma_\nu\gamma_\lambda\gamma_\sigma)$. Taking account of:

$$\left.\begin{array}{l} \gamma_\mu\gamma_\nu\gamma_\lambda\gamma_\sigma = 2g_{\mu\nu}\gamma_\lambda\gamma_\sigma - \gamma_\nu\gamma_\mu\gamma_\lambda\gamma_\sigma, \\ \gamma_\nu\gamma_\mu\gamma_\lambda\gamma_\sigma = 2g_{\mu\lambda}\gamma_\nu\gamma_\sigma - \gamma_\nu\gamma_\lambda\gamma_\mu\gamma_\sigma, \\ \gamma_\nu\gamma_\lambda\gamma_\mu\gamma_\sigma = 2g_{\mu\sigma}\gamma_\nu\gamma_\lambda - \gamma_\nu\gamma_\lambda\gamma_\sigma\gamma_\mu, \end{array}\right\} \tag{15.23}$$

we obtain:

$$\gamma_\mu\gamma_\nu\gamma_\lambda\gamma_\sigma = 2g_{\mu\nu}\gamma_\lambda\gamma_\sigma - 2g_{\mu\lambda}\gamma_\nu\gamma_\sigma + 2g_{\mu\sigma}\gamma_\nu\gamma_\lambda - \gamma_\nu\gamma_\lambda\gamma_\sigma\gamma_\mu. \tag{15.24}$$

In this way the final result is as follows:

$$\mathrm{Sp}\,(\gamma_\mu\gamma_\nu\gamma_\lambda\gamma_\sigma) = 4g_{\mu\nu}g_{\lambda\sigma} - 4g_{\mu\lambda}g_{\nu\sigma} + 4g_{\mu\sigma}g_{\nu\lambda}. \tag{15.25}$$

To find the spur of the arbitrary number of matrices, we use the equation:

$$\mathrm{Sp}\,\{\gamma_{\mu_1}\gamma_{\mu_2}.....\gamma_{\mu_{2n}}\} = 4\sum(-1)^N g_{\nu\lambda}g_{\sigma\kappa}...., \tag{15.26}$$

where $\nu, \lambda, \sigma, \kappa, ...$ is a particular combination of the indices $\mu_1, \mu_2, ..., \mu_{2n}$ with the following prescription [16]. Each matrix γ_{μ_i} is associated with a point of a circle, and the matrices have the same order as in the left-hand side of the equality (15.26). Next, all the points are pairwise joined by the straight lines, and each line joining the points, for instance, ν and λ is associated with the factor $g_{\nu\lambda}$. All possible connections of the points are taken, each connection in the sum (15.26) being associated with the term: $(-1)^N g_{\nu\lambda}g_{\sigma\kappa}....$ where N is the number of intersection points for the lines. From (15.26) we can easily derive:

$$\mathrm{Sp}\,(\gamma_5\gamma_\mu\gamma_\nu\gamma_\lambda\gamma_\sigma) = 4i\varepsilon_{\mu\nu\lambda\sigma}. \tag{15.27}$$

15.3   Relativistic covariance

Let us investigate the properties of the Dirac wavefunction $\psi(x)$ under the Poincare transformations:

$$x'_\mu = \Lambda^\nu_\mu x_\nu + a_\mu.$$

In a new reference system $\psi(x)$ is determined by the relation:

$$\psi'_r(x) = \sum_{\kappa=1}^{4} S_{r\kappa}(\Lambda)\psi_\kappa[\Lambda^{-1}(x-a)], \tag{15.28}$$

where the matrix indices are written in the explicit form. Carrying out the similar reasoning as in the case of the Duffin-Kemmer equation, we lead to the conclusion that the Dirac equation is relativistically covariant if and only if the following condition is fulfilled:

$$S^{-1}(\Lambda)\gamma_\mu S(\Lambda) = \Lambda_\mu^\nu \gamma_\nu. \tag{15.29}$$

Evidently, the matrices $\gamma'_\mu = \Lambda_\mu^\nu \gamma_\nu$ are also satisfying the relations of Clifford's algebra. A system of Eqs. (15.29) is determining the matrix S to within the numerical factor, whose value is significant for finding of the matrix that transforms the wavefunction $\overline{\psi}$. This factor is calculated as follows. Performing the Hermitian conjugation procedure for both sides of the relation (15.29), we have

$$(S^{-1}\gamma^\mu S)^\dagger = (\Lambda_\nu^\mu \gamma^\nu)^\dagger = \left(\Lambda^{\mu 0}\gamma^0 - \sum_{l=1}^{3}\Lambda^{\mu l}\gamma^l\right)^\dagger = \Lambda^{\mu 0}\gamma^0 + \sum_{l-1}^{3}\Lambda^{\mu l}\gamma^l. \tag{15.30}$$

Multiplying (15.30) by γ_0 on the left and on the right, and considering that the properties of γ^μ with respect to the Hermitian conjugation may be written in the compact form.

$$(\gamma^\mu)^\dagger = \gamma^0\gamma^\mu\gamma^0,$$

we get:

$$\gamma^0\left(\Lambda^{\mu 0}\gamma^0 + \sum_{l-1}^{3}\Lambda^{\mu l}\gamma^l\right)\gamma^0 = \Lambda_\nu^\mu\gamma^\nu = S^{-1}\gamma^\mu S \tag{15.31}$$

and

$$\gamma^0(S^{-1}\gamma^\mu S)^\dagger\gamma^0 = (\gamma^0 S^\dagger\gamma^0)\gamma^\mu(\gamma^0 S^\dagger\gamma^0)^{-1}. \tag{15.32}$$

Combining the obtained relations, as a result we have:

$$(S\gamma^0 S^\dagger\gamma^0)\gamma^\mu(S\gamma^0 S^\dagger\gamma^0)^{-1} = \gamma^\mu,$$

that is, the matrix $S\gamma^0 S^\dagger\gamma^0$ commutes with all the matrices γ^μ and hence it is divisible by the unit matrix:

$$S\gamma^0 S^\dagger\gamma^0 = aI. \tag{15.33}$$

Due to the Hermitian properties of γ^0, the constant a is real. Later it will be shown that for the matrix S the normalization det $S = 1$ is always possible. In this case $a^4 = 1$ and $a = \pm 1$. Next, we consider the identity:

$$S^\dagger S = S^\dagger\gamma^0\gamma^0 S = a\gamma^0 S^{-1}\gamma^0 S = a\gamma^0\Lambda_\nu^0\gamma^\nu = a\left(\Lambda_0^0 I - \sum_{l=1}^{3}\Lambda_l^0\gamma^0\gamma^l\right). \tag{15.34}$$

As the matrix S is nonsingular and nonzero, the matrix $S^\dagger S > 0$ and its eigenvalues are real and positive numbers, that is, Sp $(S^\dagger S) > 0$. Then, taking account of Sp $(\gamma^0\gamma^l) = 0$, we get:

$$\text{Sp}\,(S^\dagger S) = 4a\Lambda^{00} > 0. \tag{15.35}$$

As follows from (15.35), at orthochronous transformations ($\Lambda^{00} \geq 0$)

$$a = 1,$$

while at the nonorthchronous ones ($\Lambda^{00} \leq 0$)

$$a = -1.$$

Using the obtained results, we can state a law for the transformation of the conjugate bispinor $\overline{\psi}$. And taking into consideration (15.33), we have:

$$\overline{\psi}' = (\psi')^\dagger \gamma^0 = \psi^\dagger S^\dagger \gamma^0 = a\psi^\dagger \gamma^0 S^{-1} = a\overline{\psi} S^{-1}. \tag{15.36}$$

Let us demonstrate how to work the definition scheme of the matrix S for different types of the Poincare transformations. We begin with the improper transformations. According to (15.29), the matrix $S(P)$ of the space inversion should satisfy the relations:

$$S^{-1}(P)\gamma^0 S(P) = \gamma^0, \qquad S^{-1}(P)\gamma^k S(P) = -\gamma^k. \tag{15.37}$$

Consequently, for $S(P)$ we can choose $\pm\gamma^0$ or $\pm i\gamma^0$.

It is not so easy to find the matrix for the time inversion with the help of (15.35) as the time-reversal operation is described by the antiunitary operator (see, Section 5.2). The definition of this matrix, as well as the selection of phase in the space-inversion transformation, will be treated after the consideration of the charge conjugation operation.

Let us establish the form of S matrix at four-dimensional rotations. We consider the case of infinitesimal transformations:

$$x'_\mu = x_\mu + \epsilon\omega^\nu_\mu x^\mu, \tag{15.38}$$

where $\epsilon \ll 1$ and $\omega_{\nu\mu} = -\omega_{\mu\nu}$. The representation associated with the transformation (15.38) is infinitesimally different from the identity one, in the first-order with respect to ϵ being of the form:

$$S(I + \epsilon\omega) = I + \epsilon G, \qquad S^{-1}(I + \epsilon\omega) = S(I - \epsilon\omega) = I - \epsilon G. \tag{15.39}$$

Substituting (15.39) into (15.29), we obtain

$$S^{-1}\gamma_\mu S = (I - \epsilon G)\gamma_\mu(I + \epsilon G) = \gamma_\mu + \epsilon(\gamma_\mu G - G\gamma_\mu) = \Lambda^\nu_\mu \gamma_\nu = \gamma_\mu + \epsilon\omega^\nu_\mu \gamma_\nu. \tag{15.40}$$

Because of this, the matrix G should have the following property:

$$\gamma_\mu G - G\gamma_\mu = \omega^\nu_\mu \gamma_\nu. \tag{15.41}$$

Normalization det $S = 1$ requires $\det(I + \epsilon G) = 1 + \epsilon\text{Sp } G = 1$, that is, Sp $G = 0$. By the addition of a matrix being divisible by the unit matrix to G, we can always reduce the spur of G to zero. The fulfillment of (15.41) and Sp $G = 0$ is guaranteed, provided the matrix G is determined by:

$$G = \frac{i}{4}\omega^{\mu\nu}\sigma_{\mu\nu}. \tag{15.42}$$

In the case of a infinitesimal space rotation by an angle ϵ about the axis x_i, it is clear that:

$$\omega_{jk} = \epsilon\varepsilon_{jki}n_i,$$

where n_i is an unit vector. According to (15.42), the transformation generator in this case is as follows:

$$G(n_i) = \frac{i}{2}\Sigma_i n_i, \tag{15.43}$$

where

$$\Sigma_i = \frac{1}{2}\varepsilon_{ijk}\sigma_{jk} = \begin{pmatrix} \sigma_i & 0 \\ 0 & \sigma_i \end{pmatrix} \tag{15.44}$$

(note that the matrix Σ_i is diagonal in any representation of the γ-matrix). Proceeding from all the above, we can write:

$$\psi(x_j) = S(\epsilon, n_i)\psi(x_j - \epsilon\varepsilon_{jki}n_i x_k) \approx S(\epsilon, n_i)\left(1 - \epsilon\varepsilon_{jki}n_i x_k \frac{\partial}{\partial x_j}\right)\psi(x_j) =$$

$$= \left[1 + i\epsilon n_i\left(\frac{1}{2}\Sigma_i + M_i\right)\right]\psi(x_j), \tag{15.45}$$

where $M_i = -i\varepsilon_{ikj}x_k\partial_j$ is a projection of the angular momentum on axis x_i.

Let the rotation be realized about the given axis, with the direction characterized by the unit vector **n** at a finite angle θ. Then the matrix associated with a particular wavefunction transformation is given by:

$$S(\theta, \mathbf{n}) = \exp\left[i(\mathbf{n}\cdot\boldsymbol{\Sigma})\frac{\theta}{2}\right] = \cos\left(\frac{\theta}{2}\right) + i(\mathbf{n}\cdot\boldsymbol{\Sigma})\sin\left(\frac{\theta}{2}\right), \tag{15.46}$$

where (to go from the exponent to trigonometric functions) we use the Taylor series expansion and properties of the γ-matrices. From (15.46) it follows that:

$$S(\theta + 2\pi) = S(\theta),$$

that is, under the rotation by 2π a sign of the wavefunction is changing. This, in its turn, points to the fact that we are dealing with spinor objects. Since a wavefunction of the Dirac equation is a four-component quantity, this function is represented by two spinors, that is, it is a four-dimensional spinor. Similar objects will be further referred to as bispinors.

Obviously, the operator $\boldsymbol{\Sigma}$ plays a role of the Dirac-particle spin operator. Considering

$$\boldsymbol{\Sigma}^2 = \frac{3}{4}\begin{pmatrix} 1 & 0 & 0 & 0 \\ 0 & 1 & 0 & 0 \\ 0 & 0 & 1 & 0 \\ 0 & 0 & 0 & 1 \end{pmatrix}$$

and taking into account that the eigenvalues of the operators specified in the form of diagonal matrices are coincident with the values of diagonal elements, we obtain $J(J + 1) = 3/4$. Thus, the Dirac equation is really describing the particles with the spin $J = 1/2$.

Now we consider the Poincare transformations relating different inertial frames of reference. Let a relative motion of the systems be along the axis x_1, that is, we have an infinitesimal rotation through the imaginary angle $i\epsilon$ in the plane $(x_0 x_1)$. Only two components of $\omega^{\mu\nu}$, $\omega^{01} = -\omega^{10} = 1$, are different from zero, and we have

$$G_{[10]} = \frac{i}{2}\sigma_{10} = -\frac{1}{2}\gamma_1\gamma_0 = \frac{1}{2}\alpha_1.$$

The finite transformation is associated with the matrix:

$$S(\varphi, n_1) = \exp\left(\alpha_1 \frac{\varphi}{2}\right) = \cosh\left(\frac{\varphi}{2}\right) + \alpha_1 \sinh\left(\frac{\varphi}{2}\right), \tag{15.47}$$

where $\tanh\varphi = v$. Provided the motion is not along the axis x_1, but along an arbitrary direction characterized by the unit vector **n**, then we have to replace the expression (15.47) by:

$$S(\varphi, \mathbf{n}) = \exp\left[(\mathbf{n}\cdot\boldsymbol{\alpha})\frac{\varphi}{2}\right] = \cosh\left(\frac{\varphi}{2}\right) + (\mathbf{n}\cdot\boldsymbol{\alpha})\sinh\left(\frac{\varphi}{2}\right). \tag{15.48}$$

The final formula determining a law of the bispinor transformation at four-dimensional rotations is as follows:

$$\psi'(x) = S(\Lambda)\psi(\Lambda^{-1}x) = \exp\left[\frac{i}{2}\omega^{\mu\nu}\left(\frac{1}{2}\sigma_{\mu\nu} + L_{\mu\nu}\right)\right]\psi(x), \qquad (15.49)$$

where $L_{\mu\nu} = i(x_\mu\partial_\nu - x_\nu\partial_\mu)$ is the angular momentum and we have taken into consideration that $p_\mu = (E, -\mathbf{p})$.

If we use the chiral representation for the γ-matrices, in (15.46) and (15.48) we can recognize the equations for transformations of the first- and second-kind spinors, well-known from Section 3.4. Let us recall the course of our previous reasoning:

Dirac equation→relativistic covariance→transformation laws of spinors.

It goes without saying that the reverse is also possible, namely: proceeding from the transformational properties of spinors with respect to the transformations of the Poincare group, we can come to the Dirac equation. Note that this is a distinctive feature of any equation for the particles with nonzero spin as opposed to the Klein-Gordon equation that represents only the relativistic relations between the energy, momentum, and mass. One will be very instructive to illustrate such a way of obtaining an equation taking the Dirac equation as an example.

In the chiral representation the bispinor ψ comprises the spinors ξ and η, transformed by the representations $(1/2,0)$ and $(0,1/2)$, respectively. According to (15.48), the spinors $\xi(0)$ and $\xi(\mathbf{p})$ are related by the boost transformation:

$$\xi(\mathbf{p}) = \exp\left[(\mathbf{n}\cdot\boldsymbol{\sigma})\frac{\varphi}{2}\right]\xi(0) = \left[\cosh\left(\frac{\varphi}{2}\right) + (\mathbf{n}\cdot\boldsymbol{\sigma})\sinh\left(\frac{\varphi}{2}\right)\right]\xi(0) =$$

$$= \left[\sqrt{\left(\frac{\beta+1}{2}\right)} + (\mathbf{n}\cdot\boldsymbol{\sigma})\sqrt{\left(\frac{\beta-1}{2}\right)}\right]\xi(0) = \frac{p_0 + m + (\mathbf{p}\cdot\boldsymbol{\sigma})}{\sqrt{2m(p_0+m)}}\xi(0), \qquad (15.50)$$

where we introduce $\beta = p_0/m$ and allow for:

$$|\mathbf{p}|\cdot\mathbf{n} = \mathbf{p}, \qquad \cosh\left(\frac{\varphi}{2}\right) = \sqrt{\frac{\beta+1}{2}}, \qquad \sinh\left(\frac{\varphi}{2}\right) = \sqrt{\frac{\beta-1}{2}}.$$

Similarly, using the explicit-form Lorentz boost, for the spinor $\eta(\mathbf{p})$ we get:

$$\eta(\mathbf{p}) = \frac{p_0 + m - (\mathbf{p}\cdot\boldsymbol{\sigma})}{\sqrt{2m(p_0+m)}}\eta(0). \qquad (15.51)$$

Recall that in the canonical basis the representation is characterized by eigenvalues of the operators:

$$m^2, \qquad S^2, \qquad p_i, \qquad S_3 = \frac{1}{m}w^\mu n_\mu^{(3)},$$

that is, the representation basis comprises the vectors $|+, m, J; \mathbf{p}, J_3>$. It means that in a system of the particle at rest the state of the particle with the spin $J = 1/2$ may be equally well described both by the spinor $\xi(0)$ and spinor $\eta(0)$, that is, the following:

$$\xi(0) = \eta(0)$$

takes place. Acting on Eq. (15.51) by the operator:

$$\frac{p_0 + (\mathbf{p}\cdot\boldsymbol{\sigma})}{m}$$

and taking into account Eq. (15.50), we obtain:

$$\xi(\mathbf{p}) = \frac{p_0 + (\mathbf{p} \cdot \boldsymbol{\sigma})}{m} \eta(\mathbf{p}). \tag{15.52}$$

Similar operations with (15.50) lead to:

$$\eta(\mathbf{p}) = \frac{p_0 - (\mathbf{p} \cdot \boldsymbol{\sigma})}{m} \xi(\mathbf{p}). \tag{15.53}$$

From the derived equations it follows that one of the spinors $\eta(\mathbf{p})$ or $\xi(\mathbf{p})$ may be chosen at will, whereas the second is expressed in terms of the first spinor. In other words, the bispinor describing the particle with the spin 1/2 has two linearly independent components only. For the bispinor:

$$\psi(p) = \begin{pmatrix} \xi(\mathbf{p}) \\ \eta(\mathbf{p}) \end{pmatrix}$$

a system of Eqs. (15.52)–(15.53) takes the form:

$$(\hat{p} - m)\psi(p) = 0. \tag{15.54}$$

So we have demonstrated that the Dirac equation actually represents the relation between the spinors transformed in terms of the representations $(1/2, 0)$ and $(0, 1/2)$.

In the chiral representation the spinors comprising the bispinor ψ are independently transformed under the Poincare transformations, that is the main advantage of this transformation. We introduce the projective operators as follows:

$$\Lambda_R = \frac{1 + \gamma_5}{2}, \qquad \Lambda_L = \frac{1 - \gamma_5}{2}. \tag{15.55}$$

For the states:

$$\psi_{R,L} = \Lambda_{R,L}\psi,$$

we have the equations for eigenvalues of the operator γ_5 subsequently referred to as the chirality operator

$$\gamma_5 \psi_{R,L} = \pm \psi_{R,L}.$$

Using the projective operators of the right (positive) chirality Λ_R and left (negative) chirality Λ_L, from the bispinor ψ we can obtain the two-component Weyl spinors:

$$\psi_R = \begin{pmatrix} \xi \\ 0 \end{pmatrix}, \tag{15.56}$$

and

$$\psi_L = \begin{pmatrix} 0 \\ \eta \end{pmatrix}. \tag{15.57}$$

Another very popular representation of the γ-matrices, the Dirac-Pauli representation, is very useful in the analysis of the nonrelativistic limit for the Dirac equation. We show that this is actually so. Represent the bispinor $\psi(p)$ in the form:

$$\psi(p) = \begin{pmatrix} \varphi(\mathbf{p}) \\ \chi(\mathbf{p}) \end{pmatrix}$$

Besides, for definiteness, we consider the case of positive energies. Then Eq. (15.54) is broken into a system of two matrix equations:

$$(\boldsymbol{\sigma} \cdot \mathbf{p})\chi(\mathbf{p}) + m\varphi(\mathbf{p}) = E_p \varphi(\mathbf{p}), \tag{15.58}$$

$$(\boldsymbol{\sigma} \cdot \mathbf{p})\varphi(\mathbf{p}) - m\chi(\mathbf{p}) = E_p\chi(\mathbf{p}). \tag{15.59}$$

From (15.58) and (15.59) we find:

$$\chi(\mathbf{p}) = \frac{(\boldsymbol{\sigma} \cdot \mathbf{p})}{E_p + m^2}\varphi(\mathbf{p}), \tag{15.60}$$

where $E_p + mc^2 \neq 0$ and we for some time return to the normal units. Since for $v \ll c$ we have

$$\chi^\dagger(\mathbf{p})\chi(\mathbf{p}) = \frac{c^2\mathbf{p}^2}{(E_p + mc^2)^2}\varphi^\dagger(\mathbf{p})\varphi(\mathbf{p}) \approx \frac{1}{4}\left(\frac{v}{c}\right)^2 \varphi^\dagger(\mathbf{p})\varphi(\mathbf{p}),$$

in the nonrelativistic limit the components $\chi(\mathbf{p})$ in the order of magnitude are equal to $(v/c)\varphi(\mathbf{p})$ and hence are small. Because of this, the lower and upper bispinor components in the Dirac-Pauli representation are identified as big and small components, respectively.

For $m = 0$ Eqs. (15.52) and (15.53) are broken into two equations called the *Weyl equations* [17], having the following form in the nonstationary case:

$$\left.\begin{array}{l} i\partial_t\eta(\mathbf{p}) = -(\mathbf{p} \cdot \boldsymbol{\sigma})\eta(\mathbf{p}), \\ i\partial_t\xi(\mathbf{p}) = (\mathbf{p} \cdot \boldsymbol{\sigma})\xi(\mathbf{p}). \end{array}\right\} \tag{15.61}$$

Having studied the properties of Dirac's spinors under the Poincare transformations, we can define a geometric character of the bilinear combinations constructed on the basis of these spinors. So, the quantity $\overline{\psi}\psi$ represents a scalar with respect to the orthochronous transformations as we have:

$$\overline{\psi}'\psi' = \overline{\psi}S^{-1}S\psi = \overline{\psi}\psi,$$

and in the case of the nonorthochronous transformations it behaves itself as a pseudoscalar. Similarly, the bilinear combination $\overline{\psi}\gamma^\mu\psi$ is transformed as a four-dimensional vector under the orthochronous transformations:

$$\overline{\psi}'\gamma^\mu\psi' = \overline{\psi}S^{-1}\gamma^\mu S\psi = \Lambda^\mu_\nu\overline{\psi}\gamma^\nu\psi \tag{15.62}$$

and as a four-dimensional pseudoscalar in the case of the nonorthochronous transformations:

$$\overline{\psi}'\gamma^\mu\psi' = -\Lambda^\mu_\nu\overline{\psi}\gamma^\nu\psi. \tag{15.63}$$

It is easy to establish that the bilinear combinations: $\overline{\psi}\sigma^{\mu\nu}\psi$ and $\overline{\psi}\gamma^\mu\gamma^\nu\gamma^\lambda\psi$ ($\mu < \nu < \lambda$) represent the second- and third-rank tensors, respectively. Using the matrix γ_5, from the above-mentioned quantities we can compose the pseudoquantities in respect to the space reversal operation. For example, the bilinear combination $\overline{\psi}\gamma_5\psi$ is transformed as:

$$\overline{\psi}'\gamma_5\psi' = a\overline{\psi}S^{-1}\gamma_5 S\psi = -a\frac{i}{4!}\Lambda^\mu_\alpha\Lambda^\nu_\beta\Lambda^\lambda_\epsilon\Lambda^\sigma_\kappa\varepsilon_{\mu\nu\lambda\sigma}\overline{\psi}\gamma^\alpha\gamma^\beta\gamma^\epsilon\gamma^\kappa\psi = a(\det\Lambda)\overline{\psi}\gamma_5\psi,$$

where we have used

$$\gamma_5 = -\frac{i}{4!}\varepsilon_{\mu\nu\lambda\sigma}\gamma^\mu\gamma^\nu\gamma^\lambda\gamma^\sigma. \tag{15.64}$$

Knowing the transformational properties of the bilinear combinations $\overline{\psi}\Gamma_j\psi$, we can construct the invariant quantities to describe various interactions between the Dirac particles. For example, an invariant, describing the electromagnetic interaction of the Dirac particles, may be obtained through the multiplication of $\overline{\psi}\gamma^\mu\psi$ with the electromagnetic-field vector potential A_μ. Besides, interaction with an electromagnetic field may be governed by the quantity $\overline{\psi}\sigma^{\mu\nu}\psi F_{\mu\nu}$ (Pauli interaction for the magnetic moment).

15.4   Solutions of the free Dirac equation

Considering that any free particle should be described by the de Broglie wave, the solutions for the Dirac equation should be of the form:

$$\psi^{(+)}(x) = u(\mathbf{p}) \exp(-ipx) \qquad \text{for positive energy,} \tag{15.65}$$

$$\psi^{(-)}(x) = v(\mathbf{p}) \exp(ipx) \qquad \text{for negative energy.} \tag{15.66}$$

To substitute (15.65) and (15.66) into the Dirac equation gives the equations for definition of the bispinors $u(\mathbf{p})$ and $v(\mathbf{p})$:

$$(\hat{p} - m)u(\mathbf{p}) = 0, \tag{15.67}$$

$$(\hat{p} + m)v(\mathbf{p}) = 0. \tag{15.68}$$

We establish a procedure that makes it possible to distinguish the states of particles and antiparticles. It seems that this may be realized with the use of the operator:

$$\Pi_\epsilon = \frac{H}{\sqrt{H^2}},$$

that commutes with the Hamiltonian. It is evident that the operator Π_ϵ is Hermitian and unitary. In the momentum representation it has the simple form:

$$\Pi_\epsilon = \frac{(\boldsymbol{\alpha} \cdot \mathbf{p}) + m\gamma_0}{E_p}. \tag{15.69}$$

Since $\Pi_\epsilon^2 = 1$, then the eigenvalues of this operator are as follows:

$$\epsilon = \frac{p_0}{E_p} = +1$$

Then, using the operator Π_+, we could derive the electron states from the solutions of the Dirac equation, while the Π_--operator could be used for the finding of the positron components of the field function. Such a procedure is quite acceptable in the nonrelativistic theory. Unfortunately, the operator Π_ϵ could not be used because its form is not relativistically covariant. Let us try to look for the covariant projection operators instead.

Considering Eqs. (15.67), (15.68), we come to the conclusion that the projection operator to the electron states may be represented as:

$$\mathcal{P}_+ = \frac{\hat{p} + m}{2m}, \tag{15.70}$$

because, due to (15.67) and (15.68), it features the following properties:

$$\mathcal{P}_+ u(\mathbf{p}) = u(\mathbf{p}), \qquad \mathcal{P}_+ v(\mathbf{p}) = 0. \tag{15.71}$$

If we are interested in the negative-energy states, then in a similar way we can define the operator:

$$\mathcal{P}_- = \frac{-\hat{p} + m}{2m}, \tag{15.72}$$

which has the properties:

$$\mathcal{P}_- v(\mathbf{p}) = v(\mathbf{p}), \qquad \mathcal{P}_- u(\mathbf{p}) = 0. \tag{15.73}$$

Since the standard relations for the projection operators:

$$\mathcal{P}_\epsilon \mathcal{P}_{\epsilon'} = \delta_{\epsilon\epsilon'}, \qquad \mathcal{P}_+ + \mathcal{P}_- = I, \tag{15.74}$$

where

$$\mathcal{P}_\epsilon = \frac{\epsilon \hat{p} + m}{2m}, \tag{15.75}$$

are fulfilled, then $\mathcal{P}_+$ and $\mathcal{P}_-$ are actually the projection operators to be found.

Eqs. (15.67) and (15.68) give no unique definitions for the bispinors $u(\mathbf{p})$ and $v(\mathbf{p})$ as each of the equations has two linearly independent solutions. It is clear that these solutions should conform to two possible values of the electron polarization. The polarization state of a nonrelativistic electron may be characterized by setting the direction, the spin projection onto which has a well-defined value. This approach is inapplicable for a relativistic electron as the spin projection operator onto the arbitrary direction given by the unit vector $\mathbf{n}$ commutes with the Hamiltonian only in the case when $\mathbf{n} \parallel \mathbf{p}$. To describe the polarization state of a relativistic electron, we use the Pauli-Lubanski pseudovector:

$$W_\mu = -\frac{1}{4}\varepsilon_{\mu\nu\lambda\rho}\sigma^{\nu\lambda}p^\rho, \tag{15.76}$$

where we have taken into consideration that the total momentum is given by the following expression:

$$J^{\nu\lambda} = \frac{1}{2}\sigma^{\nu\lambda} + M^{\nu\lambda}.$$

Next, we introduce the four-dimensional vector $\tilde{p}$ (very useful for studies of polarization phenomena) that is characterized by

$$\tilde{p}^2 = -m^2, \qquad p_\mu \tilde{p}^\mu = 0. \tag{15.77}$$

In a system of the particle at rest the vector $\tilde{p}_\mu$ is of the form:

$$\tilde{p}_\mu = m(0, \mathbf{n}). \tag{15.78}$$

Then in the moving reference frame with $\mathbf{p} \parallel \mathbf{n}$ it has the components:

$$\tilde{p} = \left(|\mathbf{p}|, \frac{p_0 \mathbf{p}}{|\mathbf{p}|}\right). \tag{15.79}$$

For the vector $n^\mu = \tilde{p}^\mu/m$ the following relation:

$$W_\mu n^\mu = -\frac{1}{2}\gamma_5 \hat{n}\hat{p} \tag{15.80}$$

is true (recall, that $\varepsilon_{0123} = -1$). Since

$$[W_\mu n^\mu, \hat{p}] = -\gamma_5 n_\mu p^\mu \hat{p} = 0,$$

then the operators $W_\mu n^\mu$ and $\hat{p}$ have the common total eigenfunctions system. In other words, the bispinors $u(\mathbf{p})$ and $v(\mathbf{p})$, apart from Eqs. (15.67) and (15.68), must satisfy the equations:

$$\left.\begin{array}{l} -m^{-1}W_\mu n^\mu u^{(\alpha')}(\mathbf{p}) = \alpha' u^{(\alpha')}(\mathbf{p}), \\ -m^{-1}W_\mu n^\mu v^{(\alpha')}(\mathbf{p}) = \alpha' v^{(\alpha')}(\mathbf{p}). \end{array}\right\} \tag{15.81}$$

Because $(m^{-1}W_\mu n^\mu)^2 = 1/4$, then it is clear that

$$\alpha' = \pm\frac{1}{2}.$$

So, two linear independent solutions of Eqs. (15.67) and (15.68) are characterized by the eigenvalues of the operator $-W_\mu n^\mu/m$:

$$-\frac{W_\mu n^\mu}{m} u^{(\alpha)}(\mathbf{p}) = \frac{1}{2}\gamma_5 \hat{n} u^{(\alpha)}(\mathbf{p}) = \frac{1}{2}\alpha u^{(\alpha)}(\mathbf{p}), \tag{15.82}$$

$$-\frac{W_\mu n^\mu}{m} v^{(\alpha)}(\mathbf{p}) = -\frac{1}{2}\gamma_5 \hat{n} v^{(\alpha)}(\mathbf{p}) = \frac{1}{2}\alpha v^{(\alpha)}(\mathbf{p}), \tag{15.83}$$

where $\alpha = \pm 1$. As follows from the derived relations, the operator selecting the states with a defined value of α for the electrons takes the form:

$$\Lambda^{(\alpha)}(n) = \frac{1}{2}(1 + \alpha\gamma_5 \hat{n}), \tag{15.84}$$

And the corresponding operator for the positrons is derived from (15.84) with the substitution:

$$\alpha \to -\alpha.$$

Let us clarify the physical meaning of the quantum number α. Start with the simplest situation, namely, with the rest particle case. For the bispinors we accept the following orthonormalization condition:

$$w^{(\kappa)\dagger}(\mathbf{p})w^{(\kappa')}(\mathbf{p}) = \frac{E_p}{m}\delta_{\kappa\kappa'}, \tag{15.85}$$

where

$$w^{(1)}(\mathbf{p}) = u^{(+)}(\mathbf{p}), \qquad w^{(2)}(\mathbf{p}) = u^{(-)}(\mathbf{p}),$$
$$w^{(3)}(\mathbf{p}) = v^{(+)}(\mathbf{p}) \qquad w^{(4)}(\mathbf{p}) = v^{(-)}(\mathbf{p}),$$

while $u^{(+)}(0)$ $(v^{(+)}(0))$ and $u^{(-)}(0)$ $(v^{(-)}(0))$ describe the states of the electron (positron) with the spin being parallel and antiparallel to the vector $\mathbf{n}$, respectively. Under fulfillment (15.85) four linearly independent solutions of the Dirac equation in the Dirac-Pauli representation for the rest particle take the form:

$$u^{(\pm)}(0) = N_0 \begin{pmatrix} \varsigma^\pm \\ 0 \\ 0 \end{pmatrix}, \qquad v^{(\pm)}(0) = N_0 \begin{pmatrix} 0 \\ 0 \\ \varsigma^\pm \end{pmatrix}, \tag{15.86}$$

where

$$N_0 = \frac{1}{\sqrt{2(n_3+1)}}, \qquad \varsigma^+ = \begin{pmatrix} n_3+1 \\ n_1+in_2 \end{pmatrix}, \qquad \varsigma^- = \begin{pmatrix} -n_1+in_2 \\ n_3+1 \end{pmatrix}. \tag{15.87}$$

Supposing $\mathbf{p} = 0$, we obtain for the operator $-W_\mu n^\mu/m$:

$$-\frac{W_\mu n^\mu}{m} = \frac{1}{2}\gamma_5\gamma^0\boldsymbol{\gamma} = \frac{1}{2}\boldsymbol{\Sigma} = \frac{1}{2}\begin{pmatrix} \boldsymbol{\sigma} & 0 \\ 0 & \boldsymbol{\sigma} \end{pmatrix}. \tag{15.88}$$

Owing to the use of this operator for the states of (15.86), it is clear that in a primary quantization theory $\alpha/2$ has the meaning of an eigenvalue of the spin projection operator onto the direction $\mathbf{n}$.

For a moving particle we have:

$$\tilde{\hat{p}}\hat{p} = -\frac{m^2}{|\mathbf{p}|}(\mathbf{p}\cdot\boldsymbol{\gamma})\gamma_0. \tag{15.89}$$

Then, taking into account Eqs. (15.82), (15.83), we arrive at the relations:

$$S(\mathbf{p})u^{(\alpha)}(\mathbf{p}) = \frac{1}{2}\alpha u^{(\alpha)}(\mathbf{p}), \qquad S(\mathbf{p})v^{(\alpha)}(\mathbf{p}) = \frac{1}{2}\alpha v^{(\alpha)}(\mathbf{p}), \qquad (15.90)$$

where

$$S(\mathbf{p}) = \frac{(\boldsymbol{\Sigma}\cdot\mathbf{p})}{|\mathbf{p}|}$$

represents the chirality operator, which commutes with the Hamiltonian. At $\mathbf{p} \neq 0$ the number $\alpha/2$ is again having the meaning of an eigenvalue for the spin projection operator, however this time onto the direction of particle motion. In this case it is customary to speak of the state with a definite helicity, positive for $\alpha = 1$ or negative for $\alpha = -1$. Provided the particle spin is directed opposite to its momentum (similar to the left-hand screw), this particle is defined as a particle of the left-handed helicity or left-hand polarization. When the spin is in the direction of the particle momentum, we can speak about a right-hand polarized particle. Thus, the operators projecting onto the electron states with the spin projection $1/2$ and $-1/2$ are as follows:

$$\Lambda^{(+)}(n) \equiv \Lambda^{(\alpha=1)}(n) \qquad \text{and} \qquad \Lambda^{(-)}(n) \equiv \Lambda^{(\alpha=-1)}(n). \qquad (15.91)$$

By contrast, in the case of the positron the operators $\Lambda^{(-)}(n)$ and $\Lambda^{(+)}(n)$ separate the states with the spin projection onto the direction $\mathbf{n}$ that is equal to $+1/2$ and $-1/2$, respectively. For $\mathbf{p} \neq 0$ the solutions for the Dirac equation may be derived from the solutions of (15.86) with the help of Lorentz transformations to a system moving at the velocity $\mathbf{p}/E_p$. Having written the formula (15.48) in the Dirac-Pauli representation

$$S(\varphi, \mathbf{n}) = \begin{pmatrix} \cosh(\varphi/2) & (\mathbf{n}\cdot\boldsymbol{\sigma})\sinh(\varphi/2) \\ (\mathbf{n}\cdot\boldsymbol{\sigma})\sinh(\varphi/2) & \cosh(\varphi/2) \end{pmatrix} =$$

$$= \begin{pmatrix} \sqrt{(E_p + m)/2m} & (\mathbf{n}\cdot\boldsymbol{\sigma})\sqrt{(E_p - m)/2m} \\ (\mathbf{n}\cdot\boldsymbol{\sigma})\sqrt{(E_p - m)/2m} & \sqrt{(E_p + m)/2m} \end{pmatrix}, \qquad (15.92)$$

we obtain the solution sought:

$$u^{(\pm)}(\mathbf{p}) = S(\varphi, \mathbf{n})u^{(\pm)}(0) = \frac{\hat{p} + m}{\sqrt{2m(E_p + m)}}u^{(\pm)}(0) = N\begin{pmatrix} \varsigma^{\pm} \\ \frac{(\mathbf{p}\cdot\boldsymbol{\sigma})}{E_p + m}\varsigma^{\pm} \end{pmatrix}, \qquad (15.93)$$

$$v^{(\pm)}(\mathbf{p}) = \frac{-\hat{p} + m}{\sqrt{2m(E_p + m)}}v^{(\pm)}(0) = N\begin{pmatrix} \frac{(\mathbf{p}\cdot\boldsymbol{\sigma})}{E_p + m}\varsigma^{\pm} \\ \varsigma^{\pm} \end{pmatrix}, \qquad (15.94)$$

where

$$N = \sqrt{\frac{E_p + m}{4E_p(n_3 + 1)}}.$$

Thus, we have established that in a primary quantization theory the field functions, describing the states with a definite sign of the energy, and the spin projection are defined as follows:

$$u^{(\alpha)}(\mathbf{p}) = \mathcal{P}_+\Lambda^{(\alpha)}(n)\epsilon(\mathbf{p}), \qquad v^{(\alpha)}(\mathbf{p}) = \mathcal{P}_-\Lambda^{(-\alpha)}(n)\epsilon(\mathbf{p}), \qquad (15.95)$$

where $\epsilon(\mathbf{p})$ is an arbitrary bispinor satisfying only the normalization condition $\bar{\epsilon}(\mathbf{p})\epsilon(\mathbf{p}) = 1$. Accepting of the orthonormalization condition (15.85) for Dirac spinors was dictated by the requirement of relativistic invariance because in this case both the left-hand and right-hand sides of relation (15.85) are transformed as the fourth component of vector. The advantages of such an orthonormalization are quite obvious under the transition to conjugate spinors. Multiplying Eq.

$$p_0 w^{(\kappa)}(\mathbf{p}) = [(\boldsymbol{\alpha} \cdot \mathbf{p}) + \gamma_0 m] w^{(\kappa)}(\mathbf{p}) \tag{15.96}$$

by $w^{(\kappa)\dagger}(\mathbf{p})\gamma_0$ on the left, while Eq.

$$p_0 w^{(\kappa)\dagger}(\mathbf{p}) = w^{(\kappa)\dagger}(\mathbf{p})[(\boldsymbol{\alpha} \cdot \mathbf{p}) + \gamma_0 m] \tag{15.97}$$

by $\gamma_0 w^{(\kappa)}(\mathbf{p})$ on the right and summarizing the results obtained, we get

$$\overline{w}^{(\kappa)}(\mathbf{p}) w^{(\kappa)}(\mathbf{p}) = \frac{m}{p_0} w^{(\kappa)\dagger}(\mathbf{p}) w^{(\kappa)}(\mathbf{p}) \tag{15.98}$$

or

$$\overline{w}^{(\kappa)}(\mathbf{p}) w^{(\kappa')}(\mathbf{p}) = \varepsilon(p_0)\delta_{\kappa\kappa'}. \tag{15.99}$$

Using the explicit form of the Dirac equation solutions (15.93) and (15.94), one may show that they satisfy the relations:

$$\sum_{\alpha} [u_r^{(\alpha)}(\mathbf{p})\overline{u}_s^{(\alpha)}(\mathbf{p}) - v_r^{(\alpha)}(\mathbf{p})\overline{v}_s^{(\alpha)}(\mathbf{p})] = \sum_{\kappa=1}^{4} \varepsilon(p_0) w_r^{(\kappa)}(\mathbf{p})\overline{w}_s^{(\kappa)}(\mathbf{p}) = \delta_{rs}, \tag{15.100}$$

where r and s are matrix indices. The order of factors in (15.100) corresponds to the direct product of bispinors w and $\overline{w}$ ($w \otimes \overline{w}$), which represents the 4×4-matrix. Note that (15.100) may be proved without the use of the solutions in the explicit form. With this aim in view we recall that the bispinors $w^{(\kappa)}(\mathbf{p})$ form a complete system, and any bispinor $U(\mathbf{p})$ may be resolved into them, that is, we have:

$$U(\mathbf{p}) = \sum_{\kappa=1}^{4} a_\kappa w^{(\kappa)}(\mathbf{p}) = \sum_{\kappa=1}^{4} w^{(\kappa)}(\mathbf{p})a_\kappa \tag{15.101}$$

(a_κ is an ordinary c-number!). To multiply (15.101) by $\overline{w}^{(\kappa')}(\mathbf{p})$ on the left and allow for (15.99) gives

$$a_{\kappa'} = \varepsilon(p_0)\overline{w}^{(\kappa')}(\mathbf{p})U(\mathbf{p}).$$

Substituting the obtained equation into (15.101), we arrive at the relation (15.100).

With the use of relation (15.100) we can perform commonly occurring operations of the summation and averaging over the spin states of an electron or positron. Let us suppose that we are concerned with the calculation of the expression:

$$\sum_{\kappa=1}^{2} (\overline{f} M w^{(\kappa)})(\overline{w}^{(\kappa)} N g) = \sum_{\kappa=1}^{2} \left(\sum_{r,s=1}^{4} \overline{f}_r M_{rs} w_s^{(\kappa)} \right) \left(\sum_{\epsilon,\beta=1}^{4} \overline{w}_\epsilon^{(\kappa)} N_{\epsilon\beta} g_\beta \right) =$$

$$= \sum_{r,s,\epsilon,\beta=1}^{4} \overline{f}_r M_{rs} \left(\sum_{\kappa=1}^{2} w_s^{(\kappa)}\overline{w}_\epsilon^{(\kappa)} \right) N_{\epsilon\beta} g_\beta, \tag{15.102}$$

where M and N are some operators (products of the γ-matrices), f and g are bispinors and, as may be seen, the sum is taken only over the electron states. To multiply (15.100) by the operator $\mathcal{P}_+$ on the left and right gives the relation to be necessary for definition of (15.102):

$$\sum_{\kappa=1}^{2} w_r^{(\kappa)}(\mathbf{p})\overline{w}_s^{(\kappa)}(\mathbf{p}) = (\mathcal{P}_+)_{rs}. \tag{15.103}$$

When the summation of (15.102) is over $\kappa = 3, 4$ that is, over the positron states, we have to use the following equation:

$$\sum_{\kappa=3}^{4} w_r^{(\kappa)}(\mathbf{p}) \overline{w}_s^{(\kappa)}(\mathbf{p}) = -(\mathcal{P}_-)_{rs}, \qquad (15.104)$$

which also follows from (15.100).

Finally, we consider zero-mass particles. This may be of particular interest because in the ultrarelativistic limit massive particles actually behave themselves as massless objects. And for the particles of so small a mass as a neutrino the energies on the order of a few kV are considered as ultrarelativistic. Unfortunately, there is no possibility to derive a solution for the massless Dirac equation:

$$i\hat{\partial}\psi(x) = 0, \qquad (15.105)$$

from the solutions of (15.92) and (15.93) by a simple changeover $m \to 0$. This is caused by the fact that now, as distinct from the massive particles, due to commutation of the operators γ_5 and $\hat{\partial}$, the field function is an eigenfunction of the chirality operator:

$$\gamma_5\psi_q(x) = q\psi_q(x),$$

where $q = \pm 1$. Consequently, the independent solution of Eq. (15.105) must be represented in the following form:

$$\psi(x) = \begin{cases} u_q(\mathbf{p})\exp(-ipx) & \text{for positive energy,} \\ v_q(\mathbf{p})\exp(ipx) & \text{for negative energy.} \end{cases} \qquad (15.106)$$

To multiply (15.105) by $\gamma_5\gamma_0$ for the case $E > 0$ leads to the equation:

$$(\mathbf{p} \cdot \boldsymbol{\Sigma})u_q(\mathbf{p}) = q|\mathbf{p}|u_q(\mathbf{p}), \qquad (15.107)$$

from which follows that the chirality is equal to the helicity of a particle. As for the negative energy solutions, the chirality and helicity have the opposite signs. In the Dirac-Pauli representation the bispinors $u_q(\mathbf{p})$ and $v_q(\mathbf{p})$ have the form:

$$u_q(\mathbf{p}) = \frac{1}{\sqrt{2}}\begin{pmatrix} g_q(\mathbf{p}) \\ qg_q(\mathbf{p}) \end{pmatrix}, \qquad v_q(\mathbf{p}) = \frac{1}{\sqrt{2}}\begin{pmatrix} d_q(\mathbf{p}) \\ qd_q(\mathbf{p}) \end{pmatrix}, \qquad (15.108)$$

where $g_q(\mathbf{p})$ and $d_q(\mathbf{p})$ satisfy the eigenvalue equations

$$(\mathbf{n} \cdot \boldsymbol{\sigma})g_q(\mathbf{p}) = qg_q(\mathbf{p}), \qquad (\mathbf{n} \cdot \boldsymbol{\sigma})d_q(\mathbf{p}) = qd_q(\mathbf{p}) \qquad (15.109)$$

and $\mathbf{n} = \mathbf{p}/|\mathbf{p}|$. Eqs. (15.108) and (15.109) lead us to the idea that the functions $u_q(\mathbf{p})$ and $v_q(\mathbf{p})$ may be distinguished by the phase factor only. Let us show that this is the case. Rewrite the chirality operator in the spherical coordinate system

$$(\mathbf{n} \cdot \boldsymbol{\sigma}) = \sigma_x \sin\theta \cos\varphi + \sigma_y \sin\theta \sin\varphi + \sigma_z \cos\theta =$$

$$= \begin{pmatrix} \cos\theta & \sin\theta e^{-i\varphi} \\ \sin\theta e^{i\varphi} & -\cos\theta \end{pmatrix}. \qquad (15.110)$$

Using (15.110), we find the eigenfunctions of the chirality operator for $E > 0$

$$g_+(\mathbf{p}) = \begin{pmatrix} \cos(\theta/2) \\ \sin(\theta/2)e^{i\varphi} \end{pmatrix}, \qquad g_-(\mathbf{p}) = \begin{pmatrix} -\sin(\theta/2)e^{-i\varphi} \\ \cos(\theta/2) \end{pmatrix}. \qquad (15.111)$$

From the relations (15.111) follows

$$g_q(\mathbf{p}) = iq\sigma_2 g^*_{-q}(\mathbf{p}). \qquad (15.112)$$

Obviously, at the charge conjugation operation the Dirac bispinors behave as follows:

$$u_q(\mathbf{p}) = v^c_{-q}(\mathbf{p}) = C\bar{v}^T_{-q}(\mathbf{p}), \qquad v_q(\mathbf{p}) = u^c_{-q}(\mathbf{p}) = C\bar{u}^T_{-q}(\mathbf{p}).$$

Later we shall show that in the Dirac-Pauli representation the charge conjugation matrix has the form:

$$C = \begin{pmatrix} 0 & -i\sigma_2 \\ -i\sigma_2 & 0 \end{pmatrix}.$$

Then, it is easy to show the fulfillment of the following relations:

$$\left. \begin{aligned} v_+(\mathbf{p}) = C\bar{u}^T_-(\mathbf{p}) = \tfrac{1}{\sqrt{2}} \begin{pmatrix} -i\sigma_2 g^*_-(\mathbf{p}) = -g_+(\mathbf{p}) \\ i\sigma_2 g^*_-(\mathbf{p}) = g_+(\mathbf{p}) \end{pmatrix} = -u_+(\mathbf{p}), \\ v_-(\mathbf{p}) = C\bar{u}^T_+(\mathbf{p}) = -u_-(\mathbf{p}). \end{aligned} \right\} \qquad (15.113)$$

Thus, when $m = 0$ for the given momentum value, an equation for a particle with the spin of $1/2$ has just two independent solutions. In other words, in this case a fermion is described by the two-component spinors with a specific chirality value, the Weyl's spinors ψ_L and ψ_R. The Dirac equation therewith is equivalent to a pair of the Weyl equations (15.61). Recall that the necessity to use bispinors for particles with the spin of $1/2$ is determined by the requirement of the theory invariance with respect to the space inversion operation. When a particle is described by a single spinor, this symmetry is lost. And in the case of a neutrino this argument causes no concern. As is known at present, neutrinos take part in the weak interaction to be not invariant with respect to $\mathcal{P}$. The Lagrangian construction for the weak interaction is so that it involves the field function of a neutrino preceded by the operator Λ_L.

Considering that normalization is dictated by the form of an equation of motion, Lorentz-invariant normalizations of the solutions (15.87) obtained for the nonzero masses in the case of massless particles take the following form:

$$\left. \begin{aligned} \bar{u}^{(\alpha)}(\mathbf{p})\gamma_0 u^{(\beta)}(\mathbf{p}) = (2m)^{-1}\bar{u}^{(\alpha)}(\mathbf{p})\{\hat{p}, \gamma_0\}u^{(\beta)}(\mathbf{p}) = 2E_p\delta_{\alpha\beta}, \\ \bar{v}^{(\alpha)}\gamma_0(\mathbf{p})v^{(\beta)}(\mathbf{p}) = 2E_p\delta_{\alpha\beta}. \end{aligned} \right\} \qquad (15.114)$$

15.5 Once more about particles spin

The idea that an electron possesses spin (spinning electron) was first put forward by R. Kronig in 1921 for explanation of the anomalous Zeeman effect. However, as fate has willed, the first to listen to him was V. Pauli. Pauli's reaction to this hypothesis was negative. Being dispirited by the opposition of Pauli and cold reception given to the spinning electron idea by his colleagues, Kronig did not dare to publish a paper with his hypothesis. Four years later fate appeared to be more favorable to the graduates of the Leiden University S. Gaudsmit and G.E. Uhlenbeck. P. Ehrenfest was the first listener of their idea, according to which an electron represents a spinning sphere with a constant angular momentum. Ehrenfest has immediately recommended them to make a relevant publication in the journal *Naturwissenschaften*. By this concept, a spinning electron should possess an intrinsic dipole magnetic moment $\boldsymbol{\mu}_s$ and intrinsic angular momentum $\mathbf{S}$, that is called the *spin moment*, or simply the *spin*. In an atom the orbital moment of an electron may be compared to the angular momentum of the Earth upon its revolution around the Sun, accompanied by the seasonal changes. And the spin moment may be compared with the angular momentum of the Earth rotating about its axis, accompanied by the alternation of day and night.

However, the calculations of Uhlenbeck have demonstrated that for the required intrinsic mechanical moment ($\mathbf{S}^2 = 3\hbar^2/4$) a linear velocity of the rotational motion of an electron should be higher than the speed of light. Confused by the obtained result, Uhlenbeck was willing to address the editorial board and recall the paper. When he informed Ehrenfest about his decision Ehrenfest had remarked: *"It is a bit late. Besides, you both are young enough to afford stupid things."*

The concept of the electron spin was successfully used for deeper understanding of the anomalous Zeeman effect and also of numerous atomic effects. In 1928 this hypothesis was given conclusive evidence when Dirac, with the use of the relativistic quantum theory, demonstrated that particles with the mass and charge of an electron should have an intrinsic angular momentum of the value predicted by S. Gaudsmit and G.E. Uhlenbeck. Naturally, the representation of electrons in the form of charged rotating spheres is acceptable only within the scope of classical physics, having no sense in quantum theory. Nevertheless, in most cases the classical interpretation is very useful. To illustrate, we consider polarization of particles. Let us suppose that there is a certain preferential direction in space along the axis z, and the conditions of the problem are those that the eigenvalues of the operators $\mathbf{S}^2$ and S_z are simultaneously and exactly measurable, that is,

$$[\mathbf{S}^2, S_z] = 0.$$

As for the spin operators, the commutation relations:

$$[S_j, S_k] = i\epsilon_{jkl} S_l, \qquad j, k, l = x, y, z,$$

are true, the values of S_x and S_y 3 are not exactly defined. Then, similar to classical mechanics, we may speak of the precession of the spin vector $\mathbf{S}$ about the z-axis at the angle θ_s ($\cos\theta_s = J_z/\sqrt{J(J+1)}$, where J_z is the eigenvalue of the operator S_z). In other words, the statement that "the spin is along the z direction" should be understood as follows. The spin vector lies somewhere at a particular cone surface in such a manner that its component in the z direction equals $1/2$, and two other components are unknown, while it is known to do nothing but that:

$$S_x^2 + S_y^2 + S_z^2 = \frac{3}{4}.$$

The only preferential direction of a free particle is the direction of its motion. In the process the intrinsic angular momentum of the particle with zero mass is oriented parallel to the direction of its motion. If, as it has been agreed, the spin is associated with some intrinsic motion, it should be realized within the plane perpendicular to the velocity—the so-called "transverse polarization." It should be emphasized that for particles with the zero mass the helicity value, the spin projection onto the direction of motion, $\Sigma(\mathbf{p}) = \pm 1$ is a relativistically invariant quantity, in contrast to particles with the nonzero mass. Also, for a particle with the nonzero mass the spin may be parallel to the velocity, that is, for this particle the transverse polarization states are also possible. But these notions are Lorentz-noninvariant since the helicity value in this case is dependent on the fact in what coordinate system the observation is fulfilled. For example, an observer outrunning the particle will observe even such a phenomenon as the helicity overturn (spin flip). With growing the velocity of a particle, the process of its outrunning is not within the capacity of every observer, and helicity acquires the feature of stability for a greater number of the observers. In the ideal case when the particle velocity is equal to the speed of light, the helicity is a good quantum number in any reference frame. For a nonzero-mass particle in a reference frame, where it is at rest, the spin moment is not parallel to the velocity because the latter is zero. In the case of such a particle, the statement concerning the spin parallelism to the velocity is true not for every observer and hence the particle should have other polarization states as well. Then for the spin J the number of polarization states becomes $2J + 1$. On the other hand, a zero-mass particle is always moving at the speed of light. This is the reason why it has no system of reference, in respect to which this particle can be at rest. Consequently, the particles of this type have only two polarization states at all values of J. In this way the particle polarization is a characteristic of its state, associated with the spin and its direction in space. When a particle is absolutely nonpolarized, its properties are identical in all directions, similar to the particle with $J = 0$. In the general case the polarization determines a degree of symmetry (or asymmetry) of the particle's properties in space. A particle is identified as polarized if its symmetry characteristic includes the screw axis, like a rotating solid body. Polarization of particles in the general case is determined by $(2J + 1)^2 - 1$ parameters. For zero-mass particles the number of these parameters is equal to two.

Polarization of particles may be realized by various methods. For instance, for neutrons we can take an iron target magnetized up to the saturation that most actively absorbs the neutrons, whose spins are parallel to the Fe magnetization. In this case a neutron beam emerging from the target is enriched in the state with the spin that is antiparallel to the magnetization. Polarized electrons may be produced by the following methods: (i) β radioactive sources (polarization due to $V - A$-interaction); (ii) polarization owing to scattering of electrons (e.g., Mott scattering); (iii) optical pumping of alkali atoms by the polarized light; (iv) ionization of polarized metastable ^4He due to the gas discharge, and so on.

However, the most popular method of the particle polarization is transmission of the particles through the high-intensity electromagnetic fields. Relating the spin with intrinsic motion, we thereby proceed to the quasi-classical approximation. Let us find a relativistic equation determining the spin dynamics in this case. We introduce the spin four-dimensional vector s_μ. In the rest frame it coincides with the spin three-dimensional vector $\boldsymbol{\xi}$, that is, $s_\mu = (0, \boldsymbol{\xi})$. Then, since s_μ and p_μ are orthogonal in the rest reference frame they must keep this property in an arbitrary reference frame

$$s_\mu p^\mu = 0. \tag{15.115}$$

Besides, in any reference frame the relation:

$$s_\mu s^\mu = -\boldsymbol{\xi}^2. \tag{15.116}$$

takes place as well. To obtain the four-dimensional vector s_μ for the electron moving with the velocity $\mathbf{v} = \mathbf{p}/E_p$, one should use a Lorentz transformation to convert the four-dimensional vector $(m, 0, 0, 0)$ to p^μ. The matrix of such a transformation represents the product of three matrices of rotation in planes (x_0, x_i) through angles $\varphi_i = \text{arctanh}(p_i/E_p)$.

The explicit form of this matrix is as follows:

$$[L(p)]^\mu_\nu = \begin{pmatrix} \dfrac{E_p}{m} & \dfrac{p_1}{m} & \dfrac{p_2}{m} & \dfrac{p_3}{m} \\ \dfrac{p_1}{m} & 1 + \dfrac{p_1^2}{m(E_p+m)} & \dfrac{p_1 p_2}{m(E_p+m)} & \dfrac{p_1 p_3}{m(E_p+m)} \\ \dfrac{p_2}{m} & \dfrac{p_2 p_1}{m(E_p+m)} & 1 + \dfrac{p_2^2}{m(E_p+m)} & \dfrac{p_2 p_3}{m(E_p+m)} \\ \dfrac{p_3}{m} & \dfrac{p_3 p_1}{m(E_p+m)} & \dfrac{p_3 p_2}{m(E_p+m)} & 1 + \dfrac{p_3^2}{m(E_p+m)} \end{pmatrix}. \tag{15.117}$$

Thus, in a moving reference frame the components of the four-dimensional spin vector are defined by the expressions:

$$\mathbf{s} = \boldsymbol{\xi} + \frac{(\mathbf{p}\cdot\boldsymbol{\xi})\mathbf{p}}{m(E_p+m)}, \qquad s_0 = \frac{(\mathbf{p}\cdot\boldsymbol{\xi})}{m}. \tag{15.118}$$

It is obvious that the left-hand side of the spin evolution equation is nothing else but a derivative of s_μ with respect to the proper time τ ($d\tau = ds$). To have the right-hand side of the desired equation, we demand it to be linear and homogeneous both in terms of the electromagnetic field tensor $F_{\mu\nu}$ and in terms of s_μ. Also, the right-hand side may involve the four-dimensional vector of velocity $u_\mu = p_\mu/m$, for which the following:

$$s_\mu u^\mu = 0$$

is true. Then, demanding the relativistic invariance of the required equation, we arrive at

$$\frac{ds_\mu}{d\tau} = \alpha F_{\mu\nu} s^\nu + \beta u_\mu F_{\nu\sigma} u^\nu s^\sigma, \tag{15.119}$$

where α and β are constant coefficients, The constant α could be found with the help of the correspondence principle. A classical analog for the spin is a mechanical angular momentum to obey the equation:

$$\frac{d\mathbf{M}}{dt} = [\mathbf{r} \times \mathbf{F}]. \tag{15.120}$$

In the case of a magnetic field Eq. (15.120) takes the form:

$$\frac{d\mathbf{M}}{dt} = \frac{e}{m}[\mathbf{M} \times \mathbf{H}] = 2[\boldsymbol{\mu} \times \mathbf{H}], \tag{15.121}$$

where we have taken into account

$$\boldsymbol{\mu} = \frac{e}{2m}\mathbf{M}.$$

By the correspondence principle, at $v \to 0$ the spin moment should satisfy the same equation as does the mechanical angular moment. Setting in Eq. (15.119) $s^\mu = (0, \boldsymbol{\xi})$, $u^\mu = (1,0,0,0)$ and $\tau = t$, we obtain:

$$\frac{d\boldsymbol{\xi}}{dt} = \alpha[\boldsymbol{\xi} \times \mathbf{H}]. \tag{15.122}$$

Further on we take into consideration that

$$\boldsymbol{\mu}_s = \mu_s \boldsymbol{\xi},$$

where the subscript s stresses the fact that we are dealing with the magnetic moment connected with the spin. Then, to compare Eqs. (15.121) and (15.122) leads to the result:

$$\alpha = 2\mu_s.$$

Differentiating the relation

$$s^\mu u_\mu = 0$$

with respect to the proper time τ and using the classical equation for a point charge moving in a external field

$$m\frac{du^\nu}{d\tau} = eF^{\nu\lambda}u_\lambda,$$

we obtain:

$$u_\mu\frac{ds^\mu}{d\tau} = -s_\mu\frac{du^\mu}{d\tau} = -\frac{e}{m}s_\mu F^{\mu\lambda}u_\lambda = \frac{e}{m}F^{\mu\lambda}u_\mu s_\lambda. \tag{15.123}$$

Multiplying the both sides of Eq. (15.119) by u_μ and taking into account the relation

$$u_\mu u^\mu = 1,$$

we find the value of β

$$\beta = 2\left(\frac{e}{2m} - \mu_s\right) = -2\mu_s^{an},$$

where μ_s^{an} is an anomalous magnetic moment connected with the spin. Therefore, the relativistic spin motion equation, Bargmann–Michel–Telegdi (BMT) equation [18], has the following form:

$$\frac{ds_\mu}{d\tau} = 2\mu_s F_{\mu\nu}s^\nu - 2\mu_s^{an}u_\mu F_{\nu\sigma}u^\nu s^\sigma. \tag{15.124}$$

From Eq. (15.124) follows that $s_\mu s^\mu =$const, that is, even if a particle moves in a external field its polarization ξ is left untouched.

In practice it is more convenient to operate with the quantity $\boldsymbol{\xi}$, directly characterizing the particle spin direction in its instantaneous rest frame, rather than with the four-dimensional vector of the spin. For the spatial components of Eq. (15.124) we have:

$$\frac{d\mathbf{s}}{dl} = \frac{2\mu_s m}{e}\{[\mathbf{s} \times \mathbf{H}] + (\mathbf{sv})\mathbf{E}\} + \frac{2\mu_s^{an}\epsilon}{m}\{\mathbf{v}(\mathbf{v}[\mathbf{s} \times \mathbf{H}]) - \mathbf{v}(\mathbf{sE}) + \mathbf{v}(\mathbf{sv})(\mathbf{vE})\}. \tag{15.125}$$

Substituting the relation

$$\mathbf{s} = \boldsymbol{\xi} + \frac{\mathbf{p}(\boldsymbol{\xi}\mathbf{p})}{m(\epsilon + m)},$$

into (15.125) we get:

$$\frac{d\boldsymbol{\xi}}{dt} = \frac{2\mu_s m + 2\mu_s^{an}(\epsilon - m)}{\epsilon}[\boldsymbol{\xi}\times\mathbf{H}] + \frac{2\mu_s^{an}\epsilon}{\epsilon + m}(\mathbf{vH})[\mathbf{v}\times\boldsymbol{\xi}] + \frac{2\mu_s m + 2\mu_s^{an}\epsilon}{\epsilon + m}[\boldsymbol{\xi}\times[\mathbf{E}\times\mathbf{v}]]. \tag{15.126}$$

where when we differentiated $\mathbf{p}$ and ϵ the motion equations:

$$\frac{d\mathbf{p}}{dt} = e\mathbf{E} + e[\mathbf{v} \times \mathbf{H}], \qquad \frac{d\epsilon}{dt} = e(\mathbf{vE})$$

have been taken into account.

At the same time, we are interested not in a change of the absolute spin vector direction in space but in changing a spin orientation with respect to the direction of motion. We introduce the unit vector $\mathbf{n} = \mathbf{v}/|\mathbf{v}|$ and decompose $\boldsymbol{\xi}$ with its help into the perpendicular and parallel components:

$$\boldsymbol{\xi} = \boldsymbol{\xi}_\perp + \boldsymbol{\xi}_\parallel, \qquad \boldsymbol{\xi}_\parallel = \mathbf{n}(\mathbf{n}\boldsymbol{\xi}), \qquad \boldsymbol{\xi}_\perp = [\mathbf{n} \times [\boldsymbol{\xi} \times \mathbf{n}]].$$

Then, the module of the parallel component of the spin vector is defined by the equation:

$$\frac{d\xi_\parallel}{dt} = 2\mu_s^{an}(\boldsymbol{\xi}_\perp[\mathbf{H} \times \mathbf{n}]) + \frac{2}{v}\left(\frac{\mu_s m^2}{\epsilon^2} - \mu_s^{an}\right)(\boldsymbol{\xi}_\perp\mathbf{E}). \tag{15.127}$$

The applicability condition of a quasi-classical approximation calls for a relatively slow varying of the particle momentum, in turn leading to the field smallness condition (for example, in a magnetic field the Larmor radius $R = pc/eH$ should be great compared to the particle wavelength) and to the condition of small changing the fields for the distances on the order of the particle wavelength. Provided a field is rapidly changed, the BMT equation should incorporate the additional components involving the field derivatives with respect to the coordinates.

15.6 Polarization density matrix for Dirac particles

In modern practice the accelerators are operating in the incoming beam-target mode as well as in the colliding-beam mode, and usually the beams are formed of particles with the spin 1/2. And the particle scattering experiments are statistical in character. The beam and a stationary or movable target comprise a great number of the particles. An experimenter measures the average scattering over a considerable number of pair collisions between the beam particles and the target. Because of this, we can represent the beam and target as a statistical ensemble of the systems, each of which contains a pair of interacting particles. And the experimentally measured polarization is an average over the statistical ensemble. Of course, each particle is always polarized in a particular manner. When considering an aggregation of particles, however, it makes sense to look at polarization of the whole system. It is clear that two variants are possible: (i) completely polarized system with all the constituting particles having the same spin state, that enables one to describe this system in terms of the field function; and (ii) system in a mixed state when all the particles are differently polarized and associated with different field functions. In this case the system represents a statistical mixture of various pure spin states with specific weights, that is, there is no common field function.

Let us define the qualitative characteristic for system polarization degree. To this end, we consider a non-relativistic electron beam. Let the vector $\mathbf{n}$ be in the z-direction, whereas N_+ and N_- denote the number of electrons in the states with $J_z = 1/2$ and $J_z = -1/2$, respectively. If $N_+ \neq N_-$ the beam be polarized. Polarization ξ_z in a direction of the z-axis is determined be the expression:

$$\xi_z = \frac{N_+ - N_-}{N_+ + N_-}. \tag{15.128}$$

At $\xi_z = \pm 1$ the beam is completely polarized while at $\xi_z = 0$ one is completely unpolarized. In the general case the beam polarization has the longitudinal and transverse components, that is, it is characterized by a polarization vector $\boldsymbol{\xi}$. This vector defines the direction along which the electron spins are mainly orientated. The module of the vector $\boldsymbol{\xi}$ gives a numerical value of a polarization. Let us define this vector for the case of a completely polarized beam. The coordinate dependence of the electron wavefunction describing a free motion with the momentum $\mathbf{p}$ is reduced to the common factor $\exp(-ipx)$, and the amplitude $\varphi(\mathbf{p})$ (two-component spinor in the nonrelativistic case) plays a role of a spin part of the wavefunction. In such a (pure) state an electron is completely polarized. We normalize the spinor $\varphi(\mathbf{p})$

$$\varphi(\mathbf{p}) = \begin{pmatrix} c_1 \\ c_2 \end{pmatrix} \tag{15.129}$$

as follows:

$$\varphi^\dagger(\mathbf{p})\varphi(\mathbf{p}) = |c_1|^2 + |c_2|^2 = 1.$$

The electron states with $J_z = 1/2$ and $J_z = -1/2$ are described by the spinors:

$$\varphi_1 = \begin{pmatrix} 1 \\ 0 \end{pmatrix}, \qquad \varphi_2 = \begin{pmatrix} 0 \\ 1 \end{pmatrix}, \tag{15.130}$$

to form the spin basis by which any spinor could be expanded. Using $\varphi_{1,2}$ one may rewrite the electron wavefunction (15.129) in the shape:

$$\varphi(\mathbf{p}) = c_1 \varphi_1 + c_2 \varphi_2. \tag{15.131}$$

The quantities $\mid c_1 \mid^2$ and $\mid c_2 \mid^2$ are equal to the probability of finding an electron in the states with $J_z = 1/2$ and $J_z = -1/2$, respectively. Therefore, the expression (15.128) could be represented in the form:

$$\xi_z = \mid c_1 \mid^2 - \mid c_2 \mid^2 = \varphi^\dagger(\mathbf{p})\sigma_z\varphi(\mathbf{p}),$$

which is naturally generalized on the three-dimensional case:

$$\boldsymbol{\xi} = \varphi^\dagger(\mathbf{p})\boldsymbol{\sigma}\varphi(\mathbf{p}). \tag{15.132}$$

In this way for a nonrelativistic electron beam the polarization vector is an average value of the spin operator $\boldsymbol{\sigma}$. The polarization vector projections $\xi_x = c_1^* c_2 + c_1 c_2^*$, $\xi_y = i(c_1 c_2^* \quad c_1^* c_2)$, and ξ_z are called the *Stokes parameters of an electron*.

Let us analyze a mixed electron beam. We assume that $\boldsymbol{\xi}^i$ is the value of electron polarization in the pure state i, and w^i—relative probability with which this pure state enters a mixed beam. Then the polarization $\boldsymbol{\xi}$ of the mixed state is determined by the equation:

$$\boldsymbol{\xi} = \sum_i w^i \boldsymbol{\xi}^i, \tag{15.133}$$

where $\sum_i w^i = 1$. To substitute (15.131) and (15.132) into (15.133) leads to the expression:

$$\boldsymbol{\xi} - \sum_i w^i \varphi^{i\dagger}(\mathbf{p})\boldsymbol{\sigma}\varphi^i(\mathbf{p}) = \sum_{\alpha,\beta=1}^{2} \sum_i w^i c_\alpha^i c_\beta^{i*} \varphi_\beta^\dagger \boldsymbol{\sigma}\varphi_\alpha = \sum_i \rho_{\alpha\beta}\varphi_\beta^\dagger \boldsymbol{\sigma}\varphi_\alpha = \text{Sp}(\rho\boldsymbol{\sigma}). \tag{15.134}$$

The matrix

$$\rho_{\alpha\beta} = \sum_i w^i c_\alpha^i c_\beta^{i*} \equiv \overline{c_\alpha^i c_\beta^{i*}} \tag{15.135}$$

derives the name "the polarization density matrix of a mixed state." From its definition it follows that in a pure state (special case of a mixed state) $\rho_{\alpha\beta}$ is reduced to the product:

$$\rho_{\alpha\beta} = c_\alpha c_\beta^*.$$

Consequently, the matrix $\rho_{\alpha\beta}$ meets the normalization conditions:

$$\text{Sp}(\rho) = 1. \tag{15.136}$$

A polarization matrix is built of the coefficients of decomposing the wavefunction of the pure state in terms of the basis vectors and of the relative probabilities with that the pure state enters into the mixed one. From (15.134) it is clear that with the help of $\rho_{\alpha\beta}$ we can calculate the average value for the spin operator, that is, a polarization density matrix completely characterizes the spin properties of a system. We can obtain a definition of the

$\rho_{\alpha\beta}$-matrix that is more convenient for practical applications. It suffices to develop equation (15.135):

$$\rho_{\alpha\beta} = \sum_i w^i \begin{pmatrix} |c_1^i|^2 & c_1^i c_2^{i*} \\ c_1^{i*} c_2 & |c_2^i|^2 \end{pmatrix} \tag{15.137}$$

and by the direct calculations to demonstrate that the expression:

$$\frac{1}{2}[\delta_{\alpha\beta} + (\boldsymbol{\xi} \cdot \boldsymbol{\sigma})_{\alpha\beta}]$$

is exactly equal to the right-hand side of (15.137). Thus, the expression:

$$\rho_{\alpha\beta} = \frac{1}{2}[\delta_{\alpha\beta} + (\boldsymbol{\xi} \cdot \boldsymbol{\sigma})_{\alpha\beta}] \tag{15.138}$$

represents the other equivalent form of writing the polarization density matrix. Assuming that $\boldsymbol{\xi}$ is equal to $\alpha\mathbf{n}$, in the right-hand side of (15.138) we recognize the projection operator on the state with the spin projection onto the direction $\mathbf{n}$ being equal to α.

In analogy with a nonrelativistic case, the polarization density matrix of a relativistic electron is given by the following expression:

$$\rho_{r\kappa}^{(+)} = \overline{u_r(\mathbf{p})\overline{u}_\kappa(\mathbf{p})}, \qquad (r, \kappa = 1, 2, 3, 4). \tag{15.139}$$

Then, in a pure state it has the form:

$$\rho_{r\kappa}^{(+)} = u_r(\mathbf{p})\overline{u}_\kappa(\mathbf{p}), \tag{15.140}$$

and, consequently, it is normalized by the condition:

$$\mathrm{Sp}(\rho^{(+)}) = 1.$$

Using (15.95) one could rewrite the expression (15.140) as follows:

$$\rho_{r\kappa}^{(+)} = \left(\mathcal{P}_+ \Lambda^{(\alpha)}(n)\right)_{r\kappa}. \tag{15.141}$$

For the mixed state in Eq. (15.141) we need an averaging procedure over the parameters characterizing the system that incorporates this electron:

$$\rho_{r\kappa}^{(+)} = \overline{\left(\mathcal{P}_+ \Lambda^{(\alpha)}(n)\right)}_{r\kappa}. \tag{15.142}$$

The only possible result of such an averaging may be a change in the spin projection value. To take this into consideration, in Eq. (15.142) it is enough to set $\alpha = 1$ and replace the unit vector n_μ by the vector s_μ, that now takes the responsibility for the sign and degree of the beam polarization upon oneself. Thus, the polarization density matrix for a relativistic electron is determined by the expression:

$$\rho_{r\kappa}^{(+)} = \frac{1}{2}\left[\frac{\hat{p} + m}{2m}(1 - \gamma_5 \hat{s})\right]_{r\kappa}. \tag{15.143}$$

The case $s_\mu^2 = -1$ corresponds to the pure state and the one $s_\mu = 0$—to the unpolarized state

$$\rho_{r\kappa}^{(+)} = \frac{1}{2}\left(\frac{\hat{p} + m}{2m}\right)_{r\kappa}. \tag{15.144}$$

The last equation may be derived by summation with respect to the polarization states of the matrix $1/2 \sum_\alpha v_r^\alpha \bar{v}_\kappa^\alpha$. If the polarization density matrix be given, the vector s_μ would be found with help of the relation:

$$s_\mu = \mathrm{Sp}\,(\rho \gamma_5 \gamma_\mu).\tag{15.145}$$

Similarly, with the use of Eq. (15.95), we can obtain an expression for the polarization density matrix of positron:

$$\rho_{r\kappa}^{(-)} = \frac{1}{2}\left[\frac{\hat{p} - m}{2m}(1 + \gamma_5 \hat{s})\right]_{r\kappa}.\tag{15.146}$$

At the modern electron-positron colliders in use the energies in a center-of-mass system are as high as hundreds of GeV, that is, leptons are ultrarelativistic particles. The energy values of the designed muon colliders (> 400 GeV) also allow one to believe muon beams are ultrarelativistic objects too. Let us consider how the spin projection operator is modified in these conditions. In the ultrarelativistic limit we have:

$$|\,\mathbf{p}\,| \approx p_0 \left(1 - \frac{m^2}{2p_0^2}\right).$$

Then in the electron case the calculations give:

$$m^{-1}\hat{p}_\mu \gamma^\mu u(p) = m^{-1}\left[\gamma_0\,|\,\mathbf{p}\,| - \frac{p_0}{|\,\mathbf{p}\,|}(\boldsymbol{\gamma}\cdot\mathbf{p})\right]u(p) \approx$$

$$\approx m^{-1}\left[\gamma_0 p_0 - \gamma_0\frac{m^2}{2p_0} - (\boldsymbol{\gamma}\cdot\mathbf{p}) + (\boldsymbol{\gamma}\cdot\mathbf{p})\frac{m^2}{2p_0^2}\right]u(p) \approx u(p).\tag{15.147}$$

From the relation (15.147) follows

$$\Lambda^{(\alpha)}(n) \to \frac{1}{2}\left(1 - \alpha\gamma_5\right),\tag{15.148}$$

that is, at high energies the helicity projection operator passes into the chirality projection operator. Note, if for an electron the operators $(1 - \gamma_5)/2$ and $(1 + \gamma_5)/2$ are the projectors onto the states with the positive and negative chirality, respectively, then in the case of a positron these operators exchange their roles.

It should be emphasized that in the relativistic quantum theory the spin of a particle is not representing an integral of motion. As a result, in an arbitrary reference frame it is impossible to use a density matrix for characterization of a relativistic electron, the density matrix being physically meaningful in the electron rest frame only.

15.7 Dirac equation in the external electromagnetic field

In quantum electrodynamics there are some problems, whose solutions may be obtained within the scope of a single-particle theory. In these problems the number of particles is fixed, and interaction may be introduced using the notion of an external (classical) electromagnetic field. Since in certain conditions external fields may generate the electron-positron pairs from the vacuum, there is a need to know the limiting values of their strengths when the single-particle approximation is applicable.

For a while let us come back to ordinary units. Due to the uncertainty relation:

$$\Delta t \Delta E \geq \frac{\hbar}{2},$$

a virtual $e^- e^+$-pair with the lifetime:

$$\Delta t \approx \frac{\hbar}{2mc^2}$$

may appear from the vacuum. In this time the electron and positron may be separated by a distance:

$$\Delta r \approx c\Delta t \approx \frac{\hbar}{2mc},$$

that is, by a distance on the order of the Compton wavelength of the electron. If an external electric field is able to do the work of $2mc^2$ on the electron at the distance Δr, then production of a pair from the vacuum becomes real. To accomplish this the field must have the order:

$$e\mathcal{E}_c \Delta r = 2mc^2 \qquad \Rightarrow \qquad \mathcal{E}_c = \frac{4m^2c^3}{e\hbar}. \qquad (15.149)$$

At such values of the field $\mathcal{E} = \mathcal{E}_0$ ($\mathcal{E}_0$ is the critical electric field) the vacuum is unstable in respect to the production of the electron-positron pairs. The critical value of the magnetic field $\mathcal{H}_0$ is established from the equality condition of the electron rotation energy quantum $\hbar \omega_H$ (ω_H is cyclotron frequency, $\omega_H = e\mathcal{H}/mc$) and the electron rest energy:

$$\hbar \omega_H = mc^2 \qquad \Rightarrow \qquad \mathcal{H}_c = \frac{m^2c^3}{e\hbar}. \qquad (15.150)$$

But the magnetic field does no work and hence the vacuum remains stable with respect to the magnetic field. In other words, the magnetic field allows for a one-particle interpretation of the problem of charged particle motion in the field even for $\mathcal{H} > \mathcal{H}_c$.

The solutions for the Dirac equation in the external electromagnetic field based on a single-particle interpretation prove to also be relevant for the many-particle problems of the quantum field theory when the particles in the initial and/or final state are located in such a field. The Furry representation [19], the method using exact solutions of the Dirac equation, was named after its author who has shown that an apparatus of the Feynman diagrams could be generalized to the case when the electron is in the bound state rather than in the free one. This method is fairly simple. The four-dimensional potential A_μ^a is divided into two parts:

$$A_\mu^a = A_\mu^{ex} + A_\mu,$$

where A_μ^{ex} describes an external field and is precisely included into the wavefunction of the electron, and A_μ—a quantized electromagnetic field that is allowed for by the perturbation theory. The calculation of the particular physical phenomena takes place with the following modification of the Feynman diagram technique: in the initial and final states the electron is located in the external field and described by the solution for the Dirac equation in this field, the electron propagator being the Green function of this electron in the external field. However, despite the importance of exact solutions for the Dirac equation in the external field, we leave them in peace (detailed discussion of the problem may be found in Ref. [20]), focusing our attention on a more modest task—the study of the Dirac equation in the external electromagnetic field and definition of an intrinsic (spin) magnetic moment of the electron.

Consider the motion of a Dirac particle in the external electromagnetic field that is specified by the potential $A_\mu^{ex}(x)$. We introduce an electromagnetic interaction by a minimal way, that is, with the help of the recipe:

$$\partial_\mu \to D_\mu = \partial_\mu + ieA_\mu^{ex}(x). \tag{15.151}$$

Then the Dirac equations for the bispinors ψ and $\overline{\psi}$ take the form:

$$(i\gamma_\mu D^\mu - m)\psi(x) = 0, \tag{15.152}$$

$$(i\gamma_\mu^T D^{\mu*} + m)\overline{\psi}(x) = 0. \tag{15.153}$$

Such a recipe provides the invariance of equations (15.152) and (15.153) with respect to the gauge transformations:

$$\psi(x) \to \psi(x)\exp\left(ie\alpha(x)\right), \qquad A_\mu^{ex}(x) \to A_\mu^{ex}(x) - \partial_\mu\alpha(x). \tag{15.154}$$

The electron rotating in a closed orbit with the momentum $\mathbf{L}$ has the orbital magnetic moment:

$$\boldsymbol{\mu}_l = \frac{e}{2m}\mathbf{L}. \tag{15.155}$$

Believing that the Creator who had constructed our world was acting by analogy, we might expect that the value of a spin magnetic moment should be as follows:

$$\boldsymbol{\mu}_s = g_s\frac{e}{2m}\mathbf{S} \tag{15.156}$$

with $g_s = 1$ (g_s is the Lande factor or the gyromagnetic ratio). But as is evidenced by the experiments, for the electron $g_s = 2$ (or actually somewhat lower, $a^{ex} = (g_s - 2)/2 = 0.0011596521884(43)$ [21], that may be explained by the inclusion of radiative corrections taking into account the interaction between a particle and vacuum fluctuations [22]). When calculating the value of $\boldsymbol{\mu}_s$, it seems very useful to pick out a term in Eq. (15.152), which describes the interaction between a spinless particle and the electromagnetic field. This may be done by quadrating the Dirac equation, that is, by introduction of the bispinor Φ in the following way:

$$\psi = \frac{1}{m}(i\hat{\partial} - e\hat{A}^{ex} + m)\Phi. \tag{15.157}$$

Substituting (15.157) into (15.152), we obtain

$$[(i\hat{\partial} - e\hat{A}^{ex})(i\hat{\partial} - e\hat{A}^{ex}) - m^2]\Phi = 0. \tag{15.158}$$

With allowance made for

$$\gamma^\mu\gamma^\nu = \frac{1}{2}[\gamma^\mu, \gamma^\nu] + \frac{1}{2}\{\gamma^\mu, \gamma^\nu\}$$

and

$$[(i\partial_\mu - eA_\mu^{ex}), (i\partial_\nu - eA_\nu^{ex})] = -ieF_{\mu\nu},$$

we find that the bispinor Φ obeys the equation:

$$\left[D_\mu D^\mu + m^2 + \frac{e}{2}\sigma^{\mu\nu}F_{\mu\nu}\right]\Phi = 0. \tag{15.159}$$

An addition (relative to the Klein-Gordon equation) term

$$\frac{e}{2}\sigma^{\mu\nu}F_{\mu\nu} = -e[(\boldsymbol{\Sigma}\cdot\boldsymbol{\mathcal{H}}) - i(\boldsymbol{\alpha}\cdot\boldsymbol{\mathcal{E}})], \tag{15.160}$$

where $H_i = -\varepsilon_{ijk}F_{jk}/2$, $E_i = F_{0i}$, defines the interaction of the spin magnetic moment $\boldsymbol{\mu}_s$ with the magnetic field $\boldsymbol{\mathcal{H}}$ and the interaction of the spin electric moment $\mathbf{d}_s$ with the electric field $\boldsymbol{\mathcal{E}}$. The value of $\boldsymbol{\mu}_s$ will be established later, and now we discuss some mathematical details. At rising the order of the Dirac equation the number of solutions is doubled. This means that among the solutions of Eq. (15.159) some are superfluous. Because of this, we have to select the required solutions. In this case the situation is saved due to the fact that Eq. (15.158) involves only the matrices $\sigma^{\mu\nu}$ commuting with the matrix γ_5. Consequently, the solutions of Eq. (15.159) satisfy the condition:

$$\gamma_5\Phi^\pm = q\Phi^\pm, \tag{15.161}$$

where $q = \pm 1$. If we restrict ourselves to the solutions (15.159) associated with only one specific eigenvalue of q, we can establish a unique correspondence between these solutions and those for the Dirac equation. Multiplying, for example, the equality (15.157) by

$$1/2(1 - \gamma_5)$$

and using Eq. (15.161) with $q = 1$, we obtain:

$$\Phi^+ = \frac{1}{2}(1 - \gamma_5)\psi, \tag{15.162}$$

that is, each function ψ corresponds to only one Φ^+ function.

Provided for the γ-matrices the chiral representation is chosen, then from (15.162) it follows that Φ^+ is a two-component spinor. Then taking into account the relations:

$$\sigma_{10} = i\gamma_1\gamma_0 = \gamma_2\gamma_3(i\gamma_0\gamma_1\gamma_2\gamma_3) = i\sigma_{23}\gamma_5,$$

$$\sigma_{20} = i\sigma_{31}\gamma_5, \qquad \sigma_{30} = i\sigma_{12}\gamma_5$$

and using for the γ-matrices the representation (15.7), we obtain the following equation for the spinor Φ^+:

$$\left[D_\mu D^\mu + m^2 - e\boldsymbol{\sigma}\cdot(\boldsymbol{\mathcal{H}} + i\boldsymbol{\mathcal{E}})\right]\Phi^+ = 0. \tag{15.163}$$

It should be noted that because

$$\overline{\Phi}^+\Phi^+ = \frac{1}{4}\overline{\psi}(1 + \gamma_5)(1 - \gamma_5)\psi = 0,$$

then we have no possibility to construct the Lagrangian from which Eq. (15.163) might follow.

For simplicity, we assume: $\boldsymbol{\mathcal{E}} = A_0 = 0$ and come back to the ordinary units. Then Eq. (15.163) takes the shape:

$$\left[\frac{\hbar^2}{c^2}\partial_t^2 - \left(\hbar\partial_k + \frac{ie}{c}A_k\right)^2 + m^2c^2 - \frac{e\hbar}{c}(\boldsymbol{\sigma}\cdot\boldsymbol{\mathcal{H}})\right]\Phi^+ = 0. \tag{15.164}$$

Being a classical analog of the orbital moment, the spin was first introduced into nonrelativistic quantum mechanics. Because of this, the most common (and illustrative) way to determine a value of the spin magnetic moment is finding of the Hamiltonian in a nonrelativistic approximation. The reference points of the energy in the nonrelativistic theory and in the special theory of relativity are distinguished by mc^2 and hence it is more convenient to introduce for the spinor Φ^+ the following transformation:

$$\Phi^+(\mathbf{r}, t) = \Phi'(\mathbf{r}, t)\exp\left(-\frac{imc^2t}{\hbar}\right). \tag{15.165}$$

Substituting (15.165) into (15.164) and neglecting the terms higher than the first order in terms of v/c, we derive:

$$i\hbar\frac{\partial\Phi'}{\partial t} = H\Phi', \tag{15.166}$$

where

$$H = \frac{1}{2m}\left(\mathbf{p} - \frac{e}{c}\mathbf{A}\right)^2 - \frac{e\hbar}{2mc}(\boldsymbol{\sigma}\cdot\boldsymbol{\mathcal{H}}). \tag{15.167}$$

Eq. (15.166) is distinguished from the nonrelativistic Schrödinger equation by the term taking the form of the potential energy of a magnetic dipole in the external field. Thus, the electron magnetic moment is defined by the expression:

$$\boldsymbol{\mu}_s = \frac{e\hbar}{2mc}\boldsymbol{\sigma} = 2\frac{e\hbar}{2mc}\mathbf{S}, \tag{15.168}$$

from which follows that the gyromagnetic ratio g_s really equals 2.

15.8 Charge conjugation, spatial inversion, and time inversion for the Dirac field

The holes theory implies the existence of electrons and positrons satisfying the same equation and having the identical mechanical and opposite electromagnetic characteristics. Consequently, the Dirac equation should possess a symmetry about the sign changing operation for the electron charge, the charge conjugation operation. To put this another way, when in an electromagnetic field the electron is described by the equation:

$$[\gamma^\mu(i\partial_\mu - eA_\mu) - m]\psi(x) = 0, \tag{15.169}$$

then the relevant equation for the positron has the form:

$$[\gamma^\mu(i\partial_\mu + eA_\mu) - m]\psi^c(x) = 0 \tag{15.170}$$

and there is one-to-one correspondence between $\psi(x)$ and $\psi^c(x)$. Direct our efforts to determining the connection between Eqs. (15.169) and (15.170). Carrying out the complex conjugation operation in Eq. (15.169), we obtain:

$$[-\gamma^{\mu*}(i\partial_\mu + eA_\mu) - m]\psi^*(x) = 0. \tag{15.171}$$

Obviously, if one could manage to find a matrix of the shape $C\gamma_0$ with properties:

$$-(C\gamma_0)\gamma^{\mu*} = \gamma^\mu(C\gamma_0), \tag{15.172}$$

Eq. (15.171) might be represented in the form of Eq.(15.170), namely,

$$[\gamma^\mu(i\partial_\mu + eA_\mu) - m]C\gamma_0\psi^*(x) = 0.$$

Then the connection sought takes the form:

$$\psi^c(x) = C\gamma_0\psi^*(x) = C\overline{\psi}^T. \tag{15.173}$$

The matrix C is called the *charge conjugation matrix* and $\psi^c(x)$—the *charge conjugate bispinor*. With the help of the relation

$$\gamma_0\gamma^\mu\gamma_0 = \gamma^{\mu\dagger},$$

the condition determining the matrix C could be rewritten in the form that is more convenient for practical applications:

$$C^{-1}\gamma^{\mu}C = -\gamma^{\mu T}. \tag{15.174}$$

Since the matrices γ_{μ}^{T} satisfy the same anticommutation relations as the matrices γ_{μ}, then, according to the fundamental theorem about the Dirac matrices (see, Section 15.2) the matrix C really exists. Using (15.174) and taking into consideration the γ-matrix algebra, one could show that the matrix C possesses the following properties:

$$C^{\dagger} = C^{-1} = C = -C^{T} = C^{*}. \tag{15.175}$$

Let us establish how the spinor $\overline{\psi}(x)$ is transformed under the charge conjugation. Since:

$$\psi(x) \to \psi^{c}(x) = C\overline{\psi}^{T}(x) = C[\psi^{\dagger}(x)\gamma_0]^{T} = C\gamma_0^{T}\psi^{\dagger T}(x), \tag{15.176}$$

then having done the Hermitian conjugation in Eq. (15.176), we obtain:

$$\psi^{\dagger}(x) \to (\psi^{c}(x))^{\dagger} = \psi^{T}(x)\gamma_0^{\dagger T}C^{\dagger}. \tag{15.177}$$

To multiply the both sides of Eq. (15.177) by γ_0 and account for (15.174), (15.175) gives:

$$\overline{\psi}(x) \to (\psi^{c}(x))^{\dagger}\gamma_0 = \psi^{T}(x)\gamma_0^{\dagger T}C^{\dagger}\gamma_0 = \psi^{T}(x)\gamma_0(-\gamma_0 C^{\dagger}) = -\psi^{T}(x)C,$$

that allows one to conclude:

$$\overline{\psi}^{c}(x) = -\psi^{T}(x)C. \tag{15.178}$$

So, due to the invariance with respect to the charge conjugation operation, a change in sign for the four-dimensional potential of the electromagnetic field:

$$A_{\mu}(x) \to A_{\mu}^{c}(x) = -A_{\mu}(x), \tag{15.179}$$

with simultaneous antilinear transformations (15.173) and (15.178) for the bispinors $\psi(x)$ and $\overline{\psi}(x)$, leads to the equations for the bispinors $\psi^{c}(x)$ and $\overline{\psi}^{c}(x)$ in the field $A_{\mu}^{c}(x)$, which in their form are identical to the Dirac equations for $\psi(x)$ and $\overline{\psi}(x)$ in the field $A_{\mu}(x)$. This means that the total Lagrangian of this system *electromagnetic field + Dirac field* retains its form under the transformations (15.173), (15.178), and (15.179). Considering that the electromagnetic field potential is included into the Dirac equations with the coefficient e,, the transformation (15.179) should be interpreted as a change in the charge sign. Invariance of the Dirac equation with respect to the transformation of the charge conjugation points to the equal rights of two signs of the charge: for each solution $\psi(x)$ associated with a certain sign of the charge there is the solution $C\overline{\psi}^{T}(x)$ having the same properties but corresponding to the opposite-sign charge.

In the Dirac-Pauli representation the matrix C has the form:

$$C = -i\gamma_2\gamma_0 = \begin{pmatrix} 0 & -i\sigma_2 \\ -i\sigma_2 & 0 \end{pmatrix}. \tag{15.180}$$

When passing on to other representations of γ-matrices, a new matrix of the charge conjugation, C', may be obtained by means of the formula:

$$C' = UCU^{T}. \tag{15.181}$$

In the assumption of the relativistic covariance of the charge conjugation operation the bispinors $\psi^{c}(x)$ and $\psi(x)$ should have the same transformational properties under the Poincare transformations, that is,

$$\psi^{c\prime}(x') = S\psi^{c}(x), \tag{15.182}$$

$$\psi'(x') = S\psi(x). \tag{15.183}$$

Substituting Eq. (15.173) into (15.182) and allowing for

$$\overline{\psi}'(x') = a\overline{\psi}(x)S^{-1},$$

we arrive at the result:

$$S^T = \begin{cases} C^{-1}S^{-1}C, & \text{at} \quad \Lambda^{00} > 0, \\ -C^{-1}S^{-1}C, & \text{at} \quad \Lambda^{00} < 0. \end{cases} \tag{15.184}$$

Next, we show that the conditions (15.184) make it possible to decrease the number of the admissible values for the matrices performing the space-inversion and time-inversion transformations.

Before we have found out that under the space inversion the bispinor $\psi(x)$ is transformed by the following way:

$$\psi'(-\mathbf{r}, t) = P\psi(\mathbf{r}, t) = \eta_p\gamma_0\psi(\mathbf{r}, t), \tag{15.185}$$

where η_p is an intrinsic parity of a spinor particle equal either to ± 1 or to $\pm i$. It is an easy matter to convince that the fulfillment of (15.184) is guaranteed by the choice $\eta_p = \pm i$. To establish the transformation law of the Dirac's conjugate bispinor we carry out the Hermitian conjugation on (15.185) and multiply the result obtained by γ_0 on the right

$$\overline{\psi}'(-\mathbf{r}, t) = \eta_p^*\overline{\psi}(\mathbf{r}, t)\gamma_0. \tag{15.186}$$

Therefore, the intrinsic parities of the Dirac particles and antiparticles are opposite.

Consider the time-inversion transformation. From Section 5.2 we know that this transformation represents the product of the complex conjugation operation of a wavefunction and some unitary transformation which we denote by T. Thus, for a time-inverted wavefunction $\psi'(\mathbf{r}, -t)$ we have:

$$\psi'(\mathbf{r}, \quad t) = T\overline{\psi}(\mathbf{r}, t). \tag{15.187}$$

To find an explicit form of a matrix T we write the Dirac equations to be satisfied by the functions $\psi'(\mathbf{r}, t)$ and $\overline{\psi}(\mathbf{r}, -t)$: $\psi'(\mathbf{r}, t)$ and $\overline{\psi}(\mathbf{r}, -t)$

$$(i\gamma_0\partial_t - i\boldsymbol{\gamma}\nabla - m)\psi'(\mathbf{r}, t) - 0, \tag{15.188}$$

$$(-i\gamma_0^T\partial_t - i\boldsymbol{\gamma}^T\nabla + m)\overline{\psi}(\mathbf{r}, -t) = 0. \tag{15.189}$$

To multiply the last equation by $-T$ on the left gives

$$(iT\gamma_0^T\partial_t + iT\boldsymbol{\gamma}^T\nabla - mT)\overline{\psi}(\mathbf{r}, -t) = 0. \tag{15.190}$$

Now we demand the function $T\overline{\psi}(\mathbf{r}, -t)$ to obey the same equation as $\psi'(\mathbf{r}, t)$:

$$(i\gamma_0\partial_t - i\boldsymbol{\gamma}\nabla - m)T\overline{\psi}(\mathbf{r}, -t) = 0. \tag{15.191}$$

Comparing Eqs. (15.188) and (15.191), we have drawn the conclusion that the matrix T must satisfy the conditions:

$$T\gamma_0^T = \gamma_0 T, \qquad T\boldsymbol{\gamma}^T = -\boldsymbol{\gamma}T. \tag{15.192}$$

It is easy to verify that in the Dirac-Pauli and Weyl representations the matrix

$$T = \eta_t\gamma^3\gamma^1\gamma^0, \tag{15.193}$$

fulfills these conditions.* So we finally have

$$\psi'(\mathbf{r}, -t) = T\overline{\psi}(\mathbf{r}, t) = \eta_t \gamma^3 \gamma^1 \gamma^0 \overline{\psi}(\mathbf{r}, t). \tag{15.194}$$

Analogous manipulation shows that under time inversion the conjugate bispinor is transformed by the law:

$$\overline{\psi}'(\mathbf{r}, -t) = \eta_t^* \psi(\mathbf{r}, t) \gamma^0 \gamma^1 \gamma^3. \tag{15.195}$$

Using the explicit form of the spin projection operator it is easy to display that under the time inversion the helicity changes its sign.

Let us find the changeover of the field function of a spinor field under the PT and CPT operations in the Dirac-Pauli and Weyl representations. Calculations lead to the result:

$$PT\psi(\mathbf{r}, t) = \eta_p \eta_t \gamma^0 \gamma^3 \gamma^1 \psi^*(-\mathbf{r}, -t), \tag{15.196}$$

$$CPT\psi(\mathbf{r}, t) = i\gamma^2 \gamma^0 [\eta_p \eta_t \gamma^3 \gamma^1 \psi^*(-\mathbf{r}, -t)]^* = -\eta_0 \gamma_5 \psi(-\mathbf{r}, -t)], \tag{15.197}$$

where $\eta_0 = \eta_t^* \eta_p^*$.

15.9   Relativistic quantization scheme

The canonically conjugate momenta in the case of the Dirac field are determined by the following expressions:

$$\pi(x) = \frac{i}{2}\psi^\dagger(x), \qquad \pi^\dagger(x) = -\frac{i}{2}\psi(x),$$

from where it follows that they are no longer independent of the "coordinates" of the field $\psi(x)$ and $\psi^\dagger(x)$. This makes us doubtful whether the use of the canonical quantization procedure is appropriate. From the simultaneous commutation relations for $\psi(x)$ and $\pi(x')$:

$$\{\psi(x), \pi(x')\}_{t=t'} = i\delta^{(3)}(\mathbf{r} - \mathbf{r}'),$$

it follows that

$$\frac{1}{2}\{\psi(x), \psi^\dagger(x')\}_{t=t'} = \delta^{(3)}(\mathbf{r} - \mathbf{r}'). \tag{15.198}$$

On the other hand, from similar relations for the canonically conjugate pair $\psi^\dagger(x)$, $\pi^\dagger(x')$ we have

$$-\frac{1}{2}\{\psi^\dagger(x), \psi(x')\}_{t=t'} = \delta^{(3)}(\mathbf{r} - \mathbf{r}'),$$

that is contrary to Eq. (15.198).

Thus, the formalism of the canonical quantization is directly inapplicable, and we have a nice opportunity to broaden our range of interests owing to the introduction of the so-called relativistic quantization procedure. It is based on a theory of representations requiring correspondence between the transformation laws for the quantized and classical fields, both under transformations of the coordinate systems and the fields themselves. Its essence is as follows.

The transformations of the coordinates and field functions:

$$x' = \Lambda x + a, \qquad \psi'(x') = S(\Lambda)\psi(x).$$

---

*(According to (15.184) η_t must equal $\pm i$)

induce a change in the state vector of the system Φ that is determined by the relation:

$$\Phi' = U_v \Phi. \tag{15.199}$$

Invariance of the norm of the state vector requires unitarity of the operator U_v while the superposition principle imposes the linearity condition on it. We consider this situation from an experimenter's viewpoint. Let us think of some dynamic observable (for example, momentum $P_\mu(x)$) as being measured in the states described by the vectors Φ and Φ'. Then a relativistic covariance requires the following relations to be fulfilled:

$$(\Phi', P_\nu(x')\Phi') = \Lambda_\nu^\mu(\Phi, P_\mu(x)\Phi). \tag{15.200}$$

This in turn leads to the requirement that is relating expressions for the momentum operator in both systems:

$$U_v^{-1} P_\mu(x') U_v = \Lambda_\mu^\nu P_\nu(x). \tag{15.201}$$

Similarly, when computing the average over the operator field function, the resultant expression retains a geometric nature of the nonquantized field function, that is, the relation:

$$(\Phi', \psi(x')\Phi') = S(\Lambda)(\Phi, \psi(x)\Phi) = (\Phi, \psi'(x')\Phi). \tag{15.202}$$

takes place. Thus, for the field with arbitrary spin value we obtain from Eq. (15.202)

$$U_v^{-1} \psi(x') U_v = S(\Lambda)\psi(x) \tag{15.203}$$

or

$$\psi'(x') = U_v^{-1} \psi(x') U_v. \tag{15.204}$$

In the case of infinitesimal transformations of the Poincare group the transformation matrix of a state vector could be represented in the shape:

$$U_v = 1 + \delta U_v, \qquad \delta U_v = i \sum_n U_n \delta\omega^n, \tag{15.205}$$

where $\delta\omega^n$ are transformation parameters. Then for infinitesimal transformations from Eq. (15.204) follows:

$$\psi'(x') = (1 - \delta U_v)\psi(x')(1 + \delta U_v) \approx \psi(x') + [\psi(x'), \delta U_v], \tag{15.206}$$

which leads to the following expression:

$$\bar{\delta}\psi(x) = \psi'(x) - \psi(x) = [\psi(x), \delta U_v]. \tag{15.207}$$

Using the explicit form of the form variation for the field function $\bar{\delta}\psi(x)$ (1.4.17), we get the conditions on the operator field functions that at the heart of relativistic quantization procedure:

$$i\left(Y_{i(n)} - \psi_{i;k} X_{(n)}^k\right) = [U_n, \psi_i]. \tag{15.208}$$

In order to make the direct application the formula (15.208) should be rewritten for concrete transformations. In the case of space-time translation:

$$X_{(n)}^k = \delta_n^k, \qquad Y_{i(n)} = 0, \qquad U_n = P_n,$$

the relation (15.208) takes the form:

$$i\frac{\partial \psi_i(x)}{\partial x^n} = [\psi_i(x), P_n]. \tag{15.209}$$

Recall that in the nonrelativistic quantum mechanics Heisenberg's operators $N(x)$ obeyed the motion equations:

$$i\frac{\partial N(x)}{\partial t} = [N(x), H],\qquad (15.210)$$

where H is a system Hamiltonian. Since in the quantum field theory we deal with operators dependent on space-time coordinates in the same sense as the Heisenberg's operators of nonrelativistic quantum mechanics are time-dependent, then a generalization of Eq. (15.210) for the space coordinates must exist as well. Such a relativistic generalization of Eq. (15.210) is represented by Eq. (15.209) derived by us for the partial case, when the field function itself plays a role of the operator $N(x)$.

Under four-dimensional rotations the condition (15.208) leads to the relation:

$$i\left\{ \psi_{i;\sigma}(x)(x_\nu \delta_\mu^\sigma - x_\mu \delta_\nu^\sigma) - (K_{\mu\nu})_i{}^j \psi_j(x) \right\} = [\psi_i(x), M_{\mu\nu}],\qquad (15.211)$$

where $M_{\mu\nu}$ is a tensor operator of a total angular momentum and we have taken into account Eqs. (1.4.15) and (1.4.34). From obtained formulae it is also obvious that under the Poincare transformations the state vector is transformed by the law:

$$\Phi' = \exp\left(1 + iP_\mu a^\mu + \frac{i}{2}M_{\mu\nu}\omega^{\mu\nu}\right).\qquad (15.212)$$

As an example of the transformations not touching upon space-time coordinates we consider the gauge transformations of the first kind.[*] In this case the state vector transformation law has the form:

$$\Phi' = \exp(i\alpha Q)\Phi,\qquad (15.213)$$

where Q should be interpreted as an charge operator (electric, baryon, etc.) of a secondary quantized field. Now the formula (15.208) leads to the relations:

$$\psi(x) = [\psi(x), Q],\qquad \psi^*(x) = -[\psi^*(x), Q].\qquad (15.214)$$

Eqs. (15.209), (15.211), and (15.214) form the base of the relativistic quantization procedure. With their use one can establish both a physical meaning of the operator field functions and the permutation relations associated with them.

15.10 Dirac field quantization (momentum space)

Let us apply the above-mentioned procedure to the Dirac particles. The form of the permutation relations may be most conveniently determined by Eq. (15.209). In this manner we should begin with the construction of an energy-momentum tensor. The standard manipulation leads to the expression:

$$T^{\mu\nu} = \frac{i}{2}\left[\overline{\psi}(x)\gamma^\nu \partial^\mu \psi(x) - \partial^\mu \overline{\psi}(x)\gamma^\nu \psi(x)\right].\qquad (15.215)$$

Though asymmetric, this tensor results in the same value for four-dimensional vector of the field momentum as does the symmetrized tensor:

$$P^\mu = \int T^{\mu 0} d^3 r = \frac{i}{2}\int d^3 r \left[\overline{\psi}(x)\gamma^0 \partial^\mu \psi(x) - \partial^\mu \overline{\psi}(x)\gamma^0 \psi(x)\right] =$$

---

[*]Sometimes they are simply called *gradient transformations*.

$$= \frac{i}{2} \int d^3r \left\{ 2\overline{\psi}(x)\gamma^0\partial^\mu\psi(x) - \partial^\mu[\overline{\psi}(x)\gamma^0\psi(x)] \right\} = i \int d^3r[\overline{\psi}(x)\gamma^0\partial^\mu\psi(x)]. \qquad (15.216)$$

During deriving (15.216) we have used the Gauss theorem and allowed for the electric current conservation law

$$\partial^\mu j_\mu^{(em)} = 0, \qquad (15.217)$$

where

$$j_\mu^{(em)} = e\overline{\psi}(x)\gamma_\mu\psi(x). \qquad (15.218)$$

Further on we expand the operators $\psi(x)$ and $\overline{\psi}(x)$ in terms of the c-number solutions of the Dirac equation:

$$\psi(x) = \psi^{(+)}(x) + \psi^{(-)}(x) =$$

$$= \frac{1}{(2\pi)^{3/2}} \int d^3p \sqrt{\frac{m}{E_p}} \sum_{r=1}^{2} [a_r(\mathbf{p})u^{(r)}(\mathbf{p})e^{-ipx} + b_r^\dagger(\mathbf{p})v^{(r)}(\mathbf{p})e^{ipx}], \qquad (15.219)$$

$$\overline{\psi}(x) = \overline{\psi^{(+)}}(x) + \overline{\psi^{(-)}}(x) =$$

$$= \frac{1}{(2\pi)^{3/2}} \int d^3p \sqrt{\frac{m}{E_p}} \sum_{r=1}^{2} [a_r^\dagger(\mathbf{p})\overline{u}^{(r)}(\mathbf{p})e^{ipx} + b_r(\mathbf{p})\overline{v}^{(r)}(\mathbf{p})e^{-ipx}], \qquad (15.220)$$

where $a_r(\mathbf{p})$ and $b_r(\mathbf{p})$ are operators. To substitute (15.209) and (15.210) into the expression for the four-dimensional momentum vector (15.217) and take into account the orthonormalization condition for bispinors (15.99) gives the following result:

$$P^\mu = \int d^3p \, p^\mu \sum_{r=1}^{2} [a_r^\dagger(\mathbf{p})a_r(\mathbf{p}) - b_r(\mathbf{p})b_r^\dagger(\mathbf{p})]. \qquad (15.221)$$

The form of expression (15.221) is almost identical to that of the four-dimensional momentum vector of a scalar field. However, a change of the sign before the second term is very dramatic within the scope of the primary-quantized theory. The energy is not positively defined (indefinite), that is, the energy spectrum has no lower limit. Recall that in the case of a scalar field the charge is indefinite instead. In this case the situation may be improved by the use of the commutation relations capable of providing a change of the sign in front of the second term in (15.221) when taking it to the traditional form of the antiparticle number operator. Obviously, quantization with the commutators presents no definiteness of the energy, whereas the problem is completely solved due to the use of anticommutators. Proceeding further, we should remember this fact. With the use of the operator expansion of the field functions (15.219), (15.220) and with the help of (15.209) we can establish that the translational invariance is liable to take place under fulfillment of the following conditions:

$$\left.\begin{array}{ll} [P_\mu, a_s(\mathbf{p})] = -p_\mu a_s(\mathbf{p}), & [P_\mu, b_s(\mathbf{p})] = -p_\mu b_s(\mathbf{p}), \\[2mm] [P_\mu, a_s^\dagger(\mathbf{p})] = p_\mu a_s^\dagger(\mathbf{p}), & [P_\mu, b_s^\dagger(\mathbf{p})] = p_\mu b_s^\dagger(\mathbf{p}). \end{array}\right\} \qquad (15.222)$$

Insertion of the expression for P_μ into the first of the relations (15.222) yields the result:

$$\left[\sum_r \left(a_r^\dagger(\mathbf{p}')a_r(\mathbf{p}') - b_r(\mathbf{p}')b_r^\dagger(\mathbf{p}')\right), a_s(\mathbf{p})\right] = -\delta^{(3)}(\mathbf{p} - \mathbf{p}')a_s(\mathbf{p}). \qquad (15.223)$$

As the operators $a_s(\mathbf{p})$ and $b_s(\mathbf{p})$ are associated with different degrees of freedom of the field, we should have:

$$[b_r(\mathbf{p}')b_r^\dagger(\mathbf{p}'), a_s(\mathbf{p})] = 0. \qquad (15.224)$$

Then the relation (15.223) takes the form:

$$\sum_r \left(a_r^\dagger(\mathbf{p}')\{a_r(\mathbf{p}'), a_s(\mathbf{p})\} - \{a_r^\dagger(\mathbf{p}'), a_s(\mathbf{p})\} a_r(\mathbf{p}') \right) = -\delta^{(3)}(\mathbf{p} - \mathbf{p}') a_s(\mathbf{p}). \qquad (15.225)$$

The obvious solution of (15.225) is as follows:

$$\{a_r(\mathbf{p}'), a_s(\mathbf{p})\} = 0, \qquad (15.226)$$

$$\{a_r(\mathbf{p}'), a_s^\dagger(\mathbf{p})\} = \delta^{(3)}(\mathbf{p} - \mathbf{p}')\delta_{rs}. \qquad (15.227)$$

Analogously, from the remaining relations (15.223) we can obtain:

$$\{b_r(\mathbf{p}'), b_s(\mathbf{p})\} = 0, \qquad (15.228)$$

$$\{b_r(\mathbf{p}'), b_s^\dagger(\mathbf{p})\} = \delta^{(3)}(\mathbf{p} - \mathbf{p}')\delta_{rs}. \qquad (15.229)$$

If the state Φ_E is the energy operator eigenstate with the eigenvalue E, then, with allowance made for the relations (15.222), we have:

$$H a_r(\mathbf{p})\Phi_E = a_r(\mathbf{p}) H \Phi_E + [H, a_r(\mathbf{p})]\Phi_E = (E - k_0) a_r(\mathbf{p})\Phi_E, \qquad (15.230)$$

$$H a_r^\dagger(\mathbf{p})\Phi_E = (E + k_0) a_r^\dagger(\mathbf{p})\Phi_E. \qquad (15.231)$$

Similar relations takes place for the states $b_r(\mathbf{p})\Phi_E$ and $b_r^\dagger(\mathbf{p})\Phi_E$. Thus, the operators $a_r(\mathbf{p})$ and $b_r(\mathbf{p})$ have the meaning of annihilation operators because they decrease the system's energy, while $a_r^\dagger(\mathbf{p})$ and $b_r^\dagger(\mathbf{p})$—the meaning of production operators as they increase the energy of this system.

To establish a physical meaning of the operators $a_r^\dagger(\mathbf{p})$ and $b_r^\dagger(\mathbf{p})$ definitively, we should have to use the expressions (15.211) and (15.214). But since the anticommutation relations between the operators have been already found, the situation may be simplified. Using the expression for the spin moment tensor, we arrive at the spin projection operator on a motion direction:

$$S_3 = \frac{1}{2} \int d^3 p [a_1^\dagger(\mathbf{p}) a_1(\mathbf{p}) - a_2^\dagger(\mathbf{p}) a_2(\mathbf{p}) + b_1(\mathbf{p}) b_1^\dagger(\mathbf{p}) - b_2(\mathbf{p}) b_2^\dagger(\mathbf{p})]. \qquad (15.232)$$

where the vector $\mathbf{p}$ is directed along the axis z. Then, with consideration for the relations (15.227) and (15.229) we shall have

$$[S_3, a_r^\dagger(\mathbf{p})] = \begin{cases} \frac{1}{2} a_1^\dagger(\mathbf{p}), & r = 1, \\ -\frac{1}{2} a_2^\dagger(\mathbf{p}), & r = 2, \end{cases} \qquad (15.233)$$

$$[S_3, b_r^\dagger(\mathbf{p})] = \begin{cases} -\frac{1}{2} b_1^\dagger(\mathbf{p}), & r = 1, \\ \frac{1}{2} b_2^\dagger(\mathbf{p}), & r = 2. \end{cases} \qquad (15.234)$$

Using the Lagrangian (15.13), we calculate the expression for the electric charge of a field:

$$Q = \int d^3 p \sum_{r=1}^{2} e[a_r^\dagger(\mathbf{p}) a_r(\mathbf{p}) + b_r(\mathbf{p}) b_r^\dagger(\mathbf{p})]. \qquad (15.235)$$

Note that the charge is positively defined even before the application of the quantization rules. This is inferred as a partial corollary to the Pauli theorem about the connection of the spin with statistics. By the classical field theory, the energy for the particles with a half-integer spin is indefinite and for those with an integer spin the charge is indefinite instead.

To obviate these inconsistencies, fermions are quantized with the use of the anticommutators, and bosons—with commutators. Using the anticommutators, we find without trouble:

$$[Q, a_r^\dagger(\mathbf{p})] = e a_r^\dagger(\mathbf{p}), \qquad [Q, b_r^\dagger(\mathbf{p})] = -e b_r^\dagger(\mathbf{p}). \tag{15.236}$$

Our further reasoning is quite clear. We take the state with a definite value of the charge and spin projection Φ_{Q,S_3}. Acting by the operators Q and S_3 on the states $a_r^\dagger(\mathbf{p})\Phi_{Q,S_3}$ and $b_r^\dagger(\mathbf{p})\Phi_{Q,S_3}$, and using the relations (15.234), (15.235) and (15.237), we obtain the following ultimate answer. The quantities $a_r^\dagger(\mathbf{p})$ and $a_r(\mathbf{p})$ represent the production and annihilation operators for particles with the momentum p, mass m ($p^2 = m^2$), charge $-|e|$, and spin projection onto the z-axis that is equal to $1/2$ ($r = 1$) or $-1/2$ ($r = 2$). The operators $b_r^\dagger(\mathbf{p})$ and $b_r(\mathbf{p})$ are associated with antiparticles, being distinguished from the foregoing by the signs of a charge ($+|e|$) and spin projection $-1/2$ for $r = 1$ and $1/2$ for $r = 2$. Such sign reversals are in line with the predictions of Dirac's theory of holes. From this viewpoint, a positron is understood as the absence of an electron with the negative energy. If the missing electron has a positive spin projection $1/2$ and is charged negatively, its absence results in the negative spin projection $-1/2$ and the positive charge.

We introduce the operators of the particles number $N_r^{(+)}(\mathbf{p})$ and the antiparticles number $N_r^{(-)}(\mathbf{p})$:

$$N_r^{(+)}(\mathbf{p}) = a_r^\dagger(\mathbf{p}) a_r(\mathbf{p}), \qquad N_r^{(-)}(\mathbf{p}) = b_r^\dagger(\mathbf{p}) b_r(\mathbf{p}). \tag{15.237}$$

With their help an energy and a charge of a field are represented in the form:

$$H = \int d^3p \sum_{r=1}^{2} E_p [N_r^{(+)}(\mathbf{p}) + N_r^{(-)}(\mathbf{p})] + E_0, \tag{15.238}$$

$$Q = \int d^3p \sum_{r=1}^{2} e [N_r^{(+)}(\mathbf{p}) - N_r^{(-)}(\mathbf{p})] + Q_0. \tag{15.239}$$

In the formulae (15.238), (15.239) E_0 and Q_0 denote the energy and total charge of the vacuum:

$$E_0 = -2 \int d^3p E_p, \qquad Q_0 = -2e \int d^3p.$$

By the holes theory, we can assume that the observables represent the differences $H - E_0$ and $Q - Q_0$ only. But in the spirit of quantum field theory another approach (normal products method) seems to be more appropriate, being also helpful in avoidance of such infinities. As we pass to the normal products, beginning just from the Lagrangian, we get the expressions for dynamic variables involving no infinite vacuum contributions. From the anticommutation relations it follows:

$$N_r^{(+)2}(\mathbf{p}) = a_r^\dagger(\mathbf{p}) a_r(\mathbf{p}) a_r^\dagger(\mathbf{p}) a_r(\mathbf{p}) = a_r^\dagger(\mathbf{p})[1 - a_r^\dagger(\mathbf{p}) a_r(\mathbf{p})] a_r(\mathbf{p}) = a_r^\dagger(\mathbf{p}) a_r(\mathbf{p}) = N_r^{(+)}(\mathbf{p}),$$

where we have allowed for that according to (15.226) $a_r(\mathbf{p}) a_r(\mathbf{p}) = 0$. In this way for fermions the particles number in one state may be only zero or unity in agreement with the Pauli principle. We characterize the vacuum state by the conditions:

$$N_r^{(+)}(\mathbf{p})\Phi_0 = N_r^{(-)}(\mathbf{p})\Phi_0 = 0,$$

to be true for all r and $\mathbf{p}$. It is obvious that the state $a_r^\dagger(\mathbf{p})\Phi_0$ is the single particle state with the momentum p, the mass m and the spin projection $\pm 1/2$ for $r = 1, 2$; $a_{r_2}^\dagger(\mathbf{p}_2) a_{r_1}^\dagger(\mathbf{p}_1)\Phi_0$ is two particle state ..., $b_r^\dagger(\mathbf{p})\Phi_0$ is the single antiparticle state and so on.

Taking into account of Eqs. (15.254) and (15.265), in the expansion of the field function (15.219) we can go from the sum over r to the sum over the eigenvalues of the spin projection operator $J_3 = \pm 1/2$ ($\alpha = \pm$):

$$\psi(x) = \frac{1}{(2\pi)^{3/2}} \int d^3p \sqrt{\frac{m}{E_p}} \sum_{\alpha = \pm} [a_\alpha(\mathbf{p})u^{(\alpha)}(\mathbf{p})e^{-ipx} + b_\alpha^\dagger(\mathbf{p})v^{(\alpha)}(\mathbf{p})e^{ipx}], \qquad (15.240)$$

where the bispinors $u^{(\alpha)}(\mathbf{p})$ and $v^{(\alpha)}(\mathbf{p})$ are now defined the following way:

$$u^{(\pm)}(\mathbf{p}) = N \begin{pmatrix} \varsigma^\pm \\ \dfrac{(\mathbf{p} \cdot \boldsymbol{\sigma})}{E_p + m} \varsigma^\pm \end{pmatrix}, \qquad v^{(\pm)}(\mathbf{p}) = N \begin{pmatrix} \dfrac{(\mathbf{p} \cdot \boldsymbol{\sigma})}{E_p + m} \varsigma^\mp \\ \varsigma^\mp \end{pmatrix}. \qquad (15.241)$$

Using the orthonormality properties of the solutions for the Dirac equation, we can realize the inversion of Eq. (15.240). Calculations give

$$a_\alpha(\mathbf{p}) = \frac{1}{(2\pi)^{3/2}} \int d^3x e^{ipx} \overline{u}^{(\alpha)}(\mathbf{p})\gamma_0 \psi(x), \qquad (15.242)$$

$$b_\alpha(\mathbf{p}) = \frac{1}{(2\pi)^{3/2}} \int d^3x \overline{\psi}(x)\gamma_0 v^{(\alpha)}(\mathbf{p}) \qquad (15.243)$$

and so on. Based on the derived relations, we can write the state vector of this system in the Fock representation

$$\Phi_{n_{\mathbf{k}_1}^+ n_{\mathbf{k}_1'}^- \cdots} = \int \cdots \int F(x_1, ..., x_n)\psi^{b_1}(x_1)...\psi^{b_n}(x_n)d^4x_1...d^4x_n \Phi_0. \qquad (15.244)$$

where $b_i = \pm$, $\psi^+(x) \equiv \psi(x)$, $\psi^-(x) \equiv \overline{\psi}(x)$ and $F(x_1, ..., x_n)$ has a meaning of the ordinary quantum-mechanical wavefunction of the system including n particles in the configuration space.

15.11   Dirac field quantization (configuration space)

In the configuration space a role of the production and annihilation operators of the electrons (positrons) is played by the quantities $\overline{\psi^{(+)}}(x)$ ($\psi^{(-)}(x)$) and $\psi^{(+)}(x)$ ($\overline{\psi^{(-)}}(x)$), respectively. Through the use of these operators the vacuum is set by the condition:

$$\psi^{(+)}(x)\Phi_0 = \overline{\psi^{(-)}}(x)\Phi_0 = 0.$$

Nonzero anticommutation relations for the number operators have the form:

$$\{\psi_\alpha^{(+)}(x), \overline{\psi^{(+)}}_\beta(x')\} =$$

$$= \frac{1}{(2\pi)^3} \int d^3p \int d^3p' \sqrt{\frac{m^2}{E_p E_{p'}}} \sum_{r,s=1}^{2} \{a_r(\mathbf{p}), a_s^\dagger(\mathbf{p}')\} v_\alpha^{(r)}(\mathbf{p})\overline{v}_\beta^{(s)}(\mathbf{p}') \times$$

$$\times \exp[-ipx + ip'x'] = \frac{1}{2(2\pi)^3} \int \frac{d^3p}{E_p} (\hat{p} + m)_{\alpha\beta} \exp[-ip(x - x')] =$$

$$= \frac{1}{2(2\pi)^3} (i\gamma^\mu \partial_\mu + m)_{\alpha\beta} \int \frac{d^3 p}{E_p} \exp\left[-ip(x - x')\right] =$$

$$= -i(i\gamma^\mu \partial_\mu + m)_{\alpha\beta} \Delta_0^+(x - x'; m^2) = -iS_{\alpha\beta}^+(x - x'). \tag{15.245}$$

Analogous manipulations for the production and annihilation operators of the positrons lead to the result:

$$\{\psi_\alpha^{(-)}(x), \overline{\psi^{(-)}}_\beta(x')\} = -iS_{\alpha\beta}^-(x - x'), \tag{15.246}$$

where

$$S_{\alpha\beta}^-(x) = (i\gamma^\mu \partial_\mu + m)_{\alpha\beta} \Delta_0^-(x; m^2). \tag{15.247}$$

With allowance made for the definition

$$S_{\alpha\beta}(x) = S_{\alpha\beta}^+(x) + S_{\alpha\beta}^-(x) = (i\gamma^\mu \partial_\mu + m)_{\alpha\beta} \Delta_0(x; m^2), \tag{15.248}$$

we arrive at the Dirac fields anticommutator for arbitrary space-time points:

$$\{\psi_\alpha(x), \overline{\psi}_\beta(x')\} = -iS_{\alpha\beta}(x - x'). \tag{15.249}$$

The function $S_{\alpha\beta}(x)$ possesses the following properties:

$$(i\gamma^\mu \partial_\mu - m)S(x) = (i\gamma^\mu \partial_\mu - m)(i\gamma^\mu \partial_\mu + m)\Delta_0(x; m^2) = -(\Box + m^2)\Delta_0(x; m^2) = 0 \tag{15.250}$$

and

$$S(x)|_{t=0} = i\gamma_0 \delta^{(3)}(\mathbf{x}). \tag{15.251}$$

Differential equation (15.250), together with the initial condition (15.251), defines the function $S(x)$ unambiguously. This singular function plays a part of the Green function in the process of solving the problems with the initial conditions as we have at hand:

$$\psi(x) = -i \int d\sigma_\mu(x') S(x - x') \gamma^\mu \psi(x'), \tag{15.252}$$

where $t > t'$. To prove the relation (15.252) is trivial. The derivative $\delta/\delta\sigma(x')$ from the right-hand side of (15.252) goes to zero because both $S(x - x')$ and $\psi(x')$ satisfy the Dirac equation. This means that the integral in the right-hand side of (15.252) is independent on a choice of the hypersurface σ'. As a result, we can take as that the hyperplane $t' = t =$const, that, with considering the initial condition for the function $S(x)$, turns (15.252) into the identity.

Comparison between Eqs. (15.248) and (15.1) gives additional evidence that the singular functions related to the wave fields, which are described by the first-order equations, may be expressed with the use of the operator $d_{rs}(\partial)$ in terms of the singular functions characterizing the Klein-Gordon field. In particular, the causal function of a Dirac particle is given by the expression:

$$S^c(x) = (i\gamma^\mu \partial_\mu + m)_{\alpha\beta} \Delta^c(x; m^2). \tag{15.253}$$

The Fourier series expansion of the function $S^c(x)$ has the form:

$$S_{\alpha\beta}^c(x) = \frac{1}{(2\pi)^4} \int d^4 p \frac{(\hat{p} + m)_{\alpha\beta}}{m^2 - p^2 - i\epsilon} \exp\left(-ipx\right). \tag{15.254}$$

Knowing $S^c(x)$, we can calculate the convolution of the field operators:

$$\psi_\alpha^s(x)\overline{\psi}_\beta^s(x') = <0|T(\psi_\alpha(x)\overline{\psi}_\beta(x')|0> = -iS_{\alpha\beta}^c(x - x'). \tag{15.255}$$

It should be noted that in the case of the Dirac field, under the T and N-ordering due to quantization with the use of anticommutators, permutation of the two operators leads to a change of the sign. For example,

$$T\left(\psi_\alpha(x)\overline{\psi}_\beta(x')\right) = \begin{cases} \psi_\alpha(x)\overline{\psi}_\beta(x'), & t > t', \\ -\overline{\psi}_\beta(x')\psi_\alpha(x), & t < t'. \end{cases} \tag{15.256}$$

and

$$N\left(\psi_\alpha(x)\overline{\psi}_\beta(x')\right) = -N\left(\overline{\psi}_\beta(x')\psi_\alpha(x)\right).$$

It is interesting that the anticommutation relations (15.249) may be also obtained with the use of a somewhat modified canonical formalism. Recall that for the Dirac field the simultaneous anticommutation relations of the pairs $\psi(x)$, $\pi(x')$ and $\psi^\dagger(x)$, $\pi^\dagger(x')$ were at variance with each other. For this problem a Solomonian decision may be based on the principle borrowed from politics: "no person, no problem." One of the canonical momenta may be removed by changing of the Lagrangian. The only "painless" modification of the Lagrangian possible is an addition or subtraction of the terms having the form of the four-dimensional divergency. By adding of the quantity:

$$L = \frac{i}{2}\partial_\mu(\overline{\psi}\gamma^\mu\psi),$$

to the Lagrangian (15.13), $\pi^\dagger(x)$ becomes zero and $\pi(x)$ receives the value $i\psi^\dagger(x)$, that in turn leads to the following anticommutation relation for the coincident times:

$$\{\psi(x), \pi(x')\}_{t=t'} = i\{\psi(x), \psi^\dagger(x')\}_{t=t'} = i\delta^{(3)}(\mathbf{r} - \mathbf{r}'). \tag{15.257}$$

Next, into the anticommutator:

$$\{\psi(x), \psi^\dagger(x')\}$$

we substitute the function $\psi^\dagger(x')$ represented with the help of the Green function (15.252). Then, selecting the hyperplane $t = t'$=const as the hypersurface σ and using (15.257), we get the same anticommutation relations in arbitrary time as within the scope of the relativistic quantization procedure. It should be emphasized that the difficulties encountered with the canonical formalism are characteristic for the first-order equations of a particle theory for any spin. Nevertheless, the above procedure for establishing the anticommutation relations, that is (mildly speaking) somewhat inconsistent as the reason for the exclusion of $\pi(x)$ and not $\pi^\dagger(x)$ is not clear, invariably gives the correct result.

To conclude this consideration, we present a number of useful relations for the charge conjugate operator field functions. In Section 15.14 it will be shown that the effect exerted by the second-quantized charge-conjugation operator $\mathcal{C}$ on the operator field function is defined by the following relation:

$$\mathcal{C}\psi(x)\mathcal{C}^{-1} = \eta_c\psi^c(x) = \overline{\psi}^T(x), \qquad \mathcal{C}\overline{\psi}(x)\mathcal{C}^{-1} = \eta_c^*\overline{\psi}^c(x) = C^{-1}\psi(x), \tag{15.258}$$

where $|\eta_c|^2 = 1$. This in turn means that the linkage between the charge conjugate operator field functions is of the same form as in a primary quantization theory. In this case, using the expansion of $\psi(x)$ (15.219) and allowing for the properties of the charge conjugation matrix, we get

$$\{\psi_\alpha^c(x), \overline{\psi}_\beta^c(x)\} = -iS_{\alpha\beta}(x - x'), \tag{15.259}$$

$$\{\psi_\alpha(x), \psi_\beta^c(x)\} = -iC_{\tau\beta}S_{\alpha\tau}(x - x'), \tag{15.260}$$

$$\{\overline{\psi}_\alpha(x), \overline{\psi}_\beta^c(x)\} = -iC_{\beta\tau}S_{\tau\alpha}(x - x'). \tag{15.261}$$

15.12 Majorana equation

The charged leptons and quarks are described by the Dirac equation. The same equation may be used to describe electrically neutral particles with the spin 1/2, on condition that the particles and antiparticles are not identical. And for the truly neutral particles with the spin 1/2 the Dirac equation should be modified. This necessity is clear because the quantity $i\gamma^\mu\partial_\mu$ in the Dirac-Pauli and Weyl representations is not actually real. Because of this, to describe a truly neutral particle, we have to choose the representation with the property:

$$\mathrm{Re}(\gamma_\mu) = 0.$$

This requirement is met by the Majorana representation:

$$\left.\begin{array}{ll}
\gamma^0 = \begin{pmatrix} 0 & \sigma_2 \\ \sigma_2 & 0 \end{pmatrix}, & \gamma^1 = -i\begin{pmatrix} \sigma_3 & 0 \\ 0 & \sigma_3 \end{pmatrix}, \\[12pt]
\gamma^2 = \begin{pmatrix} 0 & \sigma_2 \\ -\sigma_2 & 0 \end{pmatrix}, & \gamma^3 = i\begin{pmatrix} \sigma_1 & 0 \\ 0 & \sigma_1 \end{pmatrix},
\end{array}\right\}
\tag{15.262}$$

that is related to the Dirac-Pauli representation by the unitary transformation:

$$\gamma_\mu() = \frac{1}{\sqrt{2}}(1 - \gamma_2)\gamma_\mu\frac{1}{\sqrt{2}}(1 + \gamma_2),
\tag{15.263}$$

where the γ-matrices in the right-hand side of (15.263) are given in the Dirac-Pauli representation. A truly neutral particle with an arbitrary spin is conventionally called the *Majorana particle*. The presently known particles of this class have an integer spin. We cannot rule out the existence of the truly neutral fermions in nature. A theory of truly neutral particles with the spin 1/2 has been first introduced by E. Majorana [23]. Their role may be played primarily by the neutrinos. The experiments on Super-Kamiokande [24] and Sudbury [25] have demonstrated that a neutrino is not a massless particle as it was supposed by the standard model. There is reason to believe that the neutrino mass is in the region of several tenths of eV. Naturally, during typical collider experiments the neutrino with such a small mass practically behaves as a massless particle. As will be shown later, for $m_\nu = 0$ the behavior of the Majorana and Dirac neutrinos is identical. The difference between these two types of neutrino is exhibited only in the processes, where the mass has a significant role. Recording of double neutrinoless beta-decay $(0\nu2\beta)$

$$(N, Z) \to (N - 2, Z + 2) + e^- + e^-.
\tag{15.264}$$

may be one of the possible experiments to support the Majorana nature of the neutrino. In the beginning of 2002 detecting this process was reported by the Heidelberg-Moscow (HM) Collaboration [26]. However, it is early to classify these results among the reliable data as some other collaborations have serious doubts (see, for example, Ref. [27]). We shall come back to this problem in Part 7.

If we were concerned with the particles described by the tensor quantities, the fact of true neutrality of the particles could be expressed by the apparent relation:

$$\psi(x) = \psi^*(x).
\tag{15.265}$$

But in the case of the spinor particles the behavior of $\psi(x)$ and $\psi^*(x)$ under the Poincare transformations is different. In the case when the orbital moment of momentum equals zero the corresponding transformation laws have the form:

$$\psi'(\Lambda x + a) = \exp\left(\frac{i}{4}\sigma_{\mu\nu}\omega^{\mu\nu}\right)\psi(x) \tag{15.266}$$

and

$$\psi'^{*}(\Lambda x + a) = \exp\left(-\frac{i}{4}\sigma_{\mu\nu}^{*}\omega^{\mu\nu}\right)\psi^{*}(x). \tag{15.267}$$

Certainly, if we impose the condition (15.265) in a particular Lorentz system, this condition is invalid in some other Lorentz system. As has been previously established, the field function transformation law dictates the equation form. In this manner, for the relativistically invariant definition of a truly neutral spinor particle both sides of Eq. (15.265) should involve the quantities obeying the same equation. This condition is met by $\psi(x)$ and $\psi^{c}(x)$. Really, at $e = 0$ Eqs. (15.169) and (15.170) takes the same form. Consequently, the relativistically invariant definition of a truly neutral particle is as follows:

$$\psi(x) = \lambda_{\odot}\psi^{c}(x), \tag{15.268}$$

where $\lambda_{\odot}$ is a phase factor ($|\lambda_{\odot}|^2 = 1$). The relation (15.268) is customarily called the *Majorana condition*.

First, we consider the case of massless neutrinos and demonstrate the possibility to describe the Majorana neutrino by the Weyl spinors using the chiral representation. We shall work in the chiral representation. Then with regard to:

$$\psi^{c} = \overline{\psi}^{T} = i\gamma^{2}\psi^{*}, \tag{15.269}$$

we obtain for the two-components Weyl's spinors $\psi_{R,L}$:

$$\psi_{L}^{c} \equiv (\psi_{L})^{c} = i\gamma^{2}\left\{\frac{1-\gamma_5}{2}\begin{pmatrix}0\\\eta\end{pmatrix}\right\}^{*} = \begin{pmatrix}i\sigma_2\eta^{*}\\0\end{pmatrix} = \frac{1+\gamma_5}{2}\begin{pmatrix}i\sigma_2\eta^{*}\\0\end{pmatrix}, \tag{15.270}$$

$$\psi_{R}^{c} \equiv (\psi_{R})^{c} = i\gamma^{2}\left\{\frac{1+\gamma_5}{2}\begin{pmatrix}\xi\\0\end{pmatrix}\right\}^{*} = \begin{pmatrix}0\\-i\sigma_2\xi^{*}\end{pmatrix} = \frac{1-\gamma_5}{2}\begin{pmatrix}0\\-i\sigma_2\xi^{*}\end{pmatrix}. \tag{15.271}$$

The massless neutrino field Lagrangian is determined by the expression:

$$\mathcal{L} = i\overline{\psi}\gamma_{\mu}\partial^{\mu}\psi = i[\overline{\psi}_{L}\gamma_{\mu}\partial^{\mu}\psi_{L} + \overline{\psi}_{R}\gamma_{\mu}\partial^{\mu}\psi_{R}]. \tag{15.272}$$

With the help of the Weyl's spinors we can define the Majorana spinors χ and ζ, for which the condition of (15.268) proves to be fulfilled:

$$\chi = \psi_{L} + \psi_{L}^{c}, \tag{15.273}$$

$$\zeta = \psi_{R} - \psi_{R}^{c}. \tag{15.274}$$

To find the connection between the Weyl's and Majorana spinors presents no special problems. Using Eqs. (15.273) and (15.274) we arrive at:

$$\psi_{L} = \frac{1-\gamma_5}{2}\chi, \qquad \psi_{L}^{c} = \frac{1+\gamma_5}{2}\chi, \tag{15.275}$$

$$\psi_{R} = \frac{1+\gamma_5}{2}\zeta, \qquad \psi_{R}^{c} = -\frac{1-\gamma_5}{2}\zeta. \tag{15.276}$$

The substitution of Eqs. (15.275) and (15.276) into (15.272) leads to the Lagrangian:

$$\mathcal{L} = \frac{i}{2}[\overline{\chi}\gamma_{\mu}\partial^{\mu}\chi + \overline{\zeta}\gamma_{\mu}\partial^{\mu}\zeta]. \tag{15.277}$$

To describe the Majorana neutrinos, we use the spinor χ. In the case of massive neutrinos the Lagrangian will look like:

$$\mathcal{L} = \frac{1}{2}[i\overline{\chi}\gamma_\mu\partial^\mu\chi - m\overline{\chi}\chi], \tag{15.278}$$

where m is a real quantity. Now instead of (15.273), we assume:

$$\chi = \psi_L \exp(i\beta_1) + \psi_L^c \exp(i\beta_2). \tag{15.279}$$

This, in its turn, fixes the phase in the Majorana condition:

$$\lambda_* = \exp(i\beta_1 + i\beta_2).$$

Rewriting the Lagrangian (15.278) in terms of ψ_L, we get:

$$\mathcal{L} = i\overline{\psi}_L\gamma_\mu\partial^\mu\psi_L - \frac{1}{2}M\overline{\psi}_L^c\psi_L + \text{conj.}. \tag{15.280}$$

where $M_L = m\exp[i(\beta_2 - \beta_1)]$. Note that the phase may be at will transferred from M_L to ψ_R.

The apparent generalization of the Lagrangian (15.280) on the four component spinors is given by the expression:

$$\mathcal{L} = i(\overline{\psi}_R\gamma_\mu\partial^\mu\psi_R + \overline{\psi}_L\gamma_\mu\partial^\mu\psi_L) + \mathcal{L}_{mass}^M, \tag{15.281}$$

where

$$\mathcal{L}_{mass}^M = -\frac{1}{2}\{M_R\overline{\psi}_R^c\psi_R + M_L\overline{\psi}_L^c\psi_L + \text{conj.}\}. \tag{15.282}$$

It will be recalled that for Dirac particles the mass term in the Lagrangian had the form:

$$\mathcal{L}_{mass}^D = -M_D\{\psi_L\psi_R + \text{conj.}\}. \tag{15.283}$$

Thus, in the Lagrangians for particles with the spin 1/2 two types of the mass terms are possible. And further they are referred to as the Dirac and Majorana mass, respectively. In the most general case the mass term in the Lagrangian is defined by the expression:

$$\mathcal{L}_m = -\frac{1}{2}\left(\overline{\psi}_L + \overline{\psi}_L^c, \overline{\psi}_R + \overline{\psi}_R^c\right)\begin{pmatrix} M_L & M_D \\ M_D & M_R \end{pmatrix}\begin{pmatrix} \psi_L + \psi_L^c \\ \psi_R + \psi_R^c \end{pmatrix}. \tag{15.284}$$

When introducing into the Lagrangian the Majorana-mass terms, the Lagrangian becomes noninvariant with respect to the gauge transformations providing the conservation of the lepton flavor:

$$\psi(x) \to \psi(x)\exp(iL\beta). \tag{15.285}$$

Because of this, the Majorana theory predicts the existence of the processes with the lepton flavor violation $|\Delta L| = 2$. The $0\nu2\beta$-process furnishes an excellent example of such an exotics.

Let us introduce the designations:

$$\sigma^\mu \equiv (1, -\sigma_i), \qquad \tilde{\sigma}^\mu \equiv (1, \sigma_i)$$

and pass to the two-components spinors η in the Lagrangian (15.280):

$$\mathcal{L} = i\eta^\dagger(\sigma^\mu\partial_\mu)\eta - \frac{im}{2}\left(\eta^T\sigma_2\eta - \eta^\dagger\sigma_2\eta^*\right), \tag{15.286}$$

where for the sake of simplicity we have set $\beta_1 = \beta_2$. By application of the variational principle, the Majorana equation in the form of the two-components spinor is followed from the Lagrangian (15.286)

$$\left. \begin{array}{l} \sigma^\mu \partial_\mu \eta + m\sigma_2 \eta^* = 0, \\ \tilde{\sigma}^\mu \sigma_2 \partial_\mu \eta^* - m\eta = 0. \end{array} \right\} \tag{15.287}$$

Eliminating η^* from Eq. (15.287) we arrive, as we might expect, at the Klein-Gordon equation:

$$(\Box + m^2)\eta = 0.$$

The representation of a truly neutral spinor field in terms of ψ_L or ψ_R is not the only possibility. For example, the Majorana spinors may be also represented as:

$$\phi_+ = \lambda_+ \frac{\psi + \psi^c}{\sqrt{2}}, \tag{15.288}$$

or

$$\phi_- = \lambda_- \frac{\psi - \psi^c}{\sqrt{2}}. \tag{15.289}$$

Since ψ and ψ^c have the identical transformation properties, then the Majorana condition:

$$\phi_\pm^c = \lambda_\pm \phi_\pm$$

appears to be relativistic covariant. It goes without saying that physics should be independent of the construction method for Majorana's spinors, that is, the representations used by us are unitary equivalent [28]. Within the scope of a second-quantized theory, the unitary equivalence points to the existence of the operator U that secures the fulfillment of the relations:

$$U\phi_- U^{-1} = \zeta, \qquad U\phi_+ U^{-1} = \chi. \tag{15.290}$$

With due regard for the definition of the spinors χ and ζ from Eq. (15.290) follows:

$$U^{-1}\psi_L U = \frac{1-\gamma_5}{2}\phi_+, \qquad U^{-1}\psi_R U = \frac{1+\gamma_5}{2}\phi_-, \tag{15.291}$$

$$U^{-1}\psi_L^c U = \frac{1+\gamma_5}{2}\phi_+, \qquad U^{-1}\psi_R^c U = -\frac{1-\gamma_5}{2}\phi_-. \tag{15.292}$$

We direct our attention to the fact that the fulfillment of (15.291) and (15.292) makes it possible to describe the Majorana neutrino simply with the use of the Weyl spinors ψ_L and ψ_R rather than their combinations satisfying the Majorana condition. To prove relations (15.291) and (15.292) we consider the simplest case, namely, when $m_\nu = 0$. This enables us to use the following expansion for the operator field function of the neutrino:

$$\psi(x) = \frac{1}{(2\pi)^{3/2}} \int d^3p \sum_{r=1}^{2} [a_r(\mathbf{p})u^{(r)}(\mathbf{p})e^{-ipx} + b_r^\dagger(\mathbf{p})v^{(r)}(\mathbf{p})e^{ipx}], \tag{15.293}$$

where the bispinors $u^{(r)}(\mathbf{p})$ and $v^{(r)}(\mathbf{p})$ satisfy the equations (see, (15.113))

$$\gamma_5 u^{(r)}(\mathbf{p}) = qu^{(r)}(\mathbf{p}), \qquad \gamma_5 v^{(r)}(\mathbf{p}) = -qu^{(r)}(\mathbf{p}) \tag{15.294}$$

with $q = 1$ for $r = 1$ and $q = -1$ for $r = 2$. Under the proof we also need the expansions of the field operators entering into Eqs. (15.291) and (15.292) in terms of the plane waves. With allowance made for

$$\psi^c = \psi(a_r(\mathbf{p}) \leftrightarrow b_r(\mathbf{p}))$$

and Eqs. (15.258), (15.294), we get:

$$\psi_L(x) = \frac{1}{(2\pi)^{3/2}} \int d^3p[a_1(\mathbf{p})u^{(1)}(\mathbf{p})e^{-ipx} + b_2^\dagger(\mathbf{p})v^{(2)}(\mathbf{p})e^{ipx}], \qquad (15.295)$$

$$\psi_R(x) = \frac{1}{(2\pi)^{3/2}} \int d^3p[a_2(\mathbf{p})u^{(2)}(\mathbf{p})e^{-ipx} + b_1^\dagger(\mathbf{p})v^{(1)}(\mathbf{p})e^{ipx}], \qquad (15.296)$$

$$\psi_L^c(x) = \frac{1}{(2\pi)^{3/2}} \int d^3p[b_2(\mathbf{p})u^{(2)}(\mathbf{p})e^{-ipx} + a_1^\dagger(\mathbf{p})v^{(1)}(\mathbf{p})e^{ipx}], \qquad (15.297)$$

$$\psi_R^c(x) = \frac{1}{(2\pi)^{3/2}} \int d^3p[b_1(\mathbf{p})u^{(1)}(\mathbf{p})e^{-ipx} + a_2^\dagger(\mathbf{p})v^{(2)}(\mathbf{p})e^{ipx}], \qquad (15.298)$$

$$\phi_\pm(x) = \frac{1}{(2\pi)^{3/2}\sqrt{2}} \int d^3p \sum_{r=1}^{2}\{[a_r(\mathbf{p}) \pm b_r(\mathbf{p})]u^{(r)}(\mathbf{p})e^{-ipx} + [\pm a_r^\dagger(\mathbf{p})+$$

$$+b_r^\dagger(\mathbf{p})]v^{(r)}(\mathbf{p})e^{ipx}\}. \qquad (15.299)$$

Intuitively we feel that the matrix U should take the form:

$$U = \exp[\theta \sum_r (b_r^\dagger a_r - a_r^\dagger b_r)],$$

assuming by this $\mathrm{Im}(\theta) = 0$. Carrying out simple manipulation, we obtain:

$$U^{-1}a_sU = a_s + \theta[\sum_r(b_r^\dagger a_r - a_r^\dagger b_r), a_s] + \frac{\theta^2}{2!}[\sum_r(b_r^\dagger a_r - a_r^\dagger b_r), [\sum_{r'}(b_{r'}^\dagger a_{r'} - a_{r'}^\dagger b_{r'}), a_s]]+$$

$$+... = a_s \cos\theta + b_s \sin\theta \qquad (15.300)$$

and

$$U^{-1}b_sU = a_s \cos\theta - b_s \sin\theta. \qquad (15.301)$$

When θ equals $\pi/4$, from the formulae obtained follows:

$$\left. \begin{array}{ll} \sqrt{2}U^{-1}a_1(\mathbf{p})U = a_1(\mathbf{p}) + b_1(\mathbf{p}), & \sqrt{2}U^{-1}a_2(\mathbf{p})U = a_2(\mathbf{p}) - b_2(\mathbf{p}), \\ \sqrt{2}U^{-1}b_1(\mathbf{p})U = b_1(\mathbf{p}) - a_1(\mathbf{p}), & \sqrt{2}U^{-1}b_2(\mathbf{p})U = b_2(\mathbf{p}) + a_2(\mathbf{p}). \end{array} \right\} \qquad (15.302)$$

By means of Eqs. (15.302) we can easily establish the validity of the relations (15.291) and (15.292), and our task is accomplished.

15.13 Quantization of Majorana field

At first, we carry out the quantization of the Majorana neutrino field with the use of the Weyl's spinors $\eta(x)$ and $\eta^*(x)$ [29]. In the most general case the operator expansion of the spinors $\eta(x)$ and $\eta^*(x)$ in terms of the plane waves have the form:

$$\eta(x) = \frac{1}{(2\pi)^{3/2}} \int d^3p N\{[a(\mathbf{p})\beta(\mathbf{p})+b(\mathbf{p})\alpha(\mathbf{p})]e^{-ipx} + [c(\mathbf{p})\beta(\mathbf{p})+d(\mathbf{p})\alpha(\mathbf{p})]e^{ipx}\}, \qquad (15.303)$$

$$\eta^*(x) = \frac{1}{(2\pi)^{3/2}} \int d^3p N\{[a^\dagger(\mathbf{p})\beta^*(\mathbf{p}) + b^\dagger(\mathbf{p})\alpha^*(\mathbf{p})]e^{ipx} + [c^\dagger(\mathbf{p})\beta^*(\mathbf{p}) + d^\dagger(\mathbf{p})\alpha^*(\mathbf{p})]e^{-ipx}\},$$

(15.304)

where N is a normalization factor to be determined, and the two-components spinors $\alpha(\mathbf{p})$ and $\beta(\mathbf{p})$ obey both the eigenvalue equations:

$$\frac{(\mathbf{p}\cdot\boldsymbol{\sigma})}{|\mathbf{p}|}\alpha(\mathbf{p}) = \alpha(\mathbf{p}), \qquad \frac{(\mathbf{p}\cdot\boldsymbol{\sigma})}{|\mathbf{p}|}\beta(\mathbf{p}) = -\beta(\mathbf{p}) \tag{15.305}$$

and the normalization conditions:

$$\left.\begin{array}{l} \alpha^\dagger(\mathbf{p})\alpha(\mathbf{p}) = \beta^\dagger(\mathbf{p})\beta(\mathbf{p}) = 1, \\ \alpha^\dagger(\mathbf{p})\beta(\mathbf{p}) = \beta^\dagger(\mathbf{p})\alpha(\mathbf{p}) = 0. \end{array}\right\} \tag{15.306}$$

Substitute the expressions (15.303) and (15.304) into the motion equations (15.287). These equations will be converted to identities under fulfillment of the following conditions:

$$\sigma_2\alpha^*(\mathbf{p}) = i\beta(\mathbf{p}), \qquad \sigma_2\beta^*(\mathbf{p}) = -i\alpha(\mathbf{p}), \tag{15.307}$$

$$d(\mathbf{p}) = \frac{m}{E_p + |\mathbf{p}|}a^\dagger(\mathbf{p}), \qquad b(\mathbf{p}) = -\frac{m}{E_p + |\mathbf{p}|}c^\dagger(\mathbf{p}). \tag{15.308}$$

Introducing the designations

$$a(\mathbf{p}) \equiv a_-(\mathbf{p}), \qquad c^\dagger(\mathbf{p}) \equiv a_+(\mathbf{p}),$$

we could rewrite the expansion of $\eta(x)$ in the form:

$$\eta(x) = \frac{1}{(2\pi)^{3/2}} \int d^3p N\{[a_-(\mathbf{p})\beta(\mathbf{p}) - \frac{m}{E_p + |\mathbf{p}|}a_+(\mathbf{p})\alpha(\mathbf{p})]e^{-ipx} +$$

$$+ [a_+^\dagger(\mathbf{p})\beta(\mathbf{p}) + \frac{m}{E_p + |\mathbf{p}|}a_-^\dagger(\mathbf{p})\alpha(\mathbf{p})]e^{ipx}\}. \tag{15.309}$$

Since "coordinates" and "momenta" of the Majorana field are not independent, we have to use the relativistic quantization scheme. At the first step we define the dynamical observables. The energy-momentum tensor will look like:

$$T^{\mu\nu} = i\eta^\dagger\sigma^\mu\partial^\nu\eta - g^{\mu\nu}\mathcal{L}. \tag{15.310}$$

With its help the energy-momentum vector could be found

$$P^\mu = \frac{i}{2} \int d^3x\{2\eta^*(x)\partial^\mu\eta(x) - g^{0\mu}\partial_t[\eta^*(x)\eta(x)]\}. \tag{15.311}$$

To insert the expansions of $\eta(x)$ and $\eta^*(x)$ into Eq. (15.311) gives the expression for the field energy:

$$E = N^2 \int d^3p\, p_0 \frac{2E_p}{E_p + |\mathbf{p}|}[a_-^\dagger(\mathbf{p})a_-(\mathbf{p}) - a_+(\mathbf{p})a_+^\dagger(\mathbf{p})]. \tag{15.312}$$

From Eq. (15.312) it has become obvious that N should be chosen in the form:

$$N = \sqrt{\frac{E_p + |\mathbf{p}|}{2E_p}}. \tag{15.313}$$

Further, by the standard way one may establish nonzero anticommutators:

$$\{a_\pm(\mathbf{p}), a_\pm^\dagger(\mathbf{p}')\} = \delta^{(3)}(\mathbf{p} - \mathbf{p}') \tag{15.314}$$

where $a_+(\mathbf{p})$ ($a_-(\mathbf{p})$) and $a_+^\dagger(\mathbf{p})$ ($a_-^\dagger(\mathbf{p})$) have meaning of the production and annihilation operators of particles with the positive (negative) helicity, mass m and momentum $\mathbf{p}$, respectively.

Let us proceed to the definition of singular functions of the Majorana field. From Eq. (15.305) and the orthonormalization conditions for the spinors $\alpha(\mathbf{p})$ and $\beta(\mathbf{p})$ it is easy to obtain:

$$\alpha_r(\mathbf{p})\alpha_s^*(\mathbf{p}) = \frac{1}{2|\mathbf{p}|}(|\mathbf{p}| + \sigma_j p_j)_{rs}, \qquad \beta_r(\mathbf{p})\beta_s^*(\mathbf{p}) = \frac{1}{2|\mathbf{p}|}(|\mathbf{p}| - \sigma_j p_j)_{rs}. \tag{15.315}$$

By application of the formulae (15.315) we could establish the values of the anticommutators for the operator field functions at an arbitrary instant of time:

$$\{\eta_r(x), \eta_s^*(x')\} = (\tilde{\sigma}^\mu \partial_\mu)_{rs} \Delta_0(x - x'; m^2), \tag{15.316}$$

$$\{\eta_r(x), \eta_s^T(x')\} = m(\sigma_2)_{rs} \Delta_0(x - x'; m^2). \tag{15.317}$$

This in turn allows us to find nonzero convolutions of the Majorana field operators:

$$< 0|T[\eta_r(x)\eta_s^\dagger(x')]|0 > = (\tilde{\sigma}^\mu \partial_\mu)_{rs} \Delta^c(x - x'), \tag{15.318}$$

$$< 0|T[\eta_r(x)\eta_s^T(x')]|0 > = m(\sigma_2)_{rs} \Delta^c(x - x'). \tag{15.319}$$

Again expression (15.319) supports the conclusion that the lepton flavor violation in the case of the Majorana neutrino takes place for nonzero mass of the neutrinos only.

The above formalism of the two-component spinors is convenient for the calculation of matrix elements when the Majorana neutrinos are the sole fermions involved in the interaction Lagrangian $\mathcal{L}_{int}$. Provided $\mathcal{L}_{int}$ also includes, for instance, the charged leptons, we are obliged to go to the two-component spinors for them as well. This procedure is not only devoid of aesthetic attractiveness but also results in a considerable increase of the terms in the amplitude of the corresponding reaction. Because of this reason to describe Majorana's neutrinos we shall use the most habitual formalism of the four-component spinors.

In the interaction Lagrangian of any electroweak theory, no matter the SM or its extension, the neutrino field function is accompanied by chirality projection operators. For example, in the SM the Lagrangian describing the interaction of neutrinos with gauge bosons has the form:

$$\mathcal{L}_{int}^\nu = \frac{g}{4\cos\theta_W}\bar{\nu}_l(x)\gamma_\mu(1-\gamma_5)\nu_l(x)Z^\mu(x) + \frac{g}{2\sqrt{2}}\bar{l}(x)\gamma_\mu(1-\gamma_5)\nu_l(x)W^\mu(x) + \text{conj.,} \tag{15.320}$$

where we are not concretizing the neutrino nature. In the SM for every generation we have the only left-handed neutrino (or more precisely, preferentially left-handed neutrino, but this aspect will be considered later). In some extensions of the SM to this neutrino a heavy right-handed neutrino is added. For now, let us direct our attention to the SM only. Then for the Majorana neutrino it is reasonable to select the representation:

$$\nu^M(x) = \lambda_\odot[\psi_L(x) + \psi_L^c(x)]. \tag{15.321}$$

Taking into account the formulae (15.249) and (15.259), we get:

$$\{\nu_\alpha^M(x), \bar{\nu}_\beta^M(x')\} = -iS_{\alpha\beta}(x - x'). \tag{15.322}$$

As directly follows from (15.322), convolution of the operators $\nu_\alpha^M(x)$ and $\bar{\nu}_\beta^M(x')$ for the Majorana field has the same value as for the Dirac field, that is,

$$< 0|T\left(\nu_\alpha^M(x)\bar{\nu}_\beta^M(x')\right)|0 > = -iS_{\alpha\beta}^c(x - x'). \tag{15.323}$$

But unlike the Dirac case, owing to the relation (15.260), the following quantity is also different from zero:

$$< 0|T\left(\nu_\alpha^M(x)\nu_\beta^M(x')\right)|0> = -i\lambda_\odot^2 m_{\alpha\beta}\Delta^c(x-x').\qquad(15.324)$$

It should be noted that at will we may remove the charge conjugation operator from the reaction amplitude. To this end, the operator C should be set up on the field function of the lepton forming with the neutrino a neutral or charged current (see the expression (15.320)). Obviously, it is more convenient to use the expansion of the Majorana field operator simply in terms of the solutions of the Dirac equation. Such an expression may be obtained if we recall that the constant orthonormalized spinors $\varsigma^\pm$, contained in the operator function expansion for the electron-positron field, are determined with an accuracy of the phase. Then the desired expansion for $\nu^M(x)$ could be represented in the shape:

$$\nu^M(x) = \frac{1}{(2\pi)^{3/2}}\int d^3p\sqrt{\frac{m}{E_p}}\sum_{\alpha=\pm}[a_\alpha(\mathbf{p})u^{(\alpha)}(\mathbf{p})e^{-ipx}+\lambda_\odot a_\alpha^\dagger(\mathbf{p})v_m^{(\alpha)}(\mathbf{p})e^{ipx}],\quad(15.325)$$

where the bispinor $u^{(\alpha)}(\mathbf{p})$ is defined by the formula (15.241) as before while the bispinor $v_m^{(\alpha)}(\mathbf{p})$ is given by the expression:

$$v_m^{(\pm)}(\mathbf{p}) = N\left(\begin{array}{c}\dfrac{(\mathbf{p}\cdot\boldsymbol{\sigma})}{E_p+m}\varsigma_m^\pm\\[4pt]\varsigma_m^\pm\end{array}\right),\qquad(15.326)$$

$$\varsigma_m^+ = -\begin{pmatrix}0\\1\end{pmatrix},\qquad\varsigma_m^- = \begin{pmatrix}1\\0\end{pmatrix}.$$

In Eq. (15.325) we have placed the phase factor before the Majorana particle production operator. This choice makes it possible to call $\lambda_\odot$ the phase factor of the production. It should be emphasized that nothing interferes with shifting of $\lambda_\odot$ to the annihilation operator. A direct checking demonstrates that when $\nu^M(x)$ is represented as (15.325) the Majorana condition appears to be fulfilled:

$$[\nu^M(x)]^c = \frac{1}{(2\pi)^{3/2}}\int d^3p\sqrt{\frac{m}{E_p}}\sum_{\alpha)}[a_\alpha^\dagger(\mathbf{p})C\gamma_0 u^{(\alpha)*}(\mathbf{p})e^{ipx}+\lambda_\odot^* a_\alpha(\mathbf{p})C\gamma_0 v_m^{(\alpha)*}(\mathbf{p})e^{-ipx}] =$$

$$= \lambda_\odot^*\frac{1}{(2\pi)^{3/2}}\int d^3p\sqrt{\frac{m}{E_p}}\sum_\alpha[a_\alpha(\mathbf{p})u^{(\alpha)}(\mathbf{p})e^{-ipx}+\lambda_\odot a_\alpha^\dagger(\mathbf{p})v_m^{(\alpha)}(\mathbf{p})e^{ipx}],$$

where we have taken into consideration the relations:

$$[v_m^{(\alpha)}(\mathbf{p})]^c = C[\overline{v}_m^{(\alpha)}(\mathbf{p})]^T = u^{(\alpha)}(\mathbf{p}),\qquad [u^{(\alpha)}(\mathbf{p})]^c = C[\overline{u}^{(\alpha)}(\mathbf{p})]^T = v_m^{(\alpha)}(\mathbf{p}).\qquad(15.327)$$

15.14   Second quantized representation of discrete operations C, P, and T for particles with spin $1/2$

Second quantized operators of discrete transformations are operators in the Fock space and, consequently, do not commute with production and annihilation operators. In order to distinguish them from operators transforming unquantized wavefunctions, we shall use the following designations: $\mathcal{C}$ for the charge conjugation operator, $\mathcal{P}$ for the space inversion operator, and $\mathcal{T}$ for the time inversion operator. It is obvious that the vacuum must be invariant with respect to these transformations, that is,

$$\mathcal{C}\Phi_0 = \mathcal{P}\Phi_0 = \mathcal{T}\Phi_0 = 0. \tag{15.328}$$

The action of the operator U $(U = \mathcal{C}, \mathcal{P}, \mathcal{T})$ on a system state vector that is given in the Fock representation (15.244) leads to the result:

$$U\Phi_{n^{+}_{\mathbf{p}_1} n^{-}_{\mathbf{p}'_1} \ldots} = \int \ldots \int F'(x_1, \ldots, x_n) U\psi^{b_1}(x_1) U^{-1} \ldots U\psi^{b_n}(x_n) U^{-1} d^4 x_1 \ldots d^4 x_n \Phi_0,$$

where $F' = F$ when U is unitary, and $F' = F^*$ when U is nonunitary. Thus, calling for invariance of the theory with respect to the U-transformation, we reduce our problem to establishing of the transformation laws for the operator field functions, that is, to the determination of the quantities: $U\psi(x)U^{-1}$ and $U\overline{\psi}(x)U^{-1}$. In so doing, in accordance with the correspondence principle, we have to take into account the transformation laws for the unquantized field functions stated by us in Section 15.8, on the one hand. And on the other hand, since the operators U have no effect on the spinor components of the operator field functions,* use of the motion equations (as in case of a unquantized theory) does not result in the definitive establishment of the transformation laws. In other words, in a second quantized theory, when searching for the transformation laws, we are in need of a new criterion. It is clear that only invariance of the permutation relations (PR) with respect to the transformations $\mathcal{C}$, $\mathcal{P}$ and $\mathcal{T}$ could be such a criterion.

Let us consider the charge conjugation operation. According to the results of Section 15.8, we define this operation as:

$$\mathcal{C}\psi(x)\mathcal{C}^{-1} = \eta_c \psi^c(x), \qquad \mathcal{C}\overline{\psi}(x)\mathcal{C}^{-1} = \eta_c^* \overline{\psi}^c(x). \tag{15.329}$$

Direct calculations indicate that, with such a definition of the $\mathcal{C}$-operation, invariance of the PR is provided under the condition that the $\mathcal{C}$-operator is unitary, as distinct from a primary quantized theory. The invariance with respect to the $\mathcal{C}$-operation takes place for an electrically neutral system only. In this case the state vectors of the initial system (Φ) and the transformed one (Φ^c) may be distinguished from each other only in a phase factor:

$$\Phi^c = \eta_C \Phi = \prod_i^N (\eta_c)_i \Phi.$$

To display this circumstance we have introduced the phase factor of the charge conjugation operation $|\eta_c| = 1$.

Now, by application (15.329) we establish transformation properties of the production and annihilation operators. Using the expansion of the operator field function (15.240), we get:

$$\mathcal{C}a_\alpha(\mathbf{p})\mathcal{C}^{-1} = \eta_c(-1)^{\frac{1+\alpha}{2}} b_\alpha(\mathbf{p}), \qquad \mathcal{C}b_\alpha(\mathbf{k})\mathcal{C}^{-1} = \eta_c^*(-1)^{\frac{1+\alpha}{2}} a_\alpha(\mathbf{p}), \tag{15.330}$$

where the following identities for the bispinors:

$$C\overline{u}^{(\alpha)}(\mathbf{p}) = (-1)^{\frac{1+\alpha}{2}} v^{(\alpha)}(\mathbf{p}), \qquad C\overline{v}^{(\alpha)}(\mathbf{p}) = (-1)^{\frac{1+\alpha}{2}} u^{(\alpha)}(\mathbf{p}) \tag{15.331}$$

---

*To be more specific $u^{(\alpha)}(\mathbf{p})$ and $v^{(\alpha)}(\mathbf{p})$ commute with the unitary operator U, whereas under permutation with antiunitary operator U these bispinors are replaced with complex conjugate ones.

have been taken into consideration. For the derivation of (15.331) it is convenient to test them first in the Dirac-Pauli representation, and, next, to make sure that the derived equations do not depend on the representation choice. For the production operators the corresponding transformation laws may be found from the relations (15.331) with an allowance made for the condition:

$$(UcU^{-1})^* = Uc^*U^{-1},$$

that is valid both for the unitary and antiunitary operator U as well as for an arbitrary operator c.

We may take the opposite course, namely, we define the charge conjugation operation as a transformation of the fermion with the given spin orientation into the antifermion with the same spin orientation. In this case for the fermion annihilation operators we immediately obtain the relations (15.240) with the omitted factor $(-1)^{\frac{1+\alpha}{2}}$, from where the transformation laws of the operator field function $\psi(x)$ may be established.

Reasoning in a similar way, we can determine the space and time inversion operations:

$$\mathcal{P}\psi(\mathbf{r}, t)\mathcal{P}^{-1} = \eta_p \gamma_0 \psi(-\mathbf{r}, t), \qquad \mathcal{P}\overline{\psi}(\mathbf{r}, t)\mathcal{P}^{-1} = \eta_p^* \gamma_0 \overline{\psi}(-\mathbf{r}, t), \qquad (15.332)$$

$$\left. \begin{array}{l} \mathcal{T}\psi(\mathbf{r}, t)\mathcal{T}^{-1} = \eta_t \gamma^3 \gamma^1 \psi(\mathbf{r}, -t), \\ \mathcal{T}\overline{\psi}(\mathbf{r}, t)\mathcal{P}^{-1} = \eta_t^* \gamma^1 \gamma^3 \overline{\psi}(\mathbf{r}, -t), \end{array} \right\} \qquad (15.333)$$

where $|\eta_p| = 1$, $|\eta_t| = 1$. In this case the operator $\mathcal{P}$ must be unitary and the operator $\mathcal{T}$—antiunitary. Using the expansion of $\psi(x)$ in the form (15.240), we define the behavior of the annihilation operators under the $\mathcal{P}$- and $\mathcal{T}$-transformations

$$\mathcal{P}a_\alpha(\mathbf{p})\mathcal{P}^{-1} = \eta_p a_\alpha(-\mathbf{p}), \qquad \mathcal{P}b_\alpha(\mathbf{p})\mathcal{P}^{-1} = -\eta_p^* b_\alpha(-\mathbf{p}), \qquad (15.334)$$

$$\left. \begin{array}{l} \mathcal{T}a_\alpha(\mathbf{p})\mathcal{T}^{-1} = -\eta_t(-1)^{\frac{1+\alpha}{2}} a_{-\alpha}(-\mathbf{p}), \\ \mathcal{T}b_\alpha(\mathbf{p})\mathcal{T}^{-1} = \eta_t^*(-1)^{\frac{1+\alpha}{2}} b_{-\alpha}(-\mathbf{p}). \end{array} \right\} \qquad (15.335)$$

When deriving the relations obtained the identities

$$\left. \begin{array}{ll} \gamma^3 \gamma^1 u^{(\alpha)*}(\mathbf{p}) = -(-1)^{\frac{1+\alpha}{2}} u^{(-\alpha)}(-\mathbf{p}), & \gamma_0 u^{(\alpha)}(\mathbf{p}) = u^{(\alpha)}(-\mathbf{p}), \\ \gamma^3 \gamma^1 v^{(\alpha)*}(\mathbf{p}) = (-1)^{\frac{1+\alpha}{2}} v^{(-\alpha)}(-\mathbf{p}), & \gamma_0 v^{(\alpha)}(\mathbf{p}) = -v^{(\alpha)}(-\mathbf{p}) \end{array} \right\} \qquad (15.336)$$

have been allowed for. From (15.334) and (15.335) it follows that, while the space inversion results only in the sign changing for the three-dimensional particle momentum, in the case of the time inversion the sign is changed both for the momentum $\mathbf{p}$ and spin projection α. The product of all three operators $\Theta = \mathcal{CPT}$ gives:

$$\Theta\psi(x)\Theta^{-1} = \eta_{cpt} \gamma_5 \gamma_0 \overline{\psi}(-x), \qquad \Theta\overline{\psi}(x)\Theta^{-1} = \eta_{cpt}^* \gamma_5 \gamma_0 \psi(-x), \qquad (15.337)$$

where $\eta_{cpt} = \eta_c \eta_p \eta_t$. So the $\Theta = \mathcal{CPT}$-operation leads to the replacement of a particle by the antiparticle, with the same momentum and opposite spin projection.

Let us consider Majorana neutrinos behavior under the discrete transformations. For the operator field function of the neutrino the expansion (15.325) will be used. We start with the charge conjugation operation. Now, instead of (15.329), we have:

$$\mathcal{C}\nu^M(\mathbf{r}, t)\mathcal{C}^{-1} = (\eta_c^* \lambda_\odot)^* \nu^M(\mathbf{r}, t),$$

that, in its turn, leads to the relations:

$$\mathcal{C}a_\alpha(\mathbf{p})\mathcal{C}^{-1} = (\eta_c^* \lambda_\odot)^* a_\alpha(\mathbf{p}), \qquad \mathcal{C}a_\alpha^\dagger(\mathbf{p})\mathcal{C}^{-1} = (\eta_c^* \lambda_\odot)^* a_\alpha^\dagger(\mathbf{p}). \qquad (15.338)$$

On the strength of unitarity of the operator $\mathcal{C}$ from Eqs. (15.338) one may conclude that the quantity $\eta'_c = \eta^*_c \lambda_\odot$ is real. Then, considering the $\mathcal{C}$-invariance of the vacuum, we get the relations:

$$\mathcal{C}|\mathbf{p}, \alpha > = \eta'_c|\mathbf{p}, \alpha >, \tag{15.339}$$

where $|\mathbf{p}, \alpha > \equiv a^\dagger_\alpha(\mathbf{p})|0 >$, that lead us to the conclusion: the free Majorana particle is the charge conjugation operator eigenstate. However, this statement is not valid for the physical Majorana neutrino as the weak interaction disturbs the $\mathcal{C}$-invariance.

From Eqs. (15.329) and (15.330) it follows that under the $\mathcal{CP}$-transformation the operator field function is developed by the law:

$$\mathcal{CP}\psi(\mathbf{r}, t)(\mathcal{CP})^{-1} = \eta_{cp}\gamma_0\psi^c(-\mathbf{r}, t). \tag{15.340}$$

For the Majorana neutrino the expression (15.340) is converted to the form:

$$\mathcal{CP}\nu^M(\mathbf{r}, t)(\mathcal{CP})^{-1} = (\eta^*_{cp}\lambda_\odot)^*\gamma_0\nu^M(-\mathbf{r}, t).$$

Using the relation obtained and with allowance made for the identities

$$\gamma^0 u^{(\alpha)}(\mathbf{p}) = u^{(\alpha)}(-\mathbf{p}), \qquad \gamma^0 v^{(\alpha)}_m(\mathbf{p}) = -v^{(\alpha)}_m(-\mathbf{p}), \tag{15.341}$$

one may establish the behavior of the operators $u_\alpha(\mathbf{p})$ and $a^\dagger_\alpha(\mathbf{p})$ under the $\mathcal{CP}$-transformation:

$$\left.\begin{array}{l} \mathcal{CP}a_\alpha(\mathbf{p})(\mathcal{CP})^{-1} = (\eta^*_{cp}\lambda_\odot)^* a_\alpha(-\mathbf{p}), \\ \mathcal{CP}a^\dagger_\alpha(\mathbf{p})(\mathcal{CP})^{-1} = -(\eta^*_{cp}\lambda_\odot)^* a^\dagger_\alpha(-\mathbf{p}). \end{array}\right\} \tag{15.342}$$

Compatibility of Eqs. (15.342) would be ensured when the condition:

$$(\eta^*_{cp}\lambda_\odot)^* = -\eta^*_{cp}\lambda_\odot$$

be fulfilled. In this way the free Majorana particle may be defined as the eigenstate of the operator $\mathcal{CP}$ with the eigenvalue:

$$-(\eta^*_{cp}\lambda_\odot)^* \equiv \eta'_{cp} = \pm i.$$

This is very unusual because for the well-known elementary particles the intrinsic $\mathcal{CP}$-parity is a real number. It is interesting to elucidate whether this statement is at variance with the reported experimental results. Let us consider the decay:

$$Z \to \nu_l + \nu_l \tag{15.343}$$

assuming the Majorana nature for the neutrino. We take neutrinos as nonrelativistic particles, and the $\mathcal{CP}$-invariance as an inviolable law of nature. In the reference frame of the rest Z-boson the total angular momentum of the initial state equals 1, and, as a result, the following final states are possible:

$$\left.\begin{array}{ll} {}^3S_1\ (l = 0, J = 1), & {}^1P_1\ (l = 1, J = 0), \\ {}^3P_1\ (l = 1, J = 1), & {}^3D_1\ (l = 2, J = 1). \end{array}\right\} \tag{15.344}$$

Since in the final state we have two identical fermions, the total wavefunction of the system is asymmetric. This condition is satisfied only by the ${}^3P_1$-state (the space part of the wavefunction is antisymmetric and the spin part—symmetric). Acting on the final state by the operator $\mathcal{CP}$ we get:

$$\mathcal{CP}|\nu\nu; l = 1, J = 1 > = \eta'^2_{cp}(-1)^l|\nu\nu; l = 1, J = 1 > =$$

$$= -\eta_{cp}'^2 |\nu\nu; l = 1, J = 1 > . \tag{15.345}$$

Because the $\mathcal{CP}$-parity of the initial state is equal to 1, then from (15.345) it follows:

$$-\eta_{cp}'^2 = 1,$$

that is, for the Majorana neutrino η_{cp}' may be really equal to $\pm i$.

In the case of the space inversion a sign of the helicity is changed and hence two states of the Majorana neutrino are associated with the opposite intrinsic $\mathcal{CP}$-parity. It is known that in the quark sector the $\mathcal{CP}$-invariance could be violated. There are also doubts about conservation of the $\mathcal{CP}$-invariance for leptons. Consequently, definition of the Majorana neutrino as the $\mathcal{CP}$-eigenstate is equally liable not to have a nature of the absolute statement.

Let us turn our attention to the $\mathcal{CPT}$-transformation. Allowing for the Majorana condition, we can represent the relation (15.337) in the form:

$$\Theta \nu^M(x) \Theta^{-1} = \eta_{cpt} \lambda_{\odot}^* \gamma_5 \gamma_0 C^{-1} \nu^M(-x). \tag{15.346}$$

From (15.346) one may obtain transformation rules for the production and annihilation operators of the Majorana neutrino:

$$\Theta a_\alpha(\mathbf{p}) \Theta^{-1} = -\eta_{cpt} \lambda_{\odot}^* (-1)^{\frac{1+\alpha}{2}} a_{-\alpha}(\mathbf{p}), \tag{15.347}$$

$$\Theta a_\alpha^\dagger(\mathbf{p}) \Theta^{-1} = \eta_{cpt} \lambda_{\odot} (-1)^{\frac{1+\alpha}{2}} a_{-\alpha}^\dagger(\mathbf{p}), \tag{15.348}$$

where the following identities for the bispinors $u^{(\alpha)}(\mathbf{p})$ and $v_m^{(\alpha)}(\mathbf{p})$:

$$\gamma_5 \gamma_0 C^{-1} u^{(\alpha)}(\mathbf{p}) = (-1)^{\frac{1+\alpha}{2}} u^{(-\alpha)*}(\mathbf{p}), \qquad \gamma_5 \gamma_0 C^{-1} v_m^{(\alpha)}(\mathbf{p}) = (-1)^{\frac{1-\alpha}{2}} v_m^{(-\alpha)*}(\mathbf{p}). \tag{15.349}$$

have been taken into account. Carrying out the charge conjugation operation in Eq. (15.348), we arrive at

$$\Theta a_\alpha(\mathbf{p}) \Theta^{-1} = \eta_{cpt}^* \lambda_{\odot}^* (-1)^{\frac{1+\alpha}{2}} a_{-\alpha}(\mathbf{p}).$$

Compatibility of the obtained equation with that (15.347) demands the fulfillment of the relation:

$$\eta_{cpt}^* = -\eta_{cpt},$$

that is, η_{cpt} must take the values $\pm i$. Since the vacuum is $\mathcal{CPT}$-invariant, then for one-particle state of the Majorana neutrino the equation:

$$\mathcal{CPT} |\mathbf{p}, \alpha > = \eta_*' |\mathbf{p}, -\alpha >, \tag{15.350}$$

where $\eta_{cpt}' = \eta_{cpt} \lambda_{\odot} (-1)^{\frac{1+\alpha}{2}}$, should be obeyed. Let us note the dependence of the $\mathcal{CPT}$-operator eigenvalue on the helicity of the initial state. In a system at rest the $\mathcal{CPT}$-transformation applied to the Majorana neutrino simply rotates its spin. Thus, in accordance with (15.350), we may state that the Majorana neutrino is identical to its antiparticle with due regard for the following. It passes into itself upon the $\mathcal{CPT}$-transformation accompanied by the rotation through 180^0. As at the present time the $\mathcal{CPT}$-invariance is believed to be a universal truth, Eq. (15.350) should be regarded as a definition for both the free and physical Majorana particles.

With respect to the discrete transformations, the above-mentioned properties have a number of very important consequences. We limit ourselves to the consideration of their effect only on the electromagnetic structure of the neutrino. Let us assume that the neutrino is in a homogeneous and static electromagnetic field with the magnetic induction $\mathbf{B}$ and the electric field strength $\mathcal{E}$. Provided the neutrino has a magnetic dipole moment, then the interaction energy of this neutrino with the external electromagnetic field is determined as:

$$E_{int} = -\mu_m < \mathbf{S} \cdot \mathbf{B} > -\mu_e < \mathbf{S} \cdot \mathcal{E} > . \tag{15.351}$$

Taking the neutrino as the Majorana particle, let us find the result of the $\mathcal{CPT}$-invariance of a theory. We give due consideration to the fact that, with respect to the $\mathcal{CPT}$-transformation, $\mathbf{B}$ and $\mathcal{E}$ are invariant, whereas the spin direction is changed. Then as a result of the $\mathcal{CPT}$-transformation, E_{int} changes over to $-E_{int}$. In this case we can save the $\mathcal{CPT}$-invariance only on the assumption that the Majorana neutrino has neither magnetic nor electric dipole moments. As is assumed by the $\mathcal{CPT}$-invariance of the theory, the Majorana neutrino has only one multipole moment that is referred to as an anapole moment μ_a [30]. This moment may be observed when placing the particle into an external vortex magnetic field. And the arising interaction energy of the neutrino in the nonrelativistic limit is given by:

$$E_{int} = -\mu_a < \mathbf{S} \cdot \mathrm{rot}\mathbf{H} > . \tag{15.352}$$

To cite a single example of the source for this moment, we take the toroid distribution of the electric current (toroidal winding).

As regards the Dirac neutrino, it can possess both the electric and magnetic dipole moments as well as the anapole moment.

15.15 Neutrino in the Dirac and Majorana theory

If the neutrino is a Majorana particle, in some processes involving the neutrino the lepton charge conservation law will be violated. Recall that this law was introduced not only for the successive subdivision of the particles into leptons and baryons. Also, it was called up to explain why in the process of interactions between the neutrino and matter the reactions:

$$\nu_l + (A, Z) \rightarrow l + (A, Z + 1), \tag{15.353}$$

where $l = e^-, \mu^-, \tau^-$, were allowed, whereas the reactions:

$$\overline{\nu}_l + (A, Z) \rightarrow l + (A, Z + 1) \tag{15.354}$$

were found forbidden. Taking for the leptons (ν_l and l) the lepton charge L being equal to 1, for antileptons ($\overline{\nu}_l$ and $\overline{l}$) $L = -1$, and assuming:

$$\sum_i L_i = \mathrm{const},$$

we have solved this problem without any difficulties. In so doing ν_l and $\overline{\nu}_l$ represented a particle and antiparticle, respectively, each of which may be in two states with different helicity values. In such an approach the neutrino is considered to be the Dirac particle.

But this interpretation is not the only possible one. We could find an explanation in specificity of the interaction of the neutrino with matter as well. We assume the $\mathcal{P}$-noninvariant weak interaction is so structured that each right-hand neutrino in the process of interaction with a matter produces $\overline{l}$, and left-handed neutrino—l. In the SM the Lagrangian describing the interaction of the neutrino with a matter has just this very form (see Eq. (15.320)). Using this Lagrangian, there is no need to "provide" the neutrino with any new quantum number. Of course, nothing interferes with the assumption that the neutrino is a truly neutral particle, that is, ν_l and $\bar{\nu}_l$ are simply two states of one particle with different helicity values. The lepton number of such neutrinos is zero and hence we consider them as particles having the Majorana nature. If $m_\nu = 0$ then, according to the Lagrangian (15.320), the Dirac neutrino represents the left-handed particle while the Dirac antineutrino represents the right-handed particle. The validity of this statement is obvious. A produced neutrino is described by the bispinor $\bar{u}^{(\alpha)}(\mathbf{p})$ followed by the factor $(1 + \gamma_5)$. In a massless case the bispinor $u^{(\alpha)}(\mathbf{p})$ is an eigenfunction of the chirality operator that in turn is coincident with the helicity operator. Taking account of the identity:

$$\bar{u}^{(\alpha)}(\mathbf{p})(1 + \gamma_5) = u^{(\alpha)\dagger}(\mathbf{p})(1 - \gamma_5)\gamma_0 = [\gamma_0(1 - \gamma_5)u^{(\alpha)}(\mathbf{p})]^\dagger,$$

we come to the conclusion that a helicity value of the neutrino is negative. The production of an antineutrino is associated with a factor in the matrix element $(1 - \gamma_5)v^{(\alpha)}(\mathbf{p})$. Since in the case of antiparticles the operator $(1 - \gamma_5)$ projects the states with the spin oriented in the direction of motion, the produced antineutrino proves to be a right-handed particle.

Inclusion of mass leads to the appearance of the right-handed (left-handed) helicity on the neutrino (antineutrino). We took no notice of these components, for example, in the phenomena of β-decay because the neutrino mass is extremely small. To illustrate, assuming $m/E \ll 1$ and using expansion for the neutrino operator wavefunction (15.240), in the case of the bispinor $u^{(\alpha)}(\mathbf{p})$ we get:

$$\sqrt{\frac{m}{E}}u^{(\alpha)}(\mathbf{p}) = \begin{pmatrix} \sqrt{\dfrac{E+m}{2E}}\varsigma^\alpha \\ \sqrt{\dfrac{E-m}{2E}}(\mathbf{n}\cdot\boldsymbol{\sigma})\varsigma^\alpha \end{pmatrix} \approx \sqrt{\frac{1}{2}}\begin{pmatrix} \varsigma^\alpha \\ \alpha\varsigma^\alpha \end{pmatrix} + \frac{m}{2\sqrt{2}E}\begin{pmatrix} \varsigma^\alpha \\ -\alpha\varsigma^\alpha \end{pmatrix}. \qquad (15.355)$$

Set $\alpha = -1$ and act the operator Λ_R on (15.355). Then, as expected, the first term goes to zero and the contribution of the second term is of the form:

$$\frac{m}{2\sqrt{2}E}\begin{pmatrix} \varsigma^- \\ \varsigma^- \end{pmatrix}.$$

The derived relation means that the amplitude of the "wrong" (right-handed) helicity of the neutrino is proportional to m/E. In a similar way we can demonstrate that, when the mass is included, for the antineutrino the emerging left-handed component is also proportional to m/E. Thus, in the processes described by the Lagrangian (15.320), the ultrarelativistic neutrinos are produced mainly in the left-helicity state, whereas for the produced antineutrinos the preferential helicity is right-handed. We should emphasize that this helicity character is true for the case of the Majorana neutrino too.

At first glance it seems that establishing whether a neutrino is the Majorana or Dirac particle is an easy task. To this end, it is sufficient to create a beam of, for instance, left-handed neutrinos ν_l. Next, it suffices to realize the neutrino spin-flip with the help of some mechanism, and then to direct these right-handed neutrinos to the target. If during interaction with the target the processes described by (15.354) are also detected, the neutrino has the Majorana nature; otherwise it has the Dirac nature. We emphasize this is true for the standard model supplemented with neutral left-handed leptons N_{lR} ($l = e, \mu, \tau$). In this case N_{lR} interact with the Higgs boson only and are actually sterile particles.

Unfortunately, under the state of the art in technology such experiments may be assigned to the category of mental ones only. Nevertheless, numerous experiments involving the neutrino may be currently conducted. Let us try to elucidate why the nature of the neutrino has not been established until the present time. As an example, we investigate the reaction:

$$\nu_l + l \rightarrow \nu_l + l. \qquad (15.356)$$

The corresponding Feynman diagrams are shown in Fig. 15.1, where the waved lines display the propagations of the virtual gauge bosons (W and Z). First, we consider the diagram connecting with the Z-boson exchange. The matrix element associated with it is given by

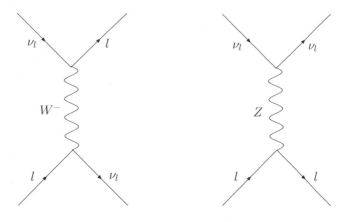

FIGURE 15.1
The Feynman diagram describing the process $\nu_l + l \to \nu_l + l$.

the expression:

$$\mathcal{M}_{\nu_l l \to \nu_l l} = <l_f; \nu_f |\mathcal{A}|l_i; \nu_i>, \tag{15.357}$$

where

$$\mathcal{A} = \frac{g^2}{16\cos^2\theta_W}\overline{\nu}_l(x)\gamma^\mu(1-\gamma_5)\nu_l(x)\left(g_{\mu\nu} + \frac{\partial_\mu\partial_\nu}{m_Z^2}\right)\Delta^c(x-y;m_Z^2)\times$$

$$\times \overline{l}(y)\gamma^\nu[\gamma_5 + 4\sin^2\theta_W - 1]l(y), \tag{15.358}$$

$$|l_i; \nu_i> = a^\dagger_{\alpha_{\nu_i}}(\mathbf{p}_{\nu_i})a^\dagger_{\alpha_{l_i}}(\mathbf{p}_{l_i})|0>, \qquad |l_f; \nu_f> = a^\dagger_{\alpha_{\nu_f}}(\mathbf{p}_{\nu_f})a^\dagger_{\alpha_{l_f}}(\mathbf{p}_{l_f})|0>.$$

and the factor $(g_{\mu\nu}+\partial_\mu\partial_\nu/m_Z^2)\Delta^c(x-y;m_Z^2)$ corresponds to the Z-boson propagator. When writing Eq. (15.357), we have taken into account the form of the Lagrangian describing the interaction of leptons with the Z-bosons (11.44).

In the matrix element (15.357) we are interested only in the part associated with the neutrino current:

$$J_\mu \equiv <l_f; \nu_f|j_\mu|l_i; \nu_i> = <l_f; \nu_f|\overline{\nu}_l(x)\gamma_\mu(1-\gamma_5)\nu_l(x)|l_i; \nu_i>. \tag{15.359}$$

First, we demonstrate that for the Majorana neutrino a vector part of the neutrino current j_μ^M becomes zero. With this in mind we consider the quantity $\overline{\nu}_l^c(x)\gamma_\mu\nu_l^c(x)$. Taking into account the γ-matrices behavior under the charge conjugation and the fact of anticommuting the operator field functions, we find:

$$\overline{\nu}_l^c(x)\gamma_\mu\nu_l^c(x) = \nu_l^T(x)\gamma_\mu^T\overline{\nu}_l^T(x) = (\nu_l)_\alpha(\gamma_\mu)_{\beta\alpha}(\overline{\nu}_l)_\beta =$$

$$= -(\overline{\nu}_l)_\beta(\gamma_\mu)_{\beta\alpha}(\nu_l)_\alpha. \tag{15.360}$$

On the other hand, due to the Majorana condition, we have

$$\overline{\nu}_l^c(x)\gamma_\mu\nu_l^c(x) = \overline{\nu}_l(x)\gamma_\mu\nu_l(x). \tag{15.361}$$

Comparison of the obtained relation with Eq. (15.360) leads to the conclusion:

$$\overline{\nu}_l(x)\gamma_\mu\nu_l(x) = 0, \tag{15.362}$$

which is what we set out to prove.

Inserting the operator field functions expansion (15.325) into (15.359) and allowing for the anticommutation relations (15.314), we get

$$(J_\mu^M)_{if} = -\overline{u}^{(\alpha_f)}(\mathbf{p}_f)\gamma_\mu\gamma_5 u^{(\alpha_i)}(\mathbf{p}_i) + \overline{v}^{(\alpha_i)}(\mathbf{p}_i)\gamma_\mu\gamma_5 v^{(\alpha_f)}(\mathbf{p}_f). \tag{15.363}$$

Because

$$v^{(\alpha)}(\mathbf{p}) = \left(u^{(\alpha)}(\mathbf{p})\right)^c = C\overline{u}^{(\alpha)T}(\mathbf{p}), \tag{15.364}$$

then

$$\overline{v}^{(\alpha)}(\mathbf{p}) = -u^{(\alpha)T}(\mathbf{p})C^{-1}. \tag{15.365}$$

Now the last term in the right-hand side of (15.363) may be rewritten in the form:

$$\overline{v}^{(\alpha_i)}(\mathbf{p}_i)\gamma_\mu\gamma_5 v^{(\alpha_f)}(\mathbf{p}_f) = -u^{(\alpha_i)T}(\mathbf{p}_i)C^{-1}\gamma_\mu\gamma_5 C\overline{u}^{(\alpha_f)T}(\mathbf{p}_f) =$$

$$= -u^{(\alpha_i)T}(\mathbf{p}_i)(\gamma_\mu\gamma_5)^T\overline{u}^{(\alpha_f)T}(\mathbf{p}_f) = -\overline{u}^{(\alpha_f)}(\mathbf{p}_f)\gamma_\mu\gamma_5 u^{(\alpha_i)}(\mathbf{p}_i),$$

which finally gives

$$(J_\mu^M)_{if} = -2\overline{u}^{(\alpha_f)}(\mathbf{p}_f)\gamma_\mu\gamma_5 u^{(\alpha_i)}(\mathbf{p}_i). \tag{15.366}$$

Having done the analogous calculations for the case of the Dirac neutrino, we obtain

$$(J_\mu^D)_{if} = \overline{u}^{(\alpha_f)}(\mathbf{p}_f)\gamma_\mu(1 - \gamma_5)u^{(\alpha_i)}(\mathbf{p}_i). \tag{15.367}$$

In the ultrarelativistic limit both the initial and final neutrinos are predominantly left-handed particle, that is, the relation:

$$\gamma_5 u^{(-)}(\mathbf{p}) \approx -u^{(-)}(\mathbf{p}) + O\left(\frac{m}{2E}\right)$$

(the same is true for the Majorana neutrino as well) takes place. Then, at $\frac{m}{E} \to 0$ we have for the Dirac current

$$\overline{u}^{(\alpha_f)}(\mathbf{p}_f)\gamma_\mu(1 - \gamma_5)u^{(\alpha_i)}(\mathbf{p}_i) \longrightarrow -2\overline{u}^{(\alpha_f)}(\mathbf{p}_f)\gamma_\mu\gamma_5 u^{(\alpha_i)}(\mathbf{p}_i), \tag{15.368}$$

that allows us to conclude

$$(J_\mu^D)_{if} \approx (J_\mu^M)_{if}. \tag{15.369}$$

Again, we see that a value of the effect enabling one to distinguish between the Majorana and Dirac neutrino is totally determined by the neutrino mass value. In other words, in experiments with the real neutrino the chances to establish the neutrino nature are growing as the energy decreases. The result of (15.369) holds not only for the exchange processes with virtual Z-bosons (processes with neutral currents) but also for those with virtual W-bosons (processes with charged currents). To prove this statement, it suffices to represent the matrix element corresponding to the diagram with the W-exchange pictured in Fig. 15.1 as a product of two currents, one of which is the neutrino current of the form given by (15.359). This may be always realized with the use of the so-called Fierz transformation

$$(\overline{u}_a Q_\alpha u_b)(\overline{u}_c Q_\alpha u_d) = -(\overline{u}_a Q_\mu u_d)(\overline{u}_c Q_\mu u_b),$$

where $Q_\alpha = \gamma_\mu(1 - \gamma_5)$, that will be proven later on.

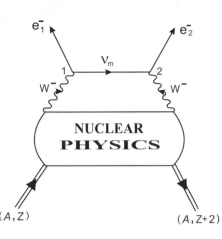

FIGURE 15.2

The diagram for the process $0\nu2\beta$.

Should the processes involving real neutrino fail to settle the dilemma with the nature of the neutrino, we have to consider the reactions involving the virtual neutrino. One of the examples is furnished by the process $0\nu2\beta$ whose diagram is shown in Fig. 15.2. Emission of two W-bosons by the initial nucleus is described by the matrix element calculated within the scope of the nuclear physics. Subsequent production of an electron pair due to the exchange of virtual neutrino is already associated with the sphere of the elementary particle physics. Proceeding further, we base our reasoning on the SM, that is, the theory comprises only the left-handed charged currents. In this case the Lagrangian form (15.320) dictates us that the neutrino emitted at the vertex 1 looks like $\bar{\nu}$ and that absorbed at the vertex 2 looks like ν. In consequence, the process proceeds only on condition that $\bar{\nu}$ and ν are identical particles. In other words, for the Dirac neutrinos it is forbidden, being allowed for the Majorana ones.

The constituent part of the diagram in Fig. 15.2 is a diagram associated with the sub-process:

$$W^{*-}W^{*-} \to e^- + e^-, \tag{15.370}$$

where W^* denotes the virtual W-boson. Note that process (15.370) in itself is of no special interest, while the time-reversal process:

$$e^- + e^- \to W^- + W^, \tag{15.371}$$

that is referred to as inverse double neutrinoless β-decay, attracts lively interest and will be studied at next-generation electron-positron colliders (for example NLC) operating in the e^-e^--mode. It is planned to study a similar process, also supporting the Majorana nature of the neutrino:

$$\mu^- + \mu^- \to W^- + W^- \tag{15.372}$$

at the muon colliders (FMC and NMC).

Let us demonstrate that in the case of $0\nu2\beta$ the effect is also determined by the neutrino mass. The neutrinos involved in the Lagrangian (15.320) belong to a definite lepton family, that is they are defined on the flavor basis. At the same time, both the Dirac and Majorana equations are written for particles with a specific mass ν_j ($j = 1, 2, 3$). Then, in the Lagrangian (15.320) we should change from the flavor basis to the physical one, that could be done with the help of the relation:

$$\nu_l(x) = U_{lj}^{NM}\nu_j(x), \tag{15.373}$$

where $U^{NM}lj$ is the neutrino mixing matrix defined by the formula (12.1) (further superscripts on this matrix will be omitted).

The matrix element of the reaction (15.370) is given by:

$$\mathcal{M}_{W^-W^-\to e^-e^-} \equiv\, <e_1^-;e_2^-|\mathcal{A}|W_1^-;W_2^->,$$

where

$$\mathcal{A} = \frac{g^2}{8}\overline{e}(x)\gamma_\mu(1-\gamma_5)U_{ej}\nu_j^{s_1}(x)W^\mu(x)\overline{e}(y)\gamma_\sigma(1-\gamma_5)U_{ej'}\nu_{j'}^{s_1}(y)W^\sigma(y) =$$

$$= \frac{g^2}{8}W^\mu(x)W^\sigma(y)L_{\mu\sigma}^{(e\nu)}, \tag{15.374}$$

and only the convolutions with $j' = j$ are different from zero:

$$\nu_j^{s_1}(y)\nu_j^{s_1}(x) =\,<0|T(\nu_j(y)\nu_j(x))|0>. \tag{15.375}$$

For (15.375) we may directly use expression (15.324). But it is difficult to keep from temptation to demonstrate a trick reducing such quantities to the calculation of the Dirac particle propagators [31]. So the problem consists in the transition from $<0|T(\nu_j(y)\nu_j(x))|0>$ to $<0|T(\nu_j(y)\overline{\nu}_j(x))|0>$. This may be performed using the charge conjugation operation and Majorana condition. The straightforward calculation yields:

$$\overline{e}\gamma_\mu(1-\gamma_5)\nu_j = \overline{(e^c)^c}\gamma_\mu(1-\gamma_5)(\nu_j^c)^c = -(e^c)^T C^{-1}\gamma_\mu(1-\gamma_5)C(\overline{\nu^c}_j)^T =$$

$$= -(e^c)^T[-\gamma_\mu(1+\gamma_5)]^T(\overline{\nu^c}_j)^T = -\overline{\nu_j^c}\gamma_\mu(1+\gamma_5)e^c = -\lambda_{\odot j}\overline{\nu}_j\gamma_\mu(1+\gamma_5)e^c. \tag{15.376}$$

Considering (15.376), the expression (15.374) takes the desired form:

$$\mathcal{A} = -\lambda_{\odot j}\frac{g^2}{8}\overline{e}(x)\gamma_\mu(1-\gamma_5)U_{ej}\nu_j^{s_1}(x)W^\mu(x)\overline{\nu}_j^{s_1}(y)\gamma_\sigma(1+\gamma_5)U_{ej}e^c(y)W^\sigma(y)$$

where

$$\nu_j^{s_1}(y)\overline{\nu}_j^{s_1}(x) =\,<0|T(\nu_j(y)\overline{\nu}_j(x))|0> = -iS_j^c(y-x).$$

Then, using the explicit form of $S_j^c(y-x)$, the orthogonality condition of the operators $(1-\gamma_5)$ and $(1+\gamma_5)$, we arrive at the same expression for $\mathcal{A}$:

$$\mathcal{A} = 2i\lambda_{\odot j}\frac{g^2}{8}\overline{e}(x)\gamma_\mu(1-\gamma_5)U_{ej}^2 m_{\nu_j}\Delta^c(y-x;m_{\nu_j}^2)W^\mu(x)\gamma_\sigma e^c(y)W^\sigma(y), \tag{15.377}$$

which would follow if the relation (15.324) is accounted for.

When calculating the matrix element $\mathcal{M}_{W^-W^-\to e^-e^-}$, we are interested just in the lepton factor, that is,

$$\mathcal{M}_{\mu\sigma}^{(e\nu)} =\,<e_1^-;e_2^-|L_{\mu\sigma}^{(e\nu)}|W_1^-;W_2^->. \tag{15.378}$$

Using the operator field function expansions of an electron (15.219), (15.220) and a neutrino (15.325), the Fourier representation for the causal function of the Dirac particle (15.254) and allowing for the anticommutation relations for the production and annihilation operators, we obtain, upon straightforward algebraic manipulation, the expression:

$$<e_1^-;e_2^-|\mathcal{A}|W_1^-;W_2^-> = M_{\mu\sigma}^{e\nu}B^{\mu\sigma}, \tag{15.379}$$

where

$$M_{\mu\sigma}^{(e\nu)} = 2\overline{u}^{(\alpha_1)}(\mathbf{p}_1)\gamma_\mu\frac{\lambda_{\odot j}U_{ej}^2 m_{\nu_j}}{k^2+m_{\nu_j}^2}\gamma_\sigma(1+\gamma_5)v^{(\alpha_2)}(\mathbf{p}_2) \tag{15.380}$$

$\mathbf{p}_{1,2}$ are three-dimensional electron momenta, k is four-dimensional virtual neutrino momentum, and $B^{\mu\sigma}$ denotes the factor that is insignificant for our line of reasoning. In this way the matrix element of the $0\nu2\beta$-process is actually determined by the neutrino mass, that is:

$$\mathcal{M}_{0\nu2\beta} \propto \sum_{j} \lambda_{\odot j} U_{ej}^2 m_{\nu_j}. \tag{15.381}$$

We can demonstrate that on the assumption of $\mathcal{CP}$-invariance the quantity $\lambda_{\odot j} U_{ej}^2$ appears to be equal to $-i\eta'_{cp}(\nu_j)|U_{ej}|^2$ and hence is a real number. All information concerning the neutrino mass obtained from a study of the $0\nu2\beta$-process or of the inverse double neutrinoless beta-decay is contained in the quantity:

$$m_{eff}^{(\nu_e)} = -i \sum_{j} \eta'_{cp}(\nu_j)|U_{ej}|^2 m_{\nu_j}. \tag{15.382}$$

If the neutrino sector were free from mixing, the right-hand side of the expression (15.382) were simply equal to m_{ν_e}. Because of this, $m_{eff}^{(\nu_e)}$ may be interpreted as the effective mass of an electron neutrino. Similarly, studying the process,

$$\mu^- + \mu^- \to W^- + W^-$$

we get information about the effective mass of a muon neutrino $m_{eff}^{(\nu_\mu)}$. Note that the theory structure may give the following variant:

$$m_{eff} \ll m_j.$$

The modern data [32] gives the following value of $m_{eff}^{(\nu_e)}$ on 99.73% CL:

$$m_{eff}^{(\nu_e)} = (0.1 - 0.9) \text{ eV}. \tag{15.383}$$

Among the available explanations to smallness of the neutrino mass with respect to the masses of charged leptons and quarks, most popular is a so-called seesaw mechanism. The idea is as follows. Recall how the energy conversion is occurring in the process of the mathematical pendulum oscillation. Its kinetic energy is maximal at some instant of time when the equilibrium point is passed through, while the potential energy is zero. As the point of a maximum deviation from the equilibrium is approached, the kinetic energy decreases and potential energy is growing. Then let us imagine a system of two particles, where a role of the energy is played by mass. In this case the heavier one of the particles, the lighter the other. With the help of the Dirac spinors $\nu_l(x)$ and $\bar{\nu}_l(x)$ in each generation two Majorana spinors are constructed, which are associated with two neutrino states. A light particle is associated with the ordinary neutrino (ν) while the term "heavy neutrino" is used for its partner. So the total number of neutrinos in such schemes equals 6. The masses of light and heavy neutrino are related by the seesaw relation:

$$m_{\nu_l} m_{N_l} = m_l^2, \tag{15.384}$$

that is a key to explanation of a small mass of the ν_l-neutrino. A representative model with such a neutrino sector is a model based on the $SU(2)_L \times SU(2)_R \times U(1)_{B-L}$-gauge group [33] (the left-right model). The lower limit for the mass of a heavy electron neutrino, obtained from analysis of the available experiments, is on the order of 1 TeV [34].

In models with additional neutrinos the chances of establishing the neutrino nature by the collider experiments, for example, inverse double neutrinoless beta-decay are drastically increased. For instance, modifying the SM so that the neutrino can be a massive Majorana particle, in the ultrarelativistic case the cross section of the process (15.371) is determined by [35]:

$$\sigma_{e^-e^- \to W^-W^-} \approx 4.16 \times 10^{-3} \left(\frac{|U_{ej}|^2 m_{\nu_j}}{\text{GeV}} \right)^2 \text{ pb}$$

and hence appears to be a small value. In the case of the left- right model the cross section of the process (15.371) is much greater. For example, when $m_{N_e} \approx 1$ TeV and $E_{cm} = 200$ GeV, it may be as much as several tenths of pb [36].

16

Massive vector field

*Bruin the second inquired if the forest hadn't
at least a university or some academy which he might burn down.
But in this respect, too, as in others, it turned out that
his predecessor had anticipated his intentions, enrolling all
university students as soldiers of remote garrisons and confining
the academicians in a hollow tree, where, submerged in
lethargy, they remain to this day.*
M.E. Saltykov Shchedrin, "Bears in Government"

16.1 Proca equation

To describe a particle with the spin 1 (vector particle), we need a field function having three independent components only. For a free-moving particle of nonzero mass the rest reference frame is always available. Obviously, just in this system the intrinsic symmetry properties of a particle itself are revealed. In this case we should consider symmetry regarding all possible rotations around the center, that is, symmetry regarding the whole group of spherical symmetry. The symmetry properties of a particle relative to this group are characterized by its spin J, determining the number $2J + 1$ of the wavefunction components transformed through each other. When the four-dimensional vector is used as a wave function, in a rest reference frame the vector particle is associated with three space components of this vector. In order to make certain that this is the case, we introduce the projection operators:

$$\Lambda^{(1)\mu}_\nu(\partial) = \delta^\mu_\nu + \frac{\partial^\mu \partial_\nu}{m^2}, \qquad \Lambda^{(0)\mu}_\nu(\partial) = -\frac{\partial^\mu \partial_\nu}{m^2}, \tag{16.1}$$

having the properties:

$$\left.\begin{aligned}
\Lambda^{(1)\mu}_\nu(\partial) + \Lambda^{(0)\mu}_\nu(\partial) = \delta^\mu_\nu, \qquad \Lambda^{(1)\mu}_\nu(\partial)\Lambda^{(1)\nu}_\sigma(\partial) = \Lambda^{(1)\mu}_\sigma(\partial), \\
\Lambda^{(0)\mu}_\nu(\partial)\Lambda^{(0)\nu}_\sigma(\partial) = \Lambda^{(0)\mu}_\sigma(\partial), \qquad \Lambda^{(0)\mu}_\nu(\partial)\Lambda^{(1)\nu}_\sigma(\partial) = 0.
\end{aligned}\right\} \tag{16.2}$$

The Pauli-Lubanski pseudovector for a field with the spin 1 is given by the expression:

$$(w_\sigma)_{\tau\kappa} = \frac{1}{2}\varepsilon_{\sigma\mu\nu\lambda}\left(\mathcal{M}^{\mu\nu}\right)_{\tau\kappa} p^\lambda, \tag{16.3}$$

where the generator of the vector representation of the proper Lorentz group $\left(\mathcal{M}^{\mu\nu}\right)_{\tau\kappa}$ coincides with that of the Lorentz group itself:

$$\left(\mathcal{M}^{\mu\nu}\right)_{\tau\kappa} = \left(I^{\mu\nu}\right)_{\tau\kappa} = \delta^\mu_\tau \delta^\nu_\kappa - \delta^\mu_\kappa \delta^\nu_\tau. \tag{16.4}$$

Acting by the square of the operator (16.3)

$$(W)^\rho_\gamma = (w^\sigma)^{\rho\tau}(w_\sigma)_{\tau\gamma}, \tag{16.5}$$

on the states

$$\psi_\rho^{(1)}(p) = \Lambda_\rho^{(1)\mu}(p)\psi_\mu(p), \qquad \psi_\rho^{(0)}(p) = \Lambda_\rho^{(0)\mu}(p)\psi_\mu(p), \qquad (16.6)$$

where

$$\Lambda_\rho^{(s)\mu}(p) = \Lambda_\rho^{(s)\mu}(\partial) \, (\partial \to -ip) \,, \qquad s = 0, 1,$$

we get

$$(W)_\gamma^\rho \, \psi_\rho^{(1)}(p) = 2m^2\psi_\gamma^{(1)}(p), \qquad (W)_\gamma^\rho \, \psi_\rho^{(0)}(p) = 0. \qquad (16.7)$$

Since the eigenvalues of the operators (16.5) are equal to $m^2 J(J+1)\delta_\gamma^\rho$, then from (16.7) it follows that $\psi_\rho^{(1)}(x)$ describes the states of a particle with the spin 1 whereas $\psi_\rho^{(0)}(x)$ does those of a particle with the spin 0. In its turn, from the expressions (16.6) it is not difficult to understand that in the rest reference frame only the space components of $\psi_\rho^{(1)}(x)$ are different from zero.

In a similar way we can show that, due to its four-dimensional nature, the field function of a vector particle may be the components of the antisymmetric second-rank four-dimensional tensor $\psi_{\mu\nu}(x)$ whose mixed components $\psi_{0k}(x)$ in a rest reference frame become zero.

Thus, the procedure of constructing an equation of motion for a vector particle is in principle clear. In the most general case we should take the quantities $\psi_\mu(x)$ and $\psi_{\mu\nu}(x)$ and then combine them in a particular manner to provide the fulfillment of both the Klein-Gordon condition and additional condition that reduces the number of independent components of the field function to three. In the case of the complex vector field the equation sought has the form [37]:

$$\left.\begin{array}{l} \partial^\nu \psi_{\mu\nu}(x) - m^2\psi_\mu(x) = 0, \\ \psi_{\mu\nu}(x) = \partial_\mu\psi_\nu - \partial_\nu\psi_\mu, \\ (+\text{conj.}) \end{array}\right\} \qquad (16.8)$$

To differentiate the first line of Eq. (16.8) with respect to x_μ and allow for a antisymmetry of $\psi_{\mu\nu}(x)$ leads to the condition:

$$\partial^\nu\psi_\nu(x) = 0. \qquad (16.9)$$

In the rest reference frame, where $\psi_\nu(x)$ does not depend on the space coordinates, we find:

$$k^0\psi_0(x) = m\psi_0(x) = 0$$

What this means is:

$$\psi_0(x) = 0,$$

that is, the equations system really describe the particle with the spin 1. Eqs. (16.8) could be derived from the Lagrangian:

$$\mathcal{L} =: \frac{1}{2}\psi_{\mu\nu}^\dagger(x)\left[\frac{1}{2}\psi^{\mu\nu}(x) - \partial^\mu\psi^\nu(x) + \partial^\nu\psi^\mu(x)\right] + \frac{1}{2}\left[\frac{1}{2}\psi_{\mu\nu}^\dagger(x) - \right.$$

$$\left. -\partial_\mu\psi_\nu^\dagger(x) + \partial_\nu\psi_\mu^\dagger(x)\right]\psi^{\mu\nu}(x) + m^2\psi^{\nu\dagger}(x)\psi_\nu(x) :, \qquad (16.10)$$

where $\psi^\nu(x)$, $\psi^{\nu\dagger}(x)$, $\psi^{\mu\nu}(x)$ and $\psi^{\mu\nu\dagger}(x)$ should be considered as the independent generalized coordinates. In this case for the description of a vector particle we use the representation that is a sum of the two irreducible representations, namely, the first: $(1/2,1/2)$ by which the four-dimensional vector $\psi_\mu(x)$ is transformed, and the second: $(0,1)\bigoplus(1,0)$ whereby the antisymmetric tensor $\psi_{\mu\nu}$ is transformed.

The components $\psi_0(x)$ are of no importance for the dynamics, and at will they may be painlessly excluded from the theory. Using Eqs. (16.8), we can express them in terms of the dynamically independent components:

$$\psi_0(x) = -\frac{1}{m^2}\partial^\mu\psi_{\mu0}(x), \qquad \psi_0^\dagger(x) = -\frac{1}{m^2}\partial^\mu\psi_{\mu0}^\dagger(x), \qquad (16.11)$$

and the derived relations may be used to determine the dynamic invariants. The momenta, canonically conjugate to $\psi_\mu(x)$ and $\psi_\mu^\dagger(x)$, are given by the expressions:

$$\pi_\mu(x) = \frac{\partial \mathcal{L}}{\partial(\partial_0 \psi^\mu(x))} = \psi^{\mu 0\dagger}(x), \qquad \pi_\mu^\dagger(x) = \frac{\partial \mathcal{L}}{\partial(\partial_0 \psi^{\mu\dagger}(x))} = \psi^{\mu 0}(x). \tag{16.12}$$

As $\pi_0(x)$ and $\pi_0^\dagger(x)$ are identically equal to zero, we have no problems with lack of the associated variables $\psi_0(x)$ and $\psi_0^\dagger(x)$. Taking energy as an example, we can demonstrate that the dynamic invariants may be expressed in terms of only the generalized "coordinates" and "momenta." We have

$$H =: \int d^3x \left[\pi^k(x)\partial_0 \psi^k(x) + \pi^{k\dagger}(x)\partial_0 \psi^{k\dagger}(x) : -\mathcal{L} \right] =: \int d^3x \left\{ (\boldsymbol{\pi}(x) \cdot \boldsymbol{\pi}^\dagger(x)) + \right.$$

$$+ (\boldsymbol{\nabla} \times \boldsymbol{\psi}(x)) \cdot (\boldsymbol{\nabla} \times \boldsymbol{\psi}^\dagger(x)) + m^2(\boldsymbol{\psi}(x) \cdot \boldsymbol{\psi}^\dagger(x)) + \frac{1}{m^2}(\boldsymbol{\nabla} \cdot \boldsymbol{\pi}(x))(\boldsymbol{\nabla} \cdot \boldsymbol{\pi}^\dagger(x)) +$$

$$\left. + \frac{1}{m^2} \boldsymbol{\nabla} \cdot \left[\boldsymbol{\pi}(x)(\boldsymbol{\nabla} \cdot \boldsymbol{\pi}^\dagger(x)) + \boldsymbol{\pi}^\dagger(x)(\boldsymbol{\nabla} \cdot \boldsymbol{\pi}(x)) \right] \right\} :=$$

$$=: \int d^3x \left[(\boldsymbol{\pi}(x) \cdot \boldsymbol{\pi}^\dagger(x)) + (\boldsymbol{\nabla} \times \boldsymbol{\psi}(x)) \cdot (\boldsymbol{\nabla} \times \boldsymbol{\psi}^\dagger(x)) + m^2(\boldsymbol{\psi}(x) \cdot \boldsymbol{\psi}^\dagger(x)) + \right.$$

$$\left. + \frac{1}{m^2}(\boldsymbol{\nabla} \cdot \boldsymbol{\pi}(x))(\boldsymbol{\nabla} \cdot \boldsymbol{\pi}^\dagger(x)) \right]. \tag{16.13}$$

From Eq. (16.13) it follows that the vector field energy H is a positive definite quantity. We can show that without ruling out the components $\psi_0(x)$ and $\psi_0^\dagger(x)$, the condition of positiveness of H were impossible.

At the same time, a vector particle is much more conveniently described by just one four-dimensional vector $\psi^\nu(x)$. In so doing, the equation:

$$(B_{\mu\lambda,\sigma\nu}\partial^\mu\partial_\lambda + m^2\delta_{\nu\sigma})\psi_\nu(x) = 0, \tag{16.14}$$

where

$$B_{\mu\lambda,\sigma\nu} = \delta_{\mu\lambda}\delta_{\sigma\nu} - \delta_{\mu\nu}\delta_{\lambda\sigma},$$

is the equivalent writing of the Proca equation. Eq. (16.14) contains the additional condition (16.9), that may be directly demonstrated by acting the operator ∂^σ on this equation and taking account of the fact that:

$$B_{\mu\lambda,\sigma\nu} = -B_{\mu\nu,\sigma\lambda} = -B_{\sigma\lambda,\mu\nu}.$$

The Lagrange function density, of which Eq. (16.14) follows, has the shape:

$$\mathcal{L} =: -B_{\mu\lambda,\sigma\nu}\partial^\mu\psi^{\sigma\dagger}(x)\partial_\lambda\psi_\nu(x) + m^2\psi_\mu^\dagger(x)\psi^\mu(x) : . \tag{16.15}$$

At first sight, it seems that the situation with the vector particle may be simplified still further. The system of Eqs. (16.8) may be replaced by the equation:

$$(\Box + m^2)\psi_\mu(x) = 0, \tag{16.16}$$

and fulfillment of the condition:

$$\partial^\mu\psi_\mu(x) = 0$$

may be required independently, or as they say, imposed manually. Of course, Eq. (16.16) together with the additional condition is equivalent to Eq. (16.14). The Lagrangian corresponding to (16.16) is given by:

$$\mathcal{L}' =: -\partial^\mu \psi^{\nu\dagger}(x) \partial_\mu \psi_\nu(x) + m^2 \psi_\mu^\dagger(x) \psi^\mu(x) : . \tag{16.17}$$

Let us pay attention to signs in the Lagrangians of a vector field. When comparing, for instance, (16.17) to the Lagrangian of a scalar field $\mathcal{L}^{J=0}$ (see, (14.4)), we can observe that the signs of $\mathcal{L}'$ and of $\mathcal{L}^{J=0}$ are the same for $\nu = 1, 2, 3$ while the signs of both Lagrangians are opposite for $\nu = 0$. Of course, a choice of the sign in the Lagrangian is not arbitrary. It is dictated by the energy positiveness requirement. Because the time components of the field function of the vector particle make no contribution to the total energy, a sign in front of the space part of the four-dimensional scalar product $\boldsymbol{\psi}^\dagger(x) \cdot \boldsymbol{\psi}(x)$ is the decisive factor in the Lagrangian.

The Lagrangian (16.17) differs from that (16.15) on the quantity:

$$\mathcal{L}' - \mathcal{L} = - : \frac{1}{2} \partial^\sigma [\partial_\mu \psi_\sigma^\dagger(x) \psi^\mu(x) + \psi_\mu^\dagger(x) \partial^\mu \psi_\sigma(x)] + \frac{1}{2} [(\partial_\mu \partial^\sigma \psi_\sigma^\dagger(x)) \psi^\mu(x) +$$

$$+ \psi_\mu^\dagger(x)(\partial^\mu \partial^\sigma \psi_\sigma(x))] : . \tag{16.18}$$

The first term in the right-hand side of (16.18) representing the four-dimensional divergence excites no apprehension, while the second one suggests an idea that both formalisms may be nonequivalent. During the calculation of dynamic variables we assume the fulfillment both of the motion equations and of consequences resulting from these equations. In this way the second term in the right-hand side of (16.18) becomes zero, and in a free case the dynamic invariants prove to be equal in the theories involving the Lagrangians $\mathcal{L}'$ and $\mathcal{L}$. This circumstance could give cause for the statement that both these theories are equivalent. This is a rather frequent error, though one may easily guess that the indicated equivalence is retained only in a free case. Indeed, let us consider the introduction of the electromagnetic interaction:

$$\partial_\mu \to D_\mu \equiv \partial_\mu - ieA_\mu.$$

In this case, (16.16) is converted into four independent Klein-Gordon equations, where naturally there is no place for interaction between the spin and electromagnetic field. Besides, it is not clear what to do with the additional condition: to keep it in the form of (16.9) or replace it by:

$$D_\mu \psi^\mu = 0.$$

It is obvious that an ultimate choice of one or another theory may be made only with the help of the experiment. Using a complex vector field, we can describe the charged carriers of the electroweak interactions, the $W^\pm$-bosons. The experimental data accumulated by the present time (measuring cross sections, determining the multipole electromagnetic moments and so on) support Eq. (16.14) (to be more specific, the Yang-Mills equation, whose linear approximation is just the Proca equation).

16.2   Dynamic invariants of the vector field

Since in a free case the Lagrangians $\mathcal{L}'$ and $\mathcal{L}$ lead to the same expressions for the field invariants, for this consideration we will use $\mathcal{L}'$. By the usual way we find the energy-momentum tensor:

$$T_{\mu\nu} = -\partial_\mu \psi_\lambda^\dagger(x) \partial_\nu \psi^\lambda(x) - \partial_\nu \psi_\lambda^\dagger(x) \partial_\mu \psi^\lambda(x) - g_{\mu\nu} \mathcal{L}', \tag{16.19}$$

the electric current vector:

$$j_\mu = ie[\partial_\mu \psi_\lambda^\dagger(x)\psi^\lambda(x) - \psi_\lambda^\dagger(x)\partial_\mu \psi^\lambda(x)], \tag{16.20}$$

and the spin moment tensor:

$$S_{\mu\nu}^\lambda = \psi_\nu^\dagger(x)\partial^\lambda \psi_\mu(x) + \partial^\lambda \psi_\mu^\dagger(x)\psi^\nu(x) - \psi_\mu^\dagger(x)\partial^\lambda \psi_\nu(x) - \partial^\lambda \psi_\nu^\dagger(x)\psi^\mu(x). \tag{16.21}$$

Recall, in order to describe spin properties of particles one uses three space components of the tensor $S_{\mu\nu}^0$ which, in its aggregate, form a three-dimensional vector $\mathbf{S}$ with the components:

$$S_m = \varepsilon_{mnl} \int d^3x S_{nl}^0, \qquad l, m, n = 1, 2, 3. \tag{16.22}$$

The expansion of the operator field functions in terms of the Proca equation solutions will look like:

$$\psi_\mu(x) = \frac{1}{(2\pi)^{3/2}} \int \frac{d^3k}{\sqrt{2k_0}} \sum_{\lambda=1}^3 \vartheta_\mu^\lambda(\mathbf{k})[a_\lambda(\mathbf{k}) \exp(-ikx) + b_\lambda^\dagger(\mathbf{k}) \exp(ikx)], \tag{16.23}$$

$$\psi_\mu^\dagger(x) = \frac{1}{(2\pi)^{3/2}} \int \frac{d^3k}{\sqrt{2k_0}} \sum_{\lambda=1}^3 \vartheta_\mu^{\lambda*}(\mathbf{k})[a_\lambda^\dagger(\mathbf{k}) \exp(ikx) + b_\lambda(\mathbf{k}) \exp(-ikx)], \tag{16.24}$$

where the polarization vectors $\vartheta_\mu^\lambda(\mathbf{k})$ may be complex quantities in the most general case. Substituting (16.23), (16.24) into (16.19), (16.20), (16.21) and carrying out the integration over the space coordinates, we derive the energy-momentum four-dimensional vector:

$$P^\mu = \int T^{0\mu} d^3x = -\int d^3k k^\mu [\psi_\nu^{(+)\dagger}(\mathbf{k})\psi^{(+)\nu}(\mathbf{k}) + \psi_\nu^{(-)\dagger}(\mathbf{k})\psi^{(-)\nu}(\mathbf{k})], \tag{16.25}$$

the electric charge:

$$Q = \int j^0 d^3x = e \int d^3k [\psi_\nu^{(+)\dagger}(\mathbf{k})\psi^{(+)\nu}(\mathbf{k}) - \psi_\nu^{(-)\dagger}(\mathbf{k})\psi^{(-)\nu}(\mathbf{k})], \tag{16.26}$$

and the spin vector:

$$\mathbf{S} = i \int d^3k \{ [\boldsymbol{\psi}^{(-)\dagger}(\mathbf{k}) \times \boldsymbol{\psi}^{(-)}(\mathbf{k})] - [\boldsymbol{\psi}^{(+)\dagger}(\mathbf{k}) \times \boldsymbol{\psi}^{(+)}(\mathbf{k})] \}, \tag{16.27}$$

where we have introduced the designations:

$$\psi_\mu^{(+)}(\mathbf{k}) = \sum_{\lambda=1}^3 \vartheta_\mu^\lambda(\mathbf{k})a_\lambda(\mathbf{k}), \qquad \psi_\mu^{(-)}(\mathbf{k}) = \sum_{\lambda=1}^3 \vartheta_\mu^\lambda(\mathbf{k})b_\lambda^\dagger(\mathbf{k}).$$

Let us consider the polarization vectors $\vartheta_\mu^\lambda(\mathbf{k})$ in greater detail. Our next objective is diagonalization of the dynamic invariants. This may be performed with the relevant choice of the vectors $\vartheta_\mu^\lambda(\mathbf{k})$. For a while, we forget an operator nature of the field function, assuming it to be an ordinary wavefunction of nonrelativistic quantum mechanics. For the solution of Eq. (16.14) we may take the quantity:

$$\psi_\mu(x) = N\vartheta_\mu(\mathbf{k}) \exp(ikx),$$

where N is a normalization constant depending on a spin projection on the chosen direction in space. Direct the particle momentum along the axis x_3. Due to the additional condition

(16.9), only three of four components involved in the polarization vector ϑ_μ are independent. Let us exclude the time component and use for the remaining space components the following two choices of three-unit vectors (local reference point):

$$\mathbf{e}^1(\mathbf{k}) = (1,0,0), \qquad \mathbf{e}^2(\mathbf{k}) = (0,1,0), \qquad \mathbf{e}^3(\mathbf{k}) = (0,0,1), \tag{16.28}$$

and

$$\mathbf{e}'^1(\mathbf{k}) = \frac{1}{\sqrt{2}}(1,i,0), \qquad \mathbf{e}'^2(\mathbf{k}) = \frac{1}{\sqrt{2}}(1,-i,0), \qquad \mathbf{e}'^3(\mathbf{k}) = (0,0,1). \tag{16.29}$$

Suppose that we work with the local reference point $\mathbf{e}^\lambda(\mathbf{k})$ and the particle polarization vector is equal to $\mathbf{e}^1(\mathbf{k})$. In this case the vector $\psi(x)$, varying in time, invariably remains to be directed along the x_1-axis. Because of this, they say that a particle with the wavefunction $\psi(x)$ is linearly polarized along the x_1-axis. And if we were concerned with photons, the essence of this statement would be more clear for a beam of photons being situated in the same state with the $\psi(x)$-function. Then in the electromagnetic field under study the electric field strength vector $\mathbf{E} = -\partial_0 \psi(x)$ would be making the oscillation along the x_1-axis, that is, we deal with the case of a field linearly polarized in the x_1 direction. However, the strength vectors $\mathbf{E}$ and $\mathbf{H}$ are observable for the electromagnetic field only. And unfortunately, the classical physics obviousness speaking to the heart, has no analogs for massive vector fields.

When the vector $\psi(x)$, varying in time, permanently remains to be directed along the x_2- or x_3-direction, we have a vector particle linearly polarized along the axis x_2 or x_3, respectively. The states with $\psi(x) \parallel \mathbf{e}^3(\mathbf{k})$ correspond to the longitudinal (relative to the momentum) polarization while those with $\psi(x) \parallel \mathbf{e}^1(\mathbf{k})$ or $\psi(x) \parallel \mathbf{e}^2(\mathbf{k})$ correspond to the transverse polarizations.

Now we assume that in the case of the local reference point (16.29) the polarization vector equals $\mathbf{e}'^1(\mathbf{k})$, that is,

$$\psi_1(x) \neq 0, \qquad \psi_2(x) \neq 0, \qquad \psi_3(x) = 0.$$

Then, in the plane $z = 0$ we arrive at the following expressions for the transverse components $\psi_1(x)$, $\psi_2(x)$:

$$\psi_1(t) = \frac{N}{\sqrt{2}} \exp{(ik_0 t)}, \qquad \psi_2(t) = \frac{N}{\sqrt{2}} \exp{\left(ik_0 t + \frac{\pi}{2}\right)},$$

i.e., for the resultant oscillation the component along the x_2-direction passes ahead of the component in the x_1-direction by $\pi/2$. All this means that the vector $\psi(t)$ is rotating, while being constant in value. This situation is referred to as the circular polarization of the field. Provided the polarization vector is $\mathbf{e}'^2(\mathbf{k})$, then the particle is again circularly polarized but this time the rotation direction of the vector $\psi(t)$ is opposite to the previous one.

Now we present this consideration in the relativistically covariant form. It is easily seen that the normalization condition and the additional condition (16.9) result in the polarization four-dimensional vectors, which should satisfy the following relation:

$$\sum_{\lambda=1}^{3} \vartheta_\mu^{\lambda\dagger}(\mathbf{k}) \vartheta_\nu^\lambda(\mathbf{k}) = -\left(g_{\mu\nu} - \frac{k_\mu k_\nu}{m^2}\right). \tag{16.30}$$

By the direct calculations we can check that at $k_\mu = (k_0, 0, 0, |\mathbf{k}|)$ the condition (16.30) is satisfied by the following two sets of the polarization vectors:

$$\left. \begin{array}{c} \vartheta_\mu^1(\mathbf{k}) = (0,1,0,0), \qquad \vartheta_\mu^2(\mathbf{k}) = (0,0,1,0), \\ \vartheta_\mu^3(\mathbf{k}) = m^{-1}(|\mathbf{k}|,0,0,k_0), \end{array} \right\} \tag{16.31}$$

and

$$\vartheta_\mu'^1(\mathbf{k}) = (\sqrt{2})^{-1}(0,1,i,0), \qquad \left.\begin{array}{l}\vartheta_\mu'^2(\mathbf{k}) = (\sqrt{2})^{-1}(0,1,-i,0), \\ \vartheta_\mu'^3(\mathbf{k}) = m^{-1}(|\mathbf{k}|,0,0,k_0),\end{array}\right\} \tag{16.32}$$

representing the four-dimensional generalization of the local reference points of (16.28) and (16.29), respectively.

In the expression for the energy the term with $\mu = 0$ proves to be negative and hence the energy is not positively defined. However, the time components of the field functions $\psi_0^{(\pm)}(\mathbf{k})$ and $\psi_0^{(\pm)\dagger}(\mathbf{k})$ are not the dynamically independent field components and should be excluded with the use of the additional conditions:

$$k^\mu \psi_\mu^{(\pm)}(\mathbf{k}) = 0, \qquad k^\mu \psi_\mu^{(\pm)\dagger}(\mathbf{k}) = 0. \tag{16.33}$$

Note that in the case of the Lagrangian $\mathcal{L}'$ the situation is somewhat burdened with the presence of nonzero momenta canonically conjugate to $\psi_0(x)$ and $\psi_0^\dagger(x)$, whereas for $\mathcal{L}$ these momenta are identically equal to zero. With the help of the expressions resulting from (16.33):

$$\psi_0^{(\pm)}(\mathbf{k}) - \frac{1}{k_0} k_n \psi_n^{(\pm)}(\mathbf{k}), \qquad \psi_0^{(\pm)\dagger}(\mathbf{k}) = \frac{1}{k_0} k_n \psi_n^{(\pm)}(\mathbf{k})$$

it is an easy matter to get

$$\psi_\nu^{(\pm)\dagger}(\mathbf{k})\psi^{(\pm)\nu}(\mathbf{k}) = -\boldsymbol{\psi}^{(\pm)\dagger}(\mathbf{k}) \cdot \boldsymbol{\psi}^{(\pm)}(\mathbf{k}) + \frac{1}{k_0^2}(\mathbf{k} \cdot \boldsymbol{\psi}^{(\pm)\dagger}(\mathbf{k}))(\mathbf{k} \cdot \boldsymbol{\psi}^{(\pm)}(\mathbf{k})). \tag{16.34}$$

Perform the expansion of the vectors $\boldsymbol{\psi}^{(\pm)}(\mathbf{k})$ in terms of the local reference point (16.31):

$$\boldsymbol{\psi}^{(+)}(\mathbf{k}) = \boldsymbol{\vartheta}^1(\mathbf{k})a_1(\mathbf{k}) + \boldsymbol{\vartheta}^2(\mathbf{k})a_2(\mathbf{k}) + \frac{\mathbf{k}}{|\mathbf{k}|}\frac{k_0}{m}a_3(\mathbf{k}), \tag{16.35}$$

$$\boldsymbol{\psi}^{(-)}(\mathbf{k}) = \boldsymbol{\vartheta}^1(\mathbf{k})b_1^\dagger(\mathbf{k}) + \boldsymbol{\vartheta}^2(\mathbf{k})b_2^\dagger(\mathbf{k}) + \frac{\mathbf{k}}{|\mathbf{k}|}\frac{k_0}{m}b_3^\dagger(\mathbf{k}). \tag{16.36}$$

Using (16.34)–(16.36), we get the following expressions for the four-dimensional momentum and the electric charge of the vector field:

$$P^\mu = \int d^3x k^\mu[(\mathbf{a}^\dagger(\mathbf{k}) \cdot \mathbf{a}(\mathbf{k})) + (\mathbf{b}^\dagger(\mathbf{k}) \cdot \mathbf{b}(\mathbf{k}))], \tag{16.37}$$

$$Q = -e \int d^3x[(\mathbf{a}^\dagger(\mathbf{k}) \cdot \mathbf{a}(\mathbf{k})) - (\mathbf{b}^\dagger(\mathbf{k}) \cdot \mathbf{b}(\mathbf{k}))]. \tag{16.38}$$

Thus, in the new variables the energy proves to be positively defined.

However, in a coordinate system with the unit vectors $\boldsymbol{\vartheta}^\lambda(\mathbf{k})$ the expression for the spin vector projection onto the direction of a particle momentum is not diagonal:

$$S_3 = i \int d^3k[a_1^\dagger(\mathbf{k})a_2(\mathbf{k}) - a_2^\dagger(\mathbf{k})a_1(\mathbf{k}) - b_1^\dagger(\mathbf{k})b_2(\mathbf{k}) + b_2^\dagger(\mathbf{k})b_1(\mathbf{k})]. \tag{16.39}$$

We can improve the situation by going to the local reference point (16.32). Inserting the expressions:

$$\boldsymbol{\psi}^{(+)}(\mathbf{k}) = \sum_\lambda \boldsymbol{\vartheta}'^\lambda(\mathbf{k})a_\lambda(\mathbf{k}), \qquad \boldsymbol{\psi}^{(-)}(\mathbf{k}) = \sum_\lambda \boldsymbol{\vartheta}'^\lambda(\mathbf{k})b_\lambda^\dagger(\mathbf{k})$$

into Eqs. (16.25)–(16.27), we must assure ourselves that in the new local reference point the quantities P^μ, Q maintain their form and the spin vector projection is brought into the diagonal form:

$$S_3 = \int d^3k [a_1^\dagger(\mathbf{k})a_1(\mathbf{k}) - a_2^\dagger(\mathbf{k})a_2(\mathbf{k}) - b_1^\dagger(\mathbf{k})b_1(\mathbf{k}) + b_2^\dagger(\mathbf{k})b_2(\mathbf{k})]. \qquad (16.40)$$

Let us establish the physical meaning of the operators: $a_n^\dagger(\mathbf{k})$, $a_n(\mathbf{k})$, $b_n^\dagger(\mathbf{k})$ $b_n(\mathbf{k})$. The form of the dynamic variables points to the fact that we have obtained the correct corpuscular structure of the vector field. Experience in quantization of the scalar field suggests that this pattern was obtained on the basis of the commutation relations for the operators. Natural generalization of the scalar field commutators in the case of a vector field is provided by the relations:

$$[a_m(\mathbf{k}), a_n^\dagger(\mathbf{k}')] = [b_m(\mathbf{k}), b_n^\dagger(\mathbf{k}')] = \delta_{mn}\delta(\mathbf{k}' - \mathbf{k}). \qquad (16.41)$$

And in the next section their validity will be proved on the basis of the canonical formalism. By application of (16.41) it is not difficult to fix:

$$[a_n(\mathbf{k}), H] = k_0 a_n(\mathbf{k}), \qquad [a_n^\dagger(\mathbf{k}), H] = -k_0 a_n^\dagger(\mathbf{k}), \qquad (16.42)$$

$$[b_n(\mathbf{k}), H] = k_0 b_n(\mathbf{k}), \qquad [b_n^\dagger(\mathbf{k}), H] = -k_0 b_n^\dagger(\mathbf{k}). \qquad (16.43)$$

If Φ_E is the energy operator eigenstate with the eigenvalue E, then:

$$Ha_n(\mathbf{k})\Phi_E = (E - k_0)a_n(\mathbf{k})\Phi_E, \qquad Ha_n^\dagger(\mathbf{k})\Phi_E = (E + k_0)a_n^\dagger(\mathbf{k})\Phi_E. \qquad (16.44)$$

and the same takes place for the states $b_n(\mathbf{k})\Phi_E$ and $b_n^\dagger(\mathbf{k})\Phi_E$. Therefore, $a_n(\mathbf{k})$ and $b_n(\mathbf{k})$ have a meaning of the annihilation operators whereas $a_n^\dagger(\mathbf{k})$ and $b_n^\dagger(\mathbf{k})$ do a meaning of the production operators.

Using Eq. (16.41), we find

$$[Q, a_n^\dagger(\mathbf{k})] = -ea_n^\dagger(\mathbf{k}), \qquad [Q, b_n^\dagger(\mathbf{k})] = eb_n^\dagger(\mathbf{k}), \qquad (16.45)$$

$$[S_3, a_n^\dagger(\mathbf{k})] = \begin{cases} a_1^\dagger(\mathbf{k}), & n = 1, \\ -a_2^\dagger(\mathbf{k}), & n = 2, \\ 0, & n = 3, \end{cases} \qquad (16.46)$$

$$[S_3, b_n^\dagger(\mathbf{k})] = \begin{cases} -b_1^\dagger(\mathbf{k}), & n = 1, \\ b_2^\dagger(\mathbf{k}), & n = 2, \\ 0, & n = 3. \end{cases} \qquad (16.47)$$

It should be emphasized that Eqs. (16.46) and (16.47) are valid only in case S_3 to be diagonal, that in turn is associated with the circular polarization of a vector particle. To act the operators Q and S_3 on the one-particle state $\Phi_\mathbf{k}^{a_n} = a_n^\dagger(\mathbf{k})\Phi_0$ leads to the equations:

$$Q\Phi_\mathbf{k}^{a_n} = Qa_n^\dagger(\mathbf{k})\Phi_0 = [Q, a_n^\dagger(\mathbf{k})]\Phi_0, \qquad (16.48)$$

$$S_3\Phi_\mathbf{k}^{a_n} = [S_3, a_n^\dagger(\mathbf{k})]\Phi_0. \qquad (16.49)$$

From Eqs. (16.44), (16.48), and (16.49) it becomes apparent that $a_n^\dagger(\mathbf{k})$ represents the production operator of the particle with the momentum $\mathbf{k}$, the energy k_0, the charge $-e$, and the spin projection onto the motion direction being equal to 1,-1,0 for $n = 1, n = 2$ and $n = 3$, respectively. The physical meaning of the operator $b_n^\dagger(\mathbf{k})$ follows from the aforementioned under changing the charge sign and the spin projection sign. We accepted the terminology, according to which particles possess the negative charge and antiparticles possess the positive charge. This is nothing else but preservation of a historic tradition as the first elementary particle discovered was the electron, the positron being the herald of the antiworld.

16.3 Commutation relations in the Proca theory

The Lagrangian $\mathcal{L}'$ helps considerably in the simplification of deriving the dynamic invariants of a vector field. But as the physical reality is approached (quantization, switching on interaction and so on), it becomes less and less applicable. This makes us to leave it and address the Lagrangians (16.10) and (16.15) instead. Both these Lagrangians give equivalent physical results and hence a choice of one of them is simply a matter of our preference. Further, we shall use the Lagrangian (16.15).

The quantization procedure with respect to the scalar field is somewhat complicated due to the presence of an additional condition. The simultaneous canonical commutation relations have the shape:

$$[\psi^m(\mathbf{r}',t),\pi^n(\mathbf{r},t)] = [\psi^m(\mathbf{r}',t),(\partial^n\psi^{0\dagger}(\mathbf{r},t) - \partial^0\psi^{n\dagger}(\mathbf{r},t))] = i\delta_{mn}\delta(\mathbf{r}'-\mathbf{r}), \qquad (16.50)$$

$$[\psi^{m\dagger}(\mathbf{r}',t),\pi^{n\dagger}(\mathbf{r},t)] = [\psi^{m\dagger}(\mathbf{r}',t),(\partial^n\psi^0(\mathbf{r},t) - \partial^0\psi^n(\mathbf{r},t))] = i\delta_{mn}\delta(\mathbf{r}'-\mathbf{r}), \qquad (16.51)$$

and all the remaining commutators containing: $\psi^m(\mathbf{r}',t)$, $\pi^n(\mathbf{r},t)$, $\psi^{m\dagger}(\mathbf{r}',t)$ and $\pi^{n\dagger}(\mathbf{r},t)$ convert to zero. Similar to the case of the scalar field, from (16.50) and (16.51) we can derive the commutation relations at an arbitrary instant of time, however, only for the space components of the operator field functions. The derived relations are not relativistically covariant, that naturally causes complications in the process of their manipulation. It is necessary to direct our efforts to establishing the commutation relations including all four components of the field function, formally considered at the same base. To that end, we previously specify the commutators involving the field function component with $\mu = 0$:

$$[\psi^l(\mathbf{r}',t),\psi_0^\dagger(\mathbf{r},t)] = \frac{1}{m^2}[\psi^l(\mathbf{r}',t),\partial_n\pi^n(\mathbf{r},t)] = \frac{i}{m^2}\partial^l\delta(\mathbf{r}'-\mathbf{r}), \qquad (16.52)$$

$$[\psi^l(\mathbf{r}',t),\psi_0(\mathbf{r},t)] = 0. \qquad (16.53)$$

With their use we can determine a set of the commutators involving the derivatives from the field:

$$[\psi^l(\mathbf{r}',t),\partial_0\psi^{n\dagger}(\mathbf{r},t)] = -i\left(g^{ln} + \frac{\partial^l\partial^n}{m^2}\right)\delta(\mathbf{r}'-\mathbf{r}), \qquad (16.54)$$

$$[\psi^l(\mathbf{r}',t),\partial^0\psi_0^\dagger(\mathbf{r},t)] = 0, \qquad (16.55)$$

$$[\psi^l(\mathbf{r}',t),\partial^n\psi_n^\dagger(\mathbf{r},t)] = 0, \qquad (16.56)$$

$$[\psi^0(\mathbf{r}',t),\partial^0\psi_0^\dagger(\mathbf{r},t)] = -\frac{i}{m^2}\partial^n\partial_n\delta(\mathbf{r}'-\mathbf{r}), \qquad (16.57)$$

where we have allowed for (16.9), (16.11) and

$$\partial_0\psi_n^\dagger(\mathbf{r},t) = \pi_n(\mathbf{r},t) + \partial_n\psi_0^\dagger(\mathbf{r},t). \qquad (16.58)$$

Substitution of the the operator field function expansions in terms of the solutions of the free Proca equation (16.23), (16.24) into the left-hand side of the relations (16.52)–(16.57) converts these relations into identities under fulfillment of the following commutation relations:

$$[\psi_\mu^{(+)}(\mathbf{k}),\psi_\nu^{(+)\dagger}(\mathbf{k}')] = [\psi_\mu^{(-)}(\mathbf{k}),\psi_\nu^{(-)\dagger}(\mathbf{k}')] = \left(g_{\mu\nu} - \frac{k_\mu k_\nu}{m^2}\right)\delta(\mathbf{k}'-\mathbf{k}). \qquad (16.59)$$

Then, with allowance made for the fact that the polarization vectors obey the condition (16.30), we find the commutation relations for the operators $a_m(\mathbf{k})$ and $b_m(\mathbf{k})$:

$$[a_m(\mathbf{k}), a_n^\dagger(\mathbf{k}')] = [b_m(\mathbf{k}), b_n^\dagger(\mathbf{k}')] = \delta_{mn}\delta(\mathbf{k}' - \mathbf{k}). \tag{16.60}$$

In its turn, Eqs. (16.60) allow us to establish the commutators for the operator field functions at an arbitrary instant of time:

$$[\psi^{(\pm)\nu}(x), \psi_\mu^{(\pm)\dagger}(y)] = i\Lambda_\mu^{(1)\nu}(\partial)\Delta_0^\pm(x - y; m^2), \tag{16.61}$$

or

$$[\psi_\nu(x), \psi_\mu^\dagger(y)] = iD_{\mu\nu}(x - y; m^2) \tag{16.62}$$

where

$$D_{\mu\nu}(x; m^2) = \Lambda_\mu^{(1)\sigma}(\partial)g_{\sigma\nu}\Delta_0(x; m^2) = \left(g_{\mu\nu} + \frac{\partial_\mu\partial_\nu}{m^2}\right)\Delta_0(x; m^2).$$

It is no trouble to verify that the commutation relations (16.61) are consistent both with the field equations and the additional condition. Also it is obvious that the operator $\Lambda_\mu^{(1)\nu}9\partial)$ in the Proca theory plays the same role as does the operator $d(\partial)$ in the formalism of the first-order equations. For example, with its help the causal Green function for the vector field could be written

$$D_{\mu\nu}^c(x; m^2) = \Lambda_\mu^{(1)\sigma}g_{\nu\sigma}(\partial)\Delta^c(x). \tag{16.63}$$

The above formalism may be easily generalized to the real vector field for $m \neq 0$ through which the neutral carrier of the electroweak interaction, the Z-boson, is described. The corresponding Lagrangian is given by the expression:

$$\mathcal{L} = -: \frac{1}{2}B_{\mu\lambda,\sigma\nu}\partial^\mu\psi^\sigma(x)\partial_\lambda\psi_\nu(x) + \frac{m^2}{2}\psi_\mu(x)\psi^\mu(x): . \tag{16.64}$$

16.4   Duffin-Kemmer equation for the vector field

The equations system (16.8) could be represented in the form of a matrix first-order equation. Let us introduce the ten-dimensional wavefunction:

$$\Psi(x) = \{\Psi_A\} = \left\{\begin{array}{c} \psi_{\nu\sigma}(x)/m \\ -i\psi_\mu(x) \end{array}\right\} = \begin{pmatrix} \psi_{01}/m \\ \psi_{02}/m \\ \psi_{03}/m \\ \psi_{23}/m \\ \psi_{31}/m \\ \psi_{12}/m \\ -i\psi_1 \\ -i\psi_2 \\ -i\psi_3 \\ -i\psi_0 \end{pmatrix}, \tag{16.65}$$

where $A = 1, 2, 3...10$. Through the use of it Eqs. (16.8) are represented in the form:

$$(i\beta^\mu\partial_\mu - m)\Psi(x) = 0, \tag{16.66}$$

where the ten-row matrices β_μ have the following nonzero components:

$$\left.\begin{array}{l}
(\beta_0)_{17} = (\beta_0)_{71} = (\beta_0)_{28} = (\beta_0)_{82} = (\beta_0)_{39} = (\beta_0)_{93} = 1, \\
(\beta_1)_{110} = -(\beta_1)_{101} = (\beta_1)_{59} = -(\beta_1)_{95} = (\beta_1)_{86} = -(\beta_1)_{68} = 1, \\
(\beta_2)_{210} = -(\beta_2)_{102} = (\beta_2)_{94} = -(\beta_2)_{49} = (\beta_2)_{67} = -(\beta_2)_{76} = 1, \\
(\beta_3)_{310} = -(\beta_3)_{103} = (\beta_3)_{48} = -(\beta_3)_{84} = (\beta_3)_{75} = -(\beta_3)_{57} = 1.
\end{array}\right\} \qquad (16.67)$$

The equations (16.66) are called the *Duffin-Kemmer equations* for vector particles [9,38]. Since the resultant representation comprises a sum of the representations $(1/2, 1/2)$, $(0, 1)$, and $(1, 0)$, then:

$$n_{max} = 2.$$

This in turn is responsible for the fact that both the expression for the operator $d(\partial)$ and commutation relations determining the β-matrices algebra are similar to those for the first-order equations of the scalar field. In this manner all the formulae derived by us for the scalar field in Section 14.11 may be used to describe the vector particle. Although, with the exception of the formula determining the spurs of the β-matrices. We address the relation (14.147) to have been obtained without using an explicit form of the matrices β_μ, P, and $\overline{P}$. In just the same way as in the case of the spin 0, we introduce the projection operators $P^{(10)}$ and $\overline{P}^{(10)}$:

$$P^{(10)} = (3 - \beta^\mu \beta_\mu)/2, \qquad \overline{P}^{(10)} = (\beta^\mu \beta_\mu - 2)/3, \qquad (16.68)$$

where $P^{(10)}$ $(\overline{P}^{(10)})$ selects the tensor (vector) part of the field function $\Psi(x)$. Then, by application of (16.67) and (16.68), it is an easy matter to verify the validity of the relation.

$$\{P^{(10)}\beta_\mu\beta_\nu P^{(10)}\}^\alpha_\tau = B_{\mu\nu}{}^{,\alpha}{}_\tau = g_{\mu\nu}\delta^\alpha_\tau - g_{\mu\tau}\delta^\alpha_\nu. \qquad (16.69)$$

With the help of Eqs. (16.69) and (3.1.147) we find that the spur of an even number of β-matrices is defined by the expression [39]:

$$\mathrm{Sp}\{\beta_{\mu_1}\beta_{\mu_2}.....\beta_{\mu_n}\} = B_{\mu_2\mu_3}{}^{,u}{}_\tau B_{\mu_4\mu_5}{}^{,\eta}{}_\rho...B_{\mu_n\mu_1}{}^{,\gamma}{}_\alpha +$$

$$+ B_{\mu_1\mu_2}{}^{,\alpha}{}_\tau B_{\mu_3\mu_4}{}^{,\tau}{}_\rho...B_{\mu_{n-1}\mu_n}{}^{,\gamma}{}_\alpha. \qquad (16.70)$$

Along similar lines it is shown that the spur of an odd number of β-matrices is equal to zero. From the general formula (16.70) we obtain, without trouble, the following expressions for the spurs from the product of two and four β-matrices:

$$\mathrm{Sp}\{\beta_\mu\beta_\nu\} = 3g_{\mu\nu}, \qquad \mathrm{Sp}\{\beta_\mu\beta_\nu\beta_\lambda\beta_\sigma\} = 3(g_{\mu\nu}g_{\lambda\sigma} + g_{\mu\sigma}g_{\nu\lambda}). \qquad (16.71)$$

17

Electromagnetic field

> *The point is that as Bruin the third lay in his*
> *den things in the forest went on as usual. It wouldn't be right*
> *to call this state of affairs welfare. But the governor's task,*
> *after all, is not to establish an ideal order, but to support*
> *and safeguard, despite its deficiencies, the existing order*
> *which has been sanctified by ages. Neither does it consist in*
> *perpetrating any manner of atrocities, whether minor, major or*
> *medium! Nay, he should be satisfied with the "natural" ones.*
> *And if ancient custom ordains that the wolves should skin*
> *rabbits and the hawks and owls pluck crows' feathers, although*
> *such order is not quite synonymous with welfare, it is,*
> *nevertheless, an order of a kind, and as such must be obeyed.*
> M.E. Saltykov Shchedrin, "Bears in Government"

17.1   Maxwell equations

An electromagnetic field theory was formulated by J. Maxwell in 1865 in the form of a system of differential equations generalizing all the fundamental laws of electromagnetic phenomena. Maxwell equations in a vacuum in the presence of the electric current with the density $\mathbf{j}$ and the electric charge with the density ρ are of the following form:

$$\left.\begin{array}{ll} \text{rot } \mathbf{E} = -\dfrac{\partial \mathbf{H}}{\partial t}, & \text{rot } \mathbf{H} = \dfrac{\partial \mathbf{E}}{\partial t} + \mathbf{j}, \\[2mm] \text{div } \mathbf{E} = \rho, & \text{div } \mathbf{H} = 0. \end{array}\right\} \tag{17.1}$$

The first equation is associated with the Faraday induction law: changing of a magnetic field leads to the beginnings of an electric field. The second equation represents the Ampere law:

$$\text{rot } \mathbf{H} = \mathbf{j},$$

where the Maxwell displacement current is introduced to take into consideration that a change of the electric field results in inducing the magnetic field. The third equation is an equivalent of the Coulomb law. The fourth equation states that there are no sources of the magnetic field other than currents, that is, magnetic charges (magnetic monopoles) in nature are nonexistent.

At first sight on the Maxwell equations it becomes apparent there is one opportunity that was either missed by nature or we are not able to detect that by our devices up till now. In a free case $\mathbf{j} = \rho = 0$ these equations are invariant with respect to the so-called dual transformations:

$$\mathbf{E} \rightarrow \mathbf{H}, \qquad \mathbf{H} \rightarrow -\mathbf{E}, \tag{17.2}$$

that is, there is the mutual symmetry of the electric and magnetic fields. Nevertheless, the presence of the electric currents and charges disturbs the equivalence of the electric and magnetic fields. Symmetry of the Maxwell equations may be restored only by the fact of existing magnetic charges with the density ρ_m and currents $\mathbf{j}_m$ connected with these charges. We denote the magnetic charge density by ρ_m and the currents associated with the magnetic charges by $\mathbf{j}_m$. In this case, the Maxwell equations would be of the form:

$$\left. \begin{array}{ll} \mathrm{rot}\ \mathbf{E} = -\dfrac{\partial \mathbf{H}}{\partial t} - \mathbf{j}_m, & \mathrm{rot}\ \mathbf{H} = \dfrac{\partial \mathbf{E}}{\partial t} + \mathbf{j}, \\ \mathrm{div}\ \mathbf{E} = \rho, & \mathrm{div}\ \mathbf{H} = \rho_m, \end{array} \right\} \tag{17.3}$$

whereas the dual transformations would be complemented by:

$$\rho \to \rho_m, \qquad \rho_m \to -\rho, \qquad \mathbf{j} \to \mathbf{j}_m, \qquad \mathbf{j}_m \to -\mathbf{j}. \tag{17.4}$$

The hypothesis that a magnetic monopole is existent in nature was put forward by P. Dirac in 1931 [40] in an effort to explain the electric charge quantization law. And since then such a hypothetic magnetic charge has been called *Dirac's monopole*. Using the Schrödinger equation for description of the interaction between the electric Q and magnetic g charges, Dirac has obtained the charge quantization condition:

$$Qg = 2\pi n\hbar c, \qquad n = 0, \pm 1, \pm 2, \pm 3, .. \tag{17.5}$$

where, for a while, we use the conventional units. In case the smallest magnetic charge g_1 is existent, from relation (17.5) it follows that the electric charge may take on only the values:

$$Q = \pm\frac{2\pi\hbar c}{g_1}, \pm\frac{4\pi\hbar c}{g_1}, \pm\frac{6\pi\hbar c}{g_1}, ... $$

On the other hand, from the charge quantization condition it follows that the magnetic charge is also quantized:

$$g_n = n\frac{2\pi\hbar c}{e} = \frac{ne}{2\alpha_{em}}. \tag{17.6}$$

As we see the minimal magnetic charge of the monopole should be equal to $\approx 68.5e$.

It appears that the condition (17.5) may be derived directly from the Maxwell equations, written in the form of Eq. (17.3), using the semiclassical and semiquantum Bohr theory [41]. Let an electric charge be slowly moving in the closed orbit and a magnetic charge be slowly moving along the loop enclosing this orbit. Motion of the magnetic charge generates the electric field, whose integral along the magnetic charge path is, in accordance with (17.3), given by the expression:

$$\oint \mathbf{E}d\mathbf{l} = -\frac{1}{c}\left(\int \mathbf{j}_m d\mathbf{S} + \frac{d\Phi}{dt}\right), \tag{17.7}$$

where

$$\Phi = \int \mathbf{H}d\mathbf{S}$$

and S is an arbitrary surface restricted by the contour $\mathbf{l}$. To find a change in the momentum $\mathbf{p}$ of the electric charge due to the electric field, Eq. (17.7) should be multiplied by Q and integrated with respect to time. As a result we have

$$\oint \Delta\mathbf{p}d\mathbf{l} = -\frac{e}{c}\left(\Delta g + \Delta\Phi\right). \tag{17.8}$$

Using the Bohr quantization rule:

$$\oint \Delta \mathbf{p} d\mathbf{l} = 2\pi n \hbar \tag{17.9}$$

and assuming that the magnetic charge comes back to the initial point ($\Delta g = g$ and $\Delta \Phi = 0$), we finally obtain

$$Qg = 2\pi n \hbar c.$$

The possibility to explain the charge quantization in a theory describing both the electric and magnetic sources still remains one of the theoretical arguments to have us treat the magnetic monopole hypothesis with the proper respect. We could list some other theoretical arguments in its support. However, a decisive role as always belongs to the experiment. But by the present time numerous efforts of experimental search for the Dirac monopole have been in vain. But from this does not follow in any way that such objects are absent in nature since so far a theoretical principle forbidding their existence has not been known.

We rewrite the Maxwell equations (17.1) in the explicit relativistically invariant form. Of course, they are Lorentz-invariant in the three-dimensional form as well. This is so because revealing the covariance of the Maxwell equations relative to the Lorentz transformations by Einstein has led to the creation of the special theory of relativity. Our objective is to go to the notation, where the covariance is expressed in the way that gives a clearer insight of the problem. We introduce the real covariant four-dimensional vector of an electromagnetic potential A_μ so that:

$$\mathbf{E} = -\text{grad } A_0 - \frac{\partial \mathbf{A}}{\partial x^0}, \qquad \mathbf{H} = \text{rot } \mathbf{A}. \tag{17.10}$$

The components of the four-dimensional rotor of the four-dimensional potential form the electromagnetic-field antisymmetric tensor:

$$F_{\mu\nu} = \partial_\mu A_\nu - \partial_\nu A_\mu, \tag{17.11}$$

the components of which are related to the components of the electric and magnetic field strength vectors as follows:

$$H_i = -\frac{1}{2}\varepsilon_{ijk}F_{jk}, \qquad E_i = F_{0i}. \tag{17.12}$$

In the free case the Maxwell equations are written by means of the tensor $F_{\mu\nu}$ in the form:

$$\partial^\mu F_{\mu\nu} = 0, \tag{17.13}$$

$$\partial^\mu F_{\nu\lambda} + \partial^\nu F_{\lambda\mu} + \partial^\lambda F_{\mu\nu} = 0. \tag{17.14}$$

Next, proceeding from $F_{\mu\nu}$ to the potential A_μ, we see that four Eqs. (17.14) are the consequence of the definition (17.11), not resulting in any equation for A_μ. Whereas four Eqs. (17.13) lead to:

$$\Box A_\mu - \partial^\nu \partial_\mu A_\nu = 0, \tag{17.15}$$

that is an analog of the Proca equation for the massless neutral vector field. The four-dimensional form of Eqs. (17.13) and (17.15) is a direct proof for the relativistic covariance of the Maxwell equations, both in terms of the electromagnetic field tensor and potential.

The procedure of introducing the potential A_μ is ambiguous. In the Maxwell theory the observables—the vectors $\mathbf{E}$ and $\mathbf{H}$, together with the motion equations and all the relations for the electromagnetic field—are invariant with respect to the second-kind gauge transformation (or second-kind gradient transformation):

$$A_\mu(x) \to A'_\mu(x) = A_\mu(x) + \partial_\mu \chi(x), \tag{17.16}$$

where $\chi(x)$ is an arbitrary function having partial derivatives of the first and second order. This ambiguity means that the dynamic description of an electromagnetic field in terms of $A_\mu(x)$ involves the degrees of freedom, nonphysical in character.

Ambiguity in a choice of the potentials allows for imposition of some additional condition. The only invariant condition, linear in $A_\mu(x)$, is the so-called Lorentz condition:

$$\partial^\mu A_\mu = 0, \tag{17.17}$$

with due regard for that Eq. (17.15) takes the form:

$$\Box A_\mu(x) = 0. \tag{17.18}$$

The principal advantage of the condition (17.17) is the fact that both in the free case and in the case with the included interaction the quantity $\partial^\mu A_\mu$ satisfies the same equation, the D'Alembert equation. Really, acting on the equation:

$$\Box A_\mu(x) = j_\mu(x) \tag{17.19}$$

by the operator ∂^μ and taking into account the electric current conservation law:

$$\partial^\mu j_\mu(x) = 0,$$

we get

$$\Box \partial^\mu A_\mu(x) = 0. \tag{17.20}$$

It is an easy matter to check that $\chi(x)$ may be always chosen so that the condition (17.17) can be fulfilled. It is also clear that the relation (17.17) and the electromagnetic field equations for the potentials (17.18) remain to be invariant with respect to the specialized gradient transformation of the second kind:

$$A_\mu(x) \to A'_\mu(x) = A_\mu(x) + \partial_\mu \chi_0(x), \tag{17.21}$$

where $\chi_0(x)$ is an arbitrary function satisfying the D'Alembert equation:

$$\Box \chi_0(x) = 0. \tag{17.22}$$

In any partial Lorentz reference frame the function $\chi_0(x)$ may be chosen in such a way that one of the components of $A_\mu(x)$, for example, the scalar potential, $A_0(x)$, becomes equal to zero. Then the Lorentz condition takes the shape:

$$\operatorname{div} \mathbf{A}(x) = 0. \tag{17.23}$$

To establish the physical meaning of the relation (17.23) we take advantage of the expansion of the four-dimensional potential in terms of the free equation solutions or, just the same, in terms of the de Broglie plane waves:

$$A_\mu(x) = \frac{1}{(2\pi)^{3/2}} \int \frac{d^3k}{\sqrt{2k_0}} [A_\mu^+(\mathbf{k}) e^{-ikx} + A_\mu^-(\mathbf{k}) e^{ikx}]. \tag{17.24}$$

To insert (17.24) into (17.23) gives

$$(\mathbf{k} \cdot \mathbf{A}^\pm(\mathbf{k})) = 0, \tag{17.25}$$

that represents the condition of the electromagnetic field transversality. Even though an electromagnetic field is described by the four-dimensional potential, only two linearly-independent components, orthogonal to the wave vector, have the physical meaning. Note that, in spite of the noncovariance of the transversality condition (17.25), one could always attain its fulfillment in any partial Lorentz system by means of the suitable choice of the function $\chi_0(x)$ in the transformation (17.21).

Thus, a massless vector field could not be considered as the limiting case of a massive vector field at $m \to 0$ because this leads to erroneous physical results. This is seen even from the manner of how mass is introduced into the particular formula for a massive vector field (for example, in the expression for the commutator function $D_{\mu\nu}(x; m^2)$). Massless fields significantly differ from massive ones. In the given case the difference is in the number of polarization states that for photons is less by one compared to the massive neutral vector particles. Such a decreased number of degrees of freedom is caused by zero mass of the photon, as only in this case the Proca equations for a neutral vector field become gradiently invariant.

The Maxwell equations (17.13) and the definition of the tensor $F_{\mu\nu}$ may be obtained from the Lagrange function density:

$$\mathcal{L} = \frac{1}{2}F^{\mu\nu}\left(\frac{1}{2}F_{\mu\nu} - \partial_\mu A_\nu + \partial_\nu A_\mu\right). \tag{17.26}$$

The wave equations for the potential follows from the Lagrange function density:

$$\mathcal{L}' = -\frac{1}{2}\partial^\nu A_\mu \partial_\nu A^\mu. \tag{17.27}$$

Now in order to derive the Maxwell equations we should add the Lorentz condition (17.17) to the wave equations. It is interesting that this condition is associated only with a theory of the classical electromagnetic field and, as will be shown later, upon quantization is replaced by the particular conditions for the system state vector, which are equivalent to the Lorentz condition exclusively from the viewpoint of the average values.

17.2 Dynamic invariants of the electromagnetic field

By application of the Lagrangian (17.26) the canonical energy-momentum tensor has the form:

$$T_\mu^{\prime\nu} = F^{\nu\sigma}\partial_\mu A_\sigma - \delta_\mu^\nu \mathcal{L}. \tag{17.28}$$

Subtracting from (17.28) the insignificant term $F^{\nu\sigma}\partial_\sigma A_\mu$ to satisfy the relation:

$$\partial_\sigma(F^{\nu\sigma}\partial_\sigma A_\mu) = 0,$$

and going to the gradiently-invariant quantities in the Lagrangian, for the metric (symmetrized) energy-momentum tensor we obtain the following expression:

$$T_{\nu\mu} = F_\nu^\sigma F_{\mu\sigma} - \frac{1}{4}g_{\nu\mu}F^{\lambda\sigma}F_{\lambda\sigma}. \tag{17.29}$$

From (17.29) it follows that the momentum and energy of the electromagnetic field are determined by the expressions:

$$\mathbf{P} = \int d^3x[\mathbf{E} \times \mathbf{H}], \tag{17.30}$$

$$H = \frac{1}{2}\int d^3x(\mathbf{E}^2 + \mathbf{H}^2). \tag{17.31}$$

With the help of (17.29) we find the total angular momentum tensor for the electromagnetic field:

$$M_{\mu\nu} = \int d^3x(x_\mu T_{\nu 0} - x_\nu T_{\mu 0}) = \int d^3x(x_\mu s_\nu - x_\nu s_\mu), \qquad (17.32)$$

where

$$s_0 = \frac{1}{2}(\mathbf{E}^2 + \mathbf{H}^2), \qquad \mathbf{s} = [\mathbf{E} \times \mathbf{H}].$$

The space components of $M_{\mu\nu}$ form the field angular momentum vector to be given by:

$$\mathbf{M} = \frac{1}{c}\int d^3x[\mathbf{r} \times [\mathbf{E} \times \mathbf{H}]]. \qquad (17.33)$$

The spin moment tensor is defined by the expression:

$$S_{\mu\nu}^\sigma = F_\mu^\sigma A_\nu - F_\nu^\sigma A_\mu.$$

Therefore, the spin vector has the form:

$$\mathbf{S} = \int d^3x[\mathbf{E} \times \mathbf{A}]. \qquad (17.34)$$

Note that the division of the photon angular momentum into the orbital and spin parts is formal. Only the total photon angular momentum has the physical meaning. This may be explained as follows. First, the ordinary definition of the spin as a moment of a particle at rest is inapplicable to a photon, since there is no rest reference frame for particle with zero mass—in any reference frame this particle is moving with the speed of light. Second, an expression for the spin proves to be gradiently-noninvariant, whereas the observables for an electromagnetic field should possess this feature. In consequence, the states with certain values of the orbital and spin moments fail to satisfy the transversality condition in the general case. But formally, the representation of the total angular momentum for the electromagnetic field as a sum of two terms appears to be very useful.

It should be noted that the photon presents another quantum-mechanical surprise. Since the photons are relativistic objects, a minimal measuring error for their coordinates, Δq, is coincident with the de Broglie wavelength. This means that it makes sense to speak of the photon coordinates only in the cases when the characteristic sizes of the problem are greater than the wavelength. In the quantum case, however, the wavelength is not considered small, so the notion of the photon coordinates becomes objectless. In other words, the wavefunction of the photon is not determined by the field value at a particular point, being rather dependent on the field distribution in the region with a size on the order of the wavelength. This suggests that localization of the photon in a smaller region is impossible and hence the notion of the probability density for the photon localization at a particular space point is absurd. Mathematically, this situation is reflected by the fact that in the coordinate representation for the photon there is no way to set up a quantity that may be interpreted as a probability density. Such a quantity should be expressed by a positive bilinear combination of the photon wavefunctions, representing the time component of the current four-dimensional vector J_μ, whose continuity equation:

$$\partial^\mu J_\mu = 0$$

just provides the particle number conservation law. It is impossible to compose a bilinear combination forming the four-dimensional vector with zero divergence from the vectors of the electromagnetic field. Such a vector could not be constructed with the use of the four-dimensional potential A_μ too. Indeed, A_μ may be introduced into the current J_ν only as a component of the $F_{\mu\nu}$-tensor due to the gauge invariance. This means that J_ν should be bilinearly formed from $F_{\mu\nu}$, F_τ^ν and four-dimensional vector k^τ. However, such a vector will deliberately be equal to zero because of the transversality condition:

$$k^\tau F_\tau^\nu = 0.$$

Recalling what stands in the gauge invariance, we come to the conclusion that the above complication is associated with the zero mass of the photon.

The Lagrangians $\mathcal{L}$ and $\mathcal{L}'$ are distinguished by the quantity that, upon integration with respect to the four-dimensional space and with due regard for the Lorentz condition, becomes zero. In a similar way the dynamic characteristics of the electromagnetic field prove to be the same for both Lagrangians with allowance made for the equations of motion and condition (17.17). A reader has the opportunity to find whether the expressions for the energy-momentum metric tensor:

$$T_{\nu\mu} = -\partial_\nu A^\sigma \partial_\mu A_\sigma + \frac{1}{2} g_{\nu\mu} \partial^\lambda A_\sigma \partial_\lambda A^\sigma \tag{17.35}$$

and the spin momentum tensor:

$$S^\sigma_{\mu\nu} = A_\nu \partial^\sigma A_\mu - A_\mu \partial^\sigma A_\nu, \tag{17.36}$$

following from the Lagrangian $\mathcal{L}'$ are equivalent to the relations (17.29) and (17.33).

In the momentum space we introduce the local reference point $e_\mu^\lambda(\mathbf{k})$ to be connected with the unit polarization vectors of the photon. Further, in Eq. (17.24) we carry out the expansion of $A_\mu^{(\pm)}(\mathbf{k})$ in terms of the vectors $e_\mu^\lambda(\mathbf{k})$:

$$A_\mu(x) = A_\mu^{(+)}(x) + A_\mu^{(-)}(x) = \frac{1}{(2\pi)^{3/2}} \int \frac{d^3k}{\sqrt{2k_0}} \sum_{\lambda=0}^{3} [e_\mu^\lambda(\mathbf{k}) a_\lambda^+(\mathbf{k}) e^{-ikx} + e_\mu^{\lambda*}(\mathbf{k}) a_\lambda^-(\mathbf{k}) e^{ikx}], \tag{17.37}$$

where $a_\lambda^\pm(\mathbf{k})$ have a sense of the de Broglie waves amplitudes in the nonquantized theory. Thus, at the given momentum value there are four linearly independent solutions to differ with polarizations. In the most simple case one may chose the vectors $e_\mu^\lambda(\mathbf{k})$ in such a way that they coincide with the unit vectors of the coordinate system n_μ, that is,

$$e_\mu^\lambda(\mathbf{k}) = \delta_\mu^\lambda. \tag{17.38}$$

Then, if one directs n_3 along $\mathbf{k}$, $\lambda = 1, 2$ will correspond to the transverse polarization, $\lambda = 3$ to the longitudinal one, and $\lambda = 0$ to the scalar (or time) one. In this coordinate system the three-dimensional vectors $\mathbf{e}^1(\mathbf{k})$, $\mathbf{e}^2(\mathbf{k})$, and $\mathbf{k}/\mid \mathbf{k} \mid$, for which the relation:

$$(\mathbf{e}^m(\mathbf{k}) \cdot \mathbf{e}^n(\mathbf{k})) = \delta_{mn}, \qquad [\mathbf{e}^m(\mathbf{k}) \times \mathbf{e}^n(\mathbf{k})] = \varepsilon^{mnk} \mathbf{e}^k,$$

is valid, form the orthonormalized system with the completeness condition:

$$e_r^1(\mathbf{k}) e_l^1(\mathbf{k}) + e_r^2(\mathbf{k}) e_l^2(\mathbf{k}) + \frac{k_r k_l}{\mid \mathbf{k} \mid^2} = \delta_{rl}, \tag{17.39}$$

where the indices $r, l = 1, 2, 3$ denote the vector components.

For a free massless particle the only direction is preferential in space, the direction of the momentum vector. It is obvious that in this case symmetry of the wavefunction of a particle with respect to the whole group of three-dimensional rotations (spherical symmetry) is lacking, and we can consider only the axial symmetry about the preferred axis. In other words, for a given $\mathbf{k}$ the photon may have only two states with different values of the spin vector projection described by the polarization vectors e_μ^1 and e_μ^2. As such we can choose:

$$e_\mu^1 = (0, 1, 0, 0), \qquad e_\mu^2 = (0, 0, 1, 0), \tag{17.40}$$

that is in accord with a linear polarization about the axis 1 or 2. Another choice is:

$$e_\mu^1 = \frac{1}{\sqrt{2}}(0, 1, -i, 0), \qquad e_\mu^2 = \frac{1}{\sqrt{2}}(0, 1, i, 0), \tag{17.41}$$

that is associated with a circular polarization of the electromagnetic field.

As we have (17.38), the total sum (and not a scalar product!) over the polarizations equals:

$$\sum_{\lambda=0}^3 e_\mu^\lambda e_\nu^{\lambda *} = \delta_{\mu\nu}. \tag{17.42}$$

At the same time, the summation should be performed only over the physical states, that is, over the transverse polarizations. This may be done using Eq. (17.39) to give the noncovariant formula:

$$\sum_{\lambda=1}^2 e_m^\lambda e_n^{\lambda *} = \delta_{mn} - \frac{k_m k_n}{|\mathbf{k}|^2}. \tag{17.43}$$

In practice it is more convenient to use a relativistically covariant expression for the summation over the photon spin states that includes all four photon polarizations possible. Using the explicit form of the vectors $e_\mu^\lambda(\mathbf{k})$, we arrive at

$$\sum_{\lambda=0}^3 e_\mu^\lambda(\mathbf{k}) e_\nu^{\lambda *}(\mathbf{k}) = -g_{\mu\nu}. \tag{17.44}$$

Now we should display that the usage of (17.44) also leads to the inclusion of only the physical photon states. Substituting the expansion (17.37) into the Lorentz condition (17.17), we get two relations:

$$|\mathbf{k}| a_3^\pm(\mathbf{k}) - k_0 a_0^\pm(\mathbf{k}) = 0, \tag{17.45}$$

to be valid for any choice of the transverse polarization vectors. Since $k_\mu k^\mu = k_0^2 - \mathbf{k}^2 = 0$, we have

$$k_0 = |\mathbf{k}|.$$

Then, from Eqs. (17.45) it follows

$$a_3^+(\mathbf{k}) a_3^-(\mathbf{k}) - a_0^+(\mathbf{k}) a_0^-(\mathbf{k}) = 0. \tag{17.46}$$

The obtained result indicates that, due to the Lorentz condition, the densities of the average numbers of longitudinal $a_3^+(\mathbf{k}) a_3^-(\mathbf{k})$ and temporal photons $a_0^+(\mathbf{k}) a_0^-(\mathbf{k})$ are equal, their contributions into the dynamic invariants being opposite in sign. Thus, longitudinal and temporal photons, hereinafter referred to as pseudophotons, are as if compensating each other. Thus, for example, using Eq. (17.44), we can show that the quadratic form involved in the definitions of the majority of dynamic variables includes no pseudophoton states:

$$-A_\mu(\mathbf{k}) A^\mu(\mathbf{k}) = -\sum_{\lambda,\lambda'=0}^3 g^{\lambda\lambda'} a_\lambda^+(\mathbf{k}) a_{\lambda'}^-(\mathbf{k}) = \sum_{\lambda=1}^2 a_\lambda^+(\mathbf{k}) a_\lambda^-(\mathbf{k}). \tag{17.47}$$

Let us define the dynamical variables in the case of the circular polarization of the photons. The energy-momentum four-dimensional vector is given by the expression:

$$P^\mu = -\int d^3k (k^\mu A_\nu^+(\mathbf{k}) A^{-\nu}(\mathbf{k})) = \sum_{\lambda=1,2} \int d^3k (k^\mu a_\lambda^+(\mathbf{k}) a_\lambda^-(\mathbf{k})). \qquad (17.48)$$

Similar to the case of a massive vector field, the energy turns to be positively defined due to the Lorentz condition only.

For the spin vector projection onto the momentum direction we find

$$S_3 = \int d^3k [a_1^+(\mathbf{k}) a_1^-(\mathbf{k}) - a_2^+(\mathbf{k}) a_2^-(\mathbf{k})]. \qquad (17.49)$$

From the expressions (17.48) and (17.49) it follows that the quantities $a_\lambda^+(\mathbf{k}) a_\lambda^-(\mathbf{k})$ represent the average number of the photons with the momentum $\mathbf{k}$, the energy k_0, possessing the spin projection onto the motion direction that is equal to $+1$ ($\lambda = 1$) and -1 ($\lambda = 2$). Note, that in the case of using the vectors of the linear photon polarization, all the dynamical invariants except the spin projection operator are reduced to the diagonal form as well.

Since the potential is a real quantity, then the relations:

$$(a_\mu^\pm(\mathbf{k}))^* - a_\mu^\mp(\mathbf{k}).$$

take place. This allows us to take the designations:

$$a_\mu^+(\mathbf{k})) = a_\mu(\mathbf{k}), \qquad a_\mu^-(\mathbf{k}) \equiv a_\mu^\dagger(\mathbf{k})$$

17.3 Electromagnetic field quantization

In the case in question quantization is performed so that the requirements of the relativistic covariance, energy positiveness, Lorentz and transversality conditions can be satisfied simultaneously. When proceeding from the Lagrangian density (17.26), the canonical quantization procedure proves to be unsuitable. Actually, as

$$\pi_\mu(x) = F_{\mu 0}(x),$$

the canonical momentum $\pi_0(x)$ is identically going to zero and the canonical quantization procedure is inapplicable to the time component of the field function. This is radically different from the situation in the case of a massive vector field, where the Lorentz condition was the operation condition included into the Lagrangian and when the component $\psi_0(x)$ could be directly eliminated from the theory. For the situation to be remedied we use the Lagrangian (17.27) and shall take the vector potential components as being the independent quantities.* Now the canonical momentum is given by:

$$\pi^\mu(x) = -\partial^0 A^\mu(x), \qquad (17.50)$$

what allows us to write the simultaneous commutation relations for all the field components

$$[\pi_\mu(\mathbf{r}, t), A_\nu(\mathbf{r}', t)] = i g_{\mu\nu} \delta^{(3)}(\mathbf{r} - \mathbf{r}') \qquad (17.51)$$

---

*It means that we discard the Lorentz condition in the operator form.

(all the remaining commutators are equal to zero). For the four-dimensional potential $A_\mu(x)$, which we consider the Hermitian operator, an expansion in terms of plane waves is of the form:

$$A_\mu(x) = \frac{1}{(2\pi)^{3/2}} \int \frac{d^3 k}{\sqrt{2k_0}} \sum_{\lambda=0}^{3} [e_\mu^\lambda(\mathbf{k}) a_\lambda(\mathbf{k}) e^{-ikx} + e_\mu^{\lambda*}(\mathbf{k}) a_\lambda^\dagger(\mathbf{k}) e^{ikx}], \qquad (17.52)$$

where $a_\lambda(\mathbf{k})$ and $a_\lambda^\dagger(\mathbf{k})$ are operator quantities. For simplicity, we take e_μ^λ in the form of δ_μ^λ to demonstrate that the canonical quantization rules may be fulfilled when the operators $a_\lambda(\mathbf{k})$ and $a_\lambda^\dagger(\mathbf{k})$ are subjected to the commutation relations:

$$[a_\mu(\mathbf{k}), a_\nu^\dagger(\mathbf{k}')] = g_{\mu\nu} \delta^{(3)}(\mathbf{k} - \mathbf{k}'). \qquad (17.53)$$

From (17.53), in its turn, one may find both the commutators for the potentials in an arbitrary instant of time:

$$[A_\mu(x), A_\nu(x')] = i g_{\mu\nu} D_0(x - x'), \qquad (17.54)$$

where

$$D_0(x) \equiv \Delta_0(x; m^2 = 0)$$

and the convolution for the potentials:

$$A_\mu^{s_1}(x) A_\nu^{s_1}(x') = i g_{\mu\nu} D_0^c(x - x'), \qquad (17.55)$$

where $D_0^c(x) = \Delta^c(x)$ at $m^2 = 0$.

Now it is high time to recall the gauge ambiguity of the potentials A_μ. Can it have an influence on the form of the operators introduced in the process of quantization of the electromagnetic field? And if so, what is this influence? We are interested in particular in the form of the photon causal function. To find the answer, we replace the Lagrangian (17.26) by the expression:

$$\mathcal{L} = -\frac{1}{4} F_{\mu\nu} F^{\mu\nu} - \frac{1}{2\xi} \partial_\mu A^\mu \partial_\nu A^\nu. \qquad (17.56)$$

The last term in (17.56) violates the gauge invariance. The quantity ξ, which may be chosen arbitrary, is called a *gauge parameter*. At $\xi = 1$ the Lagrangian (17.56) differs from that given by the expression (17.27) on insignificant four-dimensional divergence. Variation of the potentials leads to the following motion equations:

$$d_{\sigma\nu}(\partial) A^\sigma = [\partial_\mu \partial^\mu g_{\sigma\nu} - \left(\frac{\xi - 1}{\xi}\right) \partial_\nu \partial_\sigma] A^\sigma = 0. \qquad (17.57)$$

In the case under investigation the photon propagator satisfies the equation:

$$[\partial_\mu \partial^\mu g_{\sigma\nu} - \left(\frac{\xi - 1}{\xi}\right) \partial_\nu \partial_\sigma] D_\rho^{\sigma c}(x - x') = g_{\nu\rho} \delta(x - x'). \qquad (17.58)$$

It immediately follows that the propagator in the momentum representation is defined by the expression:

$$d_{\sigma\nu}(k) D_\rho^{\sigma c}(k) = [g_{\sigma\nu} - \left(\frac{\xi - 1}{\xi}\right) \frac{k_\nu k_\sigma}{k^2}] D_\rho^{\sigma c}(k) = -\frac{g_{\nu\rho}}{k^2}. \qquad (17.59)$$

Since the matrix operator $d_{\sigma\nu}(k)$ has nonzero eigenvalues on the vector k_ν:

$$[g_{\sigma\nu} - \left(\frac{\xi-1}{\xi}\right)\frac{k_\nu k_\sigma}{k^2}]k^\nu = \frac{1}{\xi}k^\nu, \qquad (17.60)$$

the matrix $d_{\sigma\nu}(k)$ possesses a reverse matrix and, as a result, Eq. (17.58) has a single-valued solution. Carrying out the division of $D_{\mu\nu}^c(k)$ on the transverse and longitudinal parts:

$$D_{\mu\nu}^c(k) = D^{(tr)}(k^2)(g_{\mu\nu} - k_\mu k_\nu/k^2) + D^{(l)}(k^2)k_\mu k_\nu/k^2 \qquad (17.61)$$

and substituting (17.61) into (17.59), we get

$$k^2\left[g^{\sigma\nu} - \left(\frac{\xi-1}{\xi}\right)\frac{k^\nu k^\sigma}{k^2}\right]D_\sigma^{\mu c} = k^2\left[g^{\sigma\nu} - \left(\frac{\xi-1}{\xi}\right)\frac{k^\nu k^\sigma}{k^2}\right][(\delta_\sigma^\mu -$$

$$-\frac{k^\mu k_\sigma}{k^2}\right)D^{(tr)}(k^2) + \frac{k^\mu k_\sigma}{k^2}D^{(l)}(k^2)\right] = \left(g^{\mu\nu} - \frac{k^\mu k^\nu}{k^2}\right)D^{(tr)}(k^2) + \frac{1}{\xi}\frac{k^\mu k^\nu}{k^2}D^{(l)}(k^2). \qquad (17.62)$$

Comparing the obtained expression with Eq. (17.59), we arrive at the expressions for the transverse and longitudinal parts of the photon propagator:

$$D^{(tr)}(k^2) = -\frac{1}{k^2}, \qquad D^{(l)}(k^2) = -\frac{\xi}{k^2},$$

that finally gives

$$D_{\mu\nu}^c(k) = -\left(g_{\mu\nu} + \frac{\xi-1}{\xi}\frac{k_\mu k_\nu}{k^2}\right)\frac{1}{k^2}. \qquad (17.63)$$

In the limit $\xi \to \infty$ the gauge-fixed term $\xi^{-1}(\partial_\mu A^\mu)^2$ disappears. This suggests the following remedy to "expulsion" ξ from the final results. In the intermediate calculations we have to keep the finite value of ξ, while in the final stage of calculations we should turn ξ to infinity. But there is no need for this procedure as, due to the gauge invariance of the theory, none of the physical results is dependent on a choice of ξ. Really, the photon propagation function is introduced into the physical quantities, scattering amplitudes, being multiplied by the transitional currents of two electrons, that is, in the combinations of the form: $j_{12}^\mu D_{\mu\nu}^c j_{34}^\nu$, where $j_{ik}^\mu = e\overline{\psi}_i(x)\gamma^\mu\psi_k(x)$. But, on the strength of the current conservation law:

$$\partial_\mu j_{ik}^\mu = 0,$$

we have

$$k_\mu j^\mu ik = 0,$$

where $k = p_i - p_k$. So, to introduce the gauge parameter into the theory is dictated by the convenience requirements only. However, as we shall see in the second volume, terms fixing the gauge are an indispensable tool under quantization of non-Abelian fields.

The limiting case $\xi \to \infty$ corresponds to the Landau gauge:

$$D_{\mu\nu}^c(k) = -\left(g_{\mu\nu} - \frac{k_\mu k_\nu}{k^2}\right)\frac{1}{k^2}. \qquad (17.64)$$

It should be noted that such a choice of the parameter ξ is analogous to the Lorentz gauge of the potentials. Further, unless otherwise specified, we shall take $\xi = 1$ (Feynman gauge):

$$D_{\mu\nu}^c(k) = -\frac{g_{\mu\nu}}{k^2}, \qquad (17.65)$$

what is associated with (17.55).

With the help of the four-dimensional vector of the energy-momentum expressed in terms of the longitudinal, transverse, and time components $a^\nu(\mathbf{k})$

$$P^\mu = -\int d^3k [k^\mu a_\nu(\mathbf{k}) a^{\nu\dagger}(\mathbf{k})], \qquad (17.66)$$

we can demonstrate that the operators $a_m(\mathbf{k})$ and $a_m^\dagger(\mathbf{k})$ ($m = 1, 2, 3$) represent the annihilation and production operators for the transverse and longitudinal photons. For the time components we have the commutation relations with the wrong sign:

$$[a_0(\mathbf{k}), a_0^\dagger(\mathbf{k}')] = \delta^{(3)}(\mathbf{k} - \mathbf{k}'). \qquad (17.67)$$

When going from $\mu = 1, 2, 3$ to $\mu = 0$ the operators $a_\mu^\dagger(\mathbf{k})$ and $a_\mu(\mathbf{k})$ as if change their roles: $a_0(\mathbf{k})$ behaves itself as $a_m^\dagger(\mathbf{k})$ rather than as $a_m(\mathbf{k})$, whereas $a_0^\dagger(\mathbf{k})$ is similar to $a_m(\mathbf{k})$. In other words, the quantization procedure is disturbed. It seems that the situation may be improved by the assumption that $a_0(\mathbf{k})$ are the production operators and $a_0^\dagger(\mathbf{k})$—the annihilation operators of temporal pseudophotons. But this situation proves to be in conflict with the fact of reality of the electromagnetic field. To show that this is the case, we take the vacuum average for expression (17.67):

$$\Phi_0^\dagger[a_0(\mathbf{k}), a_0^\dagger(\mathbf{k}')]\Phi_0 = -\Phi_0^\dagger a_0^\dagger(\mathbf{k}) a_0(\mathbf{k}')\Phi_0 = \delta^{(3)}(\mathbf{k} - \mathbf{k}'). \qquad (17.68)$$

Multiplying the left-hand side of (17.68) by $F^\dagger(\mathbf{k}')F(\mathbf{k})$ ($F(\mathbf{k})$ is an arbitrary analytical function) and integrating over $\mathbf{k}$ and $\mathbf{k}'$, we arrive at the result:

$$-\int d^3k' \Phi_0^\dagger a_0^\dagger(\mathbf{k}') F^\dagger(\mathbf{k}') \int d^3k F(\mathbf{k}) a(\mathbf{k})\Phi_0 = -\Phi_0^\dagger \mid \int d^3k F(\mathbf{k}) a(\mathbf{k}) \mid^2 \Phi_0 < 0. \quad (17.69)$$

Carrying out the same mathematical operations with the right-hand side of (17.68), we get

$$\int d^3k' F^\dagger(\mathbf{k}') \int d^3k F(\mathbf{k}) \delta^{(3)}(\mathbf{k} - \mathbf{k}') = \int d^3k \mid F(\mathbf{k}) \mid^2 > 0,$$

that contradicts the relation (17.69).

It is also easy to see that the norm of the field state vector with an odd number of scalar photons is negative. For example:

$$(\Phi_{1\mathbf{k}0}, \Phi_{1\mathbf{k}0}) = (a_0^\dagger(\mathbf{k})\Phi_0, a_0^\dagger(\mathbf{k})\Phi_0) = (\Phi_0, a_0(\mathbf{k}) a_0^\dagger(\mathbf{k})\Phi_0) = -1.$$

17.4   Quantization with use of an indefinite metric

To restore a sequence of the quantization procedure—all $a_\mu(\mathbf{k})$ and $a_\mu^\dagger(\mathbf{k})$ have the meaning of the annihilation and production operators—one could use the procedure proposed by Gupta [42] and Bleuler [43]. The content of this procedure is as follows.

To provide a physically interpreted theory, the Hilbert space $\mathcal{H}$ should include the subspace $\mathcal{H}_+$ with a positively defined metric, whose vectors describe the physical states of a system. An operator in the complete space $\mathcal{H}$ corresponds to the physical observable if it acts as a Hermitian operator on the physical states of $\mathcal{H}_+$. In the Hilbert space we introduce a linear and unitary metric operator η with the properties:

$$[\eta, a_m(\mathbf{k})] = \{\eta, a_0(\mathbf{k})\} = 0, \qquad \eta\Phi_0 = 0, \qquad \eta^2 = 1 \qquad (17.70)$$

Then, having determined the metric conjugation operation:

$$L' = \eta L \eta = qL, \qquad q = \pm 1,$$

one may speak of that the operators connected with the vector field part have the positive metric parity and those corresponding the scalar field part possess the negative metric parity. The fulfillment of (17.70) may be realized if we select η in the following form:

$$\eta = (-1)^{n_0},$$

where n_0 is a number of the scalar pseudophotons in the given state.

Conjugation of the operators by Gupta is understood as a set of the Hermitian and metric conjugations:

$$L^\star = \eta L^\dagger \eta = qL^\dagger. \tag{17.71}$$

Next we assume that, unlike other components, the scalar field component $A_0(x)$ is anti-Hermitian, that is,

$$A_\mu^\dagger(x) = -A^\mu(x),$$

and this is equivalent to going to the Hermitian component $A_4(x) = -iA_0(x)$. To realize self-conjugation (by Gupta) of the field operators, we have to redefine the scalar product of the state vectors as follows:

$$(\Phi, \Phi') = \int \Phi^\dagger \eta \Phi' d\tau = \int \Phi^\star \Phi' d\tau. \tag{17.72}$$

At $\eta = 1$ we obtain the ordinary definition of the scalar product. In the general case (17.72) may result in a positive as well as negative norm of the vector Φ, not excepting the situation when $\Phi \neq 0$ but (Φ, Φ) equals zero. Because of this, such spaces are thought of as possessing an indefinite metric. The definition (17.71) is actually providing the self-conjunction of operators as we have

$$[\int \Phi^\star \Lambda_\mu(x) \Phi d\tau]^\dagger = \int \Phi^\star \Lambda_\mu(x) \Phi d\tau.$$

The above-mentioned procedure brings about the commutation relations:

$$[a_\mu(\mathbf{k}), a_\nu^\star(\mathbf{k}')] = -\delta_{\mu\nu}\delta(\mathbf{k} - \mathbf{k}'), \tag{17.73}$$

$$[A_\mu(x), A_\nu^\star(x')] = -i\delta_{\mu\nu}D_0(x - x'). \tag{17.74}$$

Let us choose, for the sake of simplicity:

$$e_\mu^{(\lambda)}(\mathbf{k}) = \delta_\mu^\lambda,$$

that is, let the polarization index coincide with the Lorentz one. Then, the relations (17.73) allow us to treat $a_\mu(\mathbf{k})$ $(a_\mu^\star(\mathbf{k}))$ as being the annihilation (production) operators of the photons with the transverse, longitudinal, and time polarizations. Thus, the relativistic quantization scheme is ensured.

The indefinite metric owes its origin to the problems arising during quantization of the electromagnetic field. Subsequently, its sphere was widened to theories of a massive charged vector field [44–46] due to the attempts to construct a renormalizable theory of the electromagnetic interaction for this field. * The idea lay in the fact that in order to describe

---

*A theory based on the Proca equations is not renormalizable and becomes such only upon the involvement of the weak interaction and Higgs mechanism.

the particles with the spin 1, one should use a collection of fields: *vector field+auxiliary field*. Most often a role of the auxiliary field was played by a scalar field. Since the norms of the vector and scalar field states had different signs ($\mathcal{H} = \mathcal{H}_+ \oplus \mathcal{H}_-$), at calculation of the scattering cross sections the subtractive procedure was working: the contributions of scalar particles were partly canceling those of the longitudinal vector particles. As a result, a theory became renormalizable. But the existence of the negative-metric subspaces was always a source of serious trouble. The reasons for that are obvious. The transitions between the states with the different metric parity may lead to negative probabilities, that is, to the unitarity violation of the theory. True enough, if the total number of the negative-metric particles in the initial and final states is even, the probability appears to be positive for any number of the positive-metric particles. There is a simple approach to verify the unitarity of a theory with the indefinite metric. With this aim in view one should calculate the cross section $\sigma_{\mathcal{H}_\pm \to \mathcal{H}_\mp}$ for processes of the type:

$$A_+ + B_+ \to C_- + D_+,$$

where a subscript on a particle points to the fact that a particle belongs to the positive- or negative-metric space. If the theory is unitary, then the quantities $\sigma_{\mathcal{H}_\pm \to \mathcal{H}_\mp}$ must vanish. Unfortunately, all efforts to retain the unitarity in theories of a massive charged vector field with the auxiliary scalar field were unsuccessful. It seems that the situation with the indefinite-metric theories may be rescued by the presence of a symmetry that takes place both in the free case and upon inclusion of the interaction. Indeed, according to the Noether theorem, in this case a conserved quantity should be in existence. And this motion integral may forbid the transitions between the subspaces with differing metrics. However, this hope proved to be in vain because in the presence of the indefinite metric the symmetry does not lead to a motion integral [47]. It may be that a golden time of the indefinite metric is in the near future, but now it is used only as some skillful device providing the relativistic covariance of the electromagnetic field quantization.

In the case of the electromagnetic field we have another problem associated with the Lorentz condition. As for quantization we assume that all the operators $A_\mu(x)$ are independent, we have no right to require the fulfillment of the Lorentz condition directly for $A_\mu(x)$. It is an easy matter to see that the way of imposing the Lorentz condition on the allowed state vectors:

$$\left(\frac{\partial}{\partial x_\mu} A_\mu(x) \right) \Phi = 0, \tag{17.75}$$

appears to be blocked as well. Really, assuming $\Phi = \Phi_0$, we have

$$\left(\frac{\partial}{\partial x_\mu} A_\mu(x) \right) \Phi_0 = \left(\frac{\partial}{\partial x_\mu} A_\mu^-(x) \right) \Phi_0 = 0. \tag{17.76}$$

Multiplying the left-hand side of (17.76) by $A_\nu^{-\star}(x')$ we get

$$A_\nu^{-\star}(x') \left(\frac{\partial}{\partial x_\mu} A_\mu^-(x) \right) \Phi_0 = \frac{\partial}{\partial x_\mu} \left(A_\nu^{-\star}(x') A_\mu^-(x) \right) \Phi_0 =$$

$$= \frac{\partial}{\partial x_\mu} \left(A_\mu^-(x) A_\nu^{-\star}(x') \right) \Phi_0 + i\delta_{\mu\nu} \frac{\partial}{\partial x_\mu} D_0^-(x - x')\Phi_0 = i \frac{\partial}{\partial x_\nu} D_0^-(x - x')\Phi_0 \neq 0,$$

which is in contradiction with the statement (17.76).

For the agreement with the classical field, it is necessary to fulfill the Lorentz condition whether in the operator form or on the average. It appears that the latter may be realized by changing the Lorentz condition; it should concern only the part of the operator $A_\mu(x)$ to annihilate photons, that is,

$$\left(\frac{\partial}{\partial x_\mu} A_\mu^+(x)\right) \Phi = 0, \qquad \text{for all } x. \tag{17.77}$$

Since, both in the free case and at including the interaction the operator $\partial^\mu A_\mu(x)$ obeys the equation:

$$\Box \partial^\mu A_\mu(x) = 0,$$

then the division of this operator on the positive- and negative-frequency parts $\partial^\mu A_\mu^\pm(x)$ is relativistically covariant. Carrying out the Gupta conjugation of the condition (17.77), we arrive at

$$\Phi^\star \left(\frac{\partial}{\partial x_\mu} A_\mu^-(x)\right) = 0, \qquad \text{for all } x. \tag{17.78}$$

The relations (17.77) and (17.78) ensure the fulfillment of the Lorentz condition on the average:

$$\Phi^\star \left(\frac{\partial}{\partial x_\mu} A_\mu(x)\right) \Phi = 0. \tag{17.79}$$

The next problem is the energy positiveness. From the auxiliary conditions, written in the momentum representation:

$$(|\mathbf{k}| a_3(\mathbf{k}) - k_0 a_0(\mathbf{k})) \Phi = 0, \tag{17.80}$$

$$\Phi^\star(|\mathbf{k}| a_3^\star(\mathbf{k}) - k_0 a_0^\star(\mathbf{k})) = 0, \tag{17.81}$$

follows that:

$$\Phi^\star[a_3^\star(\mathbf{k}) a_3(\mathbf{k}) - a_0^\star(\mathbf{k}) a_0(\mathbf{k})]\Phi = \Phi^\star a_3^\star(\mathbf{k})[a_3(\mathbf{k}) - a_0(\mathbf{k})]\Phi = 0 \tag{17.82}$$

or

$$\Phi^\star[N_3(\mathbf{k}) - N_0(\mathbf{k})]\Phi = 0, \tag{17.83}$$

where $N_3(\mathbf{k})$ and $N_0(\mathbf{k})$ are number operators number longitudinal and temporal pseudophotons, respectively. As is seen, the Lorentz condition in a reference frame with $\mathbf{k} \parallel \mathbf{c}^3(\mathbf{k})$ permits only the states of a system that involves equal numbers of the longitudinal and temporal pseudophotons having the same momentum. The temporal pseudophotons make a negative contribution into the dynamic invariants that is canceled by the contribution from the longitudinal pseudophotons. Thus, the total energy of the field:

$$< H > = \int d^3 k k_0 < -a_\mu^\star(\mathbf{k}) a^\mu(\mathbf{k}) > = \int d^3 k k_0 < \sum_{m=1}^{2} a_m^\star(\mathbf{k}) a_m(\mathbf{k}) > \tag{17.84}$$

proves to be positive defined. Note that the concept of the electromagnetic field vacuum should be updated because of the relation (17.83). The vacuum of the electromagnetic field represents such a state of the field, where the real transverse photons are lacking. As regards the longitudinal and scalar pseudophotons, their numbers $N_3(\mathbf{k})$ and $N_0(\mathbf{k})$ are always nonzero. But, similar to any other physically realized state, the state vector for the vacuum is proportional to the value $N_3(\mathbf{k}) - N_0(\mathbf{k})$.

A final accord in history with the indefinite metric will be the proof of the fact that in concrete calculations we can do without its introduction. We show that the average values over the observables are the same both for the ordinary norm definition and for definition of the norm with the indefinite metric. To this end, we rewrite an expansion for a positive frequency part of the potential in terms of the local reference point $e_\mu^\lambda(\mathbf{k}) = \delta_\mu^\lambda$ in the form:

$$A_\mu^+(\mathbf{k}) = \sum_{m=1}^{2} e_\mu^m(\mathbf{k}) a_m(\mathbf{k}) + \left(\frac{k_\mu}{|(\mathbf{k})|} - \delta_\mu^0 \right) a_3(\mathbf{k}) + \delta_\mu^0 a_0(\mathbf{k}) = A_\mu^{tr,+}(\mathbf{k}) +$$

$$+ k_\mu \Lambda(\mathbf{k}) + \delta_\mu^0 B^+(\mathbf{k}), \tag{17.85}$$

where $A_\mu^{tr,+}(\mathbf{k})$ is a transverse component of the potential and $B^+(\mathbf{k})$ obeys the condition:

$$B^+(\mathbf{k})\Phi = 0.$$

Due to the gradient invariance the potentials $A_\mu(x)$ and $A_\mu'(x) = A_\mu(x) + \partial_\mu f(x)$ ($f(x)$) ($f(x)$ is an arbitrary function) lead to the equivalent physical results. Because of this, in the expressions for the potential we can neglect the quantities with the structure of the four-dimensional gradient or, when an analysis is performed for the momentum representation, the quantities proportional to k_μ. It should be noted that this procedure is also valid when the interaction is included as the electromagnetic interaction Lagrangian $\mathcal{L}_{em} = ieA_\mu(x)j^\mu(x)$, by virtue of the current conservation law:

$$\partial_\mu j^\mu = 0,$$

retains its form when going from $A_\mu(x)$ to $A_\mu(x)'$. Thus, we obtain

$$A_\mu^+(\mathbf{k})\Phi = A_\mu^{tr,+}(\mathbf{k})\Phi, \qquad \Phi^\star A_\mu^-(\mathbf{k}) = \Phi^\star A_\mu^{tr,-}(\mathbf{k}), \tag{17.86}$$

which is equivalent to the statement:

$$\Phi^\star A_\mu(k)\Phi = \Phi^\star A_\mu^{tr}(k)\Phi.$$

Let us show the validity of the more general statement:

$$\Phi^\star \mathcal{P}(A)\Phi = \Phi^\star \mathcal{P}(A^{tr})\Phi, \tag{17.87}$$

where $\mathcal{P}(A)$ ($\mathcal{P}(A^{tr})$) is an operator being polynomial on A (A^{tr}). Substitute the expression for $A_\mu(k)$ (17.85) into the relation (17.87). To use the Lorentz conditions (17.77), (17.78) there is a need to perform the commutation of the quantities A^{tr} and $B = B^+ + B^-$. Since:

$$[A^{tr,\pm}, B^\pm] = [A^{tr,\mp}, B^\pm] = 0,$$

then, as a result, only the quantities $A^{tr,-}$ and $A^{tr,+}$ remain under the average symbol in the left-hand side of (17.87), as was to be shown.

Now, we proceed to the main stage of our proof, that is, we are going to show the validity of of the relation:

$$\Phi^\star \mathcal{P}(A^{tr})\Phi = \Phi_{tr}^\dagger \mathcal{P}(A^{tr})\Phi_{tr}, \tag{17.88}$$

where Φ_{tr} is a vector of the pure photon state (pseudophotons are absent). A vector of an arbitrary state could be represented as a superposition of the Φ_{tr}-state and Φ_{ps}-states containing different numbers of pseudophotons. Due to the condition (17.83) Φ_{ps}-states may incorporate pseudophoton operators only in the combination B^-, that is,

$$\Phi = \{1 + \sum_n C_n \prod_{1 \le i \le n} [a_3^\star(\mathbf{k}_i) - a_0^\star(\mathbf{k}_i)]\}\Phi_{tr}, \tag{17.89}$$

$$\Phi^\star = \Phi^\star_{tr}\{1 + \sum_n C_n \prod_{1 \le i \le n} [a_3(\mathbf{k}_i) - a_0(\mathbf{k}_i)]\}. \tag{17.90}$$

To insert (17.89) and (17.90) into the left-hand side of the expression (17.88) produces the result:

$$\Phi^\star \mathcal{P}(A^{tr})\Phi = \Phi^\star_{tr}\{1 + \sum_n C_n \prod_{1 \le i \le n} [a_3(\mathbf{k}_i) - a_0(\mathbf{k}_i)]\}\mathcal{P}(A^{tr})\{1+$$

$$+ \sum_m C_m \prod_{1 \le j \le m} [a_3^\star(\mathbf{k}_j) - a_0^\star(\mathbf{k}_j)]\}\Phi_{tr} = \Phi^\star_{tr}\{1+$$

$$+ \sum_m C_m \prod_{1 \le j \le m} [a_3^\star(\mathbf{k}_j) - a_0^\star(\mathbf{k}_j)]\}\mathcal{P}(A^{tr})\{1 + \sum_n C_n \prod_{1 \le i \le n} [a_3(\mathbf{k}_i) - a_0(\mathbf{k}_i)]\}\Phi_{tr} =$$

$$= \Phi^\star_{tr}\mathcal{P}(A^{tr})\Phi_{tr} = \Phi^\dagger_{tr}\eta\mathcal{P}(A^{tr})\Phi_{tr} = \Phi^\dagger_{tr}\mathcal{P}(A^{tr})\eta\Phi_{tr} =$$

$$= \Phi^\dagger_{tr}\mathcal{P}(A^{tr})\Phi_{tr}, \tag{17.91}$$

which confirms the validity of (17.88). Because of this, for the subsequent descriptions of the electromagnetic field we can use the conventional formulae for the Hilbert space with the positive (definite) metric. Using (17.91) it is readily shown that the norm of physically allowed states proves to be positively definite. However, note that this is not a consequence of the introduction of the indefinite metric. The true reasons are: the Lorentz condition and the invariance of the electromagnetic field theory with respect to the specialized gradient transformation of the second kind.

In conclusion, we have to indicate a possibility to reformulate the Maxwell equations into the matrix Duffin-Kemmer equations in the same manner as it was done for the theory of the massive vector field. The only difference is in matching of the dimensionalities for the components of the vector A_μ and the tensor $F_{\mu\nu}$ that is realized due to the introduction of an arbitrary constant with the dimensionality of the inverse length.

17.5 Photon polarization

We start with the three-dimensional description of the nonquantized electromagnetic field. At a given momentum the photon state is characterized by the unit polarization vector $\mathbf{e}^\lambda$ that plays the role of the spin part of the photon wavefunction. An arbitrary polarization $\mathbf{e}^\lambda$ may be represented as the imposing of two mutually orthogonal polarizations $\mathbf{e}^{(1)}$ and $\mathbf{e}^{(2)}$ chosen by some particular way:

$$\mathbf{e}^\lambda = a_1^\lambda \mathbf{e}^{(1)} + a_2^\lambda \mathbf{e}^{(2)}, \tag{17.92}$$

where in the most general case a_1^λ and a_2^λ are complex quantities satisfying the normalization condition:

$$|a_1^\lambda|^2 + |a_2^\lambda|^2 = 1.$$

Since the common factor standing ahead of $\mathbf{e}^\lambda$ is arbitrary, then a vector of any polarization could be written in the form:

$$\mathbf{e}^\lambda = \mathbf{e}^{(1)}\cos\alpha + \mathbf{e}^{(2)}\sin\alpha e^{i\beta}, \tag{17.93}$$

where α and β are real numbers. Let $\mathbf{e}^{(i)} = \delta_r^i$, where $i = 1, 2$ and $r = 1, 2, 3$, then $\beta = 0$ would correspond to the linear polarization at an angle α to the x_1-axis. At $\beta = \pm\pi/2$ and

$\alpha = \pi/4$ we have the circular polarizations while the elliptic polarization correlates with arbitrary values of β and α.

Note that in the case of a three-dimensional transverse gauge and with the proviso that the z-axis has been chosen along $\mathbf{k}$, the vectors $\mathbf{e}^{(1)}$ and $\mathbf{e}^{(2)}$ involved in (17.92) have zero projection onto the z-axis, that is, they are two-dimensional objects. Proceeding from all the above, we can establish a formal analogy between the photons and massive particles with the spin 1/2. Both of them have two polarization states. True enough, in the case of the electron we consider the spin projection onto the momentum, whereas for the photons— spin projection onto the axis perpendicular to the momentum direction. Similar to the electrons, these photon states may be represented with the use of two-dimensional column vectors under the condition that the $\mathbf{k}$-direction (the z-axis) still remains the quantization axis.

The state of a photon with a definite momentum and polarization is a pure state; it is described by the wavefunction and associated with the complete quantum-mechanical description of the particle state. At the same time, there are some situations when we cannot assign a certain wavefunction to the photon. For example, at the photon scattering by the electron there exists only the wavefunction of a system *electron+photon*, whose expansion in terms of the wavefunctions of the free photon already includes the wavefunctions of the electron. In other words, there is a possibility for the mixed photon states corresponding to a less complete description using the density matrix rather than the wavefunction.

Let us consider a case when the photon momentum is given, while its polarization state is indeterminate because the factors a_1^λ and a_2^λ, or the quantities α and β, depend on the parameters characterizing another system. Such a state of the photon is called the *partial polarization state* that may be described by the density matrix. The polarization density matrix of the photon represents the second-rank tensor ρ_{mn}^γ $(m, n = 1, 2)$ specified in the plane perpendicular to the $\mathbf{k}$-vector:

$$\rho_{mn}^\gamma = \sum_\lambda g^\lambda a_m^\lambda a_n^{\lambda*} = \overline{a_m^\lambda a_n^{\lambda*}}, \qquad (17.94)$$

where g^λ is a relative probability with which the λ- polarization enters into the mixed beam, and an overline denotes averaging over the relevant parameters characterizing a total system. From the density matrix definition it follows that this matrix is Hermitian:

$$\rho_{mn}^\gamma = \rho_{nm}^{\gamma*}, \qquad (17.95)$$

and is normalized by the condition:

$$\mathrm{Sp}\,\rho^\gamma = \rho_{11}^\gamma + \rho_{22}^\gamma = 1. \qquad (17.96)$$

Thanks to (17.95) the diagonal components ρ_{11}^γ and ρ_{22}^γ are real, whereas the nondiagonal ones are complex and they satisfy the condition:

$$\rho_{12}^\gamma = \rho_{21}^{\gamma*}.$$

Thus, the density matrix is defined by three real parameters.

Let us find out the physical meaning of elements of the photon density matrix. When one chooses $\mathbf{e}^{(1)} \parallel Ox$ and $\mathbf{e}^{(2)} \parallel Oy$, then from (17.94) it follows that $\rho_{11}^\gamma = \overline{a_1^\lambda a_1^{\lambda*}}$ characterizes the probability of the photon polarization along the x-axis, while ρ_{22}^γ—along the y-axis. For the case of the circular polarization the vectors $\mathbf{e}^{(1)\prime}$ and $\mathbf{e}^{(2)\prime}$, which are connected with the vectors $\mathbf{e}^{(1)}$ and $\mathbf{e}^{(2)}$ by the relations:

$$\mathbf{e}^{(1)} = \frac{1}{\sqrt{2}}\left(\mathbf{e}^{(1)\prime} + \mathbf{e}^{(2)\prime}\right), \qquad \mathbf{e}^{(2)} = \frac{1}{i\sqrt{2}}\left(\mathbf{e}^{(1)\prime} - \mathbf{e}^{(2)\prime}\right),$$

enter into the formula (17.92). In the new unit vectors this formula takes the form:

$$\mathbf{e}^\lambda = \frac{1}{\sqrt{2}} \left(a_1^\lambda - ia_2^\lambda\right) \mathbf{e}^{(1)\prime} + \frac{1}{\sqrt{2}} \left(a_1^\lambda + ia_2^\lambda\right) \mathbf{e}^{(2)\prime} = a_1^{\lambda\prime} \mathbf{e}^{(1)\prime} + a_2^{\lambda\prime} \mathbf{e}^{(2)\prime}.$$

Then, the probability of the left and right circular polarization (complying with the vectors $\mathbf{e}^{(1)\prime}$ and $\mathbf{e}^{(2)\prime}$) is equal to

$$\rho_{11}^{\gamma\prime} = \overline{a_1^{\lambda\prime} a_1^{\lambda\prime*}} = \frac{1}{2}[\rho_{11}^\gamma + \rho_{22}^\gamma + i(\rho_{12}^\gamma - \rho_{21}^\gamma)] = [1 + i(\rho_{12}^\gamma - \rho_{21}^\gamma)],$$

$$\rho_{22}^{\gamma\prime} = \frac{1}{2}[1 - i(\rho_{12}^\gamma - \rho_{21}^\gamma)].$$

So, knowing the density matrix, we can find for the photon the probability of its having a polarization associated with the vector $\mathbf{e}^{(i)}$. This probability $dw^{(i)}$ is defined as a "projection" of the tensor ρ_{mn}^γ onto the direction of the vector $\mathbf{e}^{(i)}$, that is,

$$dw^{(i)} = \rho_{mn}^\gamma e_m^{(i)} e_n^{(i)*}. \tag{17.97}$$

In the general case a partial polarization may be conveniently described by three real Stokes parameters ξ_1, ξ_2, and ξ_3, representing the density matrix as:

$$\rho^\gamma = \frac{1}{2}\left(1 + \sum_{j-1}^{3} \xi_j \sigma_j\right) = \frac{1}{2}\begin{pmatrix} 1 + \xi_3 & \xi_1 - i\xi_2 \\ \xi_1 + i\xi_2 & 1 - \xi_3 \end{pmatrix}. \tag{17.98}$$

This formula may be converted to determine the Stokes parameters through the matrix density:

$$\boldsymbol{\xi} = \mathrm{Sp}(\rho^\gamma \boldsymbol{\sigma}). \tag{17.99}$$

The parameter ξ_3 characterizes a linear polarization along the directions of the x- or y-axis. The probabilities that the photon is linearly polarized in these axes are equal to $(1 + \xi_3)/2$ and $(1 - \xi_3)/2$ respectively, the values $\xi_3 = 1$ or $\xi_3 = -1$ corresponding to the complete polarizations in these directions.

The parameter ξ_1 characterizes a linear polarization in the directions for which $\alpha = \pi/4$ or $\alpha = \pi/4$ ($\beta = 0$). The probabilities that the photon is linearly polarized in these directions are found by the projection of the tensor ρ_{mn}^γ onto the vectors $\mathbf{e}^{(i)} = (1, \pm 1, 0)/\sqrt{2}$ and proves to be equal to $(1 + \xi_1)/2$ or $(1 - \xi_1)/2$, respectively. Again, at $\xi_1 = +1$ or $\xi_1 = -1$ we have the case of the complete polarization.

The parameter ξ_2 defines the degree of the circular polarization. The projection of the density matrix onto the circular polarization vectors gives the following expressions for the probability of the right and left polarization:

$$\frac{1}{2}\left(1 + \xi_2\right), \qquad \frac{1}{2}\left(1 - \xi_2\right).$$

As the probabilities must be positive and less than 1, then we conclude:

$$|\xi_n| \leq 1.$$

For the unpolarized photon we have

$$\xi_1 = \xi_2 = \xi_3 = 0, \qquad \rho^\gamma = 1/2.$$

In the case of the complete polarized photon the relation:

$$\xi_1^2 + \xi_2^2 + \xi_3^2 = 1. \tag{17.100}$$

should be carried out. Let us demonstrate that this is so indeed. To fulfill the relation (17.100) the Stokes parameters may be taken as:

$$\xi_1 = \sin 2\alpha \cos \beta, \qquad \xi_2 = \sin 2\alpha \sin \beta, \qquad \xi_3 = \cos 2\alpha. \qquad (17.101)$$

At these values of ξ_n the matrix ρ^γ coincides with that to be built up with the help of the polarization vector (17.93) and, consequently, be associated with the state of the complete polarization

In the theory of the quantized electromagnetic field the polarization vectors represent the factors standing ahead of the production and annihilation operators in the operator expansion of the four-dimensional potential. Similar to a case of the electron, the polarization density matrix is determined as:

$$(\rho^\gamma)_{mn} = \overline{e_m^\lambda e_n^{\lambda*}}, \qquad (17.102)$$

where $\mathbf{e}^\lambda$ ($\lambda = 1, 2$) are the three-dimensional polarization vectors. Recall that for the three-dimensional transverse gauge $(\mathbf{k} \cdot \mathbf{A}) = 0$ in the reference frame with $\mathbf{k} \parallel Oz$ the vectors $\mathbf{e}^\lambda$ are two-dimensional. Obviously, representation of the density matrix in terms of the Stokes parameters (17.98) is valid in the second quantized theory as well. This is not surprising as, unlike the electron, the photons have no nonrelativistic region, and quantum mechanics of the photon is a relativistic theory from the start.

One may represent the photon density matrix in the four-dimensional form as well. It is natural to define this matrix in the following way:

$$(\rho^\gamma)_{\mu\nu} = \overline{e_\mu^\lambda e_\nu^{\lambda*}}, \qquad (17.103)$$

where e_μ^λ are four-dimensional polarization vectors, and λ takes the values equal to 1,2 as before. When we choose the three-dimensional gauge $e^\lambda = (0, \mathbf{e}^\lambda)$, then in the reference frame with $\mathbf{k} \parallel Oz$ the nonzero components of the tensor $(\rho^\gamma)_{\mu\nu}$ will coincide with (17.102). In order to obtain the four-dimensional representation of the photon density matrix in terms of the Stokes parameters, we introduce two unit space-like real vectors $n_\mu^{(1)}$ and $n_\mu^{(2)}$, which are orthogonal both to each other and to the four-dimensional vector of the photon momentum k_μ:

$$\left. \begin{array}{l} (n^{(l)} \cdot n^{(l')}) = -\delta_{ll'}, \qquad l, l' = 1, 2, \\ (k \cdot n^{(l)}) = 0. \end{array} \right\} \qquad (17.104)$$

From (17.104) we find

$$e_\sigma^\lambda = -(e^\lambda \cdot n^{(l)}) n_\sigma^{(l)}. \qquad (17.105)$$

Since the vectors $n^{(l)}$ are space-like and orthogonal, then one may seek the reference frame in which they have the form:

$$n_\mu^{(1)} = (0, \mathbf{e}^1), \qquad n_\mu^{(2)} = (0, \mathbf{e}^2).$$

The relation (17.105) is conceptually the relativistic generalization of this fact. By application of (17.105) one may deduce the required expression for the density matrix:

$$\rho_{\mu\nu}^\gamma = \overline{(e^\lambda \cdot n^{(l)})(e^{\lambda*} \cdot n^{(l')})} n_\mu^{(l)} n_\nu^{(l')} = \frac{1}{2} \left(1 + \sum_{j=1}^3 \xi_j \sigma_j \right)_{ll'} n_\mu^{(l)} n_\nu^{(l')}. \qquad (17.106)$$

Note that the conditions (17.104) provide no unambiguous definition for the vectors $n_\mu^{(1)}$ and $n_\mu^{(2)}$. Indeed, we can add to each of them a four-dimensional vector of the form Λk_μ, where Λ is an arbitrary analytical function, to make them still meet the conditions (17.104). This arbitrariness is due to the gauge ambiguity of the photon polarization density matrix stating that matrices related by the gauge transformation are describing the same physical reality.

Problems

3.1. Let a scalar field be confined in a spatial cube with the volume $V = L^3$, where L is a cube edge length. Using the periodicity condition of a wavefunction ψ on every spatial coordinate with the period L, write down an expansion of ψ in terms of frequency components. Tending the cube size to infinity, turn to continuous representation. Write down the four-dimensional energy-momentum vector and show that now the field confined in the volume V may be considered as the harmonic oscillators totality. Find the energy and momentum of these oscillators.

3.2. Determine the charge and space parity of two scalar particles (particle and antiparticle) with the relative orbital moment equal l.

3.3. Find the explicit form of the field functions transformation matrix in the case of purely spacial rotations for the Duffin-Kemmer equation describing scalar particles.

3.4. Working within the first order equations formalism, find the values of the following quantities:

$$\varphi^{(\pm)}(k)\overline{\varphi}^{(\pm)}(k).$$

3.5. When one works within the first quantized theory, the Klein-Gordon equation leads to negative probabilities. In order to avoid the appearance of these probabilities the motion equation does not include the time derivatives higher of the first order. Proceeding from that and taking into account that the following relation

$$E = \sqrt{\mathbf{p}^2 + m^2}$$

must be fulfilled for relativistic particles, obtain the Dirac equation.

3.6. Consider the totality of transformations consisting both of purely spacial rotations and of spacial axis inversions for a particle with the spin $1/2$. Show, that the field function frequency parts $\psi^{(+)}$ and $\psi^{(-)}$ are transformed independently.

3.7. Prove the validity of the spin summation rules (15.103) and (15.104) with the help of the Green function to obey the inhomogeneous equation

$$(\hat{p} + m)G(p) = 1.$$

3.8. Show that for the particles with the spin $1/2$ the energy-momentum tensor obeys the continuity equation.

3.9. Find the values of the following expressions:

$$\mathrm{Sp}\{\hat{p}_1\hat{p}_2\}, \qquad \mathrm{Sp}\{(\hat{p}_1 + m_1)(\hat{p}_2 + m_2)\}$$

$$\mathrm{Sp}\{\hat{k}_1(\hat{p}_1 + m_1)\hat{k}_2(\hat{p}_2 + m_2)\}, \qquad \mathrm{Sp}\{\hat{k}_1\hat{k}_2(\hat{p}_1 + m_1)(\hat{p}_2 + m_2)\},$$

$$\gamma^\nu\gamma^\mu\gamma_\nu, \qquad \gamma^\nu\gamma^\mu\gamma^\sigma\gamma_\nu, \qquad \gamma^\nu\gamma^\mu\gamma^\sigma\gamma^\rho\gamma\gamma_\nu.$$

3.10. Write down the Dirac equation in such a representation where it has no imaginary coefficients.

3.11. In the rest reference frame the spin of a free Dirac particle is conserved and its wavefunction in the Dirac-Pauli representation has only two components for which the spin projections onto the given axis are equal to $\pm 1/2$. Find such a representation where the wavefunction (plane wave) possesses only two components corresponding to the definite values of the spin projection in the rest reference frame.

3.12. Consider the totality of 16 linearly independent four-row matrices Γ^A ($A = 1, 2 \ldots 16$)

$$1, \qquad \gamma^\nu, \qquad \gamma^5, \qquad i\gamma^5\gamma^\nu, \qquad \sigma^{\mu\nu}$$

and obtain the completeness condition for them. Carry out the same for 4 linearly independent two-row matrices Σ^B ($B = 1, 2, 3, 4$)

$$1, \qquad \sigma_x, \qquad \sigma_y, \qquad \sigma_z.$$

3.13. Employing the bispinors ψ_a and ψ_b, build up 16 bilinear combinations to be grouped into 5 different Lorentz-covariant quantities. Making use of the field functions transformation law, check upon their transformation properties.

3.14. Build up Lorentz scalars from four bispinors ψ_a, ψ_b, ψ_c, and ψ_d.

3.15. In calculating scattering reactions, the following formulae

$$(\overline{\psi}_a\Gamma_A\psi_b)(\overline{\psi}_c\Gamma_A\psi_d) = \sum_{A'} C_{AA'}(\overline{\psi}_a\Gamma_{A'}\psi_d)(\overline{\psi}_c\Gamma_{A'}\psi_b)$$

(Firz relations) are very useful. Find all the coefficients $C_{AA'}$.

3.16. Find the energy levels and the wavefunctions of the electron in the constant magnetic field.

3.17. Derive the Pauli equation

$$i\hbar\frac{\partial\varphi}{\partial t} = \left[\frac{1}{2m}\left(\mathbf{p} - \frac{e}{c}\mathbf{A}\right)^2 + eA_0 - \frac{e\hbar}{2mc}(\boldsymbol{\sigma}\cdot\boldsymbol{\mathcal{H}})\right]\varphi$$

from the Dirac one.

3.18. Find the Dirac equation solution in the field of the arbitrary polarized plane wave

$$A_\mu = A_\mu(\theta), \qquad \theta = k^\mu x_\mu, \qquad k_\mu A^\mu = 0,$$

where k_μ is a wave vector. Define the effective electron mass $m_*^2 = q_\mu q^\mu$, where q_μ is a time average of the kinetic momentum density Q_μ (the kinetic momentum is $p_\mu - eA_\mu$.)

3.19. Define a change of a particle polarization direction moving in the plane perpendicular to a homogeneous magnetic field ($\mathbf{v} \perp \boldsymbol{\mathcal{H}}$).

3.20. Find a change of a particle polarization direction moving in a homogeneous electric field.

3.21. In supersymmetric theories a particle $\overline{\gamma}$ with the spin $1/2$ is a superpartner of γ. At the same time $\overline{\gamma}$ possesses the Majorana nature. Assuming that in the process

$$\gamma + \gamma \to \overline{\gamma} + \overline{\gamma}$$

the C-parity is conserved, find the intrinsic parity of $\overline{\gamma}$.

3.22. In the general form the matrix element of the conserved electromagnetic current $J_\mu^{(em)}$ for the neutrino is as follows:

$$< \nu(p_f, s_f)|J_\mu^{(em)}|\nu(p_i, s_i) >= \overline{u}(p_f, s_f)\{F_Q(q^2)\gamma_\mu + F_A(q^2)(q^2 q_\mu - \hat{q}q_\mu)\gamma_5 -$$

$$-i[F_M(q^2) + F_E(q^2)\gamma_5]\sigma_{\mu\nu}q^\nu\}u(p_i, s_i),$$

where $q^\mu = p_f^\mu - p_i^\mu$, $F_n(q^2)$ are formfactors determining distributions of the electric charge ($n = Q$), axial charge ($n = A$), magnetic dipole moment ($n = M$), and electric dipole moment ($n = E$), respectively. Demanding the CPT-invariance of the theory, show that for the Majorana neutrino the single nonvanishing formfactor is the axial one (this formfactor describes an anapole moment).

3.23. For an arbitrary field with the spin $1/2$ the CP-transformation is determined by the following way:

$$CP\Psi(\mathbf{r}, \mathbf{t})(CP)^{-1} = \eta_{\mathbf{CP}}^* \gamma_4 \gamma_2 \Psi^*(-\mathbf{r}, \mathbf{t}),$$

where η_{CP} is a phase factor. Using this definition and assuming the Dirac neutrino nature, prove the validity of the relation:

$$CP|a_r(\mathbf{p}) >= \eta_{CP}|b_r(-\mathbf{p}) >,$$

where $|a_r(\mathbf{p}) > (|b_r(\mathbf{p}) >)$ designates the state to involve a particle (antiparticle) with the momentum $\mathbf{p}$ and polarization r.

3.24. If the vector field causal function is defined by the relation:

$$D_{kl}^{c\prime}(x) = \theta(x_0)D_{kl}^{(-)}(x) - \theta(-x_0)D_{kl}^{(+)}(x) \qquad (A)$$

where

$$D_{kl}^{(\pm)}(x) = \frac{\pm 1}{(2\pi)^3 i} \int e^{ikx}\theta(\pm k_0)\delta(k^2 - m^2)\left(g_{kl} - \frac{k_l k_n}{m^2}\right) dk,$$

the direct calculations of the expression (A) using the integral representation of the θ function

$$\theta(\pm x_0) = \frac{1}{2\pi i}\int_{-\infty}^{+\infty} \frac{e^{ix_0\tau}}{\tau \mp i\epsilon}d\tau,$$

leads to the expression different from

$$D_{kl}^{c}(x) = \left(g_{kl} + \frac{1}{m^2}\frac{\partial^2}{\partial r^k \partial x_l}\right) D^c(x)$$

in neighborhood of the point $x = 0$. Find $D_{kl}^{(\pm)}(x)$ and point out the reason of this contradiction.

3.25. In the photon case find the vector functions (spherical vectors) to be eigenfunctions both of the square angular momentum operator and of the angular momentum projection operator onto the given axis. [NOTE: Make use of the relation known from nonrelativistic quantum mechanics

$$[\hat{L}_k, \hat{A}_k] = i\epsilon_{klm}\hat{A}_m,$$

where $\hat{A}_m$ is some vector physical quantity to characterize a system.]

3.26. Show that the theory to be described by the Lagrangian of the problem **1.16** is quantized with the use of the indefinite metric.

References

[1] E. Schrödinger, Ann. Phys. **79**, 361 (1926).

[2] W. Gordon, Zeit. Phys. **40**, 117 (1926).

[3] O. Klein, Zeit. Phys. **37**, 895 (1926).

[4] V. Fock, Zeit. Phys. **38**, 242 (1926).

[5] J. Kudar, Ann. Phys. **81**, 632 (1926).

[6] I. S. Gradschtein, I. M. Ryzhik, *Tables of Integrals, Sums and Series*, (Fiz. Mat. Glz., Moskow, 1963).

[7] J. Schwinger, Phys. Rev. **74**, 1439 (1948); **75**, 651 (1949).

[8] H. Umezawa, Prog. Theor. Phys. **7**, 551 (1952).

[9] R. J. Duffin, Phys. Rev. **54**, 1114 (1938); N. Kemmer, Proc. Roy. Soc. **A173**, 91 (1939).

[10] P. A. M. Dirac, Proc. Roy. Soc. **A117**, 610 (1928).

[11] W. Pauli, Ann. de l'Institut Henri Poincare, **6**, 137 (1936).

[12] W. Pauli, *Handbuch der Physik*, (Bd. 24/1, Berlin, 1933).

[13] P. A. M. Dirac, Proc. Cambr. Phil. Soc. **26**, 376 (1930).

[14] J. R. Oppenheimer, Phys. Rev. **35**, 562 (1930).

[15] C. D. Anderson, Phys. Rev. **43**, 491 (1933).

[16] E. R. Caianello, S. Fubini, Nuovo Cimento **9**, 1218 (1952).

[17] H. Weyl, Zeit. Phys. **56**, 330 (1929).

[18] V. Bargman, L. Michel, V. Telegdi, Phys. Rev. Lett. **2**, 435 (1959).

[19] W. H. Furry, Phys. Rev. **81**, 115 (1951).

[20] V. B. Berestetskii, E. M. Lifshitz, and L. P. Pitaevskii, *Relativistic Quantum Theory*, (Pergamon Press, Oxford, 1971).

[21] R. S. van Dyck Jr., P. B. Schwinberg, and H. G. Dehmelt, Phys. Rev. Lett. **59**, 26 (1987).

[22] P. J. Mohr and B. N. Taylor, Rev. Mod. Phys. **72**, 351 (2000).

[23] E. Majorana, Nuova Cimento **14**, 171 (1937).

[24] S. Fukuda *et al.*, Super-Kamiokande Collaboration, Phys. Rev. Lett. **86**, 5656 (2001).

[25] Q. R. Ahmad *et al.*, SNO Collaboration, Phys. Rev. Lett. **87**, 071301 (2001).

[26] H. V. Klapdor-Kleingrothaus, A.Dietz, H.L.Harney and I.V.Krivosheina (Heidelberg-Moscow Collaboration), Mod. Phys. Lett. **A16**, 2409 (2002).

[27] C. E. Aalseth *et al.*, Mod. Phys. Lett. **A510**, 1475 (2002).

[28] B. Touschek, Zeit. Phys. **125**, 108 (1949); F. Gursey, Nuovo Cimento, Serie X **7**, 411 (1958).

[29] R. V. Case, Phys. Rev. **107**, 307 (1957).

[30] I. B. Zel'dovich, JETF **33**, 1531 (1957).

[31] B. Kayser, F. Gibrat-Debu and F. Perrier, *The physics of massive neutrinos*, (World Scientific, Singapore, 1989).

[32] H. V. Klapdor-Kleingrothaus *et al.* Phys. Lett. **A586**, 198 (2004).

[33] J. C. Pati and A. Salam, Phys. Rev. **D10**, 275 (1974); R. N. Mohapatra and G. Senjanovic, Phys. Rev. **D23**, 165 (1981).

[34] O. M. Boyarkin, G. G. Boyarkina, T. I. Bakanova, Phys. Rev. **D70**, 113010-1 (2004).

[35] D. London, G. Belanger, and J. N. Ng, Phys. Lett. **B188**, 155 (1987).

[36] O. M. Boyarkin, D. Rein, Phys. Rev. **D53**, 361 (1996).

[37] A. Proca, J. Phys. et Rad. **7**, 347 (1936).

[38] N. Kemmer, Proc. Roy. Soc. **A166**, 127 (1938).

[39] P. Roman, *Theory of elementary particles*, (N. H. Pub. Co. Amsterdam, 1964); I. Corson, *Theory of tensors, spinors and wave equations*, (Benjamin Press, NY, 1953).

[40] P. A. M.Dirac, Proc. Roy. Soc. **A133**, 821 (1931).

[41] M. G. Galkin, Phys. Lett. **A28**, 45 (1968).

[42] S. Gupta, Proc. Phys. Soc. **A63**, 681 (1950).

[43] K. Bleuler, Helv. Phys. Acta. **23**, 567 (1950).

[44] T. D. Lee, C. N. Yang, Phys. Rev. **128**, 885 (1962); T. D. Lee, G. C. Wick, Nucl. Phys. **B9**, 209 (1969).

[45] J. Hsu, Lett. Nuovo Cimento **12**, 503 (1975).

[46] O. M. Boyarkin, JETP **48**, 13 (1978).

[47] O. M. Boyarkin, J. Phys. G: Nucl. Phys. **8**, 161 (1982); O. M. Boyarkin, Sov. Phys. J. **10**, 828 (1989).

Part IV

Quantum Electrodynamics

18

S-matrix

You are not wrong, who deem
That my days have been a dream;
Yet if hope has flown away
In a night, or in a day,
In a vision, or in none,
Is it therefore the less gone?
All that we see or seem
Is but a dream within a dream.
Edgar Allan Poe, "A Dream within a Dream"

18.1 Equations and dynamic variables in the Heisenberg representation

Having used the principle of the local gauge invariance with respect to the $U(1)$-group in Section 4.5 we have established the form of the Lagrangian to describe the interaction between the electromagnetic and electron-positron fields:

$$\mathcal{L} = \frac{i}{2}\left[\overline{\boldsymbol{\psi}}(x)\gamma^\mu D_\mu\boldsymbol{\psi}(x) - D_\mu^\dagger\overline{\boldsymbol{\psi}}(x)\gamma^\mu\boldsymbol{\psi}(x)\right] - m\overline{\boldsymbol{\psi}}(x)\boldsymbol{\psi}(x) - \frac{1}{4}\mathbf{F}_{\mu\nu}(x)\mathbf{F}^{\mu\nu}(x), \qquad (18.1)$$

where $D_\mu = \partial_\mu + ie\mathbf{A}_\mu(x)$ and $\mathbf{F}_{\mu\nu}(x) = \partial_\mu\mathbf{\Lambda}_\nu(x) - \partial_\nu\mathbf{A}_\mu(x)$. Here the operators of the electromagnetic and electron-positron fields are denoted by a bold font to stress that they are set in the Heisenberg representation. Attention is drawn to the fact that the local gauge invariance is kept only when an electromagnetic field quanta mass is identical equal to zero. If the photons had a mass m, then the Lagrangian of a free electromagnetic field would include the mass term $-m^2 A_\mu A^\mu$ that is gauge noninvariant.

From the Lagrangian (18.1) follows the basic equations of the QED:

$$[\gamma^\mu(i\partial_\mu - e\mathbf{A}_\mu(x)) - m]\boldsymbol{\psi}(x) = 0, \qquad (18.2)$$

$$[\gamma^{\mu T}(i\partial_\mu + e\mathbf{A}_\mu(x)) + m]\overline{\boldsymbol{\psi}}(x) = 0, \qquad (18.3)$$

$$\partial^\mu\mathbf{F}_{\mu\nu}(x) = j_\nu(x). \qquad (18.4)$$

The electromagnetic current is defined by the expression:

$$j_\nu(x) = e\overline{\boldsymbol{\psi}}(x)\gamma_\nu\boldsymbol{\psi}(x), \qquad (18.5)$$

which formally coincides with that for the current in the case of the free electron-positron field. However, it is not exactly the same since this expression contains the quantities $\overline{\boldsymbol{\psi}}(x)$ and $\boldsymbol{\psi}(x)$ to obey the equations for interacting fields rather than those for free fields. The field operators act on the state vectors in the particles number space $\mathbf{\Phi}$. Since we work in the Heisenberg representation, then the following relation:

$$\frac{\partial}{\partial t}\mathbf{\Phi} = 0$$

is valid.

In Eq. (18.4) one may cross to the electromagnetic potentials:

$$\Box \mathbf{A}_\nu(x) = j_\nu(x), \tag{18.6}$$

In so doing, we should supplement this equation, as in the case of free fields, with the condition on the state vectors:

$$\left(\frac{\partial}{\partial x_\mu}\mathbf{A}_\mu^+(x)\right)\mathbf{\Phi} = 0. \tag{18.7}$$

Thanks to the current continuity, the operator $\partial^\mu A_\mu(x)$ satisfies one and the same equation both in a free case and in the presence of interaction:

$$\Box\frac{\partial \mathbf{A}_\mu(x)}{\partial x_\mu} = 0. \tag{18.8}$$

Therefore, the additional condition (18.7) is relativistically invariant.

Eqs. (18.2), (18.3), and (18.6) follow from the Lagrangian that one conveniently writes down as a sum of the Lagrangians of free and interacting fields:

$$\mathcal{L} = \mathcal{L}_0^{(e)} + \mathcal{L}_0^{(\gamma)} + \mathcal{L}_{int}, \tag{18.9}$$

where

$$\mathcal{L}_0^{(e)} = \frac{i}{2}\left[\overline{\psi}(x)\gamma^\mu\partial_\mu\psi(x) - \partial_\mu\overline{\psi}(x)\gamma^\mu\psi(x)\right] - m\overline{\psi}(x)\psi(x), \tag{18.10}$$

$$\mathcal{L}_0^{(\gamma)} = -\frac{1}{2}\partial_\mu\mathbf{A}^\nu\partial^\mu\mathbf{A}_\nu, \tag{18.11}$$

$$\mathcal{L}_{int} = -j_\mu(x)\mathbf{A}^\mu(x), \tag{18.12}$$

Recall, that the Lagrangians (18.1) and (18.9) differ from each other by an insignificant term having the four-dimensional divergence form. The interaction Lagrangian $\mathcal{L}_{int}$ does not contain derivatives of the field operators. We speak of such interactions as couplings without derivatives. In this case the interaction Hamiltonian density $\mathcal{H}_{int}$ is simply equal to $-\mathcal{L}_{int}$ and the interaction Hamiltonian is given by:

$$H_{int}(t) = -\int \mathcal{L}_{int}d^3x. \tag{18.13}$$

A system of differential equations (18.2)–(18.4) contains only a part of the information about the interacting fields. These equations should be amplified by permutation relations (PR) between the field operators. As evolution of the operators in time is defined by equations of motion, the PR in the case of interacting fields may be given for the initial instant of time only. It is clear that the determination of the PR at some time is equivalent to their calculation at the coincident instants of time. Having realized what PR we should find, we can easily guess the way of their searching. We remember that precisely the simultaneous PR are at the base of the canonical quantization procedure. Then, with the help of the canonical formalism one may find the following nonvanishing simultaneous PR (see, the formulae (15.257) and (17.51)):

$$[\mathbf{A}_\mu(x), \partial_0 \mathbf{A}_\nu(x')]_{t_0 = t'_0} = ig_{\mu\nu}\delta^{(3)}(\mathbf{r} - \mathbf{r}'). \tag{18.14}$$

$$\{\boldsymbol{\psi}(x), \boldsymbol{\psi}^\dagger(x')\}_{t_0 = t'_0} = \delta^{(3)}(\mathbf{r} - \mathbf{r}'). \tag{18.15}$$

As it seems, a change to the PR valid for any time intervals may be realized in the same way as for free fields. So, for example, in the free case for the electromagnetic field the solution $A_\nu(x)$ on arbitrary space-like hypersurface σ_0 could be determined by application of the Green function $D(x)$ in terms of the solution on any more early space-like hypersurface

$$A_\nu(x) = \int_{\sigma_0} d\sigma^\mu(x') \left[\frac{\partial D(x - x')}{\partial x'^\mu} A_\nu(x') - D(x - x') \frac{\partial A_\nu(x')}{\partial x'^\mu} \right]. \tag{18.16}$$

Further, selecting the hypersurface σ_0 by the corresponding way, one can obtain the PR sought. If the formula similar to (18.16) existed in the case of the interacting fields, then we would manage to define the PR at arbitrary instants of time. However to derive such a formula demands solutions knowledge of Eqs. (18.2), (18.3), and (18.6), which represent the partial differential inhomogeneous equations system. It is obvious that the task of solving this system is rather complicated. Moreover, in the majority of cases solutions may be obtained by means of approximate methods only. Because approximate PR do not suit us, we have to look for other ways to solve this problem.

18.2 Interaction representation

Having carried out the unitary transformation on the Heisenberg operators N:

$$\mathbf{N} \to \mathbf{N}^{(s)} = e^{-i\mathbf{H}t}\mathbf{N}e^{i\mathbf{H}t}, \tag{18.17}$$

where $\mathbf{H}$ is a total Hamiltonian of a system, we pass to the Schrödinger representation in which, as it is obvious from (18.17), operators are not changed as time goes on:

$$\frac{d}{dt}\mathbf{N}^{(s)} = 0. \tag{18.18}$$

The operator transformation (18.17) is associated with the state vector transformation:

$$\boldsymbol{\Phi} \to \boldsymbol{\Phi}^{(s)}(t) = e^{-i\mathbf{H}t}\boldsymbol{\Phi}. \tag{18.19}$$

In the Schrödinger representation the state vector $\boldsymbol{\Phi}^{(s)}(t)$ satisfies the Schrödinger equation:

$$i\frac{\partial}{\partial t}\boldsymbol{\Phi}^{(s)}(t) = \mathbf{H}^{(s)}\boldsymbol{\Phi}^{(s)}(t). \tag{18.20}$$

From (18.17) it follows that in the representations by Heisenberg and Schrödinger the Hamiltonians coincide:

$$\mathbf{H}^{(s)} = \mathbf{H}. \tag{18.21}$$

However, in interacting fields theory an interaction representation proves to be more convenient. This representation is sometimes named after the authors who shaped it into an explicit covariant form [1], the Tomonaga-Schwinger representation.

Let us divide the total Hamiltonian $\mathbf{H}^{(s)}$ into the free and interaction Hamiltonians:

$$\mathbf{H}^{(s)} = \mathbf{H}_0^{(s)} + \mathbf{H}_{int}^{(s)}. \tag{18.22}$$

Pass to an interaction representation connected with the Schrödinger one by a unitary transformation:

$$\mathbf{N}^{(s)} \to N = e^{i\mathbf{H}_0^{(s)}(t-t_0)} \mathbf{N}^{(s)} e^{-i\mathbf{H}_0^{(s)}(t-t_0)}, \tag{18.23}$$

$$\mathbf{\Phi}^{(s)}(t) \to \Phi(t) = e^{i\mathbf{H}_0^{(s)}(t-t_0)} \mathbf{\Phi}^{(s)}(t). \tag{18.24}$$

From Eqs. (18.23) and (18.24) it follows that both operators and state vectors belonging to the Schrödinger and new representations coincide at $t = t_0$.

Let us find out how to change operators and state vectors in the interaction representation with the passage of time. With allowance made for (18.18) from (18.23) follows:

$$\frac{\partial}{\partial t} N = i\mathbf{H}_0^{(s)} e^{i\mathbf{H}_0^{(s)}(t-t_0)} \mathbf{N}^{(s)} e^{-i\mathbf{H}_0^{(s)}(t-t_0)} - e^{i\mathbf{H}_0^{(s)}(t-t_0)} \mathbf{N}^{(s)} \mathbf{H}_0^{(s)} e^{-i\mathbf{H}_0^{(s)}(t-t_0)} = i[H_0, N], \tag{18.25}$$

where we have taken into account that $\mathbf{H}_0^{(s)} = H_0$. Analogously, for the state vectors we have:

$$i\frac{\partial}{\partial t} \Phi(t) = -\mathbf{H}_0^{(s)} e^{i\mathbf{H}_0^{(s)}(t-t_0)} \mathbf{\Phi}^{(s)}(t) + e^{i\mathbf{H}_0^{(s)}(t-t_0)} (\mathbf{H}_0^{(s)} + S\mathbf{H}_{int}^{(s)}) \mathbf{\Phi}^{(s)}(t) = H_{int}\Phi(t). \tag{18.26}$$

Thus, in the interaction representation both operators and state vectors are varied, the evolution law of operators being defined by the free fields Hamiltonian and that of state vectors—by the interaction Hamiltonian.

Now we establish a connection between the interaction representation and the Heisenberg one. Setting in (18.24)

$$\mathbf{\Phi}^{(s)}(t) = e^{-i\mathbf{H}^{(s)}(t-t_0)} \mathbf{\Phi}, \tag{18.27}$$

we obtain

$$\Phi(t) = S(t, t_0)\mathbf{\Phi}, \tag{18.28}$$

where

$$S(t, t_0) = e^{i\mathbf{H}_0^{(s)}(t-t_0)} e^{-i\mathbf{H}^{(s)}(t-t_0)}. \tag{18.29}$$

Note, that usage in the exponent of the quantity $t - t_0$ ensures the coincidence of state vectors in the interaction and Heisenberg representations at the initial moment of time. From (18.29) it follows that the operator $S(t, t_0)$ obeys the equation:

$$i\frac{\partial}{\partial t} S(t, t_0) = H_{int} S(t, t_0) \tag{18.30}$$

and the condition $S(t_0, t_0) = 1$. It is also an easy matter to verify that this operator relates operators of the Heisenberg representation with those of the interaction representation. With this aim in view one should substitute the expression

$$\mathbf{N}^{(s)} = e^{-i\mathbf{H}^{(s)}(t-t_0)} \mathbf{N} e^{i\mathbf{H}^{(s)}(t-t_0)},$$

into (18.23). As a result we get

$$N = S(t, t_0)\mathbf{N} S^{-1}(t, t_0). \tag{18.31}$$

Now we are quite ready to feel all the advantages of the interaction representation. First, we show that in this representation operators of the electromagnetic and electron-positron fields satisfy equations for the free fields. For this purpose equations for field operators in the commutator form (18.25) are needed. In so doing to find the four-dimensional vector of field momentum we shall use the expression:

$$P_\mu = \int d\sigma^\nu T_{\mu\nu} = \int d\sigma^\nu [\sum_k \frac{\partial \mathcal{L}}{\partial \psi^{k;\nu}} \frac{\partial \psi^k}{\partial x^\mu} - \mathcal{L} g_{\mu\nu}]. \tag{18.32}$$

The usefulness of such a definition implies that on the strength of the conservation law:

$$\partial^\mu T_{\mu\nu} = 0, \tag{18.33}$$

the expression (18.32) does not depend on a choice of a space-like hypersurface σ. To obtain an equation describing behavior of the operator $\psi(x)$ one must set $N(x)$ equal to $\gamma_0 \psi$ in (18.25). Then, all our efforts are reduced to the calculation of the commutator between $\gamma_0 \psi$ and $H_0 = \int d\sigma^\nu T^{(0)}_{0\nu}$ ($T^{(0)}_{\mu\nu}$ is an energy-momentum tensor for the free electromagnetic field). Since a hypersurface σ is arbitrary, there is nothing to prevent us to take it as a hyperplane with $t = t'$. In this case H_0 acquires the view:

$$H_0 = \frac{1}{2} \int_{t-t'} dr' \left[-2 \frac{\partial A^\nu(x')}{\partial t'} \frac{\partial A_\nu(x')}{\partial t'} + \frac{\partial A^\nu(x')}{\partial x^{\mu\prime}} \frac{\partial A_\nu(x')}{\partial x_{\mu\prime}} + \overline{\psi}(x') \left(i\gamma_k \frac{\partial}{\partial x'^k} + \right. \right.$$

$$\left. \left. -i\gamma_k \frac{\overleftarrow{\partial}}{\partial x'^k} + 2m \right) \psi(x') \right], \tag{18.34}$$

where the arrow, above the differential operator ∂_k, directed in the left means the operator acts on a field function standing in the left. Further on, one should take into consideration that at coinciding times operators in the interaction representation satisfy the same PR as do operators in the Heisenberg representation (18.14), (18.15). Then, allowing for the operator relation:

$$[A, BC] = \{A, B\}C - B\{A, C\},$$

we result in

$$[H_0, \gamma_0 \psi(x)] = \frac{1}{2} \int_{t=t'} dr' \left\{ \{\psi^\dagger(x'), \psi(x)\} \left(i\gamma^k \frac{\partial \psi(x')}{\partial x'_k} \right) + \right.$$

$$\left. -i\gamma^k \{ \frac{\partial \psi^\dagger(x')}{\partial x'_k}, \psi(x) \} \psi(x') + 2m \{\psi^\dagger(x'), \psi(x)\} \psi(x') \right\} =$$

$$= -\frac{1}{2} \int_{t=t'} dr' \left\{ \delta^{(3)}(\mathbf{r} - \mathbf{r}') \left(i\gamma^k \frac{\partial \psi(x')}{\partial x'_k} \right) - i\gamma^k \frac{\partial \delta^{(3)}(\mathbf{r} - \mathbf{r}')}{\partial x'_k} \psi(x') + \right.$$

$$\left. +2m\delta^{(3)}(\mathbf{r} - \mathbf{r}')\psi(x') \right\} = -\frac{1}{2} \int_{t=t'} dr' \left\{ 2\delta^{(3)}(\mathbf{r} - \mathbf{r}') \left(i\gamma^k \frac{\partial \psi(x')}{\partial x'_k} \right) - \right.$$

$$-i\gamma^k \frac{\partial}{\partial x'_k} \left(\delta^{(3)}(\mathbf{r} - \mathbf{r}')\psi(x') \right) + 2m\delta^{(3)}(\mathbf{r} - \mathbf{r}')\psi(x') \right\} = -i\gamma^k \frac{\partial \psi(x)}{\partial x_k} - m\psi(x), \tag{18.35}$$

where to reach the last line we have made an allowance for the three-dimensional Gauss theorem. Connecting (18.25) and (18.35) we get

$$[i\gamma^\mu \partial_\mu - m]\psi(x) = 0, \tag{18.36}$$

For $N(x) = \psi^\dagger(x)$ and $N(x) = \partial_0 A_\mu(x)$ analogous operations lead to the equations:

$$[i\gamma^{\mu T} \partial_\mu + m]\overline{\psi}(x) = 0. \tag{18.37}$$

$$\Box A_\mu(x) = 0. \tag{18.38}$$

So, in the interaction representation field operators do satisfy equations for free fields. In order to make the picture complete these equations should be supplemented by the general character PR, that is, by the PR at incoincident moments of time. To change from the simultaneous PR to the general ones could be realized with the help of formulae determining the operators evolution in time, that is, by means of (18.16) and

$$\psi(x) = -i \int d\sigma_\mu(x') S(x - x') \gamma^\mu \psi(x'). \tag{18.39}$$

The insertion of the expression (18.39) into an anticommutator for electron-positron field operators:

$$\{\psi(x), \overline{\psi}(x')\},$$

choice of a hypersurface in the form of a hyperplane $t = t'$, and usage of the simultaneous PR, give us the PR for an arbitrary moment of time:

$$\{\psi(x), \overline{\psi}(x')\} = -iS(x - x'), \tag{18.40}$$

in which we immediately recognize the PR for the free case. In a similar manner, by application of the formulae (18.14) and (18.16) we must assure ourselves that for the electromagnetic field the PR in the interaction representation has the same view as the Heisenberg representation for the free case:

$$[A_\mu(x), A_\nu(x')] = ig_{\mu\nu} D_0(x - x'). \tag{18.41}$$

18.3   Scattering matrix

In quantum field theory, similar to quantum mechanics, two problems are of principal importance. The first of them is reduced to finding a spectrum of the bound states (for example, spectrums of the hydrogen atom, of positronium, of muonium, and so on), and the second problem is connected with investigating both elastic and inelastic particles scattering. Here we shall centre on the latter problem.

Recall statements of the problem about particle collision in nonrelativistic quantum mechanics. It has been assumed that in the initial state at $t = -\infty$, the particles are too distant from each other so that interaction between them is absent, that is, $H_{int} = 0$. Then the particles are drawing together and coming into an interaction. Further, they again move apart, with an infinitely large distance between them, to become free at $t = \infty$. The scattering problem consists in finding the wavefunction for a system of particles at $t = \infty$ provided the wavefunction of a system of free particles at $t = -\infty$ is known. An expansion of the final state wavefunction in terms of the wavefunctions of free particles makes it possible to determine the amplitudes for different scattering processes.

When we use this procedure in the quantum electrodynamic (QED), the situation is greatly complicated. Since we take into account particle interaction effects with a vacuum, the particles are not free both before and after the collision. The only formal possibility to switch off the interaction between particles in the QED is to set a coupling constant between the electron-positron and electromagnetic fields (electron charge) equal to zero. But varying the charge of the electron, we change its other characteristics as well (mass, magnetic moment, and so on). For example, the electron mass is the lowest energy of the state with the charge e in the case of interacting fields. It is naive to think that this quantity will be the same as the bare electron mass. A particle is referred to as "bare" when it has no interaction with the vacuum of the QED whose structure is given by H_{int}. And remaining within the scope of the proposed procedure, we have a possibility to study scattering processes only of bare and not real particles. However, the situation may be improved owing to a convenient method called the *adiabatic hypothesis*. By this hypothesis, the interaction Hamiltonian should be additionally multiplied by a converging factor of the form $\exp(-\alpha|t|)$ that, at the end of our calculations, should be set equal to unity by the limiting transition $\alpha \to 0$. And the scattering process per se is represented as follows.

For $t = -\infty$ a distance between the particles is too large and there is no interaction

$$\exp(-\alpha|t|)H_{int} = 0. \tag{18.42}$$

In this manner we have a system of bare free particles, that is, the system is described by the state vector that is an eigenvector of the operator H_0. Then adiabatic switching-on of the interaction between these particles and vacuum takes place and, as a result, the particles "dress themselves" in fur coats consisting of virtual particles. Despite the fact that at this stage bare particles are transformed into the real ones, as before the particles have no interaction with each other. Following the next full switching-on of the interaction, scattering of the particles takes place. Subsequently, as the particles become more and more distant from each other, they regain their freedom, however keeping their fur coats due to the interaction with the vacuum. At $t = +\infty$ the interaction is adiabatically switched off, whereas the particles (losing their fur coats and being more and more separated) are transformed into free bare particles. Since both switching-on and switching-off of the interaction is adiabatic, the states of a system of fields are subjected only to an adiabatic change that does not lead to the appearance of new states or the disappearance of old ones. This, in turn, is indicative of a complete agreement between the states of bare and real particles and hence between the processes of their scattering. We will come back to the characteristics of real particles later (under discussion of the QED renormalizability), and now we consider real and bare particles, making no distinction between them.

Thus, the scattering problem statement in the QED proves to be the same as in nonrelativistic quantum mechanics. In the process of solving this problem we use the interaction representation. It is easily seen that a formal solution for the state vector equation (18.26) is of the form:

$$\Phi(t) = S(t, t_0)\Phi(t_0). \tag{18.43}$$

Assuming $t_0 = -\infty$ and $t = +\infty$, we can find the linkage between the initial and final state vectors of the field system we are interested in as follows:

$$\Phi(\infty) = S(\infty, -\infty)\Phi(-\infty). \tag{18.44}$$

In (18.44) the operator $S(\infty, -\infty) \equiv S$ that changes the initial state vector $\Phi(-\infty)$ to the final state vector $\Phi(\infty)$ is called the *scattering matrix* or the *S-matrix*. This matrix has been first introduced in Ref. [2] devoted to the problems of nuclei scattering. In 1943 it was again studied in detail by Heisenberg [3] but as applied to a theory of elementary particles. The S-matrix obeys an integral equation:

$$S(t, t_0) = 1 - i \int_{t_0}^{t} H_{int}(t')S(t', t_0)dt', \tag{18.45}$$

which is equivalent to the differential equation (18.30) with the boundary condition

$$S(t_0, t_0) = 1.$$

Having set
$$\Phi(-\infty) = \Phi_i,$$

where the index i denotes an initial state of a system consisting of electrons and photons, we rewrite (18.44) in the form:
$$\Phi(\infty) = S\Phi_i. \tag{18.46}$$

Now we should expand a final state in terms of an orthonormalized set of free field state vectors:
$$\Phi(\infty) = \sum_f a_f \Phi_f. \tag{18.47}$$

Using the orthonormalization condition, from (18.47) we obtain the following expression for the coefficients $a_f \equiv a_{i \to f}$:
$$a_{i \to f} = (\Phi_f, \Phi(\infty)) = (\Phi_f, S\Phi_i) \equiv\, < f|S|i > . \tag{18.48}$$

So, probability amplitudes for different scattering processes are defined by the S-matrix elements.

To determine the explicit form of the S-matrix we address Eq. (18.30) where the interaction Hamiltonian is given by:
$$H_{int}(t) = \int j_\mu(x) A^\mu(x) d\mathbf{r}. \tag{18.49}$$

We shall seek a solution of this equation in the form of series in powers of e:
$$S(t, t_0) = \sum_{n=0}^{\infty} S^{(n)}(t, t_0), \tag{18.50}$$

where the matrix $S^{(n)}$ is proportional to e^n. To substitute (18.50) into (18.30) leads to the equation:
$$i\frac{\partial}{\partial t}\left(S^{(0)} + S^{(1)} + S^{(2)} + S^{(3)} + ...\right) = H_{int}(t)\left(S^{(0)} + S^{(1)} + \right.$$
$$\left. + S^{(2)} + S^{(3)} + ...\right). \tag{18.51}$$

Allowing for that $H_{int} \sim e$, we equate terms at the same powers of e:
$$i\frac{\partial S^{(0)}}{\partial t} = 0, \qquad i\frac{\partial S^{(1)}}{\partial t} = H_{int}(t)S^{(0)}, \qquad i\frac{\partial S^{(2)}}{\partial t} = H_{int}(t)S^{(1)},$$
$$\tag{18.52}$$
$$i\frac{\partial S^{(3)}}{\partial t} = H_{int}(t)S^{(2)}, ...$$

The solutions of the system (18.52) are yielded by the expressions:
$$S^{(0)}(t, t_0) = 1, \qquad S^{(1)}(t, t_0) = -i\int_{t_0}^{t} dt_1 H_{int}(t_1), \qquad S^{(2)}(t, t_0) =$$
$$= -i\int_{t_0}^{t_1} dt_2 H_{int}(t_2)S^{(1)}(t_2) = (-i)^2 \int_{t_0}^{t} dt_1 \int_{t_0}^{t_1} dt_2 H_{int}(t_1)H_{int}(t_2),$$
$$.................. \tag{18.53}$$

$$S^{(n)}(t, t_0) = (-i)^n \int_{t_0}^{t} dt_1 \int_{t_0}^{t_1} dt_2 \int_{t_0}^{t_2} dt_3 ... \int_{t_0}^{t_{n-1}} dt_n H_{int}(t_1) H_{int}(t_2) H_{int}(t_3) ... H_{int}(t_n).$$

The upper integration limits of these integrals are different and lay in the interval (t, t_0). It would be more attractive to transform the integrals in (18.53) so that integration in each of them is carried out over one and the same interval (t_0, t) rather than over different ones $(t_0, t), (t_0, t_1), ...(t_0, t_{n-1})$. The first to make this was F. Dyson [4]. For this purpose to be achieved, it is sufficient to choose the time axis direction, that is, for example, to demand

$$t > t_1 > t_2 > ...t_{n-1} > t_n > t_0.$$

We show how that is being done by the example of the expression for $S^{(2)}(t, t_0)$.

In $S^{(2)}(t, t_0)$ the integration domain is simply the triangle, which is placed below the bisectrix of the coordinate angle in the plane (t_1, t_2) (Fig. 18.1), and for which $t_1 > t_2$. Having fulfilled replacing the variables

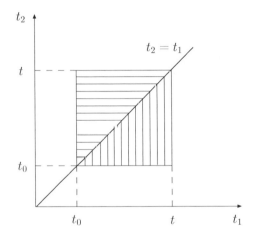

FIGURE 18.1
The integration domain for $S^{(2)}(t, t_0)$.

$$t_1 \to t_2, \qquad t_2 \to t_1,$$

we obtain for $S^{(2)}(t, t_0)$ the expression:

$$S^{(2)}(t, t_0) = -\int_{t_0}^{t} dt_2 \int_{t_0}^{t_2} dt_1 H_{int}(t_2) H_{int}(t_1). \tag{18.54}$$

Let us change the integration order in $S^{(2)}(t, t_0)$, namely, first we shall integrate over t_1. It leads to modification of the integration limits:

$$S^{(2)}(t, t_0) = -\int_{t_0}^{t} dt_1 \int_{t_1}^{t} dt_2 H_{int}(t_2) H_{int}(t_1). \tag{18.55}$$

At present, for the integration domain we have the triangle, placed above bisectrix of the coordinate angle, for which $t_2 > t_1$. To combine the transformed expression (18.55) with the original gives:

$$S^{(2)}(t_0, t) = -\frac{1}{2} \int_{t_0}^{t} dt_1 \left[\int_{t_0}^{t_1} dt_2 H_{int}(t_1) H_{int}(t_2) + \int_{t_1}^{t} dt_2 H_{int}(t_2) H_{int}(t_1) \right]. \tag{18.56}$$

In the case of the commutation of the operators $H_{int}(t_2)$ and $H_{int}(t_1)$ the expression (18.56) would be simply rewritten down in the form of the double integral:

$$S^{(2)}(t_0, t) = -\frac{1}{2} \int_{t_0}^{t} dt_1 \int_{t_0}^{t} dt_2 H_{int}(t_1) H_{int}(t_2),$$

where both integrations be already fulfilled in one interval (t_0, t), that is, over the whole quadrate pictured in Fig. 18.1. However, the operators $H_{int}(t_2)$ and $H_{int}(t_1)$ do not commute and we should think out some trick allowing us to simplify the expression (18.56). Matters could be straightened out by the use of the chronological product operator T. Since the fermion factors enter into the operator $H_{int}(t)$ by pairs in which both factors refer to the same moment of time, we have:

$$T\left(H_{int}(t_1)H_{int}(t_2)\right) = \begin{cases} H_{int}(t_1)H_{int}(t_2), & t_1 > t_2, \\ H_{int}(t_2)H_{int}(t_1), & t_2 > t_1, \end{cases} \tag{18.57}$$

or

$$T\left(H_{int}(t_1)H_{int}(t_2)\right) = \Theta(t_1 - t_2)H_{int}(t_1)H_{int}(t_2) + \Theta(t_2 - t_1)H_{int}(t_2)H_{int}(t_1). \tag{18.58}$$

Then the expression for $S^{(2)}(t, t_0)$ (18.56) could be written in the following compact form:

$$S^{(2)}(t, t_0) = -\frac{1}{2} \int_{t_0}^{t} dt_1 \int_{t_0}^{t} dt_2 T(H_{int}(t_1)H_{int}(t_2)). \tag{18.59}$$

Once we had understood the basic idea, we consider the S-matrix element of the nth-order. Allowing for the chosen time direction, we can represent the expression for $S^{(n)}(t, t_0)$ in the view:

$$S^{(n)}(t, t_0) = (-i)^n \int_{t_0}^{t} dt_1 \int_{t_0}^{t_1} dt_2 ... \int_{t_0}^{t_{n-1}} dt_n \Theta(t_1 - t_2)\Theta(t_2 - t_3)...\Theta(t_{n-1} - t_n) \times$$

$$\times H_{int}(t_1)H_{int}(t_2)...H_{int}(t_n)). \tag{18.60}$$

Assuming, next, that the integration order rearrangement is allowed, we put down $S^{(n)}(t, t_0)$ as:

$$S^{(n)}(t, t_0) = \frac{(-i)^n}{n!} \sum_{P} \int_{t_0}^{t} dt_1 \int_{t_0}^{t} dt_2 ... \int_{t_0}^{t} dt_n \Theta(t_{\alpha_1} - t_{\alpha_2})...\Theta(t_{\alpha_{n-1}} - t_{\alpha_n})H_{int}(t_{\alpha_1})...H_{int}(t_{\alpha_n}),$$

$$\tag{18.61}$$

where the symbol $\sum_P$ denotes the summation over all permutations $(t_1, t_2, ...t_n)$. Taking into account that the Hamiltonian, which contains only bilinear combinations of fermion fields, satisfies the relation:

$$T\left(H_{int}(t_1)H_{int}(t_2)...H_{int}(t_n)\right) = \sum_{P} \Theta(t_{\alpha_1} - t_{\alpha_2})...\Theta(t_{\alpha_{n-1}} - t_{\alpha_n})H_{int}(t_{\alpha_1})...H_{int}(t_{\alpha_n}),$$

$$\tag{18.62}$$

we finally get

$$S^{(n)}(t, t_0) = \frac{(-i)^n}{n!} \int_{t_0}^{t} dt_1 \int_{t_0}^{t} dt_2 ... \int_{t_0}^{t} dt_n T\left(H_{int}(t_1)H_{int}(t_2)...H_{int}(t_n)\right). \tag{18.63}$$

Thus, we have shown that the S-matrix is defined by the following series:

$$S(t, t_0) = \sum_{n=0}^{\infty} \frac{(-i)^n}{n!} \int_{t_0}^{t} dt_1 \int_{t_0}^{t} dt_2 ... \int_{t_0}^{t} dt_n T\left(H_{int}(t_1)H_{int}(t_2)...H_{int}(t_n)\right). \tag{18.64}$$

The formal summation of this series leads to the expression:

$$S(t, t_0) = T \left(\exp \left[-i \int_{t_0}^{t} dt' H_{int}(t') \right] \right).$$ (18.65)

Now, it is high time to stop and check whether in the process of finding $S(t, t_0)$ we have done some unlawful things after which $S(t, t_0)$ is not the formal solution of Eq. (18.30). With this aim in view, we differentiate the expansion (18.64) with respect to t:

$$\frac{\partial S(t, t_0)}{\partial t} = \sum_{n=1}^{\infty} \frac{(-i)^n}{n!} \int_{t_0}^{t} dt_1 \int_{t_0}^{t} dt_2 ... \int_{t_0}^{t} dt_{n-1} n H_{int}(t) T \left(H_{int}(t_1) H_{int}(t_2) ... H_{int}(t_{n-1}) \right).$$ (18.66)

Under writing the right-hand side of the equality (18.66) we have used the integrand expression symmetry and the fact that $t > t_1, ... > t_{n-1}$. All of this taken together has allowed to take out $H_{int}(t)$ for the sign of the T-product and put it on the left of the remaining factors. Thus, the equality (18.66) may be represented in the following form:

$$i \frac{\partial S(t, t_0)}{\partial t} = H_{int}(t) \sum_{n=1}^{\infty} \frac{(-i)^{n-1}}{(n-1)!} \int_{t_0}^{t} dt_1 \int_{t_0}^{t} dt_2 ... \int_{t_0}^{t} dt_{n-1}(t) \times$$

$$\times T \left(H_{int}(t_1) H_{int}(t_2) ... H_{int}(t_{n-1}) \right) = H_{int}(t) \sum_{n=0}^{\infty} \frac{(-i)^n}{n!} \int_{t_0}^{t} dt_1 \int_{t_0}^{t} dt_2 ... \int_{t_0}^{t} dt_n \times$$

$$T \left(H_{int}(t_1) H_{int}(t_2) ... H_{int}(t_n) \right) = H_{int}(t) S(t, t_0),$$ (18.67)

which is what we set out to prove.

To obtain the S-matrix in the scattering problem, in the expansion (18.65) t_0 must tend to $-\infty$ while t must tend to $+\infty$. It gives

$$S = T \left(\exp \left[-i \int_{-\infty}^{\infty} H_{int}(t) dt \right] \right)$$ (18.68)

or, after substitution of the explicit expression for $H_{int}(t)$ (18.12),

$$S = T \left(\exp \left[-i \int_{-\infty}^{\infty} N(j_\mu(x) A^\mu(x)) d^4 x \right] \right).$$ (18.69)

It is obvious that the S_n-element of the scattering matrix represents the T-product of the N-ordered factors. The T-product of such a kind will be called the *mixed T-product*.

Let us focus our attention on some mathematical aspects in calculating the S-matrix. First, transition to the limits $t = \pm\infty$ in the $S^{(n)}$-element may be realized by $n!$ different methods. Second, without the use of wavepackets for the description of the initial and final states, it is necessary to find the way for averaging of the terms periodically dependent on t_0 and t. Nevertheless, all these problems are removed due to the adopted adiabatic hypothesis that necessitates the following substitution in the S-matrix:

$$H_{int}(t) \to H_{int}(t) \exp \left[-\alpha |t| \right],$$ (18.70)

and the transition to the limit $\alpha \to 0$ only after all the integrations are performed. We shall demonstrate that this mathematical procedure also allows painlessly working with unnormalized states of the plane-wave type.

Let us take $S^{(2)}$ from the expansion (18.53) and tend t_0 to $-\infty$ while t to $+\infty$. Further on, for the obtained expression we find the matrix element between the initial and final states: $S_{i \to f}^{(2)} \equiv < f|S^{(2)}|i >$. Passing to the Schrödinger representation and introducing the total system of the eigenstates of H_0 between $H_{int}(t_1)$ and $H_{int}(t_2)$, we arrive at:

$$S_{i \to f}^{(2)} = - \sum_n \int_{-\infty}^{+\infty} dt_1 \int_{-\infty}^{t_1} dt_2 \times$$

$$\times e^{i(E_f - E_n)t_1} e^{i(E_n - E_i)t_2} < f|H_{int}^{(s)}|n >< n|H_{int}^{(s)}|i >, \qquad (18.71)$$

where for the sake of simplicity we have assumed that the Schrödinger and interaction representations coincide at $t_0 = 0$. If $|i >$ is the plane wave state, then the integral over t_2 on the lower limit represents a fast oscillating quantity. To obtain the finite result would be possible when the $|i >$-state be the wave package:

$$|i >= \int \psi_{\mathbf{p}} e^{i(Et - \mathbf{p} \cdot \mathbf{r})} d^3 p \qquad \text{with} \qquad \int |\psi_{\mathbf{p}}|^2 d^3 p < \infty.$$

Then, the contribution from the lower limit $t_2 = -\infty$ would be simply equal to zero according to the Riemann-Lebesgue lemma. The replacement (18.70) proves to lead to the finite result as well. Really, after its performance we have:

$$S_{i \to f}^{(2)} = -2\pi i \delta(E_i - E_f) \sum_n \frac{< f|H_{int}^{(s)}|n >< n|H_{int}^{(s)}|i >}{E_i - E_n + i\alpha}. \qquad (18.72)$$

Using the similar line of reasoning, one may show the validity of the following formula:

$$< f|S|i >= -2\pi i \delta(E_i - E_f) \left\{ < f|H_{int}^{(s)}|i > + \sum_n \frac{< f|H_{int}^{(s)}|n >< n|H_{int}^{(s)}|i >}{E_i - E_n + i\alpha} + \right.$$

$$\left. + ... + \sum_{n_1, n_2 ... n_k} \frac{< f|H_{int}^{(s)}|n_1 >< n_1|H_{int}^{(s)}|n_2 > ... < n_k|H_{int}^{(s)}|i >}{(E_i - E_{n_1} + i\alpha)(E_{n_1} - E_{n_2} + i\alpha)...(E_{n_k} - E_f + i\alpha)} \right\}, \qquad (18.73)$$

in which we immediately get to know the result of the "ancient" (preceded S-matrix) perturbation theory.

Let us demonstrate that the S-matrix defined by the expression (18.69) is covariant. By virtue of the fact that $j_\mu(x)A^\mu(x)$ and d^4x are scalars, and the normal product operation is covariant, the exponent in the expression (18.69) is an invariant. Since the chronological product operator is defined by the invariant way as well, then the relativistic covariance of the S-matrix becomes obvious.

From the definition (18.69) it also follows one more important property of the S-matrix, namely, its unitarity. Actually, due to hermicity of $\mathcal{L}(x)$ the matrix $S^\dagger$ is expressed in the product form of the same factors as S with only two differences, namely, the factors in $S^\dagger$ are taken with the opposite signs and installed in the reverse chronological order. By this reason, under multiplication of S and $S^\dagger$ all the factors pairwise give 1.

The S-matrix must also satisfy the causality condition that reads: "*some event having occurred in a system could have an impact on a system evolution course only in the future and could not have any influence on a system behavior in the past.*" To mathematically formulate this condition it is necessary to introduce the function $g(x)$, which has the value in the interval $(0,1)$ and characterizes intensity of switching-on an interaction in the point x. In the regions where $g(x) = 0$ an interaction is absent, where $g(x) = 1$ it is switched on completely, and under $0 < g(x) < 1$ it is switched on only in part. Replacing $\mathcal{L}_{int}$ with $\mathcal{L}_{int} g(x)$, we arrive at an interaction having been switched on with the intensity $g(x)$ and, accordingly, the scattering matrix $S(g)$ depending on $g(x)$.

Let us illustrate the idea of obtaining the causality condition in the explicit view. To do this, we shall take up the following case. The space-time region G, in which the function $g(x)$ is different from zero, breaks up into two separate subregions G_1 and G_2, all points of G_1 laying in the past as related to some moment of time $t = \tau$ and all points of G_2 laying in the future as related to $t = \tau$ (in our designations it will look like $G_2 > G_1$). In this case the function $g(x)$ is defined as a sum of two functions:

$$g(x) = g_1(x) + g_2(x), \tag{18.74}$$

one of them $g_1(x)$ not being zero only in G_1 while the other $g_2(x)$ nonvanishing only in G_2. Then, at the moment of time τ the state vector Φ_τ not depending on an interaction in the region G_2 will be defined as:

$$\Phi_\tau = S(g_1)\Phi_i, \tag{18.75}$$

where $S(g_1)$ is the scattering matrix that describes switching-on an interaction with the intensity $g_1(x)$, Φ_i is the initial state vector of the system. The final state Φ_f could be obtained from Φ_τ with the help of the operator $S(g_2)$ governing an interaction in the region G_2:

$$\Phi_f = S(g_2)\Phi_\tau. \tag{18.76}$$

Comparing Eqs. (18.74)–(18.76) with

$$\Phi_f = S(g)\Phi_i, \tag{18.77}$$

we get

$$S(g_1 + g_2) = S(g_1)S(g_2) \qquad \text{at} \qquad G_2 > G_1. \tag{18.78}$$

If all the points of the region G_1 are space-like to all the ones of the region G_2 ($G_1 \sim G_2$), then the time order of the regions can always be changed by a corresponding Lorentz transformation, that is:

$$S(g_1)S(g_2) = S(g_2)S(g_1) \qquad G_1 \sim G_2. \tag{18.79}$$

The formulae (18.78) and (18.79) are the essence of the integral principle of the causality. One may formulate the causality condition in a differential form as well. However, these formulae are not a source of any additional information about the S-matrix and in subsequent discussions we shall not use both the integral and differential formulations of the causality condition. We refer to the reader taking an active interest in this question to the book [5] where these questions are considered in detail.

18.4 Representation of the S-matrix as the sum of normal products

Since the operators $\psi(x), \overline{\psi}(x)$ and $A_\mu(x)$ represent the sum of the production and annihilation operators of the corresponding particles, then every term of the S-matrix expansion consists of the products sum of the production and annihilation operators of electrons, positrons, and photons. Let us define conditions under which products of such a kind have nonvanishing matrix elements for the process we are interested in $|i> \rightarrow |f>$. We assume that one photon is present in the initial state $|i> = |1_\gamma>$, and there are electrons and positrons in the final state $|f> = |1_{e^-}; 1_{e^+}>$. As this takes place, in the appropriate nonzero matrix element:

$$S_{i \rightarrow f} = <i|S|f> \tag{18.80}$$

one of the annihilation operators should destroy a photon in the state $|i>$, two production operators should create an electron and a positron in the state $|f>$, and all other operators should be in pairs, the operators in each pair generating and annihilating the same particle. As a source of virtual particles, these paired operators make the calculation of matrix elements for a scattering matrix too complicated. Fortunately, there is a method to take into account similar virtual processes, relating the number of paired operators to the perturbation degree in which a process under study is evaluated. The idea is fairly simple: the scattering matrix should be represented as a sum over normal products of the particle production and annihilation operators. Then in the process of calculations the annihilation operators destroy only the particles in the initial states, whereas the production operators create only the particles to be in the final states. So, our task is to exhibit the mixed product as a sum of the N-products. It could be done employing two Wick's theorems [6].

The first Wick's theorem. The T-product of n linear operators is equal to a sum of their N-products that includes all possible combinations of pairings and the term without pairings, that is,

$$
\begin{aligned}
&T(B_1 B_2 B_3 B_4 ... B_n) = \\
&= N(B_1 B_2 B_3 B_4 ... B_n) + \qquad\qquad\qquad\qquad (a) \\
&N\left(B_1^{s_1} B_2^{s_1} B_3 B_4 ... B_n\right) + N\left(B_1^{s_1} B_2 B_3^{s_1} B_4 ... B_n\right) + \Big\} \\
&+ ... + N\left(B_1^{s_1} B_2 B_3 B_4 ... B_n^{s_1}\right) + ... \qquad\qquad (b) \\
&+ N\left(B_1^{s_1} B_2^{s_1} B_3^{s_2} B_4^{s_2} ... B_n\right) + N\left(B_1^{s_1} B_2^{s_2} B_3^{s_1} B_4^{s_2} ... B_n\right) + \Big\} \\
&+ ... + N\left(B_1^{s_1} B_2 B_3^{s_2} B_4^{s_2} ... B_n^{s_1}\right) + \qquad\qquad (c) \\
&+ ....+ \\
&+ N\left(B_1^{s_1} B_2^{s_2} B_3^{s_2} ... B_k^{s_{n/2}} ... B_{n-1}^{s_{n/2}} B_n^{s_1}\right) + \Big\} \\
&+ ... + N\left(B_1^{s_1} B_2^{s_1} B_3^{s_2} ... B_k^{s_{n/2}} ... B_{n-1}^{s_1} B_n^{s_{n/2}}\right) + ..., \qquad (d)
\end{aligned}
\right\} \quad (18.81)
$$

where: (a) is a normal product without any pairings, (b) is all possible normal products with one pairing, (c) is all possible normal products with two pairings, (d) is a term with all paired operators. Here, by linear operators are meant operators being linear combinations of the production and annihilation operators. When Bose-field operators are paired, then one may place them simply alongside each other. The Fermi-field operators being paired may be placed alongside each other as well, however, in so doing, multiplication on the value $(-1)^p$, where p is a number of operator transpositions to have been carried out, must be fulfilled. One may factor out pairings from the sign of the normal product.

Let us prove the first Wick's theorem. If we perform in each term of the formula (18.81) identical simultaneous permutations of the factors, the equality is not violated. Because of this, without loss of generality, it can be assumed that the operators on the left and right sides of (18.81) are already in chronological order. And the symbol of the T-product on the left side of (18.81) may be omitted. Next, we arrange the operators so that all the annihilation operators be positioned to the right of the production ones, that is, we perform N-ordering of the expression (18.81). To this end, we sequentially rearrange the farthermost left N-unordered production operator with all the annihilation operators that are situated to the left of it. In the process some additional terms with pairing between the rearranged operators according to formula:

$$
AB = T(AB) = N(AB) + A^{s_1} B^{s_1} = \pm BA + A^{s_1} B^{s_1}, \qquad (18.82)
$$

where the sign $+$ $(-)$ is associated with Bose (Fermi) operators, will appear. A similar ordering operation is performed for other unordered production operators too. As a result, the initial T-product is represented as a sum over N-products. These N-products may have the positive or negative sign. At the same time, rearranging the factors under the N-product sign to make them T-ordered again, we obtain the positive sign for all N-products.

A sum of the derived N-products includes only pairings between the pairs of N-unordered operators rather than all the pairings possible. Considering that pairings between the N-ordered operators, being simultaneously the T-ordered ones, are equal to zero, we can state: a sum of N-products in (18.81) involves all the pairings possible. Thus, the first Wick's theorem has been proved.

The second Wick's theorem. A mixed product of the operators equals a sum of their N-products, where the operators are related by all the pairings possible, except for the pairings between the operators within one and the same N-product. The proof is found in an analogy with the first Wick's theorem. It should be remembered that there is no need to interchange the positions of the operators under the same sign of N-product as these operators are already N-ordered and hence their associated pairings are zero. Subsequently, we will use the second Wick's theorem that will be referred to simply as the Wick's theorem.

18.5   Calculation of S-matrix elements

Obviously, we should above all choose the definite normalization of state vectors just as we do in the scattering problems of quantum mechanics. The normalization on the volume unit proves to be the most convenient. To find a state vector with such a normalization we address the limiting process, namely, consider the sequence of state vectors with an unlimitedly growing ordinary norm. First, we carry out this operation for the case of one particle. Let us take the one-particle state vector in the view:

$$\Phi_1 = \int d^3p \phi_\sigma(\mathbf{p}) a_\sigma^\dagger(\mathbf{p}) \Phi_0, \tag{18.83}$$

where the symbol σ is used to set the plurality of such quantum numbers as mass, spin projection, and electric and lepton charges. It is clear that the norm of this state equals

$$\Phi_1^\dagger \Phi_1 = \int d^3p |\phi_\sigma(\mathbf{p})|^2 - N. \tag{18.84}$$

Then, assuming $N = 1$, one may state that the expression:

$$|\phi_\sigma(\mathbf{p})|^2 d^3p$$

gives the probability that the particle characterized by the quantum number σ possesses the momentum in the interval d^3p about the average value $\mathbf{p}$. It, in its turn, means that the function $\phi_\sigma(\mathbf{p})$ is a wavefunction in the momentum representation. Therefore, its Fourier image:

$$\psi_\sigma(\mathbf{r}) = \frac{1}{(2\pi)^{3/2}} \int d^3p \phi_\sigma(\mathbf{p}) e^{i\mathbf{p}\mathbf{r}} \tag{18.85}$$

is nothing but the wavefunction in the coordinate representation. The norm of $\psi_\sigma(\mathbf{r})$ appears to be equal to:

$$\int d^3x |\psi_\sigma(\mathbf{r})|^2 = \int d^3p |\phi_\sigma(\mathbf{p})|^2 = N. \tag{18.86}$$

It is clear, at $N = 1$ the quantity $dw_x = |\psi_\sigma(\mathbf{r})|^2 d^3x$ should be treated as the probability of finding a particle with quantum numbers σ in an infinitesimal element of volume d^3x. When N being equal to the particles number ($N \gg 1$), dw_x defines a mean number of particles in the volume d^3x. Now, we shall unlimitedly increase the norm N in such a way as to tend $\phi_\sigma(\mathbf{p})$ to the expression:

$$(2\pi)^{3/2}\delta^{(3)}(\mathbf{p} - \mathbf{p}_0). \tag{18.87}$$

As a result, we obtain the wave package with more and more decreasing spread of momentum about the central value $\mathbf{p}_0$. Then, from (18.85) we find:

$$|\phi_\sigma(\mathbf{p})| \to 1, \tag{18.88}$$

and as a consequence:

$$|\phi_\sigma(\mathbf{p})|^2 \to 1. \tag{18.89}$$

The relation (18.89) means that in the limit the volume unit contains one particle. Passing in (18.83) to the limit, we get the following expression for the one-particle state vector normalized by the volume unit:

$$\Phi_1 = (2\pi)^{3/2}a_\sigma^\dagger(\mathbf{p})\Phi_0. \tag{18.90}$$

In the case of particle k-kinds one should consider the expression:

$$\Phi_k = \int d^3p_1\phi_{\sigma_1}(\mathbf{p}_1)a_{\sigma_1}^\dagger ... \int d^3p_k\phi_{\sigma_k}(\mathbf{p}_k)a_{\sigma_k}^\dagger\Phi_0, \tag{18.91}$$

rather than Eq. (18.85). The norm of such a state vector represents the k product of the one-particle states norm:

$$\Phi_k^\dagger\Phi_k = \prod_{j=1}^{k}\int d^3p_j|\phi_{\sigma_j}(\mathbf{p}_j)|^2 = \prod_{j=1}^{k}N_j. \tag{18.92}$$

It is obvious that the above-stated reasoning scheme could be transferred to every factor in (18.91) and (18.92). Then the multiparticle state vector, which is normalized by the volume unit for every particle being in this state, is determined by the expression:

$$\Phi_k = (2\pi)^{3k/2}a_{\sigma_1}^\dagger(\mathbf{p}_1)a_{\sigma_2}^\dagger(\mathbf{p}_2)...a_{\sigma_k}^\dagger(\mathbf{p}_k)\Phi_0. \tag{18.93}$$

After all these preparations we could pass at once to the determination of the S-matrix elements. For this purpose we use Wick's theorem. As an example, we consider the Compton-effect on a free electron

$$e^-(p) + \gamma(k) \to e^-(p') + \gamma(k'), \tag{18.94}$$

where four-dimensional momenta of particles are denoted by letters in parentheses. The final and initial state vectors normalized by the volume unit have the form:

$$|i> = \Phi_i = (2\pi)^3 a_r^{(e)\dagger}(\mathbf{p})a_s^{(\gamma)\dagger}(\mathbf{k})|0>,$$

$$|f> = \Phi_f = (2\pi)^3 a_{r'}^{(e)\dagger}(\mathbf{p}')a_{s'}^{(\gamma)\dagger}(\mathbf{k}')|0>,$$

where the production operators of the electrons and photons are supplied by the superscripts e and γ, respectively. Thus, the matrix element is set down in the form:

$$S_{i\to f} = <f|\left(\sum_n S^{(n)}\right)|i> = I + S_{i\to f}^{(1)} + S_{i\to f}^{(2)} + S_{i\to f}^{(3)} + ... \tag{18.95}$$

In virtue of (18.69) and Wick's theorem, the expression for the matrix element in the first order of the perturbation theory will look like:

$$S^{(1)}_{i \to f} = -ie(2\pi)^6 \int d^4x < 0|a^{(e)}_{r'}(\mathbf{p}')a^{(\gamma)}_{s'}(\mathbf{k}')N\left(\overline{\psi}(x)\gamma^\mu A_\mu(x)\psi(x)\right)a^{(e)\dagger}_r(\mathbf{p})a^{(\gamma)\dagger}_s(\mathbf{k})|0>.$$

(18.96)

Taking into account the vacuum state vector normalization condition and commutativity of the operators of the electron-positron and photon fields, one could report the expression (18.96) as a product of two parts:

$$S^{(1)}_{i \to f} = -ie(2\pi)^6 \int d^4x \left[< 0|a^{(e)}_{r'}(\mathbf{p}')N\left(\overline{\psi}(x)\gamma^\mu\psi(x)\right)a^{(e)\dagger}_r(\mathbf{p})|0> \right] \times$$

$$\times \left[< 0|a^{(\gamma)}_{s'}(\mathbf{k}')A_\mu(x)a^{(\gamma)\dagger}_s(\mathbf{k})|0> \right] =$$

$$= -ie(2\pi)^6 \int d^4x \left\{ < 0|a^{(e)}_{r'}(\mathbf{p}')[\overline{\psi^{(+)}}(x)\gamma^\mu\psi^{(+)}(x) + \overline{\psi^{(-)}}(x)\gamma^\mu\psi^{(+)}(x) + \right.$$

$$\left. + \overline{\psi^{(+)}}(x)\gamma^\mu\psi^{(-)}(x) - \psi^{(-)}(x)\gamma^{\mu T}\overline{\psi^{(-)}}(x)]a^{(e)\dagger}_r(\mathbf{p})|0> \right\} \times$$

$$\times \left[< 0|a^{(\gamma)}_{s'}(\mathbf{k}')A_\mu(x)a^{(\gamma)\dagger}_s(\mathbf{k})|0> \right],$$

(18.97)

where we have allowed for both the electron-positron field operators expansion into the positive and negative frequency parts and the relation:

$$N(\overline{\psi^{(-)}}_\alpha(x)(\gamma^\mu)_{\alpha\beta}\psi^{(-)}_\beta(x)) = -\psi^{(-)}_\beta(x)(\gamma^\mu)^T_{\beta\alpha}\overline{\psi^{(-)}}_\alpha(x) = -\psi^{(-)}(x)\gamma^{\mu T}\overline{\psi^{(-)}}(x).$$

Let us calculate the individual contributions from all the four terms entering into the electron-positron part. Consider the first term. Since the relations:

$$\psi^{(+)}(x)|0> = < 0|\overline{\psi^{(+)}}(x) = 0,$$

take place, we should carry the operator $\psi^{(+)}(x)$ to the right until it acts on $|0>$ and the operator $\overline{\psi^{(+)}}(x)$ to the left up to the state $< 0|$. The calculation gets:

$$< 0|a^{(e)}_{r'}(\mathbf{p}')\overline{\psi^{(+)}}(x)\gamma^\mu\psi^{(+)}(x)a^{(e)\dagger}_r(\mathbf{p})|0> = \frac{1}{(2\pi)^3}\sum_m\sum_n\int d^3p_1\int d^3p_2\sqrt{\frac{m^2}{E_{p_1}E_{p_2}}}\times$$

$$[< 0|\left(\{a^{(e)}_{r'}(\mathbf{p}'),a^{(e)\dagger}_m(\mathbf{p}_1)\} - a^{(e)\dagger}_m(\mathbf{p}_1)a^{(e)}_{r'}(\mathbf{p}')\right)\overline{u}^{(m)}(\mathbf{p}_1)e^{ip_1x}]\gamma^\mu\times$$

$$\times [u^{(n)}(\mathbf{p}_2)e^{-ip_2x}\left(\{a^{(e)}_n(\mathbf{p}_2),a^{(e)\dagger}_r(\mathbf{p})\} - a^{(e)\dagger}_r(\mathbf{p})a^{(e)}_n(\mathbf{p}_2)\right)|0>] =$$

$$= \frac{1}{(2\pi)^3}\sqrt{\frac{m^2}{E_pE_{p'}}}e^{i(p'-p)x}\overline{u}^{(r')}(\mathbf{p}')\gamma^\mu u^{(r)}(\mathbf{p}),$$

(18.98)

where we have used the expansions of the operators $\psi(x)$ and $\overline{\psi}(x)$ into the Fourier integrals and anticommutation relations for the operators $a^{(e)}_n(\mathbf{p})$ and $a^{(e)\dagger}_r(\mathbf{p})$.

The similar manipulation leads to the conclusion that all the remaining terms in the electron-positron factor give zero contribution.

Now we calculate the photon factor entering to (18.97). Using the expansion of A_μ into the Fourier integral, we obtain

$$< 0|a^{(\gamma)}_{s'}(\mathbf{k}')A_\mu(x)a^{(\gamma)\dagger}_s(\mathbf{k})|0> = \frac{1}{(2\pi)^{3/2}}\int\frac{d^3k_1}{\sqrt{2k_{10}}} < 0|a^{(\gamma)}_{s'}(\mathbf{k}')\sum_{\lambda=0}^3\times$$

$$\times [e_\mu^\lambda(\mathbf{k}_1) a_\lambda^{(\gamma)}(\mathbf{k}_1) e^{-ik_1 x} + e_\mu^{\lambda*}(\mathbf{k}_1) a_\lambda^{(\gamma)\dagger}(\mathbf{k}_1) e^{ik_1 x}] a_s^{(\gamma)\dagger}(\mathbf{k}) |0> =$$

$$= \frac{1}{(2\pi)^{3/2}} \int \frac{d^3 k_1}{\sqrt{2k_{10}}} \sum_{\lambda=0}^{3} \left\{ e_\mu^\lambda(\mathbf{k}_1) <0| a_{s'}^{(\gamma)}(\mathbf{k}') e^{-ik_1 x} \left([a_\lambda^{(\gamma)}(\mathbf{k}_1), a_s^{(\gamma)\dagger}(\mathbf{k})] + \right. \right.$$

$$\left. + a_s^{(\gamma)\dagger}(\mathbf{k}) a_\lambda^{(\gamma)}(\mathbf{k}_1) \right) |0> + e_\mu^{\lambda*}(\mathbf{k}_1) <0| \left([a_{s'}^{(\gamma)}(\mathbf{k}'), a_\lambda^{(\gamma)\dagger}(\mathbf{k}_1)] + \right.$$

$$\left. \left. + a_\lambda^{(\gamma)\dagger}(\mathbf{k}_1) a_{s'}^{(\gamma)}(\mathbf{k}') \right) e^{ik_1 x} a_s^{(\gamma)\dagger}(\mathbf{k}) |0> \right\} = 0. \tag{18.99}$$

Thus, the matrix element of the process (18.94) in the first order of the perturbation theory becomes zero. To put this another way, in the first order of the perturbation theory the Compton-effect does not proceed.

Now we pass to the definition of the matrix element of the process (18.94) in the second order of the perturbation theory. According to Wick's theorem we have

$$< f|S_{i \to f}^{(2)}|i > = < f| \frac{(-ie)^2}{2} \int d^4 x_1 \int d^4 x_2 \left\{ N \left(\overline{\psi}(x_1) \gamma^\mu \psi(x_1) \overline{\psi}(x_2) \gamma^\nu \psi(x_2) \right) + \right.$$

$$+ N \left(\overline{\psi}^{s_1}(x_1) \gamma^\mu \psi(x_1) \overline{\psi}(x_2) \gamma^\nu \psi^{s_1}(x_2) \right) +$$

$$+ N \left(\overline{\psi}(x_1) \gamma^\mu \psi^{s_1}(x_1) \overline{\psi}^{s_1}(x_2) \gamma^\nu \psi(x_2) \right) +$$

$$\left. + \overline{\psi}^{s_2}(x_1) \gamma^\mu \psi^{s_1}(x_1) \overline{\psi}^{s_1}(x_2) \gamma^\nu \psi^{s_2}(x_2) \right\} \left\{ N(A_\mu(x_1) A_\nu(x_2)) + A_\mu^{s_1}(x_1) A_\nu^{s_1}(x_2) \right\} |i> . \tag{18.100}$$

In the factor containing only electron-positron operators the second and third terms are equivalent. To make sure in this we shall do the following operations: (i) rearrange the operators $\overline{\psi}(x_2) \gamma^\nu \psi(x_2)$ and $\overline{\psi}(x_1) \gamma^\mu \psi(x_1)$ in the second term under the sign of the normal product; (ii) carry out the replacement of the integration variables $x_1 \leftrightarrow x_2$; (iii) rename the summation indices $\mu \leftrightarrow \nu$; and (iv) take into account the relation:

$$N(A_\mu(x_1) A_\nu(x_2)) = N(A_\nu(x_1) A_\mu(x_2)).$$

As a result we get the expression to coincide with the third term.

Using the expansion of the operators of the electron-positron field and the anticommutation relations one could show that the terms

$$N(\overline{\psi}(x_1) \gamma^\mu \psi(x_1) \times \overline{\psi}(x_2) \gamma^\nu \psi(x_2)) \qquad \text{and} \qquad \overline{\psi}^{s_1}(x_1) \gamma^\mu \psi^{s_2}(x_1) \overline{\psi}^{s_2}(x_2) \gamma^\nu \psi^{s_1}(x_2)$$

give zero contribution. The term $A_\mu^{s_1}(x_1) A_\nu^{s_1}(x_2)$ entering into the photon factor yields no contribution too. Thus, all that remains is to evaluate the term:

$$S_{i \to f}^{(2)} = -e^2 (2\pi)^6 \int d^4 x_1 \int d^4 x_2 \left\{ <0| a_{r'}^{(e)}(\mathbf{p}') N \left(\overline{\psi}(x_1) \gamma^\mu \psi^{s_1}(x_1) \overline{\psi}^{s_1}(x_2) \gamma^\nu \psi(x_2) \right) \times \right.$$

$$\left. \times a_r^{(e)\dagger}(\mathbf{p}) |0> \right\} \left\{ <0| a_{s'}^{(\gamma)}(\mathbf{k}') N(A_\mu(x_1) A_\nu(x_2)) a_s^{(\gamma)\dagger}(\mathbf{k}) |0> \right\}. \tag{18.101}$$

From the calculation of the matrix element in the first order of the perturbation theory it follows that the factor $N(\overline{\psi}(x_1) \gamma^\mu \psi^{s_1}(x_1) \overline{\psi}^{s_1}(x_2) \gamma^\nu \psi(x_2))$ will produce only one nonvanishing term:

$$\overline{\psi^{(+)}}(x_1) \gamma^\mu \psi^{s_1}(x_1) \overline{\psi}^{s_1}(x_2) \gamma^\nu \psi^{(+)}(x_2).$$

To substitute the Fourier-expansions of the electron-positron field operators and the pairings between the operators leads to the following expression for the first factor in (18.100):

$$\frac{i}{(2\pi)^7}\sqrt{\frac{m^2}{E_p E_{p'}}}\overline{u}^{(r')}(\mathbf{p'})e^{ip'x_1}\gamma^\mu \int d^4 p_1 \frac{\hat{p}_1 + m}{p_1^2 - m^2}e^{-ip_1(x_1-x_2)}\gamma^\nu u^{(r)}(\mathbf{p})e^{-ipx_2}. \qquad (18.102)$$

Integration over the four-dimensional momentum is obliged by the chronological pairing of operators. In the general case of the n-pairings of arbitrary particle operators n-multiple integral over momenta of corresponding particles will appear.

Having taken advantage of the electromagnetic field operators expansion in terms of the positive and negative frequency parts, we find the expression for the photon factor:

$$A_\mu(x_1)A_\nu(x_2) = A_\mu^{(+)}(x_1)A_\nu^{(+)}(x_2) + A_\mu^{(+)}(x_1)A_\nu^{(-)}(x_2) + A_\mu^{(-)}(x_1)A_\nu^{(+)}(x_2)+$$

$$+A_\mu^{(-)}(x_1)A_\nu^{(-)}(x_2). \qquad (18.103)$$

The operators $A_\mu^{(+)}(x_1)A_\nu^{(+)}(x_2)$ and $A_\mu^{(-)}(x_1)A_\nu^{(-)}(x_2)$ do not yield any contribution to the matrix element. Allowing for the Fourier-integral expansion of $A_\mu^{(\pm)}(x)$ and the commutation relations for the production and annihilation photon operators, we obtain the contributions from the second and third terms in the photon factor (18.100):

$$< 0|a_{s'}^{(\gamma)}(\mathbf{k'})A_\mu(x_1)A_\nu(x_2)a_s^{(\gamma)\dagger}(\mathbf{k})|0 >= \frac{1}{(2\pi)^3}\frac{1}{\sqrt{2k_0 k_0'}}\left(e_\mu^{s'*}(\mathbf{k'})e^{ik'x_1}e_\nu^s(\mathbf{k})e^{-ikx_2}+\right.$$

$$\left.+e_\nu^{s'*}(\mathbf{k'})e^{ik'x_2}e_\mu^s(\mathbf{k})e^{-ikx_1}\right). \qquad (18.104)$$

Insert (18.102) and (18.104) into (18.100). As a result we arrive at the expression maintaining the coordinates x_1 and x_2 in the index of the exponent only. Then integration over the variables x_1 and x_2 leads to the appearance of the δ-functions:

$$\int d^4 x_1 \int d^4 x_2 e^{ix_1(p'+k'-p_1)}e^{ix_2(p_1-p-k)} = (2\pi)^8\delta^{(4)}(p'+k'-p_1)\delta^{(4)}(p_1-p-k), \quad (18.105)$$

$$\int d^4 x_1 \int d^4 x_2 e^{ix_1(p'-k-p_1)}e^{ix_2(p_1-p+k')} = (2\pi)^8\delta^{(4)}(p'-k-p_1)\delta^{(4)}(p_1-p+k'), \quad (18.106)$$

where the expression (18.105) arises at multiplication of the electron factor by the first term in (18.104) while the one (18.106) does at multiplication of the electron factor with the second term from (18.104). Every δ-function entering either into (18.105) or into (18.106) ensures the four-dimensional momentum conservation in the diagram vertices x_1 and x_2. To integrate over the space-time coordinates means the passage to the momentum representation. Therefore, having substituted (18.102), (18.104) into (18.100) and having integrated over the variables x_1 and x_2, we come to the expression for the Compton-effect matrix element in the second order of the perturbation theory in the momentum representation:

$$S_{i\to f}^{(2)} = -ie^2(2\pi)^4 \int d^4 p_1 \sqrt{\frac{m^2}{4E_p E_{p'}k_0 k_0'}}\left\{\overline{u}^{(r')}(\mathbf{p'})\gamma^\mu\delta^{(4)}(p'+k'-p_1)\frac{\hat{p}_1+m}{p_1^2-m^2}\gamma^\nu\times\right.$$

$$\times\delta^{(4)}(p+k-p_1)u^{(r)}(\mathbf{p})e_\mu^{s'*}(\mathbf{k'})e_\nu^s(\mathbf{k}) + \overline{u}^{(r')}(\mathbf{p'})\gamma^\mu\delta^{(4)}(p'-k-p_1)\frac{\hat{p}_1+m}{p_1^2-m^2}\gamma^\nu\times$$

$$\left.\times\delta^{(4)}(p-k'-p_1)u^{(r)}(\mathbf{p})e_\nu^{s'*}(\mathbf{k'})e_\mu^s(\mathbf{k})\right\} =$$

$$= -ie^2(2\pi)^4\sqrt{\frac{m^2}{4E_pE_{p'}k_0k_0'}}\left\{\overline{u}^{(r')}(\mathbf{p}')\left[e_\mu^{s'*}(\mathbf{k}')\gamma^\mu\frac{\hat{p}+\hat{k}+m}{(p+k)^2-m^2}e_\nu^s(\mathbf{k})\gamma^\nu+\right.\right.$$

$$\left.\left.+e_\mu^s(\mathbf{k})\gamma^\mu\frac{\hat{p}-\hat{k}'+m}{(p'-k)^2-m^2}e_\nu^{s'*}(\mathbf{k}')\gamma^\nu\right]u^{(r)}(\mathbf{p})\right\}\delta^{(4)}(p+k-p'-k'). \qquad (18.107)$$

In the obtained expression the remaining δ-function expresses the conservation law of the four-dimensional momentum. Note that in the second order of the perturbation theory one manages to get rid of the integration over the four-dimensional momentum but this is not the case when $n \geq 3$.

One may continue the process of finding the Compton-effect matrix elements in the higher-orders of the perturbation theory. However, we have already achieved the main goal. We have learned the process of obtaining the matrix elements of the scattering matrix on the basis of general analytical expression for the matrix element in the given order of the perturbation theory by application of the Wick's theorem. The scheme of finding $S_{i\rightarrow f}^{(n)}$ is as follows: (i) the initial and final state vectors ($|i>$ and $|f>$) are built; (ii) with the help of the Wick's theorem the scattering matrix $S^{(n)}$ is reduced to the normal form; (iii) the matrix element of the process is found and the terms giving nonzero contributions are selected; (iv) commutating the annihilation (production) operators entering to the S-matrix with the production (annihilation) operators belonging to $|i>$ ($<f|$) is fulfilled; (v) integration over four-dimensional coordinates is carried out (it is equivalent to the passage to the momentum representation); and (vi) to integrate over four-dimensional momenta of virtual particles is accomplished.

The above-stated procedure may be shaped to a more convenient form that allows, not addressing to Wick's theorem each time, to find the S-matrix elements at once. To accomplish this it is necessary to formulate rules that put in correspondence to each normal product a particular graph, a diagram.

18.6   Feynman rules in coordinate space

Let us set the correspondence rules between the scattering matrix elements and graphical images. In the QED the analytical expression for the scattering matrix elements $S^{(n)}$ represents a combination including functions corresponding to final and initial particles, pairings of the electron-positron and electromagnetic field operators, and some number of γ-matrices. It is clear that we ought to specify the graphical image of the above-mentioned elements.

Let us agree to correlate four-dimensional vectors $x_1, x_2, ...x_n$, on which the integration in $S^{(n)}$ is accomplished, with diagram points (diagram vertices). As far as field operators are concerned, they will be associated with lines passing through these vertices. On the strength of the local character of the interaction Lagrangian (18.12), in every diagram vertex one photon line, one incoming and one outgoing electron or positron lines come together. Since the photon field operator $A_\mu(x)$ by its geometrical nature is the four-dimensional vector, then a vertex is also characterized by a corresponding four-dimensional index μ. We agree to correlate the vertex with the factor $-ie\gamma^\mu$. Obviously, the number of vertices coincides with the order of the perturbation theory. The operator $A_\mu(x)$ will be displayed by the waved line without a definite direction with the origin in the vertex x and the index μ; the operator $\psi(x)$—by the directed solid line incoming to the vertex x; the operator $\overline{\psi}(x)$—by the directed solid line outgoing from the vertex x. All these lines are called the *external ones* and they describe real particles. Since $A_\mu(x)$ represents the sum of the production and annihilation operators for photons, then a waved external line displays a photon emitted or absorbed as a result of the scattering process. In a similar way, because $\psi(x)$ is the sum of the electron annihilation operator and the positron production operator, then the solid line directed to a vertex will denote an electron existing before scattering or a positron to come into being as a result of the scattering process. As $\overline{\psi}(x)$ comprises the sum of the electron production operator and the positron annihilation operator, then a solid line outgoing from a vertex will display either a positron in a initial state or an electron in a final state.

Let us agree to confront pairings between operators with internal lines of a diagram and consider that these lines correspond to the propagation of virtual particles. The pairing of the photon operators:

$$A_\mu(x)^s A_\nu^s(y) = ig_{\mu\nu} D_0^c(x - y)$$

will be depicted by a nondirectional waved line connecting the vertices x and y in which the matrices γ^μ and γ^ν are operating, respectively. As this takes place, we consider this line as a display of photon motion between the points x and y. Nonsymmetric pairing:

$$\psi^s(x)\overline{\psi}^s(y) = -iS^c(x - y)$$

is confronted with a solid diagram line to connect the vertices x and y. From the meaning of the operators $\psi(x)$ and $\overline{\psi}(y)$ it follows that this line could be interpreted as an electron production in the point y and its subsequent annihilation in the point x (and vice versa for a positron). Therefore, it is expedient to choose the direction of this line from the point y to the point x.

In the case of an interaction with an external (classical) electromagnetic field the Lagrangian contains an additional term:

$$\mathcal{L}_{int}^e = -e : \overline{\psi}_k(x)\,(\gamma^\mu)_{kl}\,\psi_l(x) : A_\mu^{(e)}(x). \tag{18.108}$$

Interaction with an external field will be displayed by a waved line to go out of a diagram vertex and come to an end in a shaded circle that symbolically denotes the scattering center. In the scattering matrix element the factor $A_\mu^{(e)}(x)$ (μ is a vertex index) is associated with this line. Let us consider, for example, electron scattering by an external field in the first order of the perturbation theory. The corresponding matrix element is given by the expression:

$$S_{i \to f} = -ie \int d^4x \sqrt{\frac{m^2}{E_p E_{p'}}}\, \overline{u}_{r'}(\mathbf{p}')A_\mu^e(x)\gamma^\mu u_r(\mathbf{p})e^{i(p'-p)x}. \tag{18.109}$$

If we substitute the Fourier expansion of an external potential:

$$A_\mu^e(x) = \frac{1}{(2\pi)^{3/2}} \int d^4q A_\mu^e(q)e^{-iqx}, \tag{18.110}$$

into (18.109), then, having integrated over x, we get

$$S_{i \to f} = -ie(2\pi)^{3/2}\sqrt{\frac{m^2}{E_p E_{p'}}} \int d^4q \overline{u}_{r'}(\mathbf{p}')A_\mu^e(q)\gamma^\mu u_r(\mathbf{p})\delta(p + q - p'). \tag{18.111}$$

The presence of the δ-function leads to equality $q = p' - p$. Since $q^2 = 0$ for a photon and the square $(p' - p)^2$ certainly differs from zero, then such a equality for real particles is impossible in principle. This suggests that an external field may be replaced by the effective virtual photon exchanged by the interacting particle with a source of the external field. It is important that such an interpretation for interaction with an external electromagnetic field is in line with physics of the phenomenon. An external field differs from a quantized one only in that it is fixed, its state remaining unchanged in the process of interactions. An external field may be represented as a limit of the average value for some quantized field when its intensity is infinitely increasing. Then interaction with this field and in the limit is realized by the virtual photons, for which the process of propagation in the limiting case is described by the given potential $A_\mu^e(x)$ and not by the causal photon function $D_0^c(x_c - x_d)$, where x_c (x_d) are the coordinates of the photon production (destruction) point.

The above correspondence rules between the factors of the $S^{(n)}$-matrix element and Feynman diagrams are known as the Feynman rules in the coordinate representation [7]. Using these rules and constructing all possible diagrams for the process under study in the particular order of the perturbation theory, one may obtain an analytical expression for $S^{(n)}$, and then go to the calculation of $S_{i \to f}^{(n)}$. But there are a few aspects to be taken into consideration.

First, a situation is possible when several normal products are distinguished only by the places of imposing couplings between the field operators and make identical contributions to the scattering matrix element. We shall designate them as the equivalent normal products. The number of the equivalent N-products associated with given diagram is found as follows. Vertices of the diagram are labeled with numbers 1,2,..n. Then we count the permutations number for these numbers, leaving the diagram form unaltered. Provided the count gives the value g, the number of the equivalent N-products r is determined by the relation:

$$r = n!/g. \tag{18.112}$$

Second, each normal product associated with a particular physical process may be represented as a sum of several addends, including the products of the production and annihilation operators of the particles, which take part in the process studied, and being distinguished only by the arrangement order of the operators. These addends are represented as the diagrams differing only by the positions of electron, positron, and photon lines. Such diagrams are known as the *topologically nonequivalent diagrams* (TND).

To illustrate, let us consider the Compton-effect in the second-order of the perturbation theory. As follows from (18.107), a normal product is divided into two parts associated with two diagrams, which differ from each other only by the positioning of the photon lines representing the initial and final photons. When the process involves n photons, after the normal product is divided into the addends including the photon production and annihilation operators, we have $n!$ addends corresponding to $n!$ TND.

Similarly, for the process involving several electrons and positrons its normal product is represented as a sum over the terms including the same electron and positron production and annihilation operators, being distinguished only by the arrangement order of these operators. For example, two TND displayed in Fig. 18.2, where $p_{1,2}$ ($p'_{1,2}$) are the four-dimensional momenta of the initial (final) electrons, k and k' are the four-dimensional momenta of the virtual photons, corresponds to the process:

$$e^- + e^- \to e^- + e^- \tag{18.113}$$

As distinct from the processes involving several photons, for which TND are associated with the matrix elements of the same sign, for the processes involving several electrons (positrons), TND are associated with the matrix elements that may have opposite signs. This is due to the fact that the operators of an electron-positron field satisfy the anticommutation relations.

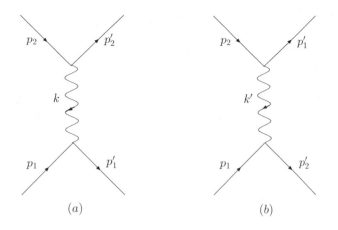

(a)   (b)

FIGURE 18.2

The Feynman diagrams (a) and (b) corresponding to the process $e^- + e^- \to e^- + e^-$.

Next, we show that matrix elements of TND (TND).for the process (18.113) have the opposite signs. The corresponding matrix element in the second order of the perturbation theory has the form:

$$S^{(2)}_{i \to f} - \frac{io^2(2\pi)^6}{2} < 0|u^{(e)}_{r'_1}(\mathbf{p}'_1)a^{(e)}_{r'_2}(\mathbf{p}'_2)[\int dx_1 \int dx_2 \; g_{\mu\nu}D^c_0(x_1 - x_2)\times$$

$$\times N\left(\overline{\psi}(x_1)\gamma^\mu\psi(x_1)\overline{\psi}(x_2)\gamma^\nu\psi(x_2)\right)]a^{(e)\dagger}_{r_2}(\mathbf{p}_2)a^{(e)\dagger}_{r_1}(\mathbf{p}_1)|0 > . \qquad (18.114)$$

Substituting the Fourier-series expansions of the electron field operators and allowing for the anticommutation relations, we arrive at:

$$S^{(2)}_{i \to f} = \frac{-ie^2}{2}\sqrt{\frac{m^4}{E_{p_1}E_{p'_1}E_{p_2}E_{p'_2}}}\int dx_1 \int dx_2 \; g_{\mu\nu}D^c_0(x_1 - x_2)\times$$

$$\times \Big\{ \overline{v}^{(r'_2)}(\mathbf{p}'_2)\gamma^\mu v^{(r_2)}(\mathbf{p}_2)\overline{v}^{(r'_1)}(\mathbf{p}'_1)\gamma^\nu v^{(r_1)}(\mathbf{p}_1)e^{i(p'_2-p_2)x_1+i(p'_1-p_1)x_2}+$$

$$+\overline{v}^{(r'_1)}(\mathbf{p}'_1)\gamma^\mu v^{(r_1)}(\mathbf{p}_1)\overline{v}^{(r'_2)}(\mathbf{p}'_2)\gamma^\nu v^{(r_2)}(\mathbf{p}_2)e^{i(p'_2-p_2)x_2+i(p'_1-p_1)x_1}-$$

$$-\overline{v}^{(r'_1)}(\mathbf{p}'_1)\gamma^\mu v^{(r_2)}(\mathbf{p}_2)\overline{v}^{(r'_2)}(\mathbf{p}'_2)\gamma^\nu v^{(r_1)}(\mathbf{p}_1)e^{i(p'_1-p_2)x_1+i(p'_2-p_1)x_2}-$$

$$-\overline{v}^{(r'_2)}(\mathbf{p}'_2)\gamma^\mu v^{(r_1)}(\mathbf{p}_1)\overline{v}^{(r'_1)}(\mathbf{p}'_1)\gamma^\nu v^{(r_2)}(\mathbf{p}_2)e^{i(p'_2-p_1)x_1+i(p'_1-p_2)x_2} \Big\} . \qquad (18.115)$$

Note that the common sign of this sum is conditional and is defined by the assignment way of the initial and final states (or, what is the same, by the arrangement order of external electron operators). So, for example, there is nothing to prevent us to take $|i >= (2\pi)^3 a^{(e)\dagger}_{r_1}(\mathbf{p}_1)a^{(e)\dagger}_{r_2}(\mathbf{p}_2)|0 >)$ instead of $|i >= (2\pi)^3 a^{(e)\dagger}_{r_2}(\mathbf{p}_2)a^{(e)\dagger}_{r_1}(\mathbf{p}_1)|0 >$. Since the matrix element enters as the square modulo into the cross section to measure in experiment, then the common sign of the matrix element for scattering of identical fermions is arbitrary at all. However, the relative sign of the addends in (18.115) does not depend on the accepted order of the external operators arrangement.

The first and second addends in (18.115) are different from each other by simultaneous rearrangement of the indices μ, ν and the arguments x_1, x_2. Because both these addends are multiplied by the expression that is symmetric both regarding μ, ν and regarding x_1, x_2, then they are equal to each other and correspond to the diagram displayed on Fig. 18.2a. The similar way one demonstrates that the third and forth addends which, in turn, are associated with the diagram of Fig. 18.2(b), are equivalent. Thus, as claimed, the matrix elements corresponding to both diagrams have the opposite signs. In other words, in the case of the presence of identical fermions in the initial or final states, the expression for the matrix element must be antisymmetric with respect to identical fermions (in the case of bosons the matrix elements should be symmetric in relation to rearrangement of identical bosons).

Generalize this result on the case when l identical fermions (electrons or positrons) participate in scattering. Let us denote momenta of particles before and after scattering by $p_1, p_2, ... p_l$ and $p'_1, p'_2, ... p'_l$, respectively. l TNDs correspond to the process. If on some two diagrams fermion lines after scattering (which are continuation of fermion lines before scattering) are designated as $p'_1, p'_2, ...$ and $p'_{j_1}, p'_{j_2}, ...$ respectively, then the relative sign of the matrix elements conforming to these diagrams is defined by the transposition parity

$$\delta_p = \begin{pmatrix} 1, & 2, & ..., & l \\ j_1, & j_2, & ..., & j_l \end{pmatrix}. \tag{18.116}$$

Since matrices enter into the interaction Lagrangian:

$$\mathcal{L}_{int} = -e : \overline{\psi}_k(x) \, (\gamma^\mu)_{kl} \, \psi_l(x) A_\mu(x) :, \tag{18.117}$$

(k and l are matrix indices), then the order of factors in $S_{i \to f}^{(n)}$ becomes to be significant as well. From the expression (18.115) we start to guess that the order of matrices placement correlates with the motion along an electron line of a diagram from the final to initial states (in the case of positrons the motion occurs in inverse direction). To ultimately become stronger in this statement, we consider a more complicated diagram with a continuous electron line. The diagram of the third order displayed on Fig. 18.3 to describe the process of the electron scattering by the external field $A_\mu^{(e)}(x)$ is a good candidate for this purpose. Note that in the QED the anomalous magnetic moment of leptons is evaluated by application

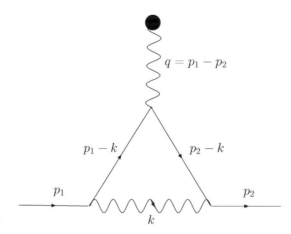

FIGURE 18.3
The process of the electron scattering by the external field.

of just these very diagrams.

The normal product corresponding to the electron scattering by the field $A_\mu^{(e)}(x)$ has the form:

$$\mathcal{A} = N\left(\overline{\psi}_m^{s_1}(x_1)(\gamma^\mu)_{mn}\psi_n(x_1)A_\mu^{s_3}(x_1)\overline{\psi}_r^{s_2}(x_2)(\gamma^\nu)_{rj}\psi_j^{s_1}(x_2)\times\right.$$

$$\left.\times A_\nu^{(e)}(x_2)\overline{\psi}_k(x_3)(\gamma^\tau)_{kl}\psi_l^{s_2}(x_3)A_\tau^{s_3}(x_3)\right). \tag{18.118}$$

Once the matrix indices to determine the matrices multiplication order have been written out, we have the right to move matrices. Using the pairing definition for the electron and photon operators, we obtain the final expression for (18.118):

$$\mathcal{A} = -i\overline{\psi}_k(x_3)(\gamma^\tau)_{kl}S_{lr}^c(x_3-x_2)(\gamma^\nu)_{rj}S_{jm}^c(x_2-x_1)(\gamma^\mu)_{mn}\psi_n(x_1)g_{\mu\tau}D_0^c(x_1-x_3)A_\nu^{(e)}(x_2),$$
$$\tag{18.119}$$

that is, the matrices placement really corresponds to the motion against the electron line direction. The correspondence rules must be supplemented by one more rule, the signs rule, which takes into account the specificity of diagrams containing even numbers of closed fermion loops. As an example, we shall deal with a diagram with one electron loop. The factor in the normal product corresponding to this loop proves to contain the minus sign:

TABLE 18.1
The Feynman rules for the QED.

No	Factor in scattering matrix	Element in Feynman diagram	State
1	Operator $\psi(x)$ under sign of normal product	External solid line incoming in vertex x	Electron (positron) in initial (final) state
2	Operator $\overline{\psi}(x)$ under sign of normal product	External solid line outgoing from vertex x	Electron (positron) in final (initial) state
3	Operator $A_\mu(x)$ under sign of normal product	External waved line at vertex x having index μ	Photon in initial or final state
4	Quantity $A_\mu^e(x)$	External waved line connecting shaded circle with vertex x having index μ	External electromagnetic field
5	Quantity $-ie\gamma^\mu$	Vertex with index μ	Interaction act
6	Pairing of electron-positron field operators $-iS^c(x-y)$	Internal solid line	Motion of virtual electron (positron) from x to y (from y to x)
7	Pairing of photon field operators $ig_{\mu\nu}D_0^c(x-y)$	Internal waved line	Motion of virtual photon between vertices x and y with indices μ and ν

$$N\left(\overline{\psi}_m^{s_1}(x_1)(\gamma^\mu)_{mn}\psi_n^{s_2}(x_1)\overline{\psi}_r^{s_2}(x_2)(\gamma^\nu)_{rj}\psi_j^{s_1}(x_2)\right) = -(\gamma^\mu)_{mn}S_{nr}^c(x_1-x_2)(\gamma^\nu)_{rj}S_{jm}^c(x_2-x_1) =$$

$$= -\text{Sp}\left(\gamma^\mu S^c(x_1-x_2)\gamma^\nu S^c(x_2-x_1)\right).$$

In the most general case the signs rule is formulated as follows: *if a diagram contains l closed fermion loops, then a matrix element must be multiplied by the factor* $(-1)^l$.

So, the matrix element conforming to the definite physical process could be presented in the form:

$$S^{(n)}_{i \to f} = \sum \mathcal{M}^{(n)}_{i \to f}, \qquad (18.120)$$

where $\mathcal{M}^{(n)}_{i \to f} \equiv\, < f|\mathcal{M}^{(n)}|i >$ and individual addends in the right-hand side differ from each other by the placement order of the operators creating and annihilating the particles to participate in the process. This expression may be obtained by two ways: from the general analytical expression for the matrix element by application of Wick's theorem, or with the help of the Feynman diagrams. To accomplish this operation within the diagram technique by Feynman we should first draw all the possible topologically different diagrams of a process in the given order of the perturbation theory. Next, to determine the form of $\mathcal{M}^{(n)}$ we must correlate every diagram element with a factor in the scattering matrix in accordance with the Feynman rules gathered in Table 18.1 and multiply the obtained expression by the quantity:

$$\eta = (-1)^l r \delta_p. \qquad (18.121)$$

Passing from $\mathcal{M}^{(n)}$ to $\mathcal{M}^{(n)}_{i \to f}$ and using the formula (18.120), we deduce the matrix element under consideration. There is no doubt that this way is more simple and more convenient than the method based on permanent usage of the Wick's theorem.

18.7   Furry's theorem

We shall prove the following theorem belonging to W. Furry [8]: *a total matrix element which corresponds to diagrams containing closed internal fermion loops to consist of odd number lines turns into zero.* To prove it, we will look at a process describing the contribution of the third order in an amplitude of a transition *vacuum→vacuum* under the influence of an external field. There are two diagrams differing only in a detour direction of a closed electron loop (Fig. 18.4). Thus, the matrix element of the reaction in question $\mathcal{M}^{(3)}_V$ includes two

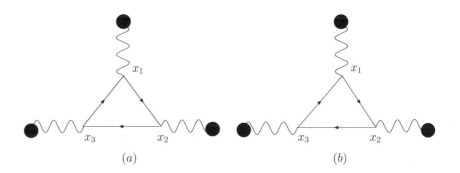

FIGURE 18.4

The Feynman diagrams (a) and (b) describing the **vacuum→vacuum**-transition under the external field influence.

addends $\mathcal{M}_a^{(3)}$ and $\mathcal{M}_b^{(3)}$ to accord with these diagrams:

$$\mathcal{M}_V^{(3)} = \mathcal{M}_a^{(3)} + \mathcal{M}_b^{(3)} = -\frac{i}{3!}(-ie)^3 \int dx_1 \int dx_2 \int dx_3 \{\mathrm{Sp}[\hat{A}^e(x_1)S^c(x_1 - x_2)\hat{A}^e(x_2)S^c(x_2 - x_3) \times$$

$$\times \hat{A}^e(x_3)S^c(x_3 - x_1)] + \mathrm{Sp}[\hat{A}^e(x_1)S^c(x_1 - x_3)\hat{A}^e(x_3)S^c(x_3 - x_2)\hat{A}^e(x_2)S^c(x_2 - x_1)]\}, \quad (18.122)$$

where $\hat{A}^e(x) = A_\mu^e(x)\gamma^\mu$.

Using the charge conjugation matrix C we rewrite the spur to enter into $\mathcal{M}_a^{(3)}$ in the following view:

$$\mathrm{Sp}[...] = \mathrm{Sp}[C^{-1}\hat{A}^e(x_1)CC^{-1}S^c(x_1 - x_2)CC^{-1}\hat{A}^e(x_2)CC^{-1}S^c(x_2 - x_3) \times$$

$$\times CC^{-1}\hat{A}^e(x_3)CC^{-1}S^c(x_3 - x_1)C]. \quad (18.123)$$

From the charge conjugation matrix property:

$$C^{-1}\gamma^\mu C = -\gamma^{\mu T},$$

follows

$$C^{-1}S^c(x_1 - x_2)C = -\frac{1}{(2\pi)^4}\int d^4p \frac{-\hat{p}^T + m}{p^2 - m^2}e^{-ip(x_1 - x_2)} =$$

$$= -\frac{1}{(2\pi)^4}\int d^4p \frac{\hat{p}^T + m}{p^2 - m^2}e^{-ip(x_2 - x_1)} = S^{cT}(x_2 - x_1). \quad (18.124)$$

This, in its turn, allows to set down (18.123) as:

$$\mathrm{Sp}[...] = (-1)^3 \mathrm{Sp}[\hat{A}^{eT}(x_1)S^{cT}(x_2 - x_1)\hat{A}^{eT}(x_2)S^{cT}(x_3 - x_2)\hat{A}^{eT}(x_3)S^{cT}(x_1 - x_3)] =$$

$$= -\mathrm{Sp}[S^c(x_1 - x_3)\hat{A}^e(x_3)S^c(x_3 - x_2)\hat{A}^e(x_2)S^c(x_2 - x_1)\hat{A}^e(x_1)] =$$

$$= -\mathrm{Sp}[\hat{A}^e(x_1)S^c(x_1 - x_3)\hat{A}^e(x_3)S^c(x_3 - x_2)\hat{A}^e(x_2)S^c(x_2 - x_1)]. \quad (18.125)$$

From (18.125) it is evident that $\mathcal{M}_a^{(3)} = \mathcal{M}_b^{(3)}$, and, as a consequence, $\mathcal{M}_V^{(3)} = 0$.

Let us interpret the obtained result. It may be assumed that the first term associated with the diagram in Fig. 18.4(a) describes propagation of an electron in one direction, whereas the second term associated with the diagram in Fig. 18.4(b) defines propagation of an electron in the reverse direction. Upon the reversal of its direction, an electron behaves like a positron. Because of this, its charge is changing the sign, and the sign of each interaction with an external electromagnetic field is also changed. Since the vertices number is odd, both diagrams compensate each other. In a similar manner, if there is a diagram that includes an internal closed electron loop with an arbitrary number of electron lines, along with this diagram, we have to consider another diagram that is distinguished from the first one only by the direction of a detour around the closed electron loop. And in case the number of vertices is odd the matrix element corresponding to the sum of these diagrams is zero.

Note that using a change of variables, we can transform the right-hand side of equality (18.125) to the initial form—the left-hand side of equality (18.123). In the process the term under consideration proves to be equal to itself with a minus sign, that is, equals zero. The same is true for the second term in (18.122). In this way each individual term proves to be zero, as one would expect based on the charge parity conservation law. A closed electron loop with k vertices, that is, with k adjoined photon lines, is associated with a neutral system of real and virtual photons, the total number of which equals k. As the photon parity is equal to -1, for odd k the initial and final states possess a different charge parity, and this is impossible as the experiment goes.

18.8   Feynman rules in the momentum representation

It is more convenient to calculate the matrix elements $S_{i \to f}^{(n)}$ in the momentum representation for the following reasons. First, in this representation the operators and the causal functions have extremely simple structures. Second, in the coordinate representation, with the use of the diagram technique, one finds the expression for an element of the scattering matrix $S^{(n)}$ and only afterwards the corresponding matrix element $S_{i \to f}^{(n)}$ may be calculated, whereas in the momentum representation $S_{i \to f}^{(n)}$ may be directly written due to the correspondence rules. To find the matrix element in the momentum representation we sufficiently carry out three operations in the scattering matrix $S^{(n)}$ written in the coordinate representation: (i) to pass on to the matrix element $S_{i \to f}^{(n)}$; (ii) to decompose the operators and the causal functions in terms of the Fourier integrals; and (iii) to integrate the obtained expressions over the space-time coordinates corresponding to the diagram vertices.

Let us fulfill these operations step-by-step. The matrix element $S_{i \to f}^{(n)}$ may be represented in the form:

$$S_{i \to f}^{(n)} = \frac{(-ie)^n}{n!} \int dx_1 \int dx_2 ... \int dx_n K_{...\mu_j...\mu_l...}(x_1,...x_n) \times$$

$$\times < f | N \left(...\overline{\psi}(x_j)\gamma^{\mu_j}...\gamma^{\mu_l}\psi(x_l)... \right) N \left(...A_{\mu_{j-1}}(x_{j-1})...A_{\mu_{l+1}}(x_{l+1})... \right) | i >, \qquad (18.126)$$

where $K_{...\mu_j...\mu_l...}(x_1,...x_n)$ comprises the product of the field operator pairings. As we have already known to calculate the expression:

$$< f | N \left(...\overline{\psi}(x_j)\gamma^{\mu_j}...\gamma^{\mu_l}\psi(x_l)... \right) N \left(...A_{\mu_{j-1}}(x_{j-1})...A_{\mu_{l+1}}(x_{l+1})... \right) | i > \qquad (18.127)$$

we should commute the production operators from the normal products (PON) with the annihilation operators involved in the states $< f |$, and also the annihilation operators from the normal products (AON) with the production operators from $|i >$ until one of PON affects $< 0 |$ or one of AON turns to be alongside $|0 >$, that gives zero as a result. The matrix element of (18.127) is nonzero only in the case when for each operator from the normal product we can find an operator in $< f |$ or $|i >$, a commutation with which results in the δ -function, that is, in the pairwise "cancelation" of both operators. In this way the expression (18.127) is nonzero when a sum of the particle numbers in each field, in the initial and final states, is precisely equal to the number of the operator functions for this field in the N-product. It should be noted that expression (18.127) proves to be nonzero when $< f |$ and $|i >$, apart from the operators "canceling" the normal-product operators, include "redundant" operators that are canceled pairwise. For this purpose it is necessary that the momenta of the "redundant" particles be equal in the $< f |$ and $|i >$ states. In this case the total number of particles in the initial and final states are exceeding the number of the operators in the N-product by some even number. But this is possible only if for whatever reasons the indicated "redundant" particles do not take place in the interaction. Because of this, such a situation is beyond our consideration. And in the experimental conditions the particles with an invariable momentum are considered as belonging to the principal initial beam, and only those changing their momenta are regarded as scattered.

So, the matrix element (18.127) is exhibited in the form of the product of the commutation results. In this case, the result of commuting $\psi(x)$ with $(2\pi)^{3/2}a_r^{(e)\dagger}(\mathbf{p})$, which is equal to:

$$\sqrt{\frac{m}{E_p}}u^{(r)}(\mathbf{p})e^{-ipx} \qquad (18.128)$$

corresponds to the electron in the initial state with the momentum $\mathbf{p}$ and polarization r, the result of commuting $\overline{\psi}(x)$ with $(2\pi)^{3/2}a_r^{(e)}(\mathbf{p})$:

$$\sqrt{\frac{m}{E_p}}\,\overline{u}^{(r)}(\mathbf{p})e^{ipx} \tag{18.129}$$

—to the electron in the final state with the momentum $\mathbf{p}$ and polarization r, the result of commuting $\overline{\psi}(x)$ with $(2\pi)^{3/2}b_r^{(e)\dagger}(\mathbf{p})$:

$$\sqrt{\frac{m}{E_p}}\,\overline{v}^{(r)}(\mathbf{p})e^{-ipx}, \tag{18.130}$$

—to the positron in the initial state with the momentum $\mathbf{p}$ and polarization r, the result of commuting $\psi(x)$ with $(2\pi)^{3/2}b_r^{(e)}(\mathbf{p})$:

$$\sqrt{\frac{m}{E_p}}\,v^{(r)}(\mathbf{p})e^{ipx}. \tag{18.131}$$

—to the positron in the final state with the momentum $\mathbf{p}$ and polarization r.

Note that in the case of the positron an incoming (outgoing) line has the arrow directed from a vertex (toward a vertex) and the momentum to point against the arrow. Analogously, we have the factor:

$$\sqrt{\frac{1}{2k_0}}\,c_\mu^\lambda(\mathbf{k})e^{-ikx} \qquad \left(\sqrt{\frac{1}{2k_0}}\,e_\mu^{\lambda*}(\mathbf{k})e^{ikx}\right) \tag{18.132}$$

for the photon in a initial (final) state with the momentum $\mathbf{k}$ and polarization λ at a vertex with an index μ.

So, the negative-frequency exponent $\exp(-ipx)$ always corresponds to particles incoming into a point x with the momentum $\mathbf{p}$ while the factor $\exp(ipx)$ is associated with particles outgoing from a point x with the momentum $\mathbf{p}$.

Substituting the integral representations of the field operator pairings into (18.126), we see that the integration over variables $x_1, x_2, ..., x_n$ is reduced to the integrals of the form:

$$\int dx_l e^{ix_l \sum_n p_n} = (2\pi)^4 \delta^{(4)}\left(\sum_n p_n\right),$$

that is, this integration leads to the appearance of the δ-functions at every vertex. To sum up the aforesaid, we come to the Feynman rules in a momentum representation.

To calculate a matrix element of a process $i \to f$ one should choose the time direction (recall, our choice is from left to right) and picture all topologically nonequivalent diagrams (diagrams distinguished only by vertex indices rearrangement are not considered to be different) in accordance with the form of the interaction Hamiltonian in the QED, that is, in every vertex one photon and two fermion lines must come across. In the vertices of the obtained diagrams one should arrange four-dimensional indices $(\mu, \nu...)$. On lines one should place momentum variables, that is, every external and internal line is assigned its four-dimensional momentum. It might be well to point out that though a photon line usually has no direction but once momentum has been assigned to it, then this line gains the momentum direction, inasmuch as the momentum comes out from one vertex and enters into the another. Then, one may get a matrix element $\mathcal{M}_{i \to f}^{(n)}$ conforming to any diagrams of nth-order using the following rules.

1) The factors:

$$\sqrt{\frac{m}{E_p}}\,\overline{u}^{(r)}(\mathbf{p}),$$

$$\sqrt{\frac{m}{E_p}}\,\overline{v}^{(r)}(\mathbf{p}).$$

correspond to external solid lines to go out of the vertices in the positive and negative direction of the time axis, respectively. The first factor describes the electron production with the momentum p and polarization r while the second one does the positron annihilation with the momentum p and polarization r.

2) The factors:

$$\sqrt{\frac{m}{E_p}}\,u^{(r)}(\mathbf{p}),$$

$$\sqrt{\frac{m}{E_p}}\,v^{(r)}(\mathbf{p}).$$

correspond to external solid lines to go into vertices in the positive and negative direction of the time axis respectively. The first factor describes the electron annihilation with the momentum p and polarization r while the second one does the positron production with the momentum p and polarization r.

3) The factor to describe the production or annihilation of a photon with the momentum k and polarization λ:

$$\sqrt{\frac{1}{2k_0}}\,e_\mu^{(\lambda)}(\mathbf{k})$$

corresponds to an external waved line at a vertex with an index μ.

4) The factor to describe either virtual electron propagation in the diagram arrow direction or virtual positron propagation in the opposite direction:

$$\frac{i}{(2\pi)^4}\frac{\hat{p}+m}{p^2-m^2},$$

corresponds to an internal solid line.

5) The factor to describe virtual photon propagation between vertices with indices μ and ν:

$$\frac{-i}{(2\pi)^4}\frac{g_{\mu\nu}}{k^2},$$

corresponds to an internal waved line.

6) The factor to describe interaction between particles:

$$-ie\gamma^\mu(2\pi)^4\delta^{(4)}(p-p'\pm k),$$

(p and p' are four-dimensional momenta of external or internal fermion lines, k is four-dimensional momentum of an external or internal photon line, and the δ-function expresses the four-dimensional momentum conservation law for all particles meeting at this vertex) corresponds to every vertex.

7) All matrices acting upon spinor indices are placed in that order as they occur at the diagram detour along an electron line from a final state to an initial one.

8) To integrate over four-dimensional momenta of internal lines displaying virtual particles is fulfilled.

9) A matrix element is multiplied by the numerical factor:

$$(-1)^l\frac{r}{n!}\delta_p.$$

In the ending we should carry out the summation over matrix elements $\mathcal{M}_{i\to f}^{(n)}$ to match all diagrams of a process under investigation. As far as an interaction with a external electromagnetic field is concerned, then, in the most general case, for it to be covered the Feynman rules in the coordinate space should be used. Feynman rules for a static external field in the momentum representation will be viewed in Section 18.11.

At the absence of an interaction, when both four-dimensional momenta and quantum numbers of particles in an initial state are equal to those of particles in a final state, the scattering matrix S evidently coincides with the identity matrix I. It is convenient to separate the identity matrix. To do it we report the scattering matrix elements in the form:

$$S_{i \to f} = \delta_{if} + iR_{i \to f}, \tag{18.133}$$

where $R_{i \to f} \equiv <f|R|i>$ are matrix elements of the so-called reaction matrix R. Beyond doubt, a phase factor in front of R is arbitrary. The choice made by us in (18.133) fits to that of the phase factor in nonrelativistic quantum scattering theory. In nondiagonal matrix elements the first term in (18.133) falls out, and thus matrix elements of the matrices S and R for a transition $i \to f$ are connected with each other by a simple relation:

$$S_{i \to f} = iR_{i \to f}. \tag{18.134}$$

In this case we have

$$R_{i \to f} = -i(2\pi)^4 \delta^{(4)} \left(\sum_i p_i - \sum_f p_f \right) \sum \mathcal{M}_{i \to f}. \tag{18.135}$$

We shall call the quantity $\sum \mathcal{M}_{i \to f}$, which enters into (18.135) by the reaction amplitude $\mathcal{A}_{i \to f}$. In the case of an interaction with an external field not all components of four-dimensional momentum will be conserved quantities (for example, only the forth component, that is, a system energy, could be conserved). This, in its turn, leads to changing a dimensionality of the δ-function in (18.135).

The rules obtained above could easily be generalized on the case of other point particles with the spin 1/2—the charged leptons μ, τ and the quarks $u_\alpha, d_\alpha, s_\alpha, c_\alpha, b_\alpha, t_\alpha$. In so doing to modify the Feynman rules resides in changing both a charge and a propagator of an electron with those of particles in question.

$$e \to q, \qquad \frac{i}{(2\pi)^4} \frac{\hat{p} + m}{p^2 - m^2} \to \frac{i}{(2\pi)^4} \frac{\hat{p} + m_f}{p^2 - m_f^2 + im_f \Gamma_f}, \tag{18.136}$$

where Γ_f is decay width of a particle with a flavor f.

18.9   Crossing symmetry

To make use of the diagram technique for the determination of scattering amplitudes, $\mathcal{A}_{i \to f}$, renders existence of a definite symmetry evident. This symmetry will be called the *crossing symmetry*.

We shall concern ourselves with binary processes with the participation of particles a, b, c, and d. Denote the particles motion direction by arrows and choose the time axis direction from left to right. In this case the following reactions are possible:[*]

$$a + b \to c + d, \tag{18.137}$$

---

[*]Recall, that one may handle reaction equations as one does with ordinary equations, with one exception only. Now, at passage from one side of an equation to another we have the changes: *particle→antiparticle*, *antiparticle→particle*.

$$a + \bar{c} \rightarrow \bar{b} + d, \qquad\qquad (18.138)$$

$$a + \bar{d} \rightarrow c + \bar{b}. \qquad\qquad (18.139)$$

and the inverse reactions as well.

The reactions (18.137)–(18.139) represent three different channels of some reaction (let us call it the generalized reaction). The definite linkage between amplitudes of these channels proves to exist. This linkage could be established by means of introducing invariant variables.

In the case of any scattering of the kind (18.137) we have two independent variables, for example, an incoming particle energy and a scattering angle. Exhibit the process amplitude, and the cross section itself, as a function of variables to be invariant with respect to the Lorentz transformation. So, there are four four-dimensional momenta of particles at our disposal. Therefore, all that could be done is to build of them scalar products:

$$(p_a p_b), \qquad (p_a p_c), \qquad (p_a p_d), \qquad (p_b p_c), \qquad (p_b p_d), \qquad (p_c p_d). \qquad (18.140)$$

Owing to the four-dimensional momentum conservation law

$$p_a + p_b = p_c + p_d \qquad\qquad (18.141)$$

and the relations $p_j^2 = m_j^2$ ($j = a, b, c, d$), only two of six scalar products in (18.140) are independent. It is commonly accepted to not use these independent scalars but Mandelstam's variables to connect with them:

$$s = (p_a + p_b)^2, \qquad t = (p_a - p_c)^2, \qquad u = (p_a - p_d)^2. \qquad (18.142)$$

It is an easy matter to verify that these variables satisfy the relation:

$$s + t + u = m_a^2 + m_b^2 + m_c^2 + m_d^2. \qquad (18.143)$$

Set down the four-dimensional momentum conservation law and the Mandelstam variables definition for the remaining two channels (18.138), (18.139)

$$\left.\begin{array}{l} p_a + p_{\bar{c}} = p_{\bar{b}} + p_d, \\ s = (p_a - p_{\bar{b}})^2, \qquad t = (p_a + p_{\bar{c}})^2, \qquad u = (p_a - p_d)^2, \end{array}\right\} \qquad (18.144)$$

$$\left.\begin{array}{l} p_a + p_{\bar{d}} = p_{\bar{b}} + p_c, \\ s = (p_a - p_{\bar{b}})^2, \qquad t = (p_a - p_c)^2, \qquad u = (p_a + p_{\bar{d}})^2. \end{array}\right\} \qquad (18.145)$$

When one fulfills the change:

$$p_a \rightarrow p_a, \qquad p_b \rightarrow p_{\bar{c}}, \qquad p_d \rightarrow p_{\bar{b}}, \qquad p_c \rightarrow p_d \qquad (18.146)$$

into (18.141), (18.142) and simultaneously passes to new designations:

$$s \rightarrow t, \qquad t \rightarrow u, \qquad u \rightarrow s, \qquad (18.147)$$

then one leads to the expressions (18.144) for the second channel. Analogously, carrying out the replacement:

$$p_a \rightarrow p_a, \qquad p_b \rightarrow p_{\bar{d}}, \qquad p_c \rightarrow p_{\bar{b}}, \qquad p_d \rightarrow p_c \qquad (18.148)$$

and passing to new designations:

$$s \rightarrow u, \qquad t \rightarrow s, \qquad u \rightarrow t, \qquad (18.149)$$

we get the expressions (18.145) for the third channel.

Thus, to describe all the three channels one may use the same variables. To put it in another way one may take any of the three reactions as an original one and thereafter passes from it to others. It is convenient to fix for every channel a definite title. One may easily be convinced that in the center-of-mass frame the quantities:

$$s = (E_a + E_b)^2, \qquad t = (E_a + E_{\bar{b}})^2 \qquad u = (E_a + E_{\bar{d}})^2$$

are total energy squares in the first, second, and third channels, respectively. For this reason the first channel is named the s-channel, the second—the u-channel, and the third—the t-channel. Now, according to accepted terminology, we denote the first process amplitude as $\mathcal{A}_s(s,t,u)$ and then keep the previous designations for the Mandelstam variables. As far as the remaining channels are concerned we rewrite their characteristics in the following view:

$$\mathcal{A}_t(s_t, t_t, u), \qquad s_t = (p_a + p_{\bar{c}})^2, \qquad t_t = (p_a - p_{\bar{b}})^2, \qquad u = (p_a - p_d)^2 \qquad (18.150)$$

and

$$\mathcal{A}_u(s_u, t, u_u), \qquad s_u = (p_b + p_{\bar{c}})^2, \qquad t = (p_a - p_c)^2, \qquad u_u = (p_a - p_{\bar{b}})^2. \qquad (18.151)$$

Now the crossing symmetry may be expressed with the help of the relations:

$$\mathcal{A}_s(s,t,u) = \mathcal{A}_t(s_t = t, t_t - s, u) - \mathcal{A}_u(s_u - u, t, u_u - s). \qquad (18.152)$$

Thanks to the crossing symmetry all three channels may be pictured by one generalized diagram (see, Fig. 18.5). Let us show that physical regions of changing the variables for

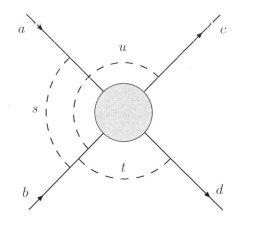

FIGURE 18.5

The s-, u-, and t-channels of the binary process.

the three processes under consideration are not mutually overlapped. To map these regions we employ a scalene coordinate system (Fig. 18.6) to be symmetric with respect to the variables s, u, t (the triangle is equilateral). Every point on this plane called *Mandelstam's plane* [9] is defined by three coordinates to be equal to the distance of the point up to the sides (or their prolongations) of the triangle abc. The positive direction for reading s, u,

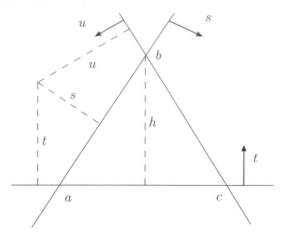

FIGURE 18.6

Mandelstam's plane.

and t is pointed by arrows guided inside the triangle. The triangle height has been chosen equal to:

$$h = s + u + t = \sum_{i=a,b,c,d} m_i^2.$$

Values that may be responsible for a physical scattering process must satisfy definite conditions being consequences of the four-dimensional momentum conservation law and the fact that momenta of initial and final particles lay on the mass shell $p_i^2 = m_i^2$.

As the product of two four-dimensional momenta obeys an inequality:

$$(p_a p_b) \geq m_a m_b, \tag{18.153}$$

then

$$(p_a + p_b)^2 \geq (m_a + m_b)^2, \qquad (p_a - p_c)^2 \leq (m_a - m_c)^2, \qquad (p_a - p_d)^2 \leq (m_a - m_d)^2. \tag{18.154}$$

From (18.154) it follows that the following inequalities take place for the s-channel:

$$\left.\begin{array}{l} (m_a + m_b)^2 \leq s \geq (m_c + m_d)^2, \\ (m_a - m_c)^2 \geq t \leq (m_d - m_b)^2, \\ (m_a - m_d)^2 \geq u \leq (m_c - m_b)^2. \end{array}\right\} \tag{18.155}$$

The similar inequalities could be obtained for the u- and t-channels.

To find remaining conditions we, following Ref. [10], introduce a four-dimensional vector:

$$K_\mu = \varepsilon_{\mu\nu\lambda\sigma} p_a^\nu p_b^\lambda p_c^\sigma. \tag{18.156}$$

In the rest frame of one of the particles, say, the particle a $(p_a = (m_a, 0, 0, 0))$ the vector K_μ has only space components $K_i = m_a \varepsilon_{i0kl} p_b^k p_c^l$, and, consequently, $K_i^2 \leq 0$, that is, the vector K_μ is space-like. The square of the vector K_μ defines the so-called boundary Kibble's function:

$$\phi = 4K_\mu^2 = \begin{vmatrix} p_a^2 & (p_a p_b) & (p_a p_c) \\ (p_b p_a) & p_b^2 & (p_b p_c) \\ (p_c p_a) & (p_c p_b) & p_c^2 \end{vmatrix} \geq 0. \tag{18.157}$$

The condition (18.157) may be expressed by way of the invariants s, t, u in the form unified for all channels:

$$stu \geq As + Bt + Cu, \qquad (18.158)$$

where

$$\left.\begin{array}{l} Ah = (m_a^2 m_b^2 - m_c^2 m_d^2)(m_a^2 + m_b^2 - m_c^2 - m_d^2), \\ Bh = (m_a^2 m_c^2 - m_b^2 m_d^2)(m_a^2 + m_c^2 - m_b^2 - m_d^2), \\ Ch = (m_a^2 m_d^2 - m_b^2 m_c^2)(m_a^2 + m_d^2 - m_b^2 - m_c^2). \end{array}\right\} \qquad (18.159)$$

As an illustration we consider the case when the s-channel describes the elastic scattering

$$m_a = m_c \equiv m, \qquad m_b = m_d \equiv \mu$$

with $m > \mu$. Since:

$$h = 2(m^2 + \mu^2), \qquad A = C = 0, \qquad B = (m^2 - \mu^2)^2,$$

then the inequality (18.158) takes the form:

$$sut \geq (m^2 - \mu^2)^2 t. \qquad (18.160)$$

The boundary of the region governed by the inequality (18.160) consists of a straight line $t = 0$ and a hyperbola

$$su = (m^2 - \mu^2)^2, \qquad (18.161)$$

which has two branches laying in sectors $u < 0$, $s < 0$ and $u > 0$, $s > 0$, and two asymptotes $s = 0$ and $u = 0$. Next, from the condition (18.155) one must additionally allow for the inequality $s \geq (m + \mu)^2$ in the s-channel and the one $u \geq (m + \mu)^2$ in the u-channel. Thereafter, all the remaining inequalities are satisfied automatically. As a result we arrive at Fig. 18.7 to display physical regions for all the three channels. Since the physical domains of variables variation for the three processes under study are not overlapping, the amplitudes may be coincident only in nonphysical domains (in Fig. 18.7 they are unshaded). In other words, the crossing relations are meaningful exclusively for the amplitudes specified over the whole domain of the variables variation, that is, they represent the analytical functions. Therefore, substitution of the variables in the relation (18.152) should be understood not formally but rather as an analytical continuation from the physical domain to the nonphysical one, for example, with respect to the variable s for the process (18.137). Then the values of the invariant amplitudes $\mathcal{A}_s(s, t, u)$, $\mathcal{A}_u(s_u, t, u_u)$, and $\mathcal{A}_t(s_t, t_t, u)$ in various physical domains are nothing else but an analytical continuation of the function specified in one of the domains. N. Bogolyubov was the first to prove a possibility for analytical continuation of the amplitude in the process of finding the dispersion relations for the πN-scattering (see, for review, the monograph [11]). Based on the microcausality principle, the existence of a unified analytical function of the complex variable s, whose boundary values constitute the cross process amplitudes, has been demonstrated. The main difficulty of the proof stems from the fact that the amplitudes include the generalized functions calling for analytical extension too. The most general and classical outcome here is a theorem (known as the "edge of the wedge" theorem) on the possible unification of the advanced and retarded Green functions into a unified analytical function. Afterwards this theorem was considered in Ref. [12]. Based on this theorem, the dispersion relations may be proved for different cases. So we come to the conclusion that a crossing symmetry follows from the scattering amplitude analyticity and hence scattering matrix analyticity.

And now let us touch upon the dispersion relations. Formally, they are Cauchy theorem corollaries from a complex variable function theory applied to the scattering amplitude

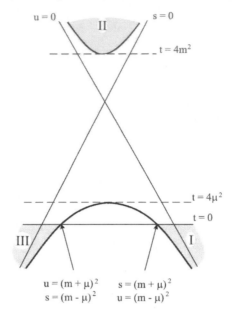

FIGURE 18.7
The physical regions corresponding to the reactions I, II, and III.

$\mathcal{A}(s)$. The dispersion relations are usually represented with integral relations of the form:

$$\text{Re } \mathcal{A}(s) = \frac{1}{\pi} \int_{-\infty}^{\infty} \frac{\text{Im } \mathcal{A}(s')}{s' - s} ds' + ..., \qquad (18.162)$$

between the real and imaginary parts of the amplitude for a two-particle forward scattering:

$$\mathcal{A}(s) = \mathcal{A}(s, \theta = 0) = \text{Re } \mathcal{A}(s) + i\text{Im } \mathcal{A}(s), \qquad (18.163)$$

each being related to the observables: total cross section $\sigma(s)$ and zero-angle differential cross section $d\sigma(s, \theta = 0)$.

R. Kronig and H. Kramers were the first to use the relations similar to (18.162) in their study of the refractive index of light in a medium [13]. With the use of (18.162) the relationship between dispersion and absorption has been established. Afterwards, the term "dispersion relations" was introduced for analytical continuations of the scattering amplitudes. The dispersion relations have been tested experimentally to reveal their applicability up to the scales of $\sim 10^{-16}$ cm. The dispersion relations formed the basis for a great number of approaches to the description of the strong interaction (see, for example, Ref. [14]). As a matter of fact, they have lost their exclusive role to a considerable degree, due to the progress in the quantum chromodynamics as the dynamic theory for the strong interaction.

Note that a crossing symmetry is a little more than a simple property of the S-matrix associated with CPT-invariance. And CPT-invariance necessitates equality of the amplitudes for the processes following from each other by permutation of the initial and final states, with substitution of antiparticles for all the particles. The cross-invariance enables such a transformation to be performed not only for all particles at once but also for any particle individually.

We shall also use the term a *cross-universality*. Its meaning is as follows. For simplicity, we consider a binary reaction, limiting ourselves to the second-order of the perturbation theory. Let us assume that it is described by two Feynman diagrams, and in these diagrams the propagators of virtual particles have denominators including the variables s and u, respectively. In this case we can state that the considered reaction proceeds through the s- and u-channels. Then, if we use a particular permutation of particles in the initial (or final) state, one diagram follows from the other. Obviously, there exists a substitution of the momenta and the polarization vectors of the particles making it possible to couple the matrix elements corresponding to both diagrams. Invariance of the amplitude with respect to such a substitution is called a *cross-universality*. This feature proves to be very useful for the calculation of scattering cross sections.

18.10 Cross section for unpolarized particles

Let us establish the relations between elements of the S-matrix and cross sections of the scattering processes. For the initial and final states normalized to the unit volume the probability of transition from the initial state to the final one is determined as follows:

$$dw'_{i \to f} = |R_{i \to f}|^2. \tag{18.164}$$

Changing one of the δ-functions in $R_{i,f}$ in accordance with the formula:

$$\left\{ \delta^{(4)}\left(\sum_i p_i - \sum_f p_f\right) \right\}_{\sum p_i = \sum p_f} = \lim_{V \to \infty} \lim_{T \to \infty} \frac{1}{(2\pi)^4} \int dt \int d^3 x \times$$

$$\times \left\{ \exp\left[i\left(\sum_i p_i - \sum_f p_f\right) x \right] \right\}_{\sum p_i = \sum p_f} = \frac{VT}{(2\pi)^4}, \tag{18.165}$$

we get

$$dw'_{i \to f} = (2\pi)^4 \delta^{(4)}\left(\sum_i p_i - \sum_f p_f\right) |\sum \mathcal{M}_{i \to f}|^2 VT. \tag{18.166}$$

Actually, the quantum-mechanical nature of the elementary particles and also the experimental conditions lead to the situation when a detector is recording the particles with, in principle, the unavoidable spread of the values measured for each of the three momentum components of particles in the final state over some intervals. As the final (as well as initial) state is associated with a continuous spectrum, of practical interest is the knowledge that the three-dimensional momenta of the final state particles $\mathbf{p}'_j$ are lying in the given intervals $d^3 p'_j$. To find this probability, Eq. (18.166) should be multiplied by the element of a phase volume (states number of particles with definite quantum numbers and momenta falling in the specified intervals):

$$\frac{d^3 p'_1 V}{(2\pi)^3} \frac{d^3 p'_2 V}{(2\pi)^3} \cdots \frac{d^3 p'_{n_f} V}{(2\pi)^3}, \tag{18.167}$$

where V is a normalization volume and n_f is a particles number in the final state. Dividing the obtained expression by VT, we get the differential probability of the process in a unit of time and in a unit of volume

$$dw_{i \to f} = \frac{dw'_{i \to f}}{VT} V^{n_f} \prod_{j=1}^{n_f} \frac{d^3 p'_j}{(2\pi)^3} =$$

$$= (2\pi)^4 \delta^{(4)} (\sum_i p_i - \sum_f p_f) |\sum \mathcal{M}_{i \to f}|^2 V^{n_f} \prod_{j=1}^{n_f} \frac{d^3 p'_j}{(2\pi)^3}. \tag{18.168}$$

Of course, it is preferable to have a quantity that is independent on the particles number in a beam or in the target, characterizing only the intensity of interaction between the colliding particles. In the lab frame the necessary features will be displayed by the quantity that is equal to the ratio of the probability $dw_{i \to f}$ to the density of the target particles ρ_t, and to the density of the incoming particles flux j_b:

$$d\sigma'_{i \to f} = (2\pi)^4 \delta^{(4)} (\sum_i p_i - \sum_f p_f) \frac{|\sum \mathcal{M}_{i \to f}|^2}{\rho_t j_b} \prod_{j=1}^{n_f} \frac{d^3 p'_j}{(2\pi)^3}, \tag{18.169}$$

where we have set the normalization volume equal to 1. However, when considering expression (18.169), a reader (even inexperienced) can raise a question: How can it be possible that the experimentally measured quantity involves the δ-function? Indeed, this function is not one of the ordinary (analytical) functions. It is classed with the generalized functions specified in terms of their effect on the ordinary ones. And the δ-function appears only at the intermediate steps of calculations, not reaching the final physical result. In Eq. (18.169) the δ-function expressing the four-dimensional momentum conservation law may be eliminated by two integrations over the three-dimensional momenta of any two particles belonging to the final state. Thus, the differential probability of scattering into the final state is given by the expression:

$$d\sigma_{i \to f} = \int' d\sigma'_{i \to f}, \tag{18.170}$$

where the prime above the integral sign denotes the twice integration eliminating the δ-function to be fulfilled. In so doing the following terminology is adopted. Provided in the final state we have two particles with the momenta p'_1 and p'_2, then we can take a set of the final states, where a three-dimensional momentum, for example, of the first particle $\mathbf{p}'_1$ is lying within the given solid angle $d\Omega$. After integrations, we obtain a quantity proportional to the element of the solid angle $d\Omega$ (beyond doubt one may transform to another variable connecting with the solid angle). This quantity is called the *differential cross section into the element of the solid angle*. The total cross section follows from the differential one after integration with respect to $d\Omega$ over the interval from 0 to 4π.

When the final state involves three particles, after the required integrations we are left with two variables, for example, y and z (and an element of the solid angle $d\Omega$ may be selected as one of them too). Then we have a double differential cross section with respect to the variables y and z. To obtain the total cross section, integration must be done with respect to y as well as z. Also, in real experiments the particles of more than three kinds may be generated. But, as a rule, we confine ourselves to detection only and hence to the description of the final states involving no more than three particles.

Next, we are constrained by discussion only situations being the most important from the physical point of view, namely, when there are only one or two particles in a initial state. In the former we are dealing with a decay while in the latter—with a particles collision. We shall address decays later, but now we proceed to the determination of a flux density of initial particles in a case of two particles collision. In the QED one could find this quantity considering a current density of the electron-positron field and the Poynting vector for the electromagnetic field. First, we define the flux density for the case when one of the initial particles is in rest (such a particle will be called a *target*). We assume that the incident flux consists of electrons. Then, the average value of the current operator in the initial state:

$$|i> = \Phi_1 = (2\pi)^{3/2} a_\sigma^\dagger(\mathbf{p}) \Phi_0, \tag{18.171}$$

is set down as:

$$< i|j_\mu(x)|i> = e(2\pi)^3 < 0|a_\sigma(\mathbf{p})\overline{\psi}(x)\gamma_\mu\psi(x)a_\sigma^\dagger(\mathbf{p})|0> = e\frac{m}{E_p}\overline{u}^\sigma(\mathbf{p})\gamma_\mu u^\sigma(\mathbf{p}) =$$

$$= e\frac{p_\mu}{E_p} = ev_\mu, \tag{18.172}$$

where v_μ is a speed of the initial electron. It is evident that we shall come to the analogous result for a positron flux too. Looking at the Poynting vector one may also establish that in the case of photons the flux density of initial particles appears to be equal to 1. Thus, we have ascertained that at the accepted normalization the flux density both for a fermion and for a photon incoming on a motionless target is simply equal to its speed.

In the general case, at the particles density of a beam and a target different from 1, we have for initial particles flux density:

$$J^{(0)} = \rho_b^{(0)}\rho_t^{(0)}v_b^{(0)}, \tag{18.173}$$

where $\rho_b^{(0)}$ and $\rho_t^{(0)}$ are flux densities of moving particles and target ones in the reference frame $K^{(0)}$ where target particles rest $v_t^{(0)} = 0$. To find an expression for the initial particles flux density in an arbitrary Lorentz reference frame K where both initial particles move, we present $\rho_b^{(0)}\rho_t^{(0)}$ and $v_b^{(0)}$ in a relativistically invariant form. Since:

$$\rho_b^{(0)}\rho_t^{(0)} = j_{b\mu}^{(0)}j_t^{(0)\mu}, \tag{18.174}$$

then in the system K the obvious relation:

$$j_{b\mu}j_t^\mu = \frac{(p_1 p_2)\rho_b\rho_t}{E_1 E_2}, \tag{18.175}$$

where $p_{1,2}$ are momenta of collided particles, takes place. Further, we have:

$$v_b^{(0)} = \frac{|\mathbf{p}_1^{(0)}|}{E_1^{(0)}} = \frac{m_2|\mathbf{p}_1^{(0)}|}{m_2 E_1^{(0)}} = \frac{\sqrt{m_2^2 E_1^{(0)2} - m_1^2 m_2^2}}{m_2 E_1^{(0)}} = \frac{\sqrt{(p_1 p_2)^2 - p_1^2 p_2^2}}{(p_1 p_2)}, \tag{18.176}$$

where we have accepted designations $p_1^{(0)} \equiv p_b^{(0)}$, $p_2^{(0)} \equiv p_t^{(0)}$. From (18.175) and (18.176) we get the expression for the flux J in an arbitrary Lorentz reference frame:

$$J = \frac{\rho_b\rho_t\sqrt{(p_1 p_2)^2 - p_1^2 p_2^2}}{E_1 E_2}. \tag{18.177}$$

Using this formula and the definitions (18.169), (18.170) and taking into account that at accepted normalization $\rho_t = \rho_t = 1$, we arrive at the following expression for the differential cross section of the scattering process:

$$d\sigma_{i\to f} = (2\pi)^4 \int{}' \delta^{(4)}(p_1 + p_2 - \sum_{j=1}^{n_f} p_j')|\sum \mathcal{M}_{i\to f}|^2 \frac{E_1 E_2}{\sqrt{(p_1 p_2)^2 - m_1^2 m_2^2}} \prod_{j=1}^{n_f} \frac{d^3 p_j'}{(2\pi)^3}. \tag{18.178}$$

We shall show that the expression for $d\sigma_{i\to f}$ may be rewritten in such a way that it will contain invariant quantities only. To do this we select noninvariant factors $1/\sqrt{E_k}$ from $\sum \mathcal{M}_{i\to f}$, that is, we exhibit $\sum \mathcal{M}_{i\to f}$ in the form:

$$\sum \mathcal{M}_{i\to f} = \mathcal{A}_{i\to f} \prod \frac{1}{\sqrt{E_k}}, \tag{18.179}$$

where multiplication should be extended over all particles both in the initial and in final states and $\mathcal{A}_{i \to f}$ is a relativistic invariant. Inserting this expression into (18.178) and allowing for the relation:

$$\int \frac{d^3 p_f}{E_f} = 2 \int d^4 p_f \delta(p_f^2 - m_f^2) \theta(E_f),$$

we obtain

$$d\sigma_{i \to f} = (2\pi)^4 \int' \delta^{(4)}(p_1 + p_2 - \sum_{j=1}^{n_f} p_j') |\mathcal{A}_{i \to f}|^2 \frac{2n_f}{\sqrt{(p_1 p_2)^2 - m_1^2 m_2^2}} \prod_{j=1}^{n_f} \left[\frac{d^4 p_j' \delta(p_j'^2 - m_j^2) \theta(E_j')}{(2\pi)^3} \right]. \tag{18.180}$$

Now the relativistic invariance of $d\sigma_{i \to f}$ will not create any doubts.

Let us come back to the expression (18.178) and eliminate the δ-functions entering in it

$$\delta^{(4)}(p_1 + p_2 - \sum_{j=1}^{n_f} p_j') = \delta^{(3)}(\mathbf{p}_1 + \mathbf{p}_2 - \sum_{j=1}^{n_f} \mathbf{p}_j') \delta(E_1 + E_2 - \sum_{j=1}^{n_f} E_j'). \tag{18.181}$$

Carry out this operation in the explicit view for the two particles state. To integrate over $d^3 p_1'$ gives rise to the disappearance of $\delta^{(3)}(\mathbf{p}_1 + \mathbf{p}_2 - \mathbf{p}_1' - \mathbf{p}_2')$. In order to remove the second δ-function we rewrite $d^3 p_2'$ in the spherical coordinate system:

$$d^3 p_2' = |\mathbf{p}_2'|^2 d|\mathbf{p}_2'| \sin\theta d\varphi d\theta = |\mathbf{p}_2'|^2 d|\mathbf{p}_2'| d\Omega = |\mathbf{p}_2'| E_2' dE_2' d\Omega, \tag{18.182}$$

where φ is an azimuthal angle of the vector $\mathbf{p}_2'$, θ is a zenith angle of the vector $\mathbf{p}_2'$ (the angle between $\mathbf{p}_2'$ and the third coordinate axis) and we have used the relation:

$$|\mathbf{p}_2'|^2 = E_2'^2 - m_2^2.$$

Integration over E_2' may be fulfilled when one replaces dE_2' with $d(E_1' + E_2')$. Since $E_1' + E_2'$ is a function of E_2', then the relation:

$$d(E_1' + E_2') = \frac{\partial(E_1' + E_2')}{\partial E_2'} dE_2' \tag{18.183}$$

takes place. With allowance made for (18.182), we obtain

$$d^3 p_2' = |\mathbf{p}_2'| E_2' d(E_1' + E_2') \left(\frac{\partial(E_1' + E_2')}{\partial E_2'} \right)^{-1} d\Omega. \tag{18.184}$$

To substitute (18.184) into (18.180) and integrate over $E_1' + E_2'$ gives the final expression for the differential cross section of the reaction:

$$d\sigma_{i \to f} = (2\pi)^{-2} \left[|\mathbf{p}_2'| E_2' \left(\frac{\partial(E_1' + E_2')}{\partial E_2'} \right)^{-1} |\sum \mathcal{M}_{i \to f}|^2 \right]_{p_1 + p_2 - p_1' - p_2'} \times$$

$$\times \frac{E_1 E_2}{\sqrt{(p_1 p_2)^2 - m_1^2 m_2^2}} d\Omega. \tag{18.185}$$

Assume that a beam of the initial state particles is not polarized, whereas devices detecting the scattered particles respond to the particles with all values of the spin projection. In other words, we are not interested in a particular polarization state of both the initial and final particles. In this case, the cross section should be summed up over all possible polarization states of the final particles and averaged over polarizations of the initial ones. We denote both operations as $\sum_{spin}$, that is, the cross section for unpolarized particles is represented as

$$\sum_{spin} d\sigma_{i \to f}.$$

Summation over the polarizations of photons is carried out with the use of Eq. (17.44), and the summation in the case of electrons and positrons—with the use of Eqs. (15.103) and (15.104). And in the case of averaging over the polarizations, we use the same equations with a preceding factor of $1/2$.

18.11 Cross section for scattering by external fields

When using the Feynman momentum-space rules in studies of the particle scattering processes by external electromagnetic fields, we need to know the Fourier images for all the "familiar" potentials. Generally, for their calculation the classical electrodynamics equations are required. In so doing, the fact whether an external field depends or not depends on time is important as well.

To illustrate the basic pattern for the description of interactions with an external field, we consider scattering of an electron by the external electromagnetic field. It should be noted that nuclear scattering of electrons used in studies of the nuclear charge distribution is a very effective instrument to reveal the structure of nuclear levels. Fig. 18.8 presents the relevant Feynman diagram in the first order of the perturbation theory. As the Fourier

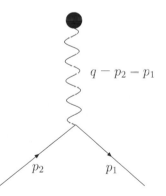

FIGURE 18.8
The Feynman diagram describing the electron scattering by the external electromagnetic field.

image for the potential of an external field is not known, we have to use the Feynman rules in the coordinate representation. In this representation the matrix element is as follows:

$$S_{i \to f} = -ie\sqrt{\frac{m^2}{E_p E_{p'}}} \int d^4x \, \overline{u}_{r'}(\mathbf{p}') A_\mu^e(x) \gamma^\mu u_r(\mathbf{p}) e^{i(p'-p)x}. \qquad (18.186)$$

The external field $A_\mu^e(x)$ may be expressed in terms of the sources density. For instance, when we have an external field created by the moving nucleus, then for the scalar potential we get:

$$\Box A_0^e(\mathbf{r}, t) = Ze\rho(\mathbf{r}, t), \tag{18.187}$$

where $\rho(\mathbf{r}, t)$ characterizes a nucleus charge distribution density. From (18.187) follows the well-known formula:

$$A_0^e(\mathbf{r}, t) = \frac{Ze}{4\pi} \int d^3 x' \frac{\rho(\mathbf{r}', t - |\mathbf{r} - \mathbf{r}'|)}{|\mathbf{r} - \mathbf{r}'|}. \tag{18.188}$$

Of course, expression (18.188) may be substituted into the matrix element (18.186), reducing our task to the computation of integrals. But these integrals are rather sophisticated to force us to use the momentum representation again.

To substitute the Fourier expansion of the nucleus potential

$$A_\mu^e(x) = \frac{1}{(2\pi)^{3/2}} \int d^3 q A_\mu^e(q) e^{-iqx}$$

into (18.186) allows to carry out integration over the variable x and this, in its turn, leads to the appearance of the δ-function. In problems of particles collision the δ-function expresses the conservation law of the four-dimensional momentum of the particles system. Because we change an external field with an effective virtual photon, a pseudophoton ($q^2 \neq 0$), herein presence of the δ-function leads to the fact that the four-dimensional pseudophoton momentum taken with the opposite sign equals the four-dimensional momentum to be transferred to the nucleus as a consequence of the electron scattering

$$-q = p - p'.$$

So, $-\mathbf{q}$ makes up the three-dimensional momentum of nucleus recoil while $-q_0$ makes up the energy imparted to the nucleus. All this means that at scattering by a fixed source the four-dimensional momentum is not conserved by this time. A source can absorb or give back a momentum without change of its state. Thus, for example, in the case of a stationary source the three-dimensional momentum is not conserved and only particles energy is conserved. Really, the variable q_0 is absent in the stationary potential expansion:

$$A_\mu^e(\mathbf{r}) = \frac{1}{(2\pi)^{3/2}} \int d^3 q A_\mu^e(\mathbf{q}) e^{i\mathbf{q}\mathbf{r}} \tag{18.189}$$

From (18.189) it is evident that in the given case, the pseudophotons transfer the three-dimensional momentum $\mathbf{q}$ and their energy q_0 equals zero. Now the matrix element (18.186) is rewritten in the following manner:

$$S_{i \to f} = -ie(2\pi)^{5/2} \int d^3 q \sqrt{\frac{m^2}{E_p E_{p'}}} \bar{u}_{r'}(\mathbf{p}') A_\mu^e(\mathbf{q}) \gamma^\mu u_r(\mathbf{p}) \delta(\mathbf{q} + \mathbf{p} - \mathbf{p}') \delta(E_{p'} - E_p) =$$

$$= -ie(2\pi)^{5/2} \sqrt{\frac{m^2}{E_p E_{p'}}} \bar{u}_{r'}(\mathbf{p}') A_\mu^e(\mathbf{p} - \mathbf{p}') \gamma^\mu u_r(\mathbf{p}) \delta(E_{p'} - E_p). \tag{18.190}$$

We shall determine the cross section of the electron on Coulomb's field of the motionless nucleus considering the nucleus field to be point charge, that is,

$$A_\mu^e(r) = \frac{Ze}{4\pi r} g_{\mu 0}$$

(In this case, one may say that unlike a real photon polarization vector, a pseudophoton polarization vector is directed along the time axis). To seek the Fourier image of the potential $A_0^e(r)$ we direct our attention to the Poisson equation for a point charge placed at the coordinates origin:

$$\Delta A_0(\mathbf{r}) = -e\delta(\mathbf{r}).$$

The consequence of this equation is the relation:

$$\Delta \frac{1}{r} = -4\pi\delta(\mathbf{r}). \tag{18.191}$$

With the help of (18.189) and (18.191) we easily obtain the required quantity:

$$A_0^e(\mathbf{q}) = \frac{Ze}{(2\pi)^{3/2}|\mathbf{q}|^2}. \tag{18.192}$$

Thus, the probability of the electron scattering by the Coulomb's field of the nucleus into one selected final state $\bar{u}_{r'}(\mathbf{p}')$ is given by the expression:

$$dW_{i\to f} = \frac{Z^2\alpha^2 m^2 (2\pi)^2 16\pi^2}{E_p E_{p'}}|\bar{u}_{r'}(\mathbf{p}')\frac{1}{|\mathbf{p}-\mathbf{p}'|^2}\gamma^0 u_r(\mathbf{p})|^2\delta(E_{p'}-E_p)\frac{T}{2\pi}, \tag{18.193}$$

where the appearance of the factor $T/(2\pi)$ is due to the δ-function

$$\delta(E_{p'} - E_p) = \lim_{E_p \to E_{p'}} \frac{1}{2\pi} \int_{-T/2}^{T/?} dt e^{i(E_p - E_{p'})t} = \frac{T}{2\pi}$$

When the initial beam of the electrons is not polarized and we are not interested in the spin states of the scattered beam, then we should sum up the obtained expression over the final spin states and average over the initial ones. Then, we arrive at the following expression for the probability of the electron scattering into the final states group $d^3p'/(2\pi)^3$ in a unit of time:

$$dw_{i\to f} = \frac{4Z^2\alpha^2 m^2}{E_p E_{p'}}\frac{1}{2}\text{Sp}\left(\frac{\hat{p}'+m}{2m}\gamma^0\frac{\hat{p}\mid m}{2m}\gamma^0\right)\frac{1}{|\mathbf{p}-\mathbf{p}'|^4}\delta(F_{p'}-F_p)d^3p'. \tag{18.194}$$

In order to get the differential cross section of scattering we should divide the expression (18.194) into initial particles flux density (in our case this quantity is simply $|\mathbf{p}|/E_p = v$) and get rid of the δ-function. Integration over $E_{p'}$ will be fulfilled without trouble if we pass in the spherical coordinate system

$$d^3p' = |\mathbf{p}'|E_{p'}dE_{p'}d\Omega.$$

Since $E_p = E_{p'}$, then $|\mathbf{p}'|$ is equal to $|\mathbf{p}|$ as well. Next, we calculate the spur of the γ-matrices entering into (18.194). Using the formulae of Section 15.2, we obtain

$$\text{Sp}\left(\frac{\hat{p}'+m}{2m}\gamma^0\frac{\hat{p}+m}{2m}\gamma^0\right) = \frac{1}{4m^2}\text{Sp}\left(\hat{p}'\gamma^0\hat{p}\gamma^0 + m^2\right) =$$

$$= \frac{1}{m^2}[2E_p E_{p'} - (p'\cdot p) + m^2] = \frac{1}{m^2}[E^2 + (\mathbf{p}'\cdot\mathbf{p}) + m^2] =$$

$$= \frac{1}{m^2}[E^2 + p^2\cos\theta + m^2] = \frac{1}{m^2}[(E^2 - m^2)(1+\cos\theta) + 2m^2] = \frac{2E^2}{m^2}\left(1 - v^2\sin^2\frac{\theta}{2}\right), \tag{18.195}$$

where θ is an angle between momenta of the incoming and scattered electron, $E = E_p = E_{p'}$ and $p = |\mathbf{p}'| = |\mathbf{p}|$. Gathering the obtained results, we find the final expression for the differential cross section of the electron scattering by the Coulomb's field of the nucleus [15]:

$$d\sigma_{i \to f} = \frac{Z^2 \alpha^2}{4 p^2 v^2 \sin^4(\theta/2)} \left(1 - v^2 \sin^2 \frac{\theta}{2} \right) d\Omega. \tag{18.196}$$

A few words about the formula (18.196). It is distinguished from Rutherford's formula by the factor $[1 - v^2 \sin^2(\theta/2)]$. This factor serves as the correction allowing for the electron spin. Really, it appears from the expression for the spur, and hence must include interaction of an electron magnetic moment with a magnetic field. In the reference frame where the electron rests ($v = 0$) this factor becomes equal to 1. As calculations show, the cross section for scattering a spinless particle by the Coulomb's field of the nucleus is defined by the Rutherford's formula. Thus, we are again convinced that the cross sections of particles possessing the spin involve the ones of scalar particles.

Computing the scattering cross section in subsequent orders of the perturbation theory leads to a divergent result to be caused by the infinite range of a Coulomb field. This problem has been solved by Dalitz [16], who has demonstrated that the scattering amplitude in a screened Coulomb field

$$A_0^e(r) = \frac{Ze}{4\pi} \frac{e^{-\eta r}}{r},$$

where η^{-1} is a screening radius ($\eta > 0$), in the nonrelativistic limit is reduced to the lower-order scattering amplitude (in the first order on $Z\eta$) multiplied by the phase factor to go to infinity as η^{-1} when $\eta \to 0$. Divergences in the higher-order of the perturbation theory are emerging just due to the infinite contribution of this phase factor. By the method proposed by Dalitz, the screened Coulomb potential $A_0^e(r)$ is used at all stages of the intermediate calculations, while at the final stage it is assumed that η is zero. In this way the final result includes no screening radius, divergences associated with a long-range interaction of the Coulomb forces being absent. This problem will be discussed in greater detail in Section 21.6 devoted to the radiation corrections for the electron scattering by the Coulomb nucleus field.

Let us proceed to the description of a particle scattering from a static external electromagnetic field in the higher-orders of the perturbation theory, that is, in the case when a matrix element involves the integrals over the virtual momenta. In Fig. 18.9 we display the Feynman diagram describing the electron scattering by a static external field in the second order of the perturbation theory. Using the Feynman rules in the coordinate representation,

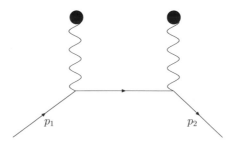

FIGURE 18.9

The Feynman diagram describing the electron scattering by the external electromagnetic field (second order of the perturbation theory).

we arrive at the expression for the corresponding matrix element:

$$S_{i \to f} = -ie^2(2\pi)\delta(E_p - E_{p'})\sqrt{\frac{m^2}{E_p E_{p'}}} u_{r'}(\mathbf{p'}) \times$$

$$\times \int d^3k A_\nu^e(\mathbf{p} - \mathbf{k})\gamma^\nu \frac{\hat{k} + m}{k^2 - m^2} A_\mu^e(\mathbf{p'} - \mathbf{k})\gamma^\mu u_r(\mathbf{p}),$$

where the integration already occurs only over the space components of the virtual electron momentum. As seen, such a disappearance of an integration with respect to the time component of the virtual particle momentum takes place only when at one of the vertices, between which this particle is moving, there is an interaction with an external field. Owing to the experience gained, the Feynman rules may be extended in the momentum representation to include an interaction with a static external electromagnetic field too. These additions are as follows.

10) The factor:

$$\frac{A_\mu^e(\mathbf{q})}{(2\pi)^{3/2}}$$

corresponds to an interaction with an external static electromagnetic field at a vertex with the index μ.

11) The factor:

$$ie\gamma^\mu(2\pi)^4\delta^{(3)}(\mathbf{p} - \mathbf{p'} - \mathbf{q})\delta(p_0 - p_0').$$

corresponds to every vertex to describe an interaction with an external static electromagnetic field.

12) Integration over variables $\mathbf{q}$ relating to external potentials is carried out.

Now we could give a general formula for the differential cross section of a particle scattering by an external static electromagnetic field. It has the following manner:

$$d\sigma_{i \to f} = \frac{2\pi}{v}\left[|\mathbf{p'_1}|E_1'|\sum \mathcal{M}_{i \to f}|^2\right]_{F_i - \sum F_j'} d\Omega \prod_{j=2}^{n_f} \frac{d^3 p_j'}{(2\pi)^3}. \tag{18.197}$$

where v and E are a speed and an energy of an incoming particle, and a matrix element $\mathcal{M}_{i \to f}$ is found by means of the Feynman rules in the momentum representation with an allowance made for the supplements done above. If scattering happens on a point Coulomb field having the charge Ze, then the external potential Fourier-images entering into $\mathcal{M}_{i,f}$ are determined by the expression (18.192).

The above-stated scheme may be applied to finding a differential cross section of scattering by any external electromagnetic field. Note, when an external field depends on time, elastic scattering of a particle occurs with changing an energy (frequency), however, when an external field is constant in time, under elastic scattering changing an energy does not take place (coherent scattering).

18.12 Decays of elementary particles

We are coming now to the analysis of elementary particle decay processes. We have the following expression for the decay probability in a unit of time:

$$d\Gamma = (2\pi)^4\delta^{(4)}(k - \sum_{j=1}^{n_f} p_j)|\sum \mathcal{M}_{i \to f}|^2 \prod_{j=1}^{n_f} \frac{d^3 p_j}{(2\pi)^3}, \tag{18.198}$$

where k is four-dimensional momentum of a decaying particle. Since $|\sum \mathcal{M}_{i \to f}|^2$ involves the factor E^{-1}, where E is an initial particle energy, then the decay probability is decreased as the speed of a decaying particle grows. The total decay probability of an unstable particle in a unit of time is found by integrating (18.198) over all momenta of secondary particles. If in this case there are k-modes (channels) of a decay, then the total probability Γ_t will be given by a sum of decay probabilities on every channel

$$\Gamma_t = \sum \Gamma_k.$$

The total decay probability in a unit of time Γ_t is connected with a lifetime of an unstable particle by the relation:

$$\tau = \frac{1}{\Gamma_t} \left(= \frac{\hbar}{\Gamma_t} \right). \tag{18.199}$$

The lifetime τ defines a time interval, during which the number of initial unstable particles is decreased by a factor equal to e ($e = 2.7183..$). The lifetime of a resting unstable elementary particle τ_0 is a physical characteristic of a particle, it being given in tables of elementary particle properties. In a reference frame where an unstable particle moves with a speed $v = p/E$, the lifetime is increased by a factor equal to E/m as compared with τ_0.

As an example, we consider a two-particle decay of the W-boson

$$W \to a + \bar{b}. \tag{18.200}$$

The differential decay probability $d\Gamma_{W \to a\bar{b}}$ in a unit of time for the resting W-boson is determined as

$$d\Gamma = (2\pi)^4 \delta^{(4)}(k - p_a - p_b) \frac{1}{m_W E_a E_b} |\sum \mathcal{M}'_{W \to a\bar{b}}|^2 \frac{d^3 p_a d^3 p_b}{(2\pi)^6}, \tag{18.201}$$

where for the sake of convenience we have selected factors m_W, E_a, and E_b, which are connected with the normalization, from $|\mathcal{M}_{W \to a\bar{b}}|^2$. Next, we assume that a Lagrangian form describing the $Wa\bar{b}$-interaction is unknown for us. Invoke the relativistic covariance for help and try to find the expression for the total decay probability. Obviously, the matrix element $\mathcal{M}_{W \to a\bar{b}}$ may be presented in the form:

$$\mathcal{M}_{W \to a\bar{b}} = g \vartheta_\mu^\lambda(\mathbf{k}) M^\mu, \tag{18.202}$$

where g is a constant of $Wa\bar{b}$-interaction and $\vartheta_\mu^\lambda(\mathbf{k})$ is a four-dimensional vector of the W-boson polarization. To average over the spin states of the W-boson leads to the result:

$$\frac{1}{3} \sum_\lambda |\mathcal{M}_{W \to a\bar{b}}|^2 = -\frac{g^2}{3} \left(g_{\mu\nu} - \frac{k_\mu k_\nu}{m_W^2} \right) M^\mu M^{\nu\dagger}. \tag{18.203}$$

From the relativistic invariance demand follows that the quantity $M^\mu M^{\nu\dagger}$ must be a four-dimensional tensor of the second rank. We can construct such a tensor using independent quantities that are at our disposal. In the case in question we have, in the capacity of those, two vectors k^μ, $(p_a - p_b)^\mu$ and the metric tensor $g^{\mu\nu}$. The vector k^μ is out, since the transversality condition:

$$k^\mu \vartheta_\mu^\lambda(\mathbf{k}) = 0$$

is valid for a massive vector field. Then, we obtain for $M^\mu M^{\nu\dagger}$

$$M^\mu M^{\nu\dagger} = A(p_a - p_b)^\mu (p_a - p_b)^\nu + B g^{\mu\nu}, \tag{18.204}$$

where in the most general case A and B are analytical functions of invariant variable $q^2 = (p_a + p_b)^2$ to be taken in the point $q^2 = m_W^2$. Substitution of (18.204) into (18.201) with allowance made for the relations

$$E_a - E_b = \frac{E_a^2 - E_b^2}{E_a + E_b} = \frac{m_a^2 - m_b^2}{m_W}, \tag{18.205}$$

$$4E_aE_b = (E_a + E_b)^2 - (E_a - E_b)^2 = m_W^2 - \frac{(m_a^2 - m_b^2)^2}{m_W^2}, \tag{18.206}$$

$$p^2 = \frac{1}{2} \left(E_a^2 + E_b^2 - m_a^2 - m_b^2 \right), \tag{18.207}$$

where $\mid \mathbf{p}_b \mid = \mid \mathbf{p}_a \mid = p$, leads to the expression:

$$d\Gamma = (2\pi)^4 \delta^{(4)}(k - p_a - p_b)\frac{g^2}{3m_W E_a E_b} \{m_W^{-2} A[w(m_W^2, m_a^2, m_b^2)]^2 -$$

$$-3B\} \frac{d^3p_a d^3p_b}{(2\pi)^6}, \tag{18.208}$$

where the function $w(A, B, C)$ being completely antisymmetric with respect to all its variables has the form:

$$w(A, B, C) = [A^2 + B^2 + C^2 - 2(AB + AC + BC)]^{1/2}. \tag{18.209}$$

To determine the total decay probability we should integrate the expression (18.208) over momenta of decay products. In this case a knowledge of the following integral:

$$I = \int \delta^{(4)}(k - p_a - p_b)\frac{d^3p_a d^3p_b}{4E_a E_b}. \tag{18.210}$$

proves to be useful. To integrate this expression over the three-dimensional momentum of the a particle gives:

$$I = \int \delta(m_W - E_a - E_b)\frac{d^3p_b}{4E_b E_a}. \tag{18.211}$$

Introduce the spherical coordinate system

$$d^3p_b = p^2 dp d\varphi d(\cos\theta)$$

and allow for that all directions of the vector $\mathbf{p}$ are equiprobable. Then (18.211) takes the view:

$$I = \pi \int \frac{p^2 dp}{E_a E_b}\delta(m_W - E_a - E_b) = \pi \int \frac{p^2}{E_a E_b} \left(\frac{d(E_a + E_b)}{dp} \right)^{-1} \delta(m_W - E_a - E_b)\times$$

$$\times d(E_a + E_b) = \frac{\pi p}{m_W} = \frac{\pi}{2}m_W^{-2}w(m_W^2, m_a^2, m_b^2), \tag{18.212}$$

where at the final stage of calculations we have taken into account the relations (18.205)–(18.207). Using (18.212) gets the following expression for the total cross section of the decay (18.200):

$$\Gamma_{W \to a\bar{b}} = \frac{g^2}{6\pi m_W^3} \{m_W^{-2} A[w(m_W^2, m_a^2, m_b^2)]^3 - 3Bw(m_W^2, m_a^2, m_b^2)\}. \tag{18.213}$$

It is clear that to define the quantities A and B we should have additional sources of information. If a successive and internally consistent theory describing the W-bosons interaction were absent, we would have to address experiments. However, nowadays we have the SM to be in perfect accord with modern experimental data. With its help we could get the explicit expression for the matrix element of the reaction (18.200). For example, when $a = l$ and $\bar{b} = \bar{\nu}_l$, A (B) proves to be equal to -1/2 -$(m_W^2 - m_l^2)/2$, where for reasons of simplicity we have neglected the antineutrino mass.

18.13 Cross section for polarized particles

In this section we take into consideration the polarization states of interacting particles. We start with a case when one electron is present both in an initial state and in a final one. Then the process amplitude is given by

$$\mathcal{M}_{i\to f} = \bar{u}_f O u_i, \qquad (18.214)$$

where $u_i \equiv u(\mathbf{p}_i)$, $u_f \equiv u(\mathbf{p}_f)$, and O is a 4×4-matrix whose concrete form is unessential in this context. The process probability is proportional to the amplitude module square:

$$dw_{i\to f} \sim |\mathcal{M}_{i\to f}|^2 = (\bar{u}_f O u_i)(\bar{u}_i \overline{O} u_f) = u_{i\alpha}\bar{u}_{i\alpha'}\overline{O}_{\alpha'\beta'} u_{f\beta'}\bar{u}_{f\beta} O_{\beta\alpha}, \qquad (18.215)$$

where $\overline{O} = \gamma_0 O^\dagger \gamma_0$ and we have set down spinor indices in the explicit form. When an electron in a initial state is completely polarized, then the formula (18.215) does not undergo any changes. In the case of a partially polarized initial electron one should carry out the following change in the formula (18.215):

$$u_{i\alpha}\bar{u}_{i\alpha'} \to (\rho_i^{(+)})_{\alpha\alpha'},$$

where $\rho_i^{(+)}$ is a polarization density matrix* of the electron in the initial state. As far as the definition of the electron polarization in a final state is concerned, then, from the point of view of experiment, two situations are possible here: (i) measurement of the probability that the electron in a final state will possess definite polarization; and (ii) finding of all probable polarization states, that is, establishing the density matrix of the electron in a final state. In the first case, a detector selects the definite polarization state of the electron. If this state is characterized by a four-dimensional vector of polarization s_d, then one must replace the bilinear combination $u_{f\beta}\bar{u}_{f\beta'}$ with the density matrix $(\rho_d^{(+)})_{\beta\beta'}$ in (18.215). Here $\rho_d^{(+)}$ represents the density matrix to define a detector and have the same structure as the matrix $\rho_i^{(+)}$: $\rho_i^{(+)}$

$$\rho_d^{(+)} = \frac{1}{2}\left[\frac{\hat{p}_f + m}{2m}(1 + \gamma_5 \hat{s}_d)\right], \qquad (18.216)$$

where $(p_f \cdot s_d) = 0$. So, if the initial electron is partially polarized and the definite polarization state d of the final electron is fixed, then the probability of the process under investigation is defined as:

$$dw_{i\to f} \sim \mathrm{Sp}(\rho_i^{(+)}\overline{O}\rho_d^{(+)}O). \qquad (18.217)$$

---

*We shall further call it by the density matrix for brevity sake.

It is obvious that interaction with a detector changes the electron state. Therefore, the matrix ρ_d and with it the probability $dw_{i \to f}$ characterizes not only the electron scattering process itself but detector properties too, that is, its capabilities for selection of one or an other electron polarization. When we are interested in a polarization state of the final electron in which the electron is brought by the interaction process itself, then we come to the second setting up a problem. The probability, that the electron described by the bispinor u_f possesses the definite polarization to be associated with the bispinor u_d, is proportional to $|\overline{u}_d u_f|^2 = u_{d\alpha} \overline{u}_{d\beta} u_{f\beta} \overline{u}_{f\alpha}$. Since the case in point is the states with partial polarization rather than a definite one, then the changes

$$u_{f\beta} \overline{u}_{f\alpha} \to (\rho_f^{(+)})_{\beta\alpha}, \qquad u_{d\alpha} \overline{u}_{d\beta} \to (\rho_d^{(+)})_{\alpha\beta} \tag{18.218}$$

must be done in the expression for the probability. Thus, the required probability proves to be proportional to the expression $\mathrm{Sp}(\rho_d^{(+)} \rho_f^{(+)})$. As the same probability is defined by the expression (18.217), then the conclusion follows

$$\mathrm{Sp}(\rho_i^{(+)} \overline{O} \rho_d^{(+)} O) \sim \mathrm{Sp}(\rho_f^{(+)} \rho_d^{(+)}). \tag{18.219}$$

So, we have for the density matrix of the electron in the final state:

$$\rho_f^{(+)} \sim' O \rho_i^{(+)} \overline{O}. \tag{18.220}$$

However, our detector has been tuned only on the electron recording, therefore, the matrix O must have somehow stored information about this fact. Accounting for:

$$\mathcal{P}_+ u_f = u_f,$$

where $\mathcal{P}_+ = (\hat{p}_f + m)/2m$ represents the projection operator onto the electron states, the matrix element (18.214) may be reported as:

$$\mathcal{M}_{i \to f} = \overline{u}_f \mathcal{P}_+ O u_i.$$

Then, the matrix $\rho_f^{(+)}$ will be given by:

$$\rho_f^{(+)} = N \mathcal{P}_+ O \rho_i^{(+)} \overline{O} \mathcal{P}_+. \tag{18.221}$$

Normalization factor N is determined by the normalization condition of the density matrix $\mathrm{Sp}(\rho_f) = 1$:

$$N^{-1} = \frac{1}{2m} \mathrm{Sp}\left[O \rho_i^{(+)} \overline{O} (\hat{p}_f + m) \right]. \tag{18.222}$$

To replace (18.222) into (18.221) gives the final expression for the density matrix of the electron in the final state:

$$\rho_f^{(+)} = \frac{1}{2m} \frac{[(\hat{p}_f + m) O \rho_i^{(+)} \overline{O} (\hat{p}_f + m)]}{\mathrm{Sp}[(\hat{p}_f + m) O \rho_i^{(+)} \overline{O}]}. \tag{18.223}$$

This matrix is defined only by the four-dimensional polarization vector of the electron in the final state s_f:

$$\rho_f^{(+)} = \frac{1}{2} \left[\frac{\hat{p}_f + m}{2m} (1 + \gamma_5 \hat{s}_f) \right], \tag{18.224}$$

where $(s_f \cdot p_f) = 0$. Knowledge of $\rho_f^{(+)}$ allows to find s_f

$$s_{f\mu} = \mathrm{Sp}(\rho_f^{(+)} \gamma_5 \gamma_\mu),$$

which, in its turn, unambiguously defines the three-dimensional polarization vector $\boldsymbol{\xi}$.

In a similar spirit we may also consider cases when initial and final states are positron or one of the states is electron while the other is positron. For example, the expression for the density matrix of the positron in the final state has the form:

$$\rho_f^{(-)} = \frac{1}{2m} \frac{[(-\hat{p}_f + m)O\rho_i^{(-)}\overline{O}(-\hat{p}_f + m)]}{\mathrm{Sp}[(-\hat{p}_f + m)O\rho_i^{(-)}\overline{O}]}. \tag{18.225}$$

The results obtained are generalized without trouble in cases of many electrons and positrons.

Now we shall pass to the investigation of processes with the participation of partially polarized photons. Let us consider a process when the initial state contains one photon and the final state contains one photon as well. In this case the matrix element of the process could be presented in the form:

$$S_{i \to f} = e_{f\mu}^{\lambda'*} O^{\mu\nu} e_{i\nu}^{\lambda}, \tag{18.226}$$

where e^{λ} are photon polarization vectors. If the initial photon is partially polarized, then in the matrix element square

$$|S_{i \to f}|^2 = e_{f\mu}^{\lambda'*} O^{\mu\nu} e_{i\nu}^{\lambda} e_{f\tau}^{\lambda'} O^{\tau\sigma*} e_{i\sigma}^{\lambda*} \tag{18.227}$$

the replacement:

$$e_{i\nu}^{\lambda} e_{i\sigma}^{\lambda*} \to (\rho_i^{\gamma})_{\nu\sigma}$$

should be fulfilled. As this takes place, when three-dimensional polarization vectors figure in the relation (18.226), then the density matrix is given by:

$$(\rho_i^{\gamma})_{ll'} = \frac{1}{2} [1 + (\boldsymbol{\xi}_i \cdot \boldsymbol{\sigma})]_{ll'}, \tag{18.228}$$

where $\boldsymbol{\xi}_i$ are Stokes's parameters for the photon in the initial state and $l, l' = 1, 2$. If four-dimensional polarization vectors enter into the relation (18.226), then the density matrix will be defined as follows:

$$(\rho_i^{\gamma})_{\mu\nu} = \frac{1}{2} \sum_{l,l'=1}^{2} [1 + (\boldsymbol{\xi}_i \cdot \boldsymbol{\sigma})]_{ll'} (n_i^{(l)})_\mu (n_i^{(l')})_\nu = \frac{1}{2} [(n_i^{(1)})_\mu (n_i^{(1)})_\nu + (n_i^{(2)})_\mu (n_i^{(2)})_\nu] +$$

$$+ \frac{\xi_1}{2} [(n_i^{(1)})_\mu (n_i^{(2)})_\nu + (n_i^{(2)})_\mu (n_i^{(1)})_\nu] - \frac{i\xi_2}{2} [(n_i^{(1)})_\mu (n_i^{(2)})_\nu - (n_i^{(2)})_\mu (n_i^{(1)})_\nu] +$$

$$+ \frac{\xi_3}{2} [(n_i^{(1)})_\mu (n_i^{(1)})_\nu - (n_i^{(2)})_\mu (n_i^{(2)})_\nu], \tag{18.229}$$

where four-dimensional vectors $(n_i^{(l)})_\mu$ $(l = 1, 2)$ satisfy the relations:

$$\left. \begin{array}{l} (n_i^{(l)} \cdot n^{(l')_i}) = -\delta_{ll'} \\ (k_i \cdot n_i^{(l)}) = 0 \end{array} \right\} \tag{18.230}$$

and k_i is a four-dimensional momentum of the photon in the initial state.

Next, we shall discuss questions of determining the photon polarization in a final state. Here, in just the same way as in the case of any particles with the spin, two setting up problems are possible: (i) finding the probability that the final photon possesses definite polarization; and (ii) determining the density matrix of the final photon. To solve the first problem is trivial and resides in the replacement of the bilinear combination $e_{f\nu}^{\lambda'} e_{f\mu}^{\lambda'*}$ involved in (18.227) with the density matrix $(\rho_d^{(\gamma)})_{\nu\mu}$ to characterize a detector:

$$\rho_d^\gamma = \rho_i^\gamma(\xi_i \to \xi_d, n_i^{(l)} \to n_d^{(l)}),$$

where $\boldsymbol{\xi}_d$ are Stokes's parameters for a detector, $n_d^{(l)}$ are two vectors satisfying the relations:

$$(n_d^{(l)} \cdot n_d^{(l')}) = -\delta_{ll'}, \qquad (k_f \cdot n_d^{(j)}) = 0$$

and k_f is a four-dimensional momentum of the photon in the final state. The probability, that the photon whose polarization in the initial state was being set by the matrix ρ_i^γ will have polarization set by the matrix ρ_d^γ, is defined by the expression:

$$dw(s_d) \sim \sum_{\lambda,\lambda'} g^\lambda g^{\lambda'} e_{d\mu}^{\lambda'*} O_{\mu\nu} e_{i\nu}^{\lambda} e_{d\mu'}^{\lambda'} O_{\mu'\nu'}^* e_{i\nu'}^{\lambda*} = \mathrm{Sp}(\rho_d^\gamma O \rho_i^\gamma O^\dagger), \qquad (18.231)$$

where g^λ is a relative probability with which λ-polarization goes into a mixed beam.

To solve the second problem we take into consideration this fact: if the photon in the final state is characterized by the four-dimensional polarization vector $e_{f\mu}^\lambda$, then the probability that the state with the four-dimensional polarization vector $e_{d\mu}^{\lambda'}$ will be recorded with the help of a detector is given by the quantity:

$$dW \sim |e_{f\mu}^\lambda e_{d\mu}^{\lambda'*}|^2 = e_{f\mu}^\lambda e_{d\mu}^{\lambda'*} e_{f\mu'}^{\lambda*} e_{d\mu'}^{\lambda'}. \qquad (18.232)$$

When we are dealing with partially polarized states rather than with pure ones, then in the expression (18.232) we should carry out the changes:

$$e_{f\mu}^\lambda e_{f\mu'}^{\lambda*} \to (\rho_f^\gamma)_{\mu\mu'}, \qquad e_{d\mu}^{\lambda'} e_{d\mu'}^{\lambda'*} \to (\rho_d^\gamma)_{\mu\mu'}, \qquad (18.233)$$

where ρ_f^γ is a density matrix of the photon in the final state. Then, for the required probability we get:

$$dw(s_d) \sim \mathrm{Sp}(\rho_f^\gamma \rho_d^\gamma). \qquad (18.234)$$

To compare (18.234) with (18.231) gives the relation:

$$\rho_f^\gamma = N(O\rho_i^\gamma O^\dagger). \qquad (18.235)$$

Having defined the coefficient N from the normalization condition, we come to the final expression for the density matrix of the photon in the final state:

$$\rho_f^\gamma = \frac{O\rho_i^\gamma O^\dagger}{\mathrm{Sp}(O\rho_i^\gamma O^\dagger)}. \qquad (18.236)$$

Substituting the definition of the matrices ρ_d^γ and ρ_f^γ, where

$$(\rho_f^\gamma)_{\mu\nu} = \frac{1}{2} \left[1 + (\boldsymbol{\xi}_f \cdot \boldsymbol{\sigma})\right]_{ll'} (n_f^{(l)})_\mu (n_f^{(l')})_\nu, \qquad (18.237)$$

and $n_f^{(l)} \equiv n_d^{(l)}$, into the relation (18.234), we arrive at:

$$dw(s_d) \sim \mathrm{Sp}(\rho_f^\gamma \rho_d^\gamma) = \frac{1}{2} \left[1 + (\boldsymbol{\xi}_f \cdot \boldsymbol{\xi}_d)\right]. \qquad (18.238)$$

The meaning of the obtained formula is as follows. If the cross section summed up over polarizations of the photon in the final state equals σ, then the production cross section of the photon with polarization parameters $\boldsymbol{\xi}_d$ will be given by the expression:

$$\sigma_d = \frac{1}{2}\sigma \left[1 + (\boldsymbol{\xi}_f \cdot \boldsymbol{\xi}_d)\right]. \qquad (18.239)$$

The reader could find more detailed information about the density matrix and its usage in calculating probabilities of different processes in elementary particles physics in the book [17].

18.14   Statistical hypothesis

As follows from the formulae (18.169) and (18.198) the differential cross section $d\sigma$ and the differential decay probability in a unit of time $d\Gamma$ are defined by the phase volume element dJ

$$dJ = (2\pi)^4 \delta^{(4)}(\sum_i p_i - \sum_f p_f) \prod_{j=1}^{n_f} \frac{d^3 p'_j}{(2\pi)^3} \qquad (18.240)$$

and several known factors. It is clear that dJ is not a relativistic invariant. However, the wavefunctions of free particles involve the normalization factors $(2E)^{-1/2}$ each of them being squared also appears in the expression describing the quantum transition. Taking into consideration this circumstance, we can always pass to the relativistically invariant phase volume element of secondary (final) particles

$$dI = (2\pi)^4 \delta^{(4)}(\sum_i p_i - \sum_f p_f) \prod_{j=1}^{n_f} \frac{d^3 p'_j}{2E_j (2\pi)^3} \qquad (18.241)$$

The formulae (18.169), (18.198) could be simplified by the assumption that the quantity $|\mathcal{M}_{i \to f}|^2$ is constant. Then $d\sigma$ and $d\Gamma$ become proportional to the relativistic phase volumes of secondary particles. This assumption is called a *statistical hypothesis* or a *phase volume approximation*. In the context of this approach, taking into consideration the four-dimensional momentum conservation laws and not concretizing process dynamics, one could estimate momentum-angle distributions of secondary particles.

The statistical hypothesis corresponds to such mechanisms of producing secondary particles in which dynamic correlations both between four-dimensional momenta of secondary particles and between four-dimensional momenta of initial and final particles are absent. Since the statistical hypothesis does not allow for interaction dynamics, then in the general case its predictions are not justified in the experiments. However, under lack of knowledge concerning a process dynamics such predictions give preliminary information about momentum-angle distributions of secondary particles. In a number of cases the similar approach allows formulation of the radically new and most effective methods: (i) for experimental search of new particles; (ii) for investigating properties of new particles and processes. Thus, for example, calculating the group spectra of secondary particles in their effective mass is realized on the basis of the statistical hypothesis, methods of experimental resonances identification extensively relying on results of these calculations. The effective mass distributions of the secondary particles group found in such a way effectively reproduce spectrum of background events or, at least, allow one to predetermine abilities of an experimental facility to be required for solving problems appearing in physics of resonances.

As an example, let us consider the effective mass distribution of two particles in a three-body system. The relativistically invariant phase volume element of a three-body state has the form

$$dI_3 \sim \delta^{(4)}(p - p_1 - p_2 - p_3)\frac{d^3p_1 d^3p_2 d^3p_3}{E_1 E_2 E_3}, \tag{18.242}$$

where p_i and E_i ($i = 1, 2, 3$) is a three-dimensional momentum and an energy of the ith particle in a final state. Since the phase volume model does not pretend to predict the absolute values of the cross sections and the decay probabilities, numerical factors in the expressions for the relativistic phase volumes are omitted. Comparison of the effective mass distribution of two secondary particles, which was obtained on the basis of Eq. (18.242), with the experimental data is put into practice only on the distribution form and not on the absolute value. By this reason, it is convenient to normalize the distribution (18.242) on a unit value of a total phase volume, that is, to calculate the quantity

$$dW_3 = \frac{dI_3}{I_3}, \tag{18.243}$$

where

$$I_3 = \int dI_3.$$

Introducing the effective mass of particles 1 and 2

$$s_{12} = \sqrt{(p_1 + p_2)^2}$$

and integrating (18.242) over d^3p_1 and d^3p_2, we obtain

$$dI_3 \sim \frac{d^3p_3}{E_3}W_2(\sqrt{s_{12}}, m_1, m_2),$$

where

$$W_2(\sqrt{s_{12}}, m_1, m_2) = \frac{2\pi\sqrt{|s_{12} - (m_1 + m_2)^2||s_{12} - (m_1 - m_2)^2|}}{s_{12}}.$$

As the phase volume element is a relativistic invariant, calculating the integral over d^3p could be fulfilled in an arbitrary reference frame. Let us work in the center-of-mass system of all three secondary particles. Inasmuch as in this system

$$E_1 + E_2 + E_3 = \sqrt{s},$$

where $\sqrt{s}$ is the effective mass of three particles system, then E_3 is defined by the equation

$$E_3 + \sqrt{E_3^2 + s_{12} - m_3^2} = \sqrt{s}, \tag{18.244}$$

whose root is as follows

$$E_3 = \frac{s - s_{12} + m_3^2}{2\sqrt{s}}. \tag{18.245}$$

Expressing components of the vector $\mathbf{p}_3$ through the spherical coordinates and changing the variables from E_3 to s_{12} with the help of Eq. (18.245), we get

$$\frac{d^3p_3}{E_3} = |\mathbf{p}_3|dE_3 d\Omega_3 = \frac{|\mathbf{p}_3|ds_{12}d\Omega_3}{2\sqrt{s}}, \tag{18.246}$$

where

$$|\mathbf{p}_3| = \sqrt{E_3^2 - m_3^2} = \frac{\sqrt{[s - (\sqrt{s_{12}} + m_3)^2][s - (\sqrt{s_{12}} - m_3)^2]}}{2\sqrt{s}} =$$

$$= \frac{\sqrt{[(\sqrt{s} + m_3)^2 - s_{12}][(\sqrt{s} - m_3)^2 - s_{12}]}}{2\sqrt{s}}.$$

To integrate over the solid angle $d\Omega_3$ leads to the following expression for the effective mass square events distribution to normalize on a unit

$$\frac{dW_3}{ds_{12}} = \sqrt{[(s_{12} - (m_1 + m_2)^2][s_{12} - (m_1 - m_2)^2]} \times$$

$$\times \frac{2\pi^2 \sqrt{[(\sqrt{s} + m_3)^2 - s_{12}][(\sqrt{s} - m_3)^2 - s_{12}]}}{s s_{12} I_3(\sqrt{s}, m_1, m_2, m_3)}, \qquad (18.247)$$

where

$$I_3(\sqrt{s}, m_1, m_2, m_3) = \frac{2\pi^2}{s} \int \frac{ds_{12}}{s_{12}} \sqrt{[(\sqrt{s} + m_3)^2 - s_{12}]} \times$$

$$\times \sqrt{[(\sqrt{s} - m_3)^2 - s_{12}][s_{12} - (m_1 + m_2)^2][s_{12} - (m_1 - m_2)^2]}. \qquad (18.248)$$

The integration limits in Eq. (18.248) are given in the following way

$$(m_1 + m_2)^2 \le s_{12} \le (\sqrt{s} - m_3)^2,$$

where the upper integration bound follows from the inequality $E_3 \ge m_3$, E_3 being determined by the expression (18.245). Normalization of (18.247) on a unit area under the distribution curve is usually performed by means of numerical calculations of the function $I_3(\sqrt{s}, m_1, m_2, m_3)$ for the specific parameter values. From (18.247) it appears that the function dW_3/ds_{12} turns into zero under the upper and lower limits of changing s_{12} and has a smooth monotonous behavior in a region placed between zeros. The maximum of dW_3/ds_{12} is situated somewhere in the middle of a possible values interval of s_{12}. As a matter of principle, in the maximum region the form of dW_3/ds_{12} distribution has little in common with the effective mass Breit-Wigner resonance distribution

$$\frac{dN}{dm^2} \sim \frac{1}{[(m^2 - m_{res}^2)^2 + (m_{res}\Gamma/2)^2]}, \qquad (18.249)$$

where $m^2 = (\sum p_f)^2$, p_f is a four-dimensional momenta of decay products (when comparing Eq. (18.249) with Eq. (18.247), one should set $m^2 = s_{12}$). Moreover, the Breit-Wigner peak always is at the same value of the effective mass equal the resonance mass $s_{12} = m_{res}$. The maximum of the phase curve dW_3/ds_{12} changes its position depending on s. In that way, there is a fundamental capability to identify resonances against a background of nonresonance statistical mass distributions.

19

Transmission of γ-radiation through matter

"Over the Mountains Of the Moon,
Down the Valley of the Shadow,
Ride, boldly ride,"
The shade replied,—
"If you seek for Eldorado!"
Edgar Allan Poe, "Eldorado"

The physics of elementary particles originally has dealt with electrons and γ-radiation, so that here the greatest amount, both of the key ideas and of technical methods, have been stored. This and the next sections will be devoted to the technical side of the QED, the calculations of particular processes. For today the QED lost the status of the precise theory in the sense that it is already not a complete theory. It, together with the weak interaction theory, forms the electroweak interaction theory. If one considers some pure electrodynamic process (there are only charged leptons and photons in the initial and final states) in the electroweak theory, then, as a rule, diagrams, additionally in relation to the QED, appear with virtual gauge bosons (W^+ and Z), neutral leptons and quarks appear. For example, in the second order of the perturbation theory, every time while the Feynman diagram has an internal photon line, then the Z-boson contribution is also possible. Such processes as

$$e^- e^+ \to e^- e^+, \mu^+ \mu^-, \tau^+ \tau^-,$$

having the s-channel Feynman diagram will feel the Z-boson effect more strongly as the energy in the center-of-mass frame is approached in the value m_Z. There are, however, restricted numbers of processes that proceed at the cost only of the electromagnetic interaction in the second order of the perturbation theory. Under transition to the third and higher orders of the perturbation theory all electrodynamic processes include contributions of the weak interaction. However, the QED is the most simple example of a gauge theory of interacting fields. Understanding its structure will help us to further advance.

19.1 Photoeffect (nonrelativistic case)

In 1887 the phenomenon of electron emission by metallic sodium irradiated with a mercury-vapor lamp was revealed by H. Hertz.* The process of light absorption with subsequent electron emission was given the name *photoeffect*. For this phenomenon A. Stoletov (1888) and F. Lenard (1902) have stated the following empirical laws: (i) a maximum kinetic energy of photoelectrons T_{max} is independent on the light intensity, (ii) T_{max} is linearly

---

*Recall, that H. Hertz has experimentally confirmed a wave nature of the electromagnetic radiation and hence the validity of Maxwell's electrodynamics.

dependent on the light frequency ω, and (iii) the photoeffect has a threshold frequency ω_0. However, these laws could not be explained within the scope of classical physics.

A theory of the photoeffect was proposed in 1905 by Einstein who had logically developed the Planck theory of light quanta. According to Einstein, light is emitted by quanta (in line with the Planck hypothesis), its propagation being due to the quanta as well. Based on a theory of the electromagnetic field energy quantization and assuming that at the photoeffect a photon is interacting with an individual electron, Einstein has put forward an explanation for the photoeffect mechanisms using a single equation:

$$\hbar\omega = \hbar\omega_0 + T, \tag{19.1}$$

where $\hbar\omega_0$ is the photoelectric work function of electrons from matter, and, for the time being, we cross to the ordinary system of units. Verifying Einstein's equation, in 1916 R. Millikan defined the quantity $\hbar$ that proved to be identical to the quantity found by Planck in a theory of thermal radiation.

A free electron is unable to absorb a photon. This transpires from the fact that for the transition of an electron from the state with the momentum $p_\mu = (E/c, \mathbf{p})$ to the state with the momentum $p'_\mu = (E'/c, \mathbf{p}')$ in the absence of the third body (for example, condensed medium, atom, etc.), the conservation laws for the energy and three-dimensional momentum:

$$E - E' = \hbar\omega, \qquad p - p' = \hbar\omega/c$$

are inconsistent at all electron speeds $v < c$. By the conventional terminology, the photoeffect in a condensed medium is called the *photoemissive effect*, and the transition of an electron from one of the atomic or molecular bound states to a continuous spectrum is referred to as *photoionization*. The energy required to carry an electron from the ground state to a continuous spectrum state with $T = 0$ is called *atom ionization energy*, and, when it is measured in eV, this energy is designated ionization potential I. In a hydrogen-like atom we have

$$I = \frac{mZ^2e^4}{2(4\pi)^2\hbar^2} = Z^2 I_0.$$

In multi electron atoms high energetic photons could knock off electrons from different electron shells. Such atoms have several ionization potentials.

Another characteristic of photoeffect is the photoionization cross section σ. By definition this is the ratio of ionization acts number per one atom in a unit of time and volume to an intensity of the coming photon flux (flux is supposed to be monochromatic). The quantity σ may be analytically calculated for the hydrogen atom and for a hydrogen-like atom with $Z \ll 137$.

First, we shall work out the cross section in a nonrelativistic case. So, in the initial state there is one electron in the bound state with the energy $E_i = -I$ and a photon with the momentum $\mathbf{k}$. The final state, belonging to a continuous spectrum, is defined by the momentum $\mathbf{p}$ and the energy $E_f \equiv E$. It is obvious that the process could be described with the help of the concept of an external field entering into the electron Hamiltonian. The electron-positron field operator may be expanded into the eigenfunctions of this Hamiltonian, the electron state concept in itself taking into account the electron interaction with environmental objects. Under such an approach investigating the processes of emitting and absorbing a photon using the scattering matrix of the first order becomes possible. It should be emphasized that only the S-matrix element of the first order differs from zero for the photoeffect. Therefore, if we use the exact wavefunctions of the electron in the nucleus field, then the photoeffect cross section obtained in the first order of the perturbation theory represents the exact expression.

Let $\psi_i(x) = \psi_i(\mathbf{r})e^{iIt}$ and $\psi_f(x) = \psi_f(\mathbf{r})e^{-iEt}$ be the wavefunctions of the initial and final electron, $A_\mu(x) = A_\mu(\mathbf{r})e^{-i\omega t}$ be the photon wavefunction (we have come back to the natural system of units). Then the matrix element of the operator $S^{(1)}$ could be represented in the form:

$$S^{(1)}_{i\to f} = -2\pi i \mathcal{A}_{i\to f}\delta(-I + \omega - E), \tag{19.2}$$

where the amplitude of the photon absorption $\mathcal{A}_{i\to f}$ is connected with the transition current:

$$j_\mu(x) = e\overline{\psi}_f(x)\gamma_\mu\psi_i(x) = j_\mu(\mathbf{r})e^{i(I+E)t} \tag{19.3}$$

by the relation:

$$\mathcal{A}_{i\to f} = \int j^\mu(\mathbf{r})A_\mu(\mathbf{r})d^3x, \tag{19.4}$$

We choose the photon wavefunction in the three-dimensional transverse gauge, that is, the four-dimensional polarization vector has the form $e_\mu^\lambda(\mathbf{k}) = (0, \mathbf{e}^\lambda(\mathbf{k}))$. Since the emitted electron belongs to a continuous spectrum, then the photoeffect cross section will be given by the formula:

$$d\sigma = 2\pi|\mathcal{A}_{i\to f}|^2\delta(-I + \omega - E)\frac{d^3p}{(2\pi)^3}, \tag{19.5}$$

where the absorption amplitude has the appearance:

$$\mathcal{A}_{i\to f} = -e\frac{1}{\sqrt{2\omega}}\int \overline{\psi}_f(\mathbf{r})(\boldsymbol{\gamma}\mathbf{e}^\lambda(\mathbf{k}))\psi_i(\mathbf{r})e^{i\mathbf{k}\mathbf{r}}d^3x. \tag{19.6}$$

Under writing of Eq. (19.5) we have also allowed for the fact that the wavefunctions, both of the electron and the photon, are normalized on one particle in the volume $V = 1$.

Let us fulfill calculations for two regions of the photon energy: for $\omega \ll m$ and for $\omega \gg I$. As $I \sim mZ^2\alpha^2 \ll m$, then these two regions are partially overlapped under $I \ll \omega \ll m$. Therefore, investigation of these regions gives the complete pattern of the photoeffect in the nonrelativistic approximation.

When $\omega \ll m$, then the electron speeds are nonrelativistic both in the initial state and in the final one. That allows us to replace the quantity $\gamma^0\boldsymbol{\gamma} = \boldsymbol{\alpha}$ with the nonrelativistic speed operator $\mathbf{v} = -i\boldsymbol{\nabla}/m$ in Eq. (19.6). Assuming that the photon wavelength is much greater than the atom size, we may change the quantity $e^{i\mathbf{k}\mathbf{r}}$ with 1. We also carry out the transition to the spherical coordinate system with $Oz\|\mathbf{k}$:

$$d^3p = \mathbf{p}^2d|\mathbf{p}|d\varphi\sin\theta d\theta = E|\mathbf{p}|dEd\Omega$$

and integrate (19.5) over E. After these operations we arrive at the expression for the differential cross section:

$$d\sigma = e^2\frac{m|\mathbf{p}|}{8\pi^2\omega}|(\mathbf{e}^\lambda(\mathbf{k})\mathbf{L}_{i\to f})|^2d\Omega, \tag{19.7}$$

where

$$\mathbf{L}_{i\to f} = -\frac{1}{m}\int(\psi_f^\dagger\boldsymbol{\nabla}\psi_i)d^3x \tag{19.8}$$

(we have saved the minus sign in (19.8) for the sake of convenience only).

We shall be constrained by investigating the photoeffect for the ground level of a hydrogen atom or a hydrogen-like ion. Then, the wavefunction of the initial electron is defined by the expression:

$$\psi_i(\mathbf{r}) = Ne^{-\eta r}, \tag{19.9}$$

where

$$\eta = \frac{Z}{a_0}, \qquad a_0 = \frac{4\pi}{me^2} = \frac{1}{m\alpha}, \qquad N = \sqrt{\frac{Z^3}{\pi a_0^3}}.$$

As far as the electron wavefunction in the final state is concerned ψ_f, then it must be the electron wavefunction in the Coulomb's nucleus field but belongs to continuous spectrum already. Recall, that in the centrally symmetric field alongside the system of the wavefunctions for the discrete spectrum:

$$\psi_{nlm} = R_{nl}(r)Y_{lm}(\theta, \varphi), \tag{19.10}$$

($R_{nl}(r)$ is the radial part of the wavefunction and $Y_{lm}(\theta, \varphi)$ is the angle part of that) which corresponds to the states with the definite energy, angular moment and its projection, we have the system of the functions for the continuous spectrum associated with the state that has the definite energy but does not possess any definite values both of the angular moment and of its projection:

$$\psi_p = \frac{1}{p}\sqrt{\frac{\pi}{2}}\sum_{l=0}^{\infty} i^l(2l+1)e^{-i\delta_l}R_{pl}(r)P_l(\mathbf{nn_1}), \tag{19.11}$$

where δ_l are scattering phases, $P_l(\mathbf{nn_1})$ are the Legendre's polynomials, $\mathbf{n} = \mathbf{p}/p$, and $\mathbf{n_1} = \mathbf{r}/r$. At infinity the system (19.11) describes the plane wave plus the convergent spherical wave. In our case, thanks to the selection rules on l, the transition from the ground state (the s-state) only to the p-state is possible. Keeping only the term with $l = 1$ and omitting insignificant phase factors, we rewrite the expression (19.11) in the form:

$$\psi_f = \frac{3}{p}\sqrt{\frac{\pi}{2}}(\mathbf{nn_1})R_{p1}(r). \tag{19.12}$$

Now the quantity $(\mathbf{e}^\lambda(\mathbf{k})\mathbf{L}_{i\to f})$ to be of interest will look like:

$$(\mathbf{e}^\lambda(\mathbf{k})\mathbf{L}_{i\to f}) = \frac{3\eta^{5/2}}{\sqrt{2}mp}\int d^3x(\mathbf{nn_1})(\mathbf{e}^\lambda(\mathbf{k})\mathbf{n_1})e^{-\eta r}R_{p1}(r) = \frac{3\eta^{5/2}}{\sqrt{2}mp}\int d\Omega'\int dr\times$$

$$\times r^2(\mathbf{nn_1})(\mathbf{e}^\lambda(\mathbf{k})\mathbf{n_1})e^{-\eta r}R_{p1}(r) = \frac{12\pi\eta^{5/2}}{\sqrt{2}pm}(\mathbf{e}^\lambda(\mathbf{k})\mathbf{n})\int drr^2e^{-\eta r}R_{p1}(r). \tag{19.13}$$

The radial wavefunctions of the continuous spectrum could be represented in terms of the degenerated hypergeometric function (see, for example, Ref. [18]). We have for $R_{p1}(r)$:

$$R_{p1}(r) = \frac{2\eta}{3}\sqrt{\frac{1+\nu^2}{\nu(1-e^{-2\pi\nu})}}pre^{-ipr}F(2+i\nu, 4, 2ipr), \tag{19.14}$$

where $\nu = \eta/p$. Taking into account the formulae

$$\int_0^\infty dze^{-\beta z}z^{\gamma-1}F(\alpha, \gamma, kz) = \Gamma(\gamma)\beta^{\alpha-\gamma}(\beta-k)^{-\alpha}, \tag{19.15}$$

$$\left(\frac{\nu+i}{\nu-i}\right)^{i\nu} = e^{-2\nu/\arctan\nu} \tag{19.16}$$

and the fact that at $n = 3, 4, \dots \Gamma(n) = (n-1)!$, we get

$$(\mathbf{e}^\lambda(\mathbf{k})\mathbf{L}_{i\to f}) = \frac{2^{7/2}\pi\nu^3(\mathbf{e}^\lambda(\mathbf{k})\mathbf{n})e^{-2\nu/\arctan\nu}}{\sqrt{p}m(1+\nu^2)^{3/2}\sqrt{1-e^{-2\pi\nu}}}. \tag{19.17}$$

Gathering the obtained results, we find that the differential cross section of the photoeffect with emitting the electron within the element of the solid angle $d\Omega$ is given by the expression:

$$d\sigma = \frac{2^7 \pi \alpha a_0^2}{Z^2} \left(\frac{I}{\omega}\right)^4 \frac{e^{-4\nu/\arctan\nu}}{1 - e^{-2\pi\nu}} (\mathbf{e}^\lambda(\mathbf{k})\mathbf{n})^2 d\Omega, \tag{19.18}$$

where we have given the account of:

$$\omega = T + I = \frac{p^2}{2m}\left(1 + \nu^2\right). \tag{19.19}$$

From Eq. (19.18) follows that the angle distribution of the photoelectrons is defined by the factor $(\mathbf{e}^\lambda(\mathbf{k})\mathbf{n})^2$. It is maximum in the directions parallel to falling photons polarization direction and turns to zero both in the directions to be perpendicular to $\mathbf{e}^\lambda(\mathbf{k})$ and in the incidence direction as well.

In the case of nonpolarized photons the formula (19.18) must be averaged over the polarization directions of falling photons that could be done with the help of the relation:

$$\frac{1}{2}\sum_{\lambda=1}^{2} e_j^\lambda e_l^\lambda = \frac{1}{2}\left(\delta_{jl} - \frac{k_j k_l}{\omega^2}\right), \tag{19.20}$$

where $j, l = 1, 2, 3$. To integrate the obtained cross section gives the total cross section of the photoeffect for unpolarized particles [19]:

$$\sigma = \frac{2^9 \pi^2 \alpha a_0^2}{3Z^2}\left(\frac{I}{\omega}\right)^4 \frac{e^{-4\nu/\arctan\nu}}{1 - e^{-2\pi\nu}}. \tag{19.21}$$

From Eq. (19.21) we notice that the photoeffect cross section tends to constant directly under threshold. At $\omega \to I$ the limiting value of σ is equal to:

$$\sigma = \frac{2^9 \pi^2 \alpha a_0^2}{3e^4 Z^2} \tag{19.22}$$

(here $e = 2.718...$).

Under small excess ω above I, when still $\omega - I \ll I$ ($\nu \gg 1$), we have:

$$\sigma \sim (I/\omega)^{8/3}.$$

When $\omega - I \sim I$ ($\nu \sim 1$):

$$\sigma \sim (I/\omega)^3.$$

Away from the red border of the photoeffect $\omega \gg I$ (however, $\omega \ll mc^2$ is still fulfilled) the Born's approximation is applied ($\nu \ll 1$) and the formula (19.21) assumes the form:

$$\sigma = \frac{2^8 \pi \alpha a_0^2 Z^5}{3}\left(\frac{I_0}{\omega}\right)^{7/2}. \tag{19.23}$$

In the case of nonrelativistic speeds of the photoelectrons the exact formulae have been obtained for σ of excited atoms (transitions from states with $n > 1$, $n1$ is the main quantum number). In the approximated calculations when $n \geq 1$ and $\omega \leq I$ one may use the quasi-classical Kramers's formula:

$$\sigma_K = \frac{2^6 \pi \alpha a_0^2}{3^{3/2} Z^2 n^5}\left(\frac{I}{\omega}\right)^3. \tag{19.24}$$

In multi electron atoms with the mean and large value of Z, the photoeffect takes place on the electrons of internal shells mainly on K-electrons, when the photons energy lays in the interval from 10^2 to 10^5 eV (roentgen radiation and γ-radiation). To get the total cross section of the photoeffect on K-shell one should multiply the expression (19.21) by 2 as this shell has two electrons. When we are increasing ω, then under reaching $\omega = I_L$ (I_L is ionization potential of L-electrons) the photoeffect starts to go on for the L-shell electrons and, as a result, σ exhibits jump-like growth.

19.2   Photoeffect (relativistic case)

We shall investigate the photoeffect in the relativistic region, that is, in the case when the photon energy is great as compared with the K-electron energy ($\omega \gg I$). Now we would take the solution of the Dirac equation in the Coulomb's nucleus field as the electron wavefunctions. However, since in the case under consideration $E_f = \omega - I \gg I$, then there is no need to solve this problem exactly and it is not unreasonable to allow for the Coulomb's field using the perturbation theory (it is valid for nuclei with $Z\alpha \ll 1$).

Because the photoelectron could be relativistic, then its unperturbed wavefunction must have the form of the relativistic plane wave, that is, ψ_f is represented by the expression:

$$\psi_f = \psi_f^{(0)} + \psi_f^{(1)} = \sqrt{\frac{m}{E_f}} \left(u_f(\mathbf{p}_f) e^{i\mathbf{P}_f\mathbf{r}} + \psi^{(1)} \right) \tag{19.25}$$

(for the sake of convenience hereafter we supply all quantities concerning the initial and final electron states by the corresponding index). In the initial state the electron is nonrelativistic, nevertheless, as will be shown later, a relativistic correction of the order $Z\alpha$ must also be included in its wavefunction ψ_i. We shall look for ψ_i in the following shape:

$$\psi_i = \psi_i^{(0)} + \psi_i^{(1)} = \left(u_i(0)\psi_S + \psi_i^{(1)} \right), \tag{19.26}$$

where ψ_S is the Schrödinger equation solution for a bound state, $u_i(0)$ is a bispinor of the view

$$u_i(0) = \begin{pmatrix} w \\ 0 \end{pmatrix},$$

and w is a bispinor to describe polarization state of an electron. Next, we set down the Dirac equation for the electron in the Coulomb's nucleus field:

$$\left[E_i + \frac{Z\alpha}{r} - m\gamma^0 + i\gamma^0(\boldsymbol{\gamma}\boldsymbol{\nabla}) \right] \psi_i = 0, \tag{19.27}$$

one acting on that by the operator:

$$\left[E_i + \frac{Z\alpha}{r} + m\gamma^0 - i\gamma^0(\boldsymbol{\gamma}\boldsymbol{\nabla}) \right].$$

The resulting equation will look like:

$$(\Delta + p_i^2 + \frac{2E_iZ\alpha}{r})\psi_i = \left[i\gamma^0(\boldsymbol{\gamma}\boldsymbol{\nabla})\frac{Z\alpha}{r} - \frac{Z^2\alpha^2}{r^2} \right] \psi_i. \tag{19.28}$$

Substituting (19.26) into (19.28), neglecting the term being quadratic on the perturbation and allowing for that ψ_S satisfies the Schrödinger equation, we arrive at the relation for $\psi_i^{(1)}$:

$$\left(\frac{1}{2m}\Delta - |E_i| + \frac{Z\alpha}{r} \right) \psi_i^{(1)} = \frac{iZ\alpha}{2m}\gamma^0(\boldsymbol{\gamma}\boldsymbol{\nabla})\frac{1}{r}u_i(0)\psi_S, \tag{19.29}$$

where E_i is the energy level of the initial state predicted by the Schrödinger theory. As is easy to see the solution of Eq. (19.29) is given by the expression:

$$\psi_i^{(1)} = -\frac{i}{2m}\gamma^0(\boldsymbol{\gamma} u_i(0)\boldsymbol{\nabla}\psi_S). \tag{19.30}$$

So, the wavefunction of the initial electron will look like:

$$\psi_i = \left[1 - \frac{i}{2m}\gamma^0(\boldsymbol{\gamma}\boldsymbol{\nabla}) \right] u_i(0)\psi_S. \tag{19.31}$$

The function ψ_i has been obtained for distances $r \sim (mZ\alpha)^{-1}$ on which the correction to it had the order of $Z\alpha$. Since the derivative of purely exponential function (19.9) (and the correction term with it in Eq. (19.31)) is always proportional to $Z\alpha$, then the formula obtained is suitable for any s-state in the case of any values of r. As will be illustrated subsequently, only the small value of r proves to be essential in the problem in question. This circumstance allows us to courageously use the formula (19.31).

Insert the obtained functions of the initial and final electrons into the transition amplitude (19.6)

$$\mathcal{A}_{i \to f} = -e\sqrt{\frac{1}{4\omega E_f}} \int d^3x \left\{ \overline{u}_f(\mathbf{p}_f) \left[(\boldsymbol{\gamma}\mathbf{e}^\lambda(\mathbf{k})) \left(1 - \frac{i}{2m}\gamma^0(\boldsymbol{\gamma}\boldsymbol{\nabla}) \right) u_i(0)\psi_S \right] \times \right.$$

$$\left. \times e^{-i(\mathbf{p}_f - \mathbf{k})\mathbf{x}} + \overline{\psi}_f^{(1)}(\boldsymbol{\gamma}\mathbf{e}^\lambda(\mathbf{k}))e^{i\mathbf{k}\mathbf{x}}u_i(0)\psi_S \right\}. \tag{19.32}$$

Recall, that we are working in the first approximation on $Z\alpha$. Then, to find the first correction in the second addend of $\mathcal{A}_{i \to f}$, one is sufficiently constrained by only the first term in the series expansion of ψ_S. That is equivalent to the replacement:

$$\psi_S \to \sqrt{\frac{Z^3}{\pi a_0^3}}. \tag{19.33}$$

Unfortunately, such a straightforward way does not achieve the object in the first addend of the expression for $\mathcal{A}_{i \to f}$. Really, in this case this addend appears to be proportional to $\delta^{(3)}(\mathbf{p}_f - \mathbf{k})$ and to turn into zero at $\mathbf{p}_f \neq \mathbf{k}$. At $v \sim 1$ the relativistic correction to be proportional to $Z\alpha$ gives the contribution to the cross section that has the same order as the next term of the ψ_S expansion in terms of $Z\alpha$. Using the Gauss theorem and passing from configuration space functions to their Fourier images, we may rewrite the expression for the photon absorption amplitude in the form:

$$\mathcal{A}_{i \to f} = -e\sqrt{\frac{1}{4\omega E_f}}\sqrt{\frac{Z^3}{\pi a_0^3}} \left\{ \overline{u}_f(\mathbf{p}_f)(\boldsymbol{\gamma}\mathbf{e}^\lambda(\mathbf{k})) [1+ \right.$$

$$\left. +\frac{1}{2m}\gamma^0\boldsymbol{\gamma}(\mathbf{p}_f - \mathbf{k}) \right] u_i(0) \left(e^{-\eta r} \right)_{(\mathbf{p}_f - \mathbf{k})} + (\overline{\psi}_f^{(1)})_\mathbf{k}(\boldsymbol{\gamma}\mathbf{e}^\lambda(\mathbf{k}))u_i(0) \bigg\}, \tag{19.34}$$

where the vector index denotes the corresponding Fourier image. From the equation:

$$(\Delta - \eta^2)\frac{e^{-\eta r}}{r} = -4\pi\delta^{(3)}(\mathbf{r}), \tag{19.35}$$

follows:

$$\left(\frac{e^{-\eta r}}{r}\right)_{\mathbf{q}} = \frac{4\pi}{\mathbf{q}^2 + \eta^2}. \tag{19.36}$$

To differentiate the obtained result with respect to the parameter η results in:

$$(e^{-\eta r})_{\mathbf{q}} = \frac{8\pi\eta}{(\mathbf{q}^2 + \eta^2)^2}. \tag{19.37}$$

To find $(\psi_f^{(1)})_{\mathbf{k}}$ we set down the Dirac equation in the Coulomb's field, substitute ψ_f into that, and restrict ourselves only by the first order of smallness

$$[\gamma^0 E_f + i(\boldsymbol{\gamma}\boldsymbol{\nabla}) - m]\psi_f^{(1)} = -\frac{Z\alpha}{r}\gamma^0 u_f(\mathbf{p}_f)e^{i\mathbf{p}_f\mathbf{r}}. \tag{19.38}$$

Acting upon both sides of this equation by the operator $[\gamma^0 E_f + i(\boldsymbol{\gamma}\boldsymbol{\nabla}) + m]$, we get:

$$(\Delta + \mathbf{p}_f^2)\psi_f^{(1)} = -Z\alpha[\gamma^0 E_f + i(\boldsymbol{\gamma}\boldsymbol{\nabla}) + m]\gamma^0 u_f(\mathbf{p}_f)\frac{e^{i\mathbf{p}\mathbf{r}}}{r}. \tag{19.39}$$

In order to transfer to the Fourier images we multiply Eq. (19.39) by $e^{-i\mathbf{k}\mathbf{r}}$ and integrate it over $\mathbf{r}$. In doing so, we use the Gauss theorem in the terms involving the operators $\boldsymbol{\nabla}$ and Δ. The result will be as follows:

$$(\mathbf{p}_f^2 - \mathbf{k}^2)\psi_f^{(1)} = -Z\alpha[2E_f\gamma^0 - (\mathbf{k} - \mathbf{p}_f)\boldsymbol{\gamma}]\gamma^0 u_f(\mathbf{p}_f)\frac{4\pi}{(\mathbf{k} - \mathbf{p}_f)^2}, \tag{19.40}$$

where we have taken into consideration that the bispinor $u_f(\mathbf{p}_f)$ obeys the equation:

$$[\gamma^0 E_f - (\mathbf{p}_f\boldsymbol{\gamma}) - m]u_f(\mathbf{p}_f) = 0.$$

From (19.40) it follows:

$$\left(\overline{\psi}_f^{(1)}\right)_{\mathbf{q}} = 4\pi Z\alpha\overline{u}_f(\mathbf{p}_f)\frac{2E_f\gamma^0 + (\mathbf{k} - \mathbf{p}_f)\boldsymbol{\gamma}}{(\mathbf{k}^2 - \mathbf{p}_f^2)(\mathbf{k} - \mathbf{p}_f)^2}\gamma^0. \tag{19.41}$$

Substituting the obtained Fourier images of the electron wavefunctions into the absorption amplitude (19.34) leads to the relation:

$$\mathcal{A}_{i\to f} = 4\sqrt{\frac{\pi}{\omega E_f}}\frac{(Z/a_0)^{5/2}}{(\mathbf{k} - \mathbf{p}_f)^2}\overline{u}_f(\mathbf{p}_f)Bu_i(0), \tag{19.42}$$

where

$$B = a(\boldsymbol{\gamma}\mathbf{e}^\lambda(\mathbf{k})) + (\boldsymbol{\gamma}\mathbf{e}^\lambda(\mathbf{k}))\gamma^0(\boldsymbol{\gamma}\mathbf{b}) + (\boldsymbol{\gamma}\mathbf{c})\gamma^0(\boldsymbol{\gamma}\mathbf{e}^\lambda(\mathbf{k})),$$

$$a = \frac{1}{(\mathbf{k} - \mathbf{p}_f)^2} + \frac{E_f}{m(\mathbf{k}^2 - \mathbf{p}_f^2)}, \qquad \mathbf{b} = \frac{\mathbf{p}_f - \mathbf{k}}{2m(\mathbf{k} - \mathbf{p}_f)^2}, \qquad \mathbf{c} = \frac{\mathbf{k} - \mathbf{p}_f}{2m(\mathbf{k}^2 - \mathbf{p}_f^2)}.$$

For the sake of simplicity we shall not be interested in the electron polarizations. Then, we should sum up the cross section over the spin states of the final electron (s_f) and average over the spin states of the initial electron (s_i). As a result of such an operation the quantity appears:

$$\frac{1}{2} \sum_{s_i, s_f} |\overline{u}_f(\mathbf{p}_f) B u_i(0)|^2 = \frac{1}{8m} \mathrm{Sp}[(\hat{p}_f + m) B (\gamma_0 + 1) \overline{B}].$$

To determine it, one must be able to find spurs of products of the γ^0- and $\boldsymbol{\gamma}$-matrices. Recall, the mechanism of deriving the four-dimensional formulae for calculating spurs is based on two facts only: (i) anticommutation of γ_5 with all the γ-matrices; and (ii) commutation relations. Using these in the three-dimensional case, we arrive at the following conclusions. Only the spurs of products with even number of the factors γ^0 and $\boldsymbol{\gamma}$ are different from zero. To calculate the spur one should collect all γ^0 in one place (that gives 1 for these products) and use the four-dimensional formulae of Section 15.2 with replacing the metric tensor g with the Kronecker symbol with the minus sign. With allowance made for these simple modifications, we get:

$$\frac{1}{2} \sum_{s_i, s_f} |\overline{u}_f(\mathbf{p}_f) B u_i(0)|^2 == \frac{1}{2m(E_f + m)} [a\mathbf{p}_f - (\mathbf{b} - \mathbf{c})(E_f + m)]^2 +$$

$$+ \frac{2(\mathbf{b}\mathbf{e}^\lambda(\mathbf{k}))}{m^2} [(E_f + m)(\mathbf{c}\mathbf{e}^{\lambda*}(\mathbf{k})) + a(\mathbf{p}_f \mathbf{e}^{\lambda*}(\mathbf{k}))], \tag{19.43}$$

where $\overline{B} = \gamma^0 B^\dagger \gamma^0$.

To define the cross section behavior character depending on the photon polarizations, we direct the axis z along $\mathbf{k}$ and choose the plane $(\mathbf{k}, \mathbf{e}^\lambda(\mathbf{k}))$ as the one xOz. Note, to transit to the cross section for unpolarized photons, to overage over the photon polarizations, is put into effect by integration over the angle φ from 0 to 2π and multiplication of the obtained result by $1/2$. Taking into consideration the relation:

$$(\mathbf{p}_f \mathbf{e}^\lambda(\mathbf{k})) - |\mathbf{p}_f| \cos\varphi \sin\theta,$$

and the fact that in the relativistic case

$$\omega = I + E_f - m \approx E_f - m,$$

we get without trouble the formulae:

$$\mathbf{k}^2 - \mathbf{p}_f^2 = -2m(E_f - m), \qquad (\mathbf{k} - \mathbf{p}_f)^2 = 2E_f(E_f - m)(1 - v\cos\theta),$$

where $v \equiv |\mathbf{v}|$ is a photoelectron speed. By application, the obtained relations we find that the differential cross section of the photoeffect for the polarized photons and unpolarized electrons is given by the expression [20]:

$$d\sigma = Z^5 \alpha^4 r_0^2 \frac{v^3(1 - v^2)^2 \sin^2\theta}{(1 - \sqrt{1 - v^2})^5 (1 - v\cos\theta)^4} \left\{ \frac{(1 - \sqrt{1 - v^2})^2}{2(1 - v^2)^{3/2}} (1 - \right.$$

$$\left. -v\cos\theta) + \left[2 - \frac{(1 - \sqrt{1 - v^2})(1 - v\cos\theta)}{1 - v^2} \right] \cos^2\varphi \right\} d\Omega, \tag{19.44}$$

where $r_0 = e^2/(4\pi m)$ is the classical radius of the electron.

In the ultrarelativistic case v tends to 1 and the cross section of the photoeffect has sharp maximum at small angles $\theta \sim \sqrt{1 - v^2}$, that is, the photoelectrons preferentially fly in the direction of photons falling.

To integrate the expression (19.44) over angles and multiply the obtained result by $1/2$ gives the total cross section of the photoeffect for unpolarized particles [20]:

$$\sigma = 2\pi Z^5 \alpha^4 r_0^2 \frac{(\gamma^2 - 1)^{3/2}}{(\gamma - 1)^5} \left[\frac{4}{3} + \frac{\gamma(\gamma - 2)}{\gamma + 1} \left(1 - \frac{1}{2\gamma\sqrt{\gamma^2 - 1}} \ln \frac{\gamma + \sqrt{\gamma^2 - 1}}{\gamma - \sqrt{\gamma^2 - 1}} \right) \right], \quad (19.45)$$

where

$$\gamma = \frac{1}{\sqrt{1 - v^2}}.$$

In the ultrarelativistic case this formula takes the simple view:

$$\sigma = 2\pi Z^5 \alpha^4 r_0^2 \gamma^{-1}. \quad (19.46)$$

In the limit of small values of $\gamma - 1 \ll 1$ we get, as was to be expected, the nonrelativistic value of the cross section (19.23).

19.3　Compton-effect (unpolarized case)

One of the first to support Einstein's idea concerning the light quanta was Compton, who in 1922 discovered a special type of electromagnetic radiation scattering. As a source, Compton has used an X-ray tube, where the thermal electrons outgoing from the cathode are accelerated up to high speeds to be decelerated when impinging the anticathode. Electron deceleration results from interactions between the electrons and a Coulomb nucleus field. And the resultant electromagnetic radiation received the name bremsstrahlung. Recall that the difference in the approaches to a nature of bremsstrahlung on the basis of the classical electromagnetic theory (CET) and the quantum theory (QT) is as follows. The CET predicts initiation of continuous radiation on each collision in the process of the electron deceleration, while the QT predicts the generation of a single photon with the energy $\hbar\omega$ that is generally different for every collision act. In Compton's experiment the radiation emitted by the molybdenum anticathode of an X-ray tube was scattered by graphite. Scattered radiation was incident upon the crystal of an X-ray spectrometer and next—upon an ionization chamber (or photographic plate). Compton has studied the intensity of scattering spectra for various angles between the direction of the initial and scattered beams θ. As it turned out, the lines of a scattering spectrum are shifted relative to the initial lines to higher values of θ, that is, a change in the wavelength of scattered beams was growing, with an increase of θ. The experimental results of Compton were first published in the *Bulletin of the National Research Council of the USA Academy of Sciences* in October 1922. The essence of this paper may be rendered as follows: X-ray beams represent a flux of photons having a particular momentum, similar to any other particles, and the scattering act is an elastic collision between a photon and an electron. Subsequently, not only is the scattering process of X-ray radiation by matter electrons, but the scattering process of a photon with any energy by an electron has been named the *Compton effect* as well.

Describing the Compton effect, for the sake of simplicity, we shall consider the electrons being in graphite to be free. It is quite natural because the binding energies of the electrons have the order of several eV, while X-ray beam photons used in the Compton experiment had the energy equal to 17.5 keV.

We now proceed to the calculation of the cross section for the Compton effect

$$e^- + \gamma \to e^- + \gamma. \quad (19.47)$$

The corresponding Feynman diagrams in the second order of the perturbation theory are

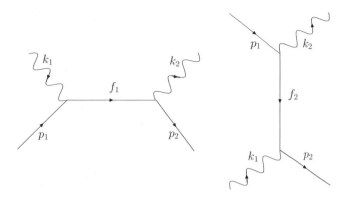

FIGURE 19.1
The Feynman diagrams corresponding to the Compton effect.

displayed in Fig. 19.1. It should be noted that in the electroweak theory any additional diagrams in relation to ones pictured in Fig. 19.1 do not appear. In the higher orders of the perturbation theory the corrections of weak interaction come about from insertions of vacuum loops both in the external and in the internal lines of the QED diagrams. The $W^-\nu_e$-, He^- and Ze^--loops represented in Fig. 19.2 may be placed in electron lines whereas the W^-W^+, $\bar{q}_i q_i$-loops displayed in Fig. 19.3—in photon lines. In accordance with the

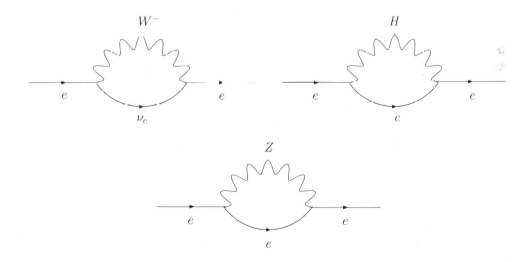

FIGURE 19.2
The loops with the $W^-\nu_e$-, He^- and Ze^--virtual pairs.

Feynman rules the amplitude of the process (19.47) takes the form:

$$\mathcal{A}_{e^-\gamma\to e^-\gamma} = -ie^2\sqrt{\frac{m^2}{4EE'\omega\omega'}}\overline{u}'(\mathbf{p}')e_\mu'^*(\mathbf{k}')O^{\mu\nu}e_\nu(\mathbf{k})u(\mathbf{p}), \qquad (19.48)$$

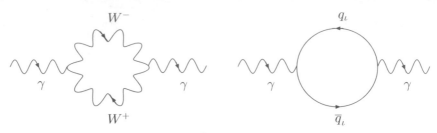

FIGURE 19.3

The loops with the W^-W^+- and $q_i\bar{q}_i$-virtual pairs.

where

$$O^{\mu\nu} = \gamma^\mu \frac{\hat{f}_1 + m}{f_1^2 - m^2} \gamma^\nu + \gamma^\nu \frac{\hat{f}_2 + m}{f_2^2 - m^2} \gamma^\mu, \tag{19.49}$$

$p = (E, \mathbf{p})$, $p' = (E', \mathbf{p}')$, $k = (\omega, \mathbf{k})$, $k' = (\omega', \mathbf{k}')$, $e_\nu(\mathbf{k})$ $(e'_\nu(\mathbf{k}'))$ is a four-dimensional polarization vector of the initial (final) photon, and

$$f_{1\mu} = (p + k)_\mu = (p' + k')_\mu, \qquad f_{2\mu} = (p - k')_\mu = (p' - k)_\mu.$$

The four-dimensional vectors $f_{1\mu}$ and $f_{2\mu}$ represent the momenta of the virtual electrons. The squares of these momenta are nothing but the Mandelstam's variables:

$$f_{1\mu}^2 = s, \qquad f_{2\mu}^2 = u. \tag{19.50}$$

To obtain the transition probability we must take the module square of the quantity $\mathcal{A}_{e^-\gamma \to e^-\gamma}$. We are constrained by the case when the initial electron is unpolarized and we are not interested in the spin projection of the final electron. Then the expression:

$$|\mathcal{A}_{e^-\gamma \to e^-\gamma}|^2$$

should be summed up over the spin states of the final electron (s') and averaged over the ones of the initial electron (s), that is, the situation is reduced to calculating the quantity:

$$\frac{1}{2} \sum_{s,s'} |\mathcal{A}_{e^-\gamma \to e^-\gamma}|^2.$$

Using the formula (15.103), we find:

$$\frac{1}{2} \sum_{s,s'} |\mathcal{A}_{e^-\gamma \to e^-\gamma}|^2 = \frac{e^4}{32EE'\omega\omega'} \mathrm{Sp}[\mathcal{M}(\hat{p}+m)\overline{\mathcal{M}}(\hat{p}'+m)], \tag{19.51}$$

where $\mathcal{M} = e'^*_\mu(\mathbf{k}')O^{\mu\nu}e_\nu(\mathbf{k})$ and $\overline{\mathcal{M}} = \gamma_0 \mathcal{M}^\dagger \gamma_0$.

With allowance made for antihermicity of γ and hermicity of γ_0, the expression under the spur sign could be represented in the view:

$$\mathcal{M}(\hat{p}+m)\overline{\mathcal{M}}(\hat{p}'+m) = \gamma^\mu \left[\frac{1}{s-m^2} e'^*_\mu(\mathbf{k}')(\hat{f}_1+m)e_\nu(\mathbf{k}) + \frac{1}{u-m^2} e_\mu(\mathbf{k})(\hat{f}_2+ \right.$$

$$\left. +m)e'^*_\nu(\mathbf{k}') \right] (\hat{p}+m)\gamma^\sigma \left[\frac{1}{s-m^2} e^*_\sigma(\mathbf{k})(\hat{f}_1+m)e'_\tau(\mathbf{k}') + \frac{1}{u-m^2} e'_\sigma(\mathbf{k}')(\hat{f}_2+m)e^*_\tau(\mathbf{k}) \right] \gamma_\tau(\hat{p}'+m). \tag{19.52}$$

We next assume that the incoming and scattered photons also are unpolarized. Then the expression (19.51) should be summed up over the final photon states (λ') and averaged over the initial ones (λ). Carrying out this operation with the help of formula (17.44), we get:

$$\frac{1}{4}\sum_{s,s',\lambda,\lambda'}|\mathcal{A}_{e^-\gamma\to e^-\gamma}|^2 = \frac{e^4}{64EE'\omega\omega'}\mathrm{Sp}\,F, \qquad (19.53)$$

where

$$F = \frac{1}{s-m^2}\left[\frac{1}{s-m^2}\gamma^\mu(\hat{f}_1+m)\gamma^\nu + \frac{1}{u-m^2}\gamma^\nu(\hat{f}_2+m)\gamma^\mu\right]\times$$

$$\times(\hat{p}+m)\gamma_\nu(\hat{f}_1+m)\gamma_\mu(\hat{p}'+m)+$$

$$+\frac{1}{u-m^2}\left[\frac{1}{s-m^2}\gamma^\mu(\hat{f}_1+m)\gamma^\nu + \frac{1}{u-m^2}\gamma^\nu(\hat{f}_2+m)\gamma^\mu\right](\hat{p}+m)\gamma_\mu(\hat{f}_2+m)\gamma_\nu(\hat{p}'+m).$$

$$(19.54)$$

It is easy to verify that the second addend in this expression follows from the first one under replacement:

$$k \leftrightarrow -k', \qquad (19.55)$$

which, in its turn, is equivalent to:

$$f_1 \leftrightarrow f_2, \qquad s \leftrightarrow u. \qquad (19.56)$$

The pointed linkage between the addends in the expression (19.54) is the crossing-universality consequence of the transition amplitude. In this context it means that in the case of unpolarized photons the matrix element corresponding to the diagram pictured in Fig. 19.1a $\mathcal{M}_a$ follows from the one pictured in Fig. 19.1(b) $\mathcal{M}_b$ under changing:

$$k \leftrightarrow -k', \qquad e_\mu(\mathbf{k}) \leftrightarrow e'_\mu(\mathbf{k}'). \qquad (19.57)$$

Having set down the amplitude square:

$$|\mathcal{A}_{e^-\gamma\to e^-\gamma}|^2 = \underbrace{\mathcal{M}_a\overline{\mathcal{M}}_a + \mathcal{M}_a\overline{\mathcal{M}}_b}_{\text{I}} + \underbrace{\mathcal{M}_b\overline{\mathcal{M}}_a + \mathcal{M}_b\overline{\mathcal{M}}_b}_{\text{II}}, \qquad (19.58)$$

we are convinced that at the replacement (19.56) the third and forth addends pass into the second and first ones, respectively. This property of the cross section still stands for identical polarized photons too. In the case when the process is described by 3 topologically nonequivalent diagrams possessing the crossing-universality, the cross section could be represented as a sum of three addends passing into each other under the particular transformation of particle momenta.

It is obvious that the both addends in (19.53) must be functions of two invariants s and u only. Having realized this circumstance, we can exhibit Sp F in the following form:

$$\mathrm{Sp}\,F = P(s,u) + P(u,s), \qquad (19.59)$$

where

$$P(s,u) = h_1(s,u) + h_2(s,u), \qquad (19.60)$$

$$h_1(s,u) = \frac{1}{(s-m^2)^2}\mathrm{Sp}[\gamma^\mu(\hat{f}_1+m)\gamma^\nu(\hat{p}+m)\gamma_\nu(\hat{f}_1+m)\gamma_\mu(\hat{p}'+m)], \qquad (19.61)$$

$$h_2(s,u) = \frac{1}{(s-m^2)(u-m^2)}\mathrm{Sp}[\gamma^\nu(\hat{f}_2+m)\gamma^\mu(\hat{p}+m)\gamma_\nu(\hat{f}_1+m)\gamma_\mu(\hat{p}'+m)]. \qquad (19.62)$$

In what follows the formulae for convolutions over four-dimensional indices will be useful:

$$\left.\begin{array}{l} \gamma^\mu \gamma_\mu = 4, \\ \gamma^\mu \gamma^\nu \gamma_\mu = -2\gamma^\nu, \\ \gamma^\mu \gamma^\nu \gamma^\tau \gamma_\mu = 4g^{\nu\tau}, \\ \gamma^\mu \gamma^\nu \gamma^\tau \gamma^\sigma \gamma_\mu = -2\gamma^\sigma \gamma^\tau \gamma^\nu, \\ \gamma^\mu \gamma^\nu \gamma^\tau \gamma^\sigma \gamma^\rho \gamma_\mu = 2(\gamma^\rho \gamma^\nu \gamma^\tau \gamma^\sigma + \gamma^\sigma \gamma^\tau \gamma^\nu \gamma^\rho). \end{array}\right\} . \tag{19.63}$$

Using (15.22), (15.25), and (19.63), we come to the expression for $h_1(s, u)$:

$$h_1(s, u) = \frac{4}{(s - m^2)^2} \mathrm{Sp} \left[\hat{f}_1 \hat{p} \hat{f}_1 \hat{p}' - 4m^2(\hat{f}_1 \hat{p} + \hat{f}_1 \hat{p}' - f_1^2 - \frac{1}{4}\hat{p}\hat{p}') + 4m^4 \right] =$$

$$= 8 \frac{4m^4 + 2m^2(s - m^2) - (s - m^2)(u - m^2)}{(s - m^2)^2}. \tag{19.64}$$

The similar calculations yield:

$$h_2(s, u) = 8m^2 \frac{2m^2 + s + u}{(s - m^2)(u - m^2)}. \tag{19.65}$$

Gathering together the necessary relations, we get the final expression for the spur F:

$$\mathrm{Sp} F = 32 \left(\frac{m^2}{s - m^2} + \frac{m^2}{u - m^2} \right) \left(\frac{m^2}{s - m^2} + \frac{m^2}{u - m^2} + 1 \right) - 8 \left(\frac{u - m^2}{s - m^2} + \frac{s - m^2}{u - m^2} \right). \tag{19.66}$$

Using the general formula (18.185), one may now write the differential cross section of the Compton effect:

$$d\sigma = (2\pi)^{-2} \frac{e^4 \omega'^2}{64 E' \omega' (pk)} \left(\frac{\partial(\omega' + E')}{\partial \omega'} \right)^{-1} \mathrm{Sp} \, F \, d\Omega, \tag{19.67}$$

where $d\Omega$ is a element of the solid angle in which the vector $\mathbf{k}'$ lays (the z-axis is directed along the vector $\mathbf{k}$).

We shall investigate the expression:

$$\frac{\partial(\omega' + E')}{\partial \omega'}.$$

Since the momentum $\mathbf{p}'$ is connected with $\mathbf{k}'$ by the conservation law:

$$E'^2 = m^2 + (\mathbf{p} + \mathbf{k} - \mathbf{k}')^2, \tag{19.68}$$

then E' is a function of ω'. Calculations are easiest fulfilled in the lab frame, in which the electron is initially at rest ($\mathbf{p} = 0$, and $E = m$). Then we have:

$$\frac{\partial(\omega' + E')}{\partial \omega'} = 1 + \frac{\omega' - \omega \cos\theta}{E'} = \frac{E' + \omega' - \omega \cos\theta}{E'} = \frac{m + \omega(1 - \cos\theta)}{E'} =$$

$$= \frac{m\omega' + \omega'\omega(1 - \cos\theta)}{\omega' E'} = \frac{m\omega' + kk'}{\omega' E'} = \frac{(pk') + (kk')}{\omega' E'} = \frac{(p'k')}{\omega' E'} = \frac{s - m^2}{2\omega' E'}. \tag{19.69}$$

To substitute (19.69) into the expression (19.67) results in:

$$d\sigma = r_0^2 \frac{\omega'^2 d\Omega}{4(s - m^2)^2} \mathrm{Sp} \, F. \tag{19.70}$$

Note, that the obtained cross section is a relativistic invariant really. The invariance of Sp F is evident while the invariance of the factor $\omega'^2 d\Omega$ follows from the identity:

$$\omega'^2 d\Omega = \int \frac{d^3 k}{\omega'} = 2 \int d^4 k \delta(k^2).$$ (19.71)

In the center-of-mass frame $E = E'$, $\omega = \omega'$ and the Mandelstam's variables takes the values:

$$\left. \begin{array}{l} s = (E + \omega)^2, \qquad t = k^2 + k'^2 - 2kk' = -2\omega^2(1 - \cos\theta), \\ u = m^2 - 2pk' = m^2 - 2\omega(E + \omega\cos\theta). \end{array} \right\}$$ (19.72)

From $s + t + u = 2m^2$ and the relations (19.72) we find without trouble:

$$\omega = \frac{(s - m^2)}{2\sqrt{s}},$$ (19.73)

that allows us to express $d(\cos\theta)$ in terms of dt

$$d(\cos\theta) = \frac{1}{2\omega^2} dt = \frac{2s}{(s - m^2)^2} dt.$$ (19.74)

As a result, we arrive at the formula for the differential cross section:

$$\frac{d\sigma}{dt} = \frac{8\pi r_0^2 m^2}{(s - m^2)^2} \left[\left(\frac{m^2}{s - m^2} + \frac{m^2}{u - m^2} \right) \left(\frac{m^2}{s - m^2} + \frac{m^2}{u - m^2} + 1 \right) - \right.$$

$$\left. - \frac{1}{4} \left(\frac{s - m^2}{u - m^2} + \frac{u - m^2}{s - m^2} \right) \right].$$ (19.75)

By application of (19.75) it is an easy matter to express the Compton effect cross section in terms of the collision parameters in any particular reference frame.

Let us find the dependence of scattered photon frequency on incoming photon frequency. Multiplying the four-dimensional momentum conservation law:

$$p_\mu + k_\mu = p'_\mu + k'_\mu,$$ (19.76)

by k_μ, we get:

$$(pk) = (p'k) + (kk').$$ (19.77)

Having rewritten (19.76) as:

$$p_\mu - k'_\mu = p'_\mu - k_\mu$$ (19.78)

and having raised to the second power both sides of the obtained expression, we get:

$$(p'k) = (pk').$$ (19.79)

Insertion of (19.79) into (19.77) leads to the Compton formula:

$$\omega' = \frac{\omega}{1 + \omega(1 - \cos\theta)/m},$$ (19.80)

where θ is an angle between $\mathbf{k}$ and $\mathbf{k}'$. At $\theta = 0$, $\omega' = \omega$ and scattering, consequently, is absent. Changing the scattered photon frequency is maximum when $\theta = \pi$, and, in accordance with (19.80), proves to be equal to:

$$\frac{\Delta\omega}{\omega'} = \frac{2\omega}{m}.$$ (19.81)

From (19.81) follows that the frequency shift becomes significant only at the photon energies to be commensurable with $m = 0.511$ MeV.

The Compton effect cross section takes the most simple view in the lab frame. Allowing for that in this case

$$s = m^2 + 2m\omega, \qquad u = m^2 - 2m\omega',$$

from (19.75) we obtain the well-known Klein-Nishina formula [21]:

$$\frac{d\sigma}{d\Omega} = \frac{r_0^2}{2} \left(\frac{\omega'}{\omega}\right)^2 \left(\frac{\omega'}{\omega} + \frac{\omega}{\omega'} - \sin^2\theta\right). \tag{19.82}$$

When $m \to \infty$ the cross section (19.82) turns into zero, that is, scattering does not happen. This situation is very instructive. If we are using this series of the perturbation theory, then we shall have to make some conclusions about the character of the physical process only after some terms of this series have been investigated. For the case under study it means just one—the photon scattering process by an infinitely heavy charge with the spin 1/2, exemplified by the atom nucleus, does not go on in the second order of the perturbation theory. The analysis shows, the scattering matrix of the forth order of the perturbation theory already contains elements to describe such a photon scattering.

If the experiment is aimed at measuring the angle distribution of scattered electrons, then the expression for the cross section (19.82) should be transformed so that it will be a function of incoming photon frequency and the photon scattering angle only. Using the Compton formula, we obtain

$$\frac{d\sigma}{d\Omega} = \frac{r_0^2}{2} \frac{1 + \cos^2\theta}{[1 + \epsilon(1 - \cos\theta)]^2} \left[1 + \frac{\epsilon^2(1 - \cos\theta)^2}{(1 + \cos^2\theta)\left[1 + \epsilon(1 - \cos\theta)\right]}\right], \tag{19.83}$$

where $\epsilon = \omega/m$. One is easily convinced that at small energies of the photon ($\epsilon \ll 1$) the formula passes to the classical expression by Thomson for the differential cross section on a free rest electron:

$$\frac{d\sigma}{d\Omega} = \frac{r_0^2}{2}(1 + \cos^2\theta). \tag{19.84}$$

One may measure the dependence of the cross section on the angle between the momenta of the scattered electron and the incoming photon (β). It is readily shown that the angles β and θ are connected by the relation:

$$\cos\beta = (1 + \epsilon)\sqrt{\frac{1 - \cos\theta}{2 + \epsilon(\epsilon + 2)(1 - \cos\theta)}}. \tag{19.85}$$

Thus, β is changed from $\pi/2$ at $\theta = 0$ up to 0 at $\theta = \pi$. By means of (19.85) the differential cross section is rewritten in the form:

$$\frac{d\sigma}{d\Omega_e} = 4r_0^2 \frac{(1 + \epsilon)^2 \cos\beta}{(1 + 2\epsilon + \epsilon^2 \sin^2\beta)^2} \left\{1 - \frac{(1 + \epsilon)^2 \sin^2 2\beta}{2[1 + \epsilon(\epsilon + 2)\sin^2\beta]^2} + \right.$$

$$\left. + \frac{2\epsilon^2 \cos^4\beta}{(1 + 2\epsilon + \epsilon^2 \sin^2\beta)[1 + \epsilon(\epsilon + 2)\sin^2\beta]}\right\}, \tag{19.86}$$

where $d\Omega_e$ is an element of the solid angle in which the scattered electron momentum lays.

To get the total cross section through invariant quantities we should integrate the differential cross section (19.75) over t from $-(s - m^2)^2/s$ to 0. We emphasize that the energy of colliding particles is considered to be given, that is, $s = \text{const.}$ Elementary integration gives the result:

$$\sigma = \frac{2\pi r_0^2}{x} \left\{ \left[1 - \frac{4}{x} - \frac{8}{x^2} \right] \ln(1+x) + \frac{1}{2} + \frac{8}{x} - \frac{1}{2(1+x)^2} \right\}, \qquad (19.87)$$

where

$$x = \frac{s - m^2}{m^2}.$$

In the ultrarelativistic region ($x \gg 1$) the cross section has the shape:

$$\sigma = \frac{2\pi r_0^2}{x} \left[\ln x + \frac{1}{2} + O\left(\frac{\ln x}{x} \right) \right]. \qquad (19.88)$$

In the nonrelativistic case ($x \ll 1$) the expression (19.87) is reduced to the classical formula by Thomson:

$$\sigma = \frac{8\pi r_0^2}{3}(1 - x). \qquad (19.89)$$

In the lab frame we have $x = 2\omega/m$ and the formulae (19.87)–(19.89) define the dependence of the Compton effect cross section on the photon energy in the case of resting electron.

The Compton effect may be investigated not only in experiments with a fixed target but in $e^-\gamma$-colliders as well, that is, in the colliding beams mode. In the first case all measurements are fulfilled in the lab frame while in the second one—in the center-of-mass frame. To be more specific, in the latter the colliding e - and γ-beams intersect at a small angle. Under data processing of such experiments all kinematic characteristics of particles are transferred by the Lorentz transformation into the center-of-mass frame where data analysis is carried out. So, we are assured once again that writing of a cross section in terms of invariant quantities is expedient. Having determined the values of s, u, t in the lab frame and in the center-of-mass frame, we obtain the cross sections both for colliding beams mode and for the fixed target mode.

Let us find out what laws of the Compton effect will be registered by observers to be situated in the lab frame and in the center-of-mass one. In this case we are limited by the ultra-relativistic region. From the formula (19.88) it follows that the behavior character of the total cross section in the lab frame and in center-of-mass one is identical. In the both frames the total cross section is decreased as the energy grows. However, in the lab frame this decrease occurs according to the law:

$$\sigma \sim \omega^{-1} \ln \frac{\omega}{m}$$

while in the center-of-mass frame ($x \approx 4\omega^2/m^2$) the cross section is decreased somewhat faster

$$\sigma \sim \omega^{-2} \ln \frac{\omega}{m}.$$

The analysis of the formula (19.75) shows that in the ultrarelativistic case the angle distribution in both of these frames has an absolutely different character. In the lab frame the differential cross section has a sharp maximum in the forward direction. In a narrow cone $\theta \preceq \sqrt{m/\omega}$ the relation:

$$\omega \approx \omega'$$

is fulfilled and the cross section goes into its maximum value $\approx r_0^2$ at $\theta \to 0$. Beyond this cone the cross section is decreased and in the region $\theta \gg \sqrt{m/\omega}$, where

$$\omega' \approx \frac{m}{1 - \cos\theta},$$

it becomes equal to:

$$\frac{d\sigma}{d\Omega} = \frac{r_0^2}{2} \frac{m}{\omega(1 - \cos\theta)}, \tag{19.90}$$

that is, it is lowered by a factor equal to $\sim \omega/m$.

In the center-of-mass frame the situation is quite the reverse. Here, the differential cross section possesses the maximum in the back direction. With the help of Eq. (19.72) one may demonstrate that at $\pi - \theta \ll 1$ the invariants entering into (19.75) are defined by the formulae:

$$\frac{s - m^2}{m^2} \approx \frac{4\omega^2}{m^2}, \qquad \frac{s + t - m^2}{m^2} \approx 1 + \frac{\omega^2}{m^2}(\pi - \theta)^2. \tag{19.91}$$

Then, we have:

$$d\sigma \approx 8\pi r_0^2 \frac{m^2 dt}{4(s - m^2)(s + t - m^2)} = \frac{r_0^2}{2} \frac{d\Omega}{1 + (\pi - \theta)^2 \omega^2/m^2}. \tag{19.92}$$

From (19.92) it is apparent that into a narrow cone $\pi - \theta \preceq m/\omega$:

$$\frac{d\sigma}{d\Omega} \sim r_0^2,$$

and beyond it the differential cross section is decreased in $\sim \omega^2/m^2$ times.

19.4   Compton-effect (polarized case)

In the case of arbitrary polarization states the matrix element square of the Compton scattering $|\mathcal{M}_{e^-\gamma \to e^-\gamma}|^2$ must be subjected to the replacement:

$$|\mathcal{M}_{e^-\gamma \to e^-\gamma}|^2 \to \frac{e^4 m^2}{4EE'\omega\omega'} \mathrm{Sp}\left[\rho^{(+)'} \rho_{\tau\mu}^{\gamma'} O^{\mu\nu} \rho^{(+)} \rho_{\nu\lambda}^{\gamma} \overline{O}^{\lambda\tau}\right]. \tag{19.93}$$

We shall use the photon density matrix in the four-dimensional form. We introduce four mutually orthogonal four-dimensional vectors:

$$P_\tau, \ N_\tau, \ K_\tau, \ M_\tau, \tag{19.94}$$

where

$$P_\tau = p_\tau + p'_\tau - K_\tau \frac{2pK}{K^2}, \qquad N_\tau = \varepsilon_{\tau\mu\nu\lambda} P^\mu M^\nu K^\lambda,$$

$$K_\tau = k_\tau + k'_\tau, \qquad M_\tau = k'_\tau - k_\tau.$$

It is easy to check that the vectors P_τ, N_τ are orthogonal to the ones K_τ, M_τ, and, as a result, they are orthogonal to k, k' as well. Since K is the time-like vector ($K^2 = 2kk' > 0$), then in the reference frame, in which $\mathbf{K} = 0$, it follows from $KP = 0$ and $KN = 0$ that

$$P_0 = 0, \qquad N_0 = 0,$$

that is, the vectors P_τ and N_τ are space-like. From P_τ and N_τ one may construct the two unit vectors:

$$n_\mu^{(1)} = \frac{N_\mu}{\sqrt{-N^2}}, \qquad n_\mu^{(2)} = \frac{P_\mu}{\sqrt{-P^2}}, \tag{19.95}$$

the former is the pseudovector while the latter does the (proper) vector. On the strength of these properties the vectors (19.95) may be used to build the four-dimensional density matrix of the initial and final photon states. It should be stressed that in the most general case there must be four such vectors rather than two ones.

Of course, one may also use the three-dimensional form of writing the photons density matrix. In this case one conveniently chooses the vectors $\mathbf{e}^1$ and $\mathbf{e}^2$ being transverse in reference to the photon momenta:

$$\mathbf{e}^1 = \frac{[\mathbf{kk'}]}{|[\mathbf{kk'}]|}, \qquad \mathbf{e}^2 = \frac{1}{\omega}[\mathbf{ke}^1], \qquad \mathbf{e}'^1 = \mathbf{e}^1, \qquad \mathbf{e}'^2 = \frac{1}{\omega'}[\mathbf{k'e}'^1]. \tag{19.96}$$

Then the vectors $\mathbf{e}^1$ and $\mathbf{e}'^1$ will be perpendicular to the scattering plane whereas the vectors $\mathbf{e}^2$ and $\mathbf{e}'^2$ will lay in this plane. However, in the three-dimensional case, which is without doubt more simple from the point of view of calculations, the final result will not have the invariant form.

To determine the photon density matrices being related to the four-dimensional unit vectors $n_\mu^{(1)}$ and $n_\mu^{(2)}$ we must first establish the linkage of these vectors with the photon polarizations. Independent directions of polarizations for each of photons are defined by the components of the three-dimensional vectors $\mathbf{n}^1$ and $\mathbf{n}^2$, which are transverse in relation both to $\mathbf{k}$ and to $\mathbf{k'}$. Both in the center-of-mass frame and in the lab one the vector $\mathbf{P}$ lays in the plane $(\mathbf{k}, \mathbf{k'})$ while the vector $\mathbf{N}$ is perpendicular to this plane. Therefore, the direction $\mathbf{n}^1$ makes sense of the polarization to be perpendicular to the scattering plane whereas the direction $\mathbf{n}^2$—of polarization in the scattering plane (see, for comparison, Eq. (19.96)). One should also take into consideration that the Stoke's parameters are defined with respect to the Cartesian coordinates system with the z-axis to be parallel to the three-dimensional photon momentum. The vectors $\mathbf{N}, \mathbf{P}_\perp, \mathbf{k}$ constitute such a system for the initial photon, and the vectors $\mathbf{N}, -\mathbf{P}'_\perp, \mathbf{k'}$ ($\mathbf{P}_\perp$ and $\mathbf{P}'_\perp$ are components of $\mathbf{P}$ to be perpendicular to $\mathbf{k}$ and $\mathbf{k'}$, respectively)—for the final photon. The sign change of $n^{(2)}$ in the photon density matrix (18.229) is equivalent to the one of ξ_1 and ξ_2. Thus, the required density matrices of the initial and final photons are given by the expressions:

$$(\rho_i^\gamma)_{\mu\nu} - \frac{1}{2}[1 + (\boldsymbol{\xi}_i \cdot \boldsymbol{\sigma})]_{ll'} n_\mu^{(l)} n_\nu^{(l')}, \qquad \boldsymbol{\xi}_i = (\xi_{i1}, \xi_{i2}, \xi_{i3}), \tag{19.97}$$

$$(\rho_d^\gamma)_{\mu\nu} = \frac{1}{2}[1 + (\boldsymbol{\xi}_d \cdot \boldsymbol{\sigma})]_{ll'} n_\mu^{(l)} n_\nu^{(l')}, \qquad \boldsymbol{\xi}_d = (-\xi_{d1}, -\xi_{d2}, \xi_{d3}), \tag{19.98}$$

where $l, l' = 1, 2$, and we have emphasized the fact that the polarization matrix of the final photon is dictated by detector characteristics ($\rho'^\gamma \equiv \rho_d^\gamma$).

We should also like to expand the second rank tensor $O^{\mu\nu}$ entering into the matrix element in terms of $n_\mu^{(1)}$, $n_\mu^{(2)}$. This tensor could always be built from products of four-dimensional vectors to govern reaction kinematics. For such vectors we can take the four mutually orthogonal vectors (19.94). However, the tensors $\rho^{\gamma'}$ and ρ^γ contain only the components N and P. Then, under construction of $O^{\mu\nu}$ it is sufficient to make use of the vectors N and P only (addends to be proportional to M and K will simply turn into zero). Sorting out all possible combinations of $n_\mu^{(1)}$ and $n_\nu^{(2)}$, we arrive at the expression for the tensor $O^{\mu\nu}$, which is similar, in its structure, to the expression for the photon density matrix in the four-dimensional form (18.229):

$$O_{\mu\nu} = O_0(n_\mu^{(1)}n_\nu^{(1)} + n_\mu^{(2)}n_\nu^{(2)}) + O_1(n_\mu^{(1)}n_\nu^{(2)} + n_\mu^{(2)}n_\nu^{(1)}) - iO_2(n_\mu^{(1)}n_\nu^{(2)} - n_\mu^{(2)}n_\nu^{(1)}) +$$

$$+O_3(n_\mu^{(1)}n_\nu^{(1)} - n_\mu^{(2)}n_\nu^{(2)}). \tag{19.99}$$

The quantity $O^{\mu\nu}$ enters into the square of the matrix element as $\bar{u}'(\mathbf{p}')O^{\mu\nu}u(\mathbf{p})$. The inclusion of this circumstance proves to be very productive under the determination of the explicit form of the quantities O_i ($i = 0, 1, 2, 3$). Since:

$$\bar{u}'(\mathbf{p}')(\hat{p} + \hat{p}')u(\mathbf{p}) = 2m\bar{u}'(\mathbf{p}')u(\mathbf{p})$$

takes place, then O_i may involve the matrices $\hat{K}$ rather than the ones $(\hat{p}+\hat{p}')$. The quantities O_0 and O_3 are scalars in the same sense as $O_{\mu\nu}$ is a four-dimensional tensor. Therefore, they have the following structure:

$$O_0 = c_0 + c_0'\hat{K}, \qquad O_3 = c_3 + c_3'\hat{K}. \tag{19.100}$$

In its turn, O_1 and O_2 must be pseudoscalars and contain the matrix γ_5 alongside the matrix $\hat{K}$:

$$O_1 = \gamma_5(c_1 + c_1'\hat{K}), \qquad O_2 = \gamma_5(c_2 + c_2'\hat{K}). \tag{19.101}$$

Taking into account orthonormalization of the vectors $n^{(1)}$ and $n^{(2)}$, we may express O_i in terms of projections of the tensor $O_{\mu\nu}$. For example,

$$O_0 = \frac{1}{2}O^{\mu\nu}(n_\mu^{(1)}n_\nu^{(1)} + n_\mu^{(2)}n_\nu^{(2)}), \qquad O_3 = \frac{1}{2}O^{\mu\nu}(n_\mu^{(1)}n_\nu^{(1)} - n_\mu^{(2)}n_\nu^{(2)})$$

an so on. In order to facilitate calculations the tensor $O_{\mu\nu}$ should be rewritten in terms of the mutually orthogonal four-dimensional vectors (19.94):

$$O^{\mu\nu} = \gamma^\mu\frac{\hat{p} + \hat{k} + m}{s - m^2}\gamma^\nu + \gamma^\nu\frac{\hat{p} - \hat{k}' + m}{u - m^2}\gamma^\mu =$$

$$= \gamma^\mu\frac{\hat{P} + 2m}{2(s - m^2)}\gamma^\nu + \gamma^\nu\frac{\hat{P} + 2m}{2(u - m^2)}\gamma^\mu - \frac{1}{t}\left(\gamma^\mu\hat{K}\gamma^\nu - \gamma^\nu\hat{K}\gamma^\mu\right). \tag{19.102}$$

After all these preliminary operations we can embark on determination of O_i. To define O_0 and O_3 we have the equations:

$$\bar{u}'(\mathbf{p}')(c_{0,3} + c_{0,3}'\hat{K})u(\mathbf{p}) = -\bar{u}'(\mathbf{p}')O^{\mu\nu}\left[N^{-2}N_\mu N_\nu \pm P^{-2}P_\mu P_\nu\right]u(\mathbf{p}), \tag{19.103}$$

where $O^{\mu\nu}$ is given by Eq. (19.102). Let us use the relations:

$$\gamma^\mu\gamma^\nu c_\mu c_\nu = c^2, \qquad \gamma^\mu\gamma^\tau\gamma^\nu c_\mu c_\nu = 2\hat{c}c^\tau - \gamma^\tau c^2, \tag{19.104}$$

where c_μ is an arbitrary four-dimensional vector, in the right-hand side of Eq. (19.103). Further, we allow for orthogonality of the four-dimensional vectors P_μ, N_μ, K_μ and the Dirac equation. Then, the comparison of the left-hand and right-hand sides of Eqs. (19.103) yields:

$$O_0 = -ma_+, \qquad O_3 = ma_+ + \frac{1}{2}a_-\hat{K}, \tag{19.105}$$

where

$$a_\pm = \frac{1}{s - m^2} \pm \frac{1}{u - m^2}.$$

The similar calculations result in:

$$O_1 = \frac{i}{2}a_+\gamma^5\hat{K}, \qquad O_2 = -ma_+\gamma^5. \tag{19.106}$$

With the help of equalities:

$$\overline{\gamma}_\mu = \gamma_\mu, \qquad \overline{\gamma_\mu\gamma_\nu...\gamma_\lambda} = \gamma_\lambda...\gamma_\nu\gamma_\mu, \qquad \overline{\gamma}_5 = -\gamma_5, \qquad \overline{\gamma_5\gamma_\mu} = \gamma_5\gamma_\mu,$$

it is easy to verify that the tensor components $\overline{O}_{\mu\nu}$ $(\overline{O}_{\mu\nu} = \gamma^0 O_{\mu\nu}^\dagger \gamma_0)$ follow from the ones $O_{\mu\nu}$ by means of the replacement $O_i \to \overline{O}_i$, where

$$\overline{O}_0 = O_0, \qquad \overline{O}_1 = -O_1, \qquad \overline{O}_2 = -O_2, \qquad \overline{O}_3 = O_3, \qquad (19.107)$$

and simultaneous rearrangement of the indices $\mu\nu$.

Usage of the expressions (19.97)–(19.99), (19.105)–(19.107) allows one to get the result:

$$|\mathcal{M}_{e^-\gamma \to e^-\gamma}|^2 = \frac{e^4 m^2}{4EE'\omega\omega'} \mathrm{Sp}\left[\rho^{(+)'}(\rho_d^\gamma)_\tau{}_\mu O^{\mu\nu}\rho^{(+)}(\rho_i^\gamma)_{\nu\lambda}\overline{O}^{T\lambda\tau}\right] =$$

$$= \frac{e^4 m^2}{4EE'\omega\omega'} \mathrm{Sp}\left\{(\rho^{(+)'}O_0\rho^{(+)}\overline{O}_0 + \rho^{(+)'}\mathbf{O}\rho^{(+)}\overline{\mathbf{O}}) + (\boldsymbol{\xi}_i + \boldsymbol{\xi}_d)(\rho^{(+)'}O_0\rho^{(+)}\overline{\mathbf{O}}+\right.$$

$$+\rho^{(+)'}\mathbf{O}\rho^{(+)}\overline{O}_0) - i(\boldsymbol{\xi}_i - \boldsymbol{\xi}_d)[\rho^{(+)'}\mathbf{O} \times \rho^{(+)}\overline{\mathbf{O}}] + (\boldsymbol{\xi}_i\boldsymbol{\xi}_d)(\rho^{(+)'}O_0\rho^{(+)}\overline{O}_0 - \rho^{(+)'}\mathbf{O}\rho^{(+)}\overline{\mathbf{O}})+$$

$$+\rho^{(+)'}(\boldsymbol{\xi}_d\mathbf{O})\rho^{(+)}(\boldsymbol{\xi}_i\overline{\mathbf{O}}) + \rho^{(+)'}(\boldsymbol{\xi}_i\mathbf{O})\rho^{(+)}(\boldsymbol{\xi}_d\overline{\mathbf{O}}) - i[\boldsymbol{\xi}_i \times \boldsymbol{\xi}_d](\rho^{(+)'}O_0\rho^{(+)}\overline{\mathbf{O}}-$$

$$\left.-\rho^{(+)'}\mathbf{O}\rho^{(+)}\overline{O}_0\right\}. \qquad (19.108)$$

Substituting (19.108) into the general expression for the scattering cross section (18.185) with $d\Omega = d\varphi \sin\theta d\theta$, where φ is a scattering plane azimuth, we deduce the formula making it possible to find the scattering cross section of polarized photons by polarized electrons. We shall use it to calculate the scattering cross section of polarized photons by unpolarized electrons. In this case we should set

$$\rho^{(+)} = \frac{\hat{p} + m}{2m}, \qquad \rho^{(+)'} = \frac{\hat{p}' + m}{2m}$$

in the expression (19.108) and double the result accordingly summation over the final electron polarizations. In the lab frame the differential cross section becomes the form [22]:

$$\frac{d\sigma}{d\varphi d\cos\theta} = \frac{r_0^2}{4}\left(\frac{\omega'}{\omega}\right)^2 \{F_0 + F_3(\xi_{i3} + \xi_{d3}) + F_{11}\xi_{i1}\xi_{d1} + F_{22}\xi_{i2}\xi_{d2} + F_{33}\xi_{i3}\xi_{d3}\}, \quad (19.109)$$

where

$$F_0 = \frac{\omega}{\omega'} + \frac{\omega'}{\omega} - \sin^2\theta, \qquad F_3 = \sin^2\theta,$$

$$F_{11} = 2\cos\theta, \qquad F_{22} = \left(\frac{\omega}{\omega'} + \frac{\omega'}{\omega}\right)\cos\theta, \qquad F_{33} = 1 + \cos^2\theta.$$

The expression (19.109) holds implicit dependence on the scattering plane azimuth through the Stokes parameters, since these parameters are defined with respect to the $x-, y-, z-$ axes (recall, that the x-axis is common for both photons and perpendicular to the scattering plane while the y-axis lays in the scattering plane). Having set $\boldsymbol{\xi}_d = 0$ and having doubled the result, we could obtain the differential cross section of scattering for the polarized initial photon (all other particles to participate in the process are unpolarized). Denoting it as $d\sigma(\boldsymbol{\xi}_i)$, we arrive at:

$$d\sigma(\boldsymbol{\xi}_i) = \frac{r_0^2}{2}\left(\frac{\omega'}{\omega}\right)^2 F d\Omega, \qquad (19.110)$$

where

$$F = F_0 + \xi_{i3}F_3 = \frac{\omega}{\omega'} + \frac{\omega'}{\omega} - (1 - \xi_{i3})\sin^2\theta. \qquad (19.111)$$

From the obtained expression follows that the photons polarized perpendicular to the scattering plane ($\xi_{i3} = 1$) are scattered stronger than the ones polarized in the scattering plane

($\xi_{i3} = -1$). Since the cross section does not include ξ_{i2}, then the dependence on the circular polarization is absent. When the linear polarization of the initial photon in relation to the x- or y-axes is missed or, again, when there is polarization in relation to directions to make the angle equal to 45^0 with these axes, then the scattering cross section coincides with the one for the unpolarized photons.

The scattering cross section of the unpolarized initial photon with detecting the final polarized photon, which follows from (19.109) at $\boldsymbol{\xi}_i = 0$:

$$d\sigma(\boldsymbol{\xi}_d) = \frac{r_0^2}{4}\left(\frac{\omega'}{\omega}\right)^2 [F_0 + \xi_{d3}F_3]d\Omega, \qquad (19.112)$$

possesses the similar properties. One may also set the task of determining the photon polarizations that arise precisely due to scattering by the electron. To do it we should first find the photon polarization matrix in the final state, and thereafter $\boldsymbol{\xi}_f$ could be obtained by application of the formula:

$$\boldsymbol{\xi}_f = \mathrm{Sp}(\rho_f^\gamma \boldsymbol{\sigma}).$$

However, one may avoid new calculations and use the available expression (19.109).

Polarization states of particles are completely defined by the Stokes parameters $\boldsymbol{\xi}$. Let us assume that we are interested in the initial photon polarization and the final photon polarization to be selected by a detector. It is obvious that the square $|\mathcal{M}_{i\to f}|^2$ is linear on each of the vectors $\boldsymbol{\xi}_i$ and $\boldsymbol{\xi}_d$. Then, $|\mathcal{M}_{i\to f}|^2$ being the function of $\boldsymbol{\xi}_d$ should have the following form:

$$|\mathcal{M}_{i\to f}|^2 = a + (\mathbf{b}\boldsymbol{\xi}_d), \qquad (19.113)$$

where a and $\mathbf{b}$, in turn, are linear functions of $\boldsymbol{\xi}_i$. On the other hand, according to (18.238) the quantity $|\mathcal{M}_{i\to f}|^2$ is represented in terms of the density matrices of the initial and final photons by the formula:

$$|\mathcal{M}_{i\to f}|^2 \sim \mathrm{Sp}(\rho_d^\gamma \rho_f^\gamma) = 1 + (\boldsymbol{\xi}_d \boldsymbol{\xi}_f). \qquad (19.114)$$

To compare the expressions (19.113) and (19.114) leads us to the conclusion:

$$\boldsymbol{\xi}_f = \frac{\mathbf{b}}{a}. \qquad (19.115)$$

Thus, having calculated the cross section as a function of the parameters $\boldsymbol{\xi}_d$, we can easily determine the photon polarization state in which the photon is brought directly by the scattering process. It should be stressed that the formula (19.115) remains valid for electrons too.

Using (19.115), we find the Stokes parameters for the final photon:

$$\xi_{f1} = \frac{2\xi_{i1}}{F}\cos\theta,$$

$$\xi_{f2} = \frac{\xi_{i2}}{F}\cos\theta\left(\frac{\omega}{\omega'} + \frac{\omega'}{\omega}\right), \qquad (19.116)$$

$$\xi_{f3} = \frac{1}{F}[\sin^2\theta + (1 + \cos^2\theta)\xi_{i3}].$$

From the obtained formulae one may make the conclusion that in the scattering result the unpolarized photon goes partially polarized. Really, setting $\xi_{i1} = 0$ into (19.116), we get:

$$\xi_{f1} = \xi_{f2} = 0, \qquad \xi_{f3} = \frac{\sin^2\theta}{\frac{\omega}{\omega'} + \frac{\omega'}{\omega} - \sin^2\theta}. \qquad (19.117)$$

As $\xi_{f3} > 0$, then the photon is polarized perpendicularly to the scattering plane. In the nonrelativistic approximation:

$$\xi_{f3} = \frac{\sin^2 \theta}{1 + \cos^2 \theta} \tag{19.118}$$

and the radiation is completely polarized at the scattering angle equal to 90^0.

From (19.116) it also follows that the circularly polarized photon is produced only in the case when the initial photon was circular polarized.

The spurs calculation in the formula (19.108) are considerably complicated for polarized electrons. Since these calculations do not involve any principal difficulties, then, when desired, the reader can independently carry them out addressing the original literature to verify obtained results (the complete summary of the results may be found in the review articles [23]).

19.5 Electron-positron pair production by the photon in the nucleus field

One more process, which happens under photon motion in matter, is photon "materialization" through e^-e^+-pair production. It is clear that the energy to be necessary for this must be no less than $2m$. One photon having the energy even greater than $2m$ is not in a position to produce the e^-e^+-pair, as the conservation laws of the momentum and energy are not being fulfilled simultaneously. To create the e^-e^+-pair by one photon the presence of an extraneous particle is needed. The role of such a particle may play an atomic nucleus. Since an action of the heavy nucleus could be allowed for as one of an external field, then the amplitude of the given process is defined by the expression (19.2) as in the case of the photoeffect. The final state is described by the product of the electron and positron wavefunctions belonging to a continuous spectrum. The electron wavefunction in its asymptotic form must contain a convergent spherical wave in parallel with a plane wave while the positron wavefunction, apart from a plane wave, must involve a divergent spherical wave. In view of the complexity of these functions one manages to exactly calculate the cross section of the pair production in two cases only: (i) in a region of small energies, when the nonrelativistic approximation is applicable; (ii) in the ultrarelativistic case.

However, if the problem conditions are such that the Born's approximation applicability criterion for the Coulomb's field is met:

$$\frac{Z\alpha}{v_{\pm}} \ll 1, \tag{19.119}$$

where v_- (v_+) is a velocity of the electron (positron), then in order to find the cross section of the e^-e^+-pair production process one sufficiently calculates the corresponding element of the scattering matrix in the second order of the perturbation theory. We shall be limited by the investigation of just that very case.

Since in our approach the electric nucleus field is considered as a pseudophotons plurality, then this process could schematically be represented in the form:

$$\gamma + \tilde{\gamma} \rightarrow e^- + e^+, \tag{19.120}$$

where hereafter the symbol $\tilde{\gamma}$ is used for designation of an external electromagnetic field. Remember that an external field is a nonquantized object, that is, it is not associated with any production and annihilation operators. The diagrams pictured in Fig. 19.4, where k is the four-dimensional photon momentum, p_- and p_+ are the four-dimensional momenta of the electron and the positron, and q is the momentum transferred nucleus, correspond to the process (19.120). Attention must be given to the minus sign in front of the four-

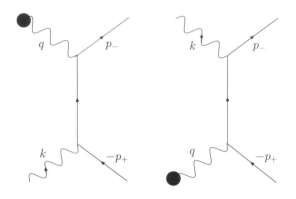

FIGURE 19.4
The Feynman diagrams describing the e^+e^--production by the photon in the nucleus field.

dimensional positron momentum. Its appearance is obligated to the sign in the exponent that accompanies the production operator of the positron in the final state. We emphasize that in the momentum representation on the Feynman diagrams a negative momentum is assigned to all antiparticles not only in the QED but also in any gauge theory.

In quantum field theory prior to calculating the cross section of some process, one must check whether this process is a crossing-channel of the other one that has been investigated before. At a single glance on the diagrams displayed in Fig. 19.4 it becomes clear that they differ from the diagrams of Fig. 19.1 describing the Compton effect by the following rearrangements only:

$$k \to k, \qquad k' \to q, \qquad p \to -p_-, \qquad p' \to p_+, \qquad (19.121)$$

where q is a four-dimensional pseudophoton momentum. Then the amplitude of the e^-e^+-pair production in the nucleus field follows from that of the Compton effect under replacement (19.121) and is defined by the expression:

$$\mathcal{M}_{\gamma\tilde{\gamma} \to e^-e^+} = -ie^2 \sqrt{\frac{m^2}{2E_-E_+\omega}} \overline{u}(\mathbf{p}_-)e^\lambda_\mu(\mathbf{k})O^{\mu\nu}A^e_\nu(q)v(\mathbf{p}_+), \qquad (19.122)$$

where

$$O^{\mu\nu} = \gamma^\mu \frac{\hat{f}_1 + m}{f_1^2 - m^2}\gamma^\nu + \gamma^\nu \frac{\hat{f}_2 + m}{f_2^2 - m^2}\gamma^\mu,$$

$A^e(q)$ is a Fourier component of the nucleus field, and

$$f_{1\mu} = (p_- - k)_\mu, \qquad f_{2\mu} = (-p_+ + k)_\mu, \qquad q = k - p_- - p_+.$$

The e^-e^+-pair production process alongside the Compton effect belong to the processes to be described by the same diagrams both in the QED and in the electroweak theory when one is constrained by the second order of the perturbation theory.

We shall neglect the recoil momentum, that is, the nucleus will be considered to be fixed ($q_0 = 0$). Then, according to (18.192), its Coulomb field is covered by the Fourier component:

$$A_0^e(\mathbf{q}) = \frac{Ze}{(2\pi)^{3/2}\mathbf{q}^2}. \tag{19.123}$$

Let us find the cross section of the e^+e^--pair production for unpolarized particles, that is, average over polarizations of the incoming photon and sum up over spin orientations of the electron and positron to be emitted. As we know, this operation is reduced to the following change in the amplitude square of the process (19.120):

$$|\overline{u}(\mathbf{p}_-)e_\mu(\mathbf{k})O^{\mu 0}A_0^e(\mathbf{q})v(\mathbf{p}_+)|^2 \rightarrow \frac{1}{8m^2}(A_0^e(\mathbf{q}))^2 \, \mathrm{Sp}\, F, \tag{19.124}$$

where

$$F = O^{\mu 0}(-\hat{p}_+ + m)\overline{O}_{\mu 0}(\hat{p}_- + m).$$

It is clear that under rearrangement of the photon and pseudophoton lines on the diagrams of Fig. 19.4 these diagrams convert to each other. Pointed rearrangement is associated with the replacement of the four-dimensional momenta of the particles:

$$p \;\leftrightarrow\; -p_+, \qquad q \leftrightarrow -q, \qquad k \leftrightarrow k. \tag{19.125}$$

The diagrams symmetry existence must lead to splitting of $\mathrm{Sp}\, F$ into two addends to transform to each other under replacement (19.124). It is easy to see that the required splitting has the view:

$$\mathrm{Sp}\, F = P_1 + P_2, \tag{19.126}$$

where

$$P_1 = \frac{1}{m^2\kappa_1}\mathrm{Sp}[O^{\mu 0}(-\hat{p}_+ + m)\gamma^0(\hat{f}_1 + m)\gamma_\mu(\hat{p}_- + m)], \tag{19.127}$$

$$P_2 = \frac{1}{m^2\kappa_2}\mathrm{Sp}[O^{\mu 0}(-\hat{p}_+ + m)\gamma_\mu(\hat{f}_2 + m)\gamma^0(\hat{p}_- + m)], \tag{19.128}$$

and we have introduced invariant quantities

$$m^2\kappa_i = f_{i\mu}^2 - m^2, \qquad i = 1, 2.$$

If one fulfils the replacement (19.124) into the second term of the expression (19.126) and take into account that in this case

$$f_1 \leftrightarrow f_2, \qquad \kappa_1 \leftrightarrow \kappa_2,$$

then, under the spur sign, we shall obtain the matrices identical to those entering in the expression for P_1 but placed in inverse order. Thus, we work out P_1 enough while the relation:

$$P_2(k, p_-, p_+) = P_1(-k, -p_+, -p_-) \tag{19.129}$$

allows us to find P_2. Using the formulae of Section 15.2, we obtain

$$\frac{1}{4}P_1 = \frac{1}{\kappa_1^2}\left[\left(\kappa_1\kappa_2 + 2\frac{\mathbf{q}^2}{m^2}\right) - \frac{4E_+^2}{m^2}(2 - \kappa_1) + \frac{4E_+E_-\kappa_1}{m^2}\right] +$$

$$+ \frac{1}{\kappa_1\kappa_2}\left[\frac{\mathbf{q}^2}{m^2}\left(\kappa_1 + \kappa_2 + \frac{\mathbf{q}^2}{m^2} + 2\right) - \frac{2E_+^2}{m^2}\left(\kappa_2 + \frac{\mathbf{q}^2}{m^2}\right) - \frac{2E_-^2}{m^2}\left(\kappa_1 + \frac{\mathbf{q}^2}{m^2}\right) -$$

$$- \frac{2E_- E_+}{m^2} (\kappa_1 + \kappa_2 - 4) \right].$$ (19.130)

By application of (19.129) we define $P_2/4$ and arrive at the following expression for Sp F:

$$\text{Sp } F = \frac{2}{\kappa_1^2 \kappa_2^2} \left\{ \frac{2\kappa_1 \kappa_2 \mathbf{q}^2}{m^2} \left[2 + \kappa_1 + \kappa_2 + \frac{\mathbf{q}^2}{m^2} - \frac{2}{m^2}(E_-^2 + E_+^2) \right] + \right.$$

$$\left. + (\kappa_1^2 + \kappa_1^2) \left(\kappa_1 \kappa_2 + \frac{2\mathbf{q}^2}{m^2} \right) - 8 \frac{(\kappa_1 E_- - \kappa_2 E_+)^2}{m^2} \right\}.$$ (19.131)

We next draw our attention to the formula (18.197) to deduce the differential cross section. Allow for that unlike the Compton effect, in the case in question the electron momentum $\mathbf{p}_-$ and the positron one $\mathbf{p}_+$ are independent, that is,

$$\left(\frac{\partial(E_- + E_+)}{\partial E_-} \right)^{-1} = 1.$$

We also make use of the relations:

$$m^2 \kappa_1 = -2p_- k = 2\mathbf{p}_- \mathbf{k} - 2E_- \omega = 2\omega(|\mathbf{p}_-| \cos \theta_- - E_-),$$

$$m^2 \kappa_2 = -2p_+ k = 2\omega(|\mathbf{p}_+| \cos \theta_+ - E_+),$$

$$q^2 = -m^2(\kappa_1 + \kappa_2 + 2) - 2(E_- E_+ + p_- p_+),$$

All this taken together allows one to get the differential cross section of the $e^+ e^-$-pair production by the photon in the Coulomb nucleus field [24]:

$$d\sigma = \frac{4Z^2 \alpha^3 |\mathbf{p}_-||\mathbf{p}_+|}{(2\pi)^2 \omega^3 \mathbf{q}^4 m^4} \left\{ -4 \left[\mathbf{k} \left(\frac{E_+ \mathbf{p}_-}{\kappa_1} + \frac{E_- \mathbf{p}_+}{\kappa_2} \right) \right]^2 + \right.$$

$$\left. + \mathbf{q}^2 \left[\mathbf{k} \left(\frac{\mathbf{p}_-}{\kappa_1} - \frac{\mathbf{p}_+}{\kappa_2} \right) \right]^2 + \frac{2\omega^2}{\kappa_1 \kappa_2} [\mathbf{k}(\mathbf{p}_- + \mathbf{p}_+)]^2 \right\} dE_+ d\Omega_- \Omega_+,$$ (19.132)

where $d\Omega_-$ and $d\Omega_+$ are elements of the solid angles in which the momenta $\mathbf{p}_-$ and $\mathbf{p}_+$ lay, E_- and E_+ are the electron and positron energies for which the relation

$$E_- + E_+ = \omega$$

takes place. The differential cross section (19.132) may be rewritten in the form to be more convenient for investigation

$$d\sigma = \frac{Z^2 \alpha^3 p_- p_+ dE_+ \sin \theta_- \sin \theta_+ d\theta_- d\theta_+ d\varphi}{(2\pi)\omega^3 \mathbf{q}^4} \left\{ \frac{p_-^2 \sin^2 \theta_-}{(E_- - p_- \cos \theta_-)^2} (4E_+^2 - \mathbf{q}^2) + \right.$$

$$+ \frac{p_+^2 \sin^2 \theta_+}{(E_+ - p_+ \cos \theta_+)^2} (4E_-^2 - \mathbf{q}^2) - 2\omega^2 \frac{p_+^2 \sin^2 \theta_+ + p_-^2 \sin^2 \theta_-}{(E_+ - p_+ \cos \theta_+)(E_- - p_- \cos \theta_-)} +$$

$$\left. + \frac{2p_- p_+ \sin \theta_+ \sin \theta_- \cos \varphi}{(E_+ - p_+ \cos \theta_+)(E_- - p_- \cos \theta_-)} (4E_- E_+ + \mathbf{q}^2 - 2\omega^2) \right\},$$ (19.133)

where $\theta_\pm$ are angles between $\mathbf{p}_\pm$ and $\mathbf{k}$, φ is an angle between the planes $(\mathbf{k}, \mathbf{p}_-)$ and $(\mathbf{k}, \mathbf{p}_+)$, $p_\pm = |\mathbf{p}_\pm|$.

As we can see, the angular distribution for the produced electron and positron has a very complicated character. It contains two high peaks in directions of the speeds $v_\pm =$

$\cos^{-1}\theta_{\pm}$ and vanishes at $\theta_- = \theta_+ = 0$. The expression for $d\sigma$ is greatly simplified in the ultrarelativistic region ($E_{\pm} \gg m$) only, when a leading role is played by small angles $\theta_{\pm}$. In this case the electron and the positron are emitted in the forward direction predominantly, that is, in the narrow cone with of the angle $\theta_{\pm} \sim m/\omega$ along the photon motion direction. Carrying out the replacements:

$$\sin\theta_{\pm} \to \theta_{\pm}, \qquad \cos\theta_{\pm} \to 1 - \frac{1}{2}\theta_{\pm}^2$$

into (19.133), we deduce the angular distribution in the small angles region:

$$d\sigma = \frac{4Z^2\alpha^3 E_-^2 E_+^2}{\pi m^2 \omega^3 \mathbf{q}^4}[\omega^2(u^2 + v^2)\xi\eta - 2E_- E_+(u^2\xi^2 + v^2\eta^2) +$$

$$+ 2(E_-^2 + E_+^2)uv\xi\eta\cos\varphi]\theta_-\theta_+ dE_+ d\theta_- d\theta_+ d\varphi, \qquad (19.134)$$

where

$$u = \frac{p_-\theta_-}{m}, \qquad v = \frac{p_+\theta_+}{m}, \qquad \xi = \frac{1}{1+u^2}, \qquad \eta = \frac{1}{1+v^2}.$$

To integrate (19.133) over the angles is a rather intricate operation. For the calculation details we refer the reader to Ref. [25] and give the final result only:

$$d\sigma = \frac{Z^2\alpha^3 p_- p_+}{\omega^3 m^2}\left\{ -\frac{4}{3} - \frac{2E_- E_+(p_-^2 + p_+^2)}{p_+^2 p_-^2} + m^2\left(l_-\frac{E_+}{p_-^3} + l_+\frac{E_-}{p_+^3} - \right.\right.$$

$$\left. - l_- l_+ \frac{1}{p_- p_+}\right) + L\left[-\frac{8E_- E_+}{3p_- p_+} + \frac{\omega^2}{p_-^3 p_+^3}(E_-^2 E_+^2 + p_-^2 p_+^2 - m^2 E_- E_+) - \right.$$

$$\left.\left. - \frac{m^2\omega}{2p_- p_+}\left(l_+\frac{E_- E_+ - p_+^2}{p_+^3} + l_-\frac{E_- E_+ - p_-^2}{p_-^3}\right)\right]\right\} dE_+, \qquad (19.135)$$

where

$$l_{\pm} = \ln\frac{E_{\pm} + p_{\pm}}{E_{\pm} - p_{\pm}}, \qquad L = \ln\frac{E_- E_+ + p_- p_+ + m^2}{E_- E_+ - p_- p_+ + m^2}.$$

Recall that all the formulae obtained above are based on the Born's approximation ($Z\alpha/v_{\pm} \ll 1$). The symmetry of the cross sections (19.131)–(19.135) with respect to the electron and the positron corresponds to just that very circumstance. This symmetry disappears in more higher orders of the perturbation theory.

In the ultrarelativistic region, when all energies are significantly larger than m ($\omega, E_{\pm} \gg m$), the expression (19.135) is appreciably simplified:

$$d\sigma = \frac{4Z^2\alpha^3}{\omega^3 m^2}\left(E_-^2 + E_+^2 + \frac{2}{3}E_- E_+\right)\left(\ln\frac{2E_- E_+}{\omega m} - \frac{1}{2}\right) dE_+. \qquad (19.136)$$

Integration of this cross section over E_+ from m to ω yields the total cross section of the $e^- e^+$-pair production by the photon:

$$\sigma = \frac{Z^2\alpha^3}{m^2}\left(\frac{28}{9}\ln\frac{2\omega}{m} - \frac{218}{27}\right), \qquad \omega \gg m. \qquad (19.137)$$

We call attention to the fact that for $\omega \to \infty$ the total cross section, though slowly varying by the logarithmic law, is still tending to infinity. Such a behavior is a symptom that a theory suffers from a serious disease: invasion of divergences. Sometimes this disease may be cured by redefinition of the finite number of the theoretical parameters (renormalizable theories), while in other cases this procedure is not a success (nonrenormalizable theories). Fortunately, the QED falls into the category of the renormalizable theories.

Close to the reaction threshold, the electron and positron velocities may be too low so that the relation (19.118) becomes invalid, making the Born's approximation inapplicable. In this case we have to apply an exact solution of the problem, that is, operate with the wavefunctions of the electron and positron in the Coulomb field of the nucleus. It is obvious that in this case there is no symmetry with respect to the electron (attracted to the nucleus) and to the positron (repelled by the nucleus).

If $v_\pm \approx \sqrt{(\omega - 2m)/\omega}$ remaining more less than 1 is of this sort that $v_\pm \gg Z\alpha$, then the Born approximation is applicable at the reaction threshold too. In the case of non-relativistic energies of the pair we have

$$\omega \approx 2m \gg p_\pm, \qquad q \approx \omega.$$

Then one may set everywhere in (19.133)

$$E_\pm = m, \qquad E_\pm - p_\pm \cos\theta_\pm = m, \qquad \omega = 2m,$$

what leads this relation to the form:

$$d\sigma = \frac{Z^2\alpha^3 p_- p_+}{64\pi^2 m^7}(p_+^2 \sin^2\theta_+ + p_-^2 \sin^2\theta_-)d\Omega_- d\Omega_+ dE_+. \tag{19.138}$$

To integrate the cross section obtained over the angles results in

$$d\sigma = \frac{Z^2\alpha^3}{6m^7}p_- p_+(p_-^2 + p_+^2)dE_+ =$$

$$= \frac{2Z^2\alpha^3}{3m^5}(\omega - 2m)\sqrt{(E_+ - m)(E_- - m)}dE_+. \tag{19.139}$$

From (19.139) one may derive the total cross section of the $e^- e^+$-pair production in the nonrelativistic region

$$\sigma = \int_0^{\omega - 2m} d\sigma = \frac{Z^2\alpha^3\pi}{12m^2}\left(\frac{\omega - 2m}{m}\right)^3. \tag{19.140}$$

If the relative velocity of the electron and the positron v_0 is small, then we must allow for their Coulomb interaction. This interaction is essential when v_0 has the same order as the velocity of a particle in the positronium

$$v_0 \leq \alpha.$$

The cross section of the $e^- e^+$-pair production having been calculated, we assumed that the nucleus field is pure Coulomb. It is true when distances from the nucleus are less than the radius of the atom K-shell. At great distances the nucleus field is partially or completely screened by the atom electrons field. As a result, the cross section of the $e^- e^+$-pair production is decreased. Let us evaluate that region of distances from the nucleus that is the most significant for the process under investigation.

From the view of the integral defining the Fourier component for the Coulomb field of the nucleus it follows that the main contribution in the integral is coming from the distances of the order of

$$R_0 \sim \frac{1}{|\mathbf{q}|}. \tag{19.141}$$

Really, the great values of r are unessential because of the fast oscillating function $\exp(i\mathbf{q}\mathbf{r})$, whereas the small ones—due to smallness of corresponding integration volume $4\pi r^2 dr$.

These circumstances allow one to consider that R_0 defines the most significant values of the impact parameter in the order of magnitude. Obviously, the screening effect will not play any role when the maximum value of R_0, which is associated with the minimum value of $q = |\mathbf{q}|$, is markedly less than the effective atom size R_a ($R_{0max} \ll R_a$). On the contrary, at fulfillment of inequality $R_{0max} \gg R_a$ the complete screening of the Coulomb nucleus field must take place.

Find q_{min} in the nonrelativistic and ultrarelativistic regions. In the former case $\omega \approx 2m$ and we get

$$q_{min} \approx 2m.$$

In the ultrarelativistic region ($\omega, E_\pm \gg m$) the result is

$$q_{min} = \omega - \sqrt{E_-^2 - m^2} - \sqrt{E_+^2 - m^2} \approx \frac{m^2 \omega}{2 E_- E_+}.$$

So, the maximum value of R_0 is given by the formulae:

$$R_{0max} \sim \begin{cases} \frac{1}{m}, & \omega \approx 2m, \qquad \sqrt{(\omega - 2m)/\omega} \ll 1, \\ \frac{2E_- E_+}{m^2 \omega}, & \omega, E_\pm \gg m. \end{cases} \tag{19.142}$$

According to the Thomas-Fermi atom model the effective atom radius is determined by the expression:

$$R_a = \frac{1}{m\alpha Z^{1/3}}. \tag{19.143}$$

From comparison of (19.142) with (19.143) one becomes clear that the condition

$$R_{0max} > R_a$$

may not be realized in the nonrelativistic region but it may take place in the ultrarelativistic region. Therefore, the inclusion of the screening effect is always needed in the high-energy region. It is obvious that the above-derived formulae for the $e^- e^+$-pair production cross section without inclusion of the screening effect in the ultrarelativistic case are valid only under realization of the condition:

$$\frac{2E_- E_+}{m\omega} \ll \frac{137}{Z^{1/3}}. \tag{19.144}$$

To take into consideration the nucleus field screening effect by external electrons one should find the Fourier component of summarized potential produced both by the Coulomb nucleus field and by the electrons field. In the stationary case we have

$$\Delta A_0^{(e)}(\mathbf{r}) = -\rho(\mathbf{r}), \tag{19.145}$$

where $\rho(\mathbf{r})$ is a charge density,

$$\rho(\mathbf{r}) = Ze\delta(\mathbf{r}) - eN(\mathbf{r})$$

and $N(\mathbf{r})$ is an electron density in an atom. By the standard way we get

$$A_0^{(e)}(\mathbf{q}) = \frac{e[Z - F(\mathbf{q})]}{(2\pi)^{3/2}\mathbf{q}^2}, \tag{19.146}$$

where $F(\mathbf{q})$ is an atom formfactor,

$$F(\mathbf{q}) = \frac{1}{(2\pi)^{3/2}} \int N(\mathbf{r})e^{-i\mathbf{q}\mathbf{r}} d^3 x.$$

So, to allow for the screening effect one should make the following replacement in the e^-e^+-pair production reaction amplitude:

$$Z \to Z - F(\mathbf{q}) = Z[1 - f(qZ^{-1/3})]. \tag{19.147}$$

In the Thomas-Fermi model $f(x)$ is an universal function that is identical for all atoms and whose values could be defined by application of numerical integration.

Introducing the atom formfactor one could report the e^-e^+-pair production cross section in the ultrarelativistic region as follows [24]:

$$d\sigma = \frac{Z^2\alpha^3}{\omega^3 m^2}\left[\left(E_-^2 + E_+^2\right)\left(\Phi_1(\zeta) - \frac{4}{3}\ln Z\right) + \frac{2}{3}E_- E_+\left(\Phi_2(\zeta) - \frac{4}{3}\ln Z\right)\right]dE_+. \tag{19.148}$$

where

$$\zeta = 200\alpha\frac{R_a}{R_{0max}} = 100\frac{m\omega}{E_- E_+ Z^{1/3}},$$

$\Phi_1(\zeta)$ and $\Phi_2(\zeta)$ are functions of ζ shown in Fig. 19.5. To be sure that the quantity ζ

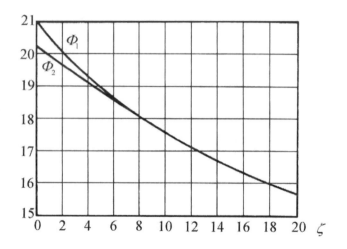

FIGURE 19.5
The graphs of the functions $\Phi_1(\zeta)$ and $\Phi_2(\zeta)$.

determines the screening effect. When ζ is decreasing (increasing) the screening effect is growing (lowering). When $\zeta = 0$ one may speak of the complete screening that obviously takes place under the fulfillment of inequality:

$$\frac{2E_- E_+}{m\omega} \gg 137Z^{-1/3}.$$

In this case

$$\Phi_1(0) = 4\ln 183, \qquad \Phi_2(0) = 4\ln 183 - \frac{2}{3}$$

and the energies distribution take the form:

$$d\sigma = \frac{4Z^2\alpha^3}{\omega^3 m^2}\left[\left(E_-^2 + E_+^2 + \frac{2}{3}E_- E_+\right)\ln(183Z^{-1/3}) - \frac{E_- E_+}{9}\right]dE_+. \tag{19.149}$$

Integrating (19.149) over E_+, we deduce the $e^- e^+$-pair production cross section in the case of the complete screening

$$\sigma = \frac{Z^2 \alpha^3}{m^2} \left(\frac{28}{9} \ln(183 Z^{-1/3}) - \frac{2}{27} \right). \tag{19.150}$$

In Fig. 19.6 we represent the plot of σ versus the photon energy for lead and aluminium with and without inclusion of the screening effect. Polarizations effects for the $e^- e^+$-pair

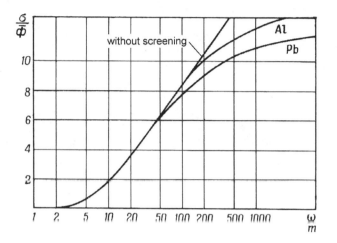

FIGURE 19.6

The total cross section of the process $\gamma \tilde{\gamma} \to e^- e^+$ versus the photon energy.

production process could be investigated by the same method to have been used in Section 19.4 under discussing the Compton effect. Here we shall not touch this problem and refer an interested reader to the original literature [26].

19.6 Resume

When γ-quanta pass through the matter (target), some quanta leave the flux due to their interaction with electrons and nuclei. The beam intensity J is decreased with the growing thickness of a target x as:

$$J = J_0 e^{-\mu x}, \tag{19.151}$$

where J_0 is a flux intensity of photons falling down on a target, μ is an absorption coefficient. In the formation of μ three processes have a decisive role: (i) photoeffect at the electron shell of an atom; (ii) Compton scattering of photons by quasi-free electrons; and (iii) the $e - e+$-pair production in the electrostatic field of the atom nucleus. When N is the number of atoms within 1 cm^3 of a target, σ_i is the cross section of the above processes per atom of the target, then we have:

$$\mu = N \sum_{i=1}^{3} \sigma_i. \tag{19.152}$$

In the case of the photoeffect its cross section is decreased with the growing photon energy: close to the red threshold it is falling as $Z^5\omega^{-7/2}$ and in the ultrarelativistic region—as $Z^5\omega^{-1}$. In the process the principal contribution of $\sim 80\%$ into the total photoeffect cross section is made by the K-shell, whereas the filled L-shell contributes $\sim 16\%$, the contribution of the M-shell being $\sim 4\%$ only.

As distinct from the photoeffect, in the case of the Compton effect a photon is not absorbed, an incident photon beam with the energy ω is transformed to the scattered beam of photons with the energy ω' that is dependent on the scattering angle θ. The energy of a scattered photon is varying within the interval from ω for $\theta = 0$ to $\omega[1+2\omega/m]^{-1}$ for $\theta = \pi$. When the energy ω is considerably higher than the binding energy of the K-electron, the total cross section of the Compton effect for an atom may be considered proportional to the electrons number, that is, in the case of neutral atoms it is proportional to the nucleus charge Z. With growing energy of the incoming photon the Compton effect cross section falls down as $\sim Z\omega^{-1}\ln(\omega/m)$.

The e^-e^+-pair production leads to the absorption of a photon, its energy is distributed between the electron and positron, some part being donated to the nucleus. With increasing photon energy the e^-e^+-pair production cross section is growing as $\sim Z^2\ln(2\omega/m)$.

A relative role of the three principal photon absorption processes in the formation of the coefficient μ is dependent on Z and on the photon energy ω. In Fig. 19.7 we display the plot of the function $\mu(\omega)$ for different substances with allowance made for the screening. The behavior of the components caused by the photoeffect, the Compton effect, and the

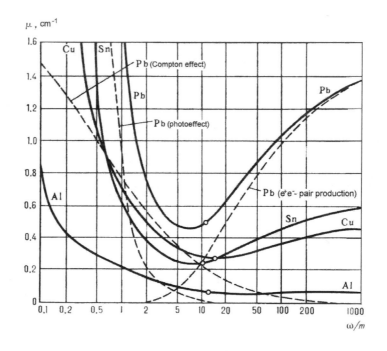

FIGURE 19.7
The plot of the function $\mu(\omega)$.

e^-e^+-pair production effect is also given for lead.

Since the photoeffect depends on Z^5 it is most important for heavy elements; that is why Pb is used for protection from X-rays and γ-radiations. The Compton effect is the most sizable for light elements. As is demonstrated by the relationship between the absorption coefficient μ, photon energy and matter charge Z, in a small-energy region $\omega < \omega_1$ the photoeffect is the basic mechanism for the interaction between a photon and matter, in the intermediate region $\omega_1 < \omega < \omega_2$ the Compton effect is more important, and in the high energy region $\omega > \omega_2$ the key role is played by the the $e^- e^+$-pair production. The boundary values of the energy separating the preferential domains of these effects for various substances are different. In the case of lead they are

$$\omega_1 = 0.5 \text{ MeV}, \qquad \omega_2 = 5 \text{ MeV}.$$

Aside from these main processes, there exist several mechanisms associated with the photon departure from the flux. The first is known as the Delbrück scattering [27], coherent photons scattering by a Coulomb nucleus field, when the $e^- e^+$-pair is formed due to an incoming photon and then annihilated to emit a photon with the initial energy

$$\gamma + \tilde{\gamma} \rightarrow \gamma + \tilde{\gamma}. \tag{19.153}$$

The lowest order of the perturbation theory in which the Delbrück scattering process proceeds is the fourth one. The corresponding Feynman diagram is shown in Fig 19 8 The

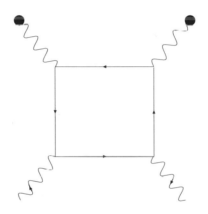

FIGURE 19.8
The Delbrück scattering process.

Delbrück scattering may be regarded as a special case of light-by-light scattering

$$\gamma + \gamma \rightarrow \gamma + \gamma. \tag{19.154}$$

However, the cross section of the process (19.153) in $Z^4 \alpha^2$ is greater than that of the process (19.154), facilitating its experimental observation.

Besides, contributions to γ-radiation absorption are given by: (i) Thomson elastic scattering by the structureless nucleus, (ii) Compton scattering by the nucleus nucleons, and (iii) absorption in the nuclear reactions of the type (γ, n),* (γ, p), and (γ, α). The latter

---

*The first symbol in parentheses denotes an incoming particle while the second one does a new outgoing particle

are important in the region of the so-called giant dipole resonance that is associated with the photon energies of the order of 10–20 MeV. In case of photons with the energies within this region the photonuclear processes may result in 5–10%-contributions into μ.

20

Scattering of electrons and positrons

Then this ebony bird beguiling my sad fancy into
smiling,
By the grave and stern decorum of the countenance
it wore,
"Though thy crest be shorn and shaven, thou," I said,
"art sure no craven,
Ghastly grim and ancient Raven wandering from
the Nightly shore—
Tell me what thy lordly name is on the Night's
Plutonian shore!"
Quoth the Raven "Nevermore."
Edgar Allan Poe, "The Raven"

20.1   Bremsstrahlung

Most important for retardation of the high-energy charged particles is a bremsstrahlung—the process accompanying every change in the state of a charged particle. According to classical electrodynamics, describing the principal features of the bremsstrahlung with a good approximation, its intensity is proportional to the squared acceleration and inversely proportional to the particle mass. As a result, within the same field the bremsstrahlung of an electron is greater than that of a proton by a factor of several millions. Therefore, the bremsstrahlung caused by scattering of electrons from the Coulomb field of the atomic nuclei and electrons is most often observed and used in practice. For example, this process is used to generate high-energy photons due to the transmission of electrons through the high-density targets.

Let us calculate the cross section of the electron bremsstrahlung by the Coulomb nucleus field. In so doing we shall assume that the field may be allowed for by the perturbation theory methods, that is, the applicability criterion of the Born approximation is fulfilled

$$\frac{Z\alpha}{v} \ll 1,$$

where v is an electron velocity. The diagrams corresponding to the process:

$$e^- + \tilde{\gamma} \to e^- + \gamma \tag{20.1}$$

are pictured in Fig. 20.1. It is clear that the process (20.1) is another cross channel of the $e^- e^+$-pair production process in the Coulomb nucleus field. Therefore, the amplitude $\mathcal{M}_{e^- \tilde{\gamma} \to e^- \gamma}$ follows from that of the $e^- e^+$-pair production by means of the replacement:

$$p_+ \to -p_1, \qquad p_- \to p_2, \qquad k \to -k. \tag{20.2}$$

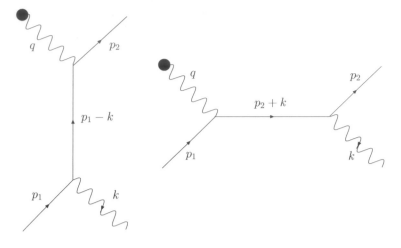

FIGURE 20.1

The Feynman diagrams for the electron bremsstrahlung by the Coulomb nucleus field.

So, we may not work out the cross section anew, and use the results that we have already obtained for the e^-e^+-pair production cross section. Carrying out the replacement (20.2) into (19.133) and taking into consideration changing both of the incoming particle flux value and of the final states number, we get the following expression for the differential cross section of the bremsstrahlung in the case of unpolarized particles [24]:

$$d\sigma = \frac{Z^2\alpha^3 p_2}{(2\pi)^2 p_1 \omega q^4}\left\{\frac{p_2^2\sin^2\theta_2}{(E_2 - p_2\cos\theta_2)^2}(4E_1^2 - q^2)+\right.$$

$$+\frac{p_1^2\sin^2\theta_1}{(E_1 - p_1\cos\theta_1)^2}(4E_2^2 - q^2) + 2\omega^2\frac{p_1^2\sin^2\theta_1 + p_2^2\sin^2\theta_2}{(E_1 - p_1\cos\theta_1)(E_2 - p_2\cos\theta_2)}-$$

$$\left.-\frac{2p_1 p_2\sin\theta_1\sin\theta_2\cos\varphi}{(E_1 - p_1\cos\theta_1)(E_2 - p_2\cos\theta_2)}(4E_1 E_2 - q^2 + 2\omega^2)\right\}d\omega d\Omega d\Omega_2, \qquad (20.3)$$

where $d\Omega$ and $d\Omega_2$ are elements of the solid angles in which the vectors $\mathbf{k}$ and $\mathbf{p}_2$ lay, φ is an angle between the planes $(\mathbf{k}, \mathbf{p}_1)$ and $(\mathbf{k}, \mathbf{p}_2)$, $p_{1,2} = |\mathbf{p}_{1,2}|$ and

$$q = |\mathbf{q}| = |\mathbf{p}_2 + \mathbf{k} - \mathbf{p}_1|.$$

Mark some typical features of the angle distribution (20.3).

(i) By virtue of the fact that the expression (20.3) is proportional to ω^{-1}, $d\sigma$ turns into infinity under $\omega \to 0$ (infrared catastrophe). This effect is caused by the zero photon mass. We shall discuss in detail the infrared catastrophe in the next Section.

(ii) In the angle distribution two peaks in the velocity directions of the initial and final electrons, which are called forth by the factors $(1 - v_1\cos\theta_1)^{-1}$ and $(1 - v_2\cos\theta_2)^{-1}$, are present. In the ultrarelativistic case ($v_1 \sim 1$ and $v_2 \sim 1$) the angle distribution has a sharp maximum near the momentum direction of the incoming electron. In this case the radiation is basically centred into a narrow cone around this direction with the angle of the order of $\theta_1 \sim m/E_1$ (the momentum of the scattered electron lays in the same cone).

To find the spectral distribution of the bremsstrahlung we carry out the replacement (20.2) in the expression (19.135). The result is

$$d\sigma_\omega = \frac{Z^2\alpha^3 p_2}{p_1\omega m^2}\left\{\frac{4}{3} - \frac{2E_1E_2(p_1^2+p_2^2)}{p_1^2p_2^2} + m^2\left(l_1\frac{E_2}{p_1^3} + l_2\frac{E_1}{p_2^3} - \right.\right.$$
$$-l_1l_2\frac{1}{p_1p_2}\right) + L\left[\frac{8E_1E_2}{3p_1p_2} + \frac{\omega^2}{p_1^3p_2^3}(E_1^2E_2^2 + p_1^2p_2^2 + m^2E_1E_2) + $$
$$+ \frac{m^2\omega}{2p_1p_2}\left(l_1\frac{E_1E_2+p_1^2}{p_1^3} - l_2\frac{E_1E_2+p_2^2}{p_2^3}\right)\right]\right\}d\omega, \tag{20.4}$$

where

$$l_{1,2} = \ln\frac{E_{1,2}+p_{1,2}}{E_{1,2}-p_{1,2}} = 2\ln\frac{E_{1,2}+p_{1,2}}{m},$$

$$L = \ln\frac{E_1E_2+p_1p_2-m^2}{E_1E_2-p_1p_2-m^2} = 2\ln\frac{E_1E_2+p_1p_2-m^2}{m\omega}.$$

The allowed values of the photon energy in the formulae obtained are limited by the condition imposed on the final electron velocity ($Z\alpha/v_2 \ll 1$), namely, the electron does not have to lose nearly all its energy

In the nonrelativistic case ($p_1 \ll m$) the spectral distribution is greatly simplified:

$$d\sigma_\omega = \frac{16Z^2\alpha^3 d\omega}{3p_1^2\omega}\ln\frac{p_1+p_2}{p_1-p_2} - \frac{8Z^2\alpha^3 d\omega}{3mT_1\omega}\ln\frac{(\sqrt{T_1}+\sqrt{T_1-\omega})^2}{\omega}, \tag{20.5}$$

where $T_1 = E_1 - m = p_1^2/(2m)$ is a kinetic energy of the initial electron. Thus, in the nonrelativistic region the radiation cross section approximately varies in inverse proportion to the photon energy and vanishes under the greatest possible photon energy $\omega = T_1$.

In the ultrarelativistic region ($E_{1,2} \gg m$) $d\sigma_\omega$ takes the form:

$$d\sigma_\omega = \frac{4Z^2\alpha^3 E_2 d\omega}{\omega E_1 m^2}\left(\frac{E_1}{E_2} + \frac{E_2}{E_1} - \frac{2}{3}\right)\left(\ln\frac{2E_1E_2}{m} - \frac{1}{2}\right) = $$
$$= \frac{4Z^2\alpha^3 d\omega}{\omega m^2}\left[1 + \left(1-\frac{\omega}{E_1}\right)^2 - \frac{2}{3}\left(1-\frac{\omega}{E_1}\right)\right]\left[\ln\frac{2E_1(E_1-\omega)}{m\omega} - \frac{1}{2}\right]. \tag{20.6}$$

From the expression (20.6) it is evident that the radiation probability by the electron of a particular part of its energy, that is, the radiation probability at the given ratio ω/m, approximately grows as $\ln(E_1/m)$. As far as the radiation intensity $J = \omega d\omega$ is concerned, it is logarithmically diverging both in the relativistic and nonrelativistic regions. However, both anomalies for the process in question are removed by means of the inclusion of the nucleus field screening by the atom electrons. To take into account the screening may be carried out in exactly in the same manner as for the case of the e^-e^+-pair production in the nucleus field. In the ultrarelativistic region we derive the formula for the bremsstrahlung cross section [24]:

$$d\sigma_\omega = \frac{Z^2\alpha^3}{\omega E_1^2 m^2}\left[(E_1^2+E_2^2)\left(\Phi_1(\zeta) - \frac{4}{3}\ln Z\right) - \frac{2}{3}E_1E_2\left(\Phi_2(\zeta) - \frac{4}{3}\ln Z\right)\right]d\omega. \tag{20.7}$$

In the case of the complete screening ($\zeta = 0$), that is, when

$$\frac{2E_1E_2}{m\omega} \gg 137Z^{-1/3}, \qquad E_1 \gg 137mZ^{-1/3},$$

the cross section (20.7) acquires the form:

$$d\sigma_\omega = \frac{4Z^2\alpha^3}{\omega E_1^2 m^2}\left[\left(E_1^2 + E_2^2 - \frac{2}{3}E_1 E_2\right)\ln(183Z^{-1/3}) + \frac{E_1 E_2}{9}\right]d\left(\frac{\omega}{E_1 - m}\right). \qquad (20.8)$$

In these conditions at the given ratio ω/E_1 the spectral distribution of the radiation does not depend on the initial electron energy E_1. Under great values of ζ (small screening) this circumstance breaks down.

Having found the bremsstrahlung cross section, one could determine the average energy loss of the electron when it moves through matter. This quantity attributed to the unit of the electron path equals

$$-\left(\frac{dE_1}{dx}\right)_r = N\int_0^{E_1-m}\omega\left(\frac{d\sigma_\omega}{d\omega}\right)d\omega = NE_1\Phi_r, \qquad (20.9)$$

where N is the number of atoms per the unit of the matter volume, and Φ_r is the cross section of the energy loss on radiation:

$$\Phi_r = \frac{1}{E_1}\int_0^{E_1-m}\omega\left(\frac{d\sigma_\omega}{d\omega}\right)d\omega. \qquad (20.10)$$

Inserting the formula (20.4) into (20.10) and fulfilling the integration, we get the expression for Φ_r:

$$\Phi_r = \overline{\Phi}\left[\frac{12E_1^2 + 4m^2}{3E_1 p_1}\ln\frac{E_1 + p_1}{m} - \frac{m^2(8E_1 + 6p_1)}{3E_1 p_1^2}\left(\ln\frac{E_1 + p_1}{m}\right)^2 - \right.$$

$$\left. - \frac{4}{3} + \frac{2m^2}{E_1 p_1}F\left(\frac{2p_1(E_1 + p_1)}{m^2}\right)\right], \qquad (20.11)$$

where $\overline{\Phi} = Z^2\alpha^3/m^2$, and the function $F(x)$ has been defined as:

$$F(x) = \int_0^x \frac{\ln(1+y)}{y}dy.$$

In order to find Φ_r in the nonrelativistic and relativistic regions we use the following expansions of the function $F(x)$:

$$F(x) = \frac{\pi^2}{6} + \frac{1}{2}\ln^2 x - F\left(\frac{1}{x}\right), \qquad \text{for any } x, \qquad (20.12)$$

$$F(x) = x - \frac{x^2}{4} + \frac{x^3}{9} - \frac{x^4}{16} + \dots \qquad x \ll 1. \qquad (20.13)$$

Allowing for (20.12) and (20.13), we arrive at:

$$\Phi_r = \frac{16}{3}\overline{\Phi}, \qquad E_1 \ll m, \qquad (20.14)$$

$$\Phi_r = 4\overline{\Phi}\left(\ln\frac{2E_1}{m} - \frac{1}{3}\right), \qquad E_1 \gg m. \qquad (20.15)$$

From the derived formulae it follows that in the nonrelativistic region the ratio of the emitted energy ω to the energy of the initial electron E_1 does not depend on E_1, while in the ultrarelativistic case Φ_r is logarithmically increased with E_1. As one would expect, this increase is removed by the inclusion of the screening. At the complete screening, in place of (20.15) we have the constant cross section:

$$\Phi_r = \overline{\Phi}\left(4\ln(183Z^{-1/3}) + \frac{2}{9}\right), \qquad E_1 \gg 137mZ^{-1/3}.$$

The energy-dependence of the cross section Φ_r with the screening inclusion in the logarithmic scale is represented in Fig. 20.2, where, for the comparison, the straight line associated with the cross section to be determined by Eq. (20.15) is also displayed. Under motion of

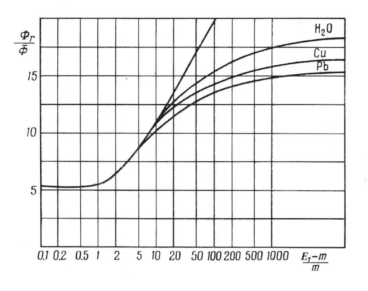

FIGURE 20.2

$\dfrac{\Phi_r}{\overline{\Phi}}$ as a function of $\dfrac{E_1 - m}{m}$.

the electrons in matter one should also bear in mind that the radiation occurs not only on the nucleus but on the electrons as well. In the ultrarelativistic case the bremsstrahlung cross section of the electron from the electron is unlike that of the electron from the nucleus by the absence of the factor Z^2 only [28]. Therefore, the presence of Z atom electrons may be approximately taken into consideration by the replacement:

$$Z^2 \rightarrow Z(Z+1).$$

There exist one more source of energy loss by the electron when it moves in matter, namely, inelastic collisions with atoms. In this case, the average lost energy along the length unit is given by the expression [29]:

$$-\left(\frac{dE_1}{dx}\right)_i = \frac{2\pi\alpha^2 NZ}{m}\ln\frac{E_1^3}{2mI^2}, \qquad (20.16)$$

where I is an average ionization potential ($I \approx 13.5Z$ eV). Thus, the ratio of the loss due to radiation to that due to collisions is equal to

$$\left(\frac{dE_1}{dx}\right)_r\left(\frac{dE_1}{dx}\right)_i^{-1} \approx \frac{1}{1600}\frac{E_1 Z}{m}, \qquad (20.17)$$

where, for the sake of simplicity, we have assumed that the logarithmic term in the formulae (20.15) and (20.16) has approximately an equal value. It is evident that the losses due to radiation and those due to collisions become equal when the relation:

$$(E_1)_0 \approx 1600mZ^{-1}$$

takes place. When $E_1 > (E_1)_0$ the losses due to radiation prevail over the ones due to collisions ($(E_1)_0$ equals 10 MeV for lead, 55 MeV—for copper, 200 MeV—for air). A reader could find the detailed discussion of the bremsstrahlung polarization in Refs. [30,31].

20.2   Radiation of long-wave photons. Infrared catastrophe

In the previous Section we have faced the problem of turning into an infinity of the probability of emitting the photons with $\omega \to 0$, the infrared catastrophe (IC). The IC proves to be caused by the fact that the ordinary perturbation theory, based on expanding the scattering matrix into a series in powers of e, is not applied to the processes in which the long wavelength photons (soft photons) participate. To make this circumstance obvious we shall investigate the inclusion of emitting the soft photons within the diagrammatic technique [32].

Let us consider an arbitrary scattering process of a charged particle that may be accompanied by the emission of any number of the ordinary photons (in the text referred to as the hard photons). Along with this process, we consider a process that is distinguished only by the emission of an additional soft photon. The cross sections of both processes $d\sigma_{ab}$ and $d\sigma_{ab\gamma}$ have a relatively simple relationship, the determination of which presents a problem at hand.

The diagrams for the process with an additional photon follow directly from those for the basic process by the "joining" of an external photon line to some electron line (external or internal), that is, by the substitution:

FIGURE 20.3
The inclusion of a soft photon in an electron line.

Obviously, the main role will be played by the diagrams, where such a joining is realized at the external electron line. In fact, if p—a momentum of the external line ($p^2 = m^2$), then at small k we also have $(p - k)^2 \approx m^2$, that is, a contribution of the electron propagator $S(p - k)$ introduced in the reaction amplitude is considerable due to the position close to its pole. This makes it possible to neglect variations in the momenta of the lines associated with the emission of a soft photon for all the remaining diagram parts.

Joining the photon line to the initial electron line p induces the following replacement in the reaction amplitude:

$$u(\mathbf{p}) \to eS(p - k)\hat{e}^{\lambda*}(\mathbf{k})u(\mathbf{p}) = e\frac{\hat{p} - \hat{k} + m}{(p - k)^2 - m^2}\hat{e}^{\lambda*}(\mathbf{k})u(\mathbf{p}) \approx$$

$$\approx -e\frac{\hat{p} + m}{2(pk)}\hat{e}^{\lambda*}(\mathbf{k})u(\mathbf{p}) = -e\frac{(pe^{\lambda*}(\mathbf{k}))}{(pk)}u(\mathbf{p}), \tag{20.18}$$

where the Dirac equation has been allowed for at the last stage of the transformations. The similar way joining the soft photon to the final electron line with p' is reduced to the replacement:

$$\overline{u}'(\mathbf{p}') \to e\overline{u}'(\mathbf{p}')\frac{(p'e^{\lambda*}(\mathbf{k}))}{(p'k)}. \tag{20.19}$$

For the sake of definiteness, we assume that the cross section $d\sigma_{ab}$ describes the electron scattering by the fixed nucleus with emitting arbitrary numbers of hard photons. The corresponding amplitude is as follows:

$$\mathcal{M}_{ab} = \overline{u}'(\mathbf{p}')\mathcal{M}u(\mathbf{p}).$$

In the expression for $\mathcal{M}_{ab}$ we first fulfill the replacement (20.18) and then the replacement (20.19). Next, summing the results obtained, we arrive at the amplitude of the bremsstrahlung for the same hard photons plus one more soft photon:

$$\mathcal{M}_{ab\gamma} = e\mathcal{M}_{ab}\left[\frac{(p'e^{\lambda*}(\mathbf{k}))}{(p'k)} - \frac{(pe^{\lambda*}(\mathbf{k}))}{(pk)}\right]. \tag{20.20}$$

We draw attention to the gauge invariance of the derived expression, namely, the reaction amplitude is not modified under changing of the four-dimensional vector of the photon polarization:

$$e_\mu^\lambda \to e_\mu^\lambda + \text{const} \cdot k_\mu.$$

To find the connection between the cross sections it is convenient to come back to the three-dimensional designation. Assuming $e^\lambda(\mathbf{k}) = (0, \mathbf{e}^\lambda(\mathbf{k}))$, we get

$$d\sigma_{ab\gamma} = e^2 d\sigma_{ab}\left|\frac{(p'e^{\lambda*}(\mathbf{k}))}{(p'k)} - \frac{(pe^{\lambda*}(\mathbf{k}))}{(pk)}\right|^2 \frac{d^3k}{2(2\pi)^3\omega} =$$

$$= e^2 d\sigma_{ab}\left|\frac{\mathbf{p}'\mathbf{e}^{\lambda*}(\mathbf{k})}{E'\omega(1 - \mathbf{v}'\mathbf{n})} - \frac{\mathbf{p}\mathbf{e}^{\lambda*}(\mathbf{k})}{E\omega(1 - \mathbf{v}\mathbf{n})}\right|^2 \frac{d^3k}{2(2\pi)^3\omega}, \tag{20.21}$$

where $\mathbf{v}$ and $\mathbf{v}'$ are velocities of the initial and final electrons, and $\mathbf{n} = \mathbf{k}/\omega$. Summing over the polarizations of the soft photon, we finally obtain

$$d\sigma_{ab\gamma} = \alpha\left(\frac{[\mathbf{v}'\mathbf{n}]}{1 - \mathbf{v}'\mathbf{n}} - \frac{[\mathbf{v}\mathbf{n}]}{1 - \mathbf{v}\mathbf{n}}\right)^2 \frac{d\omega d\Omega}{4\pi^2\omega}d\sigma_{ab}. \tag{20.22}$$

From (20.22) follows that with a precision of ω^{-1} the expression standing in front of $d\sigma_{ab}$ coincides with the radiation intensity dI calculated in the classical radiation theory, that is,

$$d\sigma_{ab\gamma} = \frac{dI}{\omega}d\sigma_{ab} \tag{20.23}$$

takes place. However, in order to consider the electron motion to be given and use the classical electrodynamics the following conditions:

$$\frac{\omega}{E} \ll 1, \qquad \frac{\omega}{|\Delta\mathbf{p}|} \ll 1, \qquad (20.24)$$

where $\Delta\mathbf{p}$ is a momentum transferred to the nucleus, must be fulfilled. Besides, the photon wave length λ must be significantly bigger than r_0. All this means that the obtained above formulae taking into account emitting a soft photon are valid under fulfillment of the conditions (20.24) only.

In a real experiment the final particles energy is known with some spreading ΔE inside that one cannot separate processes proceeding with and without emitting soft photons. Then the quantity of interest for us, $d\sigma_{ab\gamma}^{(\Delta E)}$, will be given by the expression:

$$d\sigma_{ab\gamma}^{(\Delta E)} = d\sigma_{ab} \int_{\omega_s}^{\Delta E} \frac{dI}{\omega} \sim \alpha \ln \frac{\Delta E}{\omega_s} d\sigma_{ab}. \qquad (20.25)$$

Since the values ω_s had no limitations from below, then at enough small ω_s the parameter $\alpha \ln(\Delta E/\omega_s)$ may become greater than 1 to lead to $d\sigma_{ab\gamma} \succeq d\sigma_{ab}$. But it means the inapplicability of the perturbation theory as, according to its ideology, the quantity $d\sigma_{ab\gamma}$ must be of a more higher smallness order than $d\sigma_{ab}$. To put it differently, at the investigation of the soft photon radiation processes the perturbation theory expansion parameter should be considered by the product $\alpha \ln(\Delta E/\omega_s)$ rather than α.

The formula (20.20) could be generalized without trouble in the event of simultaneously emitting several soft photons. For every one of the photon the factor having the same form as that standing at $\mathcal{M}_{ab}$ in (20.20) is added to the amplitude $\mathcal{M}_{ab\gamma}$. Let us convince ourselves by the example of emitting two soft photons that this is the case. The lines of both soft photons must be joined at external electron lines in two different sequences. Thus, for example, the diagram with a initial external line p is changed by two ones pictured in Fig. 20.4. Then the additional factors:

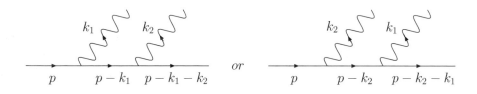

FIGURE 20.4
The diagrams with two soft photons.

$$\frac{(pe^{\lambda_1 *}(\mathbf{k}_1))(pe^{\lambda_2 *}(\mathbf{k}_2))}{2[(pk_1) + (pk_2)]2(pk_1)} \quad \text{and} \quad \frac{(pe^{\lambda_1 *}(\mathbf{k}_1))(pe^{\lambda_2 *}(\mathbf{k}_2))}{2[(pk_1) + (pk_2)]2(pk_2)}$$

appear in the amplitudes corresponding to the first and second diagrams, respectively. Further we should join first the lines k_1 and k_2 at a final electron line p', subsequently k_1—at a final electron line while k_2—at a initial electron line, and lastly k_2—at a final electron line while k_1—at a initial electron line. To sum the results obtained really leads to the formula of the kind (20.20):

$$\mathcal{M}_{ab\gamma} = e^2 \mathcal{M}_{ab} \left[\frac{(p'e^{\lambda_1 *}(\mathbf{k}_1))}{(p'k_1)} - \frac{(pe^{\lambda_1 *}(\mathbf{k}_1))}{(pk_1)} \right] \left[\frac{(p'e^{\lambda_2 *}(\mathbf{k}_2))}{(p'k_2)} - \frac{(pe^{\lambda_2 *}(\mathbf{k}_2))}{(pk_2)} \right]. \qquad (20.26)$$

The cross section of a process is factorized in a similar way. Thus, soft photons are emitted independently. The cross section of a process with emitting n soft photons could be represented in the form:

$$d\sigma_{ab\gamma} = d\sigma_{ab} dw_1 dw_2 ... dw_n, \tag{20.27}$$

where $dw_1, dw_2 ...$ are probabilities of the individual emitting of photons with the four-dimensional momentum k_1, k_2 ... Under integration of this formula over the finite interval of frequencies and directions one should introduce the factor $1/n!$, which takes into account the identity of these particles.

The formula (20.27) may be rewritten in a somewhat different form too. Since emitting the soft photon does not influence an electron motion (this is ensured by the fulfillment of the conditions (20.24)), then the processes of subsequent emitting the soft photons are statistically independent. Hence, it follows that the probability of emitting such photons is defined by the Poisson formula:

$$w(n) = (\overline{n})^n \frac{e^{-\overline{n}}}{n!}, \tag{20.28}$$

where $\overline{n}$ is an average number of the photons emitted:

$$\overline{n} - \int_{\omega_s}^{\Delta E} \omega^{-1} dI. \tag{20.29}$$

Now the cross section of the process with emitting the n photons is represented in the view:

$$d\sigma_{ab\gamma} = d\sigma_{ab} w(n). \tag{20.30}$$

If $\omega_s = 0$, then according to (20.25) and (20.29) $\overline{n}$ is equal to infinity, and, consequently, $w(n)$ vanishes. It means that the probability of electron scattering with emitting the finite number of the soft photons equals zero. In the same time, because

$$\sum_{n=0}^{\infty} w(n) = 1,$$

then

$$\sum_{n=0}^{\infty} d\sigma_{ab\gamma} = d\sigma_{ab}$$

and $d\sigma_{ab}$ comprises the cross section of scattering, which is accompanied by emitting any number of the soft photons. This cross section may be defined by application of the perturbation theory in the case of a sufficiently weak external field. On the other hand, the cross section, accompanied by emitting one soft photon or finite number of the soft photons, is incorrectly given by the perturbation theory. In accordance with the perturbation theory this cross section is equal to infinity whereas it vanishes in a real situation.

As regards the cross sections for the emission of hard photons, which is naturally calculated by the perturbation theory, then the emission cross section, for example, of a single hard photon should be interpreted as a cross section describing the emission both of the hard photon and of an arbitrary number of soft photons.

Because the perturbation theory leads to incorrect values of the cross sections for the interactions between electrons, protons and soft photons, we should pick out the dangerous border ω_{min}, below which the perturbation theory methods are inapplicable, assuming that for a real photon the following is true $\omega > \omega_{min}$. However, in practice it is more convenient to use the equivalent condition: setting no limit to the photon energy, we assume that a photon has some small mass $\lambda \neq 0$. A change from the noninvariant quantity ω_{min} to the relativistically invariant quantity (mass) is called for the maintenance of the relativistic invariance of the theory. Note that the introduction of the photon mass is nothing else but a formal approach to consideration of the infrared divergences. Of course, the photon mass must not enter into any formulae for the observables. To ensure the fulfillment of this circumstance we would simply go to the limit $\lambda \to 0$ after all the intermediate calculations. But we need not do this. As has been demonstrated previously, at sufficiently small ΔE an experimenter can hardly distinguish between the reaction associated with the emission of a soft photon and the corresponding elastic scattering reaction. Because of this, the cross section $d\sigma_{ab\gamma}$ should be considered together with the cross section $d\sigma_{ab}$ calculated in the same order of the perturbation theory. In the next section, we shall find the cross section of the pure Coulomb scattering in the second order of the perturbation theory $d\sigma_{ab}^{(2)}$ and show that it also contains the infrared divergence. In so doing the infrared divergences entering into the observed quantity:

$$d\sigma_{obs} = d\sigma_{ab}^{(2)} + d\sigma_{ab\gamma}^{(2)}$$

are canceled, that is, the infrared cut-off parameter disappears. So, the fictitious photon mass λ is required only with the aim to make the intermediate calculations meaningful. Note that, since the above procedure takes into account the experimental resolution for the energy ΔE, the quantity ΔE is involved in the final result. This, in its turn, reveals the fact that for the low-frequency region of particular importance is the total energy carried over by the emitted soft photons, as related to ΔE, rather than the number of these photons.

Our next task is to establish the linkage between ω_{min} and λ. To solve it we consider the soft photon bremsstrahlung process by the electron in the Coulomb field of the nucleus [32]. Using the replacements:

$$-2(pk) \to -2(pk) + \lambda^2, \qquad 2(p'k) \to 2(p'k) + \lambda^2,$$

we introduce the photon mass into Eqs. (20.18), (20.19). Choose the soft photon polarization vector $e_\mu^\nu(\mathbf{k})$ in the form δ_μ^ν. Then the bremsstrahlung amplitude is determined by the expression:

$$\mathcal{M}_{e^-\tilde{\gamma} \to e^-\gamma} = -ie^2 \sqrt{\frac{m^2}{2EE'\omega}} [\bar{u}(\mathbf{p}')\hat{A}^e(\mathbf{q})u(\mathbf{p})] \left[\frac{2p_\mu}{\lambda^2 - 2(pk)} + \frac{2p'_\mu}{\lambda^2 + 2(p'k)} \right]. \tag{20.31}$$

Since $E, E' \gg \lambda$, then

$$\lambda^2 - 2(pk) = \lambda^2 - 2E\sqrt{\mathbf{k}^2 + \lambda^2} + 2\mathbf{pk} \approx -2E\sqrt{\mathbf{k}^2 + \lambda^2} + 2\mathbf{pk} = -2(pk),$$

$$\lambda^2 + 2(p'k) \approx 2(p'k).$$

Further, we assume that the photon energy does not exceed the value ΔE that defines the experimental energy resolution and satisfies the demand:

$$\lambda \ll \Delta E \ll E.$$

For the sake of simplicity, we shall not be interested in the polarizations of the initial and final particles. Then, with allowance made for Eqs. (20.21) and (4.1.197), we arrive at the following expression for the differential cross section:

$$d\sigma_{e^-\tilde{\gamma} \to e^-\gamma}^{(\Delta E)} = e^2 d\sigma_0 \frac{B}{2(2\pi)^3}, \tag{20.32}$$

where

$$d\sigma_0 = \frac{Z^2\alpha^2}{4p^2v^2\sin^4(\theta/2)}\left(1 - v^2\sin^2\frac{\theta}{2}\right)d\Omega, \qquad B = \sum_{\mu=1}^{4}\int_{k_0\leq\Delta E}\left[\frac{p'_\mu}{(p'k)} - \frac{p_\mu}{(pk)}\right]^2\frac{d^3k}{\omega},$$

$d\Omega$ is a solid angle element in which the scattered electron momentum lays.

To estimate the connection between ω_{min} and λ, we should calculate the quantity B entering into (20.32) by two expedients: (i) $\lambda \neq 0$ while k_0 is changed from $k_0 = \lambda$ to $k_0 = \Delta E$; and (ii) $\lambda = 0$ while k_0 is changed from $k_0 = \omega_{min}$ to $k_0 = \Delta E$. To compare the obtained expressions that we denote as B_λ and $B_{\omega_{min}}$ will lead us to the sought-for relation. The calculation of B_λ results in

$$B_\lambda = -\int_{|\mathbf{k}|=0}^{|\mathbf{k}|=\Delta E}\frac{\mathbf{k}^2d|\mathbf{k}|d\Omega_\gamma}{\sqrt{\mathbf{k}^2+\lambda^2}}\left[\frac{m^2}{(E'\sqrt{\mathbf{k}^2+\lambda^2}-\mathbf{p}'\mathbf{k})^2} + \frac{m^2}{(E\sqrt{\mathbf{k}^2+\lambda^2}-\mathbf{p}\mathbf{k})^2} + \right.$$

$$\left. + \frac{2(pp')}{(E\sqrt{\mathbf{k}^2+\lambda^2}-\mathbf{p}\mathbf{k})(E'\sqrt{\mathbf{k}^2+\lambda^2}-\mathbf{p}'\mathbf{k})}\right] =$$

$$= \int_{|\mathbf{k}|=0}^{|\mathbf{k}|=\Delta E}\frac{\mathbf{k}^2d|\mathbf{k}|d\Omega_\gamma}{\sqrt{\mathbf{k}^2+\lambda^2}}\left[\frac{m^2}{(E'\sqrt{\mathbf{k}^2+\lambda^2}-\mathbf{p}'\mathbf{k})^2} + \frac{m^2}{(E\sqrt{\mathbf{k}^2+\lambda^2}-\mathbf{p}\mathbf{k})^2} + \right.$$

$$\left. + (pp')\int_{-1}^{1}\frac{dz}{(E_z\sqrt{\mathbf{k}^2+\lambda^2}-\mathbf{p}_z\mathbf{k})^2}\right], \tag{20.33}$$

where $d\Omega_\gamma$ is a solid angle element in which the photon momentum lays,

$$\mathbf{p}_z = \frac{1}{2}[(1+z)\mathbf{p} + (1+z)\mathbf{p}'], \qquad E_z = \frac{1}{2}[(1+z)E + (1-z)E'],$$

and we have used the identity:

$$\frac{1}{(E\sqrt{\mathbf{k}^2+\lambda^2}-\mathbf{p}\mathbf{k})(E'\sqrt{\mathbf{k}^2+\lambda^2}-\mathbf{p}'\mathbf{k})} = \frac{1}{2}\int_{-1}^{1}\frac{dz}{(E_z\sqrt{\mathbf{k}^2+\lambda^2}-\mathbf{p}_z\mathbf{k})^2}.$$

The integrals over $d\Omega_\gamma$ could easily be worked out in the spherical coordinates whose polar axis is directed first along the vector $\mathbf{p}$, then along $\mathbf{p}'$, and lastly along $\mathbf{p}_z$. Thus, for example,

$$\int\frac{d\Omega_\gamma}{(E_z\sqrt{\mathbf{k}^2+\lambda^2}-\mathbf{p}_z\mathbf{k})^2} = \frac{2\pi}{|\mathbf{p}_z||\mathbf{k}|(E_z\sqrt{\mathbf{k}^2+\lambda^2}-|\mathbf{p}_z||\mathbf{k}|\cos\theta_\gamma)}\Bigg|_{\theta_\gamma=0}^{\theta_\gamma=\pi} =$$

$$= \frac{4\pi}{E_z^2\lambda^2+\mathbf{k}^2(E_z^2-\mathbf{p}_z^2)}.$$

Taking into consideration the integral value:

$$\int_0^{\Delta E}\frac{\mathbf{k}^2d|\mathbf{k}|}{[\mathbf{k}^2(E^2-\mathbf{p}^2)+\lambda^2E^2]\sqrt{\mathbf{k}^2+\lambda^2}} = \frac{1}{E^2-\mathbf{p}^2}\ln\frac{2\Delta E}{\lambda} - \frac{E}{2|\mathbf{p}|(E^2-\mathbf{p}^2)}\ln\frac{E+|\mathbf{p}|}{E-|\mathbf{p}|},$$

we deduce the following expression for B_λ:

$$B_\lambda = 4\pi\left\{\left[-2 + (pp')\int_{-1}^{1}\frac{dz}{E_z^2-\mathbf{p}_z^2}\right]\ln\frac{2\Delta E}{\lambda} + \frac{1}{2}\left[\frac{E}{|\mathbf{p}|}\ln\frac{E+|\mathbf{p}|}{E-|\mathbf{p}|} + \right.\right.$$

$$+ \frac{E'}{|\mathbf{p}'|} \ln \frac{E' + |\mathbf{p}'|}{E' - |\mathbf{p}'|} - (pp') \int_{-1}^{1} \frac{dz}{E_z^2 - \mathbf{p}_z^2} \frac{E_z}{|\mathbf{p}_z|} \ln \frac{E_z + |\mathbf{p}_z|}{E_z - |\mathbf{p}_z|} \Bigg] \Bigg\}. \qquad (20.34)$$

For more compact writing it is convenient to introduce a new integration variable ξ:

$$\xi = \frac{|\mathbf{p}_z|}{v E_z} = \sqrt{\cos^2 \frac{\theta}{2} + z^2 \sin^2 \frac{\theta}{2}}$$

and the functions Φ and $G(v, \theta)$ determined by the relations:

$$\sinh^2 \ \Phi = \frac{\mathbf{q}^2}{4m^2} = \frac{\mathbf{p}^2 \sin^2(\theta/2)}{m^2} = \frac{v^2}{1 - v^2} \sin^2(\theta/2), \qquad (20.35)$$

$$G(v, \theta) = \int_{\cos(\theta/2)}^{1} \frac{d\xi}{(1 - v^2 \xi^2)\sqrt{\xi^2 - \cos^2(\theta/2)}} \ln \frac{1 + v\xi}{1 - v\xi}. \qquad (20.36)$$

Then B_λ takes the ultimate appearance:

$$B_\lambda = 4\pi \left[2(2\Phi \coth 2\Phi - 1) \ln \frac{2\Delta E}{\lambda} + \frac{1}{v} \ln \frac{1+v}{1-v} - \frac{1-v^2}{v \sin(\theta/2)} \cosh 2\Phi \ G(v, \theta) \right], \quad (20.37)$$

where we have allowed for the relation:

$$(pp') = m^2 \cosh 2\Phi.$$

The gained experience allows one to calculate $B_{\omega_{min}}$ without any trouble

$$B_{\omega_{min}} = 4\pi \left[-2 + (pp') \int_{-1}^{1} \frac{dz}{E_z^2 - \mathbf{p}_z^2} \right] \int_{\omega_{min}}^{\Delta E} \frac{d|\mathbf{k}|}{|\mathbf{k}|} = 8\pi (2\Phi \coth 2\Phi - 1) \ln \frac{\Delta E}{\omega_{min}}. \quad (20.38)$$

To compare the expressions (20.37) and (20.38) gives the linkage between the infrared cut-off parameter and the photon mass:

$$\ln \frac{2\omega_{min}}{\lambda} = \frac{1}{2(1 - 2\Phi \coth 2\Phi)} \left[\frac{1}{v} \ln \frac{1+v}{1-v} - \frac{1-v^2}{v \sin(\theta/2)} \cosh 2\Phi \ G(v, \theta) \right]. \qquad (20.39)$$

In the nonrelativistic case this formula results in the relation:

$$\ln \ 2\omega_{min} = \ln \ \lambda + \frac{5}{6}. \qquad (20.40)$$

Knowledge of B_λ permits one to obtain the differential cross section of the electron scattering in the Coulomb nucleus field with emitting the soft photon whose energy does not exceed ΔE [32]:

$$d\sigma_{e^- \tilde{\gamma} \to e^- \gamma}^{(\Delta E)} = \frac{Z^2 \alpha^2}{4p^2 v^2 \sin^4(\theta/2)} \left(1 - v^2 \sin^2 \frac{\theta}{2} \right) \frac{\alpha}{\pi} \Bigg[2(2\Phi \ \coth 2\Phi - 1) \ln \frac{2\Delta E}{\lambda} +$$

$$+ \frac{1}{v} \ln \frac{1+v}{1-v} - \frac{1-v^2}{v \sin(\theta/2)} \cosh 2\Phi \ G(v, \theta) \Bigg] d\Omega. \qquad (20.41)$$

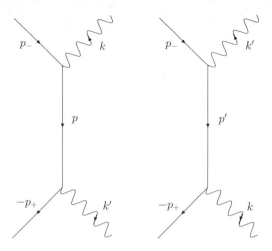

FIGURE 20.5

The Feynman diagrams describing the process $e^- e^+ \to \gamma\gamma$.

20.3 $e^- e^+$-pair annihilation into two γ-quanta. $e^- e^+$-pair production under annihilation of two γ quanta

We shall investigate the process:

$$e^- + e^+ \to \gamma + \gamma, \tag{20.42}$$

whose diagrams are displayed in Fig. 20.5. These diagrams have precisely the same form as do the Compton effect diagrams with one exception only: one fermion line and one photon line interchange their positions. Therefore, the process (20.42) and the Compton effect represent two crossing-channels of one and the same generalized reaction. In this case the Compton effect is the s-channel while the process (20.42) does the t-channel. This circumstance allows us to take advantage the of Compton effect calculation results to find the cross section of the reaction (20.42).

To match Fig. 20.5 to Fig. 19.1 shows that the diagrams of the process (20.42) follow from those of the Compton effect under replacement:

$$p \to p_-, \qquad p' \to -p_+, \qquad k \to -k_1, \qquad k' \to k_2. \tag{20.43}$$

In our case the Mandelstam variables are defined as follows:

$$\left. \begin{aligned} s_t &= (p_- - k_1)^2 = (k_2 - p_+)^2, \\ t_t &= (p_- + p_+)^2 = (k_1 + k_2)^2, \\ u &= (p_- - k_2)^2 = (k_1 - p_+)^2. \end{aligned} \right\} \tag{20.44}$$

Let us find the cross section of the process (20.42) when the initial and final particles are unpolarized. Then the amplitude square $|\mathcal{A}_{e^- e^+ \to \gamma\gamma}|^2$ being expressed in terms of the invariants s_t, t_t, and u, coincides with the similar quantity for the Compton effect under changing the significance of the invariants only (recall, that the polarization states number for the electron and the photon proves to be equal, and by this reason, the factor $1/2$ connected with averaging is conserved). Therefore, all that remains is to anew enumerate the factors standing in front of $|\mathcal{A}_{e^-e^+\to\gamma\gamma}|^2$ in the expression for the differential cross section. Allowing for

$$(p_-p_+)^2 - m^4 = \frac{1}{4}t_t(t_t - 4m^2)$$

and coming back to the standard designations:

$$s_t = t, \qquad t_t = s, \qquad u = u,$$

we get the following expression for the differential cross section of the e^-e^+-pair annihilation into two photons [33]:

$$\frac{d\sigma}{dt} = \frac{8\pi r_0^2 m^2}{s(s - 4m^2)}\left[\left(\frac{m^2}{t - m^2} + \frac{m^2}{u - m^2}\right)\left(\frac{m^2}{t - m^2} + \frac{m^2}{u - m^2} + 1\right) - \right.$$

$$\left. -\frac{1}{4}\left(\frac{t - m^2}{u - m^2} + \frac{u - m^2}{t - m^2}\right)\right]. \tag{20.45}$$

In the center-of-mass frame the expression (20.45) takes on the simple form:

$$d\sigma = \frac{m^2 r_0^2}{4pE}\left(\frac{E^2 + p^2 + p^2\sin^2\theta}{E^2 - p^2\cos^2\theta} - \frac{2p^4\sin^4\theta}{(E^2 - p^2\cos^2\theta)^2}\right)d\Omega, \tag{20.46}$$

where $E = E_- = E_+$, $p = |\mathbf{p}_-| = |\mathbf{p}_+|$ and θ is an angle between $\mathbf{p}_-$ and $\mathbf{k}_1$.

Note some inherent features of the cross section in the center-of-mass frame. As long as two identical particles are present in the final state, then the angle distribution is symmetric with respect to the angle 90^0. Under high energies the peaks caused by the presence of the denominators $E^2(1 - v\cos\theta)(1 + v\cos\theta)$ are observed both for forward-scattering and for back-scattering.

In order to evaluate the total cross section of the e^-e^+-pair annihilation one must integrate $d\sigma$ (20.46) over θ from 0 to π while over azimuth φ from 0 to π (this corresponds to that circumstance that the photons appearing in the final state are indiscernible). It is more convenient, however, to deduce the total cross section in the invariant variables. To fix the integration limits over t we find t in the center-of-mass frame:

$$t = m^2 - 2(p_-k_1) = m^2 - 2E\omega + 2p\omega\cos\theta = m^2 - \frac{s}{2} + \frac{1}{2}\sqrt{s(s - 4m^2)}\cos\theta. \tag{20.47}$$

From (20.47) it follows

$$m^2 - \frac{s}{2} - \frac{1}{2}\sqrt{s(s - 4m^2)} \le t \le m^2 - \frac{s}{2} + \frac{1}{2}\sqrt{s(s - 4m^2)}.$$

The straightforward integration gives

$$\sigma = \frac{2\pi r_0^2}{\epsilon^2(\epsilon - 4)}\left[(\epsilon^2 + 4\epsilon - 8)\ln\frac{\sqrt{\epsilon} + \sqrt{\epsilon - 4}}{\sqrt{\epsilon} - \sqrt{\epsilon - 4}} - (\epsilon + 4)\sqrt{\epsilon(\epsilon - 4)}\right], \tag{20.48}$$

where $\epsilon = s/m^2$.

In the nonrelativistic case we obtain from Eq. (20.48)

$$\sigma = \frac{\pi r_0^2}{\sqrt{\epsilon - 4}}. \tag{20.49}$$

In the ultrarelativistic region the total cross section acquires the form:

$$\sigma = \frac{2\pi r_0^2}{\epsilon}(\ln \epsilon - 1). \tag{20.50}$$

We present the total cross section value in the center-of-mass frame as well

$$\sigma = \frac{\pi r_0^2 (1 - v^2)}{4v}\left[\frac{3 - v^4}{v}\ln\frac{1 + v}{1 - v} - 2(2 - v^2)\right], \tag{20.51}$$

where $v = |\mathbf{p}|/E$ is a velocity of colliding particles.

The formulae describing the positron collision process with the resting electron are of practical interest too. In this case the invariant ϵ takes the view:

$$\epsilon = 2(1 + \gamma), \qquad \gamma = \frac{E_+}{m},$$

where E_+ is a positron energy in the reference frame where the electron is in rest, and the total cross section is given by the expression:

$$\sigma = \frac{\pi r_0^2}{\gamma + 1}\left[\frac{\gamma^2 + 4\gamma + 1}{\gamma^2 - 1}\ln(\gamma + \sqrt{\gamma^2 - 1}) - \frac{\gamma + 3}{\sqrt{\gamma^2 - 1}}\right]. \tag{20.52}$$

In the nonrelativistic region the total cross section is inversely proportional to the positron velocity:

$$\sigma = \frac{\pi r_0^2}{v_+}, \tag{20.53}$$

while the positron annihilation probability in matter w does not depend on the velocity:

$$w = Znv_+\sigma = Zn\pi r_0^2,$$

where n is a number of atoms in the volume unit. So, for the positron flying into lead the lifetime τ is equal to

$$\tau - \frac{1}{w} \approx 10^{-10} \text{ s.}$$

At high energies of the positron $(E_+ \gg m)$ we have

$$\sigma = \frac{\pi m r_0^2}{E_+}\left(\ln\frac{2E_+}{m} - 1\right), \tag{20.54}$$

the positron lifetime tending to zero when $v_+ \to 1$.

Further, without going into details we discuss some features of polarization effects for the process under investigation. To do this we should above all find out the wavefunction properties of two photon systems. It is convenient to carry out an investigation in the center-of-mass frame of the photons [34]

$$\mathbf{k}_1 = -\mathbf{k}_2 = \mathbf{k}.$$

Such a system could be determined in all cases apart from the one when two photons are moving with respect to each other in a parallel way in the same direction. Really, for such photons the summarized momentum $\mathbf{K} = \mathbf{k}_1 + \mathbf{k}_2$ and the total energy $\omega = \omega_1 + \omega_2$ must satisfy the relation:

$$\mathbf{K}^2 = \omega^2,$$

that is, there is no reference frame with $\mathbf{K} = 0$.

The wavefunction of the two photon system in the momentum representation is represented in the form of three-dimensional second rank tensor $\Phi_{im}(\mathbf{n})$. This tensor is the product of the coordinate and spin wavefunctions. The coordinate part of the wavefunction comprises the spherical function $Y_{lm}(\mathbf{n})$ $(\mathbf{n} = \mathbf{k}/k)$, which is simply a scalar from the point of view of the geometrical properties. As distinct from it, the spin part being a bilinear combination of the spin photon wavefunctions is multi component. Let the index i correspond to the first photon, whereas the one m corresponds to the second photon. Then, from the wavefunction transversality condition follows:

$$n_i \Phi_{im}(\mathbf{n}) = n_m \Phi_{im}(\mathbf{n}) = 0. \tag{20.55}$$

The mutual rearrangement of the photons means the rearrangement of the indices of the tensor $\Phi_{im}(\mathbf{n})$ along with the simultaneous change of the sign $\mathbf{n}$. Taking into consideration that the photons obey the Bose statistics, we get

$$\Phi_{im}(-\mathbf{n}) = \Phi_{mi}(\mathbf{n}). \tag{20.56}$$

Divide the tensor $\Phi_{im}(\mathbf{n})$ into the symmetric and antisymmetric parts with respect to the indices i and m:

$$\Phi_{im}(\mathbf{n}) = S_{im}(\mathbf{n}) + A_{im}(\mathbf{n}).$$

Obviously, both the symmetric and antisymmetric parts of the tensor $\Phi_{im}(\mathbf{n})$ must satisfy the relations (20.55) and (20.56). Then, from (20.56) it is evident that

$$S_{im}(-\mathbf{n}) = S_{im}(\mathbf{n}), \tag{20.57}$$

$$A_{im}(-\mathbf{n}) = -A_{im}(\mathbf{n}). \tag{20.58}$$

The space inversion on its own does not change the components sign of the second rank tensor but the one changes the sign of $\mathbf{n}$. Therefore, it follows from (20.57) that the wavefunction is symmetric with respect to the space inversion, that is, the one corresponds to the even states of the system. Analogously, it stems from (20.58) that the wavefunction $A_{im}(\mathbf{n})$ describes the odd states of the two-photons system.

The antisymmetric tensor on its transformation properties is equivalent to the axial vector $\mathbf{a}(\mathbf{n})$ with the components:

$$a_l(\mathbf{n}) = \frac{1}{2} \varepsilon_{lim} A_{im}(\mathbf{n}).$$

As

$$A_{im}(\mathbf{n}) = \varepsilon_{iml} a_l(\mathbf{n}),$$

then it follows from the orthogonality condition (20.55)

$$A_{im}(\mathbf{n}) n_m = \varepsilon_{iml} a_l(\mathbf{n}) n_m = [\mathbf{n} \times \mathbf{a}(\mathbf{n})]_i = 0,$$

that is, the vectors $\mathbf{n}$ and $\mathbf{a}(\mathbf{n})$ are parallel. Consequently, the vector $\mathbf{a}(\mathbf{n})$ may be represented in the form:

$$\mathbf{a}(\mathbf{n}) = \mathbf{n}\varphi(\mathbf{n}). \tag{20.59}$$

The relation (20.59) permits us to correlate the antisymmetric tensor $A_{im}(\mathbf{n})$ to the zero value of the system spin, that is, this tensor describes the state with $l = j$. From (20.58) it is evident that

$$\mathbf{a}(-\mathbf{n}) = -\mathbf{a}(\mathbf{n}),$$

what, in its turn, gives

$$\varphi(-\mathbf{n}) = \varphi(\mathbf{n}). \tag{20.60}$$

The equality (20.60) points to the fact that the scalar $\varphi(\mathbf{n})$ is being built from the spherical functions only of the even order l (including 0). Thus, we have shown that the tensor $A_{im}(\mathbf{n})$ accords with the odd states of the two-photons system with the even moment j.

As far as the symmetric tensor $S_{im}(\mathbf{n})$ is concerned, then, to find out its spin nature, it must be expanded into irreducible parts. This operation is reduced to the expansion into $\mathrm{Sp}[S_{im}(\mathbf{n})]$ (the scalar) and the symmetric tensor $S'_{im}(\mathbf{n})$ with the zero spur (five components). The scalar is associated with the spin zero, and, hence, it governs the states with the even value of $j = l$. It is clear that the tensor $S'_{im}(\mathbf{n})$ corresponds to the spin value to equal 2. Allowing for that in this case l must be even, we find:
(i) at even $j \neq 0$ three states are possible

$$l = j \pm 2, j;$$

(ii) at odd $j \neq 1$ two states are possible

$$l = j \pm 1.$$

The exception are provided by $j = 0$ with one state $l = 2$ and $j = 1$ with one state $l = 2$.

However, the orthogonality condition of the tensor $S_{im}(\mathbf{n})$ to the vector $\mathbf{n}$ has not been taken into account in our calculation. Because of this, one must subtract the number of the states to be associated with a symmetric tensor of the second rank, which is parallel to the vector $\mathbf{n}$ from the obtained states number. Such a tensor has the form:

$$S''_{im}(\mathbf{n}) = n_i b_m(\mathbf{n}) + n_m b_i(\mathbf{n}), \tag{20.61}$$

where $\mathbf{b}(\mathbf{n})$ is a vector which, according to (20.57), satisfies the demand.

$$\mathbf{b}(-\mathbf{n}) = -\mathbf{b}(\mathbf{n}). \tag{20.62}$$

From (20.61) and (20.62) it follows that the tensor $S''_{im}(\mathbf{n})$ is equivalent to an odd vector which, in its turn, may be expressed through the spherical functions only of the odd orders l. Inasmuch as the vector is associated with the spin 1, then one may state: (i) at even $j \neq 0$ two states with $l = j \pm 1$ are possible; and (ii) at odd j one state with $l = j$ is possible. One state with $j = 0$ and $l - 1$ is of a special case.

The derived results are gathered into the Table 20.1:

TABLE 20.1
The number of states of the two-photons system.

j	Number of even states	Number of odd states
0	1	1
1	0	0
2s	2	1
2s+1	1	0

where s is an integer positive nonzero number. Attention is drawn to the fact that the two-photons system cannot be in the state with $j = 1$ at all. Therefore, a system having the moment equal to 1 cannot decay into two photons.

The wavefunction $\Phi_{ij}(\mathbf{k})$ controls the polarizations correlation of photons. The probability that two photons simultaneously have the definite polarizations $\mathbf{e}^{\lambda_1}$ and $\mathbf{e}^{\lambda_2}$ is proportional to

$$\Phi_{ij}(\mathbf{k})e_i^{\lambda_1*}e_j^{\lambda_2*}.$$

Thus, if the polarization of one photon is given, then that of a second photon is defined as follows:

$$e_j^{\lambda_2} \sim \Phi_{ij}(\mathbf{k}) e_i^{\lambda_1 *}. \tag{20.63}$$

In odd states of the system $\Phi_{ij}(\mathbf{k})$ coincides with the antisymmetric tensor $A_{ij}(\mathbf{k})$. Then we have

$$\mathbf{e}^{\lambda_2} \mathbf{e}^{\lambda_1 *} \sim A_{ij}(\mathbf{k}) e_i^{\lambda_1 *} e_j^{\lambda_1 *} = 0,$$

that is, the polarizations of both photons are mutually orthogonal. In the case of the linear polarization it means the perpendicularity of their directions while in the case of the circular polarization it signifies the direction opposite of their rotations.

Now we can discuss the qualitative features of the polarization effects in the center-of-mass frame for the nonrelativistic and ultrarelativistic regions of the particles to collide.

At $v \to 0$ the process is proceeding only in the case when the system *electron+positron* is in the S-state. Since in this case the system possesses the negative parity, then the system parity of the photons to come into being is negative too. For the odd states, as we have learned earlier, the photon polarization vectors are mutually orthogonal.

When the electron and the positron are polarized, in the nonrelativistic region the process (20.42) will proceed if and only if their spins are antiparallel. Actually, as the annihilation happens into the S-state, then the total moment coincides with the total spin of the particles which is equal to 1 when their spins are parallel. However, the two-photons system can in no way possess the states with the total moment to equal 1.

In the ultrarelativistic case the chirality projection operator becomes the helicity projection operator (see, Section 15.6):

$$\Lambda^{(\alpha)}(n) \to \frac{1}{2} \left(1 + \alpha \gamma_5\right),$$

Then the analysis of the amplitude

$$\mathcal{A}_{e^- e^+ \to \gamma\gamma} = -ie^2 \sqrt{\frac{m^2}{4E_- E_+ \omega_1 \omega_2}} \, \overline{v}(\mathbf{p}_+) e_\nu^{\lambda_1 *}(\mathbf{k}_1) \left(\gamma^\nu \frac{\hat{p}_- - \hat{k}_1 + m}{(p_- - k_1)^2 - m^2} \gamma^\mu + \right.$$

$$\left. + \gamma^\mu \frac{\hat{p}_- - \hat{k}_2 + m}{(p_- - k_2)^2 - m^2} \gamma^\nu \right) e_\mu^{\lambda_2 *}(\mathbf{k}_2) u(\mathbf{p}_-), \tag{20.64}$$

leads us to the conclusion that in the case when one may neglect the electron mass the reaction will only proceed for the electron and the positron with the opposite signs of the helicity (in the center-of-mass frame the parallel spins are associated with the different values of the helicity).

In the ultrarelativistic case the annihilation of the electron and the positron with the same helicities is possible solely at the inclusion in the process amplitude of the terms containing m. In order of the magnitude the amplitude of this process differs from that of the annihilation of the $e^- e^+$-pair with the parallel spins by the factor m/E (a distinction between corresponding cross sections are proportional to $(m/E)^2$).

A more detail analysis of the polarization effects shows that the process in question is extremely sensitive to the positron polarization. This circumstance is used for determining the helicities of the positrons to be emitted in a β-decay.

A comprehensive investigation of the polarization effects at the $e^- e^+$-pair annihilation into the two photons could be found in Refs. [35].

Sometimes the process of the $e^- e^+$-pair annihilation into the two photons is called the *annihilation in the flight*. This process is not the dominated one under the positrons absorption by matter. In the vicinity of 80% the positrons stop to form a positronium, a nonstable bound state of the electron and the positron.

In parallel with the conversion of the $e^- e^+$-pair into the photons the process:

$$\gamma + \gamma \rightarrow e^- + e^+ \tag{20.65}$$

is also possible. The corresponding Feynman diagrams are pictured in Fig. 20.6. It is

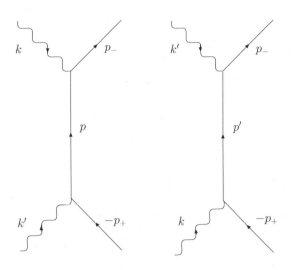

FIGURE 20.6

The Feynman diagrams for the process $\gamma\gamma \rightarrow e^- e^+$.

obvious that the diagrams of the electron-positron pair production follow from those of the process (20.42) under the time inversion. Therefore, the amplitude squares of both processes are identical and their linkage with the cross section differs only in that now we have

$$I = (k_1 k_2)^2 = \frac{t_t^2}{4} = \frac{s^2}{4}.$$

Thus, the differential cross section of the $e^- e^+$-pair production under the two-photons annihilation for the case of unpolarized particles is given by the expression [33]:

$$\frac{d\sigma}{dt} = \frac{8\pi r_0^2 m^2}{s^2} \left[\left(\frac{m^2}{t - m^2} + \frac{m^2}{u - m^2} \right) \left(\frac{m^2}{t - m^2} + \frac{m^2}{u - m^2} + 1 \right) - \right.$$
$$\left. - \frac{1}{4} \left(\frac{t - m^2}{u - m^2} + \frac{u - m^2}{t - m^2} \right) \right]. \tag{20.66}$$

Let us cross over to the center-of-mass frame. Taking into consideration the relations:

$$\omega_1 = \omega_2 = \omega_0, \qquad E_- = E_+ = \omega_0,$$

we can represent the expression for the cross section in the form:

$$d\sigma = \frac{r_0^2 m^2 \sqrt{\omega_0^2 - m^2}}{4\omega_0^3} \left[\frac{2\omega_0^2 - m^2 + (\omega_0^2 - m^2)\sin^2\theta}{m^2 \cos^2\theta + \omega_0^2 \sin^2\theta} - \right.$$
$$\left. - \frac{2(\omega_0^2 - m^2)^2 \sin^4\theta}{(m^2 \cos^2\theta + \omega_0^2 \sin^2\theta)^2} \right] d\Omega. \tag{20.67}$$

At integration of (20.67) one should allow for that because the two final particles (electron and positron) are nonidentical, the final result must not be divided into 2 as it took place in the case of the $e^- e^+$-pair annihilation. Thus, the total cross section is equal to

$$\sigma = \frac{\pi r_0^2 m^2}{\omega_0^2} \left[\left(2 + \frac{2m^2}{\omega_0^2} - \frac{m^4}{\omega_0^4} \right) \ln \left(\frac{\omega_0}{m} + \sqrt{\frac{\omega_0^2}{m^2} - 1} \right) - \sqrt{1 - \frac{m^2}{\omega_0^2}} \left(1 + \frac{m^2}{\omega_0^2} \right) \right]. \quad (20.68)$$

20.4   Positronium lifetime

In the preceding section we have considered the $e^- e^+$-pair annihilation neglecting the Coulomb force. However, the latter may manifest itself as the appearance of a neutral system (positronium), a bound state of the electron and the positron. The positronium represents a lepton analog of atomic bound states, but in contradistinction to them, the positronium is unstable. We shall solve the problem of finding the lifetime of the positronium in the ground state (S-state).

In the positronium the electron and positron spins may be parallel and antiparallel. The first case is associated with the triplet state ^3S called the *orthopositronium*, and the second case—with the singlet state ^1S referred to as the *parapositronium*. The s size of the positronium is approximately twice as big as that of a hydrogen atom (as the reduced mass of the positronium equals $m/2$, where m is the electron mass), whereas its binding energy is only one half of that for the hydrogen atom. An energy level of the ground state of the parapositronium is lower than that of the orthopositronium by 8.41×10^{-4} eV, and in a magnetic field there is a possibility for the transitions between these positronium states. It is noteworthy that the positronium is the simplest system coupled by pure electromagnetic forces and hence the properties of the free positronium are of special interest for testing the validity of the QED. As a system of two photons could not be in the state with the total momentum equal to one, the orthopositronium could not break down into two photons. Moreover, because the positronium state ^3S is an odd-charge state, we can state that breaking-down into an even number of photons for the positronium is impossible at all. On the other hand, the parapositronium is an even-charge state and hence its breaking-down into any odd number of photons proves to be forbidden. So the principal process determining the positronium lifetime is a two-photons annihilation in the case of the parapositronium and a three-photons annihilation in the case of the orthopositronium. Besides, the corresponding probabilities of the positronium decay may be associated with the annihilation cross sections of a free $e^- e^+$-pair [36].

From the uncertainty relation it follows that in order of magnitude the momenta of the electron and the positron in the positronium are

$$\sim me^2/\hbar$$

where we have gone to the ordinary system of units. By virtue of the fact that they are more less than mc, then in order to calculate the annihilation probability one may transfer to the limit of two particles resting in the origin of coordinates. Then, having multiplied the annihilation cross section σ by the flux density to be equal to $v|\psi(0)|^2$, we obtain the annihilation probability. Here v is the relative velocity of the particles and $\psi(r)$ is the wavefunction of the ground state of the positronium

$$\psi(r) = \frac{1}{\sqrt{\pi a^3}} e^{-r/a}, \qquad a = \frac{2\hbar^2}{me^2}. \quad (20.69)$$

Let us denote the decay probability of the orthopositronium and the parapositronium by w_1 and w_0 while the decay probability of the positronium averaging over the spin orientations by $\overline{w}$. Since the relative weights of the states with the spins 1 and 0 are equal to 3/4 and 1/4 respectively, then

$$\overline{w} = \frac{3}{4}w_1 + \frac{1}{4}w_0. \tag{20.70}$$

takes place. It is evident that the quantity $\overline{w}$ has the view:

$$\overline{w} = \overline{w}_{2\gamma} + \overline{w}_{3\gamma} + ..., \tag{20.71}$$

where $\overline{w}_{n\gamma}$ is the probability of converting the e^-e^+-pair into n photons, which have been averaged over the spin states. From (20.70) and (20.71) follows

$$\frac{1}{4}w_0 = \overline{w}_{2\gamma} + \overline{w}_{4\gamma} + ..., \qquad \frac{3}{4}w_1 = \overline{w}_{3\gamma} + \overline{w}_{5\gamma} + ...,$$

that is,

$$\frac{1}{4}w_0 \approx \overline{w}_{2\gamma}, \qquad \frac{3}{4}w_1 \approx \overline{w}_{3\gamma}. \tag{20.72}$$

Using the formula (20.53) for the total cross section of the e^-e^+-pair annihilation into the two photons at small velocities, one may get the lifetime of the parapositronium from (20.72)

$$\tau_0 = \frac{1}{w_0} = \frac{1}{4|\psi(0)|^2(v_+\sigma)_{v_+\to 0}} = \frac{2\hbar}{mc^2\alpha^5} \approx 1.23 \times 10^{-10} .$$

As the level width $\Gamma_0 = \hbar/\tau_0$ is small compared with the energy of the ground state

$$|E_0| = \frac{me^4}{4\hbar^2} = mc^2\frac{\alpha^2}{4},$$

then our treatment of the parapositronium as the quasistationary system is well founded.

Since the lifetime of the orthopositronium is defined by the expression:

$$\tau_1 = \frac{1}{w_1} = \frac{3}{4|\psi(0)|^2(v_+\sigma_{3\gamma})_{v_+\to 0}}, \tag{20.73}$$

our next task is to determine the cross section of the process:

$$e^- + e^+ \to \gamma + \gamma + \gamma. \tag{20.74}$$

The process is figured by six diagrams, one of them is shown in Fig. 20.7 (the remaining five are distinguished from that by rearrangements of the photon momenta k_1, k_2, and k_3 only). The corresponding amplitude is as follows:

$$\mathcal{A}_{3\gamma} = -ie^3\sqrt{\frac{m^2}{8E_-E_+\omega_1\omega_2\omega_3}}e_\mu^{\lambda_1*}(\mathbf{k}_1)e_\nu^{\lambda_2*}(\mathbf{k}_2)e_\rho^{\lambda_3*}(\mathbf{k}_3)\overline{v}(\mathbf{p}_+)O^{\mu\nu\rho}u(\mathbf{p}_-), \tag{20.75}$$

where

$$O^{\mu\nu\rho} = \sum_P \gamma^\mu\frac{\hat{k}_3 - \hat{p}_+ + m}{(k_3 - p_+)^2 - m^2}\gamma^\nu\frac{\hat{p}_- - \hat{k}_1 + m}{(p_- - k_1)^2 - m^2}\gamma^\rho$$

and the sum is taken over all rearrangements of the photon numbers 1,2,3 along with simultaneous rearrangements of the indices of the corresponding vertices μ, ν, and ρ.

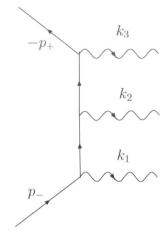

FIGURE 20.7
The Feynman diagram for the process $e^-e^+ \to \gamma\gamma\gamma$.

To sum over the spin states of the final particles and average over the spin states of the initial particles leads to the expression:

$$\frac{1}{4}\sum_{spin}|\mathcal{A}_{3\gamma}|^2 = \frac{e^6}{128E_-E_+\omega_1\omega_2\omega_3}\text{Sp}[(\hat{p}_- + m)O^{\mu\nu\rho}(-\hat{p}_+ + m)\overline{O}_{\mu\nu\rho}. \tag{20.76}$$

In the case of small velocities of the electron and the positron one may set their three-dimensional momenta equal to zero to give: $p_- = p_+ = p = (m,0,0,0)$. Then the expressions both for the propagators and for the density matrix become the simple forms:

$$\frac{\hat{k}_3 - \hat{p}_+ + m}{(k_3 - p_+)^2 - m^2} \approx \frac{\hat{k}_3 + m(1 - \gamma_0)}{-2m\omega_3},$$

$$\frac{1}{2m}(\hat{p}_- + m) \approx \frac{1}{2}(\gamma_0 + 1).$$

Adduce the scheme of further calculations. An appreciable simplification is achieved above all due to using the cross-universality. The amplitude square module involves 36 terms of the type $\mathcal{M}^*_{ijk}\mathcal{M}_{i'j'k'}$ (where the indices i, j, k take the values 1,2,3 and define the order of following of the photon vertices at the motion along the diagram from the bottom upwards). It is sufficient to calculate only six terms from 36 ones, for example, the group:

$$\mathcal{M}^*_{123}(\mathcal{M}_{123} + \mathcal{M}_{132} + \mathcal{M}_{321} + \mathcal{M}_{312} + \mathcal{M}_{231} + \mathcal{M}_{213}), \tag{20.77}$$

and, thanks to the cross-universality, all the remaining terms will follow from this group under the corresponding transformation of the photon momenta. Next, in the expression (20.77) one may separate the terms that go into each other under different rearrangements of the photons. The products (pk_i) (i=1,2,3) appearing under the taking of spurs are expressed in terms of the electron mass and the photon energies $\omega_1, \omega_2, \omega_3$. The products $(k_1k_2),...$ are defined by the conservation law of the four-dimensional momentum:

$$2p = k_1 + k_2 + k_3. \tag{20.78}$$

Thus, for example, having rewritten the expression (20.78) in the form $2p - k_3 = k_1 + k_2$ and having squared it, we get

$$(k_1k_2) = 2m(m - \omega_3).$$

It is clear that one of the validity tests of the obtained result will be the symmetry of $\sum_{spin} |\mathcal{A}_{3\gamma}|^2$ with respect to ω_1, ω_2, and ω_3. In response to straightforward but extremely cumbersome operations we get

$$\frac{1}{4} \sum_{spin} |\mathcal{A}_{3\gamma}|^2 = 16e^6 \left[\left(\frac{m - \omega_1}{\omega_2 \omega_3} \right)^2 + \left(\frac{m - \omega_2}{\omega_1 \omega_3} \right)^2 + \left(\frac{m - \omega_3}{\omega_2 \omega_1} \right)^2 \right]. \tag{20.79}$$

Making use of (20.79), we arrive at the following expression for the differential cross section of the $e^- e^+$-pair annihilation into three photons [37]:

$$d\sigma_{3\gamma} = \frac{e^6}{(4\pi)^3 \pi^2 m^2 v} \left[\left(\frac{m - \omega_1}{\omega_2 \omega_3} \right)^2 + \left(\frac{m - \omega_2}{\omega_1 \omega_3} \right)^2 + \left(\frac{m - \omega_3}{\omega_2 \omega_1} \right)^2 \right] \times$$

$$\times \delta^{(3)} (\mathbf{k}_1 + \mathbf{k}_2 + \mathbf{k}_3) \delta(\omega_1 + \omega_2 + \omega_3 - 2m) \frac{d^3 k_1 d^3 k_2 d^3 k_3}{\omega_1 \omega_2 \omega_3}. \tag{20.80}$$

The first of δ-functions in (20.80) is eliminated by means of the integration over $\mathbf{k}_3$

$$\delta^{(3)} (\mathbf{k}_1 + \mathbf{k}_2 + \mathbf{k}_3) d^3 k_3 \to 1.$$

To liquidate the second δ-function we first rewrite the differentials in the spherical system of coordinates

$$d^3 k_1 d^3 k_2 = 4\pi \omega_1^2 d\omega_1 2\pi \omega_2^2 d(\cos\theta_{12}) d\omega_2,$$

where θ_{12} is an angle between $\mathbf{k}_1$ and $\mathbf{k}_2$, and it is meant that the integrations both over the directions $\mathbf{k}_1$ and over the azimuth $\mathbf{k}_2$ with respect to $\mathbf{k}_1$ have already been fulfilled. Further, differentiating the equality

$$\omega_3 = \sqrt{\omega_1^2 + \omega_2^2 + 2\omega_1 \omega_2 \cos\theta_{12}},$$

we find

$$d(\cos\theta_{12}) = \frac{\omega_3}{\omega_1 \omega_2} d\omega_3. \tag{20.81}$$

Now the integration over ω_3 is carried out without any trouble and we arrive at the following expression for the cross section:

$$d\sigma_{3\gamma} = \frac{8e^6}{(4\pi)^3 v} \left[\left(\frac{m - \omega_1}{\omega_2 \omega_3} \right)^2 + \left(\frac{m - \omega_2}{\omega_1 \omega_3} \right)^2 + \left(\frac{m - \omega_3}{\omega_2 \omega_1} \right)^2 \right] d\omega_1 d\omega_2. \tag{20.82}$$

Each of the frequencies $\omega_1, \omega_2, \omega_3$ are changed within the limits from 0 to m (the value m is reached by two frequencies when the third is equal to 0). At a given value of ω_1 the frequency ω_2 varies between $m - \omega_1$ and m. To integrate over ω_1 and ω_2 with allowance made for the photons identity (it is taken into account by multiplying the cross section into $1/6$) yields the total cross section of the $e^- e^+$-pair annihilation into the three photons:

$$\sigma_{3\gamma} = \int_0^m d\omega_1 \int_{m-\omega_1}^m d\omega_2 \frac{d\sigma_{3\gamma}}{d\omega_1 d\omega_2} = \frac{8e^6}{(4\pi)^3 3v} \int_0^m d\omega_1 \left\{ \frac{\omega_1 (m - \omega_1)}{(2m - \omega_1)^2} + \frac{2m - \omega_1}{\omega_1} + \right.$$

$$\left. + \left[\frac{2m(m - \omega_1)}{\omega_1^2} - \frac{2m(m - \omega_1)^2}{(2m - \omega_1)^3} \right] \ln \frac{m - \omega_1}{m} \right\} = \frac{4\alpha r_0^2 (\pi^2 - 9)}{3v}. \tag{20.83}$$

Therefore, the lifetime of the orthopositronium equals

$$\tau_1 = \frac{9\pi\hbar}{2(\pi^2 - 9) mc^2 \alpha^6} \approx 1.4 \times 10^{-7} \text{ s.} \tag{20.84}$$

We call attention to the fact that the inequality $\Gamma_1 \ll |E_0|$ is fulfilled for the orthopositronium even to a greater extent than for the positronium.

Comparison of experimental and theoretical data for the positronium reveals their complete agreement, which serves as one more argument in support of the QED. By the present time, the positronium is a perfectly understood system. Clearly, its properties in a matter, specifically its lifetime, are different from the characteristics of a free positronium, being determined by the matter structure. This makes it possible to use the positronium in studies of the physical and chemical properties of the matter. For example, the positronium may be used to study high-rate chemical reactions, with the rates comparable to the positronium lifetime. To this end, one measures a change in the positronium lifetime or splitting of the energy levels in the ortho- and para states [38].

Another example of a bound leptonic system presents the muonium $(e^-\mu^+)$, the bound metastable state of the electron and the positively charged muon. As leptons take no part in the strong interaction, the muonium properties are in a very good agreement with the QED. However, it is impossible to estimate the muonium lifetime on the basis of the QED because its instability is conditioned by the muon decay due to the weak interaction

$$\mu^+ \to e^+ + \overline{\nu}_\mu + \nu_e. \tag{20.85}$$

Measurements of the muonium polarization in various substances became the basis of a new efficient method of investigating the structure of condensed media, kinetic phenomena, chemical reactions, phase changes and so on [39]. Also, an interest to muonium is caused by a search for its transitions to antimuonium, $(\mu^+e^-) \to (\mu^-e^+)$. The detection of such transitions would be evidence of the violation of the electron or muon flavor.

20.5   Möller scattering $e^- + e^- \to e^- + e^-$

We are coming now to the investigation of processes disconnected with the emission and absorption of the photons. In the electroweak theory these processes are supplemented by diagrams with exchanges of the gauge Z-boson already in the second order of the perturbation theory. In this sense, calculations of the cross sections of similar processes within the QED scopes have an approximate character.

Let us examine a reaction of the elastic scattering of the electron by the electron:

$$e^- + e^- \to e^- + e^-, \tag{20.86}$$

that was investigated for the first time in Ref. [40]. The Feynman diagrams for this process are displayed in Fig. 20.8. The corresponding amplitude is given by:

$$\mathcal{A}_{e^-e^- \to e^-e^-} = -ie^2 \sqrt{\frac{m^2}{E_1 E_2 E_1' E_2'}} \left\{ \frac{[\overline{u}(\mathbf{p}_2')\gamma_\mu u(\mathbf{p}_2)][\overline{u}(\mathbf{p}_1')\gamma^\mu u(\mathbf{p}_1)]}{(p_1 - p_1')^2} - \right.$$
$$\left. - \frac{[\overline{u}(\mathbf{p}_1')\gamma_\mu u(\mathbf{p}_2)][\overline{u}(\mathbf{p}_2')\gamma^\mu u(\mathbf{p}_1)]}{(p_1 - p_2')^2} \right\}, \tag{20.87}$$

where p_1, p_2 (p_1', p_2') is the electron momentum in the initial (final) state.

In accordance with the general rules the differential cross section of the process (20.86) for the unpolarized particles is determined by the expression:

$$d\sigma = \frac{e^2 dt}{s(s - 4m^2)} \left[P_1(t, u) + P_2(t, u) + P_1(u, t) + P_2(u, t) \right], \tag{20.88}$$

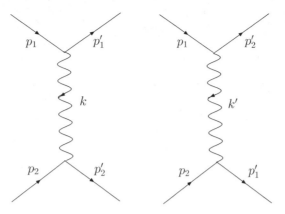

FIGURE 20.8

The Feynman diagrams for the process $e^-e^- \to e^-e^-$.

where

$$P_1(t,u) = \frac{1}{16t^2} \mathrm{Sp}[(\hat{p}_2' + m)\gamma^\mu(\hat{p}_2 + m)\gamma^\nu]\mathrm{Sp}[(\hat{p}_1' + m)\gamma_\mu(\hat{p}_1 + m)\gamma_\nu],$$

$$P_2(t,u) = -\frac{1}{16tu} \mathrm{Sp}[(\hat{p}_2' + m)\gamma^\mu(\hat{p}_2 + m)\gamma^\nu(\hat{p}_1' + m)\gamma_\mu(\hat{p}_1 + m)\gamma_\nu],$$

$$s - (p_1 \mid p_2)^2, \qquad t - (p_1 - p_1')^2, \qquad u = (p_1 - p_2')^2.$$

Simple calculations result in

$$P_1(t,u) = \frac{2}{t^2}\{(p_1 p_2)^2 + (p_1 p_2')^2 + 2m^2[m^2 - (p_1 p_1')]\} = \frac{1}{t^2}\left[\frac{s^2 + u^2}{2} + 4m^2(t - m^2)\right], \tag{20.89}$$

$$P_2(t,u) = \frac{2}{tu}[(p_1 p_2) - 2m^2](p_1 p_2) = \frac{2}{tu}\left(\frac{s}{2} - m^2\right)\left(\frac{s}{2} - 3m^2\right). \tag{20.90}$$

To substitute (20.89) and (20.90) into (20.88) yields the following expression for the differential cross section:

$$d\sigma = \frac{4\pi r_0^2 m^2 dt}{s(s - 4m^2)}\left\{\frac{1}{t^2}\left[\frac{s^2 + u^2}{2} + 4m^2(t - m^2)\right] + \right.$$

$$\left. +\frac{1}{u^2}\left[\frac{s^2 + t^2}{2} + 4m^2(u - m^2)\right] + \frac{4}{tu}\left(\frac{s}{2} - m^2\right)\left(\frac{s}{2} - 3m^2\right)\right\}. \tag{20.91}$$

In the center-of-mass frame the Mandelstam variables take the values

$$s = 4E^2, \qquad t = -4\mathbf{p}^2 \sin^2\frac{\theta}{2}, \qquad u = -4\mathbf{p}^2 \cos^2\frac{\theta}{2}. \tag{20.92}$$

Note that in this frame the momentum $\mathbf{p}$ and the energy E of the electrons are invariable. The cross section (20.91) in terms of the invariant variables becomes the form (Möller formula):

$$d\sigma = \frac{m^2 r_0^2 (E^2 + \mathbf{p}^2)^2}{4\mathbf{p}^4 E^2}\left[\frac{4}{\sin^4\theta} - \frac{3}{\sin^2\theta} + \left(\frac{\mathbf{p}^2}{E^2 + \mathbf{p}^2}\right)^2\left(1 + \frac{4}{\sin^2\theta}\right)\right] d\Omega. \tag{20.93}$$

As follows from (20.93) there are two peaks in the cross section, one peak for the forward-scattering and another for the back-scattering.

In the nonrelativistic limit ($v \ll 1$, $E/m \approx 1$) the expression (20.93) goes into the Rutherford formula with the inclusion of the exchange (in the Born approximation):

$$d\sigma = \frac{r_0^2}{16v^4} \left[\frac{1}{\sin^4(\theta/2)} + \frac{1}{\cos^4(\theta/2)} - \frac{1}{\cos^2(\theta/2)\sin^2(\theta/2)} \right] d\Omega, \qquad (20.94)$$

where $\mathbf{v} = 2\mathbf{p}/m$ is a relative velocity of the electrons. In the ultrarelativistic region we have

$$d\sigma = \frac{r_0^2 m^2}{E^2} \frac{(3 + \cos^2 \theta)^2}{4\sin^4 \theta} d\Omega. \qquad (20.95)$$

We choose, as the lab frame, the frame in which the second electron was in rest before collision. Introduce the quantity Δ

$$\Delta = \frac{E_1 - E_1'}{m} = \frac{E_2' - m}{m}, \qquad (20.96)$$

to define the energy transferred by the incoming electron in units of m. Rewriting the Mandelstam variables in the lab frame

$$s = 2m(m + E_1), \qquad t = -2m^2\Delta, \qquad u = -2m(E_1 - m - m\Delta) \qquad (20.97)$$

and inserting their values into (20.91), we obtain

$$d\sigma = \frac{2\pi r_0^2 d\Delta}{\epsilon^2 - 1} \left[\frac{\epsilon^2(\epsilon - 1)^2}{\Delta^2(\epsilon - 1 - \Delta)^2} - \frac{2\epsilon^2 + 2\epsilon - 1}{\Delta(\epsilon - 1 - \Delta)} + 1 \right], \qquad (20.98)$$

where $\epsilon = E_1/m$. This formula defines the energy distributions for the secondary electrons generated during passage of the relativistic electron through the matter. Note that the relation (20.98) is symmetric with respect to the kinetic energies of the final electrons $m\Delta$ and $m(\epsilon - 1 - \Delta)$ due to the identity of both particles.

When trying to obtain the total cross section for the reaction (20.86), we may have an unpleasant surprise: an integral with respect to the angles turns out to be divergent because of the propagator denominators. This is caused by the fact that with zero photon mass the electromagnetic forces have an infinite range of action and, no matter how far separated are the charge particles, they are scattered from each other anyway. Similar to the asymptotic properties of the field operators, the whole modern scattering theory may be formulated only in the case when the forces possess a finite range of action or when all the particles involved have no matter how small a mass that is finite anyway. But it is known that the situation may be improved by screening. In reality, experimenters never have to deal with two isolated charges because there always exist other charges causing screening of the Coulomb field at long distances.

A reader can find the discussion of the polarization effects for the process (20.86) in Refs. [41].

20.6 Processes $e^+ + e^- \to l + \bar{l}$ ($l = e^-, \mu^-, \tau^-$)

We shall consider the other cross channel of the reaction (20.86), namely, the electron-positron pair annihilation with the subsequent production of the electron-positron pair:

$$e^- + e^+ \to e^- + e^+. \qquad (20.99)$$

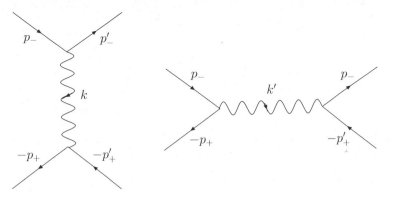

FIGURE 20.9

The Feynman diagrams for the process $e^-e^- \to l\bar{l}$.

The corresponding Feynman diagrams are pictured in Fig. 20.9. We draw attention to the different characters of both diagrams. In the former the lines of the initial and final electron are intersected in the top vertex while the ones of the initial and final positron—in the bottom vertex. In the latter the lines of the electron and the positron are intersected in every vertex; in the first vertex the e^-e^+-annihilation with emitting a virtual photon occurs (time flows from left to right), whereas in the second one the e^-e^+-production from a virtual photon takes place. By this reason the second diagram is called the *annihilation one* and the first diagram—the *scattering one*. It should be marked that for the annihilation diagram the squared four-dimensional momentum of the virtual photon $k_\mu^2 = (p_- + p_+)_\mu^2$ is greater than zero (time-like) while for the scattering diagram $k_\mu^2 = (p_- - p'_-)_\mu^2$ is less than zero (space-like). It is also clear that the scattering diagram leads to the big peak in the amplitude for the forward-scattering.

The transition from the reaction (20.86) to the reaction (20.99) is performed by the replacement:

$$p_1 \to -p'_+, \qquad p_2 \to p_-, \qquad p'_1 \to -p_+, \qquad p'_2 \to p'_-. \tag{20.100}$$

In the case of the reaction (20.99) the Mandelstam variables are defined by the expression:

$$s = (p_- + p_+)^2, \qquad t = (p_+ - p'_+)^2, \qquad u = (p_- - p'_+)^2. $$

Again, we are not interested in the particle polarizations. Then in the formula (20.91) we must fulfill the replacement:

$$s \to u, \qquad t \to t, \qquad u \to s. \tag{20.101}$$

As a result we arrive at the following expression for the differential cross section [42]:

$$d\sigma = \frac{4\pi r_0^2 m^2 dt}{s(s - 4m^2)} \left\{ \frac{1}{t^2} \left[\frac{s^2 + u^2}{2} + 4m^2(t - m^2) \right] + \right.$$

$$\left. + \frac{1}{s^2} \left[\frac{u^2 + t^2}{2} + 4m^2(s - m^2) \right] + \frac{4}{st} \left(\frac{u}{2} - m^2 \right) \left(\frac{u}{2} - 3m^2 \right) \right\}. \tag{20.102}$$

In the center-of-mass frame

$$s = 4E^2, \qquad t = -4\mathbf{p}^2 \sin^2 \frac{\theta}{2}, \qquad u = -4\mathbf{p}^2 \cos^2 \frac{\theta}{2} \tag{20.103}$$

and the cross section takes the form:

$$d\sigma = \frac{m^2 r_0^2}{16E^2} \left[\frac{(E^2 + \mathbf{p}^2)^2}{\mathbf{p}^4 \sin^4(\theta/2)} - \frac{8E^4 - m^4}{\mathbf{p}^2 E^2 \sin^2(\theta/2)} + \right.$$

$$\left. + \frac{12E^4 + m^4}{E^4} - \frac{4\mathbf{p}^2(E^2 + \mathbf{p}^2)}{E^4} \sin^2\frac{\theta}{2} + \frac{4\mathbf{p}^4}{E^4} \sin^4\frac{\theta}{2} \right] d\Omega, \qquad (20.104)$$

where E and θ are an energy and a scattering angle in the center-of-mass frame. The expression (20.104) does not already possess the symmetry with respect to the replacement $\theta \to \pi - \theta$ to be typical for scattering of identical particles.

In the nonrelativistic region this formula is reduced to the Rutherford one without exchange:

$$d\sigma = \left(\frac{\alpha}{mv^2} \right)^2 \frac{d\Omega}{\sin^4(\theta/2)}, \qquad (20.105)$$

where $\mathbf{v} = 2\mathbf{p}/m$ is a relative velocity of the electron and the positron.

It is instructive to find out the reasons of absence of the exchange addends in (20.105). In the nonrelativistic approximation the field operators of the electron-positron field may be considered by two component spinors. Since the relation

$$\bar{u}(\mathbf{p})\gamma_\mu u(\mathbf{p}) \approx \varphi_-^\dagger(\mathbf{p})\varphi_-(\mathbf{p}), \qquad (20.106)$$

where $\varphi_-(\mathbf{p})$ denotes the big components of the electron spinor, takes place (the similar relation is valid for the positron as well), then the amplitude corresponding to the scattering diagram becomes the form:

$$\mathcal{A}_{sc} \sim \frac{1}{t^2}[\varphi_-^\dagger(\mathbf{p}_-')\varphi_-(\mathbf{p}_-)][\varphi_+^\dagger(\mathbf{p}_+')\varphi_+(\mathbf{p}_+)]. \qquad (20.107)$$

Allowing for that in the center-of-mass frame $t^2 = -\mathbf{q}^2$, we come to the conclusion that the contribution from the scattering diagram exactly gives the Rutherford formula. In this approximation the contribution coming from the annihilation diagram is equal to zero. The mathematical explanation of this fact is as follows. At finding of $|\mathcal{A}_{e^-e^+ \to e^-e^+}|^2$ the terms

$$\mathcal{A}_{sc}^\dagger \mathcal{A}_{an} + \mathcal{A}_{an}^\dagger \mathcal{A}_{an}$$

where

$$\mathcal{A}_{an} \sim [\varphi_+^\dagger(\mathbf{p}_+)\varphi_-(\mathbf{p}_-)][\varphi_-^\dagger(\mathbf{p}_-')\varphi_+(\mathbf{p}_+')],$$

vanish at the cost of the orthogonality of the electron and positron spinors.

In the ultrarelativistic limit we get

$$d\sigma = \frac{r_0^2 m^2}{E^2} \frac{(3 + \cos^2\theta)^2 \cos^4(\theta/2)}{4\sin^4\theta} d\Omega, \qquad (20.108)$$

that is, this cross section differs from that of the electron-electron scattering by the factor $\cos^4(\theta/2)$ only.

Now we investigate the processes of the $\mu^-\mu^+$-, $\tau^-\tau^+$-pair production under the electron-positron annihilation. This task is analogous to the previous one but significantly simpler. Because the QED Lagrangian contains the electron, muon, and tau-lepton currents in the form of the individual addends, then the Feynman diagrams do not have any vertices in which the $\mu^- \to e^-$-, $\tau^- \to e^-$-, $\tau^- \to \mu^-$-conversions or the $\mu^- e^+$-, $\tau^- e^+$-, $\mu^-\tau^+$-pair annihilations would occur. Every chain of diagram lines (from entrance to exit or whole closed loop) conserves the lepton flavor, that is, the conservation law of the individual lepton flavor:

$$\sum_{i=1}^{N} L_{ei} = \text{const}, \qquad \sum_{i=1}^{N} L_{\mu i} = \text{const}, \qquad \sum_{i=1}^{N} L_{\tau i} = \text{const} \qquad (20.109)$$

is valid in the QED. Because of this, for example, the process of the muon pair production under the $e^- e^+$-pair annihilation in the second order of the perturbation theory is represented by only one diagram as opposite to the process (20.99) (see, Fig. 20.10). In the

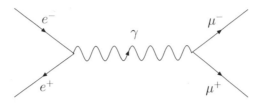

FIGURE 20.10
The Feynman diagram for the process $c^- c^+ \rightarrow \mu^- \mu^+$.

invariant variables

$$s = (p_{e^-} + p_{e^+})^2, \qquad t - (p_{e^-} - p_{\mu^+})^2, \qquad u = (p_{e^-} - p_{\mu^-})^2$$

the differential cross section for the case of the unpolarized particles is provided by the expression [43]:

$$d\sigma = \frac{e^4 dt}{4\pi s^3 (s - 4m_e^2)} \left[\frac{t^2 + u^2}{2} + (m_e^2 + m_\mu^2)(2s - m_e^2 - m_\mu^2) \right]. \qquad (20.110)$$

Neglecting the electron mass, we find the limits of the integration over t. For this purpose we carry out the transition to the center-of-mass frame and next return to the Mandelstam variables in the obtained expression for t:

$$t \approx m_\mu^2 - 2E^2 + 2\mathbf{p}_e \mathbf{p}_\mu \cos\theta = m_\mu^2 - \frac{s}{2} + \frac{1}{4}\sqrt{s(s - 4m_\mu^2)} \cos\theta. \qquad (20.111)$$

So, at a given s the value of t lays within the interval:

$$m_\mu^2 - \frac{s}{2} - \frac{1}{4}\sqrt{s(s - 4m_\mu^2)} \leq t \leq m_\mu^2 - \frac{s}{2} + \frac{1}{4}\sqrt{s(s - 4m_\mu^2)}. \qquad (20.112)$$

Fulfilling the integration over t into (20.110), we gain the total cross section of the muon pair production:

$$\sigma = \frac{4\pi\alpha^2}{3s}\sqrt{1 - \frac{4m_\mu^2}{s}}\left[1 + \frac{2m_\mu^2}{s}\right]. \qquad (20.113)$$

The cross section (20.113) goes into the maximum when $\sqrt{s} \approx 1.7\sqrt{s_0}$, where s_0 is a value of s on the reaction threshold (in the center-of-mass frame $s_0 = 4m_\mu^2$). In this case the value of the cross section proves to be 20 times as small as that of the cross section of the electron-positron annihilation into the two photons under the same energy.

As follows from (20.113), if the relative velocity of the created particles tends to zero ($s \rightarrow 4m_\mu^2$), the cross section becomes zero. However, this inference is invalid. The formula

(20.113) is no longer valid close to the reaction threshold. As in this case the Coulomb interaction becomes considerable, the created muons could not be considered as free particles. In the nonrelativistic quantum scattering theory it is demonstrated that the cross section of the reaction describing two slowly attracted particles at $v \to 0$ tends to the constant limit. By the detailed balancing principle —for an isolated system the probability of a direct transition $n \to m$ is equal to that of the reverse transition $m \to n$—the cross section of this reaction is associated with the cross section of the reverse reaction, that is, the reaction of producing the same charged particles. In consequence, when two charged particles are created, the cross section close to the threshold is tending to the constant limit. This means that upon the transition through a threshold such a reaction takes place immediately with the finite value of the cross section.

21

Radiative corrections

But the Raven, sitting lonely on the placid bust, spoke
only
That one word, as if his soul in that one word he did
outpour.
Nothing father then he uttered—not a feather then
he fluttered—
Till I scarcely more than muttered "Other friends have
flown before—
On the morrow he will leave me, as my Hopes have
flown before."
Then the bird said "Nevermore."
Edgar Allan Poe, "The Raven"

So far the calculations for scattering cross sections have been performed being restricted to the Born approximation. The matrix elements were fairly simple in form involving no integration with respect to the four-dimensional momenta of virtual particles. The corresponding diagrams, including a single internal line, looked just like the trees we have drawn in early childhood. In analogy, Born diagrams are usually called the *tree diagrams*, whereas the Born approximation is called the *tree approximation*.

In this section we investigate higher orders of the perturbation theory. This is motivated not only by the necessity to improve the accuracy of the calculation but also by our wish to make sure of the consistency and correctness of the QED. In the process we have to consider two problems: (i) elucidation of the meaning for the parameters m and e involved in the Lagrangian, and (ii) appearance of divergent integrals.

In concept, in the original Dirac equations m and e represent a mass and a charge of a free electron, i.e., the electron completely isolated from the effect of an electromagnetic field. In other words, m and e are the characteristics of a particular hypothetical object—the bare electron. It is clear that an energy spectrum of the interacting fields is no longer coincident with that of free fields. Because of this, the mass of the bare electron should be different from that of the physical electron, i.e., different from a minimum energy of the singly charged state of the interacting fields. It makes sense to assume that the electron mass is given by the quantity:

$$m_{exp} \equiv m_R = \frac{(\Psi_p, H\Psi_p)}{(\Psi_p, \Psi_p)} \mid_{\mathbf{p}=0},$$

where H is a total Hamiltonian of a system, calculated within the required accuracy in higher orders of the perturbation theory. In the case of a physical charge of the electron we proceed in a similar way. To illustrate, a physical charge may be determined in a investigation of low-frequency electromagnetic waves scattered from the electron at rest. As the process under study is purely classical, in the limiting case $\mathbf{p} \to 0$ no quantum correction should occur. Then a physical charge of the electron $e_{exp} \equiv e_R$ is determined as a factor entering into the Thomson formula for the effective scattering of the photon at

$\mathbf{p} \to 0$,

$$\sigma = \frac{8\pi}{3} \left(\frac{e_R^2}{4\pi m_R} \right)^2.$$

Thus, we should find the relations:

$$e = e(e_R), \qquad m = m(m_R)$$

and eliminate the parameters m and e having expressed them in terms of the experimental values of the mass and the charge. Such a redefinition of the theory parameters is called the *renormalization* and makes a compound part of any calculation to be compared with the experimental results.

The appearance of the divergences stems from a singular character of the Green functions at relatively short distances. In the momentum space this is equivalent to the fact that the associated Fourier images at infinity are decreasing rather slowly. We shall illustrate the divergences appearance by the example of the electron scattering from the Coulomb field of a nucleus. The Feynman diagrams describing this process in the first order and the radiation correction in the third order of the perturbation theory are shown in Fig. 21.1. The matrix

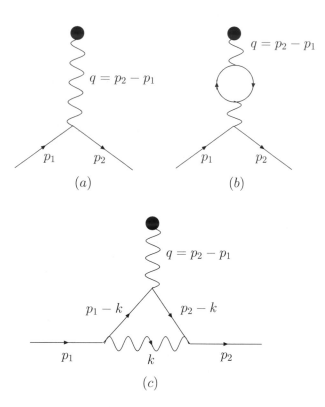

FIGURE 21.1

The Feynman diagrams (a)-(c) describing the electron scattering from the Coulomb field of the nucleus.

element corresponding to the diagrams of Fig.21.1(a) is proportional to the integral:

$$I_\nu = \int \gamma_\mu \frac{\hat{p}_2 - \hat{k} + m}{(p_2 - k)^2 - m^2} \gamma_\nu \frac{\hat{p}_1 - \hat{k} + m}{(p_1 - k)^2 - m^2} \gamma^\mu \frac{1}{k^2} d^4 k. \tag{21.1}$$

In the integrand (21.1) the power of the momentum k, with respect to which the integration is performed, is six both in the numerator and denominator. Since the integration over k is going from 0 to ∞, at the upper limit the integral as well as the matrix element is infinite or divergent, and in this case the divergence is logarithmic. The divergences deriving from great momenta are referred to as *ultraviolet*. Recall that the integrals may be divergent at small values of k as well as due to zero mass of the photon, and in this case the divergences are termed as *infrared*.

The appearance of the divergences and the results of their subsequent elimination are suggestive that quantum field theory is similar to the land described by the famous Russian tale-teller E. Schwartz where people are living according to the principle: *"No matter what happens, we always have a holiday."* The appearing divergences force us to redefine (regularize) the Green functions. On the other hand, inclusion of the interactions necessitates a change from the characteristics of bare particles to those of the physical ones. Owing to the regularized Green functions and also due to the redetermined mass and charge of the electron, ultraviolet divergences disappear from all the observables, whereas the infrared divergences, as noted above, are canceled with the allowance for higher orders of the perturbation theory. In this manner by the solution of two a priori independent problems we can derive the finite results of the QED, being in an excellent agreement with the experimental data.

21.1 Calculations of integrals over virtual momenta

Let us consider a diagram having an arbitrary number of external lines to correspond to virtual particles. In the most general case the number of the vertex δ-functions may happen to be insufficient for the convolution of all integrals. The necessary condition for this is the presence of closed loops in the diagram. So, if the number of the diagram vertices is less than the one of the internal lines, then the corresponding matrix element contains the integral over virtual four-dimensional momenta of the kind:

$$I_l = \int d^4k_1...d^4k_l F(k_1, ...k_l), \tag{21.2}$$

where l is a number of topologically independent loops, and $F(k_1, ...k_l)$ represents the product of vertex factors multiplied by the electron and photon propagators. In the QED the integral (21.2) may be reported in terms of the product of integrals, each of them has the following structure:

$$I = \int d^4k \frac{F(k)}{\alpha_1...\alpha_n}, \tag{21.3}$$

where $F(k)$ is a polynomial of some power in k, and $\alpha_1, ...\alpha_n$ are polynomials of the second power with respect to k. Therefore, our prime goal is to get acquainted with the technique of calculating the integrals of the kind (21.3).

To compute the integrals (21.3) we make use of the parametrization method by Feynman [7]. Here the starting point is the formula:

$$\frac{1}{\alpha_1...\alpha_n} = (n-1)! \int_0^1 d\xi_1 \int_0^1 d\xi_2... \int_0^1 d\xi_n \frac{\delta(\xi_1 + \xi_2 + ... + \xi_n - 1)}{[\alpha_1\xi_1 + \alpha_2\xi_2 + ... + \alpha_n\xi_n]^n} =$$

$$= (n-1)! \int_0^1 dx_1 \int_0^{x_1} dx_2... \int_0^{x_{n-2}} \frac{dx_{n-1}}{[\alpha_1 x_{n-1} + \alpha_2(x_{n-2} - x_{n-1}) + ... + \alpha_n(1 - x_1)]^n}, \tag{21.4}$$

or the equivalent one:

$$\frac{1}{\alpha_1...\alpha_n} = (n-1)! \int_0^1 u_1^{n-2} du_1 \int_0^1 u_2^{n-3} du_2... \int_0^1 du_{n-1} \times$$

$$\times \frac{1}{[\alpha_1 u_1...u_{n-1} + \alpha_2 u_1...u_{n-2}(1-u_{n-1}) + ... + \alpha_n(1-u_1)]^n}, \qquad (21.5)$$

where $x_1 = u_1$, $x_2 = u_1 u_2$, ..., $x_{n-1} = u_1 u_2...u_{n-1}$. One may differentiate the equality (21.4) with respect to α_n and gain the identity of the type:

$$\frac{1}{\alpha_1...\alpha_n^2} = n! \int_0^1 (1-x_1) dx_1 \int_0^{x_1} dx_2... \int_0^{x_{n-2}} dx_{n-1} \times$$

$$\times \frac{1}{[\alpha_1 x_{n-1} + \alpha_2(x_{n-2} - x_{n-1}) + ... + \alpha_n(1-x_1)]^{n+1}}. \qquad (21.6)$$

The advantage of the last formula implies that when there are two identical factors it takes one integration less than the factors number.

Let us prove the validity of the relation (21.5). At $n = 2$ we may write

$$\frac{1}{\alpha_1 \alpha_2} = \frac{1}{\alpha_2 - \alpha_1}\left(\frac{1}{\alpha_1} - \frac{1}{\alpha_2}\right) = \frac{1}{\alpha_2 - \alpha_1} \int_{\alpha_1}^{\alpha_2} \frac{dx}{x^2}. \qquad (21.7)$$

Transferring to new variable $x = \alpha_1 u + \alpha_2(1-u)$, we obtain

$$\frac{1}{\alpha_1 \alpha_2} = \int_0^1 du \frac{1}{[\alpha_1 u + \alpha_2(1-u)]^2}. \qquad (21.8)$$

which is what we set out to prove. Note that the equality (21.8) is true for all values of α_1 and α_2. If α_1 and α_2 have, however, the opposite signs, then u should be regarded as a complex variable and the integration contour must depart from the real axis to bypass the singularity at the point $u = \alpha_2/(\alpha_2 - \alpha_1)$.

Next, we prove that if the relation (21.5) is fulfilled under some n, it will be fulfilled under $n+1$ as well. Since the operations of integrating over different u_i in (21.5) are commutative, then

$$\frac{1}{\alpha_1...\alpha_n \alpha_{n+1}} = (n-1)! \int_0^1 u_1^{n-2} du_1 \int_0^1 u_2^{n-3} du_2... \int_0^1 \frac{du_{n-1}}{A^n \alpha_{n+1}}, \qquad (21.9)$$

where A^n is a denominator being in the right-hand side of the formula (21.5). On the other hand

$$\frac{1}{nA^n \alpha_{n+1}} = \int_0^1 \frac{u_0^{n-1} du_0}{[Au_0 + \alpha_{n+1}(1-u_0)]^{n+1}}. \qquad (21.10)$$

From (21.9) and (21.10) it immediately follows

$$\frac{1}{\alpha_1...\alpha_n \alpha_{n+1}} = n! \int_0^1 u_0^{n-1} du_0 \int_0^1 u_1^{n-2} du_1... \int_0^1 \frac{du_{n-1}}{[Au_0 + \alpha_{n+1}(1-u_0)]^{n+1}}$$

which proves our statement.

Now, by application (21.4) the integral (21.3) could be represented in the view:

$$\int d^4k \frac{F(k)}{\alpha_1...\alpha_n} = (n-1)! \int_0^1 dx_1 \int_0^{x_1} dx_2... \int_0^{x_{n-2}} dx_{n-1} \int \frac{F(k)dk}{[(k-a)^2 + b]^n}, \qquad (21.11)$$

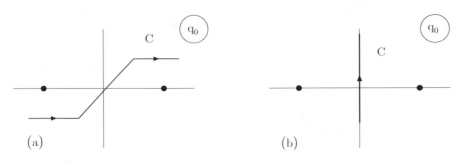

FIGURE 21.2

The rotation of the integration path in (a) and (b).

where a four-dimensional vector a and a scalar b depend on parameters $x_1, ... x_{n-1}$. Here the integration over the variable k_0 is carried out in accordance with the rule of the poles detour, that is, along the contour C shown in Fig. 21.2(a). In doing so, every of the factors α_i^{-1} in the left-hand side of (21.11) has two poles over the variable k_0 which are bypassed along the contour C. The integrand in (21.11) has $2n$ simple poles over the variable k_0 whereas, after the Feynman parametrization, it has only two nth order poles with the same detour rule along the contour C. Since the singular points of the integrand place in the real axis, then one may swing the contour C through $\pi/2$, that is, bring it into coincidence with the imaginary axis (Fig.21.2(b)). As this takes place, k_0 is changed by ik_4 (k_4 is a real quantity) and $k_0^2 - \mathbf{k}^2$ converts to $-(k_4^2 + \mathbf{k}^2)$. Thus, the rotation of the integration path around the angle $\pi/2$ counterclockwise (Wick rotation) corresponds to the transition in the k-space from the pseudo-Euclidean metric to the Euclidean one.

Using the spherical system of coordinates in the four-dimensional Euclidean space

$$
\left.
\begin{aligned}
k_1 &= k \cos\varphi \sin\theta \sin\xi, \qquad k_2 = k \sin\varphi \sin\theta \sin\xi, \\
k_3 &= k \cos\theta \sin\xi, \qquad k_4 = k \cos\xi, \\
d^4 k &= k^3 dk \sin\theta d\theta d\varphi \sin^2 \xi d\xi,
\end{aligned}
\right\}
\tag{21.12}
$$

we may get

$$
d^4 k = dk_0 d^3 k \to d_E^4 k = i dk_4 d^3 k = i k^3 dk \sin^2 \xi d\xi d\Omega,
$$

where $d\Omega$ is a four-dimensional solid angle in the Euclidean space. Having fulfilled the integration over the angles, we arrive at

$$
d^4 k \to 2i\pi^2 k^3 dk.
\tag{21.13}
$$

If the integral (21.11) is divergent, then one conveniently carries out the integration first over some finite invariant region Σ characterized by the value N, where the limit $N \to \infty$ is associated with the whole infinite four-dimensional k-space. Such a finite invariant region may be specified, for example, by the inequalities:

$$
|k^2| \le N^2, \qquad \frac{(ks)^2}{s^2} \le N^2,
$$

where s is an arbitrary time-like four-dimensional vector [44]. After the Wick rotation the region Σ is transformed to a four-dimensional sphere.

It is convenient to begin calculating the integrals of the kind (21.11) with the integral:

$$
I_2(p, l) = \int_\Sigma d^4 k \frac{1}{[k^2 - 2(pk) + l]^2}.
\tag{21.14}
$$

The shift of the origin of coordinates for divergent integrals does not leave these integrals unchanged. As a result of such a shift an additive constant is added to divergent integrals. However, as will be shown later, the similar transformation of coordinates does not lead to changing the values of the logarithmically divergent integrals. After shifting the coordinates origin $k - p \to k$, the integral (21.14) takes the form:

$$J_2(p, l) = \int_{\Sigma'} \frac{d^4 k}{[k^2 + l - p^2]^2}. \tag{21.15}$$

Carrying out the Wick rotation and realizing the integration, we get

$$J_2(p, l) = \int \frac{d_E^4 k}{[-k^2 + l - p^2]^2} = i\pi^2 \left(\ln \frac{L'^2}{p^2 - l} - 1 \right), \tag{21.16}$$

where L' is an integration sphere radius associated with the region Σ'. When we took the integral (21.15) over the region Σ, then the quantity L would appear in Eq. (21.16) rather than L'. However L' differs from L by a finite value. Therefore, if L' is sufficiently big, then ln L' is distinguished from ln L by a small value of the order of L^{-1}. Thus, we have the right to set ln $L' = \ln L$ in the formula (21.16).

Further, we compute the logarithmically divergent integral:

$$I_3^{\mu\nu}(p, l) = \int_{\Sigma} \frac{k^\mu k^\nu d^4 k}{[k^2 - 2(pk) + l]^3}.$$

After the coordinates origin shift $k - p \to k$, it becomes the form:

$$I_3^{\mu\nu}(p, l) = \int_{\Sigma'} \frac{(k^\mu + p^\mu)(k^\nu + p^\nu) d^4 k}{[k^2 + l - p^2]^3}. \tag{21.17}$$

Breaking (21.17) into parts and yielding the Wick rotation, we obtain

$$\int_{\Sigma'} \frac{k_\mu k_\nu d^4 k}{(k^2 - l')^3} = \frac{1}{4} g_{\mu\nu} \int_{\Sigma'} \frac{k^2 d^4 k}{(k^2 - l')^3} = -\frac{1}{4} g_{\mu\nu} \int \frac{k^2 d_E^4 k}{(-k^2 - l')^3} = \frac{i g^{\mu\nu} \pi^2}{4} \left(\ln \frac{L'^2}{l'} - \frac{3}{2} \right),$$

$$\int_{\Sigma'} \frac{k^\mu d^4 k}{(k^2 - l')^3} = 0, \qquad p^\mu p^\nu \int_{\Sigma'} \frac{d^4 k}{(k^2 - l')^3} = \frac{i\pi^2 p^\mu p^\nu}{2l'},$$

where $l' = p^2 - l$ and we have allowed for that integrals of an odd function over symmetric limits vanish. Thus, the integral $I_3^{\mu\nu}(p, l)$ appears to be equal to

$$I_3^{\mu\nu}(p, l) = \frac{i g^{\mu\nu} \pi^2}{4} \left(\ln \frac{L^2}{p^2 - l} - \frac{3}{2} \right) + \frac{i\pi^2 p^\mu p^\nu}{2(p^2 - l)}. \tag{21.18}$$

Integrating the formula (21.18) over p^ν, we find

$$I_2^\mu(p, l) = \int_{\Sigma} \frac{k^\mu d^4 k}{[k^2 - 2(pk) + l]^2} = i p^\mu \pi^2 \left(\ln \frac{L^2}{p^2 - l} - \frac{3}{2} \right) + g^\mu(L^2, l). \tag{21.19}$$

To determine the quantity $g^\mu(L^2, l)$ we substitute $p^\mu = 0$ into Eq.(21.19). Then, in the left-hand side we obtain the integral of an odd function over the symmetric limits and (21.19) gives the result:

$$g^\mu(L^2, l) = 0. \tag{21.20}$$

According to (21.20) the formula (21.19) acquires the final form:

$$I_2^\mu(p, l) = \int_{\Sigma} \frac{k^\mu d^4 k}{[k^2 - 2(pk) + l]^2} = i p^\mu \pi^2 \left(\ln \frac{L^2}{p^2 - l} - \frac{3}{2} \right). \tag{21.21}$$

Integrating (21.16) over l, we come to the following integral:

$$I_1(p,l) = \int_\Sigma \frac{d^4k}{k^2 - 2(pk) + l} = -i\pi^2 \left[l \ln L^2 + (p^2 - l)\ln(p^2 - l) - p^2 + g(L^2, p^2) \right], \quad (21.22)$$

where $g(L^2, p^2)$ is some function of L^2 and p^2. Differentiating (21.22) with respect to p^μ, we get

$$J_2^\mu(p,l) = \int_\Sigma \frac{k^\mu d^4k}{[k^2 - 2(pk) + l]^2} = -i\pi^2 p^\mu \left[\ln(p^2 - l) + \frac{\partial g(L^2, p^2)}{\partial p^2} \right]. \quad (21.23)$$

To compare (21.23) with (21.21) results in:

$$\frac{\partial g(L^2, p^2)}{\partial p^2} = \frac{3}{2} - \ln L^2,$$

that is,

$$g(L^2, p^2) = \frac{3}{2}p^2 - p^2 \ln L^2 + g_0(L^2).$$

To substitute the obtained value of $g(L^2, p^2)$ into (21.22) gives

$$I_1(p,l) = \int_\Sigma \frac{d^4k}{k^2 - 2(pk) + l} = -i\pi^2 \left[l \ln L^2 + (p^2 - l)\ln(p^2 - l) + \frac{1}{2}p^2 - p^2 \ln L^2 + g_0(L^2) \right].$$
$$(21.24)$$

To find $g_0(L^2)$ we set $p^2 = 0$ in the both sides of the expression (21.24). Then, with allowance made for

$$\int_\Sigma \frac{d^4k}{k^2 + l} = -i\pi^2 \left[L^2 + l \ln \frac{L^2}{-l} \right],$$

we gain finally

$$I_1(p,l) = i\pi^2 \left[L^2 + (p^2 - l)\ln \frac{L^2}{p^2 - l} - \frac{1}{2}p^2 \right]. \quad (21.25)$$

It should be noted that the integral (21.25) allows us to find practically all integrals to appear under calculation of radiative corrections (RC). Thus, for example, the elementary differentiation of (21.25) over p_μ and l leads to the following expressions (recall, the quantity $p^2 - l$ enters into (21.25) under the module sign):

$$\int_\Sigma \frac{k_\mu d^4k}{[k^2 - 2(pk) + l]^2} = \frac{1}{2}\frac{\partial I_1(p,l)}{\partial p^\mu} = i\pi^2 p_\mu \left(\ln \frac{L^2}{p^2 - l} - \frac{3}{2} \right), \quad (21.26)$$

$$\int_\Sigma \frac{k_\mu k_\nu d^4k}{[k^2 - 2(pk) + l]^3} = \frac{1}{8}\frac{\partial^2 I_1(p,l)}{\partial p^\mu \partial p^\nu} = \frac{i\pi^2}{4}g_{\mu\nu}\left(\ln \frac{L^2}{p^2 - l} - \frac{3}{2} \right) + \frac{i\pi^2}{2(p^2 - l)}p_\mu p_\nu, \quad (21.27)$$

$$\int_\Sigma \frac{k_\nu d^4k}{[k^2 - 2(pk) + l]^3} = -\frac{1}{4}\frac{\partial^2 I_1(p,l)}{\partial p^\nu \partial l} = \frac{i\pi^2}{2(p^2 - l)}p_\nu, \quad (21.28)$$

$$\int_\Sigma \frac{d^4k}{[k^2 - 2(pk) + l]^3} = \frac{1}{2}\frac{\partial^2 I_1(p,l)}{\partial l^2} = \frac{i\pi^2}{2(p^2 - l)}. \quad (21.29)$$

$$\int_\Sigma \frac{d^4k}{[k^2 - 2(pk) + l]^2} = -\frac{\partial I_1(p,l)}{\partial l} = i\pi^2 \left(\ln \frac{L^2}{p^2 - l} - 1 \right). \quad (21.30)$$

The quadratically divergent integral:

$$I_{\mu\nu}^{(2)}(p,l) = \int_\Sigma \frac{k_\mu k_\nu d^4k}{[k^2 - 2(pk) + l]^2}.$$

demands some additional efforts. To estimate it one may integrate (21.27) over l. A constant appearing under this integration could be defined with the help of the integral:

$$J_3^{\mu\nu\lambda}(p,l) = \int_\Sigma \frac{k^\mu k^\nu k^\lambda d^4 k}{[k^2 - 2(pk) + l]^3}.$$

It may be done by application of the formula:

$$\frac{1}{a^n} - \frac{1}{b^n} = \int \frac{n(b-a)dx}{[(a-b)x + b]^{n+1}},\tag{21.31}$$

where $J_3^{\mu\nu\lambda}(p,l)$ has been taken as a and the vanishing integral

$$\int_\Sigma \frac{k^\mu k^\nu k^\lambda d^4 k}{[k^2 + l]^3}$$

has been used as b. We give the final result of calculations:

$$\int_\Sigma \frac{k_\mu k_\nu d^4 k}{[k^2 - 2(pk) + l]^2} = \frac{i\pi^2}{2} g_{\mu\nu} \left[(l - p^2)\ln \frac{L^2}{p^2 - l} + \frac{5p^2 - 3l}{6} + \frac{L^2}{2} \right] +$$

$$+ i\pi^2 \left(\ln \frac{L^2}{p^2 - l} - \frac{11}{6} \right) p_\mu p_\nu.\tag{21.32}$$

21.2   Regularization methods

The appearance of divergent integrals into the matrix elements induce us to do violence to the conventional mathematical apparatus, namely, we have to sum up and multiply these matrix elements as if they are finite. Though we have succeeded in driving the divergences out of a theory at the final stages of the calculations, we are not saved from regularization of the theory at the intermediate stages. This procedure comprises in the replacement of the original expressions by more smooth ones, and, as a result, all the integrals become finite.

As has been established in Section 14.9, the causal Green functions are the generalized singular functions possessing singularities at the light cone. The divergent integrals considered in the previous section are associated with closed loops of the diagrams representing the products of the causal Green functions with the coincident arguments. These products include the squared generalized singular functions of the form $\delta(x^2)$, x^{-2} and hence are not well defined. In other words, the products of the Green functions require a supplement to the definition, as one would expect from their membership to the generalized functions. Indeed, the generalized function is specified by the integration rules for its product with sufficiently regular functions, giving no special receipt for the integration of the product of several generalized functions.

Thus, our task is to formulate the operation of replacing the causal Green functions $\Delta_c(x)$ or integrals of their products by some approximating regularized expressions not including any ultraviolet divergences or, what is the same when we work in the coordinate representation—the singularities at the light cone. As it takes place, two approaches are possible.

By the first, the so-called covariant regularization is based on the derivation of regular approximations to the causal functions themselves, i.e., on the substitution of some regularized approximations for the generalized singular functions

$$\Delta_c(x) \to \text{reg } \Delta_c(x).$$

It is obvious that the products of new propagators become regular automatically.

In the second approach the original expressions for the propagators are unchanged while both the propagator products and integrals over the virtual four-dimension momenta are redefined by means of regularized approximations. Among such a regularization kind is above all a cut-off regularization (CR). In its simplest and well-known variant, after the Wick rotation, the integration for each loop is limited to a compact sphere of radius L with $k^2 \le L^2$ that makes all the integrals convergent. One more way consists in the change-over to the discrete space-time (DR), i.e., in the assumption that x_μ takes on only discrete values associated with, for example, sites of a regular lattice. Of course, cut-off at short distances is equivalent to that at large values of the momenta. But these two approaches are not relativistically covariant, being unable to retain the initial gauge symmetry of the theory. In contrast to the CR and DR, the dimensional regularization, despite its belonging to the same class, retains the symmetry properties of the corresponding nonregularized expressions.

Covariant regularization. This regularization was suggested in Ref. [45] and has come to be known as the Pauli-Villars regularization. This scheme possesses the relativistic covariance and conserves the gauge invariance of a theory under appropriate modification. The Pauli-Villars regularization resides in the replacement of singular Green functions of a free field $G^c(x; m)$ by a linear combination:

$$G^c(x; m) \to \text{reg } G^c(x; m) = G^c(x; m) + \sum_i c_i G^c(x; M_i) \qquad (21.33)$$

The quantities $G^c(x; M_i)$ are causal functions of fictitious fields with the masses M_i and the coefficients c_i standing in front of them sorted out so that the regularized function reg $G^c(x; m)$ becomes sufficiently regular in the vicinity of the light cone or, what is the same, its Fourier image reg $G^c(p; m)$ is sufficiently fast decreasing in the region of big values of $|p^2|$.

So, it follows from the formula (14.110) that in the zero spin case the most strong singularities at the light cone $\delta(\lambda)$ and λ^{-1} enter into $\Delta^c(x; m^2)$ with coefficients not depending on the mass m. Because of this, setting c_i

$$1 + \sum_i c_i = 0, \qquad (21.34)$$

we arrive at the expression for reg $\Delta^c(x; m^2)$, which does not contain the singularities $\delta(\lambda)$ and λ^{-1}. In the region of big values of $|p^2|$ its Fourier image reg $\Delta^c(p; m^2)$ will be decreasing as $|p^2|^{-2}$. In a similar manner, the condition:

$$m^2 + \sum_i c_i M_i^2 = 0 \qquad (21.35)$$

ensures the singularities absence of the kind $\theta(\lambda)$, $\ln|\lambda|$ into reg $\Delta^c(x; m^2)$ and leads to the fact that reg $\Delta^c(p; m^2)$ is decreasing as $|p^2|^{-6}$ in the ultraviolet region. It is remarkable that the conditions (21.34) and (21.35) ensure the Green functions regularization in the QED as well.

The Pauli-Villars regularization represents a specific modification of the one-particle propagator. In the simplest variant, it is reduced to the introduction of one supplementary mass M with the factor $c = -1$. Thus, for example, in the momentum representation we have for the scalar field

$$\text{reg } \Delta^c(k) = \frac{1}{k^2 - m^2} - \frac{1}{k^2 - M^2} = \frac{m^2 - M^2}{(k^2 - m^2)(k^2 - M^2)}. \tag{21.36}$$

As follows from (21.36), the regularization procedure results in the considerably changed behavior of the propagator in the ultraviolet region or, equivalently, in the neighborhood of the light cone. This particular case reveals the essence of the regularization operation. Any elimination of infinities is related to the existence of a subtraction procedure. On the other hand, propagators are bilinear combinations of the field functions, and the negative sign appearing in the propagator results from quantization of a theory with the use of an indefinite metric. Consequently, in the case of a regularization of the photon propagator:

$$\text{reg } \Delta^c_{\mu\nu}(k) = \frac{g_{\mu\nu}}{k^2} - \frac{g_{\mu\nu}}{k^2 - M^2} \tag{21.37}$$

the second addend in (21.37) should be considered as a result of an additional interaction between the electron-positron field and vector field, whose quanta possess the mass M and whose norm is not positively defined.

After the intermediate calculations are completed, we have to get rid of fictitious fields, that is, to take away the regularization. As follows from Eq. (21.36), with M tending to infinity we arrive at the initial propagator. This allows us to state that removal of the Pauli-Villars regularization is realized when going to the limit $M_i \to \infty$. However, it is naive to believe that application of this procedure for the expressions involving the propagator products may bring back the same results as are possible without the regularization. Let us estimate the effect of the regularization taking the one-loop diagram with two scalar internal lines in Fig. 21.3 as an example. The corresponding matrix element contains the

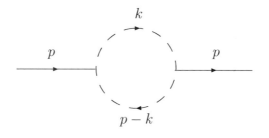

FIGURE 21.3

The loop with two virtual scalar particles.

product of two propagators with coincident arguments:

$$\mathcal{M} \sim \int d^4x e^{ikx} \int d^4y e^{iky} \Delta^c(x - y; m^2) \Delta^c(y - x; m^2).$$

After the transition to the momentum representation we obtain the integral:

$$I(p^2) = \int d^4k \frac{1}{[(k - p)^2 - m^2]} \frac{1}{[k^2 - m^2]}. \tag{21.38}$$

This logarithmically divergent integral will be the subject of our investigation. It is clear that in order to make this integral finite, one sufficiently carries out the replacement (21.36) in one of the propagators. This results in

$$\text{reg } I(p^2) = \int d^4k \frac{1}{(k-p)^2 - m^2} \left[\frac{1}{[k^2 - m^2]} - \frac{1}{[k^2 - M^2]} \right] =$$

$$= \int d^4k \frac{m^2 - M^2}{[(k-p)^2 - m^2][k^2 - m^2][k^2 - M^2]}.$$

Obviously, reg $I(p^2)$ could be expanded into such two addends, that, after tending M to the infinity, the former diverges at the integration over the four-dimensional momenta while the latter is convergent. The required splitting is given by the formula:

$$\text{reg } I(p^2) = I_D(0) + I_C(p^2), \tag{21.39}$$

where

$$I_D(0) = \int d^4k \frac{m^2 - M^2}{(k^2 - m^2)^2(k^2 - M^2)}, \tag{21.40}$$

$$I_C(p^2) = \int d^4k \frac{m^2 - M^2}{(k^2 - m^2)(k^2 - M^2)} \left[\frac{1}{(k-p)^2 - m^2} - \frac{1}{k^2 - m^2} \right]. \tag{21.41}$$

Because of the addend $I_C(p^2)$ does not contain any divergencies, in the integrand we may transform to the limit $M \to \infty$ what results in:

$$I_C(p^2) = \int d^4k \frac{2(pk) - p^2}{(k^2 - m^2)^2[(k-p)^2 - m^2]}. \tag{21.42}$$

The expressions (21.40) and (21.41) represent the first two terms of the expansion of the function reg $I(p^2)$ into the Taylor series in the vicinity of the point $p^2 = 0$. Since only the divergent part $I_D(0)$ involves the fictitious mass M, then one may connect the quantity M with the cut-off parameter.

So, we have shown that the finite part of the regularized integral $I_D(p^2)$ is given by the difference of the regularized integral and the first term of its expansion into the Taylor series. It should be emphasized, that the procedure of obtaining the finite expressions for the matrix elements by application of expanding into the Taylor series and subsequent subtraction of divergent integrals is at the heart of the foundation of the renormalization theory.

Making use of Eq. (21.6), one may present the integral (21.42) in the following form:

$$I_C(p) = 2 \int_0^1 (1-x)dx \int d^4k \frac{2(kp) - p^2}{[(k-xp)^2 - m^2 + x(1-x)p^2]^3} =$$

$$= 2 \int_0^1 (1-x)dx \int d^4k \frac{(2x-1)p^2}{[k^2 - m^2 + x(1-x)p^2]^3}. \tag{21.43}$$

At the last stage we have fulfilled the variables change-over $k \to k+xp$ and dropped the term that is linear on k because its contribution equals zero due to the oddness of the integrand. To estimate the integral over momenta we carry out the Wick rotation and transit to the spherical coordinates. As a result we get

$$\int d^4k \frac{1}{[k^2 - m^2 + x(1-x)p^2]^3} = i\pi^2 \int_0^\infty \frac{k^2 d(k^2)}{[-k^2 - m^2 + x(1-x)p^2]^3}. \tag{21.44}$$

Allowing for the well-known formula for the beta-function:

$$\int_0^\infty \frac{z^{m-1}dz}{(z+a^2)^n} = \frac{1}{(a^2)^{n-m}} \frac{\Gamma(m)\Gamma(n-m)}{\Gamma(n)}, \tag{21.45}$$

one can calculate the remaining integral

$$\int d^4k \frac{1}{[k^2 - m^2 + x(1-x)p^2]^3} = -\frac{i\pi^2}{2[m^2 - x(1-x)p^2]}. \tag{21.46}$$

Then we arrive at the following expression for $I_C(p^2)$:

$$I_C(p^2) = -i\pi^2 \int_0^\infty dx \frac{(1-x)(2x-1)p^2}{[m^2 - x(1-x)p^2]}. \tag{21.47}$$

The evaluation of the divergent part $I_D(0)$ is fulfilled in the similar way

$$I_D(0) = iM^2\pi^2 \int_0^1 \frac{x\,dx}{x(m^2 - M^2) + M^2} \approx i\pi^2 \ln \frac{M^2}{m^2} \tag{21.48}$$

(in the numerator we have neglected by the term m^2 compared with M^2). Thus, we ultimately gain

$$\text{reg } I(p^2) = i\pi^2 \ln \frac{M^2}{m^2} - i\pi^2 \int_0^\infty dx \frac{(1-x)(2x-1)p^2}{[m^2 - x(1-x)p^2]}, \tag{21.49}$$

where the finite part could be integrated to completion. Realizing the limiting transition $M \to \infty$, we do not get rid of the auxiliary mass M. To put this another way, we cannot manage to take off the regularization directly. However, the regularization enables us to perform splitting a matrix element into a finite part and a part containing the ultraviolet (UV) divergencies in the relativistically invariant form; in so doing the mass M enters into a divergent part only.

Dimensional regularization. Since the UV divergencies appear under integrating over virtual momenta in the four-dimensional space-time, then, lowering the dimensionality of the space-time one could make these integrals finite. This idea is at heart of the basis of the dimensional regularization [46]. In this case one may define the Feynman integrals as analytical functions of a parameter n, a space-time dimensionality, and the UV divergencies will be exhibited as singularities at $n \to 4$. As in the previous case, the finite part of the integral can be obtained by subtracting some first terms of the expansion into the Taylor series. The technical advantage of the dimensional regularization lies in the fact that it maintains the symmetry properties and the corresponding invariance of irregularized expressions.

We shall elucidate the details of this method by an example of calculating the integral (21.38). To define the integral over n-dimensional space we shall believe that the internal momentum has n-components—$k_\mu = (k_0, k_1, ..., k_{n-1})$—while the external one has only four nonvanishing components—$p_\mu = (p_0, p_1, p_2, p_3, 0, ...0)$. Then, in the n-dimensional space, the integral (21.38) takes the form:

$$I(p^2; n) = \int d^n k \frac{1}{[(k-p)^2 - m^2](k^2 - m^2)}. \tag{21.50}$$

It is obvious that it is converging when $n < 4$. Let us try to assign a specific meaning to this expression under non-integer n. With this in mind we first transform (21.50) using the formula (21.4), then carry out the Wick rotation and make the change-over of variables $k \to k - xp$. As a result we get

$$I(p^2; n) = \int_0^1 dx \int \frac{d^n k}{[(k - xp)^2 - m^2 + x(1-x)p^2]^2} = i \int_0^1 dx \int \frac{d^n k}{[k^2 - m^2 + x(1-x)p^2]^2}.$$

Getting over to the n-dimensional spherical coordinate system and fulfilling the integration over the angle variables, we obtain

$$I(p^2; n) = i \int_0^1 dx \int_0^\infty k^{n-1} dk \int_0^{2\pi} d\theta_1 \int_0^\pi \sin\theta_2 d\theta_2 \int_0^\pi \sin^2\theta_3 d\theta_3 ... \int_0^\pi \frac{\sin^{n-2}\theta_{n-1} d\theta_{n-1}}{[k^2 - m^2 + x(1-x)p^2]^2} =$$

$$= \frac{2i\pi^{n/2}}{\Gamma(n/2)} \int_0^1 dx \int_0^\infty \frac{k^{n-1} dk}{[k^2 - m^2 + x(1-x)p^2]^2}, \tag{21.51}$$

where we have used the formula:

$$\int_0^\pi \sin^m \theta d\theta = \frac{\sqrt{\pi}\,\Gamma\left(\frac{m+1}{2}\right)}{\Gamma\left(\frac{m+2}{2}\right)}.$$

Thus, we have extracted the dependence on n in the explicit view. The integral (21.51) is the well-defined quantity in the complex plane of n at $0 < \mathrm{Re}(n) < 4$. The limitation from the bottom is caused by the existence of the infrared (IF)-divergency that is not a source of our interest at this time. Completing the integration of the expression (21.51) by parts, we may extend the analyticity region of the function $I(p^2; n)$. So, making use of the expression:

$$n\Gamma(n) = \Gamma(n+1) \tag{21.52}$$

and accomplishing a single integration by parts, we have

$$\frac{1}{\Gamma\left(\frac{n}{2}\right)} \int_0^\infty \frac{k^{n-1} dk}{[k^2 - m^2 + x(1-x)p^2]^2} =$$

$$= -\frac{2}{\Gamma\left(\frac{n}{2}+1\right)} \int_0^\infty k^n dk \frac{d}{dk}\left(\frac{1}{[k^2 - m^2 + x(1-x)p^2]^2}\right). \tag{21.53}$$

The integral (21.53) is the analytical function even at $-2 < \mathrm{Re}(n) < 4$. Iterating the procedure of the integration by parts d times we extend the analyticity region to $-2d < \mathrm{Re}(n) < 4$ and, further, up to $-\infty < \mathrm{Re}(n)$. This, in its turn, means that the integral (21.53) represents the analytical function in the region $\mathrm{Re}(n) < 4$.

To clear up the integral behavior at $n \to 4$ we estimate it using the formula (21.45) when $n < 4$:

$$I(p^2; n) = i\pi^{n/2}\Gamma\left(2 - \frac{n}{2}\right) \int_0^1 \frac{dx}{[m^2 - x(1-x)p^2]^{2-n/2}}. \tag{21.54}$$

It follows from the relation (21.52)

$$\Gamma\left(2 - \frac{n}{2}\right) = \frac{\Gamma\left(3 - \frac{n}{2}\right)}{2 - \frac{n}{2}} \to \frac{2}{4-n} \qquad \text{when} \qquad n \to 4, \tag{21.55}$$

that is, the singularity at $n = 4$ is a simple pole. If we shall expand all the quantities into the series in the vicinity of the point $n = 4$:

$$\Gamma\left(2 - \frac{n}{2}\right) = \frac{2}{4-n} + A + (n-4)B + ..., \tag{21.56}$$

$$[m^2 - x(1-x)p^2]^{n-4} = 1 + (n-4)\ln[m^2 - x(1-x)p^2] + ..., \qquad (21.57)$$

where A and B are some constants, then we get in the limit

$$\lim_{n\to 4} I(p^2; n) = I(p^2) = \frac{2i\pi^2}{4-n} - i\pi^2 \int_0^1 dx \ln[m^2 - x(1-x)p^2] + i\pi^2 A. \qquad (21.58)$$

To single out the finite part in (21.58) it is sufficient to subtract the first term of the expansion of the function $I(p^2)$ into the Taylor series in the vicinity of the point $p^2 = 0$:

$$I(p^2) = I(0) + I_C(p^2), \qquad (21.59)$$

where

$$I(0) = \pi^2 \left(\frac{2i}{4-n} - i \ln\, m^2 + iA \right) \approx \frac{2i\pi^2}{(4-n)}, \qquad (21.60)$$

$$I_C(p^2) = -i\pi^2 \int_0^1 dx \ln\left[\frac{m^2 - x(1-x)p^2}{m^2} \right] = -i\pi^2 \int_0^1 dx \frac{(1-x)(2x-1)p^2}{m^2 - x(1-x)p^2}, \qquad (21.61)$$

and in (21.61) we have done the integration by parts. Thus, we have obtained the finite part of the integral (21.38), which has precisely the same form as in the case of the covariant regularization. Therefore, as could be expected, the finite part of the Green functions product does not depend on the regularization way. It depends on the subtraction point only. The divergency of the term $I(0)$ is related to the simple pole at $n = 4$, which is associated with the term $\ln(M^2/m^2)$ appearing at the covariant regularization.

Successive calculation of the matrix elements containing the ultraviolet (UV) divergencies consists of three stages. First, we transit to the regularized Green functions Re G^c. At the second stage, the parts G_C^c that are left finite under removing the regularization ($M \to \infty$) are extracted from the obtained expressions for Re G^c. Finally, the divergencies associated with the divergent parts of Re G^c are eliminated from the theory. The quantity G_C^c is obtained by subtracting several terms of the Taylor series from Re G^c and does not depend on the regularization way. This allows us to technically accept the most convenient strategy that provides the finite result for every diagram not demanding any regularization under an intermediate stage. This strategy is based on the regularization by means of the cut-off. To calculate divergent integrals is carried out using the Feynman technique where the integration sphere radius is identified with the boundary momentum. Further, to split divergent integrals into the finite and infinite parts, following Dyson [4], we fulfill the expansion of integrand $R(p,t)$ (p is an external momentum and t is an internal one) into the Taylor series in external momenta:

$$R(p,t) = R(0,t) + p_\mu \left(\frac{\partial R(p,t)}{\partial p_\mu} \right)_{p=0} + \frac{1}{2} p_\mu p_\nu \left(\frac{\partial^2 R(p,t)}{\partial p_\mu \partial p_\nu} \right)_{p=0} + ... \qquad (21.62)$$

Every differentiation increases the difference of the power t in the denominator and numerator by one. Because of this, if the integral is diverging logarithmically (linearly or quadratically), then only one term (two or first three terms) will be diverging under term-by-term integration of the expansion $R(p,t)$. As will be shown later, to liquidate divergent expressions, the counter-terms, whose physical meaning reside in the redefinition of particle parameters, are introduced into the theory.

21.3 Self-energy of the electron

In the interaction representation the time evolution of the state vector

$$\Phi(t_0) = \Phi(\mathbf{p}; s) = (2\pi)^{3/2} a_s^\dagger(\mathbf{p})|0>$$

to describe a free electron at the instant of time t_0 (at the remote past) is determined by the equation:

$$\Phi(t) = S(t, t_0)\Phi(t_0), \tag{21.63}$$

where

$$i\frac{\partial S(t, t_0)}{\partial t} = H_{int}S(t, t_0).$$

The energy of the state $\Phi(\mathbf{p}; s)$ is equal to E_0, where E_0 is the eigenvalue of the Hamiltonian operator for a free particle. However, the energy of an electron state is changed as the result of the interaction of the electron with the quantized electron-positron and electromagnetic fields. This, in turn, leads to modifying a phase in the amplitude of the transition from the state at the instant of time t_0 to the state with the time t:

$$(\Phi(\mathbf{p}; s), \Phi(\mathbf{p}; s))e^{-i\Delta E(t-t_0)} = \frac{(\Phi(\mathbf{p}; s), S(t, t_0)\Phi(\mathbf{p}; s))}{(\Phi_0, S(t, t_0)\Phi_0)}. \tag{21.64}$$

In Eq. (21.64) the time difference $t - t_0$ must be taken very big and the average over all oscillations from t to t_0 must be fulfilled.

Let us explain the presence of the quantity $(\Phi_0, S(t, t_0)\Phi_0)$ in Eq. (21.64). Among the addends entering into $(\Phi(\mathbf{p}; s), S(t, t_0)\Phi(\mathbf{p}; s))$, there are such ones that are associated with the diagrams for a transition: *vacuum→vacuum* (Fig. 21.4). We shall show that

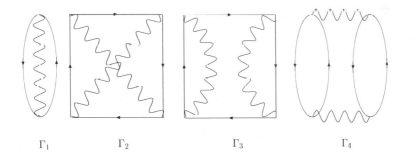

$$\Gamma_1 \qquad \Gamma_2 \qquad \Gamma_3 \qquad \Gamma_4$$

FIGURE 21.4

The Feynman diagrams describing the transition ***vacuum→vacuum***.

the total contribution from such vacuum processes is given by just the quantity: $\mathcal{A}_V = (\Phi_0, S(t, t_0)\Phi_0)$. Denote the elementary connected vacuum diagrams (diagrams that may be bypassed moving only along internal lines) by $\Gamma_1, \Gamma_2, \Gamma_3,...$ Then, in the most general case, the diagram of the vacuum fluctuations is derived from the n_1 ones of the kind Γ_1, from the n_2 ones of the kind Γ_2, and so on. The contribution of the diagram $\Gamma = n_1\Gamma_1 + n_2\Gamma_2 + ...$ into the matrix element will be equal to:

$$(\Phi_0, S_\Gamma\Phi_0) = (\Phi_0, S_{\Gamma_1}\Phi_0)^{n_1}(\Phi_0, S_{\Gamma_2}\Phi_0)^{n_2}..., \tag{21.65}$$

where $(\Phi_0, S_{\Gamma_1}\Phi_0)$ is a matrix element corresponding to the diagrams Γ_1. The fact, that the contributions of the diagrams $\Gamma_1, \Gamma_2...$ are simply multiplied, follows from the disconnectedness of the diagrams under consideration. If the diagram contained some identical parts, then family terms would not all be different, because the indices rearrangement of their vertices resulted in the same diagrams. It immediately follows that, in the most general case, the amplitude of the transition *vacuum→vacuum* is given by the expression:

$$\mathcal{A}_V = \sum_{n_1, n_2, ...} \frac{1}{n_1! n_2! ...} (\Phi_0, S_{\Gamma_1}\Phi_0)^{n_1} (\Phi_0, S_{\Gamma_2}\Phi_0)^{n_2} ... =$$

$$= e^{(\Phi_0, S_{\Gamma_1}\Phi_0)} e^{(\Phi_0, U_{\Gamma_2}\Phi_0)} ... = e^{<0|S_{\Gamma_1} + S_{\Gamma_2} + ...|0>} = e^{<0|S|0>_c} = e^{-N}, \qquad (21.66)$$

where $<0|S|0>_c$ denotes the contribution of all the connected diagrams. Inasmuch as the quantity N is purely imaginary, then the probability that the vacuum will remain the vacuum is equal to 1.

Investigate some real process described by the plurality of diagrams that are associated with an amplitude $\mathcal{A}$. In higher orders of the perturbation theory the same diagrams will appear in accompaniment of the vacuum diagrams disconnected with them. To sum over all the vacuum processes gives rise simply to the multiplication of the amplitude $\mathcal{A}$ by the phase factor $\exp(-N)$ whose modulus equals 1.

Thus, the appearance cause of the factor $(\Phi_0, S(t, t_0)\Phi_0)$ into the denominator of (21.64) is correlated to the inclusion of diagrams corresponding to the transition *vacuum→vacuum*. This factor is canceled with the analogous one, which is held in the numerator. Next, we decide not to take into account the diagrams that describe the vacuum fluctuations and, hence, the quantities of the kind $(\Phi_0, S(t, t_0)\Phi_0)$ will not appear in our consideration.

In both sides of the expression (21.64) we carry out the expansion into the series restricted to two nonvanishing terms. As a result we arrive at the equation:

$$-i\Delta E\, T = (-ie)^2 < \Phi(\mathbf{p}; s)| \int d^4x_1 \int d^4x_2 T(: \overline{\psi}(x_1)\gamma^\mu A_\mu(x_1)\psi(x_1) : \times$$

$$\times : \overline{\psi}(x_2)\gamma^\nu A_\nu(x_2)\psi(x_2) :)|\Phi(\mathbf{p}; s) >= T\frac{m}{E_0}\overline{u}_s(\mathbf{p})\Sigma^{(2)}(p)u_s(\mathbf{p}), \qquad (21.67)$$

or

$$\Delta E = \frac{im}{E_0}\overline{u}_s(\mathbf{p})\Sigma^{(2)}(p)u^s(\mathbf{p}), \qquad (21.68)$$

where m denotes the mass of the bare electron now, and

$$\Sigma^{(2)}(p) = -\frac{e^2}{(2\pi)^4} \int d^4k \gamma_\mu \frac{\hat{p} - \hat{k} + m}{(p-k)^2 - m^2}\gamma^\mu \frac{1}{k^2}. \qquad (21.69)$$

Under writing Eq. (21.69) we have allowed for that $\delta(0) = VT/(2\pi)^4$ and $V = 1$. The matrix element standing in the right-hand side of the expression (21.67) is in line with the diagram Fig. 21.5(a) to describe the emission and the subsequent absorption of the virtual photon by the free electron. The diagram of Fig. 21.5(a) determines a self-energy of the electron in the second order of the perturbation theory. It represents the simplest case of the electron self-energy diagram (ESED), which is considered to mean a block confined between two electron lines. The more complicated ESEDs of the forth order of the perturbation theory are pictured in Figs. 21.5(a)–(e).

The only measurements possible for a free particle are associated with the determination of its energy E_0 and momentum $\mathbf{p}$. Proceeding from these quantities, we can form a single invariant combination, the particle mass, $m^2 = E_0^2 - \mathbf{p}^2$. Thus, the effect of a relativistically

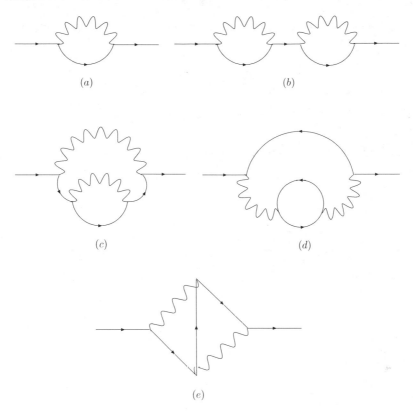

FIGURE 21.5

The Feynman diagrams (a)–(e) for the electron self-energy.

invariant interaction of a particle with the vacuum is limited to a change in its mass by some amount δm. In this case the change-over of a bare particle energy is defined by the expression:

$$E_0 + \Delta E = \sqrt{(m + \delta m)^2 + \mathbf{p}^2}. \tag{21.70}$$

Expanding the expression (21.70) into the series in δm and being constrained by the first order of smallness, we get

$$\Delta E = \frac{m \delta m}{E_0}. \tag{21.71}$$

Now it becomes obvious that the operator $\Sigma^{(2)}(p)$ has a meaning of the mass operator of the electron in the second order of the perturbation theory.

   The effect of the self-energy for a free particle may result only in changing its mass. This principle forming the basis for a renormalization theory was first used by Kramers [47] in classical electrodynamics. If one recalls the phrase of M. Twain: *"clothes adorn a person and naked people have little if any influence in the society,"* then Kramers'idea may be regarded simply as an asymptotic extension of the social observations to the phenomena of the microworld. In his work Kramers also ignores any effect of the bare mass by the argument that it could not be measured experimentally and the only observable mass is $m_R = m + \delta m = Z_m m$. In this way, from a theoretical viewpoint, the main part is played by the observable m_R rather than by the quantities m or δm taken separately. As this takes place, a change in the mass δm may prove to be infinite. But this is of no importance as in the physical observables, for example, the energy levels or the cross sections expressed in terms of the observable mass m_R are finite.

The operation of the mass redefinition is called the *mass renormalization*. It may explicitly be included into the theory by means of introducing a counter-term into the Lagrangian. The Lagrangian of a free field contains a term:

$$-m : \overline{\psi}(x)\psi(x) :,$$

which could be presented as follows:

$$-(m_R - \delta m) : \overline{\psi}(x)\psi(x) : .$$

Then the term

$$\delta m : \overline{\psi}(x)\psi(x) :$$

may be considered as a part of the interaction Lagrangian along with

$$-e : \overline{\psi}(x)\gamma^\mu A_\mu(x)\psi(x) : .$$

In so doing, the contributions coming from $\delta m : \overline{\psi}(x)\psi(x) :$ must always be examined with the ones coming from the diagrams of the electron self-energy. This statement is illustrated in Fig. 21.6, where the circles at the diagrams are associated with the contribution from the term

$$\delta m : \overline{\psi}(x)\psi(x) : .$$

The infinite quantity δm is so defined that in all orders of the perturbation theory the

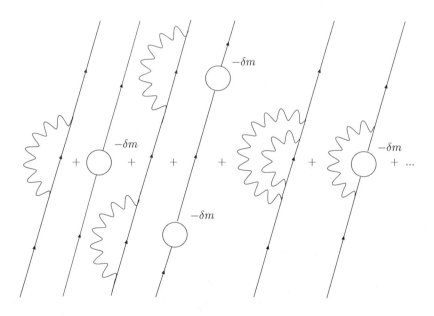

FIGURE 21.6
The Feynman diagrams illustrating the mass renormalization.

contributions of the self-energy diagrams for a free electron are mutually canceled with the ones coming from the diagrams corresponding to the counter-term $-\delta m : \overline{\psi}(x)\psi(x) :$. The amplitude that the electron being in the state $|E_p, \mathbf{p}; s >$ will be remaining in the same state is proportional to the quantity:

$$\bar{u}_s(\mathbf{p})[1 + \Sigma'(p)]u_s(\mathbf{p}),$$

where $\Sigma'(p)$ denotes the contribution of all the self-energy diagrams. The demand of the mass renormalization implies that at $\hat{p} = m_R$ the quantity $\Sigma'(p)$ must be equal to zero, because we include the contribution, coming from the term δm, in $\Sigma'(p)$.

The diagrams in Fig. 21.6 may also be considered as a modification of the internal electron line. Then the original propagator $S^c(p) = (\hat{p} - m)^{-1}$ is transmuted into

$$S^{c'}(p) = \frac{1}{\hat{p} - m} + \frac{1}{\hat{p} - m}\Sigma'(p)\frac{1}{\hat{p} - m}. \tag{21.72}$$

Now the mass renormalization principle can be formulated as the following demand: *"the pole of the modified propagator to be regarded as a function of $\hat{p}$ must be positioned at the point associated with the experimental value of the particle mass."* It is obvious that this demand fixes the value of δm in every order of the perturbation theory.

We now proceed to calculating the mass operator of the electron in the second order of the perturbation theory [7]. This operator, apart from the UV divergencies, contains the IF divergencies as well. To liquidate the latter we introduce as usual the photon mass λ into the electron propagator, that is, carry out the replacement:

$$D_0^c(k) = \frac{1}{k^2} \rightarrow \frac{1}{(k^2 - \lambda^2)}. \tag{21.73}$$

Substituting (21.73) into the expression (21.69), we obtain

$$\Sigma^{(2)}(p) = -\frac{e^2}{(2\pi)^4}\int d^4k\gamma_\mu\frac{\hat{p} - \hat{k} + m}{(p-k)^2 - m^2}\gamma^\mu\frac{1}{k^2 - \lambda^2} = \frac{\alpha}{2\pi^3}\int d^4k\frac{\hat{p} - \hat{k} - 2m}{(p-k)^2 - m^2}\frac{1}{k^2 - \lambda^2}.$$

Taking advantage of the Feynman parametrization, we present $\Sigma^{(2)}(p)$ in the form:

$$\Sigma^{(2)}(p) = \frac{\alpha}{2\pi^3}\int_0^1 dx\int d^4k\frac{\hat{p} - \hat{k} - 2m}{\{k^2 + [-2(kp) + p^2 - m^2 + \lambda^2]x - \lambda^2\}^2}.$$

At present we may use the formulae (21.26) and (21.30) to give:

$$\Sigma^{(2)}(p) = \frac{i\alpha}{2\pi}\int_0^1 dx\left\{[\hat{p}(1-x) - 2m]\ln\frac{L^2}{p^2x(x-1) + m^2x + \lambda^2(1-x)} + \right.$$

$$\left. +\hat{p}\left(\frac{3}{2}x - 1\right) + 2m\right\}. \tag{21.74}$$

To extract the finite part $\Sigma_C^{(2)}(p)$ two first terms of the expansion of $\Sigma^{(2)}(p)$ into the Taylor series in powers $(\hat{p} - m_R)$ should be subtracted from the expression for $\Sigma^{(2)}(p)$:

$$\Sigma_C^{(2)}(p) = \Sigma^{(2)}(p) - \Sigma^{(2)}(m_R) - (\hat{p} - m_R)\left(\frac{\partial\Sigma^{(2)}(p)}{\partial\hat{p}}\right)_{\hat{p}=m_R}. \tag{21.75}$$

In accordance with (21.68) and (21.71) the quantity $\Sigma^{(2)}(m_R)$ coincides with the electromagnetic mass of the electron $\delta m^{(2)}$ within the constant factor i:

$$\delta m^{(2)} = i\Sigma^{(2)}(m_R) = \frac{\alpha}{2\pi}\int_0^1 dx\left[m_R(1+x)\ln\frac{L^2}{m_R^2x^2 - \lambda^2(x-1)} - m_R\left(\frac{3}{2}x + 1\right)\right]. \tag{21.76}$$

Having designated δm as the electromagnetic mass, we emphasized its origin. It is apparent that the analogous corrections to the bare electron mass will come from the weak and gravitational interactions as well.

Having fulfilled the integration and neglecting the quantity λ as compared with m_R, we arrive at the following expression for the electromagnetic mass of the electron:

$$\delta m^{(2)} = m_R - m = \frac{3\alpha}{4\pi} m_R \left(\ln \frac{L^2}{m_R^2} + \frac{1}{2} \right). \tag{21.77}$$

Thus, in the second order of the perturbation theory the physical meaning of the electron mass is given as:

$$m_R = Z_m m, \tag{21.78}$$

where the singular factor Z_m:

$$Z_m = 1 + \frac{3\alpha}{4\pi} m_R \left(\ln \frac{L^2}{m_R^2} + \frac{1}{2} \right)$$

is logarithmically diverging under removing the regularization ($L^2 \to \infty$).

Differentiating (21.74), we gain

$$\left(\frac{\partial \Sigma^{(2)}(p)}{\partial \hat{p}} \right)_{\hat{p}=m_R} = \Sigma_1^{(2)} = \frac{i\alpha}{2\pi} \int_0^1 dx \left[(1-x) \ln \frac{L^2}{m_R^2 x^2 - \lambda^2(x-1)} + \right.$$

$$\left. + \frac{3}{2}x - 1 - \frac{2x(1-x^2)m_R^2}{m_R^2 x^2 - \lambda^2(x-1)} \right] = \frac{i\alpha}{2\pi} \left[\frac{1}{2} \ln \frac{L^2}{m_R^2} + \frac{9}{4} + \ln \frac{\lambda^2}{m_R^2} \right]. \tag{21.79}$$

To insert $\Sigma^{(2)}(m_R)$ and $\Sigma_1^{(2)}$ into (21.76) leads to the following expression for the finite part of the mass operator in the second order of the perturbation theory:

$$\Sigma_C^{(2)}(p) = \frac{i\alpha}{2\pi} \int_0^1 dx \left\{ [\hat{p}(1-x) - 2m_R] \ln \frac{m_R^2 x^2 - \lambda^2(x-1)}{p^2 x(x-1) + m_R^2 x - \lambda^2(x-1)} + \right.$$

$$\left. + (\hat{p} - m_R) \frac{2m_R^2 x(1-x^2)}{m_R^2 x^2 - \lambda^2(x-1)} \right]. \tag{21.80}$$

In the most general case, the diagram for the electron self-energy may be a part of a more complex diagram and hence the outgoing electron lines (one or both) are associated with the electron propagation functions. Then, to allow for the counter-term proportional to δm in the Lagrangian will lead to subtraction of the divergences due to the presence of $\Sigma(p)$ in matrix elements in every order of the perturbation theory. This means that, for example, during calculations of any process, the diagram of which includes the second-order diagram for the electron self-energy, we can directly use the finite expression for the mass operator (21.80), disregarding the counter-term $\sim \delta m$.

Integration of (21.80) may be carried out in the analytical form under the assumption:

$$\rho = \frac{m_R^2 - p^2}{m_R^2} \gg \frac{\lambda}{m_R} \quad \text{and} \quad \rho > 0. \tag{21.81}$$

The final result is as follows:

$$\Sigma_C^{(2)}(p) = \frac{i\alpha}{2\pi m_R} (\hat{p} - m)^2 \left\{ \frac{1}{2(1-\rho)} \left[1 - \frac{2-3\rho}{1-\rho} \ln \rho \right] - \right.$$

$$-\frac{\hat{p}+m}{\rho m_R}\left[\frac{1}{2(1-\rho)}\left(2-\rho+\frac{\rho^2+4\rho-4}{1-\rho}\ln\rho\right)+1+2\ln\frac{\lambda}{m_R}\right]\right\}. \tag{21.82}$$

At the analytical continuation of the expression (21.82) into the region $\rho<0$ the logarithm phase is defined by the replacement $m_R \to m_R - i\varepsilon$ in accordance with the poles detour rule. As this takes place, $\rho \to \rho - i\varepsilon$, so that, when $\rho<0$, $\ln\rho$ should be understood as:

$$\ln\rho = \ln|\rho| - i\pi.$$

The expression for $\Sigma_C^{(2)}(p)$ is greatly simplified in the region of big momenta $(p^2 \gg m_R^2)$:

$$\Sigma_C^{(2)}(p) = \frac{\alpha\hat{p}}{4\pi}\ln\frac{p^2}{m_R^2}. \tag{21.83}$$

21.4   Vacuum polarization

We shall examine the electron scattering by the external electromagnetic field taking into account the radiative corrections (RC). Introduce the counter-term to be connected with the mass renormalization into the interaction Lagrangian $\mathcal{L}_{int}$. Then $\mathcal{L}_{int}$ is determined by the expression:

$$\mathcal{L}_{int} = -e\{N[\overline{\psi}(x)\gamma^\mu\psi(x)(A_\mu(x)+A_\mu^e(x))]\} + \delta m N[\overline{\psi}(x)\psi(x)]. \tag{21.84}$$

In Fig. 21.7 we represent the corresponding Feynman diagrams up to the third order of the perturbation theory inclusive. Here we shall centre on the diagram shown in Fig. 21.7f_1. Let us agree to identify the block confined between two photon internal or external lines as a photon self-energy diagram (PSED). Obviously, in the case under consideration we deal with the PSED of the second order of the perturbation theory. In the diagram of Fig. 21.7f_1 the external photon line, the electron-positron loop and the internal photon line may be changed by a single effective photon line with the momentum q:

$$a_\mu^e(q) = A_\mu^e(q) + D_{\mu\nu}^c(q)\Pi^{(2)\nu\sigma}(q)A_\sigma^e(q) = A_\mu^e(q) + \delta A_\mu^e(q), \tag{21.85}$$

where

$$\Pi_{\mu\nu}^{(2)}(q) = \frac{(-ie)^2}{(2\pi)^4}\int_\Sigma d^4k \, \text{Sp}\left[\gamma_\mu\frac{1}{\hat{q}+\hat{k}-m}\gamma_\nu\frac{1}{\hat{k}-m}\right], \tag{21.86}$$

$q = p_2 - p_1$. Note that the minus sign connected with the existence of the fermion loop has been included into the definition of $\Pi_{\mu\nu}^{(2)}(q)$. The operator $\Pi_{\mu\nu}^{(2)}(q)$ is called the *polarization operator of the second order*. From (21.85) it follows that thanks to the gauge invariance the relation:

$$q^\nu\Pi_{\nu\sigma}^{(2)}(q) = 0 \tag{21.87}$$

must be fulfilled (beyond doubt this relation still stands in any order of the perturbation theory). Apparently, the tensor $\Pi_{\nu\sigma}^{(2)}(q)$ can be represented in the form:

$$\Pi_{\nu\sigma}^{(2)}(q) = g_{\nu\sigma}q^2\Pi_1(q^2) + q_\nu q_\sigma\Pi_2(q^2), \tag{21.88}$$

where owing to the gauge invariance $\Pi_{1,2}(q^2)$ are functions of q^2 only. For the relation (21.87) to be met the polarization operator must be transverse with respect to q, that is, it must have the form:

$$\Pi_{\nu\sigma}(q) = (q_\nu q_\sigma - g_{\nu\sigma}q^2)\Pi(q^2), \tag{21.89}$$

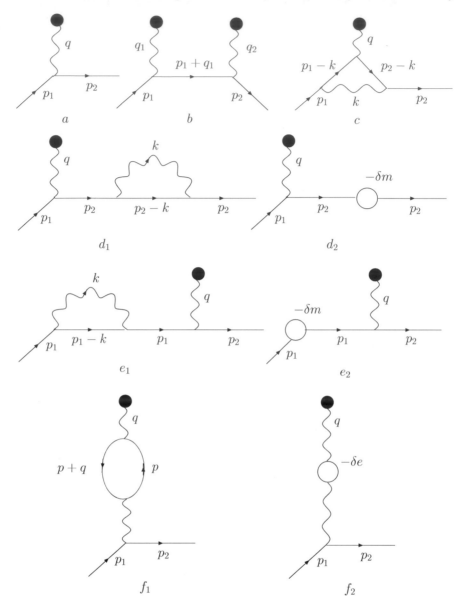

FIGURE 21.7
The Feynman diagrams describing the RC for the electron scattering by the external electromagnetic field.

where $-\Pi_2(q^2) = \Pi_2(q^2) = \Pi(q^2)$ and in order to emphasize the validity of the expansion (21.89) for the polarization tensor of any order we have omitted the superscript 2 in $\Pi_{\mu\nu}(q)$. In accordance with (21.85) the quantity $a_\mu^e(q)$ is the sum of the given external electromagnetic field $A_\mu^e(q)$ and an additive $\delta A_\mu^e(q)$, which may be interpreted as an electromagnetic field caused by a polarization of the electron-positron field vacuum under switching on the field $A_\mu^e(q)$. The field potential $A_\mu(x)$ is associated with the current $J_\mu(x)$, which creates it by means of the Maxwell equation:

$$\Box A_\mu(x) = J_\mu(x).$$

In the Fourier components this linkage looks so:

$$q^2 A_\mu(q) = -J_\mu(q). \tag{21.90}$$

Using (21.85) and (21.90), we could write

$$j_\mu^e(q) = J_\mu^e(q) + \delta J_\mu^e(q) = [1 + i\Pi^{(2)}(q^2)]J_\mu^e(q). \tag{21.91}$$

From the obtained expression it is evident that the closed electron-positron loop in this diagram induces the current $j_\mu^e(q)$ being the source of the photons with the momentum q, which influence the electron and scatter it.

We shall calculate the polarization operator in the second order of the perturbation theory. Since in the expression (21.86) the numerator contains the fourth power of the integration variable while the denominator contains the second one, then the integral is diverging quadratically. However, in reality $\Pi_{\nu\sigma}^{(2)}(q)$ possesses only the logarithmic divergency. It results from the gauge invariance. To make sure of it, first one should regularize the expression (21.86) on the electron mass and then use the transversality condition (21.87) [46,50]. On the other hand, the same result could be got by a more simple way too. Making use of the Feynman parametrization and calculating the spur of the γ-matrices product, we present $\Pi_{\mu\nu}(q)$ in the following form:

$$
\begin{aligned}
\Pi_{\mu\nu}(q) &= -\frac{\alpha}{4\pi^3} \int d^4k \frac{\mathrm{Sp}[\gamma_\mu(\hat{k} + \hat{q} + m)\gamma_\nu(\hat{k} + m)]}{[(k+q)^2 - m^2][k^2 - m^2]} = \\
&= -\frac{\alpha}{4\pi^3} \int_0^1 dx \int d^4k \frac{\mathrm{Sp}[\gamma_\mu(\hat{k} + \hat{q} + m)\gamma_\nu(\hat{k} + m)]}{[(k+qx)^2 + q^2(x - x^2) - m^2]^2} = \\
&= -\frac{\alpha}{4\pi^3} \int_0^1 dx \int d^4k \frac{\mathrm{Sp}[\gamma_\mu(\hat{k} + \hat{q}(1-x) + m)\gamma_\nu(\hat{k} - \hat{q}x + m)]}{[k^2 + q^2(x - x^2) - m^2]^2} = \\
&= -\frac{\alpha}{\pi^3} \int_0^1 dx \int d^4k \frac{(g_{\mu\nu}q^2 - 2q_\mu q_\nu)(x - x^2) + g_{\mu\nu}(m^2 - k^2/2)}{[k^2 + q^2(x - x^2) - m^2]^2}. \tag{21.92}
\end{aligned}
$$

Because, as already stated, the integral (21.92) is diverging only logarithmically, then the variable change over $k \to k - qx$ being made in (21.92) is quite legal. Acting upon the last line of (21.92) by the operator q_μ, we come to recognize that the polarization tensor will obey the transversality demand under the fulfillment of the condition:

$$\int d^4k \frac{-q^2(x - x^2) + m^2 - k^2/2}{[k^2 + q^2(x - x^2) - m^2]^2} = 0,$$

that is, the quadratically divergent part of the integral (21.92) proves to be equal to zero. With allowance made for the preceding we obtain for the polarization tensor

$$\Pi_{\mu\nu}(q) = (q_\mu q_\nu - g_{\mu\nu}q^2)\Pi(q^2), \tag{21.93}$$

where

$$\Pi(q^2) = \frac{2\alpha}{\pi^3} \int_0^1 dx(x - x^2) \int \frac{d_E^4 k}{[-k^2 + q^2(x - x^2) - m^2]^2}. \tag{21.94}$$

Carrying out the integration into (21.94), we arrive at the final expression for the polarization tensor in the second order of the perturbation theory [7,45]:

$$\Pi_{\mu\nu}^{(2)}(q^2) = \frac{i\alpha}{\pi}(q_\mu q_\nu - g_{\mu\nu}q^2)\left[\frac{1}{3}\ln\frac{L^2}{m^2} + \frac{4m^2 + 2q^2}{3q^2}(1 - \vartheta\cot\vartheta) - \frac{1}{9}\right], \tag{21.95}$$

where $q^2 = 4m^2 \sin^2 \vartheta$.

Inasmuch as the polarization tensor is diverging logarithmically, then in the expansion into the Taylor series in powers of q^2:

$$\Pi^{(2)}(q^2) = \Pi^{(2)}(0) + \Pi^{(2)\prime}(0)q^2 + ..., \tag{21.96}$$

where

$$\Pi^{(2)\prime}(0) = \left. \frac{\partial \Pi^{(2)}(q^2)}{\partial q^2} \right|_{q^2=0},$$

all terms starting with the second one are finite. Really, in subsequent terms of the expansion in powers of q^2 the exponent of the integration variable k is decreasing as much as the one of the external variable q is increasing. In the expansion of the function $\Pi^{(2)}(q^2)$ the term not depending on q^2 contains the integral over k, whose integrand behaves as dk/k. On this reason $\Pi^{(2)}(0)$ is diverging logarithmically. The next term of the expansion has the power q^2 and contains the integral of the manner $\int dk/k^3$, which is already converging under big values of k. Therefore, $\Pi^{(2)\prime}(0)$ and all the higher terms of the expansion (21.96) are reported by the expressions convergent on the variable k. Thus, the regularized operator $\Pi^{(2)}(q^2)$ is determined by the expression:

$$\Pi_C^{(2)}(q^2) = \Pi^{(2)}(q^2) - \Pi_D^{(2)}(q^2) = \frac{i\alpha}{\pi} \left[\frac{4m_R^2 + 2q^2}{3q^2} \left(1 - \vartheta \cot \vartheta \right) - \frac{1}{9} \right], \tag{21.97}$$

where $\Pi_D^{(2)}(q^2) = \Pi^{(2)}(0)$.

To define the behavior of $\Pi_C^{(2)}(q^2)$ in the limit of small momenta we are sufficiently constrained by the second term in the expansion (21.96). To calculate gives:

$$\left. \frac{\partial \Pi^{(2)}(q^2)}{\partial q^2} \right|_{q^2=0} = -\frac{4\alpha}{\pi^3} \int_0^1 dx(x - x^2)^2 \int \frac{d_E^4 k}{[-k^2 - m_R^2]^3} = \frac{i\alpha}{15\pi m_R^2},$$

that is,

$$\Pi_C^{(2)}(q^2) = \frac{i\alpha}{15\pi} \frac{q^2}{m_R^2}, \qquad \text{when} \qquad q^2 \ll m_R^2. \tag{21.98}$$

The value $q^2 = 4m_R^2$ is the threshold for the production of the single electron-positron pair by the virtual photon. At $q^2 > 4m_R^2$ the expression for $\Pi_C^{(2)}(q^2)$ becomes complex and the matrix element $\mathcal{M}^{(e)}$ has both the real and imaginary parts. The imaginary part is associated with the time decrease of the scattering amplitude without the real pair production. On the other hand, the amplitude of the transition probability into the final state, which includes specified real pairs, will accordingly be increasing in time. Due to the S-matrix unitarity the increase and accompanied decrease of these matrix elements are such that the total probability of the transition from the initial state equals to 1. To define the analytical continuation (21.96) in the region $q^2 > 4m_R^2$ one should recall the poles detour rule ($m_R \to m_R - i\varepsilon$) to lead to the following replacements:

$$[1 - q^2/(4m_R^2)]^{1/2} \to -i[q^2/(4m_R^2) - 1]^{1/2}, \qquad \vartheta \to \frac{\pi}{2} + iu, \qquad \cot \vartheta \to -i \tanh u,$$

where $\sinh u = [q^2/(4m_R^2) - 1]^{1/2}$. Then, one may show that

$$\Pi_C^{(2)}(q^2) = \frac{i\alpha}{3\pi} \left[\ln \frac{q^2}{m_R^2} - i\pi \right], \qquad \text{when} \qquad q^2 \gg m_R^2. \tag{21.99}$$

Analogously, for the case of big space-like values of q^2 we have

$$\Pi_C^{(2)}(q^2) = -\frac{i\alpha}{3\pi} \ln\left(-\frac{q^2}{m_R^2}\right), \qquad \text{when} \qquad q^2 \to -\infty. \tag{21.100}$$

Again we shall focus our attention on the process of the electron scattering by the Coulomb nucleus field, assuming, in this case, the mass renormalization to be fulfilled. The matrix element corresponding to the diagram shown in Fig. 21.7f_1 is given by the expression:

$$\mathcal{M}^{(e)} = -(2\pi)^{3/2} ie \sqrt{\frac{m_R^2}{E_{p_1} E_{p_2}}} \overline{u}_s(\mathbf{p}_2) M^{(e)} u_s(\mathbf{p}_1),$$

where

$$M^{(e)} = -i\gamma_\mu \frac{1}{q^2} \Pi^{\mu\nu}(q) A_\nu^e(q).$$

Let us analyze the sum of the matrix elements $\mathcal{M}^{(a)} + \mathcal{M}^{(e)} = \mathcal{A}$, restricted by two limiting cases: (i) $q^2 \to 0$; (ii) $q^2 \to -\infty$. Then, in the former case this sum is defined by the expression:

$$\mathcal{A} = -(2\pi)^{3/2} ie \sqrt{\frac{m_R^2}{E_{p_1} E_{p_2}}} \overline{u}_s(\mathbf{p}_2) \gamma^\mu \left[1 - \frac{e^2}{12\pi^2} \ln\frac{L^2}{m_R^2}\right] A_\mu^e(q) u_s(\mathbf{p}_1), \tag{21.101}$$

while in the latter one ($L^2 \gg -q^2$) we have:

$$\mathcal{A} = -(2\pi)^{3/2} ie \sqrt{\frac{m_R^2}{E_{p_1} E_{p_2}}} \overline{u}_s(\mathbf{p}_2) \gamma^\mu \left[1 - \frac{e^2}{12\pi^2} \ln\frac{L^2}{q^2}\right] A_\mu^e(q) u_s(\mathbf{p}_1). \tag{21.102}$$

In these expressions only the big logarithmic contributions are kept. It is clear that in terms of the initial parameters the perturbation theory does not work, since the big corrections ($\sim e^2 \ln L^2$) already appear in the approximation that follows after the Born one. Let us show that using the naive renormalization one may improve the situation. For the sake of simplicity, for the time being, we are distracted from the existence of other divergencies appearing in the QED. Further, in the formula (21.101) we redefine the charge and the external field potential:

$$e_R = Z_0^{-1/2} e, \qquad A_{R\mu}^e = Z_0^{-1/2} A_\mu^e, \tag{21.103}$$

where the singular factor Z_0 in the second order of the perturbation theory is given by the expression:

$$Z_0^{-1} = 1 + i\Pi^{(2)}(0) = 1 - \frac{e^2}{12\pi^2} \ln\frac{L^2}{m_R^2}.$$

Then, the amplitude $\mathcal{A}_{q^2 \to 0}$ expressed in terms of the variables e_R and $A_{R\mu}^e$ takes the same form as the Born one $f(e, A^e)$ in which the following replacements must be done: $e \to e_R$, $A_\mu^e \to A_{R\mu}^e$:

$$\mathcal{A}_{q^2 \to 0} = f_B(e_R, A_R^e) = -(2\pi)^{3/2} ie_R \sqrt{\frac{m_R^2}{E_{p_1} E_{p_2}}} \overline{u}_s(\mathbf{p}_2) \gamma^\mu A_{R\mu}^e(q) u_s(\mathbf{p}_1).$$

Thus, we again include the infinity into the observable quantities. True, in the successive approach resulting in the renormalization of the charge, the Green functions, the wavefunctions of the electron-positron and electromagnetic fields the infinite constants in the relations (21.103) will take an other form.

From the expression (21.91) follows that the divergent part of the induced current $j^e_\mu(q)$ is exactly proportional to the inducing external current $J^e_\mu(q)$. Because of this, in the given case the renormalization should be considered as the operation under which both the four-dimensional density of the external current $J^e_\mu(q)$ (or the potential) and the charge are renormalized. It is impossible to experimentally separate the external current $J^e_\mu(q)$ from the induced current, which is proportional to it. Just as the electron is always surrounded by the fur coat of virtual quanta (and as a result the physical mass is always measured), the external current polarizes vacuum at all times. Therefore, the observable values of the external current and the charge will always be determined by the expressions (21.103) when we carry out their measurements at big distances.

So, if one uses from the outset the quantities e_R and A^e_R as the expansion parameters, then the diagram pictured in Fig. 21.7f_1 may be excluded from consideration at all. To put it otherwise, one must "by hand" distract its contribution at the point $q^2 = 0$. One could perform the same operation in the next orders of the perturbation theory as well. Then, after its fulfillment, the scattering amplitude taken at the point $q^2 = 0$ will coincide with the Born amplitude $f_B(e_R, A^e_R)$ now in all orders of the perturbation theory. The exact amplitude $\mathcal{A}$ proves as though it were normalized on the Born one at the point $q^2 = 0$. By this reason, the value $q^2 = 0$ in the procedure in question is treated as the subtraction point or the normalization point. It should be emphasized that to introduce the renormalized quantities in accordance with the formula (21.103) makes the scattering amplitude finite at any values of q^2. Since the corresponding matrix element is logarithmically diverging, then for the matrix element to be made finite, one distraction is sufficient at an arbitrary point. In particular, after the substitution (21.103) the amplitude (21.102) is represented with an accuracy of terms $\sim e^4_R$ in the form:

$$\mathcal{A}_{q^2 \gg m^2_R} = f_B(e_R, A^e_R) \left(1 - \frac{e^2_R}{12\pi^2} \ln \frac{m^2_R}{q^2}\right).$$

It is obvious that we may introduce the counter-term connected with the charge renormalization into the starting Lagrangian just as we do in the case of the mass renormalization. The vacuum polarization produces changing the Maxwell equations for a free field, that is, changing the components $A_\mu(x)$. Hence, to take into consideration the vacuum polarization effect by the gauge invariant way, in the total Lagrangian one must carry out the replacement:

$$F_{\mu\nu}(x) \rightarrow F_{\mu\nu}(x)\sqrt{1 + \delta e}.$$

Here δe is sorted out in such a way that the polarization operator divergencies are compensated in every order of the perturbation theory. Thus, for example, for this purpose to be achieved in the second order of the perturbation theory, the fulfillment of the demand:

$$\frac{1}{2}\delta e = i\Pi^{(2)}_D(q^2)$$

is needed. Accordingly, the interaction Lagrangian is defined as follows:

$$\mathcal{L}_{int} = -e_R N[\overline{\psi}(x)\gamma^\mu \psi(x) A_\mu(x)] + \delta m N[\overline{\psi}(x)\psi(x)]+$$

$$-\frac{1}{4}\delta e F_{\mu\nu}(x) F^{\mu\nu}(x)]. \tag{21.104}$$

Then during calculations for each diagram including the irreducible self-energy photon part we add a diagram, where the ordinary self-energy part of the photon is replaced by the part given in Fig. 21.8. With the summation of these diagrams the infinities due to the vacuum polarization are canceled giving as a result a matrix element, whose factors

FIGURE 21.8

The Feynman diagram connecting with inclusion of the charge renormalization counter-term.

represent only the finite part of the polarization tensor, that is, the quantities of the kind $\Pi^{(2)}_{C\mu\nu}(q^2)$. From the qualitative point of view, varying of the mass or the charge is quite natural for the particles under the effect of interactions. A similar phenomenon has been encountered in the classical electrodynamics. In the medium a foreign charge creates around itself a polarization cloud, being partially screened by this cloud. Because of this, for long distances the effective value of the observed charge turns to be lower than the real one. Such a particle is moving together with its polarization cloud and this leads to the effective change in its inertia properties, that is, the mass. As seen, even in the absence of divergences in the theory, the renormalization is a must. This is due to the fact that in the real world it is impossible to turn off interactions between particles; we always observe the parameters of bare particles plus the corrections to them. Nobody can abandon the rule: *"observables must be expressed in terms of the renormalized constants."*

Without doubt, the finite part of the polarization tensor $\Pi_C(q^2)$ leads to the observed effects. Let us consider, for example, the Möller scattering taking into account the vacuum polarization effect in higher orders of the perturbation theory. The corresponding diagrams are displayed in Fig. 21.9. As the series of the perturbation theory are convergent, then

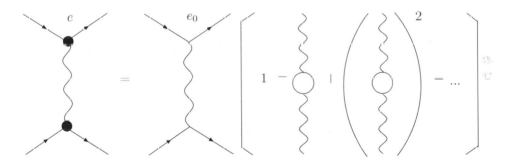

FIGURE 21.9

The Feynman diagrams illustrating the charge renormalization.

we essentially deal with the decreased geometric progression whose denominator is equal to $i\Pi_C(q^2)$ $(q = p_1 - p_1')$. Summing up the obtained series, we come to the following formula:

$$e^2(q^2) = \frac{e_R^2}{1 - i\Pi_C(q^2)}.$$

So, the experimentally defined charge depends on the value q^2 in the experiment. The fine structure constant:

$$\alpha(q^2) \equiv \frac{e^2(q^2)}{4\pi}$$

also proves to be depending on q^2. By this reason it is called the running coupling constant of the QED. By virtue of the fact that in our case $q^2 < 0$, the quantity $\Pi_C(q^2)$ is determined by the expression (21.100) in the limit of big values of $Q^2 = -q^2$. Then, the running coupling constant of the QED will be define as:

$$\alpha(Q^2) = \frac{3\pi\alpha(m_R)}{3\pi - \alpha(m_R)\ln(Q^2/m_R^2)}, \tag{21.105}$$

where $\alpha(m_R)$ is the macroscopic value of the fine structure constant ($\approx 1/137$). Special attention must be given to the following. First, the expression (21.104) maintains only observed quantities that are finite without question. Second, the quantity α exhibits the extremely moderate growth in the whole interval of the values Q^2 available.

Using the expression (21.97), we can find the effective potential of the external electromagnetic field in the second order of the perturbation theory caused by vacuum polarization:

$$[A_\mu^e(q)]_{eff} = A_\mu^e(q) + D_{\mu\nu}^c(q)\Pi_C^{(2)\nu\sigma}(q^2)A_\sigma^e(q) =$$

$$= \left\{ 1 - \frac{\alpha}{\pi} \left[\frac{4m_R^2 + 2q^2}{3q^2}(1 - \vartheta\cot\vartheta) - \frac{1}{9} \right] \right\} A_\mu^e(q). \tag{21.106}$$

At $q^2 = 0$ we get the photon field, $\delta A_\mu^e(q)$ vanishing in this case. So, a single photon moving in a space is not inducing the current. It means that the vacuum polarization effects are absent for the electromagnetic waves free to propagate. However, this statement is valid only until the effects of the lowest order is being considered. In Fig. 21.10 we present one of the diagrams describing the scattering of two photons by each other as the result of the vacuum polarization effects in the forth order of the perturbation theory. In the limit of

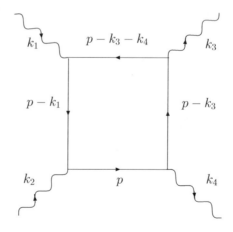

FIGURE 21.10
The Feynman diagram describing the scattering of two photons by each other.

small values of q^2 the expression (21.106) becomes the form:

$$[A_\mu^e(q)]_{eff} = \left[1 - \frac{\alpha}{15\pi m_R^2}q^2 \right) A_\mu^e(x)$$

or

$$[A_\mu^e)(x)]_{eff} = \left(1 + \frac{\alpha}{15\pi m_R^2}\Box \right) A_\mu^e(x). \tag{21.107}$$

As follows from (21.107), the vacuum polarization causes smearing of the potential over the region, the radius of which is of the order of the Compton wavelength of the electron. The physical pattern of the phenomenon is well known. Due to the interaction with an electron-positron field, the charge e becomes surrounded by a cloud of virtual electrons and positrons. Some part of them, possessing the total charge δe with the same sign as e, goes to infinity. And a part of the cloud with the total charge $-\delta e$ is concentrated around the initial charge in the region with the sizes $\lambda_c = \hbar/mc$. When a charge is observed from a distance exceeding λ_c, this charge seems to be effectively equal to the renormalized charge $e_R = e - \delta e$. At the same time, were the observations from distances considerably below λ_c possible, one could measure the bare charge e.

According to (21.107), in a hydrogen atom the effective potential for the electron, including only the vacuum polarization effects, is given by the expression:

$$-Ze^2 \left(\frac{1}{r} - \frac{4\alpha}{15m_R^2} \delta^{(3)}(\mathbf{r}) \right).$$

Uehling has been the first [48] to note that due to the vacuum polarization the electrostatic potential of a point charge in the vacuum is not described by the exact Coulomb law. Deviation from the Coulomb law has been calculated by Uehling soon after the appearance of the early Dirac [49] and Heisenberg [50] papers devoted to the vacuum polarization. It should be recorded that modifying the Coulomb law is caused not only by means of the PSEDs. The electron self-energy and vertex diagrams give contributions as well. We shall investigate this question in detail under consideration of the Lamb shift in Section 21.8.

At $q^2 = 0$ the polarization tensor $\Pi_{\mu\nu}(q^2)$ describes the photon self-energy. In the second order of the perturbation theory the photon self-energy appears as a result of the production of the virtual electron-positron pair and its subsequent annihilation (Fig. 21.11). The

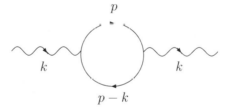

FIGURE 21.11
The photon self-energy diagram.

matrix element associated with the diagram shown in Fig. 21.11 is defined by the expression:

$$\mathcal{M} \sim e_\mu^\lambda(\mathbf{q})\Pi^{\mu\nu}(q^2)e_\nu^\lambda(\mathbf{q}) = e_\mu^\lambda(\mathbf{q})(q^\mu q^\nu - g^{\mu\nu}q^2)e_\nu^\lambda(\mathbf{q})\Pi(q^2) =$$

$$= -q^2 g_{\mu\nu}e_\nu^\lambda(\mathbf{q})e_\mu^\lambda(\mathbf{q})\Pi(q^2) = 0. \tag{21.108}$$

The vanishing of the photon self-energy is a consequence of the gauge invariance of the expression for $\Pi_{\mu\nu}$. This circumstance, in its turn, is connected with the fact that the photon mass is exactly equal to zero. Thus, owing to the gauge invariance there is no need to introduce the counter-term for the photon mass renormalization. One may relate the particle mass value with the pole position of the particle propagator. Let us define the exact photon propagator by an infinite plurality of diagrams displayed in Fig. 21.12. Then, the

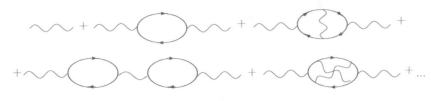

FIGURE 21.12

The Feynman diagrams defining the exact photon propagator.

zero-mass value of the photon follows from the statement: "The exact photon propagator has the pole at $q^2 = 0$."

The photon interacts not only with the electron-positron field, it interacts with fields of other particles as well. An interaction with a vacuum of these fields is described by the same self-energy diagrams whose internal lines are assigned to corresponding particles. In this case it appears that contributions coming from different particles are simply summarized [51]. Since in the second order of the perturbation theory the radiative effects caused by the vacuum polarization are inversely proportional to a mass square of a particle giving rise to a loop, then contributions from more light particles dominate. One important point to remember is that the QED allows us to calculate the contributions coming from the charged leptons only.

21.5   Vertex function of the third order

In the previous two sections we have considered the diagrams of the electron self-energy and the vacuum polarization in the second order of the perturbation theory. The divergencies corresponding to them exhaust all the possible ones of the second order scattering matrix $S^{(2)}$. Let us proceed now to the investigation of the elements matrix behavior in $S^{(3)}$. A part of divergent terms in $S^{(3)}$ incorporating the second order divergencies, such as the ones associated with the diagrams of Fig. 21.13, is compensated by the second order counter-terms simultaneously with removing the divergencies from $S^{(2)}$. By this reason, specific divergencies of the third order may be contained only in the terms of $S^{(3)}$ corresponding to the diagrams pictured in Fig. 21.14. On the strength of the Furry theorem the matrix elements appropriate to the diagrams presented in Figs. 21.14(a),(b) turn into zero. Thus, in the third order of the perturbation theory, we shall have to examine only the matrix element corresponding to the diagram shown in Fig. 21.14(c):

$$\mathcal{M} = -ie\sqrt{\frac{m^2}{E_{p_1}E_{p_2}}}\,\overline{u}(\mathbf{p}_2)\Lambda_\mu^{(3)}(p_1, p_2)u(\mathbf{p}_1)A^\mu(q), \qquad (21.109)$$

where $\Lambda_\mu^{(3)}(p_1, p_2)$ is the third-order vertex function:

$$\Lambda_\mu^{(3)}(p_1, p_2) = -\frac{ie^2}{(2\pi)^4}\int d^4k\gamma_\nu\frac{\hat{p}_2 - \hat{k} + m}{(p_2 - k)^2 - m^2}\gamma_\mu\frac{\hat{p}_1 - \hat{k} + m}{(p_1 - k)^2 - m^2}\gamma^\nu\frac{1}{k^2 - \lambda^2}, \qquad (21.110)$$

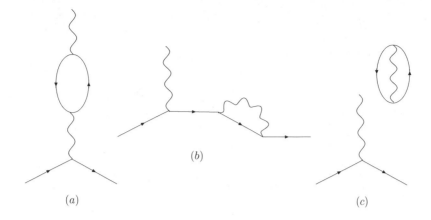

FIGURE 21.13

The Feynman diagrams (a)–(c) of the third order containing divergences.

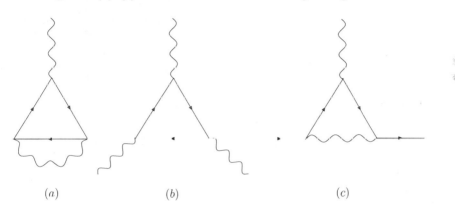

FIGURE 21.14

The vertex diagrams (a)–(c) of the third order.

and we have modified routinely the photon propagator to eliminate the infrared (IR) divergencies. It is convenient to present the expression (21.110) in the form:

$$\Lambda_\mu^{(3)}(p_1, p_2) = -\frac{ie^2}{(2\pi)^4}\{\gamma_\nu(\hat{p}_2 + m)\gamma_\mu(\hat{p}_1 + m)\gamma^\nu J_1 - [\gamma_\nu(\hat{p}_2 + m)\gamma_\mu\gamma^\sigma\gamma^\nu +$$

$$+ \gamma_\nu\gamma^\sigma\gamma_\mu(\hat{p}_1 + m)\gamma^\mu]J_\sigma + \gamma_\nu\gamma^\sigma\gamma_\mu\gamma^\tau\gamma^\nu J_{\sigma\tau}\}, \tag{21.111}$$

where

$$J(1; \sigma; \sigma\tau) = \int_\Sigma \frac{d^4k(1; \sigma; \sigma\tau)}{[k^2 - 2(p_1 k) + p_1^2 - m^2][k^2 - 2(p_2 k) + p_2^2 - m^2](k^2 - \lambda^2)} \tag{21.112}$$

in the parenthesis $(1; \sigma; \sigma\tau)$ one must maintain 1 for the integral J_1, k_σ—for the integral J_σ, and $k_\sigma k_\tau$—for the integral $J_{\sigma\tau}$. Among the three integrals only the third integral contains the UV-divergency (it is logarithmic). Therefore, under calculations of the integrals J_σ and $J_{\sigma\tau}$ one may set λ equal to zero. Further, we shall be constrained by the investigation of the case when p_1 and p_2 comprise the four-dimensional momenta of a free electron $(p_1^2 = p_2^2 = m_R^2)$.

Using the Feynman parametrization, we represent J_1, J_σ, and $J_{\sigma\tau}$ in the following form:

$$J_1 = 2 \int_0^1 dy \int_0^1 x dx \int_\Sigma \frac{d^4 k}{[k^2 - 2(p_x k) + l_x]^3}, \qquad (21.113)$$

$$J_{(\sigma;\sigma\tau)} = 2 \int_0^1 dy \int_0^1 x dx \int_\Sigma \frac{d^4 k(k_\sigma; k_\sigma k_\tau)}{[k^2 - 2(p_x k)]^3}, \qquad (21.114)$$

where

$$p_y = y p_1 + (1-y) p_2, \qquad p_x = x p_y, \qquad l_x = (x-1)\lambda^2.$$

Integrating the obtained expressions over k is accomplished by application of the formulae (21.27)–(21.29). Having done the integration over x, we come to the expressions:

$$J_1 = \frac{i\pi^2}{2} \int_0^1 \frac{dy}{p_y^2} \ln \frac{p_y^2}{\lambda^2}, \qquad (21.115)$$

$$J_\sigma = i\pi^2 \int_0^1 \frac{dy}{p_y^2} p_{y\sigma}, \qquad (21.116)$$

$$J_{\sigma\tau} = \frac{i\pi^2}{4} \int_0^1 dy \left\{ g_{\sigma\tau} \left[\ln \frac{L^2}{p_y^2} - \frac{1}{2} \right] - 2 \frac{p_{y\sigma} p_{y\tau}}{p_y^2} \right\}, \qquad (21.117)$$

where under calculation of J_1 we have neglected the term:

$$\frac{\lambda^2}{2 p_y^2} \int_0^1 \frac{dx}{x^2 p_y^2 - \lambda^2(x-1)}.$$

Let us express $q^2 = (p_2 - p_1)^2$ in terms of the angle ϑ in accordance with the relation:

$$q^2 = 4m^2 \sin^2 \vartheta. \qquad (21.118)$$

Introducing a new variable ξ instead of y

$$2y - 1 = \frac{\tan \xi}{\tan \vartheta}, \qquad (21.119)$$

we can rewrite p_y^2 and $p_{y\sigma}$ as follows:

$$p_y^2 = [p_1 + q(1-y)]^2 = m^2 - q^2 y(1-y) = m^2 [1 - 4y(1-y)\sin^2 \vartheta] =$$

$$= m^2 \cos^2 \vartheta \left(1 + \mathrm{tg}^2 \xi \right) = m^2 \frac{\cos^2 \vartheta}{\cos^2 \xi}, \qquad (21.120)$$

$$p_{y\sigma} = \frac{\tan \xi}{2 \tan \vartheta}(p_{1\sigma} - p_{2\sigma}) + \frac{1}{2}(p_{1\sigma} + p_{2\sigma}). \qquad (21.121)$$

Passing on to the new variable and integrating by parts, we get for J_1

$$J_1 = -\frac{2i\pi^2}{m^2 \sin 2\vartheta} \int_0^\vartheta \left[\ln \frac{\cos \xi}{\cos \vartheta} + \ln \frac{\lambda}{m} \right] d\xi = -\frac{2i\pi^2}{m^2 \sin 2\vartheta} \left[\int_0^\vartheta \xi \tan \xi d\xi + \vartheta \ln \frac{\lambda}{m} \right]. \qquad (21.122)$$

The similar mathematics lead to the following values of two remaining integrals:

$$J_\sigma = \frac{i\pi^2}{m^2} \frac{\vartheta}{\sin 2\vartheta}(p_{1\sigma} + p_{2\sigma}), \qquad (21.123)$$

$$J_{\sigma\tau} = \frac{i\pi^2}{2}\left[g_{\sigma\tau}\left(\ln\frac{L}{m} - \frac{1}{4}\right) - \frac{\vartheta(p_{1\sigma} + p_{2\sigma})(p_{1\tau} + p_{2\tau})}{2m^2\sin 2\vartheta} + \right.$$
$$\left. + \left(g_{\sigma\tau} - \frac{q_\sigma q_\tau}{q^2}\right)(1 - \vartheta\cot\vartheta)\right]. \tag{21.124}$$

The next step is to simplify the coefficients that accompany the integrals $J_1, J_\sigma, J_{\sigma\tau}$ in the expression for the vertex function. Since we calculate the vertex operator taken in plates between the bispinors $\overline{u}(\mathbf{p}_2)$ and $u(\mathbf{p}_1)$, then the progress in this direction should be expected under invoking the Dirac equation. Using the relations (19.63), we rewrite the coefficient standing in front of J_1 in the following view:

$$\gamma_\nu(\hat{p}_2 + m)\gamma_\mu(\hat{p}_1 + m)\gamma^\nu = -2\hat{p}_1\gamma_\mu\hat{p}_2 - 2m^2\gamma_\mu + 2m(\gamma_\mu\hat{p}_2 + \hat{p}_2\gamma_\mu + \gamma_\mu\hat{p}_1 + \hat{p}_1\gamma_\mu). \tag{21.125}$$

Next, taking into consideration the relations:

$$\hat{p}_1 = \hat{p}_2 - \hat{q}, \qquad \hat{p}_2 = \hat{p}_1 + \hat{q},$$

we arrive at

$$\overline{u}(\mathbf{p}_2)\{-2(\hat{p}_2 - \hat{q})\gamma_\mu(\hat{p}_1 + \hat{q}) - 2m^2\gamma_\mu + 2m[\gamma_\mu(\hat{p}_1 + \hat{q}) + \hat{p}_2\gamma_\mu + \gamma_\mu\hat{p}_1 +$$
$$+ (\hat{p}_2 - \hat{q})\gamma_\mu]\}u(\mathbf{p}_1) = \overline{u}(\mathbf{p}_2)(4m^2\gamma_\mu + 2\hat{q}\gamma_\mu\hat{q})u(\mathbf{p}_1) = \overline{u}(\mathbf{p}_2)(4m^2\gamma_\mu - 2q^2\gamma_\mu)u(\mathbf{p}_1), \tag{21.126}$$

where we have allowed for that

$$\overline{u}(\mathbf{p}_2)q^\nu q^\lambda\gamma_\nu\gamma_\mu\gamma_\lambda u(\mathbf{p}_1) = \frac{1}{2}\overline{u}(\mathbf{p}_2)q^\nu q^\lambda\left(\gamma_\nu\gamma_\mu\gamma_\lambda + \gamma_\lambda\gamma_\mu\gamma_\nu\right)u(\mathbf{p}_1) =$$

$$= \frac{1}{2}\overline{u}(\mathbf{p}_2)q^\nu q^\lambda\left(4g_{\mu\lambda}\gamma_\nu - 2g_{\nu\lambda}\gamma_\mu\right)u(\mathbf{p}_1) = -\overline{u}(\mathbf{p}_2)q^2\gamma_\mu u(\mathbf{p}_1).$$

The analogous mathematical manipulation allow one to present the factors standing before the integrals J_σ and $J_{\sigma\tau}$ as:

$$\gamma^\nu(\hat{p}_2 + m)\gamma_\mu\gamma_\sigma\gamma_\nu + \gamma^\nu\gamma_\sigma\gamma_\mu(\hat{p}_1 + m)\gamma_\nu = 4mg_{\sigma\mu} + 2(\hat{q}\gamma_\mu\gamma_\sigma + \gamma_\sigma\gamma_\mu\hat{q}), \tag{21.127}$$

$$\gamma^\nu\gamma_\sigma\gamma_\mu\gamma_\tau\gamma_\nu = -2\gamma_\tau\gamma_\mu\gamma_\sigma. \tag{21.128}$$

Gathering together the obtained results, we arrive at the following expression for the third-order vertex function:

$$\Lambda_\mu^{(3)}(p_1, p_2) = \frac{ie^2}{(2\pi)^4}\left\{(4m^2 - 2q^2)\gamma_\mu J_1 + [4im(p_1 + p_2)_\mu + 2(\hat{p}_1 + \hat{p}_2)\gamma_\mu\hat{q} - \right.$$

$$\left. -2\hat{q}\gamma_\mu(\hat{p}_1 + \hat{p}_2) - \frac{1}{2}(\hat{p}_1 + \hat{p}_2)\gamma_\mu(\hat{p}_1 + \hat{p}_2)\right]K_1 - 2\hat{q}\gamma_\mu\hat{q}K_2 - 4\gamma_\mu K_3\right\}, \tag{21.129}$$

where

$$K_1 = \frac{i\pi^2}{m^2}\frac{\vartheta}{\sin 2\vartheta}, \qquad K_2 = \frac{i\pi^2}{2q^2}(1 - \vartheta\cot\vartheta),$$

$$K_3 = \frac{i\pi^2}{2}\left(1 - \vartheta\cot\vartheta + \ln\frac{L}{m} - \frac{1}{4}\right).$$

Because we are dealing with the determination of the quantity:

$$\overline{u}(\mathbf{p}_2)\Lambda_\mu^{(3)}(p_1, p_2)u(\mathbf{p}_1),$$

then we may again use the Dirac equation in (21.129). Then, we gain the following expression for the vertex operator:

$$\Lambda_\mu^{(3)}(p_1, p_2) = -\frac{e^2 \gamma_\mu}{(2\pi)^2} \left[\frac{2\vartheta}{\tan 2\vartheta} \left(\ln \frac{m}{\lambda} - 1 \right) - \frac{2}{\tan 2\vartheta} \int_0^\vartheta \xi \tan \xi \, d\xi - \right.$$

$$\left. - \frac{\vartheta}{2} \tan \vartheta - \frac{1}{8} - \frac{1}{4} \ln \frac{L^2}{m^2} \right] - \frac{e^2}{32\pi^2 m} (\gamma_\mu \hat{q} - \hat{q} \gamma_\mu) \frac{2\vartheta}{\sin 2\vartheta}. \tag{21.130}$$

The derived expression is diverging logarithmically. Therefore, to extract the finite part $\Lambda_{C\mu}^{(3)}(p_1, p_2)$ it is sufficient to subtract only the first term of the expansion into the Taylor series in the vicinity of the point $q = 0$ from (21.130). Since

$$\Lambda_\mu^{(3)}(p_1, p_1) = \Lambda_\mu^{(3)}(p_1, p_2)|_{q=0} = \frac{e^2 \gamma_\mu}{(2\pi)^2} \left(\frac{1}{4} \ln \frac{L^2}{m^2} + \frac{9}{8} - \ln \frac{\lambda}{m} \right), \tag{21.131}$$

where $p_1^2 = p_2^2 = m_R^2$, then the finite value of $\Lambda_\mu^{(3)}(p_1, p_2)$ is given by the expression [7]:

$$\Lambda_{C\mu}^{(3)}(p_1, p_2) = \Lambda_\mu^{(3)}(p_1, p_2) - \Lambda_\mu^{(3)}(p_1, p_1) =$$

$$= -\frac{\alpha}{\pi} \left[(2\vartheta \cot 2\vartheta - 1) \left(\ln \frac{m_R}{\lambda} - 1 \right) - 2 \cot 2\vartheta \int_0^\vartheta \xi \tan \xi \, d\xi - \frac{\vartheta}{2} \tan \vartheta \right] \gamma_\mu -$$

$$- \frac{\alpha}{8\pi m_R} (\gamma_\mu \hat{q} - \hat{q} \gamma_\mu) \frac{2\vartheta}{\sin 2\vartheta}. \tag{21.132}$$

Attention is drawn to the fact that the dependence on the fictitious photon mass λ is conserved in $\Lambda_{C\mu}^{(3)}(q)$, and, as is easy to see, $\Lambda_{C\mu}^{(3)}(q)$ is diverging logarithmically when $\lambda \to 0$. As will be shown in Section 21.7 this divergency is fictitious and it disappears under the inclusion of the radiative corrections.

The formula (21.132) is true at $p_1^2 = p_2^2 = m_R^2$ and $q^2 \leq 4m_R^2$. When the values of the momentum q are small, the expression for the operator $\Lambda_{C\mu}^{(3)}(q)$ takes the simple form:

$$\Lambda_{C\mu}^{(3)}(q) = \frac{\alpha}{2\pi} \left[\frac{\hat{q} \gamma_\mu - \gamma_\mu \hat{q}}{4m_R} + \frac{2q^2}{3m_R^2} \left(\ln \frac{m_R}{\lambda} - \frac{3}{8} \right) \gamma_\mu \right]. \tag{21.133}$$

If $q^2 > 4m_R^2$, then the external potential can create real pairs and the denominators of Eqs. (21.113) and (21.114), containing the auxiliary variables x and y, have the poles. At the same time, these integrals are well defined because the additive to the mass $(i\varepsilon)$ gives the single-valued prescription to calculate them. At $p_1^2 \neq m_R^2$ and $p_2^2 \neq m_R^2$ the vertex operator was first obtained in Ref. [52]. For the case $p_{1,2}^2 \gg m_R^2$ and $|q^2| \gg m_R^2$ it has the form:

$$\Lambda_{C\mu}^{(3)}(q) = -\frac{\alpha}{\pi} \left[\ln \frac{|q^2|}{4m_R^2} \ln \frac{m_R}{\lambda} + \frac{1}{4} \ln^2 \frac{|q^2|}{4m_R^2} \right] \gamma_\mu. \tag{21.134}$$

21.6　Radiative corrections to scattering of an electron by the Coulomb nucleus field

So, we have studied the basic technical receptions of the calculations in higher orders of the perturbation theory. Possessing a stored arsenal of knowledge, we can rush to the storm of some particular process already not being constrained by the tree approximation. Let us deduce the cross section of the elastic scattering of the electron by the Coulomb potential with inclusion of the RC. In so doing, we pose the following goals: (i) to prove the UV-divergencies cancelation in a given order of the perturbation theory; (ii) to illustrate the technique of calculating the cross sections under inclusion of the RC; and (iii) to show, that the final result is free from the IR-divergencies and does not contain the fictitious photon mass.

The Feynman diagrams describing the electron scattering by the Coulomb nucleus field up to the third order of the perturbation theory inclusive are shown in Fig. 21.7. We call attention to the RC absence for external lines. It is caused by the definition of the regularized expressions for the mass electron operator $\Sigma_C(p)$ and the polarization tensor $\Pi_C^{\mu\nu}(q^2)$. Really, from Eqs. (21.75) and (21.97) follows

$$\frac{1}{\hat{p} - m_R}\Sigma_C(p)\bigg|_{\hat{p}=m_R} = \Sigma_C(p)\frac{1}{\hat{p} - m_R}\bigg|_{\hat{p}=m_R} = 0,$$

$$\frac{1}{q^2}\Pi_C^{\mu\nu}(q^2)\bigg|_{q^2=0} = 0,$$

ensuring the vanishing contribution from insertions of the self-energy diagrams into the electron and photon external lines.

The amplitude of the Coulomb scattering is given by the expression:

$$\mathcal{A} = -ie(2\pi)^{3/2}\frac{m}{E}\overline{u}(\mathbf{p}_2)\left[M^{a)}_{(1)} + M^{b)}_{(2)} + M^{c)}_{(3)} + M^{d_1)}_{(3)} + M^{d_2)}_{(3)} + M^{e_1)}_{(3)} + \right.$$

$$\left. + M^{e_2)}_{(3)} + M^{f_1)}_{(3)} + M^{f_2)}_{(3)}\right]u(\mathbf{p}_1), \tag{21.135}$$

where

$$M^{a)}_{(1)} = \gamma^\mu \Lambda^e_\mu(\mathbf{p}_2 \quad \mathbf{p}_1), \tag{21.136}$$

$$M^{b)}_{(2)} = -\frac{e}{(2\pi)^{3/2}}\int d^3k \gamma^{\nu} A^e_\mu(\mathbf{p}_2 - \mathbf{k})\frac{\hat{k} + m}{k^2 - m^2}\gamma^{\nu} A^e_\nu(\mathbf{p}_1 - \mathbf{k}), \tag{21.137}$$

$$M^{c)}_{(3)} = \Lambda^{(3)\mu}(p_2, p_1)A^e_\mu(\mathbf{p}_2 - \mathbf{p}_1), \tag{21.138}$$

$$M^{d_1)}_{(3)} = i\Sigma^{(2)}(p_2)\frac{1}{\hat{p}_2 - m}\gamma^\mu A^e_\mu(\mathbf{p}_2 - \mathbf{p}_1), \tag{21.139}$$

$$M^{d_2)}_{(3)} = -\delta m\frac{1}{\hat{p}_2 - m}\gamma^\mu A^e_\mu(\mathbf{p}_2 - \mathbf{p}_1), \tag{21.140}$$

$$M^{e_1)}_{(3)} = i\gamma^\mu A^e_\mu(\mathbf{p}_2 - \mathbf{p}_1)\frac{1}{\hat{p}_1 - m}\Sigma^{(2)}(p_1), \tag{21.141}$$

$$M^{e_2)}_{(3)} = -\delta m\gamma^\mu A^e_\mu(\mathbf{p}_2 - \mathbf{p}_1)\frac{1}{\hat{p}_1 - m}, \tag{21.142}$$

$$M^{f_1)}_{(3)} = -i\gamma_\nu\frac{1}{q^2}\Pi^{(2)\nu\mu}(q^2)A^e_\mu(\mathbf{p}_2 - \mathbf{p}_1), \tag{21.143}$$

$$M^{f_2)}_{(3)} = -\frac{i}{2}\delta e\gamma_\nu\frac{1}{q^2}[q^2 g^{\nu\mu} - q^\mu q^\nu]A^e_\mu(\mathbf{p}_2 - \mathbf{p}_1), \tag{21.144}$$

$q = p_2 - p_1$. We start our analysis with the addends (21.139) and (21.141). Since they contain the factor $(\hat{p}-m)^{-1}$, their denominators turn into zero when taken in plates between the bispinors $\overline{u}(\mathbf{p}_2)$ and $u(\mathbf{p}_1)$. This difficulty is connected with the limiting transition—the initial and final times are directed to $\pm\infty$ in the scattering matrix $S(t_f, t_i)$. To gain the finite result one should address to the adiabatic hypothesis, that is, one introduces the damped factors to be necessary for the right definition of the initial and final bare states [4,7]. Then the interaction Lagrangian (21.84) is replaced with the expression:

$$\mathcal{L}_{int} = -eg(t)\{N[\overline{\psi}(x)\gamma^\mu\psi(x)(A_\mu(x) + A_\mu^e(x))]\} + \delta m[g(t)]^2 N[\overline{\psi}(x)\psi(x)], \qquad (21.145)$$

where the damped function $g(t)$ is adiabatically switched on and switched off the interaction between fields and we have also allowed for that δm is caused by the second-order effect. It is assumed that the time, during which the function $g(t)$ is significantly changed, considerably exceeds the scattering process duration. When one denotes the Fourier image of the function $g(t)$ by $G(\Gamma_0)$, then

$$g(t) = \int_{-\infty}^{\infty} d\Gamma_0 G(\Gamma_0) e^{-i\Gamma_0 t} = \int_{-\infty}^{\infty} d\Gamma_0 G(\Gamma_0) e^{-i\Gamma x}, \qquad (21.146)$$

where $\Gamma_\mu = (\Gamma_0, 0, 0, 0)$. We also choose the normalization:

$$g(0) = \int_{-\infty}^{\infty} d\Gamma_0 G(\Gamma_0) = 1 \qquad (21.147)$$

and assume that the function $G(\Gamma_0)$ behaves in much the same way as the δ-function and essentially differs from zero only for the values of Γ_0 to equal T^{-1} in the order of magnitude.

When employing the interaction Lagrangian in the form (21.145), the matrix element (21.141) is substituted for the expression:

$$M_{(3)}^{e_1)} = i \int_{-\infty}^{\infty} d\Gamma_0 \int_{-\infty}^{\infty} d\Gamma_0' \gamma^\mu A_\mu^e(p_2 - p_1 - \Gamma - \Gamma') \frac{1}{\hat{p}_1 + \hat{\Gamma} + \hat{\Gamma}' - m} \times$$

$$\times \Sigma^{(2)}(p_1 + \Gamma)G(\Gamma_0)G(\Gamma_0'). \qquad (21.148)$$

When $T \to \infty$ and $\Gamma_0, \Gamma_0' \to 0$, then the integrand could be transformed. In a reconstruction process the expansion being valid for arbitrary operators A and B [7]:

$$\frac{1}{A+B} = \frac{1}{A} - \frac{1}{A}B\frac{1}{A} + \frac{1}{A}B\frac{1}{A}B\frac{1}{A} - \dots. \qquad (21.149)$$

proves to be useful. The validity of (21.149) could be proved through the iteration of the identity:

$$\frac{1}{A+B} = \frac{1}{A} - \frac{1}{A}B\frac{1}{A+B} = \frac{1}{A}(A+B)\frac{1}{A+B} - \frac{1}{A}B\frac{1}{A+B} = \frac{1}{A}(A+B-B)\frac{1}{A+B}.$$

Expanding the operator $\Sigma^{(2)}(p_1 + \Gamma)$ with the help of the relation (21.149) and keeping the terms of the first power in Γ only, we get

$$\Sigma^{(2)}(p_1 + \Gamma) = -\frac{\alpha}{4\pi^3} \int d^4k \gamma^\mu \frac{1}{\hat{p}_1 + \hat{k} + \hat{\Gamma} - m} \gamma_\mu \frac{1}{k^2 - \lambda^2} = \Sigma^{(2)}(p_1) +$$

$$+\frac{\alpha}{4\pi^3} \int d^4k \gamma^\mu \frac{1}{\hat{p}_1 + \hat{k} - m} \gamma^\nu \frac{1}{\hat{p}_1 + \hat{k} - m} \gamma_\mu \frac{1}{k^2 - \lambda^2} \Gamma_\nu = \Sigma^{(2)}(p_1) + i\Lambda^{(3)\nu}(p_1, p_1)\Gamma_\nu.$$

$$(21.150)$$

Thus, the matrix element $M_{(3)}^{e_1)}$ becomes the form:

$$M_{(3)}^{e_1)} = i \int_{-\infty}^{\infty} d\Gamma_0 \int_{-\infty}^{\infty} d\Gamma_0' \gamma^\mu A_\mu^e (p_2 - p_1 - \Gamma - \Gamma') \frac{1}{\hat{p}_1 + \hat{\Gamma} + \hat{\Gamma}' - m} \times$$

$$\times \left[\Sigma^{(2)}(p_1) + i\Lambda^{(3)\nu}(p_1, p_1)\Gamma_\nu \right] G(\Gamma_0)G(\Gamma_0'). \tag{21.151}$$

After the similar operations the matrix element $M_{(3)}^{e_2)}$ will look like:

$$M_{(3)}^{e_2)} = -\delta m \int_{-\infty}^{\infty} d\Gamma_0 \int_{-\infty}^{\infty} d\Gamma_0' \gamma^\mu A_\mu^e (p_2 - p_1 - \Gamma - \Gamma') \frac{1}{\hat{p}_1 + \hat{\Gamma} + \hat{\Gamma}' - m} G(\Gamma_0)G(\Gamma_0'). \tag{21.152}$$

Let us compare the expressions for the matrix elements $M_{(3)}^{e_1)}$ and $M_{(3)}^{e_2)}$. Since $\delta m = i\Sigma^{(2)}(p_1 = m_R^2)$, then the addition of $M_{(3)}^{c_1)}$ and $M_{(3)}^{e_2)}$ leads to the disappearance of the infinities associated with the mass renormalization:

$$M_{(3)}^{e_1)} + M_{(3)}^{e_2)} = - \int_{-\infty}^{\infty} d\Gamma_0 \int_{-\infty}^{\infty} d\Gamma_0' \gamma^\mu A_\mu^e (p_2 - p_1 - \Gamma - \Gamma') \frac{1}{\hat{p}_1 + \hat{\Gamma} + \hat{\Gamma}' - m} \times$$

$$\times \Lambda^{(3)\nu}(p_1, p_1)\Gamma_\nu G(\Gamma_0)G(\Gamma_0'). \tag{21.153}$$

Next, in order to symmetrize (21.153) we change Γ_ν with $(\Gamma_\nu + \Gamma_\nu')/2$. Since $M_{(3)}^{c_1)} + M_{(3)}^{c_2)}$ is being taken in plates between $u(\mathbf{p}_1)$ and $\bar{u}(\mathbf{p}_2)$, then the operator:

$$\frac{1}{2}(\gamma^\nu p_{1\nu} - m) \tag{21.154}$$

may be added to the expression:

$$\frac{1}{2}\gamma^\nu(\Gamma_\nu + \Gamma_\nu')$$

All this allows us to present the expression (21.153) in the following manner:

$$M_{(3)}^{e_1)} + M_{(3)}^{e_2)} = - \int_{-\infty}^{\infty} d\Gamma_0 \int_{-\infty}^{\infty} d\Gamma_0' \gamma^\mu A_\mu^e (p_2 - p_1 - \Gamma - \Gamma') \frac{1}{\hat{p}_1 + \hat{\Gamma} + \hat{\Gamma}' - m} \times$$

$$\times \Lambda^{(3)\nu}(p_1, p_1) \frac{1}{2}[\hat{\Gamma} + \hat{\Gamma}' + \hat{p}_1 - m]G(\Gamma_0)G(\Gamma_0') = -\frac{1}{2}\gamma^\mu A_\mu^e (p_2 - p_1)\Lambda^{(3)\nu}(p_1, p_1), \tag{21.155}$$

where in the final stage we have gone to the limit $\Gamma_0, \Gamma_0' \to 0$.

Applying the similar mathematical operations to the matrix elements (21.139) and (21.140), we obtain

$$M_{(3)}^{d_1)} + M_{(3)}^{d_2)} = -\frac{1}{2}\gamma^\mu A_\mu^e (p_2 - p_1)\Lambda^{(3)\nu}(p_1, p_1). \tag{21.156}$$

To summarize (21.155) and (21.156) with (21.138) results in

$$M_{(3)}^{e_1)} + M_{(3)}^{e_2)} + M_{(3)}^{d_1)} + M_{(3)}^{d_2)} + M_{(3)}^{c)} = A_\mu^e(\mathbf{p}_2 - \mathbf{p}_1)[\Lambda^{(3)\mu}(p_1, p_2) - \Lambda^{(3)\mu}(p_1, p_1)] =$$

$$= A_\mu^e(\mathbf{p}_2 - \mathbf{p}_1)\Lambda_C^{(3)\mu}(p_1, p_2), \tag{21.157}$$

that is, the mutual cancelation of the divergencies associated with the self-energy and vertex diagrams take place. Later we shall show that the similar cancelation occurs for any processes in all orders of the perturbation theory and is the consequence of the gauge invariance of the QED.

The addition of Eqs. (21.143), (21.144) and the inclusion of the relation $\delta e = 2i\Pi_D^{(2)}(q^2)$ also bring about the finite result:

$$M_{(3)}^{f_1)} + M_{(3)}^{f_2)} = -i\gamma_\nu \frac{1}{q^2}\Pi_C^{(2)\nu\mu}(q^2)A_\mu^e(\mathbf{p}_2 - \mathbf{p}_1). \qquad (21.158)$$

So, we have demonstrated that although each of the matrix elements corresponding to the diagrams displayed in Figs. 21.7(c)–(f) taken individually has the UV-divergencies, the sum of these matrix elements is finite and equals

$$\mathcal{M}_{(3)} = \sum_{j=c,d,e,f} \mathcal{M}_{(3)}^{j)} = -ie(2\pi)^{3/2}\frac{m}{E}\overline{u}(\mathbf{p}_2)[\Lambda_C^{(3)\mu}(p_1,p_2) - \frac{i\gamma_\nu}{q^2}\Pi_C^{(2)\nu\mu}(q^2)]A_\mu^e(\mathbf{p}_2 - \mathbf{p}_1)u(\mathbf{p}_1).$$
$$(21.159)$$

Now, we are coming to the determination of the cross section. According to the general rules it is defined by the expression:

$$d\sigma = \frac{2\pi|\mathbf{p}|E}{v}|\mathcal{M}_{(1)} + \mathcal{M}_{(2)} + \mathcal{M}_{(3)}|^2 d\Omega, \qquad (21.160)$$

where $|\mathbf{p}_1| = |\mathbf{p}_2| = |\mathbf{p}|$,

$$\mathcal{M}_{(1)} = -ie(2\pi)^{3/2}\frac{m}{E}\overline{u}(\mathbf{p}_2)A_\mu^e(\mathbf{p}_2 - \mathbf{p}_1)\gamma^\mu u(\mathbf{p}_1),$$

$$\mathcal{M}_{(2)} = -ie(2\pi)^{3/2}\frac{m}{E}\overline{u}(\mathbf{p}_2)M_{(2)}^{b)}u(\mathbf{p}_1).$$

As we have agreed to calculate the cross section including the third order diagrams, then we can obtain the complete picture considering the terms proportional to e^6 but no higher (thus, for example, with allowance made for the terms proportional to e^8 we would be forced to take into account the fourth-order diagrams). So, in the third order of the perturbation theory the cross section is presented by the sum:

$$d\sigma = d\sigma_{(1)} + d\sigma_{(2)} + d\sigma_{rad}, \qquad (21.161)$$

where $d\sigma_{(1)}$ is the cross section in the first Born approximation, and the quantity:

$$d\sigma_{(2)} = \frac{4\pi|\mathbf{p}|E}{v}d\Omega\,\mathrm{Re}\{\mathcal{M}_{(1)}^*\mathcal{M}_{(2)}\}, \qquad (21.162)$$

$$d\sigma_{rad} = \frac{4\pi|\mathbf{p}|E}{v}d\Omega\,\mathrm{Re}\{\mathcal{M}_{(1)}^*\mathcal{M}_{(3)}\} \qquad (21.163)$$

are the corrections to $d\sigma_{(1)}$. It is obvious that $d\sigma_{(2)}$ and $d\sigma_{rad}$ may be deduced independently from each other.

Let us consider the matrix element $M_{(2)}^{b)}$ determined by the expression (21.137). In the pure Coulomb field of the resting nucleus we have

$$A_0^e(r) = \frac{Z|e|}{4\pi r}.$$

The integral in Eq. (21.137) is logarithmically diverging for such a potential. This divergency is specific for the Coulomb field and is connected with the decrease slowness of this field at large distances. However, we have known the prescription that allows us to get rid of this difficulty. We must take into consideration the screening, that is, carry out the replacement:

$$\frac{Z|e|}{4\pi r} \rightarrow \frac{Z|e|}{4\pi r}e^{-\eta r}, \qquad (21.164)$$

where η is a positive constant, and pass on to the limit $\eta \to 0$ in the end of calculations.

In the case of the screened Coulomb potential the quantity $A_\mu^e(q)$ has the form:

$$A_\mu^e(q) = \frac{Z|e|g_{\mu 0}}{(2\pi)^{3/2}(\mathbf{q}^2 + \eta^2)}. \tag{21.165}$$

Making use of this expression, we find

$$\mathcal{M}_{(2)} = \frac{2iZ^2\alpha^2}{\pi}\frac{m}{E}\overline{u}(\mathbf{p}_2)[(\gamma^0 E + m)J_1 + (\boldsymbol{\gamma}\mathbf{J})]u(\mathbf{p}_1), \tag{21.166}$$

where we have introduced the designations:

$$J_1 = \int \frac{d^3k}{[(\mathbf{p}_2 - \mathbf{k})^2 + \eta^2][(\mathbf{p}_1 - \mathbf{k})^2 + \eta^2][\mathbf{p}_1^2 - \mathbf{k}^2 + i\epsilon]}, \tag{21.167}$$

$$\mathbf{J} = \int \frac{\mathbf{k}d^3k}{[(\mathbf{p}_2 - \mathbf{k})^2 + \eta^2][(\mathbf{p}_1 - \mathbf{k})^2 + \eta^2][\mathbf{p}_1^2 - \mathbf{k}^2 + i\epsilon]} = \frac{\mathbf{p}_1 + \mathbf{p}_2}{2}J_2 \tag{21.168}$$

and allowed for that so far as the integral $\mathbf{J}$ is symmetric with respect to $\mathbf{p}_1$ and $\mathbf{p}_2$, then it must be directed along $\mathbf{p}_1 + \mathbf{p}_2$ (the addend $i\epsilon$ has been included so as to remind the reader of the pole detour rule in the electron propagator). Eliminating the matrices γ by application of the Dirac equation

$$(\boldsymbol{\gamma}\mathbf{p})u(\mathbf{p}) = (\gamma^0 E - m)u(\mathbf{p}),$$

we rewrite (21.166) in the following form:

$$\mathcal{M}_{(2)} - \frac{2iZ^2\alpha^2}{\pi}\frac{m}{E}\overline{u}(\mathbf{p}_2)[\gamma^0 E(J_1 + J_2) + m(J_1 - J_2)]u(\mathbf{p}_1). \tag{21.169}$$

In order to calculate the integrals $J_{1,2}$ entering into (21.169), we shall employ the Feynman parametrization. With the help of Eq. (21.4) the integral J_1 is presented as:

$$J_1 = -2\int_0^1 d\xi_1 \int_0^1 d\xi_2 \int_0^1 d\xi_3 \times$$

$$\times \int \frac{d^3k\,\delta(1 - \xi_1 - \xi_2 - \xi_3)}{\{[(\mathbf{p}_2 - \mathbf{k})^2 + \eta^2]\xi_1 + [(\mathbf{p}_1 - \mathbf{k})^2 + \eta^2]\xi_2 + [\mathbf{k}^2 - \mathbf{p}_1^2 - i\epsilon]\xi_3\}^3} =$$

$$= -2\int_0^1 d\xi_1 \int_0^{1-\xi_1} d\xi_2 \times$$

$$\times \int \frac{d^3k}{[\eta^2(\xi_1 + \xi_2) + \mathbf{p}_1^2(2\xi_1 + 2\xi_2 - 1) - 2\mathbf{k}(\xi_1\mathbf{p}_2 + \xi_2\mathbf{p}_1) + \mathbf{k}^2 - i\epsilon]^3}. \tag{21.170}$$

Having introduced a new variable $\mathbf{f} = \mathbf{k} - \xi_1\mathbf{p}_2 - \xi_2\mathbf{p}_1$, we can take without trouble the three-dimensional integral:

$$\int \frac{d^3f}{(\mathbf{f}^2 - a^2 - i\epsilon)^3} = i\frac{\pi^2}{4a^3}$$

and arrive at the following value for J_1:

$$J_1 = -\frac{i\pi^2}{2}\int_0^1 d\xi_1 \int_0^{1-\xi_1} d\xi_2 \times$$

$$\times \frac{1}{[\mathbf{p}_1^2(\xi_1^2 + \xi_2^2 - 2\xi_1 - 2\xi_2 + 1) + 2\xi_1\xi_2\mathbf{p}_1\mathbf{p}_2 - \eta^2(\xi_1 + \xi_2) - i\epsilon]^{3/2}}. \qquad (21.171)$$

Next, we pass on to new variables:

$$x = \xi_1 + \xi_2, \qquad y = \xi_1 - \xi_2.$$

The integration over y from 0 to x is easily fulfilled and gives the result:

$$J_1 = -\frac{i\pi^2}{2|\mathbf{p}|^3} \int_0^1 \frac{x\,dx}{[bx^2 - 2x + 1 - \eta^2 x|\mathbf{p}|^{-2} - i\epsilon][(1-x)^2 - \eta^2 x|\mathbf{p}|^{-2} - i\epsilon]^{1/2}}, \qquad (21.172)$$

where

$$b = \frac{\mathbf{p}_1^2 + \mathbf{p}_1\mathbf{p}_2}{2\mathbf{p}_1^2} = \cos^2\frac{\vartheta}{2}.$$

To work out the integral over x at $\eta \to 0$ we divide the integration region into two parts:

$$\int_0^1 ...dx = \int_0^{1-\eta_1} ...dx + \int_{1-\eta_1}^1 ...dx,$$

where $\eta/|\mathbf{p}| \ll \eta_1 \ll 1$. Having set η equal to 0 in the first integral, we get

$$\int_0^{1-\eta_1} ...dx = \frac{1}{2(1-b)} \ln \frac{(1-x)^2}{(bx^2 - 2x + 1 - i\epsilon)}\Big|_0^{1-\eta_1} = \frac{1}{2(1-b)}\left[\ln\frac{\eta_1^2}{1-b} + i\pi\right]. \qquad (21.173)$$

Note, that the pole detour rule to be specified by the additive $-i\epsilon$ allowed us to define changing the logarithm argument under transition from 0 to $1 - \eta_1$. In the second integral one may set $x = 1$ everywhere aside from the addend $(1-x)^2$ and also put $\eta = 0$ in the first square brackets of the denominator. Then, taking into account the pole detour rule again, we obtain

$$\int_{1-\eta_1}^1 ...dx = -\frac{1}{1-b}\int_0^{\eta_1} \frac{dx'}{(x'^2 - \eta^2|\mathbf{p}|^{-2} - i\epsilon)^{1/2}} = -\frac{1}{1-b}\left[i\int_0^{\eta/|\mathbf{p}|} \frac{dx'}{(\eta^2|\mathbf{p}|^{-2} - x'^2)^{1/2}} + \right.$$

$$\left. + \int_{\eta/|\mathbf{p}|}^{\eta_1} \frac{dx'}{(x'^2 - \eta^2|\mathbf{p}|^{-2})^{1/2}}\right] = -\frac{1}{1-b}\left[\frac{i\pi}{2} + \ln\frac{2|\mathbf{p}|\eta_1}{\eta}\right]. \qquad (21.174)$$

When summarizing the both integrals, the auxiliary quantity η_1 falls out and we get

$$J_1 = \frac{i\pi^2}{2|\mathbf{p}|^3 \sin^2(\vartheta/2)} \ln\left[\frac{2|\mathbf{p}|\sin(\vartheta/2)}{\eta}\right]. \qquad (21.175)$$

The analogous mathematical manipulations yield the following expression for the integral J_2:

$$J_2 = J_1 - \frac{\pi^3[1 - \sin(\vartheta/2)]}{4|\mathbf{p}|^3 \cos^2(\vartheta/2)\sin(\vartheta/2)} - \frac{i\pi^2}{2|\mathbf{p}|^3 \cos^2(\vartheta/2)} \ln[\sin(\vartheta/2)]. \qquad (21.176)$$

It should be pointed out that the real parts of the deduced integrals do not depend on the parameter η:

$$\text{Re } J_1 = 0, \qquad \text{Re } J_2 = \frac{\pi^3[1 - \sin(\vartheta/2)]}{4|\mathbf{p}|^3 \cos^2(\vartheta/2)\sin(\vartheta/2)}. \qquad (21.177)$$

After all these preparations we can find $d\sigma_{(2)}$. Assuming the initial and final electrons unpolarized, we arrive at [53]:

$$d\sigma_{(2)} = \frac{\pi(Z\alpha)^3 E}{4|\mathbf{p}|^3 \sin^3(\vartheta/2)}[1 - \sin(\vartheta/2)]\,d\Omega. \qquad (21.178)$$

The screening constant does not enter into the final result. This is the consequence of the fact that at the intermediate stages of calculations we have repeatedly used the smallness of η. Inasmuch as the transition from the case of the electron scattering to that of the positron scattering is realized by means of the replacement $Z \to -Z$, then the similar expression for the positron has the opposite sign. Recall that in the first Born approximation the cross sections for the electron and the positron coincide.

We are coming now to the definition of $d\sigma_{rad}$. Employing Eqs. (21.96) and (21.132), one may find the matrix element $\mathcal{M}_{(3)}$. However, in so doing, we must analytically continue the expressions for $\Lambda_{C\mu}^{(3)}(p_1, p_2)$ and $\Pi_C^{(2)}(q^2)$ into a region of the negative values of q^2. Putting $\vartheta/2 \to i\Phi$, where the angle Φ is now connected with the scattering angle by the relation:

$$-q^2 = 4\mathbf{p}^2 \sin^2 \frac{\theta}{2} = 4m^2 \sinh^2 \Phi, \qquad (21.179)$$

in Eqs. (21.96), (21.132) and allowing for (21.165) (here the screening constant may be set equal to zero), we find

$$\mathcal{M}_{(3)} = \frac{iZe^2\alpha}{\pi|\mathbf{q}|^2} \frac{m}{E} \overline{u}(\mathbf{p}_2)\gamma_0 \left\{ \left[(1 - 2\Phi \coth 2\Phi) \left(\ln \frac{m_R}{\lambda} - 1 \right) + \right. \right.$$

$$+ 2\coth 2\Phi \int_0^\Phi (u \tanh u)du + \frac{1}{2}\Phi \tanh \Phi \right] - \frac{\hat{q}}{2m_R} \frac{1\Phi}{\sinh 2\Phi} -$$

$$\left. - \left[\left(1 - \frac{1}{3}\coth^2 \Phi \right)(1 - \Phi \coth \Phi) - \frac{1}{9} \right] \right\} u(\mathbf{p}_1). \qquad (21.180)$$

Further, we substitute $\mathcal{M}_{(3)}$, $\mathcal{M}_{(1)}$ into (21.163) and carry out the summation and average over the electron spin states. As a result we derive [54]

$$d\sigma_{rad} - -\frac{Z^2\alpha^3 E}{4\pi|\mathbf{p}|^3 \sin^4(\theta/2)}[1 - v^2 \sin^2(\theta/2)] \left[2(1 - \Phi \coth \Phi)\left(1 - \frac{1}{3}\coth^2 \Phi\right) - \right.$$

$$- \frac{2}{9} + \Phi \tanh \Phi + 2(1 - 2\Phi \coth 2\Phi)\left(1 + \ln \frac{\lambda}{m_R}\right) + 4\coth 2\Phi \int_0^\Phi u \tanh u \, du +$$

$$\left. + \frac{2\Phi v^2 \sin^2(\theta/2)}{\sinh 2\Phi[1 - v^2 \sin^2(\theta/2)]} \right]. \qquad (21.181)$$

The last thing that remains to be done is to get rid of the fictitious photon mass λ. As follows from Section 20.2 we must be considering the pure elastic scattering along with the scattering with emitting a soft photon whose energy ΔE is significantly less than that of the electron. We add the cross section accompanied by emitting a soft photon (see Eq. (20.41)) to the cross sections sum $d\sigma_{(1)} + d\sigma_{(2)} + d\sigma_{rad}$ and take into account that the integral entering into (21.181) may be converted to the form:

$$\int_0^\Phi u \tanh u \, du = \frac{(1 - v^2)\sinh 2\Phi}{4\sin(\theta/2)} \int_{\cos(\theta/2)}^1 \frac{\xi \ln(1 - v^2\xi^2)d\xi}{(1 - v^2\xi^2)\sqrt{\xi^2 - \cos^2(\theta/2)}} +$$

$$+ \frac{\Phi}{2}\ln \frac{1}{1 - v^2}. \qquad (21.182)$$

Then, as will readily be observed, the addends containing $\ln \lambda$ are canceled and we come to the final formula for the cross section of the electron scattering by the Coulomb nucleus field in the e^6 approximation [55]:

$$\frac{d\sigma}{d\Omega} \bigg/ \left(\frac{d\sigma}{d\Omega}\right)_M = [1 - \delta(E, \theta, \Delta E)], \qquad (21.183)$$

where $\delta(E, \theta, \Delta E) = \delta_{rad} - \delta_{(2)}$, $(d\sigma/d\Omega)_M$ is the Mott effective cross section,

$$\left(\frac{d\sigma}{d\Omega}\right)_M = \frac{Z^2\alpha^2}{4|\mathbf{p}|^2 v^2 \sin^4(\theta/2)}[1 - v^2\sin^2(\theta/2)]$$

and the quantities $\delta_{(2)}$ and δ_{rad} are determined by the expressions:

$$\delta_{(2)} = \pi Z\alpha v \sin(\theta/2)\frac{1 - \sin(\theta/2)}{1 - v^2\sin^2(\theta/2)}, \tag{21.184}$$

$$\delta_{rad} = \frac{\alpha}{\pi}\left\{2(1 - 2\Phi\coth 2\Phi)\left(1 + \ln\frac{2\Delta E}{m_R}\right) + \Phi\tanh\Phi + 2(1 - \Phi\coth\Phi) \times\right.$$

$$\times\left(1 - \frac{1}{3}\coth^2\Phi\right) - \frac{2}{9} + \frac{1}{v}\ln\frac{1-v}{1+v} + 2\Phi\coth 2\Phi\ln\frac{1}{1-v^2} +$$

$$+\frac{2\Phi v^2\sin^2(\theta/2)}{\sinh 2\Phi[1 - v^2\sin^2(\theta/2)]} + \frac{(1-v^2)\cosh 2\Phi}{v\sin(\theta/2)}\int_{\cos(\theta/2)}^1\left[\frac{\ln(1+v\xi)}{1-v\xi} - \right.$$

$$\left.\left. - \frac{\ln(1-v\xi)}{1+v\xi}\right]\frac{d\xi}{\sqrt{\xi^2 - \cos^2(\theta/2)}}\right\}. \tag{21.185}$$

In a similar way the photon mass is removed in the cross section to be calculated in more higher orders of the perturbation theory. So, considering any scattering process (as complex as is wished), being the effect of the nth order of the perturbation theory, along with its RC that are associated with the $(n + 2)$th order of the perturbation theory, we must take into account the process of emitting a real soft photon as well. And in this case the summary cross section does not include the photon mass [56]. From Eqs. (21.184) and (21.185) it is evident that in the nonrelativistic region the quantities $\delta_{(2)}$ and δ_{rad} tend to zero when $v \to 0$. Of course one may use the formulae (21.184) and (21.185) only in the case when the quantities $\delta_{(2)}$ and δ_{rad} are rather small. In particular, q^2 must not be very large and the nucleus must be light enough (the second Born approximation gives the right results at $Z < 10$—15 only). From the formula (21.185) it also follows that with decreasing ΔE, that is, with improving the energy detector resolution, the quantity δ_{rad} becomes large and, as a result, the expression (21.183) breaks down. Under these circumstances, the RC of higher orders become important. In Ref. [57] it was shown that the cross section may be represented in the form (21.183) under inclusion of the RC of a more higher order as well. The expression standing in the right-hand side of (21.183) resembles expanding the exponent into the series. By this reason J. Schwinger made the assumption [58] that, in the most general case, the formula:

$$\frac{d\sigma}{d\Omega}\left/\left(\frac{d\sigma}{d\Omega}\right)_M\right. = e^{-\delta(E,\theta,\Delta E)}. \tag{21.186}$$

takes place. Some time later it was demonstrated [59] that under small values of ΔE ($\Delta E \ll E$) the Schwinger assumption is asymptotically valid. In order to gain some insight into the effect of the RC, in Table 21.1 we display the $\delta_{(2)}$ and δ_{rad} values expressed in percent versus E, Z, θ, and ΔE. We refer the reader, who wishes to gain more detailed information concerning the RC calculation in the QED, to the excellent paper by Brawn and Feynman [60] devoted to computing the cross section of the Compton scattering in the e^6 approximation.

TABLE 21.1

The $\delta_{(2)}$ and δ_{rad} values versus E, Z, θ, and ΔE.

E	2.5 MeV			4 MeV			9.5 MeV		
θ	45^0	90^0	135^0	45^0	90^0	135^0	45^0	90^0	135^0
$\delta_{(2)}/Z$	0.61	0.89	0.86	0.62	0.92	0.99	0.63	0.94	1.07
δ_{rad} at ΔE=10 kEV	4.8	7.4	8.7	6.9	9.9	11.3	12.4	15.9	17.5
δ_{rad} at ΔE=25 keV	3.9	6.0	7.1	5.7	8.1	9.3	10.5	13.5	14.8
δ_{rad} at ΔE=50 keV	3.2	5.0	5.9	4.7	6.8	7.9	9.0	11.7	12.8

21.7 Anomalous magnetic moment of leptons

It is known that when a nonrelativistic particle with the magnetic moment is placed in an external magnetic field, then it acquires an additional energy $E_{int} = -(\boldsymbol{\mu}\mathbf{H})$. Therefore, for the intrinsic magnetic moment of a particle to be measured, one must investigate its interaction with a constant or slowly varying electromagnetic field and extract in the interaction energy a term proportional to the magnetic field.

We now turn to the electron scattering by an external field taking into account the diagrams up to the third-order inclusive (Fig. 21.8). In this case one may understand the electron scattering as that by the effective potential $(A^e_\mu)_{eff}$. Assuming this field weak, we can be limited by an approximation linear in $(A^e_\mu)_{eff}$. Then, the scattering amplitude including the contributions coming from the diagrams a, b, d, e, and f pictured in Fig. 21.8 is given by the expression:

$$\mathcal{A} = -ie(2\pi)^{3/2}\sqrt{\frac{m^2}{E_{p_1}E_{p_2}}}\overline{u}(\mathbf{p}_2)\Lambda^\mu(q)u(\mathbf{p}_1)A^e_\mu(p_2 - p_1), \tag{21.187}$$

where

$$\Lambda^\mu(q) = [\gamma^\mu + \gamma^\nu D^c_{\nu\tau}(p_2 - p_1)\Pi^{(2)\tau\mu}_C(p_2 - p_1) + \Lambda^{(3)\mu}_C(p_2, p_1)]. \tag{21.188}$$

It should be emphasized that, in accordance with Eqs. (21.96) and (21.132) at small values of q,[*] only the quantity $\Lambda^{(3)\mu}_C(p_2, p_1)$ contains addends linear with respect to q.

The most general structure of the matrix element that follows from the relativistic and gauge invariance (see, for example, deducing the formula (10.53)) is as follows:

$$\mathcal{A} = -ie(2\pi)^{3/2}\sqrt{\frac{m^2}{E_{p_1}E_{p_2}}}\overline{u}(\mathbf{p}_2)\left[F_E(q^2)\gamma^\mu + \frac{i}{2m}F_M(q^2)\sigma^{\nu\mu}q_\nu\right]u(\mathbf{p}_1)A^e_\mu(p_2 - p_1). \tag{21.189}$$

Thus, the electron vertex function Λ_ν is represented in terms of two form factors $F_{E,M}$ to describe electromagnetic properties of the electron. If we examine the vertex in the first order of the perturbation theory, then

$$F^0_E(q^2) = 1, \qquad F^0_M(q^2) = 0. \tag{21.190}$$

It means that deviations of the form factors $F_{E,M}(q^2)$ from $F^0_{E,M}(q^2)$ should be interpreted as electromagnetic corrections of higher orders. To give the physical explanation of these corrections we address the nonrelativistic approximation.

---

[*]Small values of q correspond to a slowly varying field.

In the lowest order of the perturbation theory the relation:

$$e \lim_{p_1, p_2 \to 0} \left[< p_2 | \int d^3 x \overline{\psi}(x) \Lambda_\mu(x) \psi(x) | p_1 > \right] = \lim_{p_1, p_2 \to 0} \left[< p_2 | \int d^3 x j_\mu(x) | p_1 > \right] \quad (21.191)$$

takes place. Obviously, with including higher orders of the perturbation theory, the left-hand side of (21.191) becomes equal to

$$e \overline{u}(0) \Lambda_\mu(0) u(0) = e \overline{u}(0) \gamma_\mu F_E(0) u(0). \quad (21.192)$$

Next, taking into account the definition of the total measured charge e_{exp}:

$$e_{exp} = \lim_{p_1, p_2 \to 0} \left[< p_2 | \int d^3 x j_0(x) | p_1 > \right] \quad (21.193)$$

and allowing for that $\overline{u}(0) \boldsymbol{\gamma} u(0) = 0$, we arrive at

$$F_E(0) = \frac{e_{exp}}{e}. \quad (21.194)$$

Due to Eq. (21.194) the quantity $F_E(q^2)$ is sometimes called the *charge form factor*. However, this title is not absolutely exact because at $q^2 \neq 0$ the magnetic moment is also connected with the matrix element $e \overline{u}(\mathbf{p}_2) \gamma_\mu u(\mathbf{p}_1)$. To make sure that it is the case we again access to the nonrelativistic approximation. Expand the matrix element of the current into the series in powers of v/c and, then, multiply every term of the series by the potential of the external electromagnetic field:

$$e \mathbf{A} \overline{u}(\mathbf{p}_2) \boldsymbol{\gamma} u(\mathbf{p}_1) \approx -i e \overline{\varphi}(\mathbf{p}_2) \left\{ \frac{\mathbf{A}(\mathbf{q}) \cdot (\mathbf{p}_1 + \mathbf{p}_2)}{2m} + i \frac{\boldsymbol{\sigma} \cdot [\mathbf{A} \times \mathbf{q}]}{2m} \right\} \varphi(\mathbf{p}_1), \quad (21.195)$$

where $\mathbf{q} = \mathbf{p}_2 - \mathbf{p}_1$, $\varphi(\mathbf{p})$ is the big component of the bispinor $u(\mathbf{p})$ and we have accepted the three-dimensional gauge of the potential

$$(\mathbf{A}(\mathbf{q}) \cdot \mathbf{q}) = 0.$$

Evidently, the first addend in (21.195) corresponds to the linkage of the current $e\mathbf{p}/m$ with the potential $\mathbf{A}$ while the second one governs the linkage of the electron with the external magnetic field $\mathcal{H}$

$$-i e \overline{\varphi}(\mathbf{p}_2) \frac{\boldsymbol{\sigma} \cdot [\mathbf{A} \times \mathbf{q}]}{2m} \varphi(\mathbf{p}_1) = \mu_0 \overline{\varphi}(\mathbf{p}_2) \boldsymbol{\sigma} \cdot \mathcal{H} \varphi(\mathbf{p}_1),$$

where $\mu_0 = e/(2m)$. In the Dirac theory the electron automatically has the "normal" magnetic moment $e g_e / 2m$ with $g_e = 2$. Thus, the title "Dirac form factor" is best suited to the form factor $F_E(q^2)$.

The form factor $F_M(q^2)$ may be named for Pauli. Really, by virtue of the relation:

$$\frac{1}{2} \sigma^{\mu\nu} F_{\mu\nu} = -[(\boldsymbol{\Sigma} \cdot \mathcal{H}) - i(\boldsymbol{\alpha} \cdot \boldsymbol{\mathcal{E}})],$$

one may state that the RC make the electron behave in such a way as if it possesses distribution of an anomalous dipole (magnetic and electric) moment to be characterized by $F_M(q^2)$. So, $F_M(0)$ measures contribution to the (intrinsic) magnetic moment of the electron caused by electromagnetic corrections of higher orders. In the Bohr magneton units the total (intrinsic) magnetic moment of the electron μ_e may be written as

$$\mu_e = \left[1 + \frac{(g-2)_e}{2} \right] \mu_B, \quad (21.196)$$

where 1 standing in the parenthesis results from the term $F_E^0(q^2)\gamma_\nu$ in Eq. (21.189), and $a_e = (g-2)_e/2$ is an anomalous magnetic moment (AMM).

Inserting (21.132) and (21.96) into (21.188), we find

$$F_E(q^2) = 1 + \frac{\alpha}{\pi}\left[(1 - 2\vartheta\cot 2\vartheta)\left(\ln\frac{m_R}{\lambda} - 1\right) + 2\vartheta\cot 2\vartheta\int_0^\vartheta \xi\tan\xi d\xi + \frac{\vartheta}{2}\tan\vartheta\right] -$$

$$-\left[\frac{4m_R^2 + 2q^2}{3q^2}(1 - \vartheta\cot\vartheta) - \frac{1}{9}\right], \tag{21.197}$$

$$F_M(q^2) = \frac{\alpha}{2\pi}\frac{2\vartheta}{\sin 2\vartheta}. \tag{21.198}$$

Since

$$F_M(q^2)|_{q^2=0} = \frac{\alpha}{2\pi},$$

then we can conclude that owing to the RC in the second order of the perturbation theory the electron placed in a slowly varying external electromagnetic field behaves as a particle having the magnetic moment

$$\mu_e = \left(1 + \frac{\alpha}{2\pi}\right)\mu_B, \tag{21.199}$$

that is,

$$\left[\frac{(g-2)_e}{2}\right]_{theor} - \frac{\alpha}{2\pi}. \tag{21.200}$$

We call attention to the fact that the AMM is directly related to the vertex function. By this reason, computing corrections to the magnetic moment it is sufficient to consider the vertex diagrams only. For example, in order to compute the electron AMM with a precision of terms of the order of α^3, we must deal with diagrams shown in Fig. 21.15. Calculations lead to the result [61]:

$$\frac{(g-2)_e}{2} = \frac{\alpha}{2\pi} - 0.328\left(\frac{\alpha}{\pi}\right)^2. \tag{21.201}$$

At present, within the QED there are the calculations of the electron AMM in higher orders of the perturbation theory

$$(a_e)^{QED} = \frac{(g-2)_e}{2} - \sum_n a_n\left(\frac{\alpha}{\pi}\right)^n. \tag{21.202}$$

For example, for $n = 4$ the result is [62]:

$$(a_e)^{QED} = \frac{\alpha}{2\pi} - 0.32847844400\left(\frac{\alpha}{\pi}\right)^2 + 1.181234017\left(\frac{\alpha}{\pi}\right)^3 - 1.5098(384)\left(\frac{\alpha}{\pi}\right)^4. \tag{21.203}$$

Emphasize a promptly increasing complexity of calculations under transition to more higher orders of the perturbation theory. With computing a_3 (the contribution is accounted for of sixth-order diagrams) the diagrams number is equal 72. In the standard model the contributions to the AMM coming from the electroweak sector a_e^{EW} and those caused by corrections to the photon propagator from virtual hadrons a_e^{had} have also been calculated [62]:

$$a_e^{EW} + a_e^{had} = 1.66(3) \times 10^{-12}. \tag{21.204}$$

Now we address to experiments on observing the electron AMM. Nafe, Nelson, and Rabi were the first [63] who detected the AMM existence investigating the hyperfine splitting of the energy levels for the atomic hydrogen and deuterium. The idea of these experiments is as follows.

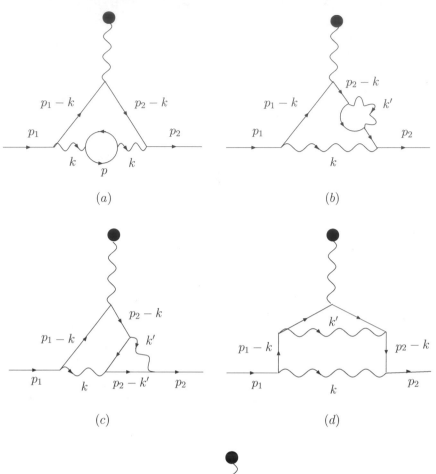

(a)   (b)

(c)   (d)

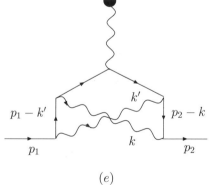

(e)

FIGURE 21.15
The Feynman diagrams (a)–(e) contributing to the electron AMM.

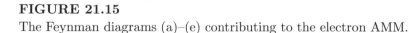

The hyperfine energy levels structure of atoms and molecules is due both to the interaction of a magnetic nucleus field with a magnetic field created by electrons and to that of a quadrupole nucleus moment with a nonhomogeneous intra-atomic electric field. In atoms and ions the magnetic interaction plays the main role. Let us find the corresponding energy for the hydrogen atom being constrained by the nonrelativistic case [64].

If the hydrogen atom nucleus has the magnetic moment $\boldsymbol{\mu}_p = \mu_p \boldsymbol{\sigma}_p$, where $\boldsymbol{\sigma}_p$ are the Pauli matrices acting upon the nucleus wavefunctions, then it produces the magnetic field:

$$\mathbf{A} = \text{rot } \frac{\mu_p \boldsymbol{\sigma}_p}{r}, \qquad \boldsymbol{\mathcal{H}} = \text{rot } \mathbf{A}.$$

The magnetic field of the nucleus must act on the electron magnetic moment $\boldsymbol{\mu}_e = -\mu_e \boldsymbol{\sigma}_e$ ($\boldsymbol{\sigma}_e$ has an effect upon the electron wavefunctions), with the result that an additional interaction between the nucleus and the electron appears. This interaction leads to the hyperfine structure of the energy levels:

$$H_{int} = -(\boldsymbol{\mu}_e \boldsymbol{\mathcal{H}}) = \mu_e \mu_p \left(\boldsymbol{\sigma}_e \text{rot rot } \frac{\boldsymbol{\sigma}_p}{r} \right) =$$

$$= \mu_e \mu_p \left[(\boldsymbol{\sigma}_e \nabla)(\boldsymbol{\sigma}_p \nabla) \frac{1}{r} - (\boldsymbol{\sigma}_e \boldsymbol{\sigma}_p) \nabla^2 \frac{1}{r} \right]. \tag{21.205}$$

In the first approximation one may assume that there are no chosen directions, and, hence, the relations:

$$(\boldsymbol{\sigma}_e \nabla)(\boldsymbol{\sigma}_p \nabla) = \frac{1}{3}(\boldsymbol{\sigma}_e \boldsymbol{\sigma}_p) \Delta, \tag{21.206}$$

$$\Delta \frac{1}{r} = -4\pi \delta(\mathbf{r}). \tag{21.207}$$

take place. Then, we get

$$H_{int} = \frac{8\pi}{3} \mu_e \mu_p (\boldsymbol{\sigma}_e \boldsymbol{\sigma}_p) \delta(\mathbf{r}). \tag{21.208}$$

From (21.208) it follows that in the first approximation the interaction of the magnetic moments has an impact on the S-state only.

To find $(\boldsymbol{\sigma}_e \boldsymbol{\sigma}_p)$ we make use of the eigenvalues equation for the operator of the total spin:

$$\frac{1}{4} \hbar^2 (\boldsymbol{\sigma}_e + \boldsymbol{\sigma}_p)^2 = \hbar^2 I(I+1), \tag{21.209}$$

where I is an eigenvalue of the hydrogen atom spin ($I - 0, 1$). Since the left-hand side of the expression (21.209) may be rewritten in the view:

$$\frac{1}{4} \hbar^2 [\sigma_e^2 + \sigma_p^2 + 2(\boldsymbol{\sigma}_e \boldsymbol{\sigma}_p)] = \frac{1}{4} \hbar^2 [6 + 2(\boldsymbol{\sigma}_e \boldsymbol{\sigma}_p)], \tag{21.210}$$

we get

$$(\boldsymbol{\sigma}_e \boldsymbol{\sigma}_p) = 2I(I+1) - 3. \tag{21.211}$$

Allowing for Eqs. (21.208) and (21.211) we could determine the average of the operator H_{int}

$$\Delta E_I = \int d^3 x \psi^\dagger(\mathbf{r}) H_{int} \psi(\mathbf{r}) = \frac{8\pi}{3} \mu_p \mu_e |\psi(0)|^2 [2I(I+1) - 3]. \tag{21.212}$$

So

$$\Delta E_{I=0} = -8\pi \mu_n \mu_e |\psi(0)|^2, \tag{21.213}$$

when the electron and proton spins are antiparallel, and

$$\Delta E_{I=1} = \frac{8\pi}{3} \mu_n \mu_e |\psi(0)|^2, \tag{21.214}$$

when the electron and proton spins are parallel. For the hydrogen atom in the S-state we have

$$|\psi(0)|^2 = \frac{1}{\pi n^3 a_0^3},$$

where

$$a_0 = \frac{\hbar^2}{me^2}.$$

Then, the difference between the levels that defines S-therm splitting caused by the electron interaction with the magnetic moment of the nucleus is governed by the expression:

$$\Delta\omega = \frac{\Delta E_{I=1} - \Delta E_{I=0}}{\hbar} = \frac{32\mu_e\mu_n}{3\hbar} \frac{1}{n^3 a_0^3}, \tag{21.215}$$

At $n = 1$ it follows from the formula (21.215)

$$\Delta\omega_{th} = 1417 \text{ MHz},$$

while in the experiments by Nafe, Nelson, and Rabi the obtained result was:

$$\Delta\omega_{exp} = 1420 \text{ MHz},$$

what has corresponded to 0.26% deviation from the theoretical result under condition $(g - 2)_e/2 = 0$. It has been found that the frequency $\Delta\omega_{th}$ was not growing to $\Delta\omega_{exp}$ with allowance for the relativistic corrections or nucleus mass finiteness. Since by that time a magnetic moment of the proton has been measured within a sufficient accuracy, the only doubts were about a magnetic moment of the electron. Because of this, an assumption put forward by Breit [65]—the electron possesses not only the kinematic magnetic moment equal to $e\hbar/mc$ but also the intrinsic magnetic moment the existence of which is not described by the conventional Dirac theory—looked quite naturally.

Using the magnetic resonance method, in 1948 P. Kusch and H. Foley [66] performed high-precision measurements for the frequencies of the Zeeman splitting in a spectrum of gallium. Their direct measurements of the electron g_e-factor provided support for Breit's assumption. Schwinger was the first to calculate the quantity of the AMM in the second order with respect to e [67]:

$$a_e^{(2)} = \frac{\alpha}{2\pi}$$

(sometimes the additive to the AMM equal to $\alpha/(2\pi)$ is called the *Schwinger term*). To substitute the value $\mu_e = (1 + a_e^{(2)})\mu_B$ into (21.215) gives the result being in the excellent agreement with the experiment.

The $(g - 2)_e/2$-anomaly considerably changes the spin dynamics of the Dirac electron. To demonstrate this we consider the electron motion in a magnetic field that is constant in time and homogeneous in space. In this case the Bargmann–Michel–Telegdi equation [68] describing the spin evolution has the form:

$$\frac{d\boldsymbol{\xi}}{dt} = \frac{ec}{E}(1 + \gamma a_e)[\boldsymbol{\xi} \times \boldsymbol{\mathcal{H}}] + \frac{ea_e(\mathbf{v}\boldsymbol{\mathcal{H}})}{Ec(1 + \gamma)}[\mathbf{v} \times \boldsymbol{\xi}], \tag{21.216}$$

where $\gamma = E/(mc^2)$ and we have turned to the ordinary system of units. As a rule, of particular importance is a change in the absolute spin direction relative to the direction of motion rather than its changing in space. Present $\boldsymbol{\xi}$ in the view:

$$\boldsymbol{\xi} = \mathbf{n}\xi_{\parallel} + \boldsymbol{\xi}_{\perp},$$

where $\mathbf{n} = \mathbf{v}/v$ and investigate the case when the magnetic field is directed perpendicular to the electron motion orbit plane $\mathbf{v}\perp\boldsymbol{\mathcal{H}} = 0$. In the right-hand side of (21.216) only the first term is left and, hence, the particle spin is rotating about the direction $\boldsymbol{\mathcal{H}}$ with the frequency:

$$\Omega_s = \Omega\left(a_e + \frac{1}{\gamma}\right),$$

where $\Omega = e\mathcal{H}/(mc)$, whereas the electron rotation frequency proves to be less

$$\Omega_e = \frac{ec\mathcal{H}}{E} = \frac{\Omega}{\gamma}.$$

The difference of these frequencies is proportional to the AMM value:

$$\Omega_s - \Omega_e = a_e\Omega \tag{21.217}$$

and does not depend on a particle motion speed. Inasmuch as the perpendicular projection of the spin $\boldsymbol{\xi}_\perp$ is the motion integral

$$\frac{d\boldsymbol{\xi}_\perp}{dt} = 0, \tag{21.218}$$

then only $\xi_\|$ can precess with respect to the direction $\mathbf{v}$. Really, allowing for the motion equations

$$\frac{d\mathbf{p}}{dt} = \frac{ec}{E}[\mathbf{p} \times \mathcal{H}],$$

we get from (21.216)

$$\frac{d}{dt}(\boldsymbol{\xi}\mathbf{p}) = -\Omega a_e[\boldsymbol{\xi} \times \mathbf{p}]\mathbf{h} \tag{21.219}$$

($\mathbf{h} = \mathcal{H}/\mathcal{H}$), that is, $\xi_\|$ oscillates with the frequency Ωa_e. Exploring the variation in time of the longitudinal polarization for an ensemble of polarized particles just comprises the essence of a precession method used to find the AMM.

Besides, there exists an other method, a spin-resonance one. It is based on the spin-resonance experiments conducted to study frequencies of the transitions between the Landau-Rabi levels for the lepton in a homogeneous magnetic field. It is clear that a significant progress in measurements of the AMM should be expected due to experiments with free electrons. For the first time the AMM value for a free electron was measured in 1957 by Dehmelt (see the review paper [69]) who, using a high-frequency field, has succeeded in the induction and observation of the spin-resonance transitions of the electrons moving in a magnetic field. Let us explain the essence of the spin-resonance method.

We take care of the Pauli equation

$$\left[\frac{1}{2m}\left(\mathbf{p} - \frac{e}{c}\mathbf{A}\right)^2 + \frac{g_e\mu_B}{2}(\boldsymbol{\sigma}\mathcal{H})\right]\psi(\mathbf{r}) = E\psi(\mathbf{r}) \tag{21.220}$$

where we are using the ordinary units as before. In a homogeneous magnetic field directed along the axis z, the electron energy spectrum has the form (Landau-Rabi levels):

$$E_{n,m_s} = \hbar\Omega\left(n + \frac{1}{2} + \frac{m_s g_e}{2}\right) + \frac{p_z^2}{2m}, \tag{21.221}$$

where $m_s = \pm 1/2$ is associated with the spin projection onto the field direction. If we had $g_e = 2$, then the electron energy would have a spin degeneracy (thus, for example, $E_{1,1/2} = E_{2,-1/2}$ and so on). However, as $g_e = 2(1 + a_e)$, this degeneracy is absent and we come to two sets of the energy levels in accordance with the spin projection onto the field direction $\mathcal{H}$:

1. $m_s = \frac{1}{2}$, $E_1 = \hbar\Omega\left(n + 1 + \frac{a_e}{2}\right)$;

2. $m_s = -\frac{1}{2}$, $E_2 = \hbar\Omega\left(n - \frac{a_e}{2}\right)$,

where we are interested in only the cyclotron orbital motion, that is, $p_z = 0$. The quantum transitions between the adjacent levels in each series are realized without a spin flip and are characterized by the frequency Ω. The transitions with a spin flip are corresponding to the transitions from one series to another. In the process the frequency of these transitions in combination with a change of the orbital quantum number turns to be proportional to $a_e \Omega$. Observation using the spin-resonance method is based on using two magnetic fields: a constant magnetic field to fix the projection of the particle spin moment, and an alternating magnetic field to generate the spin-flip transitions. Creation of the Penning trap has improved the potentialities of the spin-resonance method considerably. In this trap, combination of stationary electric and magnetic fields provide a finite motion of the electrons in the finite region of space. To illustrate, an isolated electron may be trapped for a period of several days, offering a unique possibility for continuous observation of its characteristics.

The electric potential of the Penning trap is chosen in the form:

$$\varphi = -\frac{e^2 a(x^2 + y^2 - 2z^2)}{2},$$

where a is an constant, and a homogeneous magnetic field $\mathcal{H} = (0, 0, \mathcal{H})$ is specified by a vector potential

$$\mathbf{A} = (-\mathcal{H}y/2, \mathcal{H}x/2, 0).$$

Since in a homogeneous magnetic field the space and spin parts of the wavefunction are separated, then the Pauli equation for the electron put in the Penning trap could be easily solved. The finiteness demand of the obtained solution leads to the following expression for the electron energy spectrum:

$$E = \hbar \left[\omega_c' \left(n + \frac{1}{2} \right) + \omega_z \left(n_z + \frac{1}{2} \right) - \omega_m \left(n_m + \frac{1}{2} \right) + \frac{g_e m_s \Omega}{2} \right], \tag{21.222}$$

where $n, n_z, n_m = 0, 1, 2, 3...$. From the viewpoint of physics, this spectrum is associated with cyclotron orbital motion at the modified frequency $\omega_c' = \Omega - \omega_m$, axial harmonic oscillations at the frequency $\omega_z = \sqrt{2e^2 a/m}$, magnetron oscillations at the frequency $\omega_m = \Omega(1 - \sqrt{1 - 2\omega_z^2/\Omega^2})/2$, and the last term in (21.222) defines the energy of interaction between the spin and magnetic field. The radial and vertical motion is simultaneously stable at the condition that

$$(1 - 2\omega_z^2/\Omega^2)^2 > 0.$$

Hence, it follows that, when

$$0 < a < \frac{\mathcal{H}^2}{4me^2},$$

the electron is remaining inside the trap under $\varphi_0 = e^2 a/2$. Since the formula:

$$a_e = \frac{\omega_a - \omega_m}{\omega_c' + \omega_m}, \tag{21.223}$$

where $\omega_a = \Omega g_e/2 - \omega_c'$, takes place, the AMM definition is reduced to measuring the frequencies in the case of the spin-resonance method too. In so doing, it is enough to find only three frequencies since all the frequencies are coupled with the relation:

$$\omega_z^2 = 2\omega_m \omega_c'.$$

Note that both precession and spin-resonance methods enable one to measure the AMM directly, independently of measuring the g-factor that specifies the gyromagnetic ratio.

Up to now, the electron AMM has most exactly been measured by application of the spin-resonance method in experiments at the University of Washington [70]

$$a_e = 0.0011596521884(43).$$

The experiments were realized using the Penning trap with a single electron during some days to have ensured the high accuracy of measurements.

Our consideration of the AMM has concerned its kinematic part only. No doubt that the AMM of any particle has a dynamic nature as well. The influence of an external magnetic field on the electron AMM has been investigated in Ref. [71]. It was shown that the additive to the electron MM induced by a homogeneous magnetic field is defined as

$$\delta a_e(E, \mathcal{H}) = \frac{\alpha}{2\pi} F(E, \mathcal{H}),$$

where $\mathcal{H}$ is a magnetic field strength, E is an electron energy, $F(E, \mathcal{H})$ is a function of a parameter $\kappa = (\mathcal{H}/\mathcal{H}_0)(E/mc^2)$, and $\mathcal{H}_0$ is the critical value of the magnetic field. In the case $\kappa \ll 1$, $F(E, \mathcal{H})$ is given by the expression:

$$F(E, \mathcal{H}) = 12\kappa^2 \left(\frac{37}{12} - \ln \frac{1}{\kappa} - C - \frac{1}{2} \ln 3 \right), \qquad (21.224)$$

where $C = 0.577....$ In ordinary conditions of experiment the values of $F(E, \mathcal{H})$ are vanishingly small. However, for the electrons with the energy 10 MeV in the field with $\mathcal{H} \approx 4 \times 10^4$ Gs,

$$F(E, \mathcal{H}) \approx 4 \times 10^{-8}$$

which is available to observe.

We are coming to the investigation of the muon AMM. In this case the correction in the second order on e $a_\mu^{(2)}$ has the same form as that for the electron AMM, that is,

$$a_\mu^{(2)} = a_e^{(2)}(m_e \to m_\mu) = \frac{\alpha}{2\pi} \mu_0, \qquad (21.225)$$

where $\mu_0 = e\hbar/(2m_\mu c)$ is the normal magnetic moment of the muon. However, this is not the case for the correction in the fourth order on e. The reason is that the vertex function includes a polarization operator $\Pi_{\mu\nu}^{(2)}$ to be determined by contributions coming from virtual $e^- e^+$-, $\mu^+ \mu^-$- and $\tau^+ \tau^-$ pairs for any charged leptons. Since these contributions are inversely proportional to the mass square of particles forming a loop, then those of more lighter particles dominate. It means that for the muon the contribution coming from the electron-positron loop will be dominated. By now computing the muon AMM has been fulfilled within the QED in the fifth order on α [62]:

$$a_\mu^{QED} = \frac{\alpha}{2\pi} + 0.765857376(27) \left(\frac{\alpha}{\pi} \right)^2 + 24.05050898(44) \left(\frac{\alpha}{\pi} \right)^3 +$$

$$+126.07(41) \left(\frac{\alpha}{\pi} \right)^4 + 930(170) \left(\frac{\alpha}{\pi} \right)^5. \qquad (21.226)$$

Growing the coefficients in the (α/π) expansion reflects the presence of large quantity $\ln(m_\mu/m_e) \approx 5.3$, which comes from electron loops. The most exact determination of the fine structure constant at the scale $E = m_e c^2$ has been done in Ref. [72]

$$\alpha^{-1}(a_e) = 137.03599958(52). \qquad (21.227)$$

To substitute (21.227) into (21.226) results in

$$a_\mu^{QED} = 116584705.7(2.9) \times 10^{-11}. \qquad (21.228)$$

Up to date the value of the muon AMM measured at Brookhaven National Laboratory (BNL) on Alternating Gradient Synchrotron with the help of the precession method [73] is given by the expression:

$$a_{\mu^-}^{exp} = 11\ 659\ 214(8)(3) \times 10^{-10}\mu_0, \qquad \text{(BNL'01)}. \qquad (21.229)$$

The uncertainty of BNL 2001 result is equal to 0.7 ppm (ppm = parts per million or 1 ppm $= 10^{-6}$). In case of the electron the attained precision is much higher, being at a level of ppb units. Nevertheless, the measurements of $a_{\mu^-}^{exp}$ give us more information about the short-range phenomena as the value of such effects proves to be proportional to m_l^2 [74]. In this way we have the amplification factor of $m_\mu^2/m_e^2 \approx 40000$ fully compensating for the inadequate accuracy in the measurement of $a_{\mu^-}^{exp}$. All this results in the sensitivity of the muon anomaly to hadron and electroweak corrections, providing a reliable instrument for observation of the effects caused by still undiscovered particles having masses on the order of several hundred GeV. As a consequence, with calculating the muon AMM contributions coming from all model sectors must be taken into account. It means that in the SM, a_μ^{SM} is defined as:

$$a_\mu^{SM} = a_\mu^{QED} + a_\mu^{EW} + a_\mu^{had}, \qquad (21.230)$$

where the electroweak contribution is given by the expression [62]:

$$a_\mu^{EW} = 152(4) \times 10^{-11}\mu_0. \qquad (21.231)$$

The largest theoretical ambiguities in a_μ are governed by the quantity a_μ^{had}, which is mainly defined by contributions of virtual hadrons to a photon propagator in the fourth and sixth orders. Lower corrections cause vacuum polarization by hadrons $a_\mu^{had}(VP1)$, while corrections of the sixth order along with vacuum polarization by hadrons $a_\mu^{had}(VP2)$ include in itself light-by-light scattering $a_\mu^{had}(LbyL)$ as well. Since a_μ^{had} contains integrals over virtual particle energies including a region of soft energies, a_μ^{had} could not be reliably calculated within the perturbative QCD. Only the existing value of the quantity $a_\mu^{had}(VP2)$ cause no doubts. As for $a_\mu^{had}(LbyL)$ all its estimations made so far are model dependent. The calculations are based on the chiral perturbation or on the extended Nambu-Jona-Lasinio model. Also the vector meson dominance is assumed and the phenomenological parametrization of the pion form factor $\pi\gamma^*\gamma^*$ is introduced for regularizing the divergences. The value of $a_\mu^{had}(VP1)$ can be calculated on the basis of experimental data concerning the investigations of the cross section $e^+e^- \to hadrons$ and/or on the basis of the analysis of data concerning a τ-decay into hadrons. However, the inclusion of the latter introduces systematic uncertainties originating from isospin symmetry breaking effects that are difficult to estimate. A recent analysis, performed in Ref. [75] showed, that $a_\mu^{had}(VP1)$, obtained exclusively on the basis of e^+e^- data differs a lot from $a_\mu^{had}(VP1)$, whose calculation is based on the data, concerning $\tau \to hadrons$. Since all data concerning the QED effects are in excellent agreement with the theoretical predictions, then only the corrections coming from the electroweak and hadron sectors are under suspicion. We shall come back to the detailed discussion of this problem in the second volume.

21.8   Lamb shift

The Dirac equation for the electron in the Coulomb field of a point nucleus predicts degeneracy of the energy levels for the bound states, possessing the same values of the main quantum number n and the quantum number of the total moment j and differing in the values of the quantum numbers for the orbital moment $l = j \pm 1/2$. Thus, for example, the states $2S_{1/2}$ and $2P_{1/2}$ should have the same energy. With the use of radio spectroscopic methods in 1947 Lamb and Retherford [76] demonstrated that in a hydrogen atom the level $2S_{1/2}$ is higher than the level $2P_{1/2}$ approximately by 1000 MHz. This discovery was an impetus for all further development of the QED finally leading to the understanding of this phenomenon. Removal of the degeneracy from the levels $2S_{1/2}$ and $2P_{1/2}$ is caused by quantum fluctuations of the vacuum of the electromagnetic and electron-positron fields modifying the potential energy $\varphi(r) = -Ze^2/r$ associated with the electron-nucleus interactions. Also, this process named the *Lamb shift* may be observed between other levels in hydrogen or hydrogen-like atoms. This term is frequently used in a more wide sense as a level shift of the bound electron states in an external field caused by the RC.

On examination of the Lamb shift we shall follow Ref. [77] in which the Dirac theory is combined with the RC. Let us consider the electron scattering process in an external field $A^e(x)$. The corresponding Feynman diagrams are represented in Fig. 21.7. From the expression (21.159) it is evident that the cumulative effect of the RC corresponding to the diagrams pictured in Figs. 21.7(a)–(f) can be regarded as changing the external field potential. Then, the effective potential for the electron proves to be equal to

$$\gamma^\mu [A^e_\mu(q)]_{eff} = [\gamma^\mu + \gamma^\nu D^c_{\nu\tau}(p_2 - p_1) \Pi^{(2)\tau\mu}_C (p_2 - p_1) + \Lambda^{(3)\mu}_C (p_2, p_1)] A^e_\mu(q), \quad (21.232)$$

rather than $\gamma^\mu A^e_\mu(q)$. It is convenient to split the change of $\gamma^\mu \delta A^e_\mu$ into three parts

$$\gamma^\mu \delta A^e_\mu(q) - \gamma^\mu [\delta A^{(\gamma)}_\mu(q) + \delta A^{AMM}_\mu(q) + \delta A^{(V_E)}_\mu(q)], \quad (21.233)$$

where a source of the correction $\delta A^{(\gamma)}_\mu(q)$ is a vacuum polarization, $\delta A^{AMM}_\mu(q)$ is associated with the AMM of the electron, and the correction $\delta A^{(V_E)}_\mu(q)$ appears due to the Dirac form factor in the expression for the vertex function. In the case of small transferred momenta q^2 (see Eqs. (21.133) and (21.98)) these quantities are as follows:

$$\gamma^\mu \delta A^{(\gamma)}_\mu(q) = -\gamma^\mu \frac{\alpha q^2}{15\pi m_R^2} A^e_\mu(q). \quad (21.234)$$

$$\gamma^\mu \delta A^{AMM}_\mu(q) = \frac{\alpha}{2\pi} \left[\frac{\hat{q}\gamma_\mu - \gamma_\mu \hat{q}}{4m_R} \right] A^e_\mu(q), \quad (21.235)$$

$$\gamma^\mu \delta A^{(V_E)}_\mu(q) = \gamma_\mu \frac{\alpha q^2}{3\pi m_R^2} \left(\ln \frac{m_R}{\lambda} - \frac{3}{8} \right) A^e_\mu(q) \quad (21.236)$$

Thus, in the coordinate representation the Lagrangian, describing the interaction between the electron and an external field slowly varying in space and time, will be given by the expression:

$$\mathcal{L}_{int} \approx e \int d^3 x \overline{\psi}(x) \gamma^\mu \left[A^e_\mu(x) + \delta A^{(\gamma)}_\mu(x) + \delta A^{AMM}_\mu(x) + \delta A^{(V_E)}_\mu(x) \right] \psi(x), \quad (21.237)$$

where

$$\gamma^\mu \delta A^{(\gamma)}_\mu(x) = \frac{\alpha}{15\pi m_R^2} \Box A^e_\mu(x), \quad (21.238)$$

$$\gamma^\mu \delta A^{(V_E)}_\mu(x) = -\gamma^\mu \frac{\alpha}{3\pi m_R^2} \left(\ln \frac{m_R}{\lambda} - \frac{3}{8} \right) \Box A^e_\mu(x), \quad (21.239)$$

$$\gamma^\mu \delta A^{AMM}_\mu(x) = \frac{\alpha}{4\pi m_R} \sigma^{\nu\mu} \partial_\nu A^e_\mu(x). \quad (21.240)$$

Using (21.237), we arrive at

$$\left\{\gamma^\mu[i\partial_\mu - eA_\mu^e(x) - \delta A_\mu^e(x)] - m\right\}\psi(x) = 0, \tag{21.241}$$

where

$$\gamma^\mu \delta A_\mu^e(x) = \gamma^\mu[\delta A_\mu^{(\gamma)}(x) + \delta A_\mu^{(V_E)}(x) + \delta A_\mu^{AMM}(x)].$$

The equation obtained is nothing more nor less than the Dirac equation written with an accuracy of the third order on e, which includes the mass operator in an external field with inclusion of higher approximation effects [78]. We shall return to the deduction of this equation in Section 22.2, and now we are coming to the investigation of (21.241) for the hydrogen atom case assuming the interaction between the electron and the proton to be purely Coulomb.

The quantity $\psi(x)$ should be interpreted as the matrix element of the field operator between the vacuum and the one-electron state in the presence of an external field. In accordance with the perturbation theory rules the additional term in Eq. (21.241) must be taken into account in the first order only. Unperturbed states are solutions of the Dirac equation for a particle in an external field:

$$\left\{\gamma^\mu[i\partial_\mu - eA_\mu^e(r)] - m_R\right\}\psi(x) = 0, \tag{21.242}$$

where

$$A_\mu^e(r) = -\frac{Ze}{4\pi r}g_{\mu 0}.$$

According to the general rules, the shift of the energy level caused by H_{int} is yielded by:

$$\delta E = e\int d^3x \psi^\dagger(\mathbf{r})\left[\delta A_0^{(\gamma)}(r) + \delta A_0^{(V_E)}(r) + \delta A_0^{AMM}(r)\right]\psi(\mathbf{r}), \tag{21.243}$$

where

$$\delta A_0^{(\gamma)}(r) = -\frac{\alpha}{15\pi m_R^2}\Delta A_0^e(r), \tag{21.244}$$

$$\delta A_0^{(V_E)}(r) = \frac{\alpha}{3\pi m_R^2}\left(\ln\frac{m_R}{\lambda} - \frac{3}{8}\right)\Delta A_0^e(r), \tag{21.245}$$

$$\delta A_0^{AMM}(r) = \frac{i\alpha}{4\pi m_R}(\boldsymbol{\gamma}\cdot\boldsymbol{\mathcal{E}}(r)). \tag{21.246}$$

To find δE one may make use of the solutions of Eq. (21.242), which have been obtained by application of the perturbation theory. We take the Schrödinger equation solution for the hydrogen atom:

$$\psi_{n,l,m_l}(\mathbf{r}) \equiv \psi_{n,l,j}(\mathbf{r}) = N_{nl}e^{-\rho/2}\rho^l L_{n+l}^{2l+1}(\rho)Y_l^{|m_l|}(\theta,\varphi), \tag{21.247}$$

where

$$\rho = \frac{2r}{na}, \qquad a = \frac{1}{m_R e^2},$$

N_{nl} is a normalization coefficient, $L_{n+l}^{2l+1}(\rho)$ is the generalized Laguerre polynomial, and $Y_l^{|m_l|}(\theta,\varphi)$ are spherical functions, as the zeroth order approximation. Allowing for the Coulomb character of the field and using

$$\Delta\left(\frac{1}{r}\right) = -4\pi\delta^{(3)}(\mathbf{r}),$$

we deduce the following expression for the energy shift initiated by the first two addends in Eq. (21.243):

$$(\delta E_1)_{n,l,j} = \frac{Ze^2\alpha}{3\pi m_R^2} \int d^3x\, \psi_{n,l,j}^\dagger(\mathbf{r}) \left[\ln\frac{m_R}{\lambda} - \frac{3}{8} - \frac{1}{5}\right] \delta^{(3)}(\mathbf{r})\psi_{n,l,j}(\mathbf{r}) =$$

$$= \delta_{l,0} \frac{4m_R\alpha}{3\pi n^3}(Z\alpha)^4 \left(\ln\frac{m_R}{\lambda} - \frac{3}{8} - \frac{1}{5}\right). \tag{21.248}$$

So, in the nonrelativistic approximation the addends (21.244) and (21.247) influence the S-state only.

The expression for $(\delta E_1)_{n,l,j}$ includes the fictitious photon mass, therefore, we have to gain an understanding of why it has happened. In the description of free-particle scattering the IR-divergences are due to the ability of a charged lepton to emit and absorb soft photons without much shifting from the mass surface $p^2 = m_R^2$. As a result, our lepton propagators have very small denominators and together with the photon propagators lead to the IR divergences. But during the calculation of the Lamb shift no IR divergences appear as in this case the four-dimensional momentum of the electron is beyond the mass surface. In this way even at very low photon energies the denominator of the electron propagator is $m_R^2 - p^2$ and not zero. The derived vertex operator is logarithmically dependent on this value. Because of this, the factor $2\ln(\lambda/m_R)$ associated with the scattering of a free particle in the case of the Lamb shift is replaced by $\ln(m_R^2 - p^2)/m_R^2$. For the bound hydrogen-like state we have

$$p_0 - m_R - E_n,$$

where E_n is a binding energy in the state n and $|\mathbf{p}| \approx Z\alpha m_R$, to give

$$\frac{1}{m_R^2}(m_R^2 - p^2) \approx \frac{2}{m_R}\left(E_n + \frac{\mathbf{p}^2}{2m_R}\right).$$

Then, in the case in question the factor $\ln\lambda^2/m_R^2$ is replaced with

$$\ln\frac{2m_R E_n + \mathbf{p}^2}{m_R^2}.$$

Due to this factor the main contribution to changing the position of the level $2S_{1/2}$ is given by the addend $\delta A^{(V_E)}$. It shifts the level $2S_{1/2}$ approximately in 1010 MHz with respect to that $2P_{1/2}$. The vacuum polarization effects provide the additive to the shift $E_{2S_{1/2}} - E_{2P_{1/2}}$ equal to -27 MHz.

Now we consider the contribution to the level shift induced by the electron AMM

$$(\delta E_2)_{n,l,j} = \frac{ie\alpha}{4\pi m_R} \int d^3x\, \psi_{n,l,j}^\dagger(\mathbf{r})(\boldsymbol{\gamma}\cdot\boldsymbol{\mathcal{E}}(r))\psi_{n,l,j}(\mathbf{r}). \tag{21.249}$$

If one neglects the Coulomb potential, then the big $\varphi(\mathbf{r})$ and small $\chi(\mathbf{r})$ components of $\psi(\mathbf{r})$ will be coupled by the approximate relation:

$$\chi(\mathbf{r}) = -i\frac{(\boldsymbol{\sigma}\cdot\boldsymbol{\nabla})}{2m_R}\varphi(\mathbf{r}).$$

Inasmuch as $e\boldsymbol{\mathcal{E}} = -Z\alpha\mathbf{r}/r^3$ and the relations:

$$\psi^\dagger(\mathbf{r})(\boldsymbol{\gamma}\cdot\mathbf{r})\psi(\mathbf{r}) = \varphi^\dagger(\mathbf{r})(\boldsymbol{\sigma}\cdot\mathbf{r})\chi(\mathbf{r}) - \chi^\dagger(\mathbf{r})(\boldsymbol{\sigma}\cdot\mathbf{r})\varphi(\mathbf{r}) \approx$$

$$\approx \frac{1}{2im_R} \left[\varphi^\dagger(\mathbf{r})(\boldsymbol{\sigma} \cdot \mathbf{r})(\boldsymbol{\sigma} \cdot \boldsymbol{\nabla})\varphi(\mathbf{r}) + \text{conj.} \right]$$

are valid, then after integration by parts the integral in (21.249) becomes the view:

$$\int d^3x \psi^\dagger(\mathbf{r}) \frac{(\boldsymbol{\gamma} \cdot \mathbf{r})}{r^3} \psi(\mathbf{r}) \approx \frac{i}{2m_R} \int d^3x \varphi^\dagger(\mathbf{r})(\boldsymbol{\sigma} \cdot \boldsymbol{\nabla}) \left[\frac{(\boldsymbol{\sigma} \cdot \mathbf{r})}{r^3} \varphi(\mathbf{r}) \right]. \tag{21.250}$$

Taking into consideration the identity:

$$[\sigma_j A_j, \sigma_k B_k] = \delta_{jk}(A_j B_k - B_k A_j) + i\varepsilon_{jkl}\sigma_l(A_j B_k + B_k A_j) \tag{21.251}$$

and the relation

$$\boldsymbol{\nabla}\frac{1}{r} = -\frac{\mathbf{r}}{r^3},$$

we arrive at

$$(\delta E_2)_{n,l,j} = \frac{Z\alpha^2}{8\pi m_R^2} \int d^3x \varphi_{n,l,j}^\dagger(\mathbf{r}) \left[4\pi\delta^{(3)}(\mathbf{r}) + 4\frac{(\mathbf{LS})}{r^3} \right] \varphi_{n,l,j}(\mathbf{r}), \tag{21.252}$$

where $\mathbf{L} = -i[\mathbf{r} \times \boldsymbol{\nabla}]$ and $\mathbf{S}$ is the spin moment operator. The eigenvalues of the operator $(\mathbf{LS})$ are given by:

$$(\mathbf{LS}) = \frac{1}{2}(1 - \delta_{l,0}) \left[j(j+1) - l(l+1) - \frac{3}{4} \right]. \tag{21.253}$$

To compute the average value for r^{-3} results in

$$< r^{-3} >_{n,l,j} = \frac{2(Zm_R\alpha)^3}{l(l+1)(2l+1)n^3}, \qquad l \geq 1. \tag{21.254}$$

With the help of (21.252), (21.253), and (21.254) we get the final result for $(\delta E_2)_{n,l,j}$

$$(\delta E_2)_{n,l,j} = \frac{\alpha(Z\alpha)^4 m_R}{2\pi n^3(2l+1)} C_{j,l}, \tag{21.255}$$

where

$$C_{j,l} = \delta_{l,0} + (1 - \delta_{l,0})\frac{j(j+1) - l(l+1) - 3/4}{l(l+1)} =$$

$$= \begin{cases} (l+1)^{-1}, & \text{when} \quad j = l + 1/2, \\ -l^{-1}, & \text{when} \quad j = l - 1/2, \ l \geq 1. \end{cases}$$

The electron AMM leads to the levels shift approximately in 68 MHz. Thus, the predicted value of the Lamb shift is in the neighborhood of 1052 MHz with an accuracy of $\alpha(Z\alpha)^4$.

In summary, we have obtained the approximate value of the Lamb shift using the average values of the operators $\Lambda_C^\mu(p_1, p_2)$ and $\Pi_C^{\mu\nu}(q^2)$ for hydrogen-like states under consideration. Strictly speaking, the expression (21.132) for the vertex operator $\Lambda_C^\mu(p_1, p_2)$ is inapplicable because it was deduced for a free electron ($p_1^2 = p_2^2 = m_R^2$) rather than for a bound electron ($p^2 \neq m_R^2$). However, in our case $|\mathbf{p}| \ll m_R$ and the deviations from the formula (21.132) are extremely small.

To upgrade the calculations, first we should go to an exact electron propagator in the Coulomb field. As there is no closed expression for the electron propagator available, this approach presents serious difficulties during the calculations and is still incompletely realized. Besides, it is necessary to take into consideration all the corrections associated with the nucleus: due to its magnetic moment, recoil, form factors, radius and mass finiteness, polarizability, etc. Moreover, it is required to take into account the metastability of the excited levels, that is, the natural width of the levels $2S_{1/2}$ and $2P_{1/2}$.

By the present time, allowances have been made for the leading higher-order corrections to the coupling constant $Z\alpha$, second-order corrections to α in the electron self-energy, for the electron AMM and the vacuum polarization, and also for the effects due to the proton radius and mass finiteness. All this yields the following theoretical value for the Lamb shift in the hydrogen [79]:

$$E_{2S_{1/2}} - E_{2P_{1/2}} = 1057.8640(140) \text{ MHz.} \tag{21.256}$$

When calculating these values, the main uncertainty results from the higher orders on coupling $(Z\alpha)$ and the second order (α^2) for the electron self-energy. The discrepancies between the theoretical and experimental measurements for the Lamb shift are presently insignificant. So, the experimental values obtained in Refs. [80,81] are

$$E_{2S_{1/2}} - E_{2P_{1/2}} = 1057.8450(90), \tag{21.257}$$

$$E_{2S_{1/2}} - E_{2P_{1/2}} = 1057.8514(19). \tag{21.258}$$

The survey of the consecutive theory of the Lamb shift in higher orders of the perturbation theory could be found in Ref. [82]. A reader interested in details and calculations technique is recommended to refer to the books [83].

An interest in the high-precision calculation and measurements of the Lamb shift in the hydrogen atom is caused not only by the existing disparity between the theoretical results and experimental data but also by the possibility of obtaining information concerning the structure and properties of the corrections having no direct relation to the QED. The Lamb shift defines the properties of the nucleus-bound electron, taking into consideration both the QED effects and those due to the nucleus structure. For example, the Lamb shift correction caused by a finite size of the proton δ_p:

$$\delta_p = \frac{m^3 (Z\alpha)^4}{12} < r_p^2 >,$$

where $< r_p^2 >$—the root mean square proton radius, is within the accuracy limits of a modern experiment. In the case of $< r_p^2 >^{1/2} = 0.862$ fm the correction amounts to 0.146 MHz. In principle, the experimental accuracy of ~ 2 kHz attained in the estimation of the Lamb shift by the use of an atomic interferometer [81] makes it possible to measure the proton radius with an error of 0.007 fm, that is nearly one half of the experimental error in the case of the elastic e^-p-scattering.

22

Renormalization theory

"Prophet!" said I, "thing of evil!—prophet still,
if bird or devil!—
Whether Tempter sent, or whether tempest tossed thee
here ashore
Desolate yet all undaunted, on this desert land
enchanted—
On this home by Horror haunted—tell me truly, I
implore—
Is there—is there balm in Gilead?—tell me—
tell me, I implore!"
Quoth the Raven "Nevermore."
Edgar Allan Poe, "The Raven"

With calculating the elements of the S-matrix in the QED, we may face three types of divergences: the UV-divergences, the IR-divergences, and singularities due to the coincidence of two or more poles in the integrand. Divergences of the last type may appear, for example, when for some special values of the particle momenta a multiparticle process may be subdivided into independent processes involving independent particle groups. Consideration of such divergences is beyond the scope of this book, and an interested reader may find the relevant information in Ref. [84]. The IR-divergences are also out of consideration because they have put emphasis in the previous Sections. Now our attention is focused on the UV-divergences, presenting much trouble indeed.

According to the terminology of quantum field theory (QFT), renormalization is understood as a procedure aimed precisely at the elimination of the UV-divergences. Such divergences are known to appear, when integrals are evaluated over the momenta in the matrix elements associated with the closed-loop Feynman diagrams. As it has been shown earlier, some the UV-divergences in the QED may be removed by the redefinition of such parameters as a mass and an electric charge of the electron. Here it will be demonstrated that this procedure is sufficient to exclude all the singular components of the matrix elements and Green functions available in the QED. In other words, all the UV-divergences may be reduced to the field additives δm and δe to the electron mass and charge. Formally, these additives may be expressed in terms of the singular factors Z_m and Z_e to the bare (initial) mass m and charge e:

$$m + \delta m = Z_m m, \qquad e + \delta e = Z_e e. \tag{22.1}$$

The quantities measured in experiments include only the functions of $Z_m m$ and $Z_e^{1/2} e$ products. The observables involve no bare mass or charge and no UV-singularities, taken separately or in combination. This makes it possible to identify the products of (22.1) with the physical meaning of the electron mass and charge:

$$m_R = Z_m m, \qquad e_R = Z_e e. \tag{22.2}$$

An operation of redefining the physical parameters:

$$m \to e_R, \qquad e \to e_R, \tag{22.3}$$

by means of their multiplication into singular factors, completely excluding the UV-divergences from the observables, is referred to as the renormalization operation. The full description of such an approach to renormalization of the QED has been first proposed by Dyson [4]. This method has been further developed in Refs. [85–88].

The operation used to cancel the UV-divergences may be formalized without the relations similar to (22.2) as well. The idea is as follows. The RC to the matrix elements are expressed in terms of the propagator products. In the coordinate representation the propagators have the light-cone singularities giving rise to the UV-divergences. Mathematically, this amounts to the problem of finding the operation of multiplication for the generalized functions. The theory of multiplying the generalized functions involved in quantum-field calculations has been developed by Bogoliubov. By application of this theory, Bogoliubov and Parasyuk have solved the problem of removing the UV-divergences [89]. They have proved a theorem (Bogoliubov–Parasyuk theorem) enabling the derivation of unambiguous expressions for the scattering matrix elements, within the scope of the perturbation theory but without the recourse to the intermediate regularization, counter-terms, and singular expressions similar to those involved in (22.2). A reader could find the extensive description of this procedure in the book [90].

It is naive to suppose that a renormalization theory should be necessarily based on a series expansion of the perturbation theory. The first attempts to formulate this theory without any approximations on the basis of an axiomatic approach have been made in Refs. [91]. A complete description of a renormalization theory in the axiomatic QFT may be found in [92]. More detailed references to the theory are given in the monograph [93] that is a good guide on the problem.

When considering a quantitative renormalization theory, we shall hold to historic sequence, that is, follow the method developed by Dyson [4] and Ward [94].

22.1 Primitive divergent diagrams

Let us review the types of UV-infinities found in the QED. To this end, we consider the general-form diagram including internal and external electron and phonon lines. A matrix element associated with this diagram involves integration over all the internal momenta p_k. As is known, such an integration may be performed conveniently after the Wick's rotation. According to Dyson [95], in the process no extra divergences are brought about. With using such transformations of an integral, a study of the integral and its divergence is simplified. When the momentum degree in the numerator is lower than that in the denominator, the integral is convergent. At the same time, when the momentum degree in the numerator is equal to or higher than that in the denominator, the integral may be divergent but not necessarily in all cases. The integral dimension D is introduced as a difference in the total degree of momenta in the numerator and denominator of the integrand. Then a sufficient convergence condition may be written as:

$$D < 0, \tag{22.4}$$

while the condition at which the divergences are possible takes the form:

$$D \geq 0, \tag{22.5}$$

$D = 0$ being understood as a logarithmic divergence. To characterize diagrams in the QED we shall use the following notations: F_e is a number of external electron lines, F_i is a number of internal electron lines, B_e is a number of external photon lines, and B_i is a number of internal photon lines, n is a number of vertices.

Specifying the number and type of external lines, we concretize the process. And assuming the values of F_i and B_i, we define an approximation for the calculation of this process. The momenta degree in the denominator and numerator is calculated in the following way:

(i) Every internal electron or photon line enters the factor d^4p into the numerator. It gives the contribution in D equal to $4(F_i + B_i)$.

(ii) In every vertex the four-dimensional δ-function appears after integration over the whole space-time. Its presence is equivalent to the addition of the fourth power of a momentum into the denominator. Since one of δ-functions is reserved for realization of the total four-dimensional momentum conservation law, then the overall contribution of vertices in D is equal to $-4(n - 1)$.

(iii) Every internal electron line introduces the first power of a momentum into the denominator to contribute $-2B_i$ to D.

(iv) Every internal photon line introduces the second power of a momentum into the denominator to give the contribution in D equal to $-2B_i$.

Therefore, the dimensionality of D will be defined by the equation:

$$D = 3F_i + 2B_i - 4(n - 1). \tag{22.6}$$

We try to express this quantity in terms of only the external lines number. If we shall manage to do it, then the divergences number in the theory proves to be finite.

Because two electron lines are converging at each vertex, a number of the electron line ends is equal to the doubled number of the vertices. As each internal electron line has two ends and each external line has only one, we get:

$$2F_i + F_e = 2n. \tag{22.7}$$

Similarly, each vertex is the end of a photon line and hence a number of the photon line ends equals n

$$2B_i + B_e = n. \tag{22.8}$$

To substitute (22.7), (22.8) into (22.6) leads to the encouraging result:

$$D = 4 - \frac{3}{2}F_e - B_e. \tag{22.9}$$

The developed approach to the evaluation of the divergences order for the integrals is, however, applicable only to matrix elements involving integration with respect to a single four-dimensional momentum. In case of a higher-multiplicity integral the calculation of degrees is insufficient to determine the integral divergence or convergence. This is illustrated in Fig. 22.1. For both diagrams pictured in this figure we have $D = -1$. Nevertheless, the diagram a is finite, while the diagram b is divergent. Because of this, the inequalities of (22.4), (22.5) are directly applicable, as a divergence or convergence criterion, exclusively to the diagrams associated with a single integration. Provided a diagram includes l internal lines, the conditions of (22.4) and (22.5) are valid only in the case when a subintegration result over an arbitrary selected $(l - 1)$th internal four-dimensional momentum is finite. In other words, in the case of a divergent diagram its divergence should appear at the final integration with respect to the four-dimensional momentum. Divergent diagrams featuring

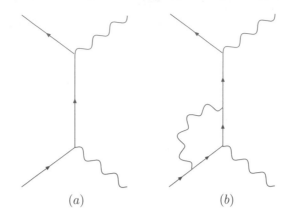

FIGURE 22.1

The diagrams (a) and (b) describing the process $e^- e^+ \to \gamma\gamma$ in the second and fourth order.

this property are referred to as primitively divergent diagrams (PDD). Such diagrams may be characterized as those leading to the finite matrix element if any of the internal lines of a diagram is broken and replaced by two external lines. Consideration of the PDD is of particular importance because they give the basic RC enabling the calculation of all other RC by the appropriate embedding. This procedure has been demonstrated during the calculation of the RC for electron scattering by an external field. At the same time, since the RC are the source of divergences, for analysis of divergences in the theory it is sufficient to take into consideration only the PDD and to study the properties of matrix elements to be associated with them.

Let us determine a degree of the PDD. By definition, the integrals associated with such diagrams are convergent when one of the integration variables is fixed, and hence divergent integrals appear in the case when in the integrand the numerator degree $4(F_i + B_i - n + 1)$ is equal to or higher the denominator degree $2B_i + F_i$, that is,

$$4(F_i + B_i - n + 1) \geq 2B_i + F_i. \tag{22.10}$$

With allowance made for (22.7) and (22.8) we arrive at the condition:

$$D = 4 - \frac{3}{2}F_e - B_e \geq 0. \tag{22.11}$$

Inasmuch as D does not contain n, the divergence degree of the PDD does not depend on the diagram vertices number. It means, that there are a finite number of the PDD in the QED. So, if by application of renormalization we manage to do all matrix elements of the PDD finite, then we lead to the finite (renormalizable) theory.

Using the formula (22.11), we define all kinds of the PDD:

(i) $F_e = B_e = 0$, $D = 4$—divergence of the fourth order;
(ii) $F_e = 0$, $B_e = 4$, $D = 0$—logarithmic divergence;
(iii) $F_e = 0$, $B_e = 3$, $D = 1$—linear divergence;
(iv) $F_e = 0$, $B_e = 1$, $D = 3$—cubic divergence;
(v) $F_e = 0$, $B_e = 2$, $D = 2$—squared divergence;
(vi) $F_e = 2$, $B_e = 0$, $D = 1$—linear divergence;
(vii) $F_e = 2$, $B_e = 1$, $D = 0$—logarithmic divergence.

The diagram of Fig. 22.2 describing the transition *vacuum→ vacuum* corresponds to the first case. This graph is eliminated from consideration on the basis of the reasons to

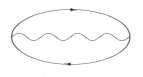

FIGURE 22.2

The Feynman diagram associated with the transition **vacuum→vacuum**.

have been brought forward in Section 20.3. An analysis of divergences in any QFT should always be performed with due regard for the symmetry properties. Due to invariance of a theory, some PDD are actually finite or zero. In the QED light-by-light scattering diagrams ($F_e = 0$, $B_e = 4$; Fig. 22.3), triangular diagrams ($F_e = 0$, $B_e = 3$; Fig. 22.4), and "tailless fish" diagrams ($F_e = 0$, $B_e = 1$; Fig. 22.5) are referred to such PDD. In case

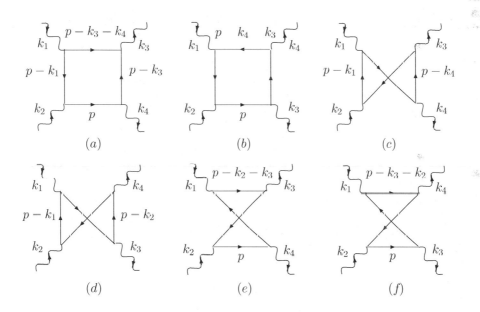

FIGURE 22.3

The Feynman diagrams (a)–(f) for the light-by-light scattering.

of the light-by-light scattering Eq. (22.11) gives $D = 0$ but in fact there is no logarithmic divergence at all, and the corresponding matrix element is finite. Let us demonstrate that this circumstance is caused by the gauge invariance of the QED.

Since the amplitude of the light-by-light scattering $\mathcal{A}(k_1, k_2, k_3, k_4)$ is logarithmically diverging, then in order to get its finite part $\mathcal{A}_R(k_1, k_2, k_3, k_4)$ one sufficiently subtracts from $\mathcal{A}_R(k_1, k_2, k_3, k_4)$ only the first term of the expansion into the Taylor series, that is

$$\mathcal{A}_R(k_1, k_2, k_3, k_4) = \mathcal{A}(k_1, k_2, k_3, k_4) - \mathcal{A}(0,0,0,0). \tag{22.12}$$

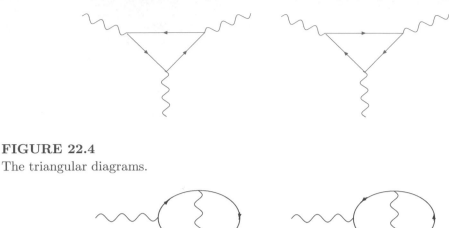

FIGURE 22.4
The triangular diagrams.

FIGURE 22.5
The "tailless fish" diagrams.

Owing to the gauge invariance observables depend on only the electromagnetic field tensor

$$F_{\mu\nu}(k) = k_\mu A_\nu(k) - k_\nu A_\mu(k),$$

rather than on the potential $A_\mu(k)$. By this reason, observables must turn into zero at $k_\mu = 0$, and, hence, $\mathcal{A}(0,0,0,0) = 0$. This result is confirmed by direct calculations [96] yielding the following expression for the cross section in the ultrarelativistic limit:

$$\sigma = \frac{\alpha^2 r_0^2}{\pi^2} \left(\frac{m}{\omega}\right)^2 \left(\ln \frac{\omega}{m}\right)^4. \tag{22.13}$$

Note that the four-photon interaction has no classical analog. It results from quantum fluctuations of the virtual pairs of charged particles. When the photons energy becomes equal to $2m$ a real pair is produced, and the cross section has the order of several μb. Despite the fact that by modern standards the cross section is so great, this effect is hardly observable due to a significant background. For example, in the interstellar space the radiation scattering by galactic hydrogen (and dust) is much greater. But as has been already mentioned in Section 19.6, another electromagnetic process with $B_e = 4$, $F_e = 0$ is our object of experimental observation, light scattering from an electrostatic field in the second order of the perturbation theory (Delbrück scattering).

As for the diagrams with $F_e = 0$, $B_e = 3$ and $F_e = 0$, $B_e = 1$, their matrix elements vanish due to the Furry theorem. Recall, that this theorem is a consequence of the QED invariance with respect to the charge conjugation operation.

Now we proceed to the discussion of the case $F_e = 0$, $B_e = 2$. The only primitively-divergent photon self-energy diagram (PSED) is shown in Fig. 22.11. Earlier it has been demonstrated that, owing to the gauge invariance, in fact this diagram has merely a logarithmic rather than quadratic divergence.

At $F_e = 2$, $B_e = 0$ we have a single primitively-divergent electron self-energy diagram (ESED) (see, Fig. 21.5(a)). According to Eq. (22.11) this diagram should have the linear divergence. However, the divergence proves to be logarithmic only (see, Section 21.3). It should be noted that in case of the odd dimensionality of D, as a rule, the closest lower even-order divergence is realized.

The last case with $F_e = 2$, $B_e = 1$ corresponds to a vertex diagram (VD). The simplest example of such a diagram is represented in Fig. 22.6(a). In the given case we have the

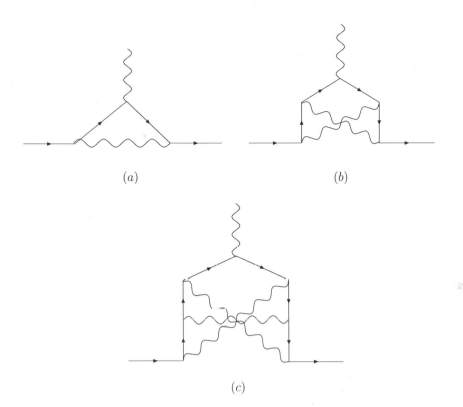

FIGURE 22.0
An example of vertex diagrams (a)–(c).

logarithmic divergence. But as shown in Section 21.6, these divergences are compensated by those existing in the self-energy electron and photon diagrams. Note that this reduction is due to the gauge invariance of the QED as in the case of a successive renormalization, only the gauge-invariant diagram sets should be always considered. The number of primitively-divergent VDs is infinite. Some of these diagrams are shown in Fig. 22.6(b),(c).

In this way the foregoing results may be defined as follows: *any primitively-divergent diagram should be the 5−, 6−, or 7−type diagram.*

22.2 Effective elements of Feynman diagrams

Such notions as compact, noncompact, reducible, and irreducible diagrams are introduced for the self-energy and vertex diagrams.

An ESED is compact when it could not be subdivided into parts connected only by one electron line, being noncompact otherwise. In Fig. 21.5 all the diagrams are compact while the *a*-diagram is noncompact.

In a similar way, it is impossible to divide a compact PSED into parts connected by a single photon line, whereas the situation is opposite in case of a noncompact PSED. In Fig. 22.7 the *b*-diagram is noncompact and the remaining ones are compact. A VD is called

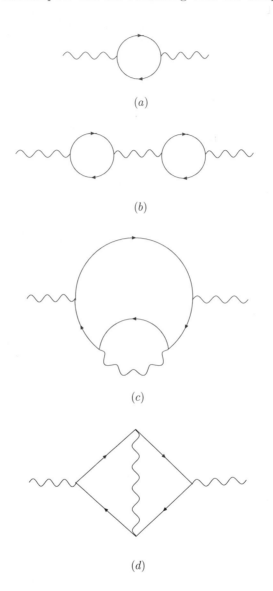

(a)

(b)

(c)

(d)

FIGURE 22.7
The PSED diagrams (a)–(d).

compact when it could not be subdivided into parts connected with each other only by an electron or photon line, and otherwise a VD is noncompact. All VDs pictured in Fig. 22.6 are compact. An example of a noncompact VD is given in Fig. 22.8. If all the self-energy and vertex parts are excluded from some diagram G, we have a reduced diagram referred to as a skeleton of the diagram G. A diagram coincident with its skeleton is called *irreducible*.

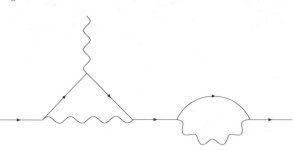

FIGURE 22.8
An example of the noncompact VD diagram.

The ESED and PSED of the second order shown in Figs. 21.5(a) and 22.7(a) are the single examples of the irreducible self-energy parts. All higher-order self-energy parts are reducible and may be obtained by inserting the PSED, ESED, and VD into the lines and vertices of the diagrams displayed in Figs. 21.5(a) and 22.7(a). There are an infinite number of irreducible VDs. Among the diagrams pictured in Fig. 22.6 the first two diagrams are irreducible.

Let us consider an arbitrary diagram containing an internal electron line. In the higher-order approximations this diagram is completed by diagrams with various ESEDs included in the line. Then the initial electron line, together with all the included ESEDs, is graphically represented as a heavy solid line called the effective internal electron line. The quantity associated with this line is called the *electron Green function* $G^{(e)}(p)$ (for the sake of simplicity, hereafter we drop spinor indices). In Fig. 22.9 an infinite sum of diagrams corresponding to the effective electron line $G^{(e)}(p)$ is represented. It is obvious that the following relation

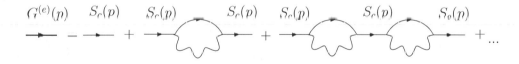

FIGURE 22.9
The diagrams corresponding the effective electron line.

takes place

$$\frac{G^{(e)}(p)}{i} = \frac{S^c(p)}{i} + \frac{S^c(p)}{i}\frac{\Sigma^*(p)}{i}\frac{S^c(p)}{i} + \frac{S^c(p)}{i}\frac{\Sigma^*(p)}{i}\frac{S^c(p)}{i}\frac{\Sigma^*(p)}{i}\frac{S^c(p)}{i} + ... =$$

$$= \frac{S^c(p)}{i} + \frac{S^c(p)}{i}\frac{\Sigma(p)}{i}\frac{S^c(p)}{i}. \tag{22.14}$$

A graphical linkage between $\Sigma^*(p)/i$ and $\Sigma(p)/i$ is displayed in Fig. 22.10, where $\Sigma^*(p)/i$ denotes a sum of quantities associated with only compact ESEDs, and an operator $\Sigma(p)/i$ describes a sum of all possible compact and noncompact ESEDs with the given momentum p (recall, the appearance of the factor i in $\Sigma^*(p)$ is produced by the Wick rotation). From Fig. 22.10 it is evident that

$$\frac{\Sigma(p)}{i} = \frac{\Sigma^*(p)}{i} + \frac{\Sigma^*(p)}{i}\frac{S^c(p)}{i}\frac{\Sigma^*(p)}{i} + \frac{\Sigma^*(p)}{i}\frac{S^c(p)}{i}\frac{\Sigma^*(p)}{i}\frac{S^c(p)}{i}\frac{\Sigma^*(p)}{i} + ... =$$

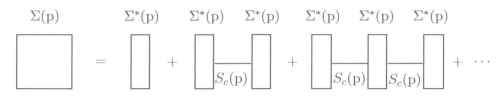

FIGURE 22.10

$\Sigma(p)$ as a sum of $\Sigma^*(p)$.

$$= \frac{\Sigma^*(p)}{i}\left[1 + \Sigma^*(p)S^c(p) + \Sigma^*(p)S^c(p)\Sigma^*(p)S^c(p) + ...\right] = \frac{\Sigma^*(p)}{i}\left[1 - S^c(p)\Sigma^*(p)\right]^{-1}.$$

$$(22.15)$$

To insert this expression into (22.14) leads to the result

$$G^{(e)}(p) = S^c(p) + S^c(p)\Sigma^*(p)[1 - S^c(p)\Sigma^*(p)]^{-1}S^c(p) = [1 - S^c(p)\Sigma^*(p)]^{-1}S^c(p), \quad (22.16)$$

from which the Dyson equation for the electron Green function follows

$$G^{(e)}(p) = S^c(p) + S^c(p)\Sigma^*(p)G^{(e)}(p). \tag{22.17}$$

It is obvious that the graphical representation of this equation has the form pictured in Fig. 22.11. And by the presentation of $G^{(e)}(p)$ in the form of (22.14) it is assumed that from the

FIGURE 22.11

The graphical representation of Eq. (22.17) in the coordinate space.

Feynman graphs we can select more simple blocks to be calculated using the general rules of the diagram technique. Combining these blocks with each other, we can subsequently obtain the proper expression for the graphs as a whole. The possibility of such a selection is an important feature of the diagram technique. This is connected with the fact that a numerical factor in the diagram is independent of its order. The same feature makes it possible to use the function $G^{(e)}$ to simplify the calculation of the RC for the amplitudes of different processes. Instead of considering the graphs with the necessary corrections to the internal electron lines every time anew, we have a possibility to replace these lines by the heavy ones, that is, we simply correlate them with the $G^{(e)}$ propagators in the required approximation.

In a similar way one can determine an effective internal photon line as an internal photon line with all possible PSEDs included in it. In a figure the internal effective line will be displayed by the heavy waved line. The quantity associated with this line is called the *photon Green function* $G^{(\gamma)}(k)$ (for simplicity, hereafter, we omit tensor indices). The graphic presentation of $G^{(\gamma)}(k)$ is given in Fig. 22.12. From this figure it is obvious that the formula:

$$\frac{G^{(\gamma)}(k)}{-i} = \frac{D^c(k)}{-i} + \frac{D^c(k)}{-i}\frac{\Pi(k)}{i}\frac{D^c(k)}{-i} \tag{22.18}$$

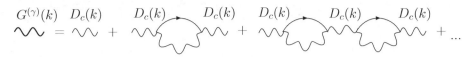

FIGURE 22.12
The graphic presentation of $G^{(\gamma)}(k)$.

takes place.

Introducing a polarization operator $\Pi^*(k)$ to be a sum of all compact PSEDs with the given momentum k, $\Pi(k)$ can graphically be displayed by Fig. 22.13. From this figure it

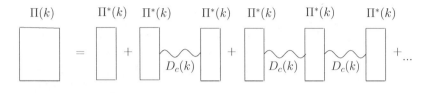

FIGURE 22.13
$\Pi(k)$ as a sum of $\Pi^*(k)$.

follows that

$$\Pi(k) = \Pi^*(k)[1 - D^c(k)\Pi^*(k)]^{-1}. \tag{22.19}$$

Substitution of the obtained expression into (22.18) results in the Dyson equation for the photon Green function

$$G^{(\gamma)}(k) = D^c(k) + D^c(k)\Pi^*(k)G^{(\gamma)}(k), \tag{22.20}$$

whose graphical form is given in Fig. 22.14. Note, the relations (22.17) and (22.20) are the

FIGURE 22.14
The graphical representation of Eq. (22.20).

integral equations. For example, in the coordinate representation the relation (22.17) looks like:

$$G^{(e)}(x,y) = S^c(x-y) + \int d^4x' \int d^4y' S^c(x-x')\Sigma^*(x',y')G^{(e)}(y',y). \tag{22.21}$$

Next, we define the effective external electron and photon lines as external electron and photon lines with all possible self-energy diagrams included in these lines. Their graphic presentation is shown in Figs. 22.15 and 22.16. According to these figures we can write

FIGURE 22.15
The effective external electron line.

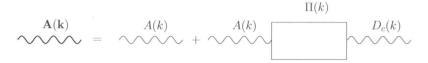

FIGURE 22.16
The effective external photon line.

$$\boldsymbol{\psi}(p) = \psi(p) + S^c(p)\Sigma(p)\psi(p), \tag{22.22}$$

$$\mathbf{A}_\mu(k) = A_\mu(k) + D^c_{\mu\nu}(k)\Pi^{\nu\sigma}(k)A_\sigma(k). \tag{22.23}$$

Analogously, an effective line for an external electromagnetic field is defined (in this figure it is displayed by the heavy waved line with the shaded circle at the end)

$$\mathbf{A}^e_\mu(k) = A^e_\mu(k) + D^c_{\mu\nu}(k)\Pi^{\nu\sigma}(k)A^e_\sigma(k). \tag{22.24}$$

Prior to the definition of the effective vertex, note that since noncompact VDs represent collections both of compact VDs and of the effective electron and photon lines, then using the effective electron and photon lines allows us to omit noncompact VDs from consideration. In this case a sum of the quantities corresponding to all possible compact VDs with the specified momenta p_1, p_2, and k on the electron and photon lines represents the vertex function $\Gamma_{\mu,\alpha\beta}(p_1, p_2; k)$ (μ is a vector index while α and β are spinor indices). In diagrams the vertex function is pictured by a heavy point, known as the effective vertex. Since p_1, p_2, and k are related by the four-dimensional momentum conservation law, in what follows k will be omitted from the argument of the vertex function. It is convenient to extract the ordinary vertex from the vertex function

$$\Gamma_\mu(p_1, p_2) = \gamma_\mu + \Lambda_\mu(p_1, p_2), \tag{22.25}$$

where we have ignored spinor indices. Using the effective lines and effective vertices, we can replace complex diagrams encountered in calculations of the RC by the effective ones, the skeleton diagrams, representing irreducible diagrams with the effective lines (instead of the ordinary lines) and effective vertices (instead of the ordinary vertices). To calculate the required matrix element, we have to know the vertex function and Green functions. The Green functions may be found provided the mass and polarization operators are known. Let us consider the way of finding these quantities.

Since only one irreducible ESED and a single irreducible PSED exist, the mass and polarization operators have one skeleton diagram each (see, Fig. 22.17). It should be noted that all the reducible self-energy parts are exhausted by the insertion of the vertex parts into one vertex rather than into both vertices. Otherwise, the same self-energy part would be taken into consideration more than once. The explanation is as follows. Consider the internal structure of the graph Σ^* in the coordinate space (Fig. 22.18). By the inclusion of all self-energy parts into the internal electron and photon lines of the irreducible diagram

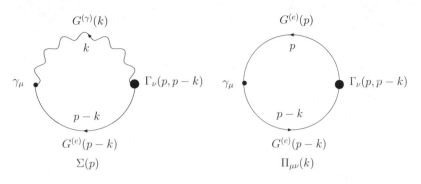

FIGURE 22.17
The skeleton diagrams for the mass and polarization operators.

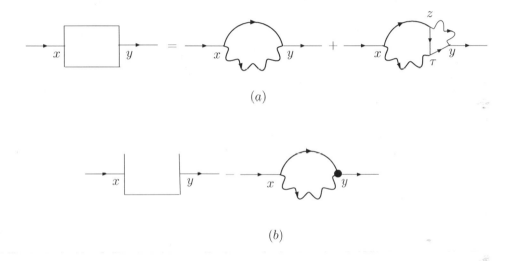

FIGURE 22.18
The internal structure of the graph Σ^* in the coordinate space in (a) and (b).

we have two heavy lines (solid and waved) outgoing from the point x. As this takes place, two variants are possible: both branches are going, without meeting, to the point y (the first term in the right-hand side of Fig. 22.18(a)) or only to some points z and τ (the second term in the right-hand side of Fig. 22.18(a)). Then the remaining part of the graph in the second term represents what we call the *vertex part* $\Lambda_\mu(p_1, p_2)$. A sum of both terms on the right-hand side of Fig. 22.18(a) just gives the desired result (Fig. 22.18(b)). Applying the rules for writing the matrix elements to the skeleton diagrams associated with the mass and polarization operators, we get

$$\Sigma^*(p) = -\frac{e^2}{(2\pi)^4} \int \gamma^\mu G^{(e)}(p-k)\Gamma^\nu(p, p-k)G^{(\gamma)}_{\mu\nu}(k)d^4k, \tag{22.26}$$

$$\Pi^*_{\mu\nu}(p) = \frac{e^2}{(2\pi)^4} \int \gamma_\mu G^{(e)}(p)\Gamma_\nu(p, p-k)G^{(e)}(p-k)d^4p. \tag{22.27}$$

To insert the expressions obtained to the Dyson equation gives the integral equation for determining the electron and photon Green functions

$$G^{(e)}(p) = S^c(p) - \frac{e^2}{(2\pi)^4} S^c(p) \int \gamma^\mu G^{(e)}(p-k) \Gamma^\nu(p, p-k) G^{(\gamma)}_{\mu\nu}(k) G^{(e)}(p) d^4k, \quad (22.28)$$

$$G^{(\gamma)}_{\mu\nu}(k) = D^c_{\mu\nu}(k) + \frac{e^2}{(2\pi)^4} D^c_{\mu\lambda}(k) \int \mathrm{Sp}\left[\gamma^\lambda G^{(e)}(p) \Gamma^\sigma(p, p-k) G^{(e)}(p-k)\right] G^{(\gamma)}_{\sigma\nu}(k) d^4p.$$
$$(22.29)$$

Our next step is finding the vertex function $\Gamma_\mu(p_1, p_2)$. But it turns out that this function could not be represented in the closed form. Indeed, the number of irreducible VDs is countless. So, to derive an integral equation for $\Gamma_\mu(p_1, p_2)$, we have to draw all the irreducible VDs and for each of them replace ordinary lines and vertices by the effective ones. The whole series for Γ_μ is generated by just this very way

$$\Gamma_\mu(p_1, p_2) = \gamma_\mu - \frac{e^2}{(2\pi)^4} \int \Gamma^\sigma(p_1, p_1 - k) G^{(e)}(p_1 - k) \Gamma_\mu(p_1 - k, p_2 - k) \times$$

$$\times G^{(e)}(p_2 - k) \Gamma^\nu(p_2 - k, p_2) G^{(\gamma)}_{\sigma\nu}(k) d^4k + \dots \quad (22.30)$$

In such a manner an equation for the vertex function and hence equations for the electron and photon Green functions could not be represented in the closed form using only differentiation and integration operations with respect to the space-time coordinates. At the same time, equations for these three functions may be obtained in the closed form when we use the functional differentiation operation [78,97]. Since the main question that we are interested in here is that of the existence of such equations rather than the technique of their derivation, then we are not going to reproduce their deduction and refer an interested reader to the monograph [98].

In the presence of an external electromagnetic field the internal electron line includes, apart from the ESED, particular parts associated with the vacuum polarization by the external field, that is, the parts joined to the remaining graph by a photon line. In this case Fig. 22.11 should correspondingly be supplemented and we come to Fig. 22.19. Then the

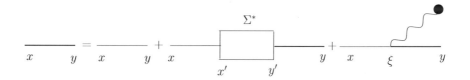

FIGURE 22.19
$G^{(e)}(x, y)$ in the presence of an external electromagnetic field.

integral equation (22.21) is replaced by:

$$G^{(e)}(x, y) = S^c(x - y) + \int d^4x' \int d^4y' S^c(x - x') \Sigma^*(x', y') G^{(e)}(y', y) +$$

$$+ e\gamma^\mu \int d^4z S^c(x - z) \mathbf{A}^e_\mu(z) G^{(e)}(z, y). \quad (22.31)$$

Here $\mathbf{A}^e_\mu(x)$ is the sum of the original external field $A^e_\mu(x)$ and the RC to it $\delta A^e_\mu(x)$. In the coordinate representation the operator $\Sigma^*(x, y)$ is given by:

$$\Sigma^*(x, y) = -e^2 \gamma^\mu \int d^4z \int d^4\tau G^{(e)}(x, z) \Gamma_\mu(z, y; \tau) G^{(\gamma)}(x, \tau), \quad (22.32)$$

where

$$\Gamma_\mu(z, y; \tau) = \gamma_\mu \delta^{(4)}(z - \tau)\delta^{(4)}(y - \tau) + \Lambda_\mu(z, y; \tau). \tag{22.33}$$

Acting by the operator $i\hat{\partial} - m$ on Eq. (22.31), we arrive at the Schwinger equation [78]:

$$[i\hat{\partial} - e\hat{\mathbf{A}}^e(x) - m]G^{(e)}(x, y) + \int d^4 y' \Sigma^*(x, y')G^{(e)}(y', y) = \delta^{(4)}(x - y), \tag{22.34}$$

where we have allowed for

$$(i\hat{\partial} - m)S^c(x - y) = \delta^{(4)}(x - y).$$

One can pass from the equation for the Green function to that for the electron wavefunction in an external field with the RC inclusion. For this purpose it is enough to omit the δ-function in the right-hand side of Eq. (22.34) to give:

$$[i\hat{\partial} - e\hat{\mathbf{A}}^e(x) - m]\psi(x) + \int d^4 y' \Sigma^*(x, y')\psi(y') = 0. \tag{22.35}$$

It should be recorded that the majority of the above manipulations are concerned with infinite quantities. Obviously, a mathematical rigor necessitates regularization of all the quantities prior to studying their properties and before the derivation of the equations they are satisfying. With the use of the Pauli-Villars regularization the quantities, which were finite without the regularization, remain unchanged, while, for example, linearly divergent quantities change to the finite linear functions of the regularizing masses.

22.3 Ward identity

As follows from Eqs. (22.26) and (22.27), relations between the vertex function and the electron as well as photon Green functions are complex in nature. But when the arguments of the vertex function are coincident, this relation becomes quite simple. We can find it for this particular case.

Having differentiated an identity

$$S^c(p)[S^c(p)]^{-1} = 1,$$

with respect to p_μ, we obtain

$$\frac{\partial S^c(p)}{\partial p^\mu} = -S^c(p)\frac{\partial [S^c(p)]^{-1}}{\partial p^\mu}S^c(p). \tag{22.36}$$

Substitution of the value

$$[S^c(p)]^{-1} = m - \hat{p}$$

into Eq. (22.36) results in

$$\frac{\partial S^c(p)}{\partial p^\mu} = S^c(p)\gamma_\mu S^c(p). \tag{22.37}$$

The right-hand side of the derived relation, to within a constant factor, is equal to the matrix element associated with a part of the diagram, where an external photon line is joined to the internal electron line (Fig. 22.20). Thus, formal differentiation of the electron

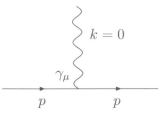

FIGURE 22.20
The graphical representation of the right-hand side of Eq. (22.37).

propagator with respect to p^μ corresponds to the situation when a photon line with zero momentum is joined to the electron line.

Next, we consider a diagram W and insert an irreducible ESED in its internal electron line, that is, carry out the replacement:

$$S^c(p) \rightarrow S^c(p)\Sigma(W,p)S^c(p). \tag{22.38}$$

Hereinafter, we introduce the diagram symbol into the argument of the self-energy operators to emphasize that now they are the components of the diagram W. It is clear that $\Sigma(W,p)$ differs from the operator $\Sigma^{(2)}(p)$

$$\Sigma^{(2)}(p) = -\frac{e^2}{(2\pi)^4} \int d^4k \gamma^\nu S^c(p-k)\gamma_\nu D^c(k) \tag{22.39}$$

by the factor i. Differentiating (22.39) and making use of the relation (22.37), we arrive at the equation

$$\frac{\partial \Sigma(W,p)}{\partial p^\mu} = -\frac{ie^2}{(2\pi)^4} \int d^4k \gamma^\nu S^c(p-k)\gamma_\mu S^c(p-k)\gamma_\nu D^c(k). \tag{22.40}$$

The expression in the right-hand side of Eq. (22.39) is equal to the lowest (third) order vertex operator at the zero-momentum transfer $\Lambda^{(3)}(p,p)$. So, we can state that the performed substitution (22.38) and subsequent differentiation of the obtained diagram with respect to p_μ is equivalent to the inclusion of the vertex into the internal electron line or replacement:

$$\gamma_\mu \rightarrow \Lambda^{(3)}(p,p)$$

at the vertex of the initial diagram. It is seen that from the compact ESED associated with the operator $\Sigma'(W,p)$ we can obtain a compact VD by joining the external photon line to some internal electron line. Formally, an insertion of the zero-momentum photon line into the electron line implies taking a derivative with respect to the external momentum from the propagator associated with this electron line. Consequently, due to differentiation of the operator $\Sigma'(W,p)$ with respect to p^μ, a photon line is joined to each electron line transferring the momentum p. Provided the ESED contains a closed electron line as in Fig. 22.21, in case of the photon line joining to the electron loop we obtain a collection of diagrams whose sum is equal to zero in accordance with the Furry theorem. Because of this, the electron loop takes no part in the differentiation with respect to the momentum of an external electron line.

If the sum of all the compact ESED's associated with the operator $\Sigma^*(W,p) \equiv \Sigma^*(p)$ is differentiated with respect to p^μ, then we shall lead to the identity

$$\frac{\partial \Sigma^*(p)}{\partial p^\mu} = \Lambda_\mu(p,p), \tag{22.41}$$

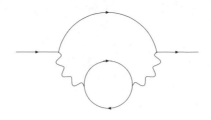

FIGURE 22.21

The ESED containing a closed electron line.

where $\Lambda_\mu(p,p)$ denotes the set of all the compact VDs, Ward has obtained for the first time [99]. Using (22.17), one can write

$$[G^{(e)}(p)]^{-1} = [S^c(p)]^{-1} - \Sigma^*(p) = [m - \hat{p} - \Sigma^*(p)]. \tag{22.42}$$

Then, differentiating (22.42) with respect to p^μ yields another writing of the Ward identity:

$$\frac{\partial[G^{(e)}(p)]^{-1}}{\partial p^\mu} = \frac{\partial}{\partial p^\mu}[m - \hat{p} - \Sigma^*(p)] = -[\gamma_\mu + \Lambda_\mu(p,p)] = -\Gamma_\mu(p,p). \tag{22.43}$$

Let us show that the Ward identity is a consequence of the gauge invariance of the QED. Under an infinitesimal gauge transformation we have

$$A_\mu(x) \to A_\mu(x) - \frac{\partial}{\partial x^\mu}\delta\phi(x) = A_\mu(x) - \delta A_\mu(x), \tag{22.44}$$

$$\left.\begin{array}{l} \psi(x) \to \psi(x)\exp[-ie\delta\phi(x)] \approx [1 - ie\delta\phi(x)]\psi(x), \\ \overline{\psi}(x) \to \overline{\psi}(x)\exp[ie\delta\phi(x)] \approx [1 + ie\delta\phi(x)]\overline{\psi}(x). \end{array}\right\} \tag{22.45}$$

In this case the change-over of the electron Green function is

$$\delta G^{(e)}(x,y) = ieG^{(e)}(x-y)[\delta\phi(y) - \delta\phi(x)]. \tag{22.46}$$

Emphasize, the gauge transformation (22.45) violates the space-time homogeneity and the function $\delta G^{(e)}$ depends on the arguments x and y individually rather than on the difference $x - y$. Introducing the Fourier representation for $\delta G^{(e)}(x,y)$ and $\delta\phi(x)$

$$\delta G^{(e)}(x,y) = \int d^4p_1 \int d^4p_2 \delta G^{(e)}(p_1,p_2)e^{i(p_1y - p_2x)}, \tag{22.47}$$

$$\delta\phi(x) = \int d^4p\,\delta\phi(p)e^{-ipx} \tag{22.48}$$

and substituting them into Eq. (22.46), we gain

$$\delta G^{(e)}(p+q,p) = ie\delta\phi(q)[G^{(e)}(p) - G^{(e)}(p+q)]. \tag{22.49}$$

Because the theory is gauge invariant, then the obtained value of the electron propagator must coincide with the value an infinitesimal external field $\delta A_\mu(x)$ generates. In the momentum representation $\delta A_\mu(x)$ takes the form:

$$\delta A_\mu(q) = -iq_\mu\delta\phi(q). \tag{22.50}$$

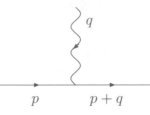

FIGURE 22.22
The graphical representation of $\delta G^{(e)}(p_1, p_2)$.

When one is limited by the first order of smallness in terms of $\delta\phi$, then the quantity $\delta G^{(e)}(p_1, p_2)$ is graphically represented by the skeleton diagram of Fig. 22.22. However, in this figure the effective external photon line associated with the factor

$$\delta A_\mu(q) + D^c_{\mu\sigma}(q)\Pi^{\sigma\nu}(q)\delta A_\nu(q) \tag{22.51}$$

must be changed by the ordinary line as the second term of (22.51) vanishes. Then, from Fig. 22.22 it follows

$$\delta G^{(e)}(p+q, p) = eG^{(e)}(p+q)\Gamma^\mu(p+q, p; q)G^{(e)}(p)\delta A_\mu(q). \tag{22.52}$$

Inserting $\delta A_\mu(q)$ defined by (22.50) into Eq. (22.52) and comparing the obtained expression with (22.49), we arrive at the relation:

$$G^{(e)}(p+q) - G^{(e)}(p) = q_\mu G^{(e)}(p+q)\Gamma^\mu(p+q, p; q)G^{(e)}(p), \tag{22.53}$$

which may also be represented in the form:

$$[G^{(e)}(p+q)]^{-1} - [G^{(e)}(p)]^{-1} = -q_\mu\Gamma^\mu(p+q, p; q). \tag{22.54}$$

Expanding both sides of the expression (22.54) into a series in the infinitesimal quantity q_μ and transiting to the limit $q \to 0$, we again get the Ward identity in the form (22.43)

$$\frac{\partial}{\partial p^\mu}[G^{(e)}(p)]^{-1} = -\Gamma_\mu(p, p), \tag{22.55}$$

where $\Gamma(p, p; 0) \equiv \Gamma(p, p)$.

22.4   Extraction of divergences

There are three kinds of the PDDs in the QED, namely, the self-energy and vertex diagrams. In Sections 21.3–21.5 we have already investigated the extraction of the divergences in the lowest orders of the perturbation theory for the matrix elements corresponding to these diagrams. In the process, we have used the invariant method by Dyson [4] based on expanding the integrand associated with a divergent diagram into the Taylor series in terms of the external momentum (see Eq. (21.62)).

We have found that the ESED of the second order is described by the expression:

$$\Sigma^{(2)}(p) = \Sigma^{(2)}(m_R) + (\hat{p} - m_R)\left(\frac{\partial\Sigma^{(2)}(p)}{\partial\hat{p}}\right)_{\hat{p}=m_R} + \Sigma_C^{(2)}(p), \tag{22.56}$$

where the first two terms are diverging linearly and logarithmically, respectively. As in this case the second-order diagram is simultaneously an irreducible diagram, Eq. (22.56) is valid for the divergences separation of the irreducible ESED, actually giving a contribution of all the irreducible ESEDs. Introducing the designations

$$\Sigma^{(2)}(m_R) = A^n, \qquad \left(\frac{\partial \Sigma^{(2)}(p)}{\partial \hat{p}}\right)_{\hat{p}=m_R} = B^n,$$

we rewrite (22.56) in the view:

$$\Sigma^n(p) = A^n + B^n(\hat{p} - m_R) + \Sigma^n_C(p), \tag{22.57}$$

where $\Sigma^n_C(p)$ vanishes for a free electron. The mass renormalization is determined in such a way that the free electron mass ($\hat{p} = m_R$) is exactly equal to m_R, that is,

$$\overline{\psi}(p)\Sigma^n(p)\psi(p) = 0 \qquad \text{at} \qquad \hat{p} = m_R, \tag{22.58}$$

to give

$$A^n = -i\delta m. \tag{22.59}$$

The polarization tensor $\Pi(W, q^2)$ appearing after replacement of the internal photon line:

$$D^r_{\mu\nu}(q) \rightarrow D^{\lambda r}_\mu(q)\Pi_{\lambda\sigma}(W, q)D^{\sigma r}_\nu(q), \tag{22.60}$$

is distinguished from $\Pi(q^2)$ by the constant factor i. This is easily seen when we take into account the explicit form of the operator $\Pi_{\lambda\sigma}(q)$ and use the Feynman gauge for the propagator of an electromagnetic field. Investigating the PSED of the second order, we have established that splitting the polarization tensor into the finite and divergent parts has the form:

$$\Pi^{(2)}(q^2) = \Pi^{(2)}(0) + \Pi^{(2)}_C(q^2), \tag{22.61}$$

where $\Pi^{(2)}(0)$ contains the logarithmic divergence and the finite quantity $\Pi^{(2)}_C(q^2)$ equals zero at $q^2 = 0$. Since the diagram under consideration is the single irreducible PSED, then the contribution of all irreducible PSEDs into $\Pi(q^2)$ is given by the expression:

$$\Pi^n(q^2) = C^n + \Pi^n_C(q^2), \tag{22.62}$$

where we have taken the designation $C^n = \Pi^n(0)$.

In the case of the VDs the situation is somewhat different. Here, apart from the third order diagram considered in Section 21.5, we have an infinite set of irreducible diagrams. In this way the separation procedure for the divergences should be extended to all the irreducible VDs.

A matrix element corresponding to some irreducible VD could be represented by an integral

$$\Lambda^{V_i}_\mu(p_1, p_2) = \int d^4k R^{V_i}_\mu(p_1, p_2, k). \tag{22.63}$$

As the contribution of the vertex parts has a logarithmic divergence, in the Taylor expansion for the integrand (22.63) the terms with the derivatives are not required. As a result we get:

$$\Lambda^{V_i}_\mu(p_1, p_2) = \Lambda^{V_i}_\mu(p_1, p_1) + \Lambda^{V_i}_{C\mu}(p_1, p_2) \tag{22.64}$$

where $\Lambda^{V_i}_\mu(p_1, p_1)$ is logarithmically diverging and the finite quantity $\Lambda^{V_i}_{C\mu}(p_1, p_2)$ turns into zero at $p_1 = p_2$. The relativistic invariance demands

$$\Lambda^{V_i}_\mu(p_1, p_1) = L(V_i)\gamma_\mu. \tag{22.65}$$

At $\hat{p}_1 = \hat{p}_2 = m_R$ and $p_1 = p_2$ the operator $\Lambda_{C\mu}^{V_i}(p_1, p_2)$ is reduced to the matrix γ_μ multiplied by a constant value. This value may be included into the term $L(V_i)\gamma_\mu$. Then the operator $\Lambda_{C\mu}^{V_i}(p_1, p_2)$ will be equal to zero at $\hat{p}_1 = \hat{p}_2 = m_R$ and $p_1 = p_2$, rather than at $p_1 = p_2$. By the way, such a choice is more justified from the physical viewpoint. Indeed, the quantity $\Lambda_{C\mu}^{V_i}(p_1, p_2)$ makes no contribution when the electromagnetic potential is constant, because such a potential is transferring zero momentum. Very well, this is clear as a constant electromagnetic potential may be eliminated any time with the use of the gauge transformation.

Having summarized (22.64) over all the irreducible VDs, we obtain the required relation:

$$\sum_{V_i} \Lambda_\mu^{V_i}(p_1, p_2) = L^n \gamma_\mu + \Lambda_{C\mu}^n(p_1, p_2), \tag{22.66}$$

where

$$L^n = \sum_{V_i} \Lambda(V_i). \tag{22.67}$$

To demonstrate a possibility for renormalization of a theory as a whole, we have to consider diagrams of all types and all orders, both reducible and irreducible. Thus, the next step necessitates demonstration of the divergences separation in the case of reducible diagrams. It should be noted that separation of the matrix elements into finite and infinite parts must be performed in the relativistically invariant and unambiguous way. Logically, it may be assumed that the formulae providing the basis for separation of divergences in the expressions associated with reducible diagrams are of the same form as for irreducible diagrams. This statement may be supported if we consider the procedure of separating the finite and infinite expressions in the general form for the ESEDs.

Let us look at a reducible electron self-energy part $\Sigma^r(W, p)$, which is a compound element of some diagram. It is clear on relativistic invariance grounds that this operator must have the form:

$$\Sigma^r(W, p) = A' + B'_\mu p^\mu + \Sigma_C^r(p), \tag{22.68}$$

where the matrix operators A' and B'_μ are diverging linearly and logarithmically, respectively. Inasmuch as $\Sigma^r(W, p)$ represents a 4×4- matrix, then it may be expanded in terms of 16 linearly independent matrices Γ_j (see, Eq. (15.11)):

$$\Sigma^r(W, p) = \Sigma_S(p) + \Sigma_{V\mu}(p)\gamma^\mu + \frac{1}{2}\Sigma_{T\mu\nu}(p)\sigma^{\mu\nu} + \Sigma_P(p)\gamma_5 + \Sigma_{A\mu}(p)\gamma^\mu\gamma_5. \tag{22.69}$$

The parity conservation provides

$$\Sigma_P(p) = \Sigma_{A\mu}(p) = 0. \tag{22.70}$$

Because the vector p_μ is the single one being in our disposal, then, due to the relativistic invariance, the vector and tensor coefficients in (22.69) are of the form:

$$\Sigma_{V\mu} = p_\mu \Sigma_V(p^2), \qquad \Sigma_{T\mu\nu} = p_\mu p_\nu \Sigma_T(p^2). \tag{22.71}$$

Note, the relativistic invariance also demands $\Sigma_S(p) \equiv \Sigma_S(p^2)$. Thus, the expression (22.69) is represented in the form:

$$\Sigma^r(W, p) = \Sigma_S(p^2) + \gamma^\mu p_\mu \Sigma_V(p^2). \tag{22.72}$$

Comparing the expressions (22.69) and (22.72), we find

$$A' = \Sigma_S(0), \qquad B'_\mu = \gamma_\mu \Sigma_V(0) = B\gamma_\mu, \tag{22.73}$$

to give

$$\Sigma^r(W,p) = A^r + B^r(\hat{p} - m_R) + \Sigma_C^r(p),$$

where $A^r = A' + B^r m_R$. So, the contribution of the reducible ESEDs into the mass operator

$$\sum_i \Sigma(W_i, p) = \Sigma(p)$$

has the same structure as in the case of the irreducible ESEDs, and we have finally

$$\Sigma(p) = A + B(\hat{p} - m_R) + \Sigma_C(p). \tag{22.74}$$

In a similar manner, one could show that the contributions of the reducible diagrams into the total operators

$$\sum_i \Pi(W_i, q^2) = \Pi(q^2), \qquad \sum_i \Lambda_\mu(p_1, p_2, V_i) = \Lambda_\mu(p_1, p_2),$$

are governed by the expressions having the same form as in the case of irreducible diagrams. All this results in

$$\Pi(q^2) = C + \Pi_C(q^2), \tag{22.75}$$

$$\Lambda_\mu(p_1, p_2) = L\gamma_\mu + \Lambda_{C\mu}(p_1, p_2), \tag{22.76}$$

where C and L are logarithmically divergent constants. It should be stressed that now all the constants entering into Eqs. (22.7), (22.75) are real. The relations (22.74)–(22.76) form the basis for the divergences separation in expressions associated both with the irreducible and reducible divergent diagrams. Note, the relation:

$$B = L, \tag{22.77}$$

which connects the divergent quantities contained in Eqs. (22.74) and (22.76), is the direct consequence of the Ward identity. This is a proof that in the QED in all orders of the perturbation theory the vertex divergence is canceled with the divergence that remains in the electron self-energy after the electron mass renormalization.

But we face unpleasant surprises under more detailed analysis of the divergences in reducible diagrams. It appears that the G diagrams, resultant from the irreducible diagram G_0 by insertion of the PSEDs, ESEDs, and VDs into various lines and vertices of G_0, may be subdivided into following three classes: (i) diagrams with absolutely non-overlapping insertions; (ii) diagrams where the insertions are totally contained within each other; and (iii) diagrams with overlapping insertions. The first two classes of divergences may be separated by the invariant Dyson method (see, Eq. (21.62)). When the insertions are non-overlapping, separation is performed independently for each insertion. Divergences of the second class are separated step-by-step, beginning from divergences of the inner insertion and ending with the divergence of the diagram G as a whole. To illustrate, in the case of a diagram given in Fig. 22.23 the divergences associated with integration with respect to virtual momenta are separated in the following sequence: k_1, k_2, and k_3. A combination of divergences of the first two types is demonstrated in Fig. 22.24. Here the sequence of separation is as follows: first we extract the divergence due to integration over k_1, then do divergence of the self-energy photon part in line k_2, and finally do divergence associated with integration over k_3.

The method of Dyson is inapplicable for overlapping insertions because in this case the integrals are divergent with respect to more than one variable. They are divergent even if integration is over one of the variables, whereas all others are fixed. As an example, we

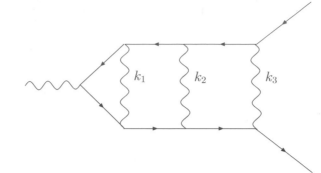

FIGURE 22.23

An example of the diagram with overlapping insertions.

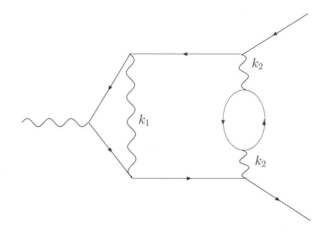

FIGURE 22.24

An example of the diagram with divergences of the (i) and (ii) type.

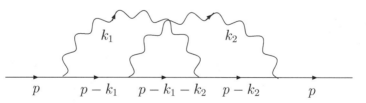

FIGURE 22.25

The overlapping ESED.

consider an overlapping ESED (Fig. 22.25), which may be obtained by insertion of the VD into any of the vertices of the second-order ESED. It is associated with the matrix element

$$\Sigma^{(4)}(p) = \frac{ie^4}{(2\pi)^8} \int d^4k_1 \int d^4k_2 \frac{1}{k_1^2} \gamma^\mu \frac{\hat{p} - \hat{k}_1 + m}{(p - k_1)^2 - m^2} \times$$

$$\times \gamma^\nu \frac{\hat{p} - \hat{k}_1 - \hat{k}_2 + m}{(p - k_1 - k_2)^2 - m^2} \gamma_\mu \frac{1}{k_2^2} \frac{\hat{p} - \hat{k}_2 + m}{(p - k_2)^2 - m^2} \gamma_\nu. \tag{22.78}$$

From (22.78) it is evident that the double integral over k_1 and k_2 is linearly diverging. To fix one of variables does not save the situation. Fixing k_2 (k_1), we get the integral logarithmically divergent with respect to k_1 (k_2).

In much the same manner, overlapping divergences are found on insertion of the VD into the vertices of the PSED. An example of such a diagram, where the insertion of the VD has been made into one of the vertices of the second-order PSED, is given in Fig. 22.26. In

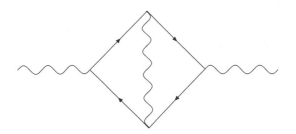

FIGURE 22.26
The PSED obtained by the insertion of the VD.

the general case a contribution made by such reducible diagrams into Π^* or Σ^* represents an integral that possesses the divergences corresponding to all the diagram construction procedures possible with the use of the VD insertion into one or both vertices of the initial irreducible diagram. The number of possible methods to construct a reducible diagram from the irreducible components in such a way is growing with the diagram order. Since for each construction method divergences are introduced independently, then investigating the appeared divergences becomes very complicated. Note that, when we look for an adequate approach to the divergence separation of the overlapping diagrams, our main difficulty is a correct enumeration of all the divergent parts, enabling interpretation of subtraction as a renormalization and leading to an unambiguous separation of the finite parts. A method enabling unambiguous separation of covariant and absolutely convergent residual terms in the case of the overlapping divergent parts has been proposed in Refs. [100]. With its help one can deduce the formulae (22.74)–(22.76) in the case of any diagrams.

22.5   QED renormalizability

Renormalizability of the QED may be proved without going into the problem of overlapping divergences. Again, the main hero of this history is the gauge invariance, or more precisely, one of its consequences, the Ward identity. We shall give the proof of the QED renormalizability following the papers by Ward [94].

First we show the way to obviate the problem of overlapping divergences in case of the ESED. We take a look at the mass operator $\Sigma(W_1, p)$ corresponding to the diagram W_1 with overlapping divergences (see, Fig. 22.25). Computing the derivative $(\partial\Sigma(W_1, p)/\partial p^\mu)$, we get three addends associated with the diagrams displayed in Fig. 22.27. These three diagrams are obtained upon inserting the zero-momentum external photon line to three internal electron lines of the diagram in Fig. 22.25. Note that insertion of the external

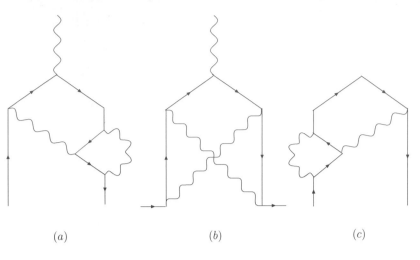

FIGURE 22.27
The diagrams (a)–(c) associated with $(\partial \Sigma(W_1, p)/\partial p^\mu)$.

photon line is (i) primarily lowering a degree of divergence (the diagram of Fig. 22.25 is diverging linearly while that of Fig. 22.27 is diverging only logarithmically), and (ii) most important, in the diagrams found in Fig. 22.27 we have no overlapping divergences. Let us denote the operators correlated with the diagrams pictured in Fig. 22.27 as $\Lambda(p_1, p_2, V_j^{(1)})$ $(j = 1, 2, 3)$. Non-overlapping divergences in $\sum \Lambda_\mu(p_1, p_2, V_j^{(1)})$ could be extracted by application the Dyson method. With allowance made for

$$\frac{\partial}{\partial p^\mu} \Sigma(W_1, p) = \sum_{j=1}^{3} \Lambda_\mu(p, p, V_j^{(1)}) = \Lambda_\mu(p, p, V^{(1)}), \qquad (22.79)$$

we carry out integration (22.79) and arrive at

$$\Sigma(W_1, p) - \Sigma(W_1, p') = \int_{p'}^{p} dq^\mu \Lambda \mu(q, q, V^{(1)}). \qquad (22.80)$$

Making use of the relation:

$$\Sigma(W_1, p') \Big|_{\hat{p}'=m_R} = -i\delta m(W_1),$$

one can present the formula (22.80) in the form:

$$\Sigma(W_1, p) + i\delta m(W_1) = \int_{p'}^{p} dq^\mu \Lambda_\mu(q, q, V^{(1)}), \qquad (22.81)$$

where $\hat{p}' = m_R$. In such a manner, if the expression for the operator $\Lambda_\mu(q, q, V^{(1)})$ (corresponding to the compact VD) with already separated divergences is used in Eq. (22.81), then divergences of the operator $\Sigma(W_1, p)$ also prove to be isolated.

In the most general case, we should go to the sum over all compact diagrams. Then, integration of the Ward identity yields

$$\Sigma^*(p) - \Sigma^*(p') = \int_{p'}^{p} dq^\mu \Lambda_\mu(q, q) = \int_{0}^{1} dx(p^\mu - p'^\mu)\Lambda_\mu(p^x, p^x), \qquad (22.82)$$

where p' is a free-electron momentum, and in order to reach the last line we have done the replacement

$$q = p^x = px + (1 - x)p'.$$

The term $\Sigma^*(p')$ allows us to automatically take into account the mass renormalization because we have by definition

$$\Sigma^*(p')|_{\hat{p}'=m_R} = -i\delta m.$$

From this point on we shall omit the term $\Sigma^*(p')$ supposing that δm has been included into the definition of $\Sigma^*(p)$.

As follows from Section 22.2, an exact value of the operator Λ_μ is found when in the expressions associated with all the irreducible VDs we perform the following substitutions:

$$\gamma_\mu \rightarrow \Gamma_\mu, \qquad S^c \rightarrow G^{(e)}, \qquad D^c \rightarrow G^{(\gamma)} \qquad (22.83)$$

and summarize the whole series. The operator Λ_μ has no overlapping as it is constructed from irreducible diagrams. Consequently, separation of divergences may be carried out without fail.

A similar procedure is available for the PSED as well. Here the problem of overlapping divergencies may be obviated by reducing analysis of the PSED to an analysis of the diagrams resultant from the differentiation over the external photon momentum. For example, let us consider the diagram W_2 shown in Fig. 22.28. The operator associated with this di-

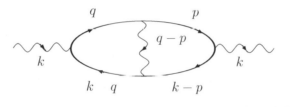

FIGURE 22.28
The PSED of the fourth order.

agram is denoted as $\Pi(W_2, k^2)$ while the symbol $\Delta_\mu(W_2, k, k)$ will be used for the operator resulting from differentiation of $\Pi(W_2, k^2)$ with respect to k_μ. Since

$$\frac{\partial}{\partial k_\mu} \left(\frac{1}{\hat{k} - \hat{q} - m} \right) = -\frac{1}{\hat{k} - \hat{q} - m} \gamma^\mu \frac{1}{\hat{k} - \hat{q} - m}, \qquad (22.84)$$

then the operator $\Delta_\mu(W_2, k, k)$ could be constructed by application of the diagrams of Figs. 22.29 obtained by means of insertions of an external photon line with the zero-momentum to those electron lines of the diagram of Fig. 22.28 that transfer an external momentum. It should be noted that the diagrams shown in Figs. 22.29 are nonzero, despite the Furry theorem, as here a sum of the diagrams is not considered. Obviously, the divergence of the obtained diagrams becomes weaker because differentiation leads to an increase in the denominator degree in the integrand by unity. In more higher degrees the PSEDs may emerge in which the external photon momentum is transferred by some of the internal photon lines (see, for example, Fig. 22.30). In this case

$$\frac{\partial}{\partial k_\mu} \left[\frac{1}{(k - q)^2} \right] = -4 \frac{k_\mu - q_\mu}{(k - q)^3}$$

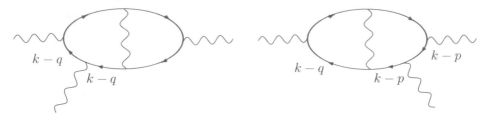

FIGURE 22.29

The diagrams obtained by means of insertions of an external photon line to the diagram pictured in Fig. 22.28.

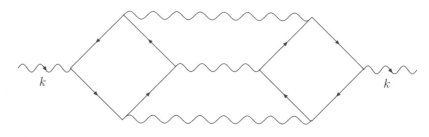

FIGURE 22.30

An example of higher degrees the PSED.

and the integral convergence is again improved (the term being proportional to q_μ vanishes in accordance with the Lorentz invariance). But in both cases the principal feature is the elimination of overlapping divergences due to differentiation with respect to k_μ. Actually, when a zero-momentum photon line is connected to some part having an overlapping divergence, the diagram element with a joined up photon line ceases to be divergent, and only one of the overlapping divergences is retained.

Let us define an operator Δ_μ (it is analogous to the operator Λ_μ) with the help of the relation:

$$\frac{\partial}{\partial k_\mu}\Pi^*(k^2) = \Delta_\mu(k,k). \tag{22.85}$$

Then, taking into account that the photon self-energy equal to zero ($\Pi^*(0) = 0$), we find from (22.85)

$$\Pi^*(k^2) = \int_0^k dq^\mu \Delta_\mu(q,q) = \int_0^1 dy k^\mu \Delta_\mu(yk,yk). \tag{22.86}$$

The definition of Δ_μ as the derivative of Π^* is in fact an implicit definition enabling one to find the functions Π^* and Δ_μ with the help of a certain class of the specially selected irreducible diagrams. Naturally, a question arises: if Δ_μ is an analog of Λ_μ, what quantity will be associated with Γ_μ? This quantity is designated by W_μ and given by the relation:

$$W_\mu(k) = -2k_\mu + \Delta_\mu(k). \tag{22.87}$$

Really, as

$$\frac{\partial[D^c(k)]^{-1}}{\partial k_\mu} = 2k^\mu,$$

then

$$\frac{\partial}{\partial k_\mu}[G^{(\gamma)}]^{-1} = \frac{\partial}{\partial k_\mu}\{[D^c(k)]^{-1} - [\Pi^*(k)]\} = 2k^\mu - \Delta_\mu(k,k) = -W_\mu(k). \tag{22.88}$$

By application of Eq. (22.88), we find

$$[G^{(\gamma)}]^{-1} = -\int_0^1 dy k^\mu W_\mu(ky). \qquad (22.89)$$

In the formalism of Feynman diagrams, the function W_μ represents a contribution of the irreducible diagrams following from all the PSEDs by connection of an external photon line to the electron lines transferring an external photon momentum (see, for example, Fig. 22.31). To obtain an exact value of the operator W_μ, we have to make substitutions (22.83)

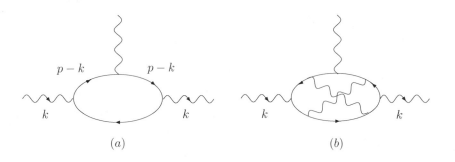

FIGURE 22.31
An example of the irreducible diagrams (a) and (b) following from the PSEDs.

in the expressions associated with the starting Feynman diagrams and to sum the derived series. This gives the result:

$$W_u = -2k_\mu - \frac{2\alpha}{3(2\pi)^3} \int \mathrm{Sp}\{\Gamma^\nu(p, p+k)G^{(e)}(p+k)\Gamma_\mu(p+k, p+k)\times$$

$$\times G^{(e)}(p+k)\Gamma_\nu(p+k, p)G^{(e)}(p)\}d^4p + ..., \qquad (22.90)$$

where we have written only the term related with the diagram shown in Fig. 22.31(a). Since, in the most general case, the operator is diverging quadratically, then the operator Δ_μ may have linear divergence and no higher. Therefore, its finite value is given as:

$$\Delta_{C\mu}(yk, yk) = \Delta_\mu(yk, yk) - \Delta_\mu(0,0) - \left(\frac{\partial\Delta_\mu(yk, yk)}{\partial(yk)_\nu}\right)_{k=0}(yk)_\nu. \qquad (22.91)$$

Actually, due to the Lorentz invariance, this divergence is reduced to the logarithmic one as there is no invariant four-dimensional vector, that is, $\Delta_\mu(0,0) = 0$. Recall, the vertex operator playing the similar role for the ESED, is logarithmically diverging too, that is

$$\Gamma_{C\mu}(p_1, p_2) = \gamma_\mu + \Lambda_\mu(p_1, p_2) - \Lambda_\mu(p_1', p_1') = \gamma_\mu + \Lambda_{C\mu}(p_1, p_2), \qquad (22.92)$$

where p_1' is a four-dimensional momentum for a free electron.

Demonstration of the QED renormalizability may be performed in two stages. At the first stage the subtraction procedure is performed, making all the elements of the S-matrix finite. The second stage is a proof that this subtraction is equivalent to renormalization, that is, dropping of the infinite terms is equivalent to redefinition of the mass and charge values for a bare electron.

First, we define the finite parts of the functions Γ_μ, $G^{(e)}$ and $G^{(\gamma)}G$, which should be inserted into skeleton diagrams during the calculation of exact matrix elements. By the appropriate subtractions, the kernels in the Dyson equations (22.17) and (22.20) are made finite

$$G^{(e)}(p) = S^c(p) + S^c(p) \int_0^1 dx (p^\mu - p'^\mu)[\Lambda_\mu(p^x, p^x) - \Lambda_\mu(p', p')]G_C^{(e)}(p), \tag{22.93}$$

$$G^{(\gamma)}(k) = D^c(k) + D^c(k) \int_0^1 dy k^\mu [\Delta_\mu(yk, yk) - \left(\frac{\partial \Delta_\mu(yk, yk)}{\partial (yk)_\nu} \right)_{k=0} (yk)_\nu] G_C^{(\gamma)}(k). \tag{22.94}$$

Next, in the modified equations (22.92)–(22.94) the constant e is replaced by the quantity $e_R(e, m)$ selected so that the modified functions:

$$G_R^{(e)}(p; e_R) = G^{(e)}(p) \Big|_{e=e_R}, \qquad G_R^{(\gamma)}(k; e_R) = G^{(\gamma)}(k) \Big|_{e=e_R},$$

$$\Gamma_{R\mu}(p_1, p_2; e_R) = \Gamma_{C\mu}(p_1, p_2) \Big|_{e=e_R}$$

are different from the initial diverging functions $G^{(e)}(p; e), G^{(\gamma)}(k; e)$ and $\Gamma_\mu(p_1, p_2; e)$ only by the common factors. This means that modification of the initial integral equations (22.17), (22.20), and (22.25) by subtraction of the divergent terms is reduced to renormalization of the charge. Thus, we should show the validity of the equations:

$$G^{(e)}(p; e) = Z_2(e_R)G_R^{(e)}(p; e_R), \tag{22.95}$$

$$G^{(\gamma)}(k; e) = Z_3(e_R)G_R^{(\gamma)}(k; e_R), \tag{22.96}$$

$$\Gamma_\mu(p_1, p_2; e) = Z_1^{-1}(e_R)\Gamma_{R\mu}(p_1, p_2; e_R), \tag{22.97}$$

where $Z_{1,2,3}$ is a divergent constant and e_R is a renormalized charge

$$e_R = Z_1^{-1} Z_2 Z_3^{1/2} e. \tag{22.98}$$

Since these equations make it possible to interpret subtraction of infinite constants as a separation of infinite constant factors, they testify to a multiplicative character of the renormalization.

Let us prove the relations (22.95)–(22.97). To this end we substitute $G^{(e)}(p; e)$, $G^{(\gamma)}(k; e)$ and $\Gamma_\mu(p_1, p_2; e)$ obtained from these relations into Eqs. (22.17), (22.20), and (22.25). It is evident that in (22.25) an every term with the coefficient $(e^2)^n$ contains n functions $G^{(\gamma)}$, $2n$ functions $G^{(e)}$ and $2n + 1$ functions Γ_μ. Then, carrying out the replacement

$$\left. \begin{array}{ll} G^{(e)}(p; e) \to G_R^{(e)}(p; e_R), & G^{(\gamma)}(k; e) \to G_R^{(\gamma)}(k; e_R), \\ \Gamma_\mu(p_1, p_2; e) \to \Gamma_{R\mu}(p_1, p_2; e_R) & e \to e_R, , \end{array} \right\}, \tag{22.99}$$

in the expression for Γ_μ and using the relations (22.95)–(22.97), we lead to the result:

$$\Lambda_\mu(e, \Gamma_\mu, G^{(e)}, G^{(\gamma)}) = Z_1^{-1} \Lambda_\mu(e_R, \Gamma_{R\mu}, G_R^{(e)}, G_R^{(\gamma)}). \tag{22.100}$$

Next, we take into consideration that in (22.90) an every term with the coefficient $(e^2)^n$ incorporates $n-1$ functions $G^{(\gamma)}$ and $2n+1$ functions $G^{(e)}$ and Γ_μ. Fulfilling the change-over (22.99) into (22.90), we come to the conclusion:

$$W_\mu(e, \Gamma_\mu, G^{(e)}, G^{(\gamma)}) = Z_1^{-1} Z_3^{-1} Z_2 W_\mu(e_R, \Gamma_{R\mu}, G_R^{(e)}, G_R^{(\gamma)}),$$

or

$$\Delta_\mu(e, \Gamma_\mu, G^{(e)}, G^{(\gamma)}) = Z_1^{-1} Z_3^{-1} Z_2 \Delta_\mu(e_R, \Gamma_{R\mu}, G_R^{(e)}, G_R^{(\gamma)}). \qquad (22.101)$$

By application of the obtained expressions, we can rewrite the integral equations for the mass and polarization operators (22.82), (22.86) as:

$$\Sigma^*(p; e_R) = Z_1^{-1} \int_0^1 dx (p^\mu - p'^\mu) \Lambda_\mu(p^x, p^x; e_R, \Gamma_{R\mu}, G_R^{(e)}, G_R^{(\gamma)}) \qquad (22.102)$$

$$\Pi^*(k; e_R) = Z_1^{-1} Z_2 Z_3^{-1} \int_0^1 dy k^\mu \Delta_\mu(yk, yk; e_R, \Gamma_{R\mu}, G_R^{(e)}, G_R^{(\gamma)}). \qquad (22.103)$$

When substituting the expressions (22.102) and (22.103) the integral equations (22.17), (22.20), and (22.25) become the form:

$$Z_2 G_R^{(e)}(p; e_R) = S^c(p) + S^c(p) Z_1^{-1} Z_2 \times$$

$$\times \int_0^1 dx (p^\mu - p'^\mu) \Lambda_\mu(p^x, p^x; e_R, \Gamma_{R\mu}, G_R^{(e)}, G_R^{(\gamma)}) G_R^{(e)}(p; e_R), \qquad (22.104)$$

$$Z_3 G_R^{(\gamma)}(k; e_R) = D^c(k) + D^c(k) Z_1^{-1} Z_2 \int_0^1 dy k^\mu \Delta_\mu(yk, yk; e_R, \Gamma_{R\mu}, G_R^{(e)}, G_R^{(\gamma)}) G_R^{(\gamma)}(k; e_R), \qquad (22.105)$$

$$Z_1^{-1} \Gamma_{R\mu}(p_1, p_2; e_R) = \gamma_\mu + Z_1^{-1} \Lambda_\mu(p_1, p_2; e_R, \Gamma_{R\mu}, G_R^{(e)}, G_R^{(\gamma)}). \qquad (22.106)$$

These equations must be compatible with the integral ones (22.92)–(22.94). In this case the compatibility conditions enable us to determine the divergent constants $Z_{1,2,3}$.

At the fulfillment of the equality

$$Z_1 \gamma_\mu = \gamma_\mu - \Lambda_\mu(p', p'; e_R), \qquad (22.107)$$

where $\Lambda_\mu(p_1, p_2; e_R) \equiv \Lambda_\mu(p_1, p_2; e_R, \Gamma_{R\mu}, G_R^{(e)}, G_R^{(\gamma)})$, Eqs. (22.106) and (22.92) identically coincide. To compare (22.107) with the relations (22.76) and (22.77) permits one to express the constant Z_1 in terms of divergent parts of the vertex and mass operators

$$Z_1 = 1 \mid B = 1 \mid L. \qquad (22.108)$$

The compatibility of Eqs. (22.93) and (22.104) results in

$$Z_2 - 1 = S^c(p) \int_0^1 dx (p^\mu - p'^\mu)[Z_1^{-1} Z_2 \Lambda_\mu(p^x, p^x; e_R) - \Lambda_\mu(p^x, p^x; e_R) + L\gamma_\mu], \quad (22.109)$$

where we have allowed for $\Lambda_\mu(p', p'; e_R) = L\gamma_\mu$. Integrating the third addend in the right-hand side of Eq. (22.109) and considering (22.108) and $\hat{p}' = m_R$, we get

$$Z_2 - 1 = Z_1 - 1 + S^c(p) \int_0^1 dx (p^\mu - p'^\mu)[Z_1^{-1} Z_2 \Lambda_\mu(p^x, p^x; e_R) - \Lambda_\mu(p^x, p^x; e_R)]. \quad (22.110)$$

From (22.110) it immediately follows

$$Z_1 = Z_2. \qquad (22.111)$$

In a similar way, the identical coincidence condition of Eqs. (22.105) and (22.94) gives

$$(Z_3 - 1)k^2 = k^\mu k^\nu \int_0^1 dy \, y \left(\frac{\partial \Delta_\mu(yk, yk)}{\partial(yk_\nu)} \right)_{k=0}.$$

The solution of this equation is

$$Z_3 = 1 + k^{-2} k^\mu k^\nu C'_{\mu\nu}, \tag{22.112}$$

where

$$C'_{\mu\nu} = \int_0^1 dy\, y \left(\frac{\partial \Delta_\mu(yk, yk)}{\partial(yk_\nu)} \right)_{k=0}. \tag{22.113}$$

From the relativistic invariance follows

$$C'_{\mu\nu} = C_0 g_{\mu\nu},$$

where $_0 = C_{\sigma\sigma}$. Taking into account the linkage between the operator $\Delta_\mu(yk, yk)$ and the polarization tensor, we arrive at

$$C_0 = C = i\Pi(0).$$

Thus, Eq. (22.112) takes the final form

$$Z_3 = 1 + C = 1 + i\Pi(0). \tag{22.114}$$

So, the relations (22.95)–(22.97) are really valid when the renormalized charge e_R is given by the expression (22.98) while the divergent constants $Z_{1,2,3}$ are defined by Eqs. (22.108), (22.111), and (22.114).

From the Dyson equation for the electron Green function we have

$$[G^{(e)}(p; e)]^{-1} = [S^c(p)]^{-1} - \Sigma^*(p). \tag{22.115}$$

After the mass renormalization (including the term $: \delta m \overline{\psi}(x)\psi(x) :$ into the interaction Lagrangian), the operator $\Sigma^*(p)$ becomes the view:

$$\tilde{\Sigma}^*(p) = B(\hat{p} - m_R) + \Sigma_C(p), \tag{22.116}$$

where $\Sigma_C(p)$ obeys the relation:

$$\left. \frac{1}{\hat{p} - m_R} \Sigma_C(p) \right|_{\hat{p} = m_R}. \tag{22.117}$$

Employing (22.108) and (22.111), one could present (22.115) as follows:

$$G^{(e)}(p; e) = \frac{1}{\hat{p} - m_R} \left[Z_2 + \frac{\Sigma_C(p)}{\hat{p} - m_R} \right]^{-1}. \tag{22.118}$$

Eq. (22.118) gives the result

$$\lim_{\hat{p} \to m_R} (\hat{p} - m_R) G_R^{(e)}(p; e_R) = 1. \tag{22.119}$$

One could analogously demonstrate that

$$\lim_{k^2 \to 0} k^2 G_R^{(\gamma)}(k) = 1. \tag{22.120}$$

$$\left. \Gamma_{R\mu}(p_1, p_2) \right|_{\hat{p}_1 = \hat{p}_2 = m_R} = \gamma_\mu. \tag{22.121}$$

As follows from the relations (22.119), (22.120), and (22.121), for free motion of an electron or photon there are no any observable RCs in all orders of the perturbation theory. As, in essence, the relations (22.119) and (22.120) determine the poles of the renormalized Green functions, these relations may be considered as a requirement for the physical electron to have an experimental mass value m_R, and for the photon to have zero mass.

To this point our consideration has been concerned only with the internal effective lines and vertices of the Feynman diagrams. Now we are coming to a discussion of external electron and photon lines. Accomplishing the mass renormalization in (22.22) and allowing for the relation

$$\tilde{\Sigma}(p) = (Z_2 - 1)(\hat{p} - m_R), \tag{22.122}$$

which is valid at $\hat{p} \to m_R$, we could represent (22.22) as:

$$\boldsymbol{\psi}(p) = \psi(p) + (Z_2 - 1)S^c(p)(\hat{p} - m_R)\psi(p). \tag{22.123}$$

Unfortunately, it is impossible to take the derived relation as a renormalization condition for the electron function. The expression (22.123) is ambiguous because it is dependent on the action direction of the operator $(\hat{p} - m_R)$. When the operator $(\hat{p} - m_R)$ effects upon $\psi(p)$, we have zero, and in case of its action upon $S^c(p)$ the result is -1. A similar situation has been encountered in Section 21.6 when the indicated ambiguity has been attributed to the limiting process under which the initial and final states of the scattering matrix are directed to $t = \pm\infty$. There is no point in repeating the procedure of adiabatic interaction switching on and off that allows one to come to the correct result. Instead, we demonstrate that the proper renormalization of the electron and photon wavefunctions is of the form [52]:

$$\boldsymbol{\psi}(p) = Z_2^{1/2}\psi_R(p), \tag{22.124}$$

$$\overline{\boldsymbol{\psi}}(p) = Z_2^{1/2}\overline{\psi}_R(p), \tag{22.125}$$

$$\boldsymbol{\Lambda}_\mu(k) = Z_3^{1/2}A_{R\mu}(k). \tag{22.126}$$

The relation similar to (22.126) exists for the potential of an external electromagnetic field. Note that, since the divergent constant Z_2 governs renormalization of the wavefunction, the equality $Z_1 = Z_2$ should be interpreted as a cancellation in all orders of the vertex divergence by the wavefunction renormalization divergence. As shown in Section 22.2, inclusion of higher orders of the perturbation theory for a particular process necessitates replacement of the ordinary Feynman diagram elements in the associated irreducible diagrams by the effective elements. Thus, when considering some nth order irreducible diagram, we can schematically represent the corresponding element of the S-matrix as follows:

$$\mathcal{M} \sim e^n \int (\Gamma)^n (G^{(e)})^{F_i} (G^{(\gamma)})^{B_i} (\boldsymbol{\psi})^{F_e} (\mathbf{A})^{B_e}, \tag{22.127}$$

where $(X)^n$ denotes that the quantity X enters the integrand n-times.

The matrix element $\mathcal{M}$ is expressed in terms of the nonrenormalized quantities. Having done the transition to the renormalized quantities in accordance with the formulae:

$$e = Z_3^{-1/2}e_R, \qquad m_R = m + \delta m, \qquad \psi = Z_1^{1/2}\psi_R, \tag{22.128}$$

$$\mathbf{A} = Z_3^{1/2}A_R, \qquad G^{(\gamma)} = Z_3 G_R^{(\gamma)}, \qquad G^{(e)} = Z_1 G_R^{(e)}, \qquad \Gamma_\mu = Z_1^{-1}\Gamma_{R\mu}, \tag{22.129}$$

we obtain

$$\mathcal{M} \sim Z_1^{-n+F_i+F_e/2} Z_3^{B_i+B_e/2-n/2} e_R^n \int (\Gamma_R)^n (G_R^{(e)})^{F_i} (G_R^{(\gamma)})^{B_i} (\psi_R)^{F_e} (A_R)^{B_e}. \tag{22.130}$$

Each vertex is a joining place of two electron lines and one photon line, an internal line entering into two vertices and an external line entering into a single vertex. Therefore,

$$n = F_i + \frac{1}{2} F_e = 2 B_i + B_e, \tag{22.131}$$

and, as a consequence, the exponents of Z_3 and Z_1 proves to be equal to zero

$$\mathcal{M} \sim e_R^n \int (\Gamma_R)^n (G_R^{(e)})^{F_i} (G_R^{(\gamma)})^{B_i} (\psi_R)^{F_e} (A_R)^{B_e}. \tag{22.132}$$

In this way the matrix elements are identically expressed in terms of nonrenormalized quantities $(e, m, \Gamma, G^{(e)}, G^{(\gamma)}, \boldsymbol{\psi}, \mathbf{A})$ and renormalized ones $(e_R, m_R, \Gamma_R, G_R^{(e)}, G_R^{(\gamma)}, \psi_R, A_R)$. As the matrix element $\mathcal{M}$ is assumed to be perfectly general, renormalizability of the QED is completely proved. It should be noted that the proof is based on the relation (22.131), connecting the vertex and line numbers in irreducible diagrams, and also on the gauge invariance (to be exact, on its consequence, the Ward identity). We can construct the proof of renormalizability for any other gauge theory in much the same manner.

It is significant that to keep the physical meaning of the renormalized mass and the renormalized charge, arbitrary renormalization constants must satisfy the particular conditions. To illustrate, the requirements for the finiteness and positivity of mass as well as finiteness and reality of charge lead to the following restrictions:

$$0 < m + \delta m < \infty, \qquad 0 < Z_3 < \infty.$$

22.6 Renormalization group

Let us consider another important aspect of the problem involving a choice of subtraction points in the process of renormalization. When calculating a physical (renormalized) charge in Section 21.4, we have made the matrix element of the diagram in Fig. 21.7(e) finite by subtraction of its contribution at the point $q^2 = 0$. From the viewpoint of experiment, this means that we determine a physical charge as a parameter that figures in the electron scattering cross section at an external electromagnetic field for $q^2 = 0$. Obviously, the adopted approach is rather ambiguous. A charge might be determined for scattering of the electron with some nonzero value $|q^2| = m^*$. This new charge $e(m^*)$ in its value is differing from the conventional one. Expansion of the S-matrix in this case is of the same form but with respect to the new charge $e(m^*)$. According to the renormalization procedure, the contributions of the diagrams involving the PSEDs should be subtracted at the point m^*. A similar selection of a subtraction point occurs for renormalization of mass as well. By the invariant Dyson method, the subtraction points are nothing else but the Taylor series expansion points for the divergent expressions. The use of different methods of selection for the series expansion points results in differing theoretical definitions of physical parameters. At the same time, all subtraction points are equivalent as physics must be independent on the selection of renormalization conditions. In other words, the renormalization procedure should be self-consistent. On the mathematical language it means as follows. If one transits from the quantities $e_1, G_1^{(e)}, G_1^{(\gamma)}, \Gamma_{1\mu}$ to the ones $e_2, G_2^{(e)}, G_2^{(\gamma)}, \Gamma_{2\mu}$ by means of transformation:

$$\left.\begin{array}{ll} e_2 = z_0^{1/2} e_1, & G_2^{(e)} = z^{-1} G_1^{(e)}, \\[2mm] G_2^{(\gamma)} = z_0^{-1} G_1^{(\gamma)}, & \Gamma_{2\mu} = z \Gamma_{1\mu}, \end{array}\right\} \tag{22.133}$$

where z, z_0 are arbitrary quantities, then matrix elements $\mathcal{M}$ expressed in terms of the first and second sets of the quantities will be identical

$$\mathcal{M}_1 = \mathcal{M}_2.$$

Assuming in (22.133) that $z = z_3$, $z_0 = z_2 = z_1$ and taking the quantities with indices 1 and 2 as renormalized and bare ones respectively, we see that Eq. (22.133) represents a multiplicative part of the renormalization procedure. It is expedient to recall our steps. First, it has been demonstrated that in order to eliminate the divergences in the theory we must introduce into the Lagrangian a finite number of counter-terms. Then we have proved that, generally speaking, this is equivalent to some renormalization of the Green functions, vertex function, and charge. In the process the counter-term associated with the mass renormalization does not result in multiplicative renormalizations.

The transformations of the multiplicative renormalization (22.133) are clearly possessing the group property. The corresponding group is identified as the multiplicative renormalization group or simply renormalization one (RG). By definition, the RG in the QED represents a group of the multiplicative transformations that act in a abstract space with the Green functions, vertex functions and charge as "coordinates," and cause no changes in the observed effects. It is clear that the presence of the RG is not peculiar to the QED and may be set in any renormalized theory. To this end, among all the acceptable counter-terms of the present theory, we should separate only those leading to multiplicative renormalization of the Green functions, vertex functions and charges, to obtain the transformation relations similar to (22.133). Moreover, such a group also may be found in other fields of theoretical physics associated with critical phenomena: theory of turbulence, physics of polymers, transport theory, magnetic hydrodynamics, etc. In the most general case, the RG appears when the mathematical description of a physical problem involves selection of a particular solution satisfying the boundary condition for some value of the argument (at a certain renormalization point). And invariance with respect to the RG transformations (renorm-invariance) indicates that the physical meaning of the theory is independent on the choice of the renormalization point.

In the QFT the RG was first investigated in Ref. [101]. Later this group has been used to obtain the concrete information about UV asymptotic form of the Green functions in the QED [102]. It has been found that the renorm-invariance of the Green functions enables derivation of their associated equations, with the solutions that allow one to carry out the calculations beyond the applicability domain of the perturbation theory. A theory of the RG has been completely developed in the works by Bogolyubov and Shirkov (see, Ref. [90]). We make no efforts to study the RG equations for the vertex function or Green functions, restricting ourselves to the consideration of an equation for the running coupling constant α.

In the QED the RG transformation can be written as:

$$R(t) = \{m^2 \to m^{*2} = m^2 t, \alpha \to \alpha' = \overline{\alpha}(t, \alpha)\}, \tag{22.134}$$

where t is a continuous transformation parameter changing the scale of the variable m^2, and $\overline{\alpha}$ satisfies the functional equation:

$$\overline{\alpha}(t, \tau) = \overline{\alpha}[t, \overline{\alpha}(\tau, \alpha)]. \tag{22.135}$$

On the strength of Eq. (22.135) the transformations $R(t)$ possess the group property:

$$R(t)R(\tau) = R(t\tau)$$

and form the continuous group, that is, the Lie group. Thus, the RG could completely be characterized by its infinitesimal element. Therefore, in place of functional equations, we may examine differential ones (they are called the *Lie equations*) corresponding to the transformations $R(t)$ with t being close to 1. Such an equation for $\overline{\alpha}$ could be set down in the form:

$$t\frac{\partial\overline{\alpha}(t,\alpha)}{\partial t} = \beta[\overline{\alpha}(t,\alpha)]. \tag{22.136}$$

Here the function $\beta(\alpha)$ representing the group generator is prescribed according to the relation:

$$\beta(\alpha) = \frac{\partial\overline{\alpha}(\xi,\alpha)}{\partial\xi}\bigg|_{\xi=1}. \tag{22.137}$$

To calculate the β-function, we use the formula (22.140), on the right-hand side of which an approximate value $\overline{\alpha}$ obtained from the perturbation theory is substituted. Let us constraint ourselves by the one-loop approximation. Then, in accordance with (21.100), the relation:

$$\overline{\alpha}(t,\alpha) = \alpha\left[1 + \frac{\alpha}{3\pi}\ln t\right], \tag{22.138}$$

holds for the ultraviolet region, and with the help of (22.137) we get

$$\beta(\alpha) = \frac{\alpha^2}{3\pi}. \tag{22.139}$$

Inserting the deduced expression $\beta(\alpha)$ into (22.136) and integrating the obtained equation, we come to the already familiar result:

$$\overline{\alpha}(t,\alpha) = \frac{\alpha}{1 + b\alpha\ln t}, \tag{22.140}$$

where $b = -1/(3\pi)$ and $\alpha = \overline{\alpha}(0,\alpha)$. This expression is an exact solution for a differential equation (22.136) or, equivalently, for the functional group equation (22.135). At the same time, upon a series expansion in terms of α it is in line with the approximate expression (22.138). Consequently, it can be stated that a method of the RG offers synthesis of the perturbation theory and renorm-invariance. The derived expression (22.140) includes a sum over all the principal logarithmic contributions of the kind $\alpha(\alpha\ln t)^n$ and may be used up to the infinitely great values of t. It is apparent that the parameter t should be identified with a ratio Q^2/m^{*2}, where Q^2 is the square of the four-dimensional momentum. By this reason, the limit $t \to \infty$ is associated with the UV asymptotic form $Q^2 \to \infty$. Recall that the formulae of the kind (22.140) for the effective coupling constants of the electroweak and strong interactions are basic for the formulation of the GUTs. A mathematical apparatus of the GUTs relies on a system of coupled Lie equations for several effective coupling constants that is a generalization of Eq. (22.136).

22.7 Consistency problems of the QED

The fact, that in the QED infinite quantities are appearing, does not matter as long as the observables are finite. The appearance of divergent quantities at the intermediate stage again may be considered as an introduction of some new class of the mathematical quantities disappearing from the end formulae for the physical observables. Examples are found in complex numbers, the δ-functions, and so on. Complex numbers have been traditionally used by physics for a long time. The sewing together of the wave functions in a quasi-classical approximation, based on the introduction of complex space coordinates, leads to the Bohr-Sommerfeld quantization rule whose validity is unquestionable. In other words, a problem associated with the infinities, appearing in the course of events of the "scenario" and happily disappearing in the denouement, may be apprehended as the a happy end of some melodrama. Unfortunately, the list of the QED problems is far from being exhausted.

Under studying the QED renormalizability we have used the S-matrix series expansion in terms of the coupling constant powers. In the process it has silently been implied that the series is convergent after renormalization. But it is known that, provided a possibility for trapping into the bound stationary states exists in the theory, after renormalization the series are divergent [103]. Recall that as this takes place, a series of the perturbation theory is divergent even in a nonrelativistic quantum mechanics [104]. In this way, even when we leave a problem of convergence alone, the renormalization pattern used is applicable only in the case of the scattering processes of free particle, when there are no bound states.

As for the series convergence in the perturbation theory, the situation is also not so good. The amplitude of an arbitrary process may be represented as a sum over the terms with increasing powers of the quantity α, every term of the series being lower that the preceding one by about a factor of 100. Because of this, it seems that a contribution made by the diagrams with higher α^n-powers should be infinitesimal so they could be omitted. But we have to point out that the number of such diagrams has a factorial growth ($\sim n!$). Then for a sufficiently high order the diagrams number is so great that it exceeds a small factor of α^n due to the high power $n!$. As a result, the higher-order corrections are growing, and a sum over the whole series is infinite. Such a series are known in mathematics, where they are referred to as asymptotic. The asymptotic series

$$F(p; \alpha) = \sum_{n=0}^{\infty} \alpha^n f_n(p) \qquad (22.141)$$

may be used for the description of a behavior exhibited by the function F, with a very high but always finite accuracy. As distinct from the convergent series, the terms of the asymptotic series $\alpha^n f_n$ are first decreased with the growing number n and then, beginning from some n_c, they are increased, generally without bound. At that, a maximum accuracy, with which the function F may be approximated, is determined by the quantity $f_{n_c}(p)$. The lower is $f_{n_c}(p)$, the higher the accuracy.

Physically, this aspect of the problem has been first explained in Ref. [105]. An asymptotic character of a series for the S-matrix in the QED has been inferred with the following reasoning. Let us suppose that we calculate a physical observable $F(p; \alpha)$ ($\alpha = e_R^2/(4\pi)$) in the form of a power series for the running coupling constant e_R^2. When this series is convergent at a particular positive value of e_R^2, it should be convergent in some circle of radius e_R^2 with a center at the origin of a complex plane $Z = e_R^2$. Consequently, the series should be also convergent for $e_R^2 = 0$ and, besides, on some interval of a real negative semiaxis, that is, at negative values of e_R^2. But $F(p; -e_R^2)$ allows for a simple physical interpretation. The quantity $F(p; -e_R^2)$ would represent the quantity F under study if the interaction between two charges were determined by the function $-e_R^2 D^c(x)$ rather than $e_R^2 D^c(x)$. Meaning that negative values of e_R^2 are associated with the pseudoworld, where like charges are attracted to each other, while unlike charges repel one another. Then the conventional definition of the vacuum as a state with the lowest energy is invalid. Indeed, imagine the situation when N electron-positron pairs are produced. In the process, all electrons are concentrated in one region of the space, whereas positrons occupy the other. If these regions are small enough and sufficiently separated from one another, in the case of great N a negative Coulomb energy of the attracted like charges is higher than the rest energy of particles and their kinetic energy. We call these bound states *pathological*. Apart from this state, let another normal state be existent that involves a number of particles. This state is separated from the pathological one of the same energy by a potential barrier, the height of which is equal to the energy required to form N pairs $(2Nm_e)$. And the tunneling effect leads to the transitions from the normal to the pathological state. Note that a pathological state, wherein the ordinary state is transferred, is nonstationary as particles are generated in ever-growing numbers. In other words, the vacuum disintegration occurs. Due to these effects, we cannot suppose that the QED with the interaction function $-e_R^2 D^c(x)$ leads to the analytical S-matrix. As a result, at $e_R^2 \neq 0$ a series for $S(e_R^2)$ is asymptotic and not convergent. However, this difficulty is rather mathematical in nature since in Ref. [105] it has been demonstrated that $n_c \approx 4\pi e_R^{-2} = 137$ and the associated maximum accuracy, to within the series (22.141) is close to reality, equals $\exp(-4\pi/e_R^2)$. Evidently, this is more than enough for the QED calculations in practice.

Another problem of the QED is related with a behavior of the effective electron charge $e_{eff} \equiv e(Q^2)$. As follows from:

$$e^2(Q^2) = \frac{3\pi e^2(m_R)}{3\pi - e^2(m_R)\ln(Q^2/m_R^2)},$$ (22.142)

for the transferred momentum:

$$Q^2 = m_R^2 \exp\left(\frac{3\pi}{e^2(m_R)}\right)$$ (22.143)

the denominator of the expression (22.142) goes to zero, the charge itself being infinitely large. As a consequence, a physically meaningless restriction is imposed on the values of the four-dimensional momentum transfer. Substituting the numerical values into (22.143), we find that the effective charge goes to zero at the fantastically high energies $\sim 10^{566}$ eV exceeding the energy of the universe as a whole. This circumstance is somewhat soothing because there is a chance to attribute the revealed inconsistency of the QED to its openness. With such high energies it is necessary to take into consideration the effects due to all other interactions, the gravitational one including.

It turned out so that at our climbing the third stage of the Quantum Stairway the QED was to be responsible for the matter structure, just as it became the first working model of the quantum field theory. Its basic properties—the relativistic covariance, the local gauge invariance, the convergence of perturbation theory series, changing the interaction constant with a momentum transferred—got those strings from which the Gobelin tapestry of modern physics was weaved. It is difficult to foresee what aspect would assume the elementary particles theory if the carriers of interactions between the electrons were massive particles with the spin 3/2 rather than the photons. We finish this volume by the words of the poet Robert Burns:

> *Sad words:*
> *"Might be,"*
> *Can't be forgotten*
> *By mice*
> *And people too.*

Problems

4.1. Free-fields operators to be entered into normal products and ordinal pairings obey the free equation of motion. So, for a scalar field we have

$$(\Box - m^2) : \psi(x)\psi(y) := 0, \qquad (\Box - m^2)\psi^s(x)\psi^s(y) = 0.$$

Find the equations to be valid for $\psi(x)\psi(y)$ standing both under the sign of chronological product and under that of the T-product.

4.2. Prove that the vacuum expectation value (VEV) of the chronological product of linear operators $A, B_1, B_2, ... B_n$ is equal to the sum of n VEVs taking from the same chronological products with all the possible pairings between one of these operators (for example, A) and all other operators, that is,

$$< 0|T(AB_1...B_n)|0 > = \sum_i < 0|T(A^s B_1...B_i^s...B_n)|0 >$$

(the third Wick's theorem).

4.3. For charged scalar particles the Lagrangian taking into account the electromagnetic interaction is given by the expression

$$\mathcal{L}^{em} = -ie[\varphi^*(x)\partial_\mu\varphi(x) - \partial_\mu\varphi^*(x)\varphi(x)]A^\mu(x) - e^2\varphi^*(x)\varphi(x)A_\mu(x)A^\mu(x).$$

Direct calculations gives the following results for the chronological pairings of scalar field operators

$$< 0|T(\partial_\mu\psi(x)\psi^*(x')|0 > = -i\partial_\mu\Delta^c(x - x'),$$

$$< 0|T(\partial_\mu\psi(x)\partial'_\nu\psi^*(x')|0 > = -i\partial_\mu\partial'_\nu\Delta^c(x - x') + 2in_\mu n_\nu\delta^{(4)}(x - x').$$

The terms depending on $\delta^{(4)}(x - x')$ also appear in $H_{int}^{(em)}$. However, as computations show, in the matrix elements the mutual cancellation of the similar terms takes place and the rectilinear procedure is reduced to completely neglecting these terms (see discussion in Chapter 1.1). With allowance made for this circumstance, build up the Feynman rules in the coordinate and momentum space.

4.4. Will Furry's theorem be valid in the scalar electrodynamics?

4.5. Calculate the decay probability of a muon through the channel

$$\mu^- \to e^- + \overline{\nu}_e + \nu_\mu,$$

setting the matrix element square equal to $8G_F^2 m_\mu^4/3$.

4.6. The muon decay amplitude through the channel

$$\mu^- \to e^- + \overline{\nu}_e + \nu_\mu$$

to be summed up over particle spins is given by the expression

$$\sum |\mathcal{M}|^2 = 128G_F^2(p_e \cdot p_{\nu_\mu})(p_\mu \cdot p_{\nu_e}). \tag{IV.1}$$

Thinking the amplitudes of the decay (IV.1) and the reaction

$$\nu_\mu + e^- \to \overline{\nu}_e + \mu^- \qquad\qquad (IV.2)$$

to obey the crossing-symmetry condition, find the differential and total cross sections of the reaction (IV.2). Represent the differential cross section as a function of the energy portion transferred to the recoil lepton: $y = E_\mu / E_{\nu_\mu}$. For the sake of simplicity, set the neutrino masses equal to zero.

4.7. Make the same as in the previous problem for the reaction

$$\overline{\nu}_e + e^- \to \overline{\nu}_\mu + \mu^-.$$

4.8. Using the Feynman rules, write down the expression for the invariant amplitude of the process

$$e^+ + e^- \to \gamma + \gamma.$$

Check up the gauge invariance of the expression obtained.

4.9. Making use of the result of the problem **4.4.**, ascertain to which energies the Born's approximation of the four-fermion contact interaction is correct (unitary bound). [NOTE: In the general case the scattering amplitude could be represented in the form

$$f(\theta) = \frac{1}{k} \sum_{l=0}^{\infty} (2l+1) f_l P_l(\cos\theta),$$

where f_l is a partial wave amplitude.]

4.10. Find physically acceptable value regions of the variables u, s, and t for the reaction

$$K_S^0 + \pi^0 \to \pi^0 + \pi^0.$$

4.11. Make the same as in the previous problem for the reaction

$$p + \gamma \to p + \pi^0.$$

4.12. The relativistically invariant phase volume element for secondary particles is as follows

$$d\Phi = \Pi_f \frac{d^3 \mathbf{p}_f}{2E_f} \delta^{(4)} \left(\sum_i p_i - \sum_f p_f \right),$$

where p_i and p_f are four-dimensional momenta of initial and final particles. Find the total phase volume

$$Phi = \int d\Phi$$

for pair of particles with the masses m_1, m_2 and the effective mass equal to $\sqrt{s}$ ($s = (p_1 + p_2)_\mu (p_1 + p_2)^\mu$).

4.13. Using the formula of the problem **4.8.** and the particle mass values, compare the phase volumes of the following decays:

$$\left.\begin{array}{l} K_1^0 \to \pi^- + \pi^+ \\ K^+ \to \pi^+ + \pi^+, \end{array}\right\} \qquad\qquad (IV.3)$$

$$\left.\begin{array}{l} K^+ \to \mu^+ + \nu_\mu \\ K^+ \to e^+ + \nu_e \end{array}\right\} \qquad\qquad (IV.4)$$

4.14. A resting particle a is decaying into particles 1, 2, and 3. Find events distribution density in variables E_1^{cm} and E_2^{cm} (E_{cm} is an energy in the center-of-mass frame), assuming that to be given by the relativistically invariant phase volume.

4.15. Using the results of the problem **4.13.** and taking into consideration experimental values of decay branchings $\mathrm{Br}(K^+ \to \pi^+ + \pi^0) = 0.21$ and $\mathrm{Br}(K_1^0 \to \pi^+ + \pi^-) = 0.684$, obtain the ratio of matrix element module squares for the following decays

$$K^+ \to \pi^+ + \pi^0, \qquad K_1^0 \to \pi^+ + \pi^-.$$

4.16. Find the isotopic relations between probabilities of the $\phi(1020)$-decays through the channels:

$$\phi \to K^+ + K^-, \qquad \phi \to K^0 + \overline{K}^0.$$

Having introduced a correction for a phase volume value, compare the obtained value with the experimental data.

4.17. The isospin is not conserved in the weak interaction. However, in most cases nonlepton decays of strange particles (decays in which a final state isospin is not defined exactly) are under the selection rule

$$\Delta I = \frac{1}{2}.$$

Using this rule, find a ratio of probabilities for the Λ-hyperon decays through the channels:

$$\Lambda \to p + \pi^-, \qquad \Lambda \to n + \pi^0.$$

With allowance made for correction for the phase volume, define a ratio

$$\frac{\Gamma(\Lambda \to n\pi^0)}{\Gamma(\Lambda \to n\pi^0) + \Gamma(\Lambda \to p\pi^-)}.$$

Compare the obtained value with the experimental one. [NOTE: The rule $\Delta I = 1/2$ is easiest to be applied assuming that some fictitious particle (spurion) with the isospin ΔI is added to the left side of equations describing decay processes (in our case $I_{sp} = 1/2$). After that it can consider these decays as processes in which the isospin is conserved.]

4.18. Calculate the differential and total cross sections of the Compton effect for a scalar particle.

4.19. Find the differential cross section of the process

$$\gamma + \tilde{\gamma} \to h^- + h^+,$$

where h^- (h^+) is a negative (positive) charged scalar boson, in the ultrarelativistic case.

4.20. Calculate the differential cross section of the bremsstrahlung for an ultrarelativistic scalar particle.

4.21. Find the differential cross section of the reaction

$$h^- + h^+ \to \gamma + \gamma,$$

where h^- (h^+) is a negative (positive) charged scalar boson.

4.22. Determine the fine structure of the parapositronium levels.

4.23. Find the differential cross section of an electron scattering by an electron in the nonrelativistic case. Consider electrons to be polarized.

4.24. Find the spin flip probability of a complete polarized electron under its scattering by a nonpolarized electron in the nonrelativistic case.

4.25. Find the differential cross section of an elastic electron scattering by a muon in the center-of-mass frame.

4.26. Find the self-energy of a charged scalar particle.

4.27. Determine the vacuum polarization in quantum electrodynamics of scalar particles.

4.28. Using the method of counting the momenta power in integrands, prove the scalar electrodynamics renormalizability.

References

[1] S. Tomonaga, Progr. Theor. Phys. **1**, 27 (1946); J. Schwinger, Phys. Rev. **74**, 1439 (1948).

[2] J. A. Wheeler, Phys. Rev. **52**, 1107 (1937).

[3] W. Heisenberg, Zeit. Phys. **120**, 513 (1943); **120**, 673 (1943).

[4] F. J. Dyson, Phys. Rev. **75**, 486 (1949); **75**, 1736 (1949).

[5] N. N. Bogoliubov, D. V. Shirkov, *Introduction to the Theory of Quantized Fields*, (Interscience Publishers Inc., New York, 1959).

[6] G. Wick, Phys.Rev. **80**, 268 (1950).

[7] R. Feynman, Phys. Rev. **76**, 749 (1949); **76**, 769 (1949).

[8] W. H. Furry, Phys. Rev. **51**, 125 (1937).

[9] S. Mandelstam, Phys. Rev. **112**, 1344 (1958).

[10] T. W. B. Kibble, Phys. Rev. **117**, 1159 (1960).

[11] J. G. Taylor, *Lectures on Dispersion Relations in Quantum Field Theory and Related Topics*, (New issue, Univrsity of Maryland, 1959).

[12] H. Bremerman, R. Oehme, J. G. Taylor, Phys. Rev. **109**, 2178 (1958).

[13] R. Kronig, J. Opt. Soc. Amer. **12**, 547 (1926); H. Kramers, Atti. Congr. Intern. Fisici Como. **2**, 545 (1927).

[14] G. Barton, *Introduction to Dispersion Techniques in Field Theory*, (W. Benjamin. Inc., New York, 1965).

[15] N. F. Mott, Proc. Roy. Soc. **A124**, 425 (1929).

[16] R. H. Dalitz, Proc. Roy. Soc. A **206**, 509 (1951).

[17] K. Blum, *Density Matrix Theory and Applications*, (Plenum Press, New York and London, 1981).

[18] L. D. Landau, E. M. Lifshitz, *Quantum Mechanics. Nonrelativistic Theory*, (Pergamon Press, London, 1959).

[19] A. Zommerfeld, *Atombau und Spectrallinlen*, (Brauschweig, 1939).

[20] F. Sauter, Ann. d. Phys. **9**, 217 (1931); **11**, 454 (1931).

[21] O. Klein, Y. Nishina, Zeit. Phys. **52**, 853 (1929).

[22] U. Fano, Phys. Rev. **93**, 121 (1954).

[23] H. A. Tolhoek, Rev. Mod. Phys. **28**, 277 (1956); H. Mc Master, Rev. Mod. Phys. **33**, 8 (1961).

[24] H. Bethe, W. Heitler, Proc. Roy. Soc. **146**, 83 (1934).

[25] R. L. Gluckstern, M. H. Hull, Phys. Rev. **90**, 1030 (1953).

[26] R. I. Gluckstern, M. H. Hull, G. Breit, Phys. Rev. **90**, 1026 (1953); K. Mc Voy, Phys. Rev. **106**, 828 (1957); H. Banerjee, Phys. Rev. **111**, 532 (1958).

[27] M. Delbrück, Zeit. Phys. **84**, 144 (1933).

[28] V. A. Baier, V. S. Fadin, V. A. Khoze, **50**, 1611 (1966); **51**, 1135 (1966).

[29] G. Breit, E. Teller, Astrophys. J. **91**, 215 (1940).

[30] K. Mc Voy, Phys. Rev. **110**, 484 (1958).

[31] J. M. Jauch, F. Rohrlich, Helv. Phys. Acta **27**, 613 (1954).

[32] R. P. Feynman, Phys. Rev. **74**, 1430 (1953).

[33] P. A. Dirac, Proc. Cambr. Phil. Soc. **26**, 361 (1930); I. Tamm, Zeit. Phys. **62**, 545 (1930).

[34] L. D. Landau, Dokl. Akadem. Nauk: Ser. Fiz. **60**, 207 (1948) (Also published in *Collected Papers of L. D. Landau* (Edited by D. Ter Haar, Pergamon Press, 1965).

[35] L. A. Page, Phys. Rev. **106**, 394 (1957); H. Mc Master, Rev. Mod. Phys. **33**, 8 (1961).

[36] I. Ya. Pomeranchuk, Dokl. Akadem. Nauk: Ser. Fiz. **60**, 218 (1947).

[37] A. Ore, J. Powell, Phys. Rev. **75**, 1696 (1949).

[38] V. I. Goldansky, V. G. Firsov, Uspekhi Khimii, **40**, 1353 (1971).

[39] A. Schenk, *Muon Spin Rotation Spectroscopy*, (Bristol, 1985).

[40] S. Möller, Ann. Phys. **14**, 568 (1932).

[41] A. M. Bincer, Phys. Rev. **107**, 1434 (1957); K. Böckman, G. Kramer, W. R. Theis, Zeit. Phys. **150**, 201 (1958).

[42] H. J. Bhabha, Proc. Roy. Soc. **154**, 195 (1936).

[43] V. B. Berestetski, I. Ya. Pomeranchuk, JETP, **29**, 864 (1955).

[44] A. I. Akhiezer, R. V. Polovin, Uspekhi Fiz. Nauk **51**, 3 (1953).

[45] W. Pauli, F. Villars, Rev. Mod. Phys. **21**, 434 (1949).

[46] G. t'Hooft, M. Veltman, Nucl. Phys. **B44**, 189 (1972); C. G. Bollini, J. J. Giambiagi, Phys.Lett. **B40**, 566 (1972).

[47] H. A. Kramers, Rapports du 8e Conseil Solvay 1948, 241, Bruxelles, 1950.

[48] E. A. Uehling, Phys. Rev. **48**, 55 (1935).

[49] P. A. M. Dirac, Proc. Cambr. Phil. Soc. **30**, 150 (1934).

[50] W. Heisenberg, Zeit. Phys. **90**, 209 (1934); **92**, 692 (1934).

[51] D. Feldman, Phys. Rev. **76**, 1369 (1949); H. Umezawa, R. Kawabe, Progr. Theor. Phys. **4**, 443 (1949).

[52] R. Karplus, N. M. Kroll, Phys.Rev. **77**, 536 (1950).

[53] W. A. McKinley, H. Feshbach, Phys. Rev. **74**, 1759 (1948).

[54] R. Dalitz, Proc. Roy. Soc. **A206**, 509 (1951).

[55] J. Schwinger, Phys. Rev. **75**, 1912 (1949); L. Elton, H. Robertson, Proc. Roy. Soc. **A65**, 145 (1952).

[56] J. Jauch, F. Rohrlich, Helv. Phys. Acta **27**, 613 (1954).

[57] H. Suura, Phys. Rev. **99**, 1020 (1955).

[58] J. Schwinger, Phys. Rev. **76**, 790 (1949).

[59] D. R. Yennie, H. Suura, Phys. Rev. **105**, 1738 (1957); Yung Su Tsai, Phys. Rev. **120**, 269 (1960); S. N. Gupta, Phys. Rev. **98**, 1507 (1955).

[60] L. M. Brawn, R. P. Feynman, Phys. Rev. **85**, 231 (1952).

[61] C. M. Sommerfield, Phys. Rev. **107**, 328 (1957); A. Peterman, Helv. Phys. Acta **30**, 409 (1957).

[62] A. Czarnecki and W. J. Marciano, Nucl. Phys. (Proc. Suppl) **76**, 245 (1999); P. J. Mohr and B. N. Taylor, Rev. Mod. Phys. **72**, 351 (2000).

[63] J. E. Nafe, E. B. Nelson, I. I. Rabi, Phys. Rev. **71**, 914 (1947).

[64] E. Fermi, Zeit. Phys. **60**, 320 (1930).

[65] G. Breit, Phys. Rev. **72**, 984L (1947).

[66] H. M. Foley and P. Kusch, Phys. Rev. **73**, 412 (1948).

[67] J. Schwinger, Phys. Rev. **73**, 416L (1948).

[68] V. Bargmann, L. Michel, V. Telegdi, Phys. Rev. Lett. **2**, 435 (1959).

[69] J. Field, E. Picasso, F. Combley, Uspekhi Fiz. Nauk, **127**, 553 (1979).

[70] R. S. van Dyck Jr., P. B. Schwinberg, and H. G. Dehmelt, Phys. Rev. Lett. **59**, 26 (1987).

[71] I. M. Ternov, V. R. Khalilov, V. N. Radionov, Yu. I. Klimenko, JETP, **74**, 1301 (1978); I. M. Ternov, V. R. Khalilov, Yad. Fiz. **35**, 328 (1982).

[72] T. Kinoshita, Rept. Prog. Phys. **59**, 1459 (1996).

[73] Muon g-2 Collaboration, G. W. Bennett at al., Phys. Rev. Lett. **92**, 161802 (2004).

[74] T. Kinoshita and W. J. Marciano, in *Quantum Electrodynamics*, (edited by T. Kinoshita, World Scientific, Singapore, 1990), 419–478.

[75] K. Hagiwara, A. Martin, D. Nomura, and T. Teubner, Phys. Rev. **D69**, 093003 (2004).

[76] W. E. Lamb, R. C. Retherford, Phys. Rev. **72**, 241 (1947).

[77] H. A. Bethe, Phys. Rev. **72**, 339 (1947).

[78] J. Schwinger, Proc. Nat. Acad. Sci. USA **37**, 452 (1951).

[79] P. J. Mohr, Phys. Rev. Lett. **34**, 1050 (1975).

[80] S. R. Lundeen, F. M. Pipkin, Phys. Rev. Lett. **46**, 232 (1981).

[81] Yu. L. Sokolov, V. P. Yakovlev, JETP **83**, 15 (1982).

[82] G. W. Erickson, D. R. Yennie, Ann. Phys. **35**, 271 (1965); **35**, 447 (1965).

[83] H. A. Bethe, E. Salpeter, *Quantum Mechanics of One and Two Electron Atoms*, (Springer-Verlag, Berlin, 1957); T. Kinoshita, *Perspective of Quantum Physics*, (Iwanami Shoten, Tokyo, 1978).

[84] R. I. Eden, Proc. Roy. Soc. **A215**, 133 (1952).

[85] P. T. Mathews, A. Salam, Rev. Mod. Phys. **23**, 311 (1951).

[86] G. Takeda, Prog. Theor. Phys. **7**, 359 (1952).

[87] J. C. Ward, Phys. Rev. **84**, 497 (1951).

[88] J. Schwinger, *Qauntum Electrodynamics*, (Dower, New York, 1958).

[89] N. N. Bogoliubov, O. S. Parasiuk, Acta Math. **97**, 227 (1958).

[90] N. N. Bogoliubov, D. V. Shirkov, *Introduction to the Theory of Quantized Fields*, (John Wiley, 1982).

[91] G. Källen, Helv. Phys. Acta **25**, 417 (1952); H. Lehmann, Nuovo Cimento **11**, 342 (1954); H. Lehmann, K. Symanzik, W. Zimmermann, Nuovo Cimento **6**, 319 (1957); A. S. Wightman, Phys. Rev. **101**, 860 (1956).

[92] N. N. Bogoliubov, A. A. Logunov, I. T. Todorov, *Introduction to an Axiomatic Quantum Field Theory*, (Benjamin, 1975); H. Epstein, V. Glaser, Annales deL'Institute Poincare **XIX**, 211 (1973).

[93] K. Hepp, *Theorie de la Renormalisation*, (Springer-Verlag, Berlin, 1969).

[94] J. C. Ward, Phys.Rev. **73**, 182 (1950); Proc. Phys. Soc. **A64**, 54 (1951).

[95] F. I. Dyson, Phys. Rev. **82**, 428 (1951).

[96] A. Akhiezer, Phys. Zs. d.Sowjetunion **11**, 263 (1937); R. Karlplus, M. Neuman, Phys. Rev. **80**, 380 (1950).

[97] J. Schwinger, Proc. Nat. Acad. USA **37**, 455 (1951).

[98] A. I. Akhiezer, V. B. Berestetski, *Quantum Electrodynamics*, (Inter., New York, 1965).

[99] J. C. Ward, Phys. Rev. **77**, 293L (1950); **78**, 182L (1950).

[100] A. Salam, Phys. Rev. **82**, 217 (1951); **84**, 426 (1951).

[101] E. Stückelberg, A. Petermann, Helv. Phys. Acta **26**, 499 (1953).

[102] M. Gell-Mann, F. Low, Phys. Rev. **95**, 1300 (1954).

[103] B. Ferretti, Nuovo Cimento **8**, 108 (1951); K. Nishijima, Progr. Theor. Phys. **6**, 37 (1951).

[104] R. Jost, A. Pais, Phys. Rev. **82**, 840 (1951).

[105] F. J. Dyson, Phys. Rev. **85**, 631 (1952).

Appendix: Natural system of units

Typical velocities of elementary particles are close to the light velocity, angular momenta represent multiples of $\hbar/2$, while energies even if they reach the order of 10^{-5} J, are ranked among a category of superhigh ones. Consequently, when we are trying to imagine the elementary particles world visually, one of the troubles in our consciousness is caused by the fact that constants, whose values cannot be laid in macroworld standards, present in the microworld physics formulae. On the other hand, it is easy to appreciate their values compared with each other. So, as the natural step, we should break off the connection with macroworld. It is achieved by the transition to the system of units where the fundamental physical constants of microworld are used as the basic units. Such systems are called *natural system of units* (NSU).

The first to suggest one of NSU kinds was M. Plank. He chose $\hbar$, c, G, and k (the Boltzmann constant) as the basic units. Under the NSU construction the fundamental constants, taking as the basic units, are formally assumed to be equal to 1. So, the Plank NSU is defined by the relation:

$$\hbar = c = G = k = 1.$$

Because the quantum field theory (QFT) is symbiosis of the special theory of relativity and nonrelativistic quantum mechanics, then the NSU of the QFT should be defined as follows:

$$\hbar = c = 1.$$

In this system the dimensionality of any dynamical observable O is connected with the mass dimensionality, that is,

$$[O] = [m^n] \qquad (n \in N).$$

For example, for velocity, action and angular momentum n is equal to 0 while for energy and momentum we have $n = 1$. From the Heisenberg uncertainty relations:

$$\Delta p_i \cdot \Delta x_i \geq \frac{\hbar}{2}, \qquad (A.1)$$

$$\Delta E \cdot \Delta t \geq \frac{\hbar}{2}, \qquad (A.2)$$

it follows that for coordinate and time n equals -1. In this system the electric charge is a dimensionless quantity as its linkage with the fine structure constant α is given by:

$$\frac{e^2}{4\pi\hbar c} = \alpha.$$

Using the definition of the Lorentz force

$$\mathbf{F} = e\mathcal{E} + \frac{e}{c}[\mathbf{v} \times \mathbf{H}],$$

one may be convinced that the strengths of the electric $\mathcal{E}$ and magnetic $\mathcal{H}$ fields have the dimensionality of m^2. In the NSU, the value "eV" (1 eV=1.6021892·10^{-19} J) and derivatives from it (keV, MeV, GeV, TeV, etc.) are used as a mass unit.

To pass on to some ordinary system of units, we must have formulae connecting the basic units of both systems. Let us chose "GeV" as a basic unit in the NSU. Then, for the CGS system the definitions of "g," "cm," and "s" may be found from the relations:

$$1\text{GeV} = 1.6021892 \cdot 10^{-10}\text{J} = 1.7826759 \cdot 10^{-24}\text{g} \cdot c^2, \qquad (A.3)$$

$$\hbar c = 1.9732858 \cdot 10^{-14}\text{GeV} \cdot \text{cm}, \qquad (A.4)$$

$$\hbar = 6.582173 \cdot 10^{-25}\text{GeV} \cdot \text{s}. \qquad (A.5)$$

Using Eqs. (A.3)–(A.5), one could express any derivative CGS unit in terms of "GeV." For example, in the CGS the force dimensionality is given by

$$1\text{dyn} = \frac{\text{g} \cdot \text{cm}}{\text{s}^2} = \frac{\cdot 10^{-12}\text{Gev}^2}{\hbar c},$$

that allows us to state 1 dyne is equal to GeV^2 in the CGS.

To compare some quantities, it is necessary that they have identical dimensionalities. One of the advantages of the NSU implies that some quantities with incoincident dimensionalities in ordinary system of units have one and the same dimensionality in the NSU. As an example, it is instructive to consider the coupling constants G_F and G defining the intensity of the weak and gravitational interactions, respectively. The calculations give

$$G_F \sim 1.4 \cdot 10^{-49}\text{Erg} \cdot \text{cm}^3 = 1.2 \cdot 10^{-5}\hbar^3 c^3 \text{GeV}^{-2} \xrightarrow{\text{NSU}} 1.2 \cdot 10^{-5}\text{GeV}^{-2},$$

$$G \sim 6.7 \cdot 10^{-8}\text{cm}^3 \cdot \text{g}^{-1} \cdot \text{s}^{-2} = 6.7 \cdot 10^{-39}\hbar c^5 \text{GeV}^{-2} \xrightarrow{\text{NSU}} 6.7 \cdot 10^{-39}\text{GeV}^{-2}.$$

The obtained result corresponds to the well known fact—in the region of energies reached to date, the intensity of the weak interaction is as great as 10^{33} times than that of the gravitational interaction.

After the formula for observable quantity O has been obtained, we should make the transition from the NSU to some ordinary system of units, to say, to the CGS. For this purpose, we must rebuild the right dimensionalities of all the physical quantities entering into O (aside from energy, of course) with the help of the following replacements:

$$[m]_E \to [m]c^2, \qquad [l]_E \to [l](\hbar c)^{-1}, \qquad [t]_E \to [t](\hbar)^{-1}, \qquad (A.6)$$

where the subscript E shows that the quantity dimensionality is taken in the NSU.

Index